AF544886

STRUCTURE OF TURBULENCE IN HEAT AND MASS TRANSFER

PROCEEDINGS OF THE INTERNATIONAL CENTRE FOR HEAT AND MASS TRANSFER
ZORAN P. ZARIĆ, EDITOR
C. L. TIEN, SENIOR CONSULTING EDITOR

---

1. Afgan and Schlünder **Heat Exchangers: Design and Theory Sourcebook** 1974
2. Afgan and Beer **Heat Transfer in Flames** 1974
3. Blackshear **Heat Transfer in Fires: Thermophysics, Social Aspects, Economic Impact** 1974
4. deVries and Afgan **Heat and Mass Transfer in the Biosphere: Transfer Processes in Plant Environment** 1975
5. Denton and Afgan **Future Energy Production Systems: Heat and Mass Transfer Processes** 1976
6. Hartnett **Alternative Energy Sources** 1976
7. Spalding and Afgan **Heat Transfer and Turbulent Buoyant Convection: Studies and Applications for Natural Environment, Buildings, Engineering Systems** 1977
8. Zarić **Thermal Effluent Disposal from Power Generation** 1978
9. Hoogendoorn and Afgan **Energy Conservation in Heating, Cooling, and Ventilating Buildings: Heat and Mass Transfer Techniques and Alternatives** 1978
10. Durst, Tsiklauri, and Afgan **Two-Phase Momentum, Heat and Mass Transfer in Chemical, Process, and Energy Engineering Systems** 1979
11. Spalding and Afgan **Heat and Mass Transfer in Metallurgical Systems** 1981
12. Jones **Nuclear Reactor Safety Heat Transfer** 1981
13. Bankoff and Afgan **Heat Transfer in Nuclear Reactor Safety** 1982
14. Zarić **Structure of Turbulence in Heat and Mass Transfer** 1982
15. Taborek, Hewitt, and Afgan **Heat Exchangers: Theory and Practice** 1982
16. Metzger and Afgan **Heat and Mass Transfer in Rotating Machinery** 1983

STRUCTURE OF TURBULENCE IN HEAT AND MASS TRANSFER

Edited by

Zoran P. Zarić

Boris Kidrič Institute, Belgrade, Yugoslavia

In cooperation with the International Union of Theoretical and Applied Mechanics

HEMISPHERE PUBLISHING CORPORATION

Washington New York London

DISTRIBUTION OUTSIDE THE UNITED STATES

McGRAW-HILL INTERNATIONAL BOOK COMPANY

Auckland Bogotá Guatemala Hamburg Johannesburg Lisbon London
Madrid Mexico Montreal New Delhi Panama Paris
San Juan São Paulo Singapore Sydney Tokyo Toronto

STRUCTURE OF TURBULENCE IN HEAT AND MASS TRANSFER

1 2 3 4 5 6 7 8 9 0 B C B C 8 9 8 7 6 5 4 3 2

Library of Congress Cataloging in Publication Data
Main entry under title:

Structure of turbulence in heat and mass transfer.

(Proceedings of the International Centre for Heat and Mass Transfer ; 14)

Proceedings of the Symposium on Heat and Mass Transfer and the Structure of Turbulence, held Oct. 6–10, 1980, in Dubrovnik, Yugoslavia, sponsored by the International Union of Theoretical and Applied Mechanics and the International Centre for Heat and Mass Transfer.

"In cooperation with the International Union of Theoretical and Applied Mechanics."

Bibliography: p.

Includes index.

1. Heat—Transmission—Congresses. 2. Mass transfer—Congresses. 3. Turbulent boundary layer—Congresses. I. Zarić, Zoran. II. Symposium on Heat and Mass Transfer and the Structure of Turbulence (1980 : Dubrovnik, Croatia) III. International Union of Theoretical and Applied Mechanics. IV. International Center for Heat and Mass Transfer. V. Series.

QC319.8.S84 536'.2 81-23730
ISBN 0-89116-233-X AACR2
ISSN 0272-880X

Contents

Preface

The symposium on Heat and Mass Transfer and the Structure of Turbulence was the first joint venture of the International Union of Theoretical and Applied Mechanics (IUTAM) and the International Centre for Heat and Mass Transfer (ICHMT), which has been affiliated with the IUTAM since 1972. The initiative for the symposium came from E. A. Brun, F. N. Frenkiel, and Z. Zarić. The ICHMT made a formal proposal for the joint organization of the symposium to the IUTAM in 1978, the General Assembly of the IUTAM approved the venture, and the Bureau of IUTAM appointed the scientific committee of the symposium late in 1978. The scientific committee consisted of A. J. Favre (France), W. Fiszdon (Poland), F. N. Frenkiel (USA), L. Fulachier (France), T. Mizushina (Japan), P. R. Owen (UK), D. B. Spalding (UK), and Z. Zarić (Yugoslavia), chairman.

In the last decade fundamental research in turbulent flows has shifted from a pure statistical approach to investigations of the physics of the phenomena involved. As a result, the existence of intermittent, quasi-deterministic structures, coherent for a detectable period of time, which appear at random but with a statistically defined periodicity, has been experimentally proved. Moreover, it has been found that these structures play a crucial role in all turbulent processes, including momentum and heat and mass transfer. On the other hand, research in heat and mass transfer in turbulent flows, including the very important field of environmental flows, is still mostly in the phenomenological stage, making very little, if any, use of the results achieved so far in the physics of turbulence.

The principal idea behind the symposium was, therefore, to gather together researchers in the physics of turbulence with those in the heat and mass transfer and environmental flows, with the aim of contributing to the clarification of heat and mass transfer processes by the results achieved so far through the research in the physics of turbulence. With this aim the symposium addressed the following topics: (1) Latest results of the investigations in the physics of turbulent shear flows, (2) Better understanding of heat and mass transfer phenomena in turbulent flows and possibilities of its modelling, and (3) Understanding of the processes in environmental flows.

The Symposium was held October 6–10, 1980, in the Hotel Libertas, Dubrovnik, Yugoslavia. It was attended by 46 participants from 12 countries. The scientific committee selected 35 papers for presentation and 3 visualisation films; 32 papers and 3 films were actually presented at the symposium's 7 sessions. Ample time was provided for discussions following each paper. In addition, 3 general discussions were held on the three main topics, given in the previous paragraph chaired by J. Laufer, L. Fulachier, and E. J. Hopfinger, respectively. A book of extended abstracts of all the admitted papers was prepared and distributed at the symposium.

This volume contains 34 papers selected by the scientific committee and 3 papers related to the visualisation films presented at the symposium as well as summaries of general discussions on heat and mass transfer in turbulent and environmental flows. The material is divided into 5 chapters, the first 3 of which correspond to the 3 main topics of the symposium, the fourth containing papers on turbulence modelling, and the fifth containing papers related to the visualisation studies. A list of participants and a subject index are also included.

One half of this volume is concerned with the physics of turbulent flows, while the other half is divided between various aspects of heat and mass transfer processes and environmental flows. This confirms the fact that the penetration of the physical approach, as opposed to statistical and phenomenological approaches, in the research of heat and mass transfer phenomena is a very slow process. Nevertheless, the very lively discussions at the symposium, between those involved in the physics of turbulent flows and those interested in solving heat and mass transfer problems, as well as between experimentalists and those involved in prediction models, showed there could be no real breakthrough in the theory of turbulent flows and, therefore, no satisfactory mathematical models of turbulence, without a more detailed insight into the physics involved. It is believed that this symposium has made a contribution towards this goal and that future similar meetings will proceed further in this direction.

The organization of the symposium was made possible by the allocations provided by IUTAM for travel supports of some of the participants, by the financial support provided by the Yugoslav National Commission for UNESCO, and by the efforts of the ICHMT Secretariat. The scientific committee wishes to acknowledge Hemisphere Publishing Corporation for preparing the book of extended abstracts and for making possible the publication of the proceedings. The assistance of Mrs. Stanislava Djurić and Mr. Ilija Ilić in the smooth operation of the symposium is also acknowledged.

Zoran Zarić

PHYSICS OF ISOTHERMAL FLOWS

Flow Instability and Turbulence

JOHN LAUFER*
Department of Aerospace Engineering
University of Southern California
Los Angeles, California 90007

1. INTRODUCTION

Traditionally, problems concerning flow instability have been treated quite independently from the turbulence problem. There is, of course, a well founded reason for this point of view. The physical perception and the mathematical treatment of the two problems have been quite different; a time dependent, linear, small perturbation theory was considered the basis of the stability problem while a time averaged, non-linear set of equations were usually the starting point of the turbulence problem. There was, and is, no viable formulation of the so-called transition from laminar to turbulent flows.

This problem is reexamined for the case of free mixing layers and jets. It has been shown in the past that the initial unstable shear layer develops into a row of vortices which subsequently interact, coalesce and become turbulent. The paper will argue that some rather simple stability considerations can describe the flow development whether the flow is fully turbulent or not.

2. THE INITIAL INSTABILITY

The behaviour of a mixing layer or jet just downstream of the trailing edge of a splitter plate or nozzle can be satisfactorily predicted by the Orr-Sommerfeld linear perturbation theory. Provided the plate thickness is less than the separating b.l. thicknesses, the wake does not have a detectable effect on the flow development [1]. However, there is considerable evidence that external perturbation near the origin play an important role on the dynamics of the layer [1], [2], [3]. In particular, the layer is most sensitive to disturbance frequencies equal or lower than the theoretical most amplified frequency. Of special relevance is the work of Wygnanski and Oster who have shown that the spreading rate of a shear layer can be increased considerably over a limited downstream distance by introducing perturbations in the initial shear layer [2]. In addition, Ho and Huang gave a more quantitative account concerning the receptivity and reaction of the shear layer to disturbances in the various frequency ranges [1]. They have shown, for instance, that the ratios of the perturbation frequency to the theoretical, most amplified frequency is an important parameter.

---

* Professor and Chairman

3. SECONDARY INSTABILITY

In their classical paper, Winant and Browand have pointed out that the pairing process between neighboring vortices plays an essential role in the development of a shear layer. Specifically, two neighboring vortices through their induced field tend to amalgamate. This process is initiated by an instability process that is not well understood yet. One possibility, suggested by the theoretical work of Kelly [4] and by numerical simulation [5] is a so-called subharmonic instability. Accordingly, a disturbance that has a wave length twice the separation distance between two vortices and in addition, has the proper phase relationship, induces the pairing process. It is conjectured that these disturbances are most likely to originate from downstream through the induced field of an already coalesced vortex pair.

4. THE SUBHARMONIC FEEDBACK LOOP

The presence of a feedback was first suggested by Dimotakis and Brown [6] in analyzing autocorrelation measurements in a mixing layer. Further evidence was found by Laufer and Monkewitz in a turbulent jet [7]. A qualitative picture of the subharmonic feedback mechanism is described by Laufer [8]. It is based on the results obtained by Ho and Nosseir in the jet impingement problem [9] and can be summarized as follows: the pressure field induced by a pairing process feeds back to the origin of the shear layer (the region most sensitive to outside disturbances) thus perturbing the layer in the form of a subharmonic (or multiple subharmonic) to the self-excited frequency of the initial layer. For a resonance feedback to occur it is required that the feedback loop period be an integer number times the resonance period. Thus,

$$\frac{x_p}{U_c} + \frac{x_p}{a} = \frac{N}{f_r}$$

where x_p is the pairing location, U_c and a the convection velocity of the turbulent vortex structures and the sonic velocity, respectively, f_r the resonance frequency and N an integer number.

Comparisons have been made with available experimental data with encouraging results [7].

It is of interest to note that in the above rather simplistic treatment no consideration has been given to the actual transition process. It seems that as far as the basic mechanism of interaction between the large scale structures is concerned, the presence or absence of small scale turbulence is of secondary importance.

ACKNOWLEDGEMENT

The author wishes to acknowledge the financial support of AFOSR under Contract No. F49620-48-C-0060 and of ONR under Contract No. N00014-76-C-0211.

REFERENCES

1. Ho, Chih-Ming and Huang, Lein-Saing 1980. Subharmonic and Vortex Merging in Mixing Layers, (submitted for publication).

2. Wygnanski, I. and Oster, D. 1980. The Forced Mixing Layer Between Parallel Streams, (submitted for publication).

3. Hussain, A.K.M.F. 1980. Perturbed and Unperturbed Turbulent Jets, (to appear in Lecture Notes in Physics, Springer).

4. Kelly, R.E. 1967. On the Stability of an Inviscid Shear Layer Which is Periodic in Space and Time, J. Fluid Mech., Vol. 27, p. 657.

5. Riley, J.J. and Metcalfe, R.W. 1980. Direct Numerical Simulation of a Perturbed, Turbulent Mixing Layer, AIAA Preprint No. 80-0274.

6. Dimotakis, P.E. and Brown, G.L. 1976. The Mixing Layer at High Reynolds Number: Large-Structure Dynamics and Entrainments, J. Fluid Mech., Vol. 78, p. 535.

7. Laufer, J. and Monkewitz, P. 1980. On Turbulent Jet Flows: A New Perspective, AIAA Preprint 80-0962.

8. Laufer, J. 1980. Deterministic Versus Stochastic Turbulence, (submitted for publication).

9. Ho, Chih-Ming and Nosseir, N.S. 1980. Dynamics of an Impinging Jet Part I: The Feedback Phenomenon, (to be published in J.F.M.).

Conditionally Averaged Patterns of Coherent Events in a Wall-bounded Turbulent Flow

ZORAN ZARIĆ
Boris Kidrič Institute of Nuclear Sciences
Beograd, Yugoslavia

ABSTRACT

Conditionally averaged probability density distributions of the axial velocity fluctuations during intermittent sweeps and ejections in a channel flow of air were determined previously. With the aim of detecting characteristic velocity patterns within these events a new analysis based on the same conditional averaging procedure and using the same digital data was perfprmed. The presented results show that intermittent events account for about 50% of the time throughout the wall layers. It is found that sweeps and ejections occuring separately, as single events, account for 35 to 45% of the time while ejections followed by a sweep account at most for about 15% of the time. These findings are in disagreement with some previous concepts in which a S shaped pattern is believed to be characteristic for coherent events in the wall layers. They are, however, in good agreement with the existing evidence of latterally shifted occurence of sweeps and ejections, most of the time.

1. INTRODUCTION

The present state of the art of the physics of turbulent shear flows is marked by the discovery of a definite coherence within the flow, detected by numerous different experimental means. It is believed that these coherent events are caused by some, not yet identified, quasi-deterministic structures embeded in the background turbulence. The importance of the full description of these coherent structures lays in the experimentally proven fact that during the events caused by these structures most of the turbulent energy is generated and most of the turbulent transport is occuring. Whatever the form of the coherent structures might be the events caused by them are detected by a stationary probe as a characteristic pattern in the signal. It is not too probable that the study of these patterns in the Eulerian frame of reference would lead to the full description of these structures. However, a hypothesis on a definite form of a structure could be checked by the recognition of its characteristic pattern. Also, although the Lagrangian frame of reference is probably more suitable for the identification of a structure, information gained in the Eulerion frame of reference is more quantitative and what is very important, more adequate for the study of the transport mechanism at the wall.

Various pattern detection schemes have been used so far in the study of the inner layers of wall bounded turbulent flows. They are reviewed elsewhere /1,2/. A technique of the pattern recognition in the signal was applied by Wallace, Brodkey and Eckelmann /3/ in a channel flow. The pattern sought was a weak deceleration of the flow, followed by a period of low streamwise velocity and then a strong acceleration leading to a period of relatively high velocity. The only condition applied, then, was that a negative streamwise velocity time derivative is followed by a stronger positive time derivative, irrespective of the velocity level. Such a S shaped signature in the signal was in fact recognized as a characteristic feature of the bursting phenomena by the pattern detection scheme applied by Blackwelder and Kaplan /4/. Ansemble averaged patterns, recognized by Wallace et.al. at different nondimensional distances from the wall (x_2^+), are shown in Fig. 1, reproduced from /3/. Patterns were recognized in the streamwise velocity (u_1') signals while normal to the wall velocity (u_2') and the instantaneous shear stress ($u_1' u_2'$) fluctuations, where recorded simultaneously and ansemble averaged in the periods recognized in the u_1 signals. More details on the technique are found in /3/. At $x_2^+ = 15$ the accepted pattern shown in Fig. 1 accounts for over 65% of the total signal /3/. A similar pattern at the same x_2^+ is found to account for only about 25% of the time by Blackwelder and Kaplan /4/. In performing sensitivity analysis

Wallace et.al. /3/ have found that if an additional condition is applied that the positive velocity gradient is twice or 3 times stronger than the negative velocity gradient, in absolute value, than the given pattern accounts for only 18% and 5% of the time, respectivelly.

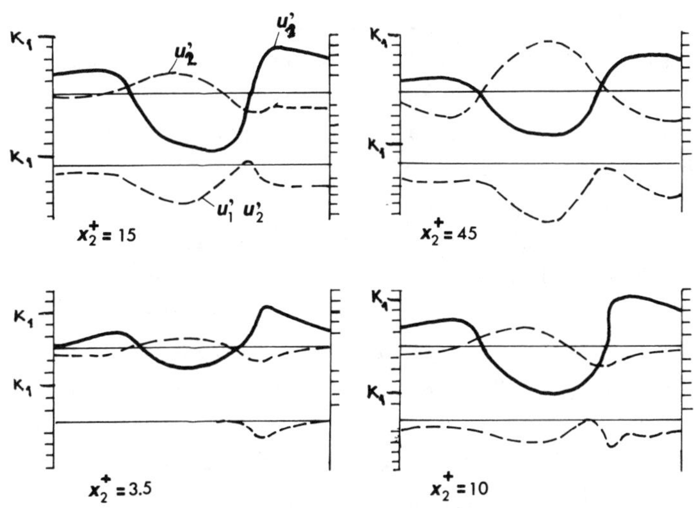

Fig. 1. Patterns recognized by Wallace et.al. /3/.

It is important to note that Wallace et.al. /3/ as well as Blackwelder and Kaplan /4/ have programmed their pattern detection schemes for the S shaped patterns shown in Fig. 1. In fact in the scheme applied in /4/ values of the short time average variance above a certain level are used as detection criteria, and the S shaped pattern with a strong positive temporal gradient leads to high values of the variance. Strong positive (du_1/dt) is an indication of an inrush of the high momentum fluid into the wall layers. This is confirmed by the result of Chen and Blackwelder /5/, obtained in a boundary layer with wall heating and a pattern detection scheme of the temperature signal similar to the one in /4/. It is concluded in /5/ that a strong temperature decrease, corresponding to an inrush, is the most obvious manifestation of a coherent structure. This is also confirmed by the shape of the u_2' pattern in Fig.1. All the mentioned authors /3,4/ have also found that an inrush of high momentum, low temperature fluid is preceeded by a period of relatively low velocity, corresponding to an ejection, or a burst as seen again from the u'_2 pattern in Fig. 1.

Not all the evidence accumulated so far in numerous studies of coherent structures in the wall layers is in accordance with the sequence of events in which an inrush, or a sweep is always preceeded by an ejection, or a burst. Early visualisation studies of the Stanford group indicate that the low velocity streaks, which are bursting, are laterally intercepted, not always axially followed by regions of faster flowing fluid /6/. The results of the visualisation study by Grass /7/ emphasized important contribution of the inrush phase close to the wall, which was previously neglected. It is also noted that the typical inrush phase is violently splaying low velocity fluid close to the wall forwards and to either side as it progresses along the wall. In the visualization study by Oldaker and Tiederman /8/ it is found that a pair of low speed streaks is formed on other side of an oval depression in the flow layer, persumably caused by a sweep. Lateral interception of high speed wallward regions and low speed bursting regions is readily observed in the photographs taken simultaneously from downstream and from above in a boundary layer by Hirata and Kasagi /9/.

The results of more quantitative measurements in the Eulerian frame of reference are also in agreement with the hypothesis that a stationary probe would more often feel a bursting or a sweep phase one by one, rather than in an sequence. Falco /1o/ recognized two different types of patterns in the wall layers with an equal probability, one with a zone average greater than the local mean, and the other less than the local mean. The results of the triple spatio-temporal correlations of temperature signals in a boundary layer with wall heating by Elena, Fulachier and Dumas /11/ show clearly that the warm fluid of the ejections is surrounded by thicker zones of colder inrushing fluid, not followed by a sweep.

Our previous work was concerned with the identification of intermittent, persumably coherent, events in the inner layers by an original conditional averaging technique which distinguishes between the sweep and the ejection events /12, 13, 14/. In what follows a short description of the experimental technique and the conditional averaging procedure applied will be presented, followed by the results of a new analysis of the same data made with the aim of recognizing the velocity patterns within a certain event.

2. CONDITIONAL PATTERN RECOGNITION

Axial velocity measurements were made in an air flow through a rectangular channel of 4o x 3oo mm^2 cross-section. The maximum velocity was u_o = 1o.2 m/s so that the Re number based on this velocity and the half-width of the channel (b) was about 12,ooo and the friction velocity amounted to u_* = o.48 m/s. Velocity measurements were made with a 5 μm hot wire and a DISA constant-temperature anemometer. Signals of 15 seconds duration were digitalized with a sampling frequency of 8,ooo values per second and stored on digital tapes in form of sets of 12o,ooo values each. Further details on the experimental setup and measuring procedure were reported in /13, 14/.

Complete axial velocity probability density distributions (PDD) were determined for 2o distances from the wall ranging from x_2^+ = 1.6, to the center of the channel (x_2^+ = 555). Selected PDDs are shown in Fig. 2. The agreement in shapes of those PDDs on a given x_2^+ with those of other authors obtained at similar relatively low Re numbers is excellent /15, 16/. We have suggested /17/ that the evolution in shape of PDDs across the wall layer is a consequence of intermittent ejection and sweep events. Close to the wall, up to x_2^+ about 15, for relatively low Re number flows, PDDs have positive skewness and high flatness factors, supposedly as a consequence of the intermittent arrival of vigorous, high-momentum sweeps. Further from the wall at these Re numbers PDDs have negative skewness and low flatness factors, as a consequance of the ejection events becoming more vigorous.

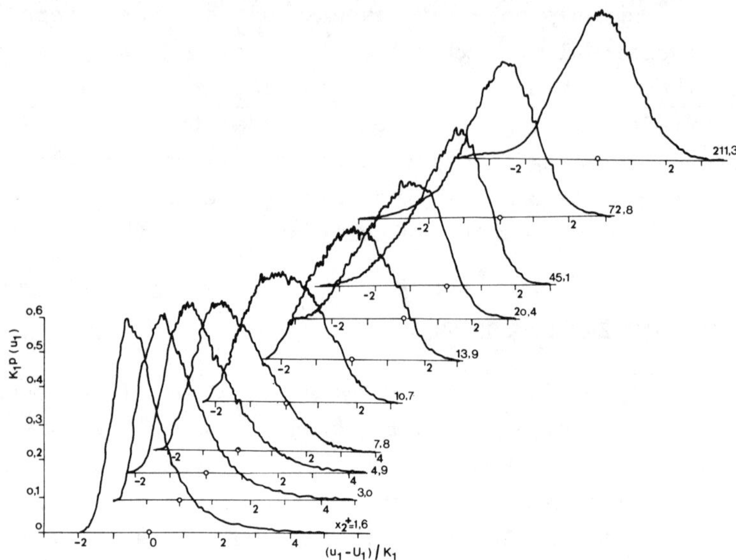

Fig. 2. Probability density distributions of axial velocity /13/.

In order to prove this hypothesis an original conditional averaging analysis was performed. The procedure is explained in detail in /12, 13, 14/, so that a brief description only is given here as far as it is necessary for the understanding of the further analysis. A three-value intermittency function of the type:

$$I(t) = \begin{cases} 1 & Z > A \quad \text{SWEEP} \\ 0 & B < Z < A \quad \text{QUIESCENT PERIOD} \\ -1 & Z < B \quad \text{EJECTION} \end{cases}$$

is introduced, where Z is in fact a combination of a number of criteria. A sweep should be associated with high, positive amplitude fluctuations, u_1 , and in addition an arrival of a sweep is associated with a high positive value of the velocity time derivative (du_1/dt). An ejection should be associated with a relatively high negative amplitude of u_1 , and in analogy with sweeps, its arrival should be associated with a relatively high negative value of the time derivative, not necessarily as high in absolute value as the one associated with the arrival of sweeps. Our criteria for the switch on of the $I(t) = 1$, or the arrival of a sweep were, then, a high positive amplitude of u_1 , or a high value of the product $Z = u_1\ (du_1/dt)$. Similarly, criteria for the arrival of an ejection were a high negative amplitude of u_1 OR a high Z value. It is, however, allowed that an ejection may end - up in a sweep, or vice versa, depending on how strong is positive or negative value of the time derivative, respectively.

The threshold values of the u_1 amplitudes and of the Z have to be chosen. Our results, as well as the results of other investigations show that the velocity time derivative has a positive skewness factor throughout the inner layers. We have however, performed a conditional averaging of the du_1/dt, the only condition being its sign. The resulting distributions of the r.m.s. values of positive ($\langle K' \rangle^+$) and negative ($\langle K' \rangle^-$) du/dt normalized by the mean r.m.s. value of the time derivative (K'), are presented across the wall layers in Fig. 3. It is seen that the value of $\langle K' \rangle^+/K$ is much higher than the value of $\langle K' \rangle^-/K$ in the innermost layers (closer than $x_2^+ = 30$) but that further from the wall they are almost of the same intensity, with the latter being even somewhat higher. We have also performed a similar conditional averaging of the product Z. The results for the r.m.s. values of the Z, $\langle K_z \rangle^+$ and $\langle K_z \rangle^-$, normalized by the product of the mean r.m.s. values of velocity (K) and its time derivative (K') are also shown in Fig. 3. It is seen that the value of $\langle K_z \rangle^+$ is many times higher than the corresponding value of $\langle K_z \rangle^-$ for $x_2^+ < 15$ and that further from the wall $\langle K_z \rangle^-$ values become higher. The conclusion is that the thresholds should depend on the type of the intermittent event as well as on the nondimensional distance from the wall. Our thresholds values were taken as proportional to the conditionally averaged r.m.s. values just described. Further details are given in /13, 14/.

The resulting conditionally averaged PDDs of the u_1 fluctuation during sweep and ejection events are shown in Figs. 4 and 5. In Fig. 4, it is shown how the PDD of the total u_1 signal (1), at $x_2^+ = 4.9$, is decomposed into three PDDs corresponding to the sweep (2), ejection (3) and quiescent (4) phase. In Fig. 5

the decomposition of the PDD with negative skewness at $x_2^+=28.2$ is shown.

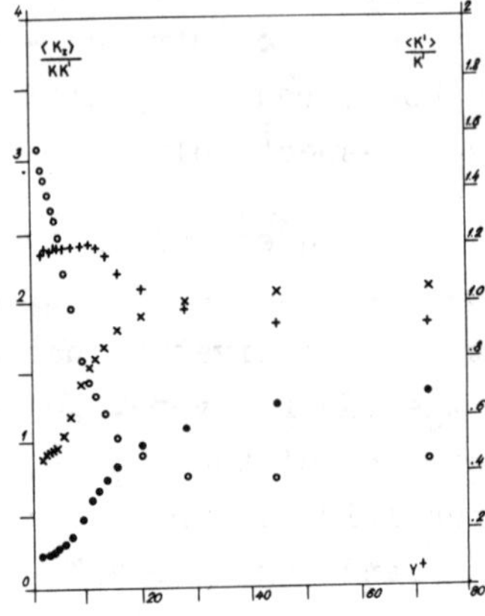

Fig. 3. Normalized values of conditionally averaged r.m.s. values of (du_1/dt) and the product $Z=u_1\cdot(du_1/dt)$.
o – $\langle K_z\rangle^+/KK'$; • – $\langle K_z\rangle^-/KK'$, + – $\langle K'\rangle^+/K'$; X – $\langle K'\rangle^-/K'$

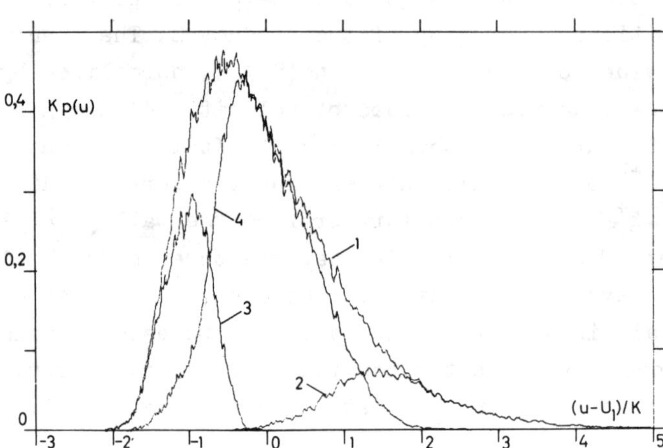

Fig. 4. Probability density distributions p(u), normalized with the local r.m.s value, K, for the total signal (1) sweep phase (2), ejection phase (3) and quiescent phase (4) at $x_2^+ = 4.9$.

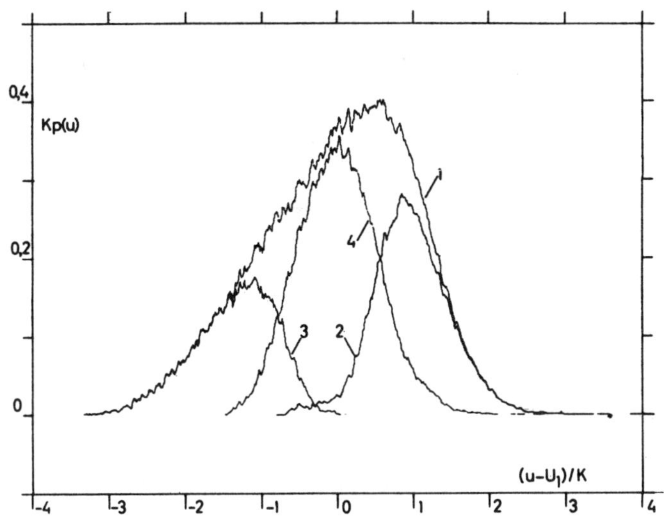

Fig. 5. Conditionally averaged probability density distributions at x_2^+ = 28.2. Notations as in Fig. 4.

The described conditional averaging analysis provides an explanation for the evolution of the PDD shapes across the wall layers. It does not give information on the velocity patterns during the intermittent events. Also, it does not provide information whether the events appear in sequence or one at a time. In order to obtain this kind of information a new analysis is made using the data registered on tapes some years ago.

The original conditional averaging procedure is kept with the same criteria and the same treshold levels. The computer is programmed to register the velocities in sequence during the sweep (S) and ejection (E) events. As it is assumed that the duration of separate events vary in a statistical manner, velocities are sorted into loo unit time intervals and then ansemble averaged at each of the time intervals, so that the longer events are compressed and the shorter streched to an ensemble averaged time interval. The procedure is similar to the one used by Wallace et. al. /3/. It is also registered if a sweep event ends up into an ejection event or vice versa. These composed events are called (S + E) and (E + S), respectively and they are ensemble averaged separately. Velocities in three time intervals of o.125 ms, or in total for a $\Delta t^+ = tu_*^2/\nu = 5$, before and after a single or a composed event, are also registered and ensemble averaged. Finally, if the time interval between an ejection and a sweep or between a sweep and an ejection was less or equal to $\Delta t^+ = 5$, it is taken that a composed (E + S) or (S + E) event is occuring and the velocities in between were included into the ensemble average of such an event.

Axial velocity patterns recognized by such a conditional averaging procedure during (E + S),(S+E) and (S) events are presented in Figs. 6, 7, 8 and 9, respectively, for 6 different distances from the wall. The patterns are presented in actual values of the ensemble averaged velocities u_1. Scales on the side give the values normalized by the mean r.m.s. value at a given x_2^+. Scales on the bottom give the ensemble average duration of an event at a given x_2^+ in miliseconds (t) and in $t^+ = t\, u_*^2/\nu$ values. The vertical lines delimit the event with the patterns immediately preceeding and following the event, also shown.

The (E + S) patterns recognized and shown in Fig. 6, have a S shape similar to the one recognized by Wallace et.al. /3/ and Blackwelder and Kaplan /4/ as being the characteristic pattern of the bursting event in the wall layers. Comparing Fig. 1 with Fig. 8, however, it is seen that the patterns recognized in this study correspond to much more vigorous events in which maximum and minimum values of the amplitudes are at least equal to K, while the maximum values of the patterns recognized in /3/ are less than the corresponding r.m.s. value. In Fig. lo percentages of the time of the whole signal (T_t) occupied by different events, S, E, E+S, S+E and Q (quescent periods) are given in function of the x_2^+. It is seen that the (E+S) event accounts at the most for about 14% of the time at about $x_2^+ = 10$ and much less than that closer and further from the wall. In /3/ and /4/ it was found that the S shaped pattern accounts for 65% and 25% of the time, respectively.

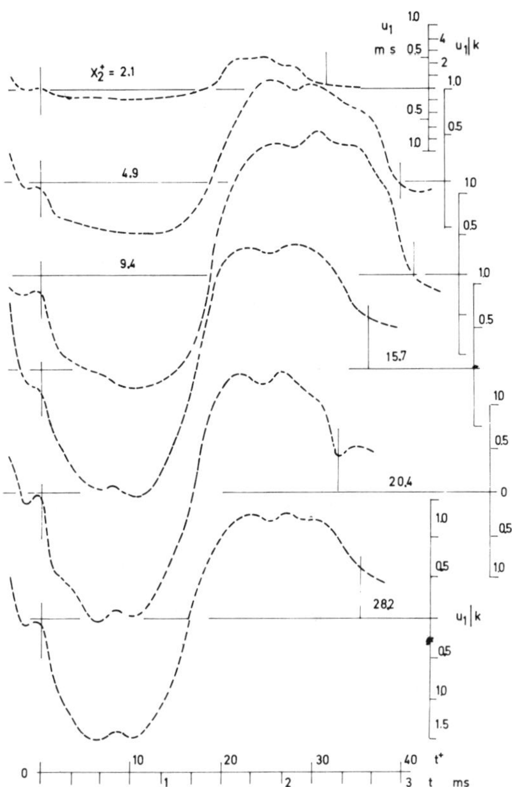

Fig. 6. Patterns recognized during a composed event in which an ejection ends up into a sweep (E+S) at different x_2^+.

From Fig. 7 it is seen that in this study patterns of the (S+E) type, in which a sweep ends up in an ejection, are also recognized. This kind of a pattern would be rejected by the procedures employed in /3/ and probably in /4/, also. From Fig. 1o it is seen, however, that these composed events are rare, accounting at the most for about 8% of the time.

In Fig. 8, patterns recognized during isolated ejection events (E) are shown at different x_2^+. It is seen that these events are very vigorous. It is also seen that these events develop from a quiescent period and end up into another quiescent period. It is also seen that for most distances from the wall the patterns shown have a stronger time gradient at the beginning of the event than at the end of it. It seems threfore, that these patterns would be at least most of tho time, rejected by the procedure employed in /3/ although they have a deformed S shape.

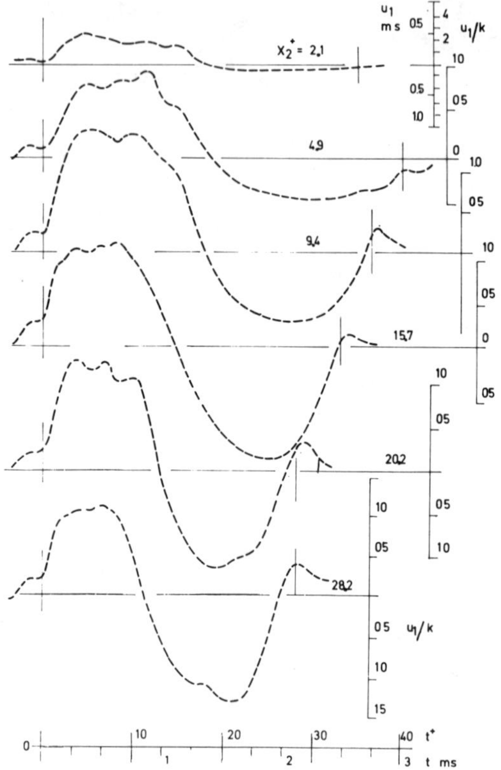

Fig. 7. Patterns recognized during a composed event in which a sweep ends up in an ejection (E+S) at different x_2^+.

In Fig. 9, patterns recognized during isolated sweep events (S) are shown at different x_2^+. It is seen that these patterns indicate very vigorous events at least close to the wall, and that they all develop at an already higher level of the velocity. An S shape could be recognized at the beginning of the event characterized by a high positive value of the velocity derivative. These patterns would most probably and at least partly be accepted by the procedures employed in both /3/ and /4/.

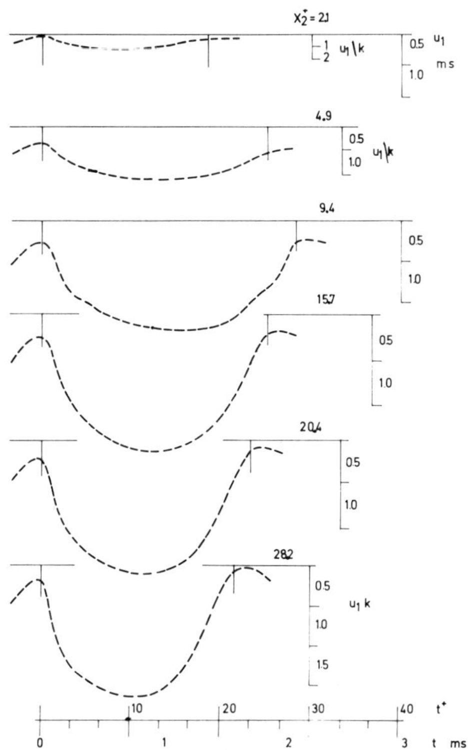

Fig. 8. Patterns recognized during an ejection event (E) at different x_2^+.

Resulting average pattern lenghts of different events expressed in real time (t_i) as well as in nondimensional time units (t^+) are plotted in Fig. 11, in function of x_2^+. Pattern lenghts of composed events (E + S) and (S + E) are given separately for the ejection part E + (S) and (S) + E, and for the sweep part, (E) + S and S + (E), respectively. It is seen that the ejection events last longer than the sweep events, except in the immediate vicinity of the wall. Esspecially long ejection events are noted around $x_2^+ = 10$. Duration of sweep events is increasing slowly and gradually wallwards.

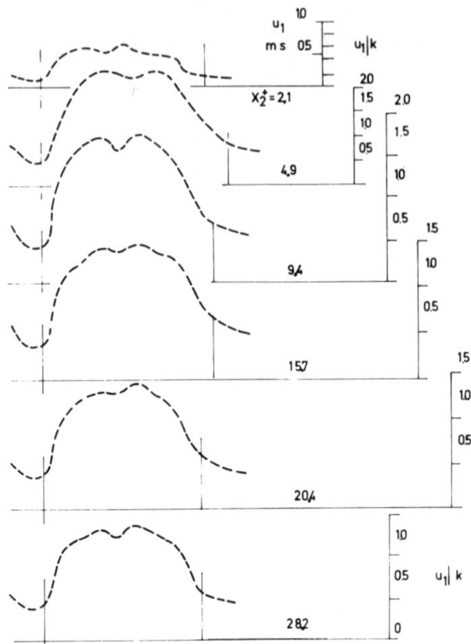

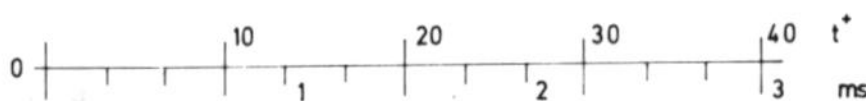

Fig. 9. Patterns recognized during a sweep event (S) at different x_2^+.

As assumed, pattern lenghts vary considerably from one to another event of the same kind. In Fig. 12, nondimensional frequencies of occurence fb/ u_o , of a given pattern lenght, at a given x_2^+, are presented for the ejection and the sweep events, separately. These pattern lenght probability distributions are similar to the ones observed by Wallace et. al. /3/. It is seen that while average pattern lenghts for an ejection event are from 17 to 27 in t^+ (fig. 11) some of the observed patterns are from 100 to 200 in lenght, expressed in t^+. Probability distributions of the pattern lenghts of sweep events are even more skewed.

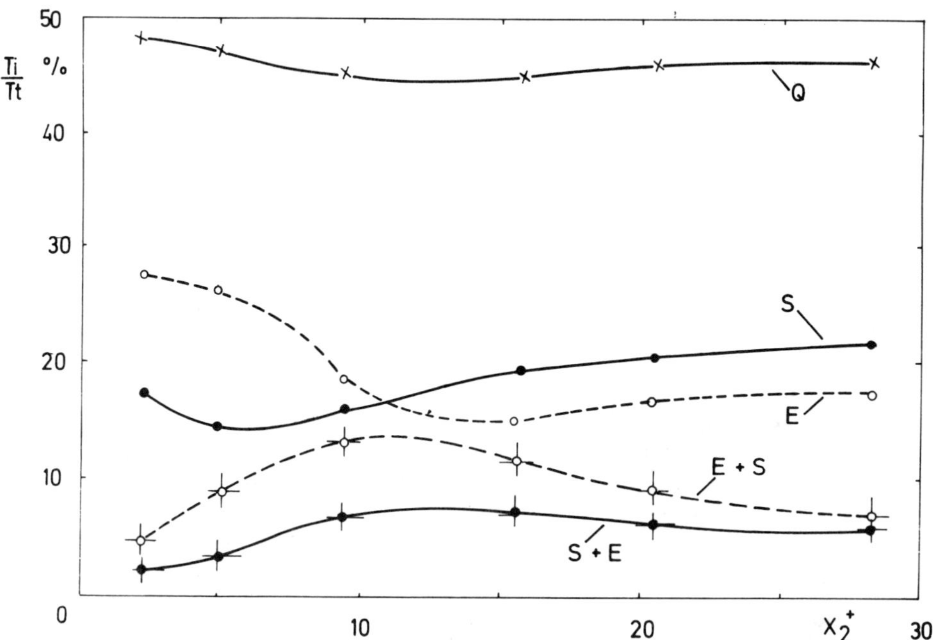

Fig. 1o. Percentages of the total time of the signal (Ti/Tt) corresponding to different events. E - isolated ejections; S - isolated sweeps; E + S - ejections ending up into sweeps; S + E - sweeps ending up into ejections; Q - quiescent periods.

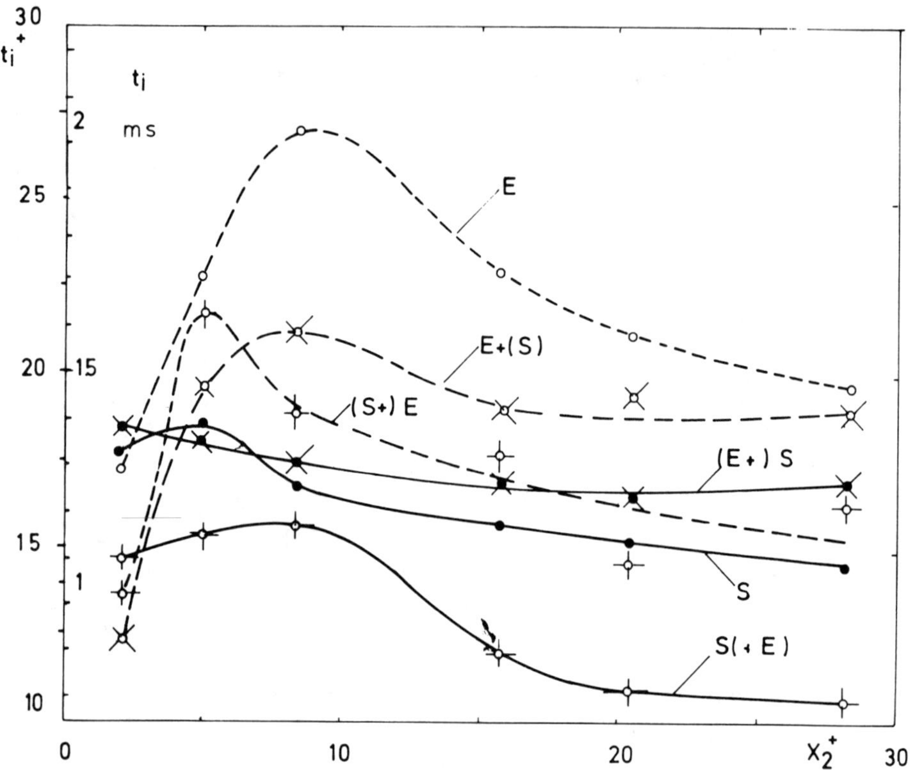

Fig. 11. Pattern lengthts of different events. E - ejection; S - sweep; E + (S) and (S) + E - ejection parts of the ejection - sweep and the sweep - ejection event; (E)+S and S + (E) - sweep part of the ejection - sweep and the sweep - ejection event.

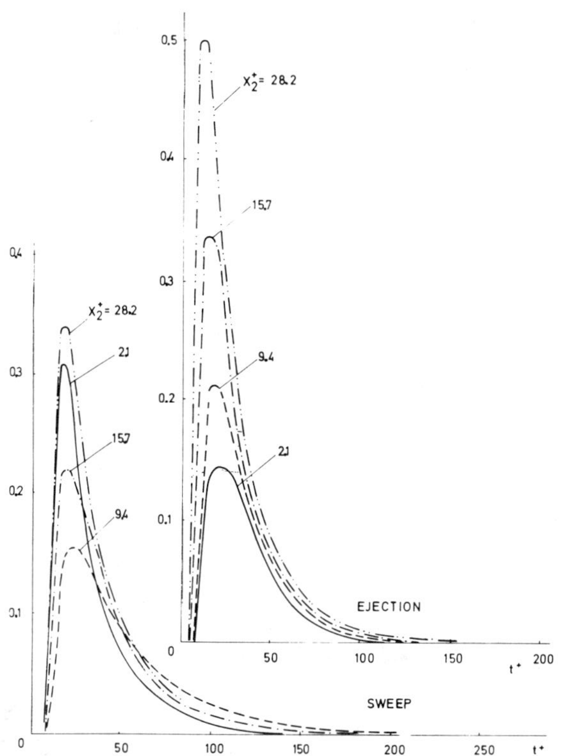

Fig. 12. Probability distribution of pattern lenghts (t^+) of sweep and ejection events.

3. DISCUSSION

The results regarding observed patterns during the bursting events obtained in this study differ significatly from those obtained by Wallace et.al. /3/ and by Blackwelder and Kaplan /4/. It has to be noted that the pattern detection procedure in this study differs in a number of aspects from those employed in /3/ and in /4/. First, it differs from procedures in both of these studies in looking separately for an ejection and for a sweep event and not for a single pattern during a burst. Then, it differs from the procedure employed in /3/ in applying a certain treshold level to the detection criteria. Finally, it differs from the procedure employed in /4/ in choosing different treshold levels in function of the type of event and on the distance from the wall.

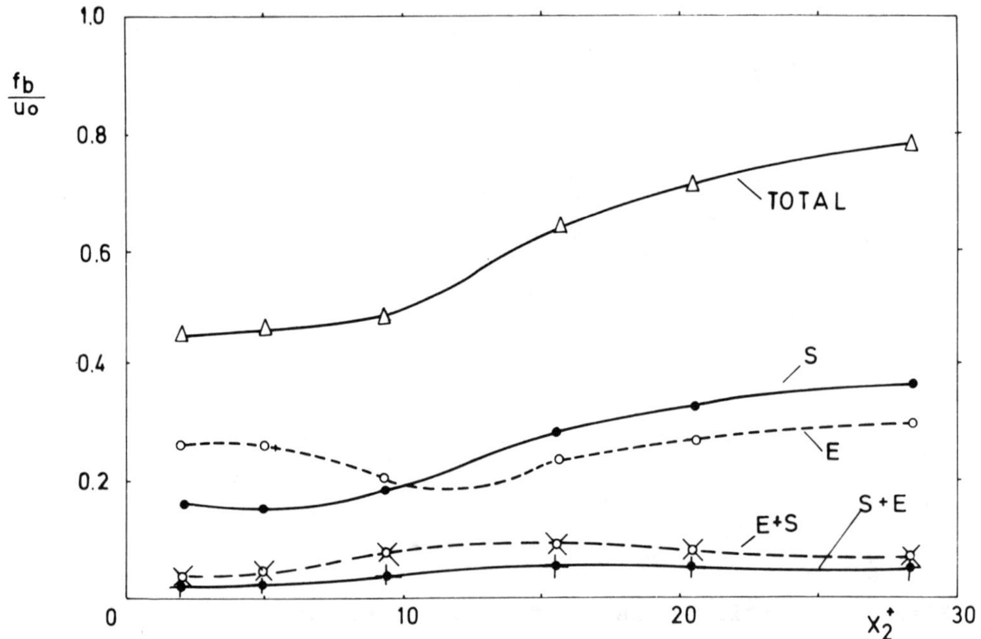

Fig. 13. Frequencies of occurrence of different intermittent events as well as of the sum of all of them in function x_2^+. Notations as in Fig. 1o.

Patterns recognized by Wallace et.al. /3/, and shown in Fig. 1, are very similar in shape to the (E + S) patterns recognized in this study. Maximum amplitudes of the patterns are, however, significantly lower than those observed here (Fig. 6). Patterns recognized in /3/ were observed to account for 65% of the time of the signal. However, under the condition that positive time gradient is in absolute value twice as strong as the negative, those patterns account for only about 18% of the time at $x_2^+ = 15$ and 5% of the time at $x_2^+ = 3o$ /3/. In this case the agreement with the lenght of the (E + S) pattern recognized in this study is fair (see Fig. 1o). It seems, then, that the patterns recognized in /3/ with no restrictions regarding the intensity of the time derivative include also some of the (S) and even some of the (E) patterns recognized in this study. However, apparently not all of them since (S) parts of the (S + E) patterns, some of the (E) patterns having stronger negative than positive time derivative, and most probably, second parts of the (S) events, are rejected by the criteria in /3/ (see Figs. 7 - 9). This is also sustained by the fact that patterns in /3/ (Fig.1) have in general lower amplitudes than the patterns recognized in this study which might be explained as being a consequence of ensemble averaging some events and some non-events, as defined by the criteria in this study.

Vigorous events leading to S shaped patterns detected by Blackwelder and Kaplan /4/ as well as by Chen and Blackwelder /5/ correspond most probably to the (E + S) events as well as to the (S) events detected in this study (Figs. 6 and 9). It is seen from Fig. 1o, that these two events account for about 3o% of the time at $x_2^+ = 15$ while it is estimated in /4/ that S shaped events account for about 25% of the total time. Amplitude intensities of the S shaped events in /4/ are somewhat lower than those in Fig. 9, probably because of the ensemble averaging (E+S)and(S) events together. The average frequency of occurrence of S shaped events, normalized by boundary layer thickness and the maximum velocity, is estimated in /4/ to amount to o.1. Frequencies of occurrence of different events as detected in this study are presented in Fig. 13m normalized by the half width of the channel (b) and the maximum velocity (U_o). It is seen that the normalized frequency of occurrence of (E + S) events alone, detected in this study is about o.1 at $x_2^+ = 15$ and of (S) and (E + S) events together amounts to between o.2 and o.4, depending on x_2^+. Chen and Blackwelder /5/ are also estimated frequencies of detection of what they call "internal temperature fronts", corresponding to the inrushes of the cold fluid wallwards. Normalized frequencies detected in /4/ were about o.25 at $x_2^+ = 35$ and up to o.4 further from the wall. This is in better agreement with the findings in this study (Fig. 13). Finally, it should be noted that sharp temperature increases corresponding to ejections of warm fluid, were also detected in /5/, but since they were not spatially correlated they were rejected as random fluctuations

On the other hand, it appears that the results of this study are in good agreement with the concept of latteraly shifted occurrence of sweeps and ejections, as indicated by a number of visualization studies mentioned in the introduction /6,7,8,9/ A stationary probe with sufficiently small latteral dimensions, would then at a time encounter either a sweep or an ejection. In this respect it should be noted from Fig. 13 that very close to the wall, ejections are detected almost twice as often than sweeps. This is in very good agreement with the notion of vigorous sweeps close to the wall pushing slow fluid sidewise and forming low velocity streaks which are eventually rized outwards. Outside the viscous sublayer sweeps become more frequent, in agreement with the findings in /5/. As for the appearance of (E + S) and (S + E) events, these could be explained in two different ways. All the events are apparently three-diemensional so that a probe detecting an ejection might in the next instance detect a sweep because of the latteral shift of the structures. It might well be also, that a new sweep encounters an already developing ejection, or vice versa. This is sustained by the finding that the composed events occur most often around $x_2^+=15$, where the turbulence production is at a maximum /Fig. 13/. Also comparing Figs. 6 and 7, corresponding to composed events, with Figs. 8 and 9, corresponding to single (E) and (S) events it is seen that in the case of (E + S) event the pattern lenght of the ejection part is shorter than the pattern lenght of a single (E) event at the same x_2^+, and in the case of a (S + E) event the pattern lenght of the sweep part is shorter than the corresponding pattern lenght of a single (S) event. This is also seen from Fig. 11. Thus, it appears that the ejection part in case of the (E + S) event, as well as the sweep part in case of the (S + E) event, did not have time to develop because of a new sweep or ejection, respectively, arriving at the probe.

It could be concluded, then, from the evidence presented, that the conditional pattern recognition procedure introduced in this study, represents a valuable tool for quantitative studies of the intermittent events in turbulent shear flows, presumably caused by coherent structures.

REFERENCES

1. Blackwelder, R.F., 1978. Pattern recognition of coherent eddies. Proc. Dynamic Flow Conf. 1978. pp. 173-19o.

2. Zarić,Z. Physical evidence on coherent structures in the wall layers of turbulent flows. Turbulent Forced convection in channels and bundles. Hemisphere. pp. 4o3-44o.

3. Wallace, J.M. Brodkey, R.S. and Eckelmann,H. 1977. Pattern recognized structures in bounded turbulent shear flows. J.Fluid Mech. Vol. 83, pp. 673-693.

4. Blackwelder,R.F. and Kaplan,R.E. 1976. On the wall structure of the turbulent boundary layer. J.Fluid Mech. Vol. 76, pp.89-112.

5. Chen,C-H.P. and Blackwelder,R.F. 1978. Large-scale motion in a turbulent boundary layer: a study using temperature contamination. J.Fluid Mech. Vol. 89, pp.1-31.

6. Kline,S.J.,Reynolds,W.C.,Schraub,F.A. and Runstadler,P.W. 1967. The structure of turbulent boundary layers. J.Fluid Mech. Vol.3o, pp. 741-775.

7. Grass,A.J. 1971. Structural features of turbulent flows over smooth and rough boundaries. J.Fluid Mech. Vol. 5o, pp. 233-255.

8. Oldaker,D.K. and Tiederman,W.G. 1977. Spatial structure of the viscous sublayer in drag-reducing channel flows. Phys. Fluids. Vol. 2o, pp. S133-S144.

9. Hirata,M. and Kasagi,N. 1979. Studies of large-eddy structures in turbulent shear flows with aid of flow-visualization techniques. Studies in heat transfer. Hemisphere-McGraw Hill,N.York. pp. 145-164.

1o. Falco,R.E. 1977. Coherent motions in the outer region of turbulent boundary layers. Phys. Fluids. Vol.2o,pp.S124-S132.

11. Elena,M.,Fulachier,L. and Dumas,R. 1979. Etude Experimentale des apports et des ejections de fluide dans une couche limite turbulentte. AGARD Conference preprint No 271.

12. Zarić,Z. 1974. Statistical analysis of wall turbulence phenomena. Advances in Geophysics. Vol. 18A, pp. 249-261. Academic Press.

13. Zarić,Z. 1974. Etude statistique de la turbulence parietale. Boris Kidrič Inst. Beograd.

14. Zarić,Z. 1975. Wall turbulence structure and convective heat transfer. Int.J.Heat Mass Transfer. Vol. 18, pp. 831-842.

15. Khabakpasheva,E.M. and Mihailova,E.S. 1976. Issledovanie turbulentnosti pri techenii vodi v ploskom kanale. Eksperimental'$_{noe}$ issledovanie strukturi turbulentnosti. Novosibirsk. pp. 33-57.

16. Kreplin,H-P. 1976. Experimentelle der Längsschwankungen und der wandparalellen Querschwankungen der Geschwindigkeit in einer turbulenten Kanalstömung. Thesis. Univ. Göttingen.

17. Zarić,Z. 1972. Distribution de probabilite des vitesses et des tempoeratures pres d'une paroi. C.R. Acad. Sci.Paris. Vol. A275, pp.459-462.

Measurement of the Structure of the Streamwise Vorticity Field in a Turbulent Boundary-Layer

J. M. WALLACE and P. VUKOSLAVČEVIĆ*
Department of Mechanical Engineering
The University of Maryland
College Park, Maryland 20742, USA

ABSTRACT

A survey of measurements of vorticity in bounded turbulent shear flows is presented. It is shown that the measured rms streamwise vorticity is systematically too large by as much as 45% when measured with a four-sensor Kovasznay-type array. Instantaneous values can be much more in error. Therefore this probe should not be used to detect instantaneous vortical structures. New attempts to measure vorticity are described.

1. INTRODUCTION

For a turbulent flow, the time averaged equations of motion, of energy and of continuity of a species in a binary system contain additional terms which are introduced by using Reynolds decomposition of the variables of the equations into their mean and fluctuating parts. These terms, the turbulent momentum, energy and mass fluxes, require additional equations relating them to the mean variables of the original equations in order to have a determinate system. This, of course, is the well known closure problem of turbulence which has engendered so much effort over the last century. Unfortunately the physical laws which would establish these relations are unknown. Theories have been developed, analogous to molecular diffusion processes, for the turbulent diffusion of momentum, heat and mass. These utilize the so-called eddy viscosity, eddy thermal conductivity, and eddy diffusivity and are widely used for flows of relatively simple geometry in engineering design. The problem with these attempts are, of course, that these diffusivities are not properties of the fluid but vary over its spatial extent and from one flow geometry to another.

A renewed effort to understand the physical laws governing the dynamics of these flows was stimulated by visual observations of both bounded and free turbulent shear flows, made in the 1960's--most notably by research groups at Stanford and Ohio State Universities and at the California Institute of Technology. They clearly established that turbulent flows, far from being completely random, show a peculiar superposition of randomness and order. In the region near solid boundaries, where turbulence dissipation and generation is most intense, the flows exhibit ordered sequences of coherent motion which can be related to the creation of high Reynolds stresses and to the production and dissipation of turbulent kinetic energy. In free turbulent shear flows large scale rotating and coalescing eddies have been visually observed and studied as possible sources of acoustic noise.

*Present address: University "Veljko Vlahović," Titograd, Yugoslavia.

The ordered structures in bounded flows occur, however, at random locations in space and at random times as seen by an observer at a fixed Eulerian position. Thus a fixed point flow property measurement using hot-wire or hot-film anemometry probes will generate a signal which appears to have the characteristics of a random function because the probe is continually intersected by these flow structures at random spatial locations within their region of coherency and at random times in their evolution. On the other hand, visual data has been difficult to quantify in any systematic manner. Thus research groups have been motivated to develop detection schemes to educt from these fixed point signals the quantitative characteristics of the flow structures. The hope has been that enough information can be obtained about the structures so that a physical model of these dynamical processes can be developed which would lead to a mathematical description to be used in modelling the unclosed terms in the mean transport equations described above. This has proved to be a demanding task, particularly for bounded flows, i.e. boundary layers and pipe and channel flows.

A physical model which appears to incorporate most of the experimental evidence in bounded flows is the stretched, hairpin-like vortex near the wall. This has been suggested by Theodorsen[1], Willmarth and Tu[2], Bakewell and Lumley[3], Kline, Reynolds, Schraub and Runstadler[3a] and by Townsend[4]. In this model, vortex lines generated at the wall are lifted away from it and stretched downstream by the mean shear. The initial spanwise orientation of the vortex lines is locally reoriented, principally in the streamwise direction, resulting in a counter-rotating pair of vortices which mutually induce their motion further outward. The vortices tend to gather low momentum fluid near the wall and lift it outward producing high momentum flux and thus, high Reynolds stress. Experiments by Blackwelder and Eckelmann[5] and by Kastrinakis, Wallace, Willmarth, Ghorashi and Brodkey[6] have added strong direct evidence for the existence of these counter-rotating pairs of vortices. If this physical model is conclusively established, it should give us a much clearer understanding of the processes which transport momentum, heat and mass.

A new turbulence closure developed by Bernard[7], which is based on a vorticity description of the flow field, has provided further motivation for developing better means of measuring the vorticity field in turbulent flows. The method used to close the mean and covariance vorticity equations can be extended to the energy and mass concentration equations. The experimental verification of the closure approximations requires measurements of fluctuating vorticity components and velocity gradient terms. This work is currently being carried out at the University of Maryland.

2. SURVEY OF PREVIOUS VORTICITY MEASUREMENTS

Thirty years ago Kovasznay[8,9] proposed a four-sensor hot-wire probe which would be sensitive to the streamwise velocity and vorticity components. A sketch of this probe is shown in Figure 1 which illustrates that the wires of the probe form the legs of a Wheatstone Bridge when operated with a constant current. The wires are supported by four prongs so that adjacent wires have a common prong. The voltage difference measured between points A-C is sensitive to the streamwise velocity and that measured between points B-D is sensitive to the streamwise vorticity. Probes of this configuration have known infrequent use because of the difficulty in constructing them small enough to resolve the smallest scales of vorticity in the flow while maintaining the geometrical and electrical symmetry required to prevent spurious measurement of vorticity.

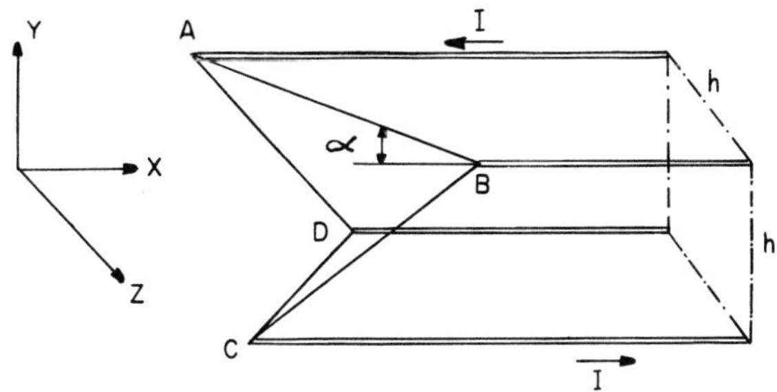

Fig. 1. Sketch of the Kovasznay-type Vorticity Probe

Two attempts have been made to measure the root-mean-square distribution of the streamwise vorticity across bounded shear flows utilizing probes of this type. The earliest was that of Corrsin and Kistler[10] in a turbulent boundary layer with "fully rough" walls. A later attempt was made by Kastrinakis[11] in a low Reynolds number turbulent channel flow with turbulent scales large enough to allow measurements as near to the wall as the edge of the viscous sublayer. Both these sets of measurements are shown in Figure 2 along with values determined by Lighthill[12] from measurements of the dissipation in isotropic turbulence by Schubauer and Klebanoff[13]. The mean squared vorticity was apparently determined from the relation $\overline{\omega^2} = \varepsilon/\nu$. The Corrsin and Kistler data appear to extend the range of the Kastrinakis data and there is remarkable agreement for $y^+ \gtrsim 20$ between the Kastrinakis data and the values determined by Lighthill. This agreement should probably not be taken too seriously in view of the rough nature of Lighthill's estimate and also of the potential sources of error in the other two sets of data to be considered shortly. The dotted line in the figure indicates the trend of the data as it approaches the wall where Kreplin[14] has measured the asymtotic limit of the rms ω_x value with a V-shaped, flush wall element hot-film probe.

A pair of Kovosznay-type probes were employed in the same channel flow used by Kastrinakis to attempt to detect the presence of counter-rotating vortex pairs. The data obtained with the pair of probes located at $y^+ = 15$ and separated by a variable spanwise spacing were analyzed with a pattern recognition algorithim developed by Wallace, Brodkey and Eckelmann[15]. The results were reported by Kastrinakis et al.[6] and are shown in Figure 3. An ensemble averaged time sequence is obtained when the streamwise velocity exhibits a form characterized by a gradual deceleration followed by a strong acceleration. When a pattern of this sort was found in the U-signal, the streamwise vorticity signal at various spanwise separations was simultaneously ensembled averaged over the same intervals of time. The figure shows two ω_x patterns at streamwise separations of 25 and 50 viscous lengths. At a separation of 25 the vorticity pattern is clearly out of phase with the velocity pattern and at 50 its ensemble averaged value is almost zero over the entire time sequence. This result is consistent with the hypothesized spacing of the vortices in the hairpin vortex model described above.

Measurements of the spanwise and normal vorticity components in a turbulent oil channel flow with very large viscous scales have been made by Eckelmann,

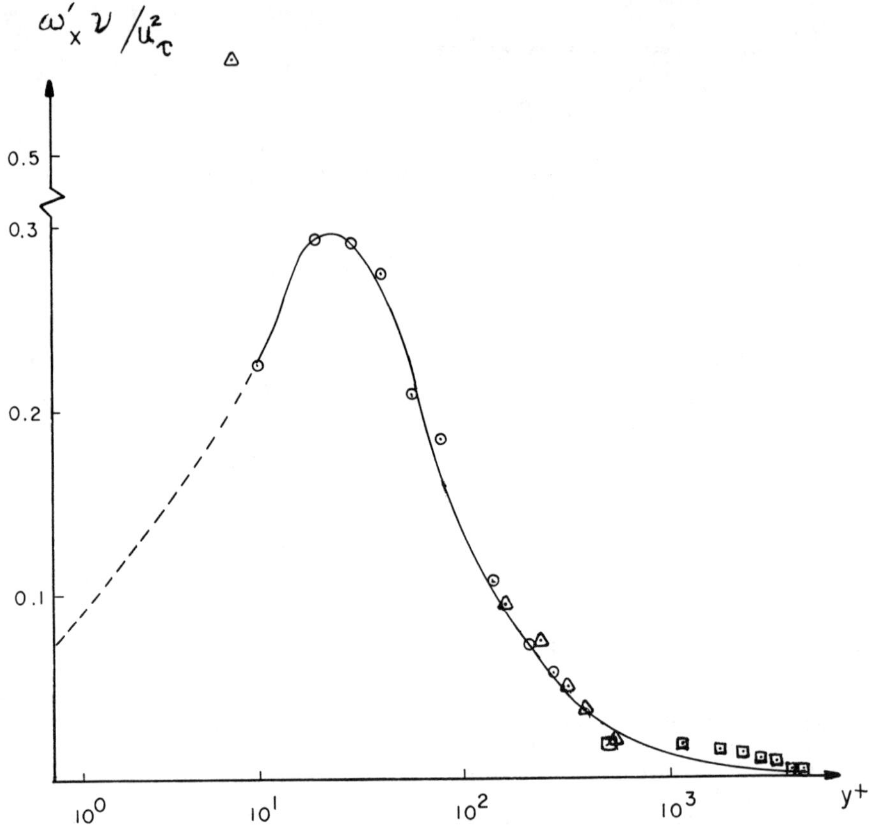

Fig. 2. The rms Distribution of the Streamwise Vorticity. Data of Corrsin and Kistler ⊡ in a "Fully Rough" Turbulent Boundary Layer, Kastrinakis ⊙ in a Turbulent Channel Flow, and Lighthill △ Estimated from Dissipation Measurements in Isotropic Flow.

Nychas, Brodkey and Wallace[16] using five hot-film sensors in another configuration. The cross-stream velocity gradients, $\partial U/\partial y$ and $\partial U/\partial z$, are measured directly and the streamwise gradients, $\partial v/\partial x$ and $\partial w/\partial x$, were approximated using Taylor's hypothesis. In this work the pattern recognition algorithim was again used, and the characteristic vorticity patterns of ω_y and ω_z obtained simultaneously with the detection of the U-pattern described above are shown in Figure 4.

Another probe utilizing a somewhat similar sensor configuration has been developed by Foss. He has utilized it to measure the spanwise vorticity in axisymmetric jet flows.

3. ANALYSIS AND MEASUREMENTS OF ERRORS IN VELOCITY AND VORTICITY MEASUREMENTS WITH MULTI-SENSOR PROBES

All of these probes are subject to serious errors in the flow quantities they purport to measure. Average measurements of odd-powers of the streamwise

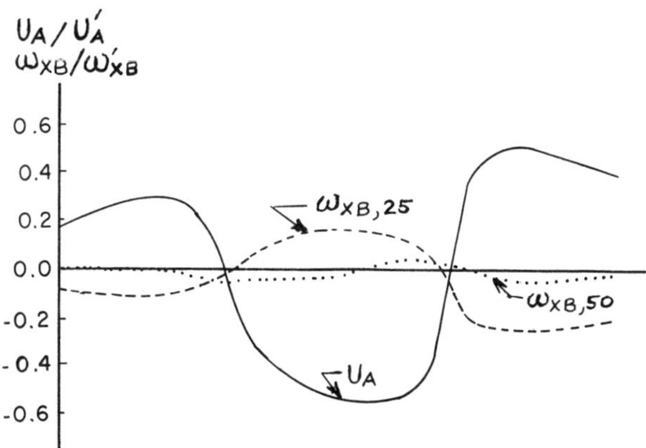

Fig. 3. Simultaneously Ensemble Averaged Pattern Recognized ω_x (Normalized with rms Value); Δz Separation: Solid Line, U Recognized on Probe A; Dashed Line, ω_x with Δz of 25; Dotted Line Same with Δz of 50. Measurements Made at $y^+ = 15$ in a Turbulent Channel Flow.

vorticity with the Kovasznay-type probe should in principle be relatively error free since the error has equal probability of having either sign for all amplitudes. This fact was exploited by Kastrinakis et al.[6]. Instantaneous values and average even powers can be badly in error however. No correction of the error caused by the cross-stream velocity components are possible to make because the probe is incapable, in principle, of measuring these components. Kastrinakis et al.[17] suggest that a way around this dilemma would be to use the same wire configuration but with each wire held by a separate pair of prongs (a total of eight) and to electrically operate each wire independently. In effect, the probe would be two X-arrays in perpendicular planes which are parallel to

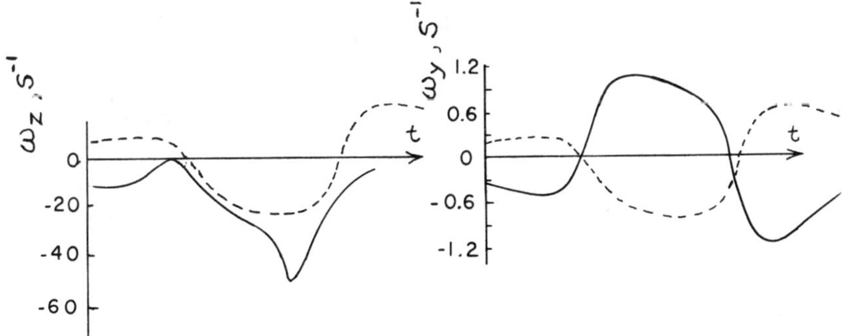

Fig. 4. Simultaneously Averaged Pattern Recognized U, ω_y and ω_z at $y^+ = 30$ in a Turbulent Channel Flow. Dotted Lines are U Pattern.

component normal to the planes containing the wires. These equations can be combined by adding equation (1.3) to (1.1) and (1.4) to (1.2) and subtracting the sums to obtain an expression for the streamwise component of vorticity as given in equation (2).

$$\omega_x = \frac{1}{2h \sin \alpha} \left[\sqrt{U_{N_3}^2 - v_o^2} + \sqrt{U_{N_1}^2 - v_o^2} - \sqrt{U_{N_4}^2 - w_o^2} - \sqrt{U_{N_2}^2 - w_o^2} \right] \quad (2)$$

where, terms containing h^2 are neglected and those containing $\partial v/\partial y$ and $\partial w/\partial y$ approximately cancel when expanded using the binomial theorem by putting the radicals in the form $(1+x)^n$.

If the cross-stream velocity components, v_o and w_o, are neglected in equation (2), the equation reduces to the expression used by the previous workers with the Kovasznay-type probe. By keeping these terms in the equations for streamwise vorticity, their effects can be included if they can be measured with the probe. Equations (1.1) - (1.4) can also be combined by subtracting equation (1.4) from (1.2) and (1.3) from (1.1) to obtain

$$w = \frac{1}{2 \sin \alpha} \left[\sqrt{U_{N_1}^2 - v_o^2} - \sqrt{U_{N_3}^2 - v_o^2} - h \frac{\partial U}{\partial y} \cos \alpha \right] \quad (3)$$

$$v = \frac{1}{2 \sin \alpha} \left[\sqrt{U_{N_4}^2 - w_o^2} - \sqrt{U_{N_2}^2 - w_o^2} - h \frac{\partial U}{\partial z} \cos \alpha \right] \quad (4)$$

It is seen that the expressions for v and w contain, in addition to the normal cooling components, in the form in which they normally appear for X-arrays, the v and w components themselves as well as gradients of the streamwise velocity components. If these additional terms are neglected, as is normally done for X-arrays, equations (3) and (4) become the difference equations for the cross-stream velocity components. This is not necessary, however, since it is possible to solve this pair of equations by an iterative process which is quite convergent.

The probe was tested in a uniform irrotational flow by pitching and yawing it through a range of angles, and by varying the flow speed. If the cross-stream velocity components are neglected in equation (2), a spurious vorticity is found as had previously been shown by Kastrinakis et al.[17] If the cross-stream velocities, which for this gradient free flow are very accurately measured, are included in the determination of ω_x from equation (3), the spurious vorticity measured is reduced to that due to random measurement error alone.

Unfortunately in a turbulent shear flow the gradients appearing in equations (3) and (4) can be expected to be substantial. We have made estimates, based on measurements reported by Eckelmann[16] et al. in a similar low Reynolds number flow, that maximum values of the gradients of the order of the mean stress at the wall can occur. That these gradients, which normally are neglected in the operation of X-arrays, have a serious effect has been indicated by Willmarth and Bogar[19]. To investigate this further we built a two sensor gradient probe with parallel wires spaced the same distance apart as the wires of the four sensor probe. The maximum gradients of the fluctuating streamwise velocity components, $\partial U/\partial y$ and $\partial U/\partial z$, at various y^+ locations across the boundary layer were obtained [see Vukoslavčević and Wallace[20]] by observing the maximum fluctuating voltage difference between the two wires on a storage oscilloscope. Indeed, as had been estimated, the maximum gradients are of the order of the wall shear stress; $\partial U/\partial y$ has a maximum value of about $0.9\ \partial U/\partial y|_{wall}$

the mean flow. Under the conventional assumptions used for the operation of X-arrays, the three velocity components could then be measured.

Cleveland[18] built a probe of this type, shown schematically in Figure 5, and gave an expression for the streamwise vorticity sensitivity of the probe under the conditions that the cross-stream velocity components are neglected. This expression has been tested in a uniform irrotational flow by spinning the probe at various angular velocities when aligned with the mean flow direction and comparing these induced vorticity values to those calculated from the expression he derived. The results confirmed the analysis.

We have built another probe, of this type with extremely good geometrical symmetry and with dimensions that satisfy Wyngaard's[19] criteria for good spatial resolution, to investigate the effects of cross-stream velocity components. An expansion of the velocity about the geometric center of the four sensors in a Taylor's series, truncating second and higher order terms, leads to the expressions of equations (1.1) - (1.4) giving the components of the total velocity vector normal to each wire in terms of the three cartesian velocity components at the array center, their gradients, and the angle α and the length h shown in Figure 5.

$$U_{N_1}^2 = \left[\left(U_o + \frac{\partial U}{\partial y}\frac{h}{2} \right) \cos\alpha + \left(w_o + \frac{\partial w}{\partial y}\frac{h}{2} \right) \sin\alpha \right]^2 + \left(v_o + \frac{\partial v}{\partial y}\frac{h}{2} \right)^2 \quad (1.1)$$

$$U_{N_2}^2 = \left[\left(U_o - \frac{\partial U}{\partial z}\frac{h}{2} \right) \cos\alpha - \left(v_o - \frac{\partial v}{\partial z}\frac{h}{2} \right) \sin\alpha \right]^2 + \left(v_o - \frac{\partial w}{\partial z}\frac{h}{2} \right)^2 \quad (1.2)$$

$$U_{N_3}^2 = \left[\left(U_o - \frac{\partial U}{\partial y}\frac{h}{2} \right) \cos\alpha - \left(w_o - \frac{\partial w}{\partial y}\frac{h}{2} \right) \sin\alpha \right]^2 + \left(v_o - \frac{\partial v}{\partial y}\frac{h}{2} \right)^2 \quad (1.3)$$

$$U_{N_4}^2 = \left[\left(U_o + \frac{\partial U}{\partial z}\frac{h}{2} \right) \cos\alpha + \left(v_o + \frac{\partial v}{\partial z}\frac{h}{2} \right) \sin\alpha \right]^2 + \left(w_o + \frac{\partial w}{\partial z}\frac{h}{2} \right)^2 \quad (1.4)$$

These are exact expressions for the normal cooling velocity components on each wire including (the last term in each expression) the effect of the velocity

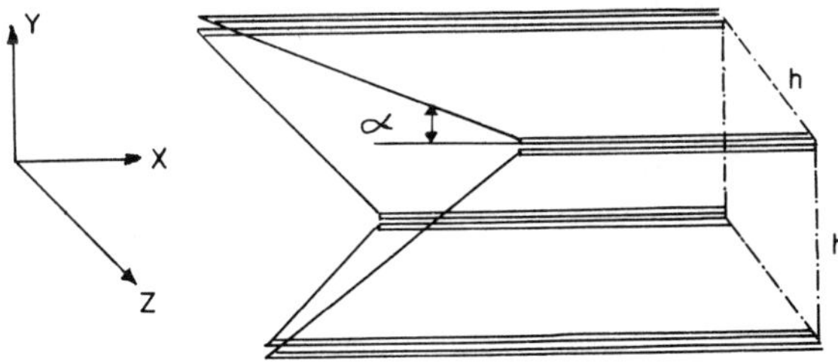

Fig. 5. Sketch of a Probe to Measure Velocity and Streamwise Vorticity. Each Wire is Supported by a Separate Pair of Prongs and Is Operated at a Constant Temperature.

at $y^+ = 20$ and $\partial U/\partial z$ is about $0.75\ \partial U/\partial y|_{wall}$. We made an estimate of the effects these gradients can have on the measured vorticity for worse test cases. The cases chosen for calculation of the maximum errors correspond to instances when w has its maximum instantaneous value, v, ω_x and $\partial U/\partial z$ are zero, the instantaneous streamwise velocity has its mean value, and $\partial U/\partial y$ is varied up to its maximum at various distances from the wall. The maximum values of w used in the calculation were taken from Kreplin and Eckelmann[21]. If v is exchanged with w and $\partial U/\partial z$ with $\partial U/\partial y$, similar large errors will be calculated for v as is easily seen by considering equations (3) and (4).

The error in measured values of ω_x which would result from these cases is plotted in Figure 6 as a function of the maximum instantaneous wall gradient. The curves indicate that the maximum instantaneous error, at locations near the wall, can be considerably greater than the maximum root-mean-square values of ω_x shown in Figure 2. We also measured the maximum values of streamwise vorticity occurring in the boundary layer at various y^+ locations by measuring the combination of instantaneous maximum values of the bridge voltages of the

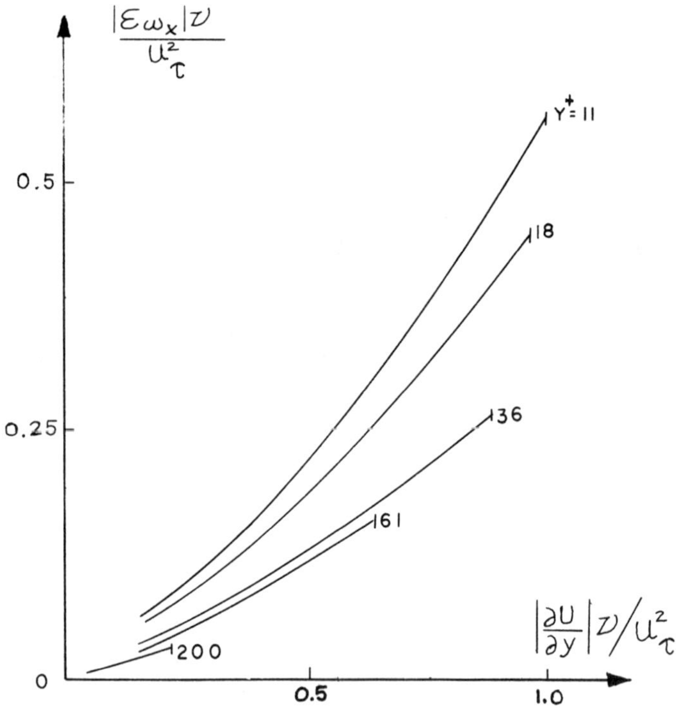

Fig. 6. Calculated Error in the Streamwise Vorticity, ε_{ω_x}, Due to Neglecting the Effect of the Gradient $\partial U/\partial y$ in Determining v and w, as a Function of $\partial U/\partial y$ at Various y^+ Positions. $R_\theta = 2{,}810$.

four wires

$$\Delta E = E_3 + E_1 - E_4 - E_2 \tag{5}$$

on a storage oscilloscope. For the cases where v and w are equal, ΔE and the calibration relations can be used to find values for the normal cooling velocity components in equation (2). Thus, for this case, the maximum of ω_x are measured fairly accurately. If v and w are not equal, two error bands can be drawn about the measured maximum streamwise vorticity. One of these, shown in Figure 7, represents the maximum errors which occur due to neglecting v and w in equation (2). If they are not neglected, the second error band shown represents the maximum errors due to the neglect of $\partial U/\partial y$ in determining v and w as previously discussed. It is clear from the figure that for $y^+ \lesssim 100$, the errors in the maximum values can be very large indeed. Of course, if the instantaneous vorticity is less than the maximum, these relative errors become even larger.

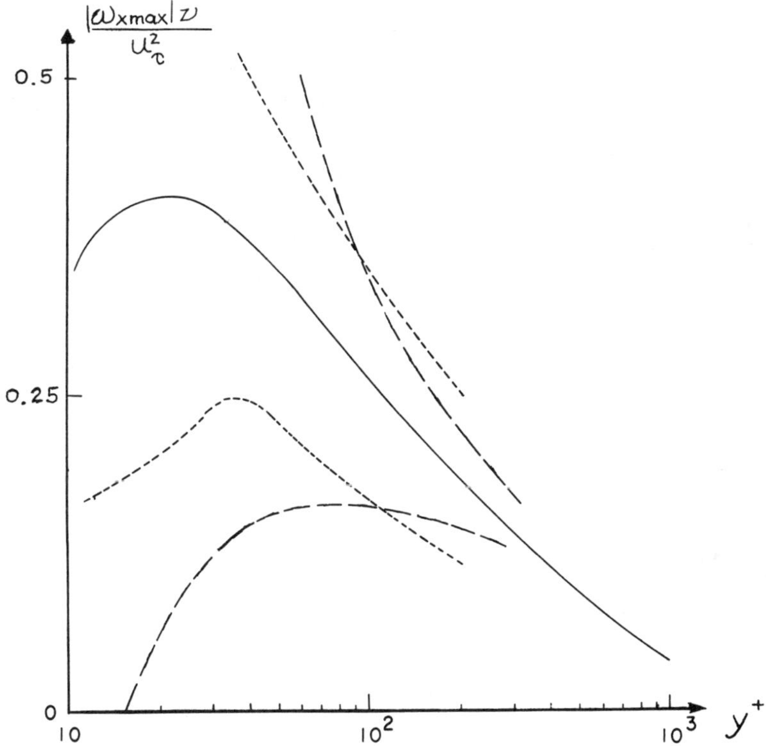

Fig. 7. Measured Maximum Values of the Streamwise Vorticity, ω_x, if v and w Are Equal, as a Function for y^+ (x). Error Band if w is Maximum And v is Zero but they are Neglected (---). Error Band if w is Maximum and v is Zero and They Are Not Neglected but $\partial U/\partial y$ is Maximum And is Neglected in Determining v and w (— —). R_θ = 2,810.

4. MEASUREMENTS WITH A THREE-SENSOR HOT-WIRE ARRAY

We wished to assess the effect of these instantaneous errors due to neglecting the cross-stream velocity components on the measured values of the root-mean-square ω_x. A new probe utilizing three sensors was constructed and is shown schematically in Figure 8. Two of the sensors are in the horizontal plane and the third is in the vertical plane. The probe was tested in a similar manner to the four-sensor probe described above, and data was recorded on a FM tape recorder and subsequently digitized at a rate of 1280 Hz. Again, the tests showed that the velocity components could be measured very accurately in a gradient free flow with this probe; however, it is subject to the same errors due to velocity gradients as any other probe which cannot measure these gradients and account for their affects on the velocity. In this case, however, the sensor wires are approximately 0.6 mm long, i.e. somewhat shorter than those of the four-sensor array, and the distance between the wire centers is considerably shorter. A binomial expansion of the terms under the radicals in equations (2) yields an expression for the streamwise vorticity in a form which gives explicitly the error due to the neglect of the cross-stream components and is given by

$$\omega_x = \frac{1}{2h \sin \alpha}\left[\left(U_{N_1} + U_{N_3} - U_{N_2} - U_{N_4}\right) + \frac{1}{2}\left(\frac{w^2}{U_{N_2}} + \frac{w^2}{U_{N_4}} - \frac{v^2}{U_{N_1}} - \frac{v^2}{U_{N_3}}\right)\right] \quad (6)$$

If the vertical wire and one of the horizontal wires are treated as two adjacent legs of a geometrically and electrically symmetric four-sensor array and the velocity components measured with the probe are used, an expression for the error, ε_{ω_x}, can be obtained in terms of the normal cooling components on the wires and the measured velocities. This expression is given as

$$\varepsilon_{\omega_x} = \frac{1}{4h \sin \alpha}\left[\frac{w^2}{U_{N_2}} - \frac{w^2}{U_{N_2} - 2v \sin \alpha} - \frac{v^2}{U_{N_3}} - \frac{v^2}{U_{N_3} + 2w \sin \alpha}\right] \quad (7)$$

The probability density distribution of this error was determined at $y^+ \sim 20$

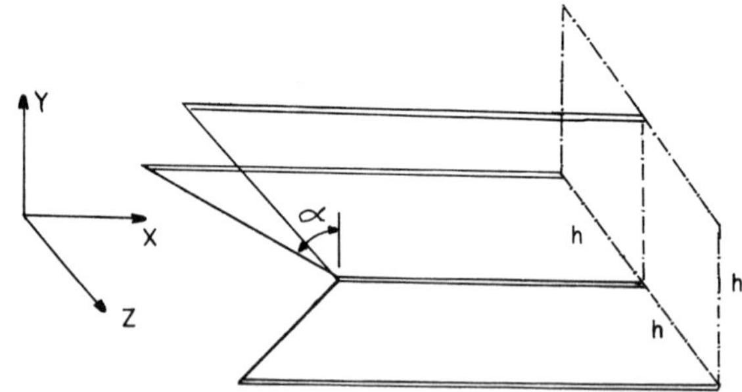

Fig. 8. Three-sensor Probe to Measure Three Velocity Components Simultaneously

and is shown in Figure 9. It is seen that the error is usually very small. However, as expected, errors even larger than the mean shear at the wall can infrequently occur, and they accumulate in the rms measurement. At this location the root-mean-square error, $\sqrt{\overline{\varepsilon_{\omega_x}^2}}\ \nu/U_\tau^2$, is about 0.13 $\partial\overline{U}/\partial y|_{wall}$, i.e about 45% of the measured value in Figure 2.

It is clear then, that the values shown in Figure 2 of the root-mean-square streamwise vorticity are systematically too large. We plan to measure this distribution with the four-sensor probe described above and correct the measurements with the measured error term described here. The four-sensor probe should not be used in experiments where conclusions about instantaneous flow structures are hoped to be drawn. The instantaneous errors in vorticity measured with this probe can be, albeit infrequently, very much in error, and there is no way to determine when this occurs.

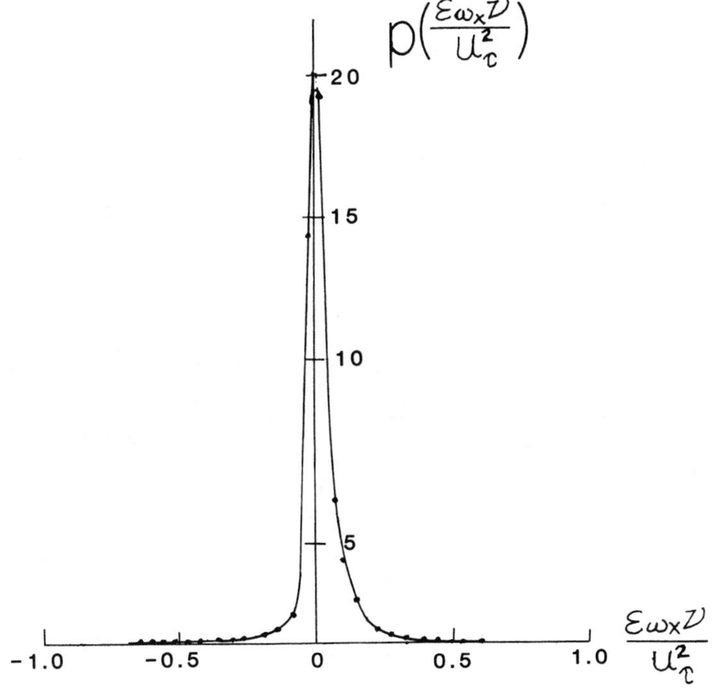

Fig. 9. Probability Density Distribution of the Error in Measuring Streamwise Vorticity with a Kovasznay-type Vorticity Probe Due to the Influence of Cross-stream Velocity Components. a) Smaller Scale. R_θ = 3,970

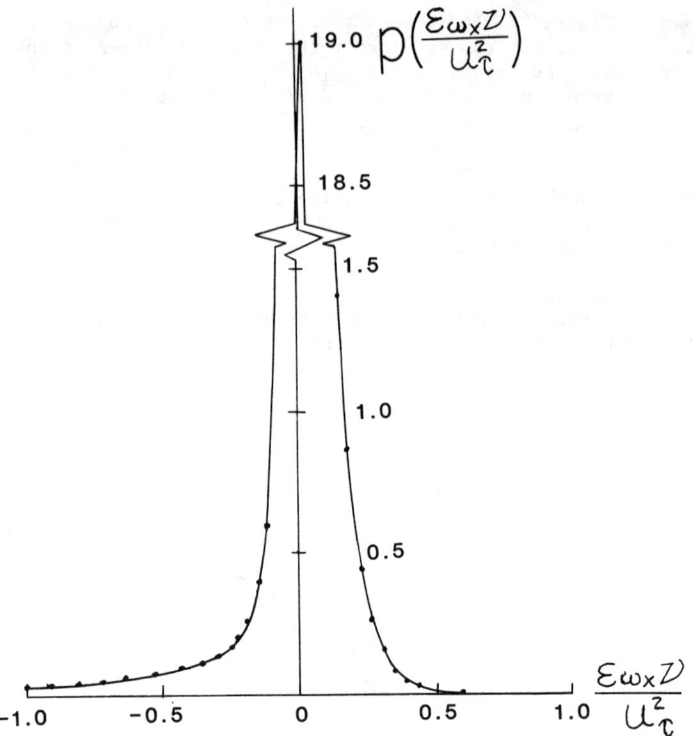

Fig. 9. b) Larger Scale

5. CONCLUSIONS

Although the need for measuring vorticity components in turbulent shear flows has long been seen as an experimental goal of very high priority, there is a paucity of actual results. This is largely due to the great technical difficulty of building probes small enough to adequately resolve the scales of vorticity in turbulent flows and which produce a measurable signal which can be shown to only be sensitive to vorticity. Probes small enough to resolve these scales have been built and are described in this paper, but inherent errors in their measurements have been uncovered and evaluated.

Other means of measuring the components of vorticity are currently being developed. Wassman and Wallace[22] have constructed a miniature probe utilizing nine hot-wire sensors so that the first-order differences in velocity can be measured. In principle, this probe should be capable of obtaining all the information necessary to unambiguously resolve the instantaneous velocity and vorticity fields. It is currently being tested. Problems with hydrodynamic interference with the flow have been encountered, but it appears that they can be overcome. A completely new method which makes use of tiny mirrors embedded in microscopic polymer spheres suspended in the flow has also been developed by Frish and Webb[23]. This optical method shows great promise of yielding usable vorticity measurements. Until such measurements are possible, it is likely that a deeper understanding of the mechanisms of momentum, heat, and mass transfer will be delayed.

ACKNOWLEDGEMENTS

Support of this work by NSF Grant CME-7919835 and NASA Grant NAG1-24 is gratefully acknowledged as is support for Mr. Vukoslačević as a Fulbright Scholar by the Institute of International Education. Special thanks are due Judy Clark for typing the manuscripts under tight time constraints.

REFERENCES

1. Theodorsen, T. (1952), Proc. 2nd Midwestern Conf. Fluid Mech.

2. Willmarth, W.W. and Tu, B.J. (1967), Phys. Fluids 10, S134.

3. Bakewell, H.P. and Lumley, J.L. (1967), Phys. Fluids 10, 1880.

3a. Kline, S.J., Reynolds, W.C., Schraub, F.A. and Runstadler, P.W. (1967), J. Fluid Mech. 30, 741.

4. Townsend, A.A. (1970), J. Fluid Mech. 41, 13.

5. Blackwelder, R. and Eckelmann, H. (1979), J. Fluid Mech., 94, 577.

6. Kastrinakis, E.G., Wallace, J.M., Willmarth, W.W., Ghorashi, B. and Brodkey, R.S. (1978), Lecture Notes in Physics 75, Springer-Verlag, 175.

7. Bernard, P. (1980), SIAM J. Appl. Math., 38, 81.

8. Kovasznay, L.S.G. (1950), Quart. Prog. Rep. Cont. NORD-8036-JHB-3D, The Johns Hopkins Univ.

9. Kovasznay, L.S.G. (1954), Art. F-2 in High Speed Aerodynamics and Jet Propulsion, Princeton Univ. Press, IX, 227.

10. Corrsin, S. and Kistler, A.L. (1955), NACA Rep. 1244.

11. Kastrinakis, E.G. (1977), Bericht 5/1977 MPI f. Strömungsforschung, Göttingen.

12. Lighthill, M.J. (1963), Chap. 2 of Laminar Boundary Layers ed. by L. Rosenhead, Oxford Univ. Press.

13. Schubauer, G.B. and Klebanoff, P.S. (1956), NACA Rep. 1289.

14. Kreplin, H.-P. (1976), Mitt. MPI f. Strömungsforschung u. Aerodyn. Versuchsanstadlt, Göttingen, 63.

15. Wallace, J.M., Brodkey, R.S. and Eckelmann, H. (1977), J. Fluid Mech. 83, 673.

16. Eckelmann, H., Nychas, S.G., Brodkey, R.S. and Wallace, J.M. (1977), Phys. Fluids 20, S225.

17. Kastrinakis, E.G., Eckelmann, H. and Willmarth, W.W. (1979), Rev. Sci. Instruments 50, 759.

18. Cleveland, W.G. (1979), M. S. Thesis, the Univ. of Maryland.

19. Wyngaard, J.C. (1969), J. Sci. Instruments 2, 983.

20. Vukoslavčević, P. and Wallace, J.M. (1980), submitted to Rev. Sci. Instruments.

21. Kreplin, H.-P. and Eckelmann, H. (1979), Phys. Fluids 20, 1210.

22. Wassmann, W.W. and Wallace, J.M. (1979), Bull. Am. Phys. Soc. 24, 1142.

23. Frisch, M.B. and Webb, W.W. (1979), Bull. Am. Phys. Soc. 24, 1143.

A Synthesis and Model of Turbulence Structure in the Wall Region

R. E. FALCO
Department of Mechanical Engineering
Michigan State University
East Lansing, Michigan 48824

ABSTRACT

A major fraction of the turbulence produced in the wall region of a turbulent boundary layer is the consequence of a well defined sequence of events, which we call the "pocket flow module". It is basically the interaction of Taylor microscale sized eddies of the outer region (which move towards the wall) with the wall region flow. The "footprint" of this module has length scales the order of 100 wall layer units in both the streamwise and spanwise directions. The pocket flow module furthermore extends the order of $100y^+$ above the wall. Its occurrence is distributed in an apparently random fashion over the wall, occurring both on, and in between, the longer streaks. The frequency of pocket formation is approximately four times the bursting frequency. The evolution of a single pocket contains within it all of the structural features described by other investigators. The purpose of this paper is to indicate how these features are phased to the evolution of the pocket, and to show that the average pocket evolution can be accurately modeled by the interaction of a vortex ring with a wall, when the ring's line of flight intersects the wall at an acute angle.

1. INTRODUCTION

It is generally agreed that the wall region of a turbulent boundary layer develops a structure which results in the redistribution of a passive contaminant into elongated streamwise orientated streaks. Observations by Kline et al. (1967), have shown that the streaky structure lifts-up, oscillates and breaksdown. The breakdown results in fluid moving out across the layer. Kim, Kline and Reynolds (1971) have shown that this process is responsible for a large fraction of the turbulence produced near the wall. Grass (1971) and later Lu and Willmarth (1973) and others showed that another event was also contributing a large fraction to the Reynolds stress. It was characterized by high speed fluid moving towards the wall, and was called a sweep, a term first used by Corino and Brodkey (1969). Offen and Kline (1974) noticed that there was a close correspondence between sweeps and ejections. Falco (1978, 1979) noticed that the sweep and ejection events, which were viewed in an observation plane parallel to the mean flow and perpendicular to the wall, were strongly coupled with the occurrence of a well defined visual pattern in observation planes parallel to the wall (when the planes were in the wall region). Combining anemometry and flow visualization,

Falco determined that the sweep and ejection events were indeed coupled in an event he called the "pocket flow module".

Furthermore, as part of the evolution of the pocket flow module, streamwise vortices formed (Kim et al. 1971), hairpin-like vortices formed (Willmarth and Tu 1967, Kline et al. 1967, many others), lifted fluid was apparently oscillated both up and down and laterally (Kline 1978), sharp shear layers formed (Blackwelder and Kaplan 1976), and a rapid breakdown into both larger and smaller scales resulted (see Offen and Kline 1973).

Although all of these features have been identified before, they have not yet been put together into an ordered picture. Indeed it has certainly not been clear that they should fit into a single picture which does not include either the long streaky structure, or motions which scale on the overall boundary layer thickness. However, this is the case. We will briefly summarize our new findings, and put forth a model of the flow module which contains all of the features described above.

2. EXPERIMENTAL TECHNIQUES

Experiments were conducted in a continuously running boundary layer flow visualization tunnel. The flow was made visible by contaminating it with tiny oil droplets. This marked fluid has been visualized with both flood lighting, and laser light which was expanded into a sheet. Details of both the tunnel and flow visualization technique are described by Falco (1980a). For the present experiments, two additional techniques have been used to visualize the flow. One was the oil fog "wash out" technique, in which smoke is introduced into the entire layer, and then turned off. For a short period after turning off the marker the slow moving wall layer structure comes out in sharp relief. This technique allows an observable concentration of marker to be present in the wall region over several boundary layer thicknesses, without the problems associated with marker introduction. The other technique used is a hybrid, consisting of flood illuminated marker introduced through a slit in the wall, and a laser sheet illuminated slice normal to the wall intersecting the marked sublayer. A second slit in which smoke was introduced into the entire boundary layer (further upstream, before the trip), allowed us to correlate the sublayer structure with the outer layer structure. This technique is called the laser sheet/flood light technique later in the text.

Hot-wires were placed in the flow in several of the experiments. Details of the probes, calibration techniques, and data acquisition, are described by Falco (1979, 1980a). The system allowed simultaneous sampling of all wires used. Probes consisting of u-wires, x-wires, and parallel u-wires, were all used. At $R_\theta = 1068$, the parallel wires and x-wires were separated by $y^+ = 3.6$.

3. RESULTS

As indicated above, all of the structural features found to be associated with the production of turbulence are associated with the evolution of the pocket flow module. The flow module's life time is, on average, about 30 wall layer time scales ($tu^{*2}/\nu = t^+$).

Examination of the evolution indicated that by breaking it down into 5 stages, the essential features mentioned in the introduction could be isolated. These stages were determined by visual inspection of the concentration patterns of the "footprint" in the smoke marked wall region. Our schematic impression of the pocket patterns during each stage is shown in Figure 1. The essential features are highlighted, in particular, regions in which lift-up is observed are shaded.

Figure 1 also shows the ensemble averaged streamwise velocity fluctuations at $y^+ = 16$, conditionally sampled by observing the times when pockets passed over the wire during the various stages. Furthermore, we only sampled data for "hits" that were close to the center line of the pockets. The ensemble sizes are 16, 7, 6, 9 and 17 for stages one through five respectively. The duration of the stages was referenced to the first indication of the pocket as seen in plan view.

Stage 1 ($0 < t^+ < 8$), was defined as that period during which marker was moved away from a local region over the wall, the marker moving roughly radially away from the center of the convecting pattern. We see that the streamwise velocity was high over the entire region cleared of marker. Additional measurements have shown that the normal velocity was towards the wall. The highest uv signals of the entire record were found during this stage of evolution of the pocket flow module. The impression is gained that a sweep (defined as high speed fluid moving towards the wall) of scale the order of 100 wall layer units has come to the wall. Using our laser sheet/flood light technique we have sometimes observed vortex ring-like eddies just above the wall during this period (see Figure 2a).

Stage 2 ($4 < t^+ < 12$), was characterized by the formation of a characteristic crescent shape. Flow visualization at this stage indicated the lift-up of sublayer fluid around the crescent. This fluid moved out to y^+ of less than 16 on average (because the wire did not record the presence of low speed fluid during this time interval), and then moved back down to the wall. The ensemble averaged u signature (see Figure 1) showed that the magnitude of the high speed fluid increased, and x-wire data indicated continued movement towards the wall. Thus, the pocket flow module continued to be characterized as a strong sweep. Our additional data from the laser sheet/flood light technique sometimes showed a vortex ring-like eddy becoming rapidly distorted during this stage. An example of this in a fully turbulent boundary layer is shown in Figure 2, where two frames which are approximately $5t^+$ apart, have been taken from a movie in which the laser sheet/flood light technique was used. We can see a vortex ring-like eddy immediately above the incipient pocket (stage 1) in Figure 2a, and can observe the eddy rapidly incline as stage 2 of the pocket develops (Figure 2b).

Between t^+ of 8 and 16 (stage 3), the crescent elongated in the stream direction, and its upstream boundary started to become diffused. Fluid continued to lift-up from the sublayer and come back towards the wall. It appeared that a pair of counter-rotating streamwise vortices formed. These vortices were joined around the upstream end of the pocket pattern, and appeared just as the vortex which was created around an obstacle placed in the center of the pocket. The way in which these vortex lines closed was not observable with the techniques used. The streamwise orientated portions of the vortex should have a strength proportional to the mean vorticity of the layer over that height. Figure 1 shows that u remained high in the upstream "half" of the pocket. Our x-wire data indicated that over this portion of the pocket the uv contribution also remained high, and thus the sweep continued to occur in the upstream part. However, near the downstream boundary of the

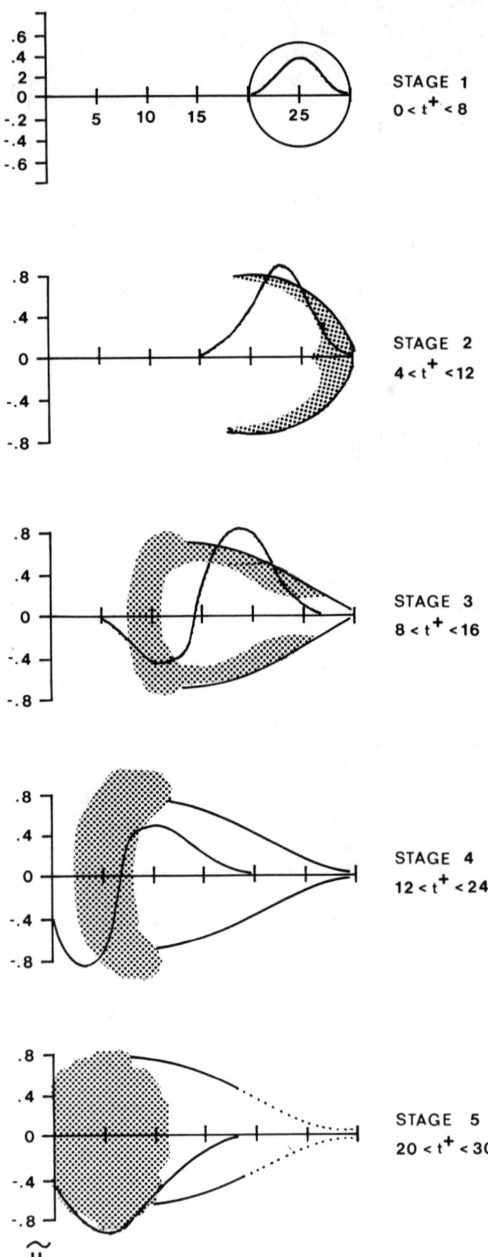

Figure 1. Sketches of the evolution of the footprint of the pocket flow module. Shaded regions indicate lifted wall region fluid. The ordinate is $(\langle u\rangle - \tilde{u})/\tilde{u}$, and the abscissa is in intervals of t^+. The flow direction is from right to left.

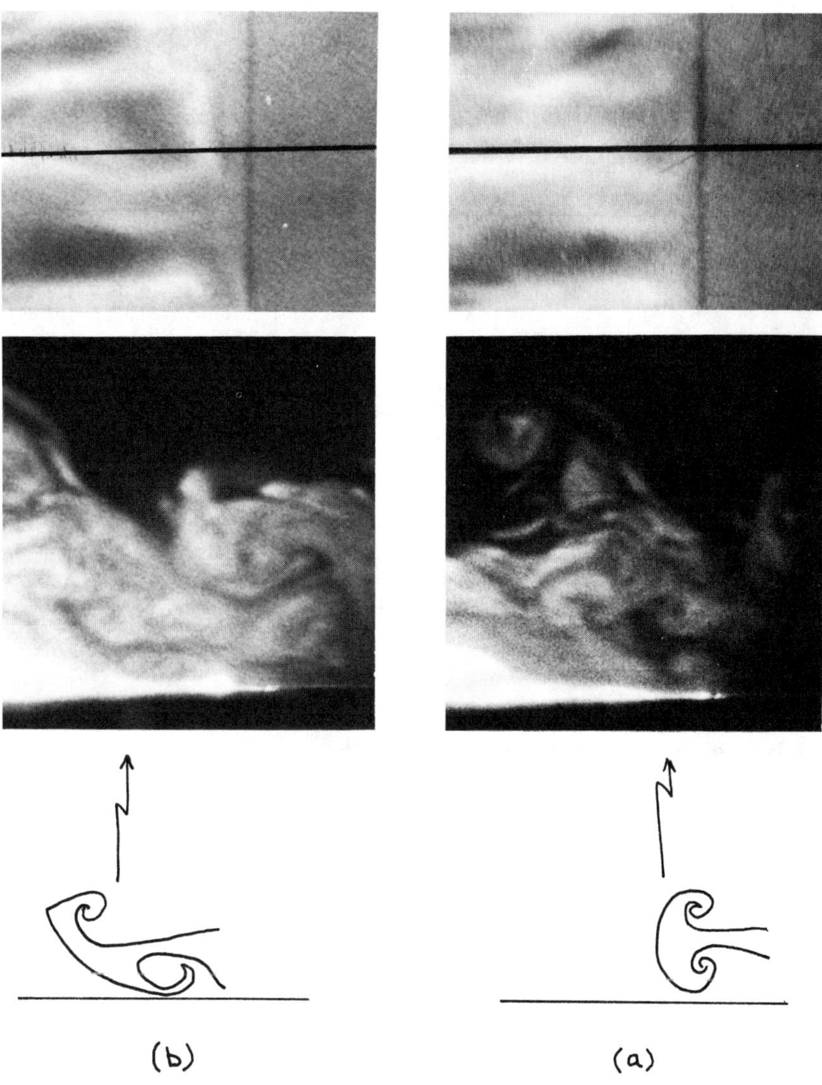

Figure 2. Flood illuminated plan views (top photos) and simultaneous laser sheet illuminated side views of a pocket flow module, showing the formation of a pocket pattern by a vortex ring-like eddy. (a) Stage one: (b) Stage two, five t^+ later. Note the laser sheet side views show the vortex ring-like eddy (see accompanying sketches) in part (a) incline rapidly in part (b). The lines drawn in the plan views indicate the position of the laser sheet in which the side views were photographed. Flow is from right to left. This evolution should be compared with that in Figure 5.

pocket pattern evidence of the lift-up of sublayer fluid existed in the visualizations. The hot-wire data showed that <u> was less than the long time average, and <v> was positive. Thus, a lift-up at least to the level of the wire had been initiated in the downstream portion.

By stage 4 ($12 < t^+ < 24$), the streamwise orientated vortices which were connected around the upstream boundary of the pocket had lost their strength. This was evidenced in our visualizations by the cessation of lift-up along the sides and upstream boundary. However, a strong new feature appeared. This was the continued outward movement of fluid from the downstream end of the pocket which started during stage three. This fluid went beyond the buffer layer, and is properly called an "ejection". Our visualization with two views has clearly shown a vortex with the same sign as the mean shear, rising away from the wall. This vortex had the appearance of a "hairpin" vortex. There is weak visual evidence of a vortex of opposite sign immediately upstream of this "hairpin". Figure 3 shows an oblique view of a "hairpin" emerging from a pocket. The streamwise velocity perturbation (see Figure 1) showed that the region of velocity defect which formed in stage three had increased in intensity, while the region of velocity excess had diminished somewhat. The uv signature due to the ejection was now very strong, while the uv contribution due to the sweep decreased.

Stage 5, which had been observed to occur for pockets between 20 and 30 t^+ old, is characterized by a wider, even more intense region of low speed fluid. No trace of the high speed fluid remains. The <uv> signals indicated a continued intense Reynolds stress. Visual observations in plan view showed that the pocket patterns had become difficult to distinguish (and would probably not be distinguishable if not for the patterns seen in previous frames of the movies). The simultaneous side views of the flow in a laser sheet showed that the lifted smoke marked sublayer fluid had lost its coherence, and the "hairpin" vortex of stage four was no longer identifiable. The lifted fluid appeared to have broken down into a range of scales, both smaller and larger than those described above.

Additional details about <uv> and <du/dy> are discussed by Falco (1980b).

4. COMPARISON WITH STRUCTURAL FEATURES OBSERVED BY OTHERS

The evolution of the pocket flow module appears to an observer moving along with it to consist on the whole as a sweep of scale proportional to the Taylor microscale, which interacts with the wall region flow and causes an ejection. The physical features that are observed to occur during this evolution are:

(1) streamwise orientated vortices just above the wall, which are observable only during stage 2, i.e. for a duration of at most $8t^+$;
(2) a "hairpin" vortex whose vorticity has the same sign as the mean vorticity, this forms during stage three and continues to exist throughout stage four, thus its duration is at most $16t^+$;
(3) a sharp shear layer which develops during stage three and intensifies during stage four, its lifetime is at most $16t^+$;
(4) rapid breakdown occurs during stage five.

Figure 4 shows a sketch of how these features are orientated with respect to the pocket pattern. From measurements of the width and length scales of the pocket patterns we can obtain an estimate of the scale of these features.

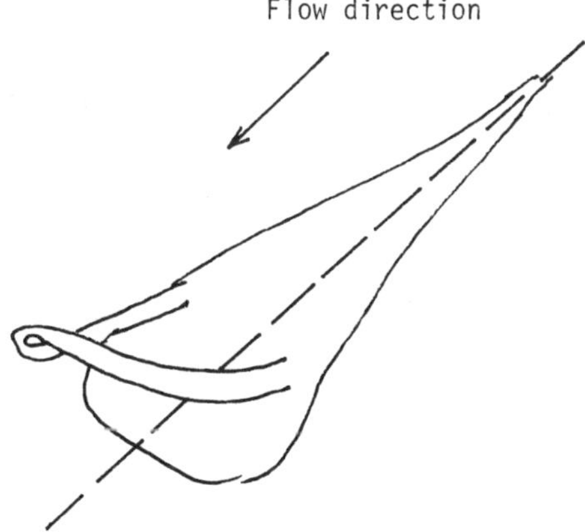

Figure 3. Oblique view of a "hairpin" vortex arising out of a pocket in stage 4.

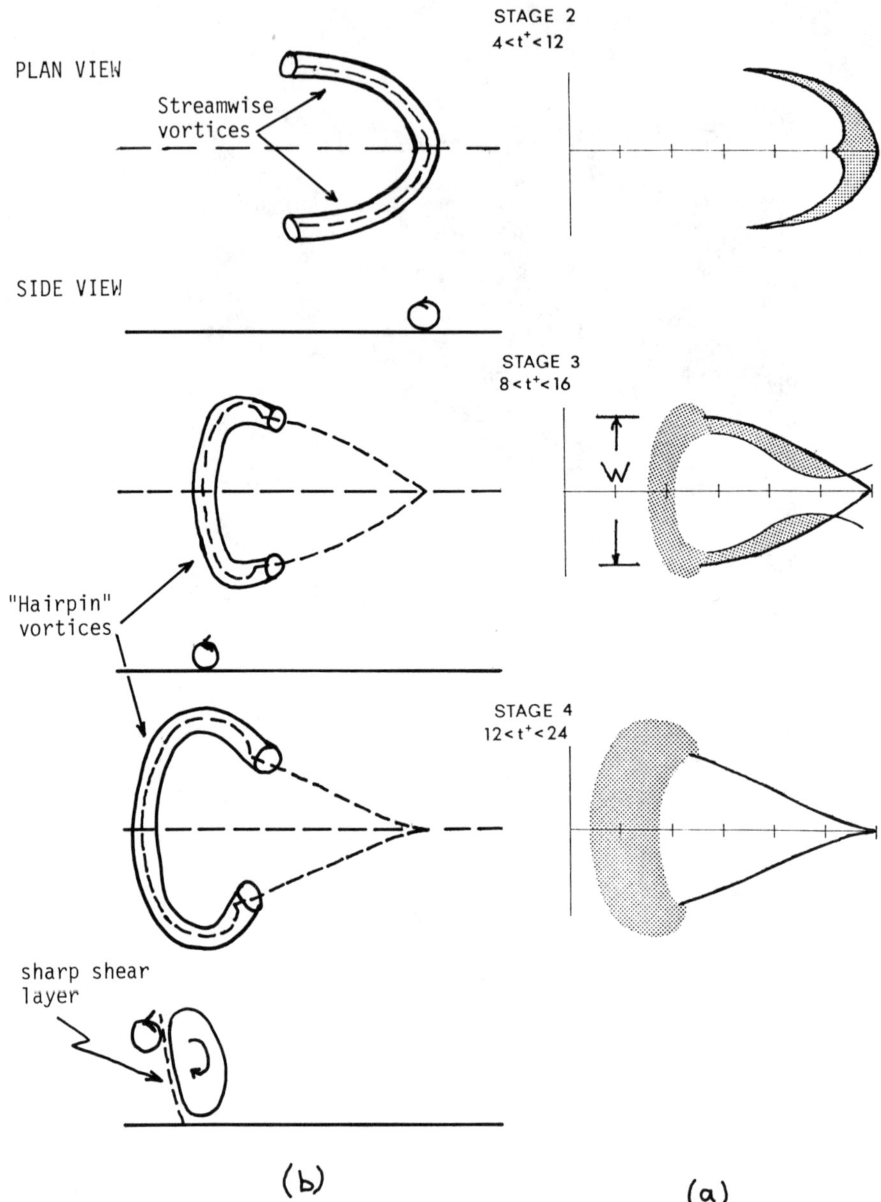

Figure 4. Sketch showing how streamwise vortices, "hairpin" vortices, and strong shear layers are orientated with respect to the pocket pattern during its evolution. (a) Plan views: (b) Pictures reconstructed from simultaneous plan and side view data. On the left, the pocket pattern is indicated by dashed lines to emphasize the vortex structure.

The width of the pattern is a function of Reynolds number:

$$\langle W\rangle^+ = \langle W\rangle u^*/\nu \sim R_\theta^{1/3}$$

at $R_\theta = 1000$, $\langle W\rangle^+ = 57$. The streamwise length scale measured at stage three is approximately 125% larger. Thus we see that the streamwise vorticity associated with the pocket pattern is both local in spatial extent, and of short duration. Near the upstream boundary of the pocket, during stage three, when there is noticable lift-up, the effects of this lift-up are not sensed by the wire. This suggests that the extent of the vortex is less than $y^+ = 16$ (the wire position) in the direction normal to the wall. All of these results are consistent with the visual observations of Kim et al. (1971).

A number of workers have suggested that "hairpin" vortices are responsible for the lift-up of wall layer fluid, while others have suggested that vortices of this type (i. e. vortices with their "legs" anchored on the wall) are responsible for the cascade process in general. Although it is kinematically inadmissible for a vortex tube which exists in an incompressible fluid to end, it is possible that a bent vortex tube could look like a "hairpin" vortex. Kline et al. (1967) gave a detailed description of how "hairpin" vortices could cause the lift-up, oscillation and breakdown observed in the dye marked sublayer. We have definitely established that hairpin-like vortices exist in the wall region of a turbulent boundary layer, and that they are formed at the downstream end of the pocket pattern (see Figure 3 and Figure 6). We have observed that these "hairpins" lift up, along with other sublayer fluid and then breakup. Thus the fluid in the "hairpin" itself forms part of the ejection. They are observed to lift up to y^+ the order of 50, in approximately $16t^+$, which is remarkably rapid. We believe that this lift-up and ejection is too rapid to be due to stretching by the mean shear with possible consequential roll around or pairing. The scales of the "hairpin" can be roughly estimated from Figure 4.

During stages two and three a sharp shear layer of duration approximately $10t^+$ forms at the probe height. A shear layer of this type has been observed by Offen and Kline (1973). During stage three another shear layer forms. This one is at the downstream end of the pocket pattern and it is considerably narrower. This shear layer persists though stage four during which it amplifies, while remaining thin. Hot-wire signals show that both the streamwise and normal velocity gradients are large at the downstream end but that only du/dy is large at the upstream end during these times. Thus we conclude that the downstream shear layer, which develops later, is inclined at a significant angle to the wall. Blackwelder and Kaplan (1976) used a burst detection technique that keyed upon an inclined shear layer. The ensemble averaged streamwise velocity signatures they recovered look like an average of those presented in Figure 1 for stages three and four. Furthermore, our $\langle uv\rangle$ signatures also look similar. Blackwelder (1978) has suggested that this sharp shear layer rapidly becomes unstable resulting in the ejection. We now see that it is part of the pocket flow module evolution, and that it is a consequence of the ejection, rather than a cause of it.

During stage five we get a rapid breakdown of the lifting hairpin and surrounding fluid. We have no indication of growing oscillations in the shear layer prior to the breakdowm. Rapid breakdown, into both smaller and larger scales, is a feature of the bursting process emphasized by Kline (1978).

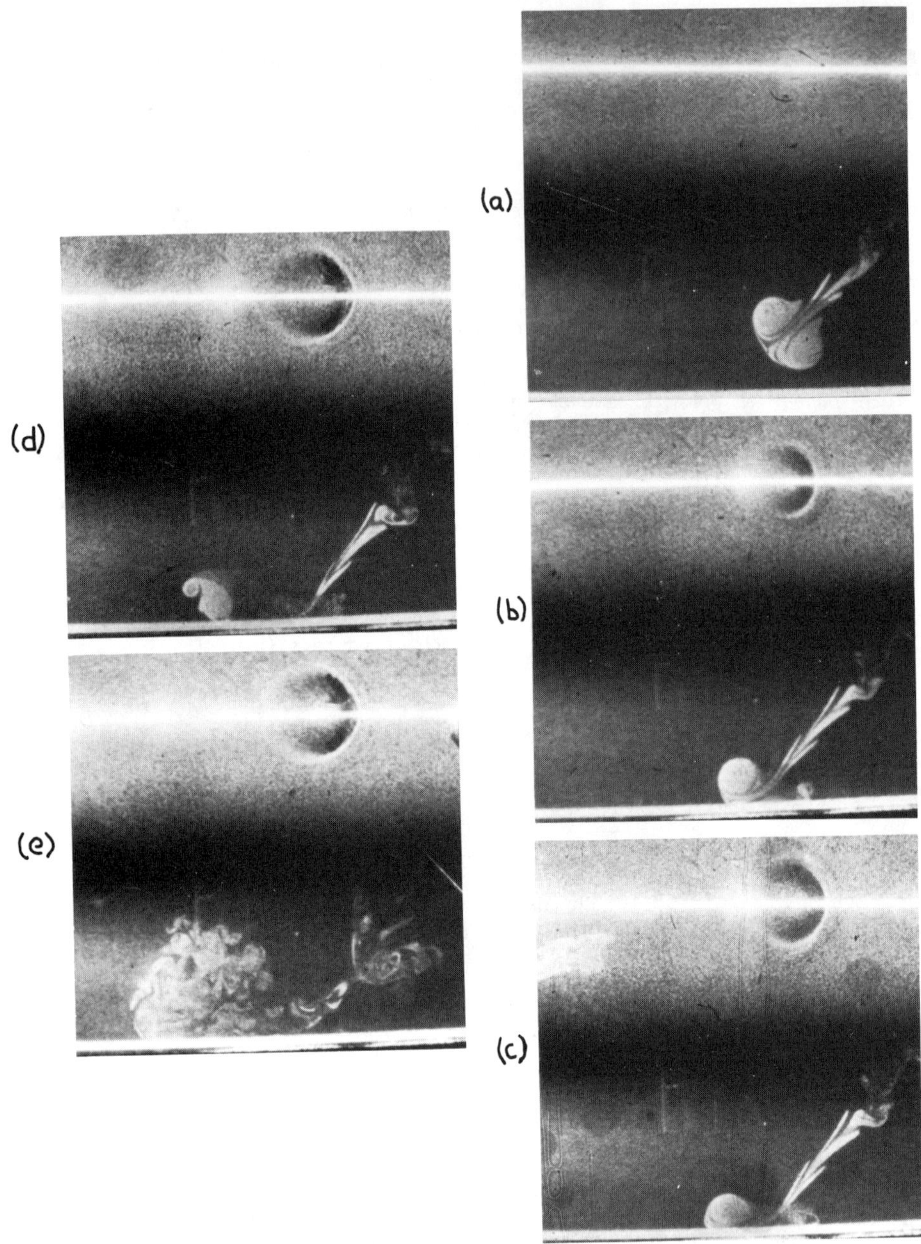

Figure 5. The interaction of a vortex ring with a wall. Five stages in the evolution are shown, in both plan and side view. (a) The ring just above the wall: (b) Back and sides of the ring have interacted, and begun to redistribute marker that has been uniformly distributed over the wall: (c) The front of the ring has come to the wall, and a "pocket" pattern has formed: (d) A "hairpin" has formed and been induced off the wall by the ring vortex: (e) The lifted fluid breaksdown.

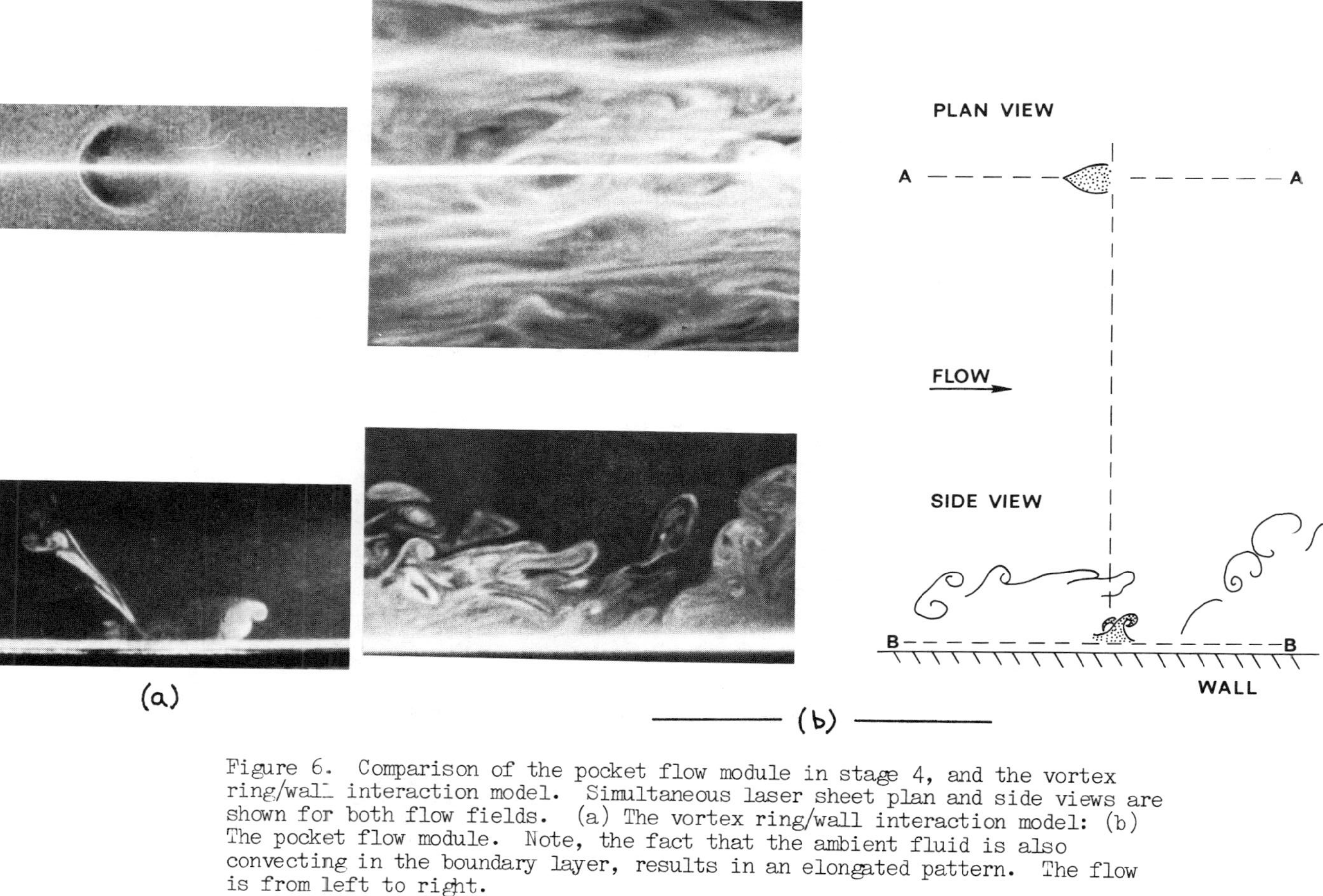

Figure 6. Comparison of the pocket flow module in stage 4, and the vortex ring/wall interaction model. Simultaneous laser sheet plan and side views are shown for both flow fields. (a) The vortex ring/wall interaction model: (b) The pocket flow module. Note, the fact that the ambient fluid is also convecting in the boundary layer, results in an elongated pattern. The flow is from left to right.

5. MODEL OF THE TURBULENCE PRODUCTION MECHANISM

It is clear that mechanisms proposed in the past cannot adequately explain the events described above. It appears, however, that a single interacting flow field can produce all of the evolutionary features described above. It is the interaction of a vortex ring with a wall. This mechanism has been suggested by the turbulent boundary layer data of the type shown in Figure 2 as well as observations made further away from the wall, Falco (1977a), which show that vortex ring-like eddies that have the same size as the pockets, are an important part of the overall structural picture. Falco (1977b) first proposed this possibility.

As a vortex ring induces itself towards a wall at some arbitrary acute angle, it appears as a three-dimensional sweep. Experiments performed with rings generated in an initially still ambient, using mutually orthogonal laser sheets, with fine particles on the wall and florescent dye in the ring, showed structural features which appeared similar to those described for the pocket flow module. Figure 5 shows plan and side views of the vortex ring/wall interaction. As the ring comes closer to the wall, it is stretched by inviscid "mirror image" effects (see Falco and Wiggert 1980, for quantitative estimates), which cause it to incline in the flow direction. At the same time vorticity is created in the fluid near the wall by the stagnation point flow field that the wallward moving ring produces. This newly created vorticity, which has the opposite sign of rotation from the portion of the ring closest to it, interacts most strongly with the closest portion of the ring. This results in fluid which was initially on the wall being induced away from the wall, i.e., lifted-up from the wall. Because the ring's vorticity is being amplified by stretching, it induces this vortical wall layer fluid back down to the wall around portions of the ring that first get closest to the wall. This stage in the ring/wall interaction results in a movement of marker very similar to stage two of the pocket evolution, with the portion of the ring that has "laid down" on the wall appearing as a pair of upward tilted streamwise vortices connected around their upstream end. Of course, they are connected at the downstream end (the vortex ring is still whole) but an observer who only marks the wall layer fluid would not see the connection.

As more of the ring comes close to the wall, the pattern created by the ring in the marked wall region flow elongates, and the upstream side of the ring appears to lose its coherence. This is reminiscent of stage three of the pocket flow module evolution. The downstream portion also causes the vorticity generated by the stagnation point flow field to be redistributed into a vortex, and this vortex (which appears similar to the "hairpin" vortex of stages 3 and 4 of the pocket flow module evolution), and the part of the ring vortex in closest proximity induce themselves off the wall. These vortices lift up a distance about equal to the original ring core diameter and then go unstable. This appears as an ejection of fluid away from the wall, followed by a breakup (see stages 4 and 5 of the pocket flow module). We expect a sharp shear layer to form between the ring and newly generated vortex. The breakup that ensues is very rapid, leading to the generation of a turbulent plume. The plume has both larger and smaller scales, a characteristic of the breakup of the bursting process noted by Kline (1978). Although marker that has been laid on the wall has been redistributed and lifted off the wall at the upstream end as well as along the sides of the pattern created by the vortex ring, most of the marker that gets away from the wall does so at the downstream end, just as observed for the pocket flow module in the boundary layer.

The details of the model process are very close to the real one. The shape of the pattern in which marker has been removed and/or concentrated, with its well defined upstream end (this is, of course, destroyed by the flow in the boundary layer), and its ragged downstream end is clearly similar to the pocket pattern of the boundary layer flow. Figure 6 shows a comparison of the pocket flow module in stage four, with the vortex ring/wall interaction

model during its downstream lift-up and ejection phase. Figure 2 accurately indicates the rapid tilting a vortex ring undergoes when it comes in contact with the wall (compare with Figure 5a and 5b).

6. DISCUSSION

It is possible to infer from the data described above the mechanism of small scale wall pressure fluctuations and shear stress fluctuations. The highly curved, stretched portion of the sweep vortex which forms around the upstream part of the pocket pattern, will be a region of low pressure, and high shear stress. On the other hand, the incoming sweep, and the ejection, will both produce regions of high pressure. As noted, the scale of these regions is consistent with the scale of the highest intensity wall pressure perturbations measured by Dinkelacker, Hessel, Meier and Schewe (1977). We believe that the intensity of the vortex at the upstream end of the pocket, which induces the lifted sublayer fluid back to the wall, is responsible for the fact that the most intense wall pressure signals are low pressure ones. Furthermore, the shape of the vortex at the upstream end of the pocket during this stage (see Figure 4) is elongated in the spanwise direction, another characteristic found by Dinkelacker (see his contribution in this symposium). We have measured the convection velocities of the pocket patterns, to further quantify the comparison. The upstream end of the pattern (where we expect low pressure) convects, on average, at $.53U_\infty$, while the downstream end convects at $.63U_\infty$. These values are close to the values found by Schewe (1979). We predict that the low pressure peaks will have a duration of approximately $3t^+$, while peaks of positive pressure should be slightly wider, approximately $6t^+$.

Our observations also allow us to predict that intense periods of streamwise vorticity will be measured, with magnitudes aproximately twice the mean vorticity of the layer at $y^+ = 16$, which are of short duration, the order of $t^+ = 5$–10.

Finally, the development of the pocket flow module is consistent with log law scaling, which requires the motions which produce most of the transport to scale proportional to y, the distance from the wall. Observations indicate that the core of the incoming vortex ring decreases in cross-section as its vortex is stretched by the presence of the wall. There is a limited region over which this decrease in scale is a linear function of the distance from the wall. Thus it is possible for the coherent sweep to satisfy log law scaling. On the other hand, the fluid that is ejected from the downstream end of the pocket (or the vortex ring/wall interaction model), breaks up into a turbulent plume. Turbulent plumes grow linearly away from their source. Thus the ejection of fluid can also satisfy log law scaling.

ACKNOWLEDGEMENTS

This work was jointly sponsored by the Air Force Office of Scientific Research, and by the Office of Naval Research, Fluid Dynamics Branch. Their financial assistance is greatly appreciated.

REFERENCES

Blackwelder, R. F. 1978. "The bursting process in turbulent boundary layers", in Coherent Structure of Turbulent Boundary Layers, ed. C.R. Smith and D.E. Abbott, AFOSR/Lehigh, pp. 48-94.

Blackwelder, R. F. and Kaplan, R. E. 1976. "On the bursting phenomena near the wall in bounded turbulent shear flows", J. Fluid Mech., Vol. 76., pp. 89-112.

Corino, E. R. and Brodkey, R. S. 1969. "A visual investigation of the wall region in turbulent flow", J. Fluid Mech., Vol. 37., pp. 1-30.

Dinkelacker, A., Hessel, M., Meier, G. E. A. and Schewe, G. 1977. "Investigation of pressure fluctuations beneath a turbulent boundary layer by means of an optical method", Phys. Fluids, Vol. 20, Pt. II, pp. S216-S224.

Falco, R. E. 1977a. "Coherent motions in the outer region of turbulent boundary layers", Phy. Fluids, Vol. 20, Part II, pp. S124-S132.

Falco, R. E. 1977b. "Vortex ring/wall interactions in still and moving ambients", Bulletin Am. Phy. Soc. Series II, Vol. 22, pg. 1277.

Falco, R. E. 1978. "The role of outer flow coherent motions in the production of turbulence near a wall", in Coherent Structure of Turbulent Boundary Layers, ed. C. R. Smith and D. E. Abbott, AFOSR/Lehigh. pp. 448-461.

Falco, R. E. 1979. "Structural aspects of turbulence in boundary layer flows", to be published in Turbulence in Liquids, ed. G. K. Patterson and J. L. Zakin.

Falco, R. E. 1980a. "Combined simultaneous flow visualization/hot-wire anemometry for the study of turbulent flows", J. Fluids Engr., Vol. 102, pp. 174-183.

Falco, R. E. 1980b. "The production of turbulence near a wall", AIAA 13th Fluids and Plasma Dynamics Conference, Snowmass, Colorado, Paper 80-1356.

Falco, R. E. and Wiggert, D. C. 1980. "The effects of dilute polymer solutions on vortex ring/wall interactions-a mechanism for drag reduction", to be published in AIAA Progress in Aeronautics and Astronautics.

Grass, A. J. 1971. "Structural features of turbulent flow over smooth and rough surfaces", J. Fluid Mech., Vol. 50, pp. 233-254.

Kim, H. T., Kline, S. J. and Reynolds, W. C. 1971. "Production of turbulence in a boundary layer", J. Fluid Mech., Vol. 50, pp. 133-160.

Kline, S. J., Reynolds, W. C., Schraub, F. A. and Runstadler, P. W. 1967. "The structure of turbulent boundary layers", J. Fluid Mech., Vol. 30, pp. 741-774.

Kline, S. J. 1978. "The role of visualization in the study of the structure of the turbulent boundary layer," in Coherent Structure of Turbulent Boundary Layers, ed. C. R. Smith and D. E. Abbott, AFOSR/Lehigh, pp. 1-26.

Lu, S. S. and Willmarth, W. W. 1973. "Measurements of the structure of the Reynolds stress in a turbulent boundary layer", J. Fluid Mech., Vol. 60, pp.

481-511.

Offen, G. R. and Kline, S. J. 1973. "Experiments on the velocity characteristics of 'Bursts' and on the interactions between the inner and outer regions of a turbulent boundary layer," Report MD-31, Dept. of Mech. Engrg., Stanford Univ., Stanford, CA.

Offen, G. R. and Kline, S. J. 1974. "Combined dye-streak and hydrogen-bubble visual observations of a turbulent boundary layer", J. Fluid Mech., Vol. 62, pp. 223-239.

Schewe, G. 1979. "Untersuchung von wanddruckund wanddruckgradientenschwankungen unter einer turbulenten grenzschichtstromung", Doctoral Diss., Max-Planck-Institut fur Stromungsforschung, Gottingen, West Germany.

Willmarth, W. W. and Tu, B. J. 1967. "Structure of turbulence in the boundary layer near the wall," Phy. Fluids, Supplement to Vol. 10, pp. S134-S137.

Do Tornado-like Vortices Play a Role in Turbulent Mixing Processes?

A. DINKELACKER
Max-Planck-Institut für Strömungsforschung
D 3400 Göttingen, BRD

ABSTRACT

At the Max-Planck-Institut für Strömungsforschung several investigations on wall pressure fluctuations under turbulent flows have been performed. Some of the results of these investigations led to the idea, that a part of the observed wall pressure patterns might be caused by "tornado-like" vortex structures occurring in the turbulent flow. Some observations leading to this idea will be reported and a flow-model will be discussed which seems to fit the observed pressure patterns. With respect to the process of turbulent mixing the proposed flow model could explain intense momentum transfer in the boundary layer.

1. INTRODUCTION

Investigations of wall pressure fluctuations under turbulent flows are of interest mainly for two reasons: Firstly because these fluctuations can cause flow noise and flow induced vibrations; secondly because these fluctuations can give information on turbulent flows. At the Max-Planck-Institut für Strömungsforschung several investigations on turbulent wall pressure fluctuations have been performed. These investigations include measurements using an optical measuring method [1], [2], [3] and measurements using electro-mechanical wall pressure transducers [4], [5], [6]. The evaluation of these measurements has been done with respect to the study of time and space histories of individual pressure events, as well as with respect to statistical properties of wall pressure fluctuations. Some of the results of these investigations led to the idea, that a part of the observed wall pressure patterns might be caused by "tornado-like" vortex structures, which might be occurring in the turbulent flow. In the following some observations leading to this idea will be briefly reported and a flow model will be discussed which seems to fit the observed wall pressure patterns.

2. EXPERIMENTAL OBSERVATIONS

2.1 Some Remarks on the Experimental Arrangement

The experiments were performed in a specially designed low noise, low vibration wind tunnel, in which the pressure transducers were built in flush with one of the tunnel walls. The flow was air flow of the type of "tunnel inlet flow", i.e. turbulent boundary layers were developing along the tunnel walls. The flow velocities used for the investigations were between 6 m/s and 9 m/s, the Reynoldsnumber Re_θ based on the momentum thickness θ was of the order of 2000 (details are given in [1], [2] and [4]).

The optical pressure transducer (Fig. 1) consists of several hundred small elastic membranes, the displacements of which can be measured by interferometric means and photographed with a high speed camera. With this method the instantaneous pressure can be measured simultaneously at many points on the wall (here 198 membranes). From these measurements one can draw maps of the instantaneous wall pressure distribution. These maps are comparable to pressure fields in meteorological weather maps (details are given in [1] and [2]).

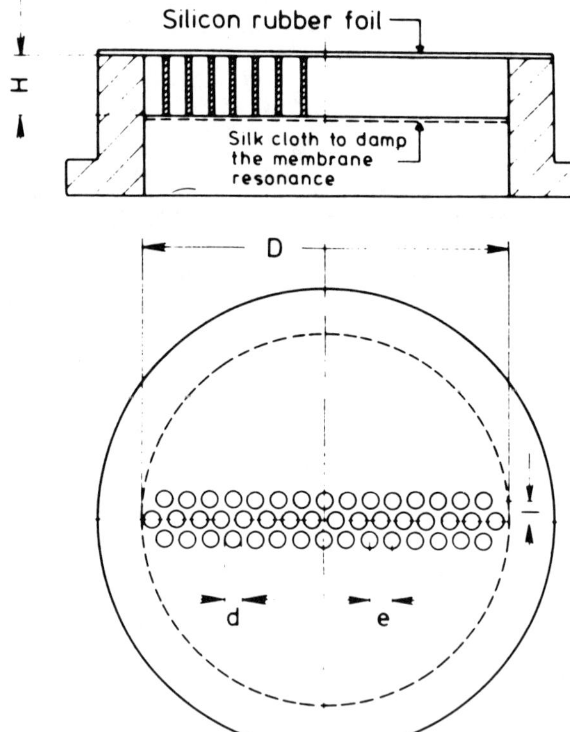

Fig. 1. Pressure transducer used for the measurements with the optical method (D = 97 mm, H = 6 mm, d = 2, 5 mm, e = 3 mm, l = 2, 6 mm)

The electro-mechanical pressure transducers used consisted of condensor microphones of various sizes with diameters down to d = 1 mm or $d^+ = 19$. Small diameters are important with respect to the spatial resolution of the wall pressure field ("+" means: made nondimensional with the characteristic sublayer properties ν and u_τ; ν = kinematic viscosity; u_τ = shear velocity; details are given in [4]).

2.2 Some Experimental Results

In this context the following experimental results seem to be especially relevant (extracted from [1] to [6]):

(a) Investigations with the optical method showed that small areas of rather low pressure occur in surroundings of comparatively high pressure. No comparable areas of high pressure were observed in low pressure surroundings. The diameter of the small low pressure areas was usually not larger (may be smaller) than the diameter of the membranes used for the measurements (here d = 2.5 mm or $d^+ = 57$). An example of such a pressure field is given in Fig. 2.

In order to investigate the observed pressure patterns in more detail the following analysis was performed: With the help of a computer program for each pressure reading p at a membrane (more exactly: for each pressure

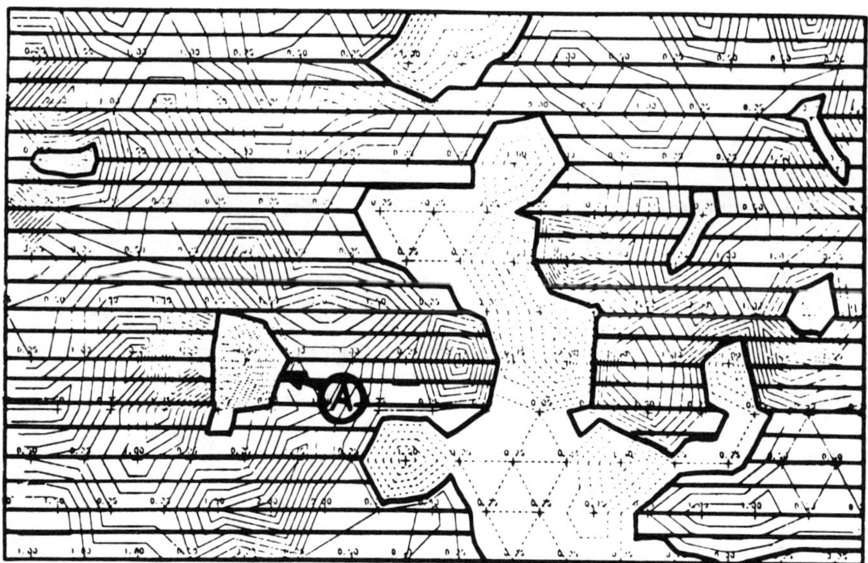

Fig. 2. Wall pressure distribution (flow velocity: U_∞ = 8.5 m/s, flow direction left to right, size of observation field: 50 mm x 30 mm, shaded area = high pressure, "A" = small area of very low pressure)

reading at a membrane which was not at the boundary of the observation field) the instantaneous value of the mean pressure p_s at the six membranes surrounding the first one was calculated. From these values the mean value $\overline{p_s}$ was calculated for the different amplitude ranges of p. Fig. 3 shows the results of these calculations. Here one sees that the distribution of $\overline{p_s}(p)$ is asymmetric. Furthermore areas of very low pressure ($p \leqslant 4.5\ p_{RMS}$) tend to be surrounded by areas of comparatively high pressure.

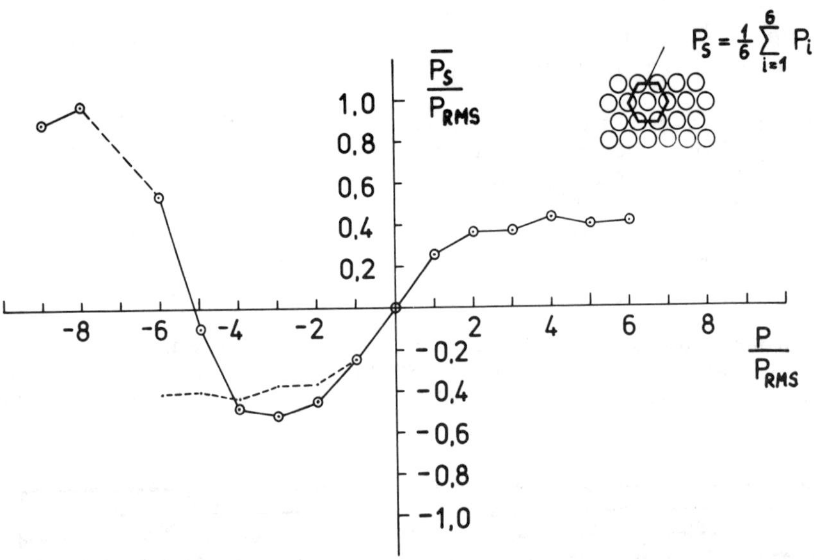

Fig. 3. Mean pressure $\overline{p_s}$ in the "surroundings of a membrane" as a function of the pressure p at the membrane.

(b) The small low pressure areas could be followed downstream over some distance. These distances, however, were considerably shorter than the distances observed for larger pressure areas - both low and high. The convection velocities u_c were also smaller than the average convection velocities (typical values $l^+ = 100$, $u_c^+ = 10$). In Fig. 4 an example for the convection effects, observed with four small electromechanical transducers, aligned in flow direction, are given.

(c) Probability analysis of the amplitudes of wall pressure fluctuations measured with small electromechanical transducers shows an asymmetry. More events occur with high amplitude and negative sign than with high amplitude and positive sign (comp. Fig. 5).

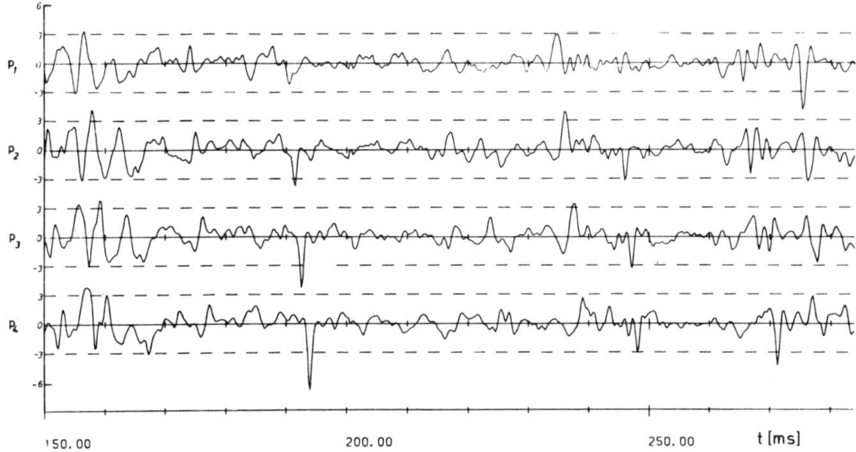

Fig. 4. Time functions of the wall pressure p measured at four consecutive pressure transducers aligned in flow direction (flow velocity $U_{\infty} = 6.3$ m/s, transducer diameter $d = 1$ mm or $d^+ = 19$, distance between adjacent transducers $l = 4$ mm or $l^+ = 75$, dashed lines mark the amplitudes of $\pm 3\ p_{RMS}$)

p_{sw} [p_{rms}]	$p < -p_{sw}$ [%]	$p > p_{sw}$ [%]
0	49.1	50.9
1	13.5	13.5
2	2.7	2.3
3	0.64	0.4
4	0.19	0.09

Fig. 5. Probability analysis of amplitudes of wall pressure fluctuations (microphone diameter $d^+ = 19$, p_{SW} = threshold pressure in multiples of p_{RMS}, here $p_{RMS} = 0.010$ q, q = dynamic pressure at centerline of tunnel)

(d) The 3rd moment (skewness) of the wall pressure fluctuations is zero for "large" transducers and becomes negative for "small" transducers, which again indicates the existence of small low pressure areas (comp. Fig. 6).

(e) Measurements with combinations of several small transducers indicate that the small areas of low pressure have a spanwise (z-direction Fig. 9) extent which is considerably larger than their extent in flow direction. For these measurements four small transducers were built flush with the tunnel wall, aligned either in flow direction or in spanwise direction. The signals of n of these four transducers (n = 1, 2, 3, 4) were added and analysed with respect to higher statistical moments of the time functions. In this way a transducer sensitivity was generated which had some similarity with that of line transducers (breadth of the transducers $b^+ = 19$, length of the transducers $l^+ = 19$, 95, 170 and 245). The results are shown in Fig. 7. As one can see the second moment (= normalized RMS values) is higher for the spanwise

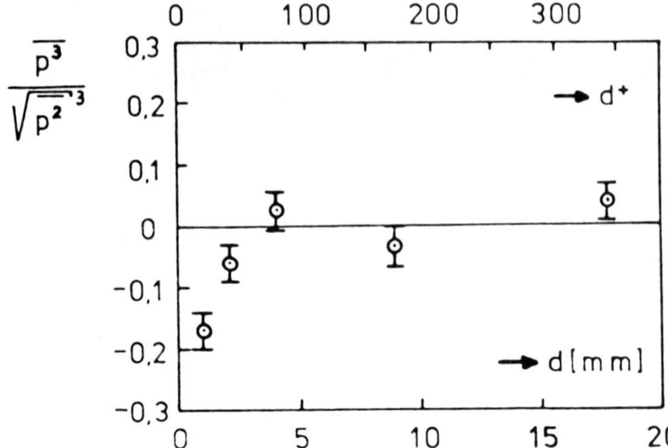

Fig. 6. Third moment (= skewness) of wall pressure fluctuations as a function of the microphone diameter

transducer arrangement than for the arrangement in flow direction. This indicates that the pressure patterns generating these fluctuations must have a dominating orientation in spanwise direction. As one also can see the 3rd moment (=skewness) is lower for the spanwise transducer arrangement than for the other one, and with the spanwise arrangement the 3rd moment remains negative for all n. This indicates that the low pressure patterns especially must have a spanwise orientation.

3. FLOW MODEL

From the observations reported in the previous section the question arises: What sort of flow structure can be connected with the observed small low pressure pattern? As the flow tube model given in [2] can not explain small low pressure patterns - on the contrary according to this model small high pressure patterns should be expected - the idea arose, that the small low pressure areas might be footprints of small tornados occurring in the turbulent boundary layer. Following this idea a flow model was proposed [7] in which pairs of "tornado-like" vortices are generated. The words "tornado-like" are used because certainly not all features of a real tornado can be expected to occur. Important features seem to be that relatively small areas of low pressure are generated and that - because of the no slip condition at the wall - the rotational movement in the tornado-like vortex decreases to zero at the wall. Hence in the vicinity of the wall the low core pressure is not compensated by centrifugal forces, and as a consequence, there will be an inflow of low momentum fluid into the vortex core. The vortex therefore will act as a "suction tube" sweeping along the wall (Fig. 8).

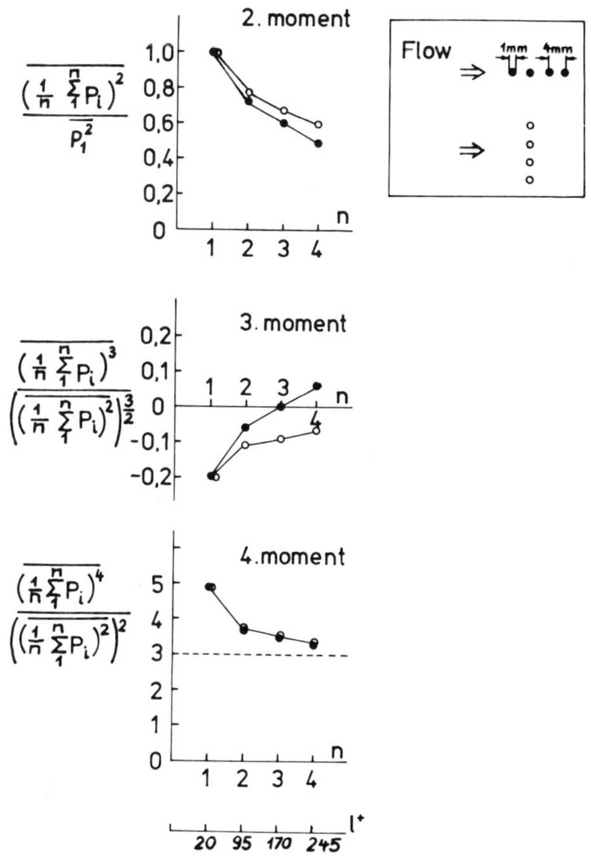

Fig. 7. Statistical moments of wall pressure fluctuations measured with different transducer arrangements (o transducers arranged in spanwise direction, • transducers arranged in flow direction, n = number of transducers used, p_i = pressure at the i th transducer, p_1 = RMS-value at the 1st transducer, l^+ = "length" of the transducer arrangement, flow velocity: U_∞ = 6.3 m/s)

The flow model proposed for the possible generation of pairs of counter-rotating tornado-like vortices has some similarity with the "horseshoe model" suggested by Theodorsen [8], [9] and with the vortex model discussed by Tu and Willmarth [10]. The essential features of the model are explained in Fig. 9. In Fig. 10 a variation of the process is shown. In this figure the process starts with a disturbance in downstream direction and hence not one but two pairs of counter rotating vortices are generated.

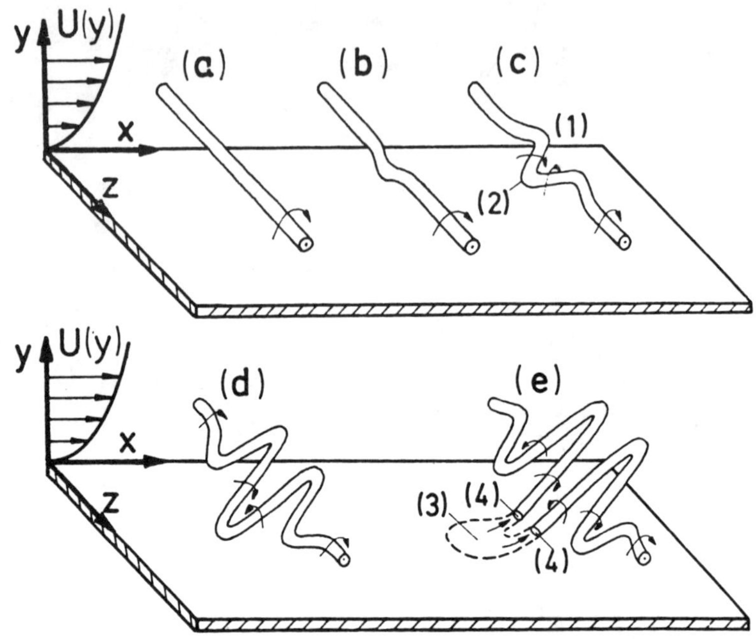

Fig. 9. Flow model for the generation of pairs of "tornado-like" vortices in a shear layer

(a) Vortex tube in spanwise direction at some distance to the wall.

(b) Small disturbance in the vortex tube.

(c) Development of disturbance: Under the action of induction and mean shear flow the front parts (1) move upwards i.e. into areas of higher flow velocity and the rear parts (2) move downwards i.e. into areas of lower flow velocity. Furthermore, the disturbance also develops in spanwise direction. Vortex stretching occurs, connected with an increase in vorticity and a decrease in core pressure.

(d) Continuation of (c).

(e) The parts of the vortex tube coming into the vicinity of the wall become distorted (action of high shear stress and high turbulence) which leads to a spreading of the vorticity (3) and to an increase of core pressure. As a consequence two counter rotating "tornado-like" vortices are formed with flow into the "open ends" (4) of the vortex tube.

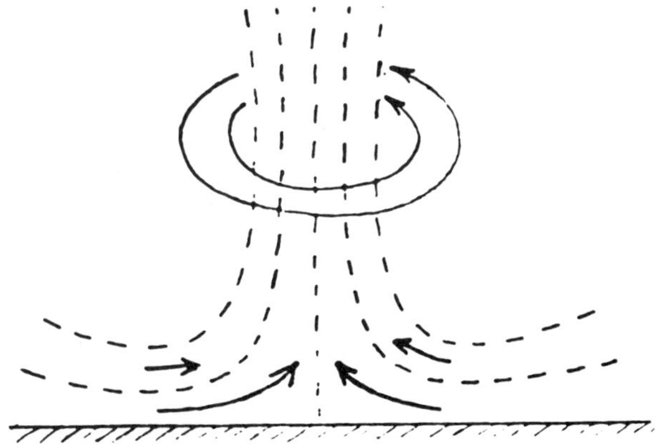

Fig. 8. Tornado-like vortex (——— flow lines, - - - - vorticity lines)

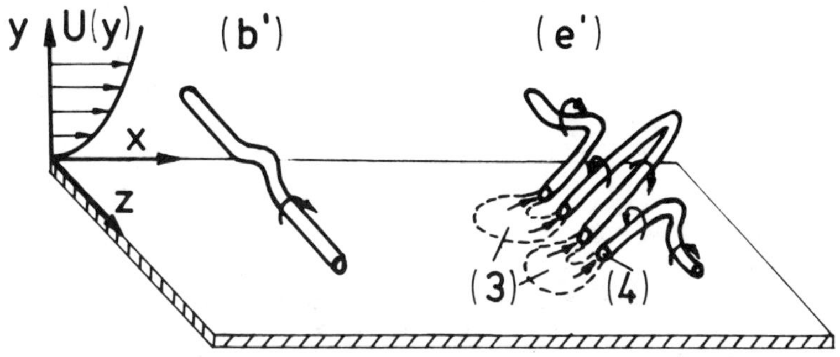

Fig. 10. Variation of the process sketched in Fig. 9

(b) Small disturbance in downstream direction.

(e) Two vortex loops reach the wall area - hence generation of two pairs of counter rotating tornado-like vortices.

As the induction strength between parts of a vortex tube depends on the curvature of the vortex tube it can be expected that elongated vortex loops (see Fig. 9) would not stay in one plane, but would deform in the way as sketched in Fig. 11. For diagram simplicity this was not included in Figs. 9 and 10.

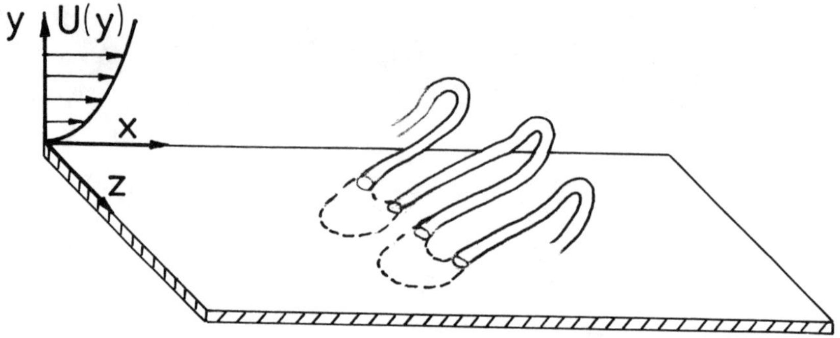

Fig. 11. Expected deformation of elongated vortex loops under the action of induction

If one considers how the history of the vortex tube model sketched in Fig. 9 might continue, the following could be expected: Firstly, more loops will reach the wall area and hence more pairs of counter-rotating tornado-like vortices will be generated. This process may occur simultaneously for several loops. Secondly, it seems not unlikely that the "open ends" of the vortex tube join with one another in the course of time. Depending on whether they join with their "right- or left-sided" partner either closed vortex loops are formed, or the old arrangement is restored and the game begins again. Thirdly, interaction with preceeding or succeeding loops of the same type may occur.

The question as to what size the proposed flow structures might have, cannot be answered clearly. However, from the fact that the skewness of the wall pressure fluctuations decreases strongly (see Fig. 6) if one replaces a pressure transducer of 2.1 mm diameter (corresponding to $d^+ = 40$) by a transducer of 1 mm diameter (corresponding to $d^+ = 19$) it can be concluded that the extension of the "open ends" of the vortex tubes is of the order of $l^+ = 20$ or smaller. From the fact that the 3rd moments remain negative even if the signals of 4 transducers aligned in spanwise direction are added (breadth of the transducer $l^+ = 245$, see Fig. 7) it can be concluded that the spanwise extension of the structure generating the low pressure patterns is of the order of $l^+ = 200$ or larger. As this extension of the pressure pattern in spanwise direction may be caused by several pairs of counter-rotating vortices it seems to be not unlikely that the individual loops have a spanwise extension of the order of $l^+ = 100$. This order of magnitude has been found for "longitudinal streaks" in turbulent boundary layers by several investigators.

With the question as to how often tornado-like vortex structures occur in a flow, it is clear that the answer depends on the definition of the structures. If one takes only the very pronounced low pressure events into account the numbers are small. (In the 130 ms of the time functions plotted in Fig. 4, five events with $p \leq 3\, p_{RMS}$ were found. The duration of these five events corresponds to about 5 % of the total time.) If one takes also smaller pressure minima into account the number of events becomes considerable larger.

4. DISCUSSION

The existence of such flow structures as those discussed in the previous section in developed turbulent boundary layers is still hypothetical. In the transition flow regime, however, flow structures which look very similar to those in Fig. 9 have been observed by many investigators, e.g. by Hama, Long and Hegarty [11], by Matsui [12] and by Knapp and Roache [13].

Breidenthal [14] recently investigated flow structures in the mixing layer between two parallel water streams of different flow velocities. With the help of a very sensitive visualization technique he was able to visualize details of the mixing process between the two streams. Fig. 12 shows a plan view of the mixing layer. As can be seen there is considerable similarity between the vortex tube picture given in Fig. 9, and the flow structures visualized in Fig. 12.

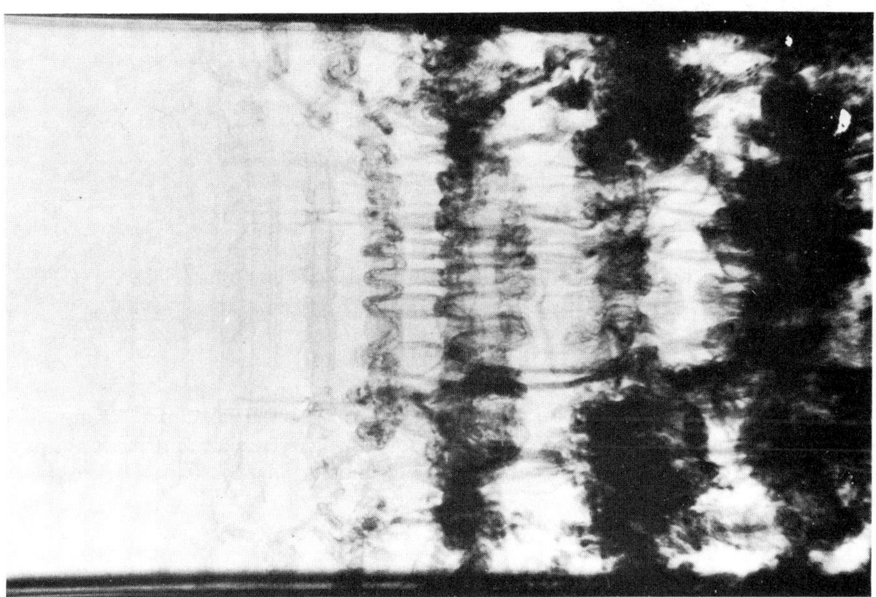

Fig. 12. Plan view of the mixing layer between two streams of water flowing with different flow velocities. (Investigation of Breidenthal, Ref. 14. Flow from left to right. Flow visualization with the help of the reaction between dilute phenolphthalein (stream 1) and dilute sodium hydroxide (stream 2). Both reactants are transparent, and only the product generated in the mixing layer between the two streams is visible.)

Furthermore the flow model discussed in the previous section seems to fit very well with results of many other investigations on turbulent boundary layers - for example:

- with the existence of pairs of counter-rotating longitudinal vortices in the vicinity of the wall as suggested by Bakewell and Lumley [15],
- with the existence of low speed and high speed streaks as observed by Kline, Reynolds, Schraub and Runstadler [16],
- with the occurrence of "ejection events" as observed by Corino and Brodkey [17] (comp. Fig. 13),
- with the occurrence of high spanwise velocity gradients in the vicinity of the wall, as observed by Willmarth and Bogar [18].

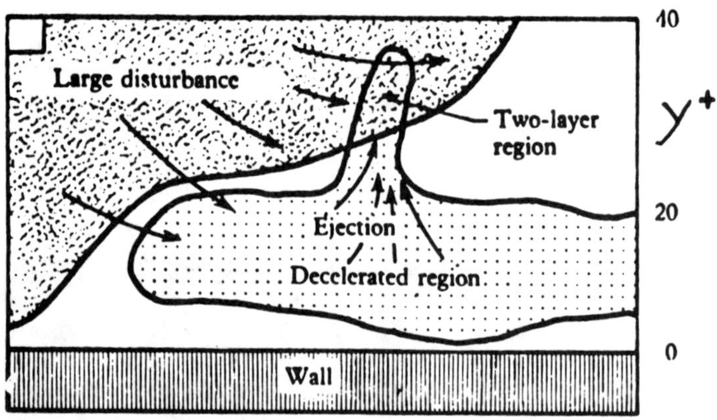

Fig. 13. "Ejection event" from a sequence of flow events observed in a moving frame of reference. (Investigation of Corino and Brodkey, Ref. 17, flow from left to right.)

From the previous discussion it seems highly probable that flow structures with pairs of counter-rotating "tornado-like" vortices occur in turbulent boundary layers and that these flow structures do play a role in the process of turbulent mixing.

REFERENCES

1. Emmerling, R. 1973. Die momentane Struktur des Wanddruckes einer turbulenten Grenzschichtströmung. Mitteilungen aus dem MPI für Strömungsforschung und der AVA, Göttingen, Nr. 56.

2. Dinkelacker, A., Hessel, M., Meier, G.E.A. und Schewe G. 1977. Investigation of pressure fluctuations beneath a turbulent boundary layer by means of an optical method. Phys. Fluids Vol. 20, No. 10,

Pt. II, October 1977, S216-S224.

3. Hessel, M. 1978. Untersuchungen von Wanddruckfeldern unter turbulenten Grenzschichten. MPI für Strömungsforschung, Göttingen, Bericht 11/1978.

4. Schewe, G. 1979. Untersuchung von Wanddruck- und Wanddruckgradientenschwankungen unter einer turbulenten Grenzschichtströmung. Mitteilungen aus dem MPI für Strömungsforschung, Göttingen, Nr. 68 A.

5. Langeheineken, Th. 1979. Experimentelle Untersuchung der Wanddruckschwankungen einer ausgebildeten turbulenten Rohrströmung. MPI für Strömungsforschung, Göttingen, Bericht 106/1979.

6. Schewe, G. 1980. Über charakteristische Strukturen in Zeitfunktionen von turbulenten Wanddruckschwankungen. MPI für Strömungsforschung, Göttingen, Bericht 1/1980.

7. Dinkelacker, A. 1979. Play tornado-like vortices a role in the generation of flow noise? Mechanics of Sound Generation in Flows, Editor: E.-A. Müller, Springer Verlag Berlin Heidelberg New York 1979, pp. 217-223.

8. Theodorsen, Th. 1952. Mechanism of turbulence. Proceedings of the Second Midwestern Conference on Fluid Mechanics, Ohio Stat University.

9. Theodorsen, Th. 1955. The structure of turbulence. 50 Jahre Grenzschichtforschung (Ed. H. Görtler u. W. Tollmien), Braunschweig: Vieweg und Sohn, pp. 55-62.

10. Tu, B.J. and Willmarth, W.W. 1966. An experimental study of the structure of turbulence near the wall through correlation measurements in a thick turbulent boundary layer. University of Michigan, Technical Report, 02920-3-T.

11. Hama, F.R., Long, J.D. and Hegarty, J.C. 1957. On transition from laminar to turbulent flow. Journ. of Applied Physics. Vol. 28, No. 4, pp. 388-394.

12. Matsui, T. 1963. A study on transition in a boundary layer by flow visualization. Research Report No. 13, Faculty of Engineering, Gifu University, Gifu Japan.

13. Knapp, C.F. and Roache, P.J. 1968. A combined visual and hot-wire anemometer investigation of boundary-layer transition. AIAA Journal Vol. 6, No. 1, pp. 29-36.

14. Breidenthal, R.1980. Reacting flow visualization of a turbulent shear layer and wake. International Symposium on Flow Visualization, Bochum, 9.-12. Sept. 1980.

15. Bakewell, H.P. and Lumley, J.L. 1967. Viscous sublayer and adjacent wall region in turbulent pipe flow. Phys. of Fluids Vol. 10, pp. 1880-1889.

16. Kline, S.J., Reynolds, W.C., Schraub, F.A. and Runstadler, P.W. 1967. The structure of turbulent boundary layers. Journ. Fluid Mech. Vol. 30, pp. 741-773.

17. Corino, E.R. and Brodkey, R.S. 1969. A visual investigation of the wall region in turbulent flow. J. Fluid Mech., vol. 37, part 1, pp. 1-30.

18. Willmarth, W.W. and Bogar, T.J. 1977. Survey and new measurements of turbulent structure near the wall. The Physics of Fluids, Vol. 20, No. 10, Pt II, October 1977, S 9-S21.

The Structure of a Turbulent Boundary Layer and Its Relation to the Wall Pressure Fluctuation

YASUJIRO KOBASHI, MAKOTO ICHIJO, and SHOJI SAKURAI
Faculty of Engineering
Hokkaido University
Sapporo 060, Japan

ABSTRACT

The wall pressure fluctuation and its relation to the turbulence structure of a boundary layer have been investigated experimentally. The effect of common mode noise of the pressure which are known to be caused by the vibration of the wall and the fan were carefully removed and various correlation analyses together with ensemble averagings which refer either to the bulge or the burst were carried out. The correlation measurements show that the wall pressure fluctuation is very much related to the flow fluctuation whole through and beyond the boundary layer. The ensemble average of the pressure fluctuation reveals that the bulge and the burst are both responsible to the wall pressure fluctuation, although the latter is more direct and effective. Various correlations suggest a strong causality between the bulge and the burst.

1. INTRODUCTION

Owing to the works of Kaplan and Laufer (1) and Kovasznay, Kibens and Blackwelder (2) , the flow structure of the outer intermittent region of a turbulent boundary layer has been made clear. It has a shape just like a chain of bumpy lumps of fluid, which they called as the bulge, and moves a little slower than the outer flow (at $0.94U_0$, where U_0 is the outer flow speed). The mean streamwise scale of the bulge is about 2.5 times of the boundary layer thickness and within which there is a slowly circulating flow in the direction of the mean vorticity. The fluid outside of the bulge flows around it and is entrained somewhere between the adjoining ones.

As to the flow structure of the wall region, Kline, Reynolds, Schraub and Runstadler (3) discovered by the visual studies the so called burst phenomenon which is accompanied by the ejection and sweep events. Since then, the phenomenon has been studied by many authors such as Wallace, Eckelmann and Brodkey (4) and Lu and Willmarth (5) as a subject of the pattern recognition study and its structure and contribution to the origin of the Reynolds stress have been clarified. Blackwelder and Kaplan (6) developed an ingineous technique which is called as the variable interval time averaging (VITA) method to adjudge the burst event quantitatively and succeeded in finding the flow pattern associated with the burst.

There is no doubt now that there are coherent structures inside of the boundary layer. However, there still remains a question if there is any relation between the bulge and the burst. Rao, Narasimha and Badri Narayanan (7) , and Ueda and Hinze (8) measured the mean period of the burst occurence and found that it is about two times of that of the bulge. Sakurai, Ishiguro, Ichijo and Kobashi (9) made the measurement of the space-time correlation both in the outer and inner layers of a turbulent boundary layer and found that the iso-correlation contours are very similar in shape and has almost the same size in streamwise direction although the contour in the inner layer is much slenderer than that of the outer layer. These facts seem to suggest the causality between them but are not enough to establish the relation. In this connection, Offen and Kline (10) proposed a hypothetical model of the burst structure to explain its quasi-cyclic nature.

Brown and Thomas (11) have proposed a structural model of a turbulent boundary layer based on their correlation measurements and Falco's visualization results (12). The adequacy of the sample they employed was, however, questioned by Falco himself.

The systematic measurements of the wall pressure fluctuation beneath the turbulent boundary layer were made by Willmarth and Wooldridge (13). Later, Willmarth and Tu (14), Bradshaw (15), Willmarth (16), Panton, Goldman, Lowery and Reischman (17) and others worked on the relation between the wall pressure fluctuation and the flow structure of the boundary layer. They found that the wall pressure fluctuation is closely correlated with the flow fluctuation and travels downwards as a group nearly at a constant speed. Non of them, however, could not realize whether it is caused by the bulge or the burst. Blackwelder and Woo (18) tried to find the dependence of the burst structure on the pressure fluctuation with a negative result.

The present experiment was carried out with the aim to find the relation, if there is any, between the bulge and the burst with the aid of the wall pressure information. The space-time correlations between the velocity component and the wall pressure and their ensemble averages refered to the bulge and the burst were measuered for this purpose.

For the measurements of the wall pressure, a condenser microphone type pressure transducer combined with a pin-hole tube was used in order to satisfy the contradicting demands of the sensitivity and the resolvability. Such a system of a pressure pick-up is known to deserve the penalty of the Helmholtz-resonance, but it was improved by reducing the cavity volume and the length of the tube as small as possible and using a damper and an electronic compensation network. The frequency response of the apparatus was flat up to 4 kHz, which was good enough for the present measurement.

Another problem we encounter when we measure the wall pressure fluctuation in a wind tunnel is the noise inherent to the wind tunnel system such as the vibration of the wall and disturbances caused by the fan. These are vital for the correlation measurements even if they are small in magnitude. In this experiment, they were compensated by using a properly located dummy probe, because they are a kind of common mode noise which prevail through the test section almost simultaneously.

The measurements were carried out at a constant flow speed, and from the simultaneous signals of the velocity and pressure fluctuations, various correlations and ensemble averages were analyzed.

The space-time correlations between the wall pressure and the velocity fluctuations at various distances from the wall show the existence of a coherent flow structure whole through and far beyond the boundary layer, which is responsible to the pressure fluctuation at the wall.

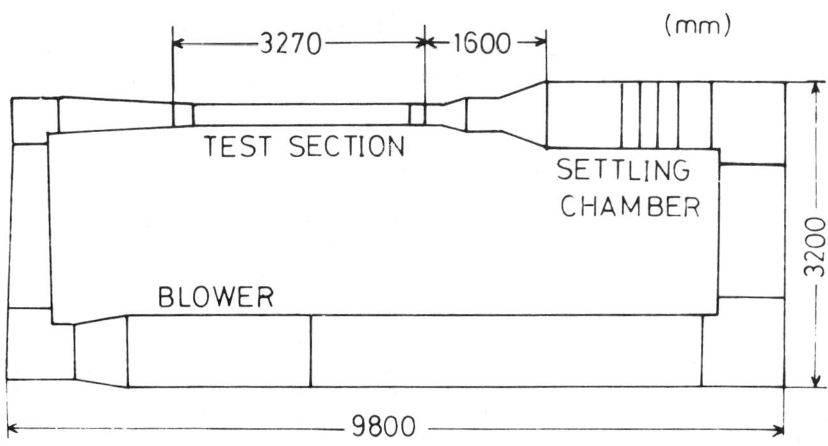

Fig.1 Wind Tunnel

The ensemble averages refered to the bulge and burst seem to suggest that both are related in their occurence. The burst appears between the valley and the peak of the bulge where the pressure induced by the main flow fluctuation is increasing.

2. APPARATUS AND PROCEDURE

2.1 Wind Tunnel

The experiments were carried out in the low turbulence wind tunnel of a return duct type whose test section has a 20cmx60cm cross section and is 3m long (Fig.1).

The measurements were made in the boundary layer of the side wall which was artificially developed by a two-dimensional roughness of 1mm diameter mounted at 15cm from the entrance of the test section.

Most of the data were taken at the center section of the wall and 140cm downstream of the roughness element, where the effect of the roughness is completely vanished. The flow conditions at this point were

main flow speed	$U_0 = 11.4$ m/s
main flow turbulence intensity	$u'/U_0 = 0.0002$
boundary layer thickness	$\delta = 23$ mm
friction velocity	$u_\tau = 0.46$ m/s
pressure fluctuation intensity	$p'/\ \rho U_0^2/2 = 12 \times 10^{-3}$

2.2 Hot-Wire Anemometer

For the flow measurements, the conventional hot-wire anemometer of a constant temperature type was used together with the linearizing unit. The U-type (single) and X-type probes were employed for the streamwise (u-) and normal (v-) flow components.

The probes were mounted on a traversing mechanism so that they could be moved separately in streamwise (x-), normal (y-) and spanwise (z-) directions.

The distance of the hot-wire from the wall surface was carefully calibrated by applying the heat-loss method.

2.3 Pressure Apparatus

In order to satisfy the contradicting demands of the pressure pick-up, the sensitivity and the resolvability, a condenser microphone type pressure transducer of 10mm diameter combined with a pin-hole tube of 1mm inner diameter which was mounted flush in the wall was employed.

The frequency response of the pressure pick-up, which otherwise is characterized by the Helmholtz-resonance when it is combined with a pressure tube, was improved up to 4 kHz by reducing the cavity volume and the length of the tube as small as possible and using a damper and an electronic compensation network.

2.4 Common Mode Noise Rejection

It is known that the wall pressure of a wind tunnel is superposed by undesirable disturbances such as the vibration of the wall and the noise caused by the fan which are inherent to the wind tunnel system and the operation condition.

These disturbances are observed in the main flow ,too, as a part of the free stream turbulence. They are small in magnitude but can not be neglected as far as the correlations of the fluctuating quantities are concerned.

These disturbances appear almost simultaneously whole through the test section, which therefore we call as the common mode noise. They can be compensated by using a properly located dummy probe. The location of the probe must be carefully selected, but in our case it was found to be good enough if it was located four times of the boundary layer thickness appart spanwisely from the measuring probe.

2.5 Measurements and Data Analyses

The velocity and the wall pressure fluctuation were measured at the fixed point (x=140cm, z=0cm) under the constant flow speed condition.

All the data were stored on the TEAC R-210A Data Recorder and were analyzed by the MELCOM 70/25 Mini Computer into the forms of space-time correlations and the ensemble averages refering to the bulge and the burst. The sampling rate was mostly fixed at 0.1ms per each channel.

3. EXPERIMENTAL RESULTS

3.1 Mean Flow Characteristics of Boundary Layer

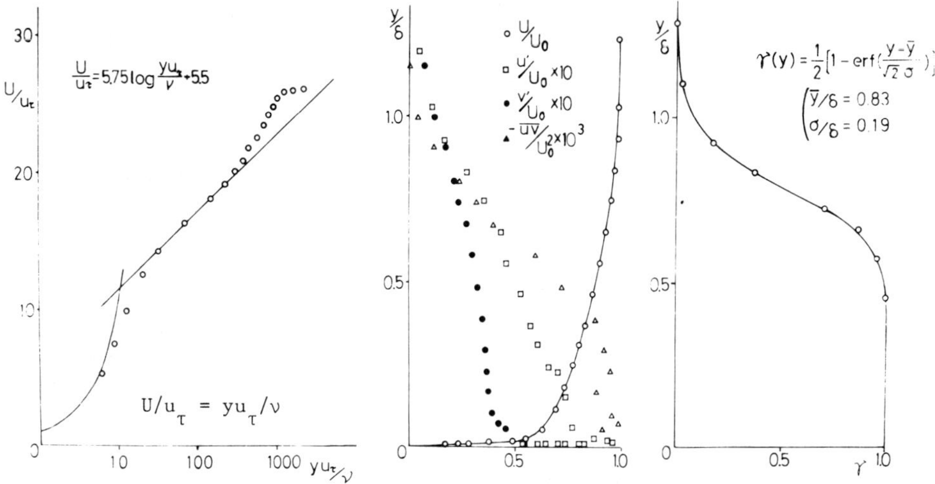

Fig.2 Mean Characteristics of the Turbulent Boundary Layer

The mean velocity, the turbulence intensity and the Reynolds stress are shown in Fig.2 together with the intermittency factor (γ) against the non-dimensional distance from the wall ($y^+=yu_\tau/\nu$, and $Y=y/\delta$). They show the typical features of the boundary layer of a flat plate in a uniform flow.

3.3 Spectra of Velocity and Pressure Fluctuations

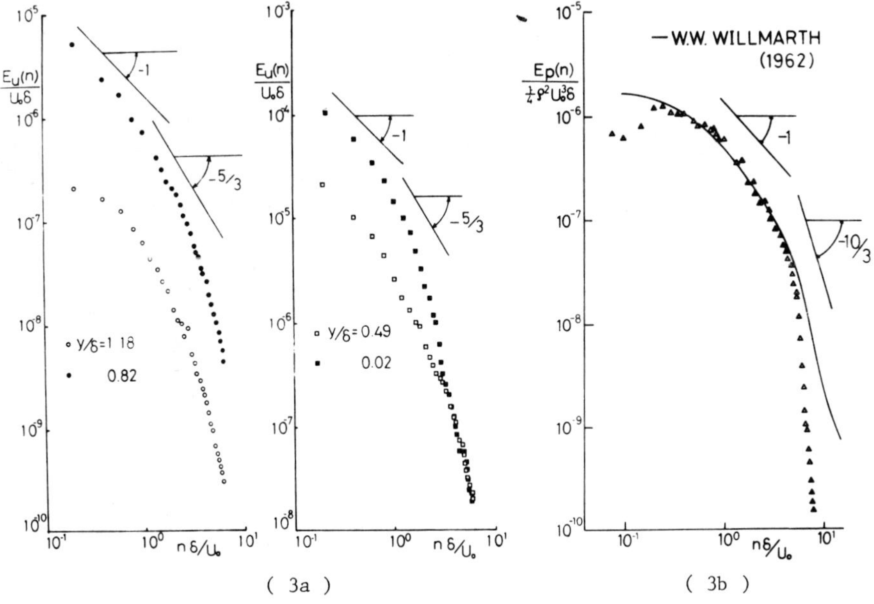

Fig.3 Spectra of Velocity and Wall Pressure

The spectra of the streamwise velocity fluctuation ($E_u(n)/U_0\delta$) at various distances from the wall surface and that of the wall pressure fluctuation ($E_p(n)/\rho^2U_0^3\delta/4$) are shown in Fig.3 against the nondimensional frequency ($n\delta/U_0$). Here, n is the frequency of the spectral component.

The specrum of the pressure fluctuation free from the common mode noise is similar to that of Willmarth and Wooldridge which is shown in the figure for comparison exept that the high frequency component decreases a little faster in our case.

3.3 Correlation between Velocity and Wall Pressure Fluctuations

In order to find the relation between the wall pressure fluctuation and the turbulence of a boundary layer, the space-time correlations of the velocity components to the wall pressure

$$R_{pu_i} = \frac{\overline{p(X_0,Y_0,Z_0,T_0)\ u_i(X_0,Y_0+Y,T_0+T)}}{p'(X_0,Y_0,Z_0)\ u_i'(X_0,Y_0,Z_0)} \qquad (1)$$

are calculated at various distances from the wall ($Y=y/\delta$). Here $X = x/\delta$, $Z = z/\delta$ and $T = U_0t/\delta$ is the nondimensional time. The sufix 0 represents the reference position of the probe and the time. u_i denotes the velocity components u (streamwise) and v (normal) corresponding to i = 1 and 2. ( )' means the r.m.s. value of the fluctuating quantity.

The correlation between the wall pressure and the Reynolds stress were also calculated. The results are shown in Figs.4a, 4b and 4c, all of which show the evidence that there exists a coherent structure whole through and far beyond the boundary layer.

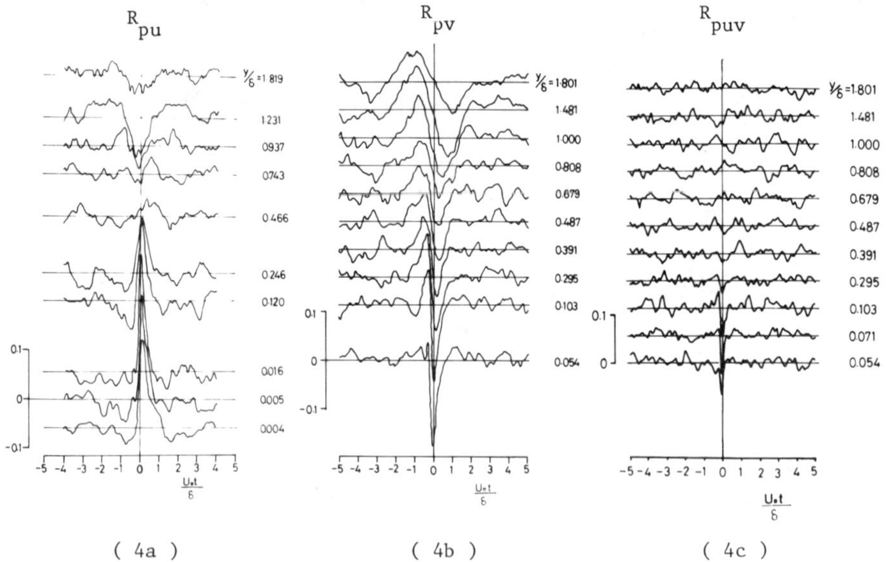

Fig.4 Correlation between Velocity and Wall Pressure

The results agree with those of other authors, but it is to be mentioned here that these correlations, except that of the Reynolds stress, are independent of the sign of the pressure fluctuation. The fact is important to understand the flow structure of the boundary layer, especially of the inner layer.

3.4 Ensemble Averages refered to Bulge

Figs.5a and 5b show the streamwise velocity fluctuation at various distances from the wall and that of the wall pressure which are sampled and averaged by the bulge. The bulge detection was made by the reference hot-wire placed outside of the turbulent region (Y=1.7) where the flow is known to fluctuate according to the bulge passage (Phillips (19)). The triggering time or the start time of sampling was given by the positive and negative peaks of the hot-wire signals or more practically by the zero-crossings from positive to negative and those from negative to positive of the differentiated signals. For the latter case, however, the low frequency electric noise (hum) must be carefully removed because the flow fluctuation is very small and the shift of the zero-crossing becomes vital. < > in the figure represents the ensemble average and the sufix 1 denotes the position of the detection probe.

The mean time interval of the bulge detections was 4.5ms or T =2.3, which is nearly same as the average period of the bulge reported so far.

The figures show that there exists a deterministic flow structure associated with the bulge, and that the outer flow fluctuation is responsible to the pressure fluctuation at the wall. The positive velocity fluctuation in the outer flow causes the negative wall pressure and the negative flow velocity produces the positive wall pressure. This is the direct effect of the outer flow fluctuation on the wall pressure, and is consistent with the negative value of the correlation between the external velocity and the wall pressure shown in Fig.3a.

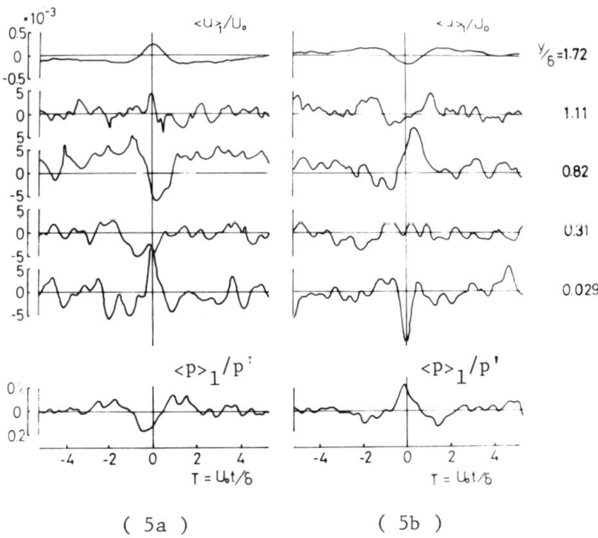

Fig.5 Ensemble Average of Velocity and Wall Pressure Fluctuations refered to Bulge

As to the positive value of the pressure-velocity correlation in the inner layer, however, the present result can not offer any reasonable explanation.

3.5 Ensemble Averages refered to Burst

In order to find the reason of the confusing results between the correlation R_{pu} and the ensemble averaged quantities in the inner layer which are refered to the bulge, the ensemble averages of the burst refered velocity and wall pressure fluctuations were calculated and the results are shown in Fig.6. The detection of the burst was made by using the variable interval time averaging (VITA) method proposed by Blackwelder et al. In our experiments, the burst detecting probe was placed at $y^+ = 25$ or $Y = 0.03$, which is denoted by the sufix 2 in the figure. Here $y^+ = yu_\tau/\nu$ is the nondimensional distance from the wall, and u_τ and ν are the friction velocity and the kinematic viscosity of the air.

The figure shows that the typical flow structure of the burst which is characterized by the ejection and the sweep events is limited within the wall region, but that it is strongly related to the flow structure all through the boundary layer. It is remarkable that the ensemble averaged flow outside of the turbulent region is nearly the same both in shape and magnitude as that refered to the bulge itself. This fact seems to suggest that the burst is synchronized to the the bulge passage, though the mean period of the burst detection was T = 4.57, which is just twice of that of the bulge. We can see from the figure that the burst occurs nearly at the mid point between the valley and the peak of the bulge.

As long as the effect on the wall pressure is concerned, the burst is more direct and effective than the bulge. The ensemble average of the wall pressure fluctuation resembles the flow structure related to the burst except that it has a sharp positive peak at the moment of ejection event. This resemblance and almost antisymmetrical character of the burst and the wall pressure can explain

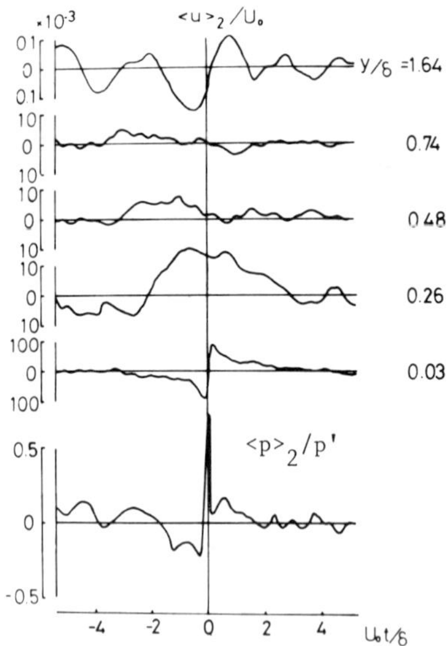

Fig.6 Ensemble Average of Velocity and Wall Pressure Fluctuations refered to Burst

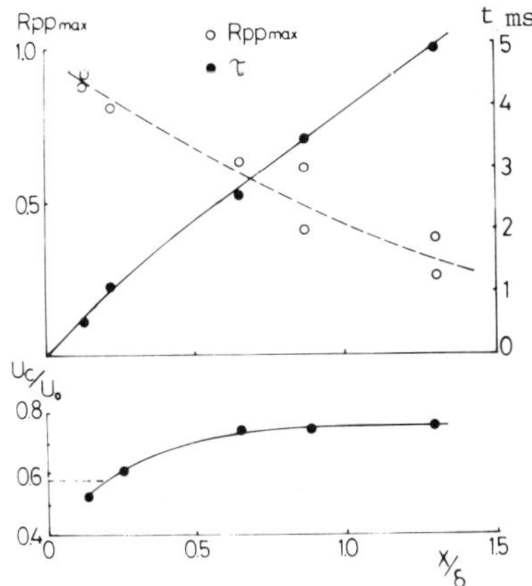

Fig.7 Decay and Convection Velocity of Wall Pressure (Calculated from Rpp)

the positive correlation between the wall pressure and the velocity near the wall independently of the sign of the pressure fluctuation.

3.6 Movement of Coherent Structure

From the simultaneous measurement of the wall pressure fluctuation at various streamwise separations, the decay and the convection velocity of the pressure were calculated and the results are shown in Fig.7. Because of the lousiness of the correlations the decay curve is not so smooth, from which, however, we can see that the wall pressure damps gradually and disappears within two or three times of the boundary layer thickness. The convection velocity (U_c) varies from the lower value ($0.53U_0$) at the begginning to $0.7\ U_0$ at its end.

The results agree with those reported by Willmarth et al and others except that the terminal convection velocity in our case is a little smaller.

In order to make clear the reason of this curious value of the convection velocity of the pressure fluctuation, the movement of the wall pressure fluctuations refered separately to the bulge and the burst were measured.

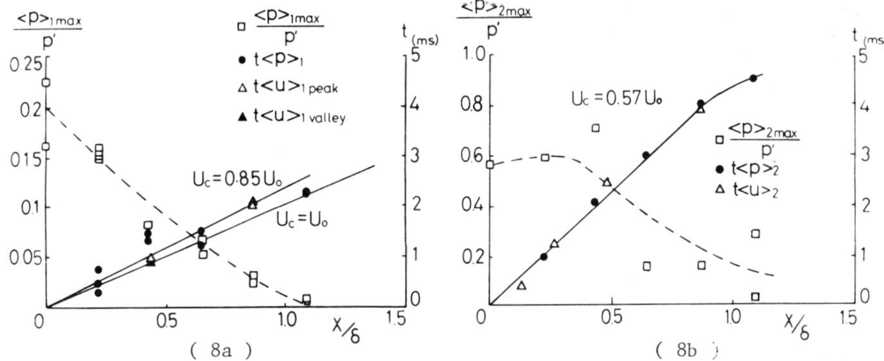

Fig.8 Decay and Convection Velocity of Wall Pressure (Calculated from Ensemble Average)

The results are shown in Figs.8a and 8b, in which the peak values of the ensemble averaged wall pressure $<p>_{1max}$, $<p>_{2max}$ and the delay time after the detection are shown aganist the nondimensional streamwise distance (X=x/δ). In the figures the traveling velocities of the bulge and the burst are also shown. Again, the decay curves are not smooth, but they show that the bulge and the pressure refered to it travel at nearly the same speed as the main flow, while the burst refered one is convected at half speed of the main flow and exactly the same as that of the burst itself. It is clear that the lower velocity of the wall pressure of Fig.7 is caused by the burst whose effect is more predominant there. The terminal velocity, however, can be reasoned from the present data.

4. CONCLUDING REMARKS

The wall pressure fluctuation has been studied in connection with the flow structure of the turbulent boundary layer.

The common mode noise inherent to the wind tunnel measurements were removed by using a dummy probe properly located in the test section.

The space-time correlations of the velocity components with respective to the wall pressure fluctuations calculated from the simultaneous measurements show that the wall pressure is very much correlated with the velocity fluctuation all through and far beyond the boundary layer, suggesting that there exists a coherent flow structure in the boundary layer which affects the pressure fluctuation of the wall.

The flow structure and related wall pressure fluctuation were measured separately refering to the bulge and the burst. The bulge was detected by the peaks of the flow fluctuation outside of the turbulent region and the burst detection was made by the VITA method proposed by Blackwelder et al.

The mean streamwise scale or the mean wave length of the bulge is 2.25 times of the boundary layer thickness. The wall pressure and the flow inside of the bulge, except the region very near the wall, fluctuate inversely to the outer flow, which shows that the pressure change caused by the outer flow fluctuation prevails all through the boundary layer. The inversion of the phase of the flow fluctuation is caused by the upward movement of the low momentum fluid in the boundary layer. The flow near the wall, on the other hand, is governed by the pressure field and varies in phase with the outer flow.

The mean period of the burst appearance is two times of that of the bulge. The flow field related to the burst shows that its typical structure is limited within the wall region, but that it is closely related to the flow structure of the outer layer and to the outer flow fluctuation. As to the pressure of the wall, the burst is more effective than the bulge. It has a sharp positive peak at the moment when the burst (the interface of the ejction and the sweep) appears, but otherwise it has a very similar feature as the burst. The related flow structure outside and inside of the bluge is nearly the same as that refered to the bulge. These facts have led us to conclude that the burst is the passive phenomenon subordinating to the bulge.

The guess that the burst is the subordinating phenomenon to the bulge seems to be correct as far as its dependency on the outer flow fluctuation and its correlation to the wall pressure are concerned, but some of the results cast a doubt on this conclusion especially when we think of the structural model of the boundary layer. The most serious problem is the difference of the velocities which were detected from the wall pressure movement refering to the bulge and the burst separately. It must cause the random phase relationship between the bulge and the burst, contradictingly to the result obtained from the burst triggered flow measurements. If we assume that the burst is an phenomenon independent of the bulge, then it is easier to model the flow structure except that the outer flow fluctuation, which is related both to the bulge and the burst, travels at nearly the same speed as the main flow. But, in this case, the question arises why the flow structure, including the size and the period of occurence, is so much related with each other.

The flow structres related to the bulge and the burst have been made clear, to some extent, with the aid of the wall pressure information, but the causality between them is not uncovered yet. For this purpose, it seems to be necessary to improve the bulge detection technique so that the outer flow fluctuation caused by the bulge can be distinguished from that related to the burst.

REFERENCES

1. Kaplan, R.E. and Laufer, J. 1969. The intermittently turbulent region of the boundary layer. Proc. 12th. Int. Congr. Mech., pp. 236-245.

2. Kovasznay, L.S.G., Kibens, V. and Blackwelder, R.F. 1970. Large-scale motion in the intermittent region of a turbulent boundary layer. J. Fluid Mech., Vol. 41, pp. 283-325.

3. Kline, S.J., Reynolds, W.C., Schraub, F.A. and Runstadler, P.W. 1967. The structure of turbulent boundary layers. J. Fluid Mech., Vol. 30, pp. 741-773.

4. Wallace, J.M., Eckelmann, H. and Brodkey, R.S. 1972. The wall region in turbulent shear flow. J. Fluid Mech., Vol. 54, pp. 39-48.

5. Lu, S.S. and Willmarth, W.W. 1973. Measurements of the structure of the Reynolds stress in a turbulent boundary layer. J. Fluid Mech., Vol. 60, pp 481-511

6. Blackwelder, R.F. and Kaplan, R.E. 1976. On the wall structure of the turbulent boundary layer. J. Fluid Mech., Vol. 76, pp. 89-112.

7. Rao, K.N., Narasimha, R. and Badri Narayanan, M.A. 1971. The bursting phenomenon in a turbulent boundary layer. J. Fluid Mech., Vol. 48, pp. 339-352.

8. Ueda, H. and Hinze, J.O. 1975. Fine-structure turbulence in the wall region of a turbulent boundary layer. J. Fluid Mech., Vol. 67, pp. 125-143.

9. Sakurai, S., Ishiguro, O., Ichijo, M. and Kobashi, Y. 1980. Organized motion near the wall in adverse pressure gradient turbulent boundary layer (in Japanese). J. Japan Soc. Aeron. Space Sci., Vol. 28, pp. 334-342.

10. Offen, G.R. and Kline, S.J. 1975. A proposed Model of the bursting process in turbulent boundary layers. J. Fluid Mech. Vol. 70, pp.209-228.

11. Brown, G.L. and Thomas, A.S.W. 1977. Large structure in a turbulent boundary layer. Phys. Fluids, Vol. 20, pp. S 243-252.

12. Falco, R.E. 1977. Coherent motions in the outer region of turbulent boundary layers. Phys. Fluids. Vol. 20, pp. S 124-132.

13. Willmarth, W.W. and Wooldridge, C.E. 1962. Measurements of the fluctuating pressure at the wall beneath a thick turbulent boundary layer. J. Fluid Mech.

Vol. 14, pp. 187-210.

14. Willmarth, W.W. and Tu, B.J. 1967. Structure of turbulence in the boundary layer near the wall. Phys. Fluids, Vol. 14, pp. 187-210.

15. Bradshaw, P. 1967. Inactive motion and pressure fluctuations in turbulent boundary layers. J. Fluid Mech. Vol.30, pp. 241-258.

16. Willmarth, W.W. 1975. Pressure fluctuations beneath turbulent boyndary layers. Ann. Rev. Fluid Mech. Vol. 7, pp. 13-38.

17. Panton, R.L., Goldman, A.L., Lowery, R.L. and Reischman, M.M. 1980. Low frequency pressure fluctuations in axisymmetric turbulent boundary layers. J. Fluid Mech., Vol. 97, pp. 299-319.

18. Blackwelder, R.F. and Woo, H.H.W. 1974. Pressure fluctuation of a turbulent boundary layer. Phys. Fluids, Vol. 17, pp.515-519.

19. Phillips, O.M. 1955. The irrotational motion outside a free turbulent boundary. Proc. Camb. Phil. Soc., Vol.51, pp. 220-229.

Reynolds Number Dependence of the Large Structure in a Slightly Heated Turbulent Boundary Layer

R. A. ANTONIA, C. S. SUBRAMANIAN, S. RAJAGOPALAN, and A. J. CHAMBERS
Department of Mechanical Engineering
University of Newcastle
New South Wales 2308 Australia

ABSTRACT

Visual identification and conditional measurements of the coherent large structure were obtained using simultaneous temperature traces in a slightly heated boundary layer with zero pressure gradient. A combined X-wire and single cold wire arrangement was used in conjunction with a rake of cold wires to obtain ensemble averages of the products uv, uθ and vθ (u, v are the longitudinal and normal velocity fluctuations, θ is the temperature fluctuation). Before examining the influence of Reynolds number on these ensemble averages, a comparison is made between the visual identification and existing detection methods that have been developed for application at a single location in the flow. None of the detection methods is in good agreement (the best correspondence is about 42%) with the visual method, although all signatures are in qualitative agreement with each other. Using the visual method, the mean period normalised by outer layer parameters, and inclination of the back of the large structure are approximately independent of R_m, the momentum thickness Reynolds number, when R_m is greater than about 1000. The width of the region near the back of the large structure that is associated with a sudden change in temperature is equal to about 0.042δ, and approximately independent of R_m. The contribution of the large structure to the Reynolds shear stress decreases as R_m decreases, and represents about 20% at $R_m \simeq 1000$.

1. INTRODUCTION

Measurements of mean velocity and various turbulence quantities in a turbulent boundary layer with low Reynolds number and in zero pressure gradient have been made and analysed in the literature. Coles (1962) found that the velocity defect in the outer part of the boundary layer was a function of R_m when $R_m \lesssim 5000$. Huffman & Bradshaw (1972) suggested that a breakdown of the velocity defect law may be attributed to the viscous superlayer at the turbulent /non-turbulent interface. Falco (1977) used combined simultaneous flow visualisation and hot-wire anemometry to determine the characteristics of the large scale motion and highly coherent three-dimensional "typical eddies" which were found to be strongly Reynolds number dependent. At $R_m \simeq 600$, the length scale of these eddies was of the order of the boundary layer thickness δ but at $R_m \simeq 6000$, the length scale was one order of magnitude smaller. Falco also found that these eddies, which formed on the upstream side or back of large scale motions at all values of R_m, produced most of the Reynolds shear stress in the outer half of the layer at $R_m = 1200$. Murlis (1975) found that, as R_m was decreased, turbulence measurements revealed a marked change in the structure of the turbulent field; for example, in the outer layer, the ratio of rms values of u and v varied with R_m. As a proportion of the wall shear stress τ_w, the kinematic Reynolds shear stress $-\overline{uv}$ decreased with decreasing R_m, a result also obtained by Wood (1976). Murlis also explored Huffman & Bradshaw's suggestion regarding the importance of the superlayer at small R_m by using conditional sampling based on the detection of the turbulent/non-turbulent interface. The statistics of turbulent bulges revealed, in agreement with Falco's observations, that bulges were longer, up to order δ, at low R_m. Unlike high R_m bulges which are solitary, low Reynolds number bulges are more split, "like a series of hands, fingers outstretched". Smoke-flow visualisation results of Head & Bandyopadhyay (1978) seemed to indicate an even more detailed difference in the appearance of the smoke filled layer at various R_m. At $R_m < 1000$, coherent motions with length scale of order δ had "a more or less" regular appearance. For $1000 < R_m < 7000$, the most characteristic feature of the layer is not the existence of large-scale coherent motions but the random amalgamation of structures, small in the streamwise direction, but highly elongated at about 40°-45° to the wall. These structures, perhaps analogous to Murlis' "fingers", were inferred to represent hairpin vortices. At larger R_m, these structures did not disappear but the large scale motion was found to be more regular.

While there is substantial agreement between the previous, mainly qualitative, observations of the effect of R_m on the motion, there have been few attempts to quantify this effect. In particular reliable contributions of the large structure to the Reynolds shear stress have not been obtained. In the present paper ensemble averaged signatures of velocity and temperature, associated with the large structure, are compared for different values of y/δ and R_m. A preliminary investigation of the R_m dependence of the large structure was made by Subramanian & Antonia (1979) who used an array of four cold wires to identify the upstream interface of the large structure in a slightly heated boundary layer. The use of temperature as a passive marker of the fluid had been explored earlier by Chen & Blackwelder (1978) who used an array of 10 cold wires. In the present investigation, a rake of 11 cold wires is used to permit a more representative coverage of the spatial extent of the large structure. However, as a conditional sampling technique is used to generate ensemble averaged characteristics of the large structure and since (e.g. Antonia, 1980, 1981) quantitative agreement between signatures obtained using different detection methods is relatively poor, it was felt important to compare the present detection method with those already available in the literature. Such a comparison is given in Section 3. Results obtained using a conditional sampling technique based on the visual identification of the large structure from the rake traces are presented and discussed in Section 4.

2. EXPERIMENTAL ARRANGEMENT AND CONDITIONS

The boundary layer developed, with zero pressure gradient, over the smooth wall of the working section (0.38 m × 0.23 m) of the wind tunnel. The maximum speed attainable in the working section is about 20 ms^{-1} and the free stream turbulence level is less than 0.5% for the range of speeds considered here. The first 3 metres of the working section floor consists of 20 identical heated Iconel strips (30 cm × 15 cm × 0.46 cm) supported by a Sindanyo (hard asbestos board) base plate. The strips are insulated from each other and connected in series to provide a uniform wall heat flux distribution. A.C. heating is provided by a transformer operated at 60V, and between 45 and 90 amps current. The last 1.83 metres of the working section floor are unheated and consist of epoxy coated Sindanyo board. A 3 mm dia. wire was placed on the heated wall at 4 cm from the entrance of the working section to trip the boundary layer. This location of the trip was determined by trial and error so that the origins of the aerodynamic and thermal layers were effectively coincident for all free stream speeds considered.

All temperature measurements were made with a 0.6 μm Pt-10% Rh cold wire (temperature coefficient of 1.5×10^{-3}°C^{-1}) operated by a constant current (0.1 mA) anemometer. Mean velocity measurements were made with a 5 μm Pt-10% Rh wire, operated by a DISA 55M10 constant temperature anemometer at an overheat ratio of 1.8. A rake of 11 cold wires (0.6 μm Pt-10% Rh) was constructed using double sided printed circuit board. Each probe was made by soldering 0.5 mm dia. × 5 mm long brass pins to the copper tracks on each side of a narrow strip (2 × 1 × 65 mm) of printed circuit board. The probes were rounded, coated with electrical insulating material and supported on one side of a dual in line edge connector. A triangular section of circuit board was clamped to the other side of the edge connector to strengthen and stabilise the rake and to which the leads, connecting each probe to a constant current bridge, were joined. The probes were arranged so that the distance between wires was about 6 mm. Wires closest and farthest from the wall were at nominal values of $y/\delta = 0.007$ and 0.93 respectively. Simultaneous measurements of u, v, θ were made with an X-wire and a cold wire located 1.23 mm upstream of the geometrical centre of the X-array and in a direction perpendicular to the plane of the X-array. The X-wire/cold wire arrangement was traversed across the boundary layer at a lateral separation of 13 mm from the rake. The X-wire/cold wire arrangement was initially calibrated for velocity and temperature in the heated core of a plane jet. The X-probe (the hot wire resistances were matched to within 1%) was operated by two DISA 55M10 anemometers and DISA 55M25 linearisers. The contamination of the velocity signal by temperature was removed using a technique similar to that suggested by Champagne (1979).

The instantaneous linearised voltage E of a single hot wire is $E = E(U,S)$ where U and S are instantaneous velocity and temperatures respectively. The fluctuating component can be written as

$$e = \alpha u + \beta \theta$$

where $\beta = \partial\overline{E}/\partial\overline{S}$ and $\alpha = \partial\overline{E}/\partial\overline{U}$.

By analogy, the X-wire voltages may be written

$$e_1 = \alpha_1 (u + v \cot \psi_1) + \beta_1 \theta$$

$$e_2 = \alpha_2 (u - v \cot \psi_2) + \beta_2 \theta \quad .$$

The constants α_1, α_2, ψ_1 and ψ_2 (effective inclinations of the hot wires to the free stream direction) are found by calibrating the hot wires in the isothermal stream for different speeds, and different yaw angles, while β_1 and β_2 are found by varying the temperature of the jet for a given speed and with zero yaw. In the present case, $\alpha_1 = \alpha_2$ and $\beta_1 = \beta_2$. The velocity contamination of the temperature signal was found to be approximately 0.006°C/ms^{-1} and no correction was necessary. A correction was however applied to the digital time series corresponding to the fluctuating temperature sensor voltage to compensate for the longitudinal separation between the X-wire and the cold wire. For the pre-ent sampling frequencies, a delay corresponding to $0.8\ \nu/U_\tau^2$ and $3.3\ \nu/U_\tau^2$ at the lowest and highest R_m respectively was applied to the temperature signal.

Signals from the rake and X-wire/temperature arrangement were recorded on an 8 channel FM tape recorder (HP3968A) at a speed of 38.1 cms^{-1} and subsequently digitised at roughly twice the Kolmogorov frequency at each value of flow speed

U_1, using suitable reduction in playback speed of the analogue tape recorder. Digitised record lengths were between 100 to 200 s duration. Signals proportional to u and v were obtained from the X-wire/cold wire voltages using analogue computer elements with allowance for the temperature contamination of the velocity. Digital records were stored on digital magnetic tapes and magnetic discs before processing on the PDP 11/34 computer.

Although measurements were made at various values of the free stream speed, the main experimental results presented in this paper were obtained at values of U_1 equal to 2.1, 8.4 and 18.8 ms^{-1}, corresponding to values of R_m equal to 990, 3100 and 7100. All measurements were obtained at a distance from the trip of 2.24 m.

3. COMPARISON BETWEEN VISUAL, IDENTIFICATION AND OTHER DETECTION METHODS

Although flow visualisation studies have provided useful, but usually qualitative, information about coherent structures in turbulent shear flows, most of the available quantitative information has been obtained with the use of the conditional sampling and averaging technique. For the present purpose, conditional sampling is defined as the eduction of information about a coherent structure with respect to the time reference provided by the detection method used to recognise the structure. While conditional sampling has helped (e.g. Kaplan, 1973; Antonia, 1980, 1981) to bridge the gap between data collected in Eulerian and Lagrangian reference frames, differences exist between quantitative results obtained by different detection methods which supposedly focus on the same structure. The onus is therefore on the experimentalist to demonstrate that the structure recognised is not an artifact of the detection method used. Most, if not all, investigators who have used conditional sampling and averaging have ensured that their particular detection method does not educe a structure when it is applied to a pseudo-turbulent structureless signal. This test, although important, is not sufficient and there is clearly a need to establish the validity and accuracy of the quantitative data obtained from conditional sampling before the contribution of the coherent structure to the momentum or heat flux can be trusted. There is also a need for a critical comparison of quantitative data obtained by different detection methods. Offen & Kline (1973) used simultaneous hot-film and dye visualisation measurements in a boundary layer to compare five kinds of conditional methods, including VITA (variable interval time averaging), a detection scheme used by Blackwelder & Kaplan (1976). The combined anemometer/dye experiments indicated that none of the detection schemes correlated very well with the visual indications of bursting or with any other scheme. These authors stated that there remained serious doubts or questions about what events were measured by each detection scheme. Nonetheless, they found that most of the methods detected at least some portion of 43 - 59% of the visual ejections when the schemes were adjusted to generate about the same total number ($\simeq$ 80) of ejections. Offen & Kline also correlated the on-off time records, with a value of 1 during an event and zero otherwise, from the various schemes, and found that the maximum correlation coefficients from any pair of these records were in the range 0.35 - 0.42. Despite the relatively poor correlation, these authors emphasised that the schemes agreed with each other, to a certain extent in their relationship to the visual data, and generally produced conditional averages which were in qualitative agreement with the expected results, many of them being as effective as the visual data at detecting periods of high uv.

Although Chen & Blackwelder (1978) used a rake of wires to identify the upstream interface of the large structure, they obtained conditional ensemble averages of θ, u, v and their products by applying VITA to the temperature signal from a single cold wire located next to a X-wire. The X-wire/cold wire combination was traversed across the boundary layer but the rake was not used for those experiments. The present simultaneous temperature traces show, like those of Chen & Blackwelder, repetitive patterns characterised by a slow increase in temperature followed by a relatively sudden decrease. This decrease is identified with the upstream interface of the large structure. This interface is inclined with the shear, the convection speed at any location on the interface being approximately equal to the local mean velocity. This latter result was obtained by Subramanian & Antonia (1979). Although the interface often extends over the entire boundary layer, its three-dimensionality is evident by the fact that it is not always detected by all sensors in the rake. Although not examined here, the three-dimensionality of the large scale motion and its dependence on Reynolds number are clearly aspects worthy of future investigation. That the interface usually occurs at all values of y and does not seem to be associated with any particular part of the boundary layer led Chen & Blackwelder (1978) to conjecture that this interface might provide the dynamical link between the large scale motion and the events observed near the wall.

While a visual method is probably the most reliable scheme for detecting bursts in the wall region of the boundary layer, the visual identification of the large structure using temperature traces obtained over a significant spatial region would appear to be more reliable than detection methods based on information at one point in space. However, in view of the obvious tediousness associated with the visual identification scheme, it was hoped that comparison of the visual scheme with other one-point detection schemes, easily adaptable to the computer, might indicate one or more one-point schemes to be satisfactory, at least after judicious tampering with the detection criteria.

The present comparison is between ensemble averaged signatures of u, v, θ and of the products uv, vθ and uθ. As the output of each detection method consists of a series of time points τ_n, each corresponding to the detected occurrence of that structure, the ensemble average of a function is defined as

$$\langle f(t)\rangle = \sum_{n=1}^{N} f(\tau_n + t)/N$$

where N is the total number of structures detected. In the VISUAL method, the reference point is first identified approximately by eye and subsequently determined accurately using method (iii) below and a very small value of K. Other methods used are:

(i) VITA, with structure detected when $\widetilde{\theta^2} - \tilde{\theta}^2 > K\theta'^2$ and $\dot{\theta} < 0$ (the dot denotes differentiation with respect to time) where

$$\tilde{\theta} = \frac{1}{T}\int_{t-\frac{1}{2}T}^{t+\frac{1}{2}T} \theta(t)\ dt$$

and prime denotes the rms value, and K (here equal to about 0.6) is a threshold parameter. A typical value of T, given by Blackwelder & Kaplan (1976) is T^+ ($\equiv TU_\tau^2/\nu$) $\approx$ 10.

(ii) A technique employed (here denoted BT) by Brown & Thomas (1977), in which the temperature (velocity in their case) signal is first low-pass filtered before forming a "high frequency" component by subtracting the filtered signal from the total signal. The output of the detection scheme consists of the times when the rectified and smoothed "high frequency" component exceeds a threshold level.

(iii) A detection method based on $\dot{\theta}^3$. The reference time of the structure is given whenever $\dot{\theta}^3 > K(\dot{\theta}^3)'$ with $K \simeq 3.2$. This method has been called RA as it is essentially that used by Rajagopalan & Antonia (1980a,b).

(iv) HOLE, a technique used by Lu & Willmarth (1973) in conjunction with the quadrant-analysis approach. These authors identified strong ejections when the magnitude of uv ($u < 0$, $v > 0$) exceeded a certain threshold, which in their case defined a hole size $H = |uv|/u'v'$. Lu & Willmarth set H equal to about 4.5 (i.e. $|uv| \simeq 10\ \overline{uv}$). It should be noted that the quadrant-analysis approach was introduced by Willmarth & Lu (1972) and Wallace et al (1972) to sort out the contributions to $\overline{uv}$ from the various observed events in the wall region of the layer. Application of the hole technique to identify contributions to $\overline{uv}$ from "bursts" or "ejections" and "sweeps" in the outer intermittent layer is, in principle, not physically meaningful unless some evidence is provided for the existence of these events in this region of the flow. It has been suggested (e.g. Van Atta, 1980) that the lack of phase information eliminates the possibility of defining the geometry of coherent structures from information obtained from the quadrant analysis decomposition. Reference times τ_n corresponding to times when $|uv|$ exceeds $H\,u'v'$ in the second quadrant, can be used as inputs to a conditional sampling technique. This is the approach used here to determine ensemble averages of velocity and temperature for comparison with averages provided by other detection methods. It must also be noted that Comte-Bellot et al (1979) used such an approach to obtain $\langle u\rangle$, $\langle v\rangle$ and $\langle uv\rangle$ in a pipe flow.

The main results of the comparison are shown in Table 1 and Fig. 1. None of the detection methods [(i) - (iv)] could be considered to be in good agreement, with respect to either the frequency of structure detection or the detailed behaviour of conditional signatures, with VISUAL. By analogy with the

TABLE 1 COMPARISON BETWEEN DIFFERENT DETECTION METHODS AT $y^+ = 40$, $R_m = 3100$. A TOTAL OF 100 EVENTS ARE CONSIDERED

Method	Reference	Parameters Used Here	$U_1\overline{T}/\delta$	Percentage Correspondence Relative to VISUAL
VISUAL	Present method: visual detection of nearly simultaneous sudden changes in θ in all traces	----	2.63	100
VITA	e.g. Blackwelder & Kaplan (1976)	$TU_\tau^2/\nu = 4.6$, $K = 0.6$ (-ve slope only)	2.43	42
RA	Rajagopalan & Antonia (1980b)	$K = 3.2$ (-ve slope only)	2.64	42
HOLE	Lu & Willmarth (1973)	$H = 1.9$ ($u < 0$, $v > 0$)	2.78	15
BT	Brown & Thomas (1977)	$K = 2.7 \times$ rms value of rectified "high frequency" signal	2.86	32

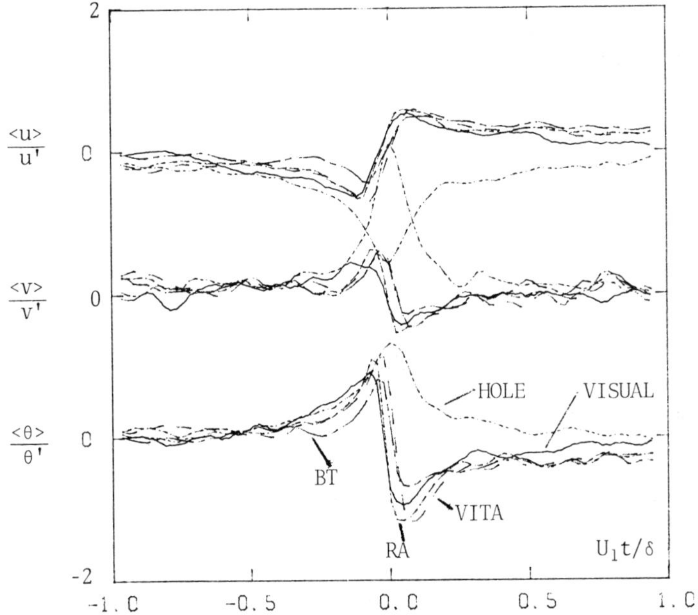

Fig. 1 Comparison of Ensemble Averages of u, v, θ Obtained by Different Methods at y^+ = 40, R_m = 3100.

comparison by Offen & Kline, the various threshold parameters in the four detection methods were adjusted to force approximate agreement (see Table 1) between the mean period $\overline{T}$ between structures detected by each method with the period obtained by VISUAL. Despite this agreement, the best correspondence (Table 1) was achieved by RA (~ 42%) while the worst correspondence was provided by HOLE (~ 15%). Of the 100 events considered in the Table, 42 were common to VISUAL, VITA and RA and only as few as 7 were common to all methods.

Ensemble averages $\langle u \rangle$, $\langle v \rangle$, $\langle \theta \rangle$, obtained at y^+ = 40 and R_m = 3100, are shown in Fig. 1, normalised by rms values. All results shown here were obtained for N ≃ 100. A study of the effect of the size of N on the ensemble average was made. It was found that N ≃ 100 was a satisfactory compromise as practically no change could be detected for larger N (values as high as 500) were considered). In this and other similar figures, t = 0 denotes the location where all reference points τ_n are brought into alignment. The time is normalised by the scale δ/U_1. There is generally poor quantitative agreement between signatures (Fig. 1) obtained from different methods. While all signatures are in qualitative agreement with each other and with those already reported, at a similar value of R_m, by Blackwelder & Kaplan (1976) and Chen & Blackwelder (1978), there are significant differences in the magnitude and location of the relatively sudden change of the ensemble average near t = 0. While VITA and BT detect the structure at approximately the same time, this time occurs later than that detected by RA or VISUAL. The amplitudes of the signatures obtained by RA are however slightly larger than those inferred from VISUAL. Perhaps not surprisingly, the signatures obtained from HOLE do not indicate the change in the sign of $\langle u \rangle$, $\langle v \rangle$

or <θ> near t = 0. Results obtained (not shown here) for <uv>, <uθ>, <vθ> are in general support with those inferred from the comparison of <u>, <v> and <θ>, except that the signatures obtained using BT are in slightly better agreement than VITA with those pertaining to VISUAL or RA.

4. RESULTS AND DISCUSSION

Conditional averages of u, v, θ, normalised by rms values, are shown in Figs. 2 and 3 for R_m of 990 and 3100 respectively. Comparison of the signatures, shown at several values of y/δ, at the two Reynolds numbers reveals several interesting features. Firstly, it is clear that, for R_m = 990, there is relatively poor collapse, near t = 0, of signatures (especially <v>) at different y values, quite in contrast with the results of Fig. 3. The sudden decrease in temperature near t = 0 is observed at an earlier time for y/δ = 0.037 than for y/δ ≡ 0.7. The delay seems to depend systematically on the particular y location. Accompanying the decrease in <θ> is a corresponding decrease in <v> and an increase in <u>. The magnitude of the slopes, near t = 0, of <θ> and <u> show a small, but systematic, increase with increasing y/δ. The magnitude of the decrease of <v> is rather small compared with that of the changes of <u> or <θ>. This is especially noticeable at y/δ = 0.037, which corresponds to $y^+ \simeq 15$. Another noticeable feature of Figs. 2 and 3 is the relatively wider temporal extent of the conditional averages at R_m = 990 relative to those at R_m = 3100. In particular, most if not all signatures at R_m = 3100 have crossed the zero axis while only <v> exhibits this behaviour for R_m = 990. The extent of <v> is smaller than that of <u> or <θ> at both Reynolds numbers. It is also of interest

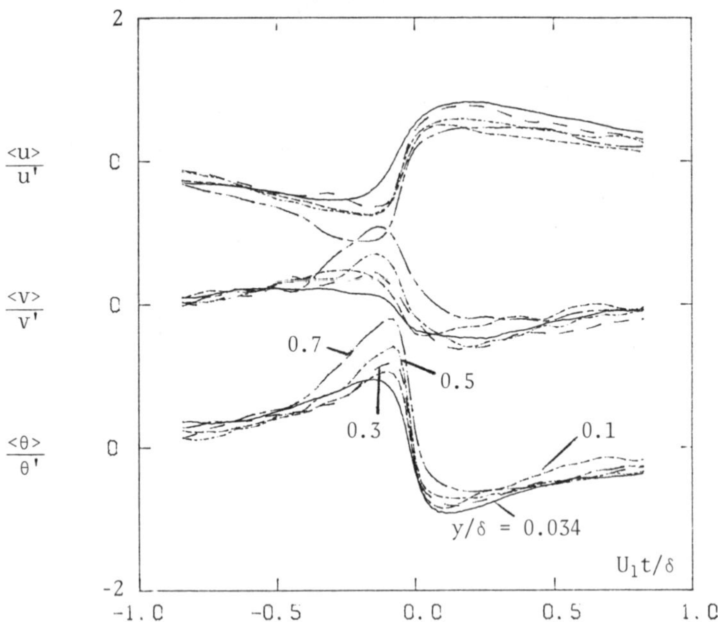

Fig. 2 Ensemble Averages of u, v, θ at Different Values of y/δ for R_m = 990.

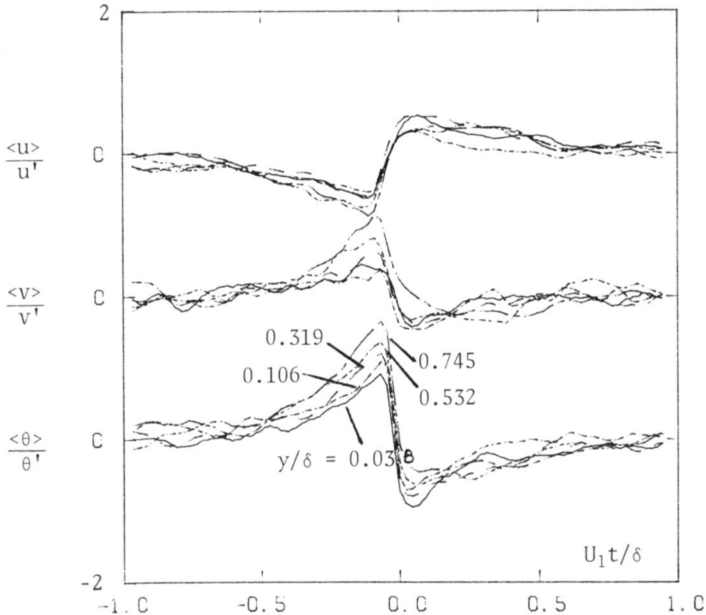

Fig. 3 Ensemble Averages of u, v, θ at Different Values of y/δ for R_m = 3100.

to note that $\langle v\rangle$, at y/δ = 0.7 (R_m = 990) or y/δ = 0.74 (R_m = 3100), crosses zero near t = 0 with considerable delay. This trend may perhaps reflect the increasing influence of the irrotational flow field fluctuations as y/δ increases. The $\langle v\rangle$ signature at y/δ = 0.7 for R_m = 7100 shows the same trend. The signatures for R_m = 7100 show little deviation from those at R_m = 3100 for corresponding values of y/δ. At smaller values of y/δ, $\langle v\rangle$ is approximately 180° out of phase with $\langle u\rangle$ while $\langle v\rangle$ and $\langle\theta\rangle$ are almost in phase. $\langle u\rangle$ and $\langle\theta\rangle$ are strongly but negatively correlated. At small values of y/δ, the conventional correlation coefficient $\overline{u\theta}/u'\theta'$ approaches the value of -0.9.

As y/δ decreases, $\langle\theta\rangle$ becomes increasingly negatively skewed. At $y^+ \simeq 15$ (Fig. 4; these results have been obtained using the rake signals as it was not possible to achieve this y^+ location with the X-probe at the higher R_m). As all signatures are relative to conventional local mean values, the mainly negative excursion in $\langle\theta\rangle$ for t > 0 emphasises the arrival at the wall of relatively cooler fluid from larger values of y. It seems reasonable to speculate that this fluid is associated with the sweep event since $\langle u\rangle$ and $\langle v\rangle$ show dominant positive and negative excursions respectively (especially Fig. 2 at y^+ = 15) at positive values of t. Unfortunately, for reasons previously given, we were not able to obtain $\langle u\rangle$ and $\langle v\rangle$ signatures at y^+ = 15 to substantiate this speculation for larger values of R_m. It is clear from Fig. 4 that little difference exists between $\langle\theta\rangle$ at the two larger R_m values. At R_m = 990, $\langle\theta\rangle$ is larger in amplitude and extends further for both positive and negative values of t. At y/δ = 0.14, the signatures (Fig. 5) at the larger R_m are in close agreement with each other but at the lower R_m, the deviation of $\langle\theta\rangle$ from the higher R_m behaviour

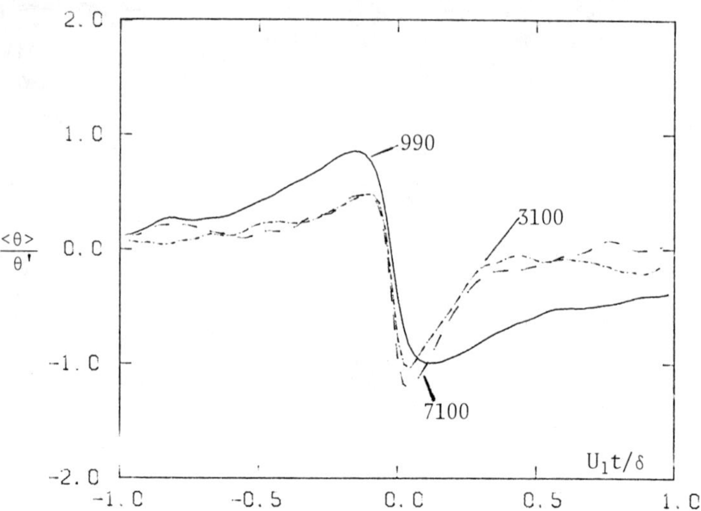

Fig. 4 Ensemble Averages of θ at $y^+ \simeq 15$ and Different Values of R_m.

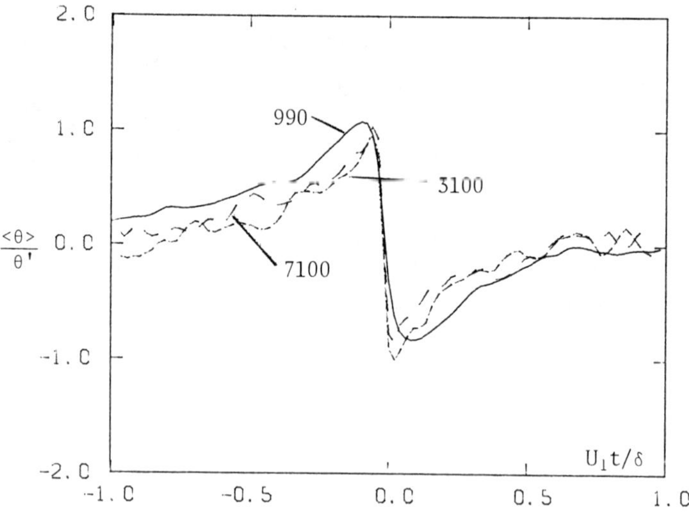

Fig. 5 Ensemble Averages of θ at $y/\delta = 0.14$ and Different R_m.

is smaller than the corresponding deviation at $y^+ = 15$. The deviation remains evident up to much larger values of y/δ but only at negative values of t. For $t > 0$, $\langle\theta\rangle$ increases relatively slowly, which is consistent with the expected return to the ambient temperature of the free stream upstream of the turbulent/non-turbulent interface.

Comparison at different R_m of conditional averages of the products uv and $v\theta$ is shown in Fig. 6 for $y/\delta \simeq 0.3$. The magnitudes of $\langle uv\rangle$ and $\langle v\theta\rangle$ (and $\langle u\theta\rangle$, not shown here) exhibit a minimum near $t = 0$ and local peaks (not well defined) on either side of $t = 0$, located at roughly $|t|U_1/\delta \simeq 0.1$. The signatures at $R_m = 990$ are different from those at larger R_m which are in fairly reasonable agreement. In particular, the magnitude of the peak at positive t for $R_m = 990$ is larger than that at negative t, while the opposite behaviour is observed at larger R_m. At smaller values of y/δ, the peak for $t > 0$ at $R_m = 3100$ and 7100 is larger in magnitude than that at $t < 0$. This seems compatible with our previous speculation, supported by measurements in other shear flows (e.g. Wallace et al's, 1977, results in a fully developed turbulent channel flow with a relatively thick viscous sublayer), of the importance of the sweep near the wall. Available experimental evidence for the existence of a double peak structure in $\langle uv\rangle$ has been reviewed by Antonia (1980) and will not be discussed here. The relatively large peak (especially in $\langle v\theta\rangle$) at $t < 0$ that occurs for large R_m in Fig. 6 becomes more pronounced at larger values of y/δ. Both $\langle uv\rangle$ and $\langle v\theta\rangle$ make a small excursion, at $y/\delta > 0.7$, towards positive and negative values respectively near $t = 0$.

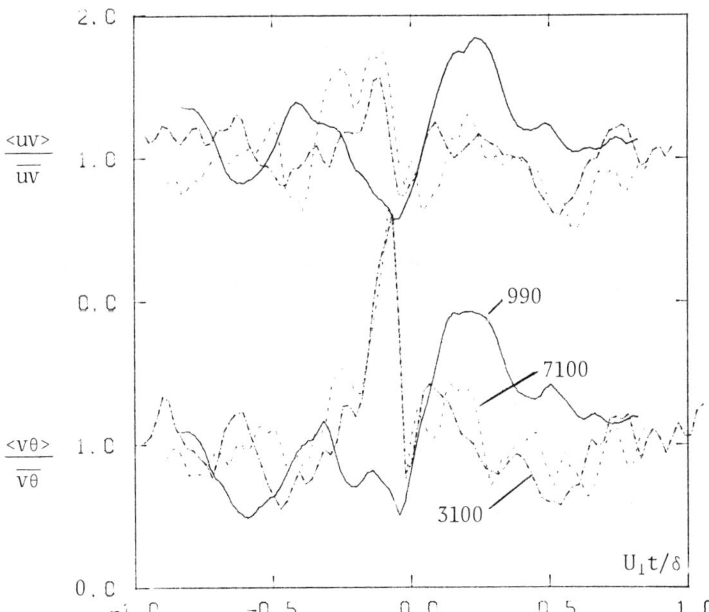

Fig. 6 Ensemble Averages of uv and $v\theta$ at $y/\delta \simeq 0.3$ and Different R_m.

Chen & Blackwelder (1978) found that, at small values of y/δ, the zero crossings of $\langle u \rangle$ and $\langle v \rangle$ occurred at earlier times than zero crossings of $\langle \theta \rangle$. As their cold wire was, as in the present arrangement, upstream of the X-probe, the phase difference could not be explained in terms of the spatial separation of velocity and temperature sensors. The authors accepted their phase difference as genuine and raised the possibility that the internal shear layer and upstream interface may not be coincident. They also suggested that a more plausible explanation for the phase difference may be the existence in the sublayer of longitudinal streamwise vortices. The present conditional averages of u, v, θ near the wall (results at $y^+ = 15$ are shown in Fig. 2) indicate that there is no phase difference between $\langle u \rangle$ or $\langle v \rangle$ and the $\langle \theta \rangle$ signature. This zero phase shift was noted at all values of R_m and seems to support Chen & Blackwelder's conjecture that the internal shear layer, associated with the temperature jump at the back of the large structure, extends down to the wall. Whether this shear layer provides, as suggested by Chen & Blackwelder, a dynamical link between the bursting phenomenon near the wall and the outer intermittent region remains a matter for further investigation.

The mean period $\overline{T}$ between occurrences of the upstream interface of the large structure was determined by counting the number of visual detections of the interface over the record examined. When $\overline{T}$ is normalised by the time scale δ/U_1, the ratio is found to be equal to about 2.5[§] when R_m is greater than about 2000. At smaller values of R_m, $\overline{T}U_1/\delta$ increases (or, equivalently, the frequency $\underline{f}\delta/U_1$ decreases), to a value of about 5.0 at $R_m \simeq 1000$. This latter value of $\overline{T}U_1/\delta$ corresponds to the "bursting" period measured by Narahari Rao et al (1971). It also corresponds approximately (and perhaps, as mentioned by Willmarth, 1975, fortuitously) with the mean period between "bursts" or "sweeps" obtained by Lu & Willmarth (1973), using the hole technique and specific values of H to characterise "bursts" and "sweeps". These authors found, at $R_m = 4230$, that $U_1\overline{T}_B/\delta^* \simeq 30$ ($\overline{T}_B$ is the mean period between "bursts", as defined in their paper and δ^* is the displacement thickness) or $U_1\overline{T}_B/\delta \simeq 5$ when H was equal to about 4.5. However, as noted earlier, a value of H of about 1.9 leads to approximate agreement in frequency of detection between HOLE and VISUAL. For H = 1.9, the results of Lu & Willmarth (1973) indicate that $U_1\overline{T}_B/\delta \simeq 2.5$, in reasonable agreement with the present value. It should also be noted that the interface crossing frequency obtained by Chen & Blackwelder (1978) at $R_m = 2800$, is equivalent to $U_1\overline{T}/\delta \simeq 2.63$. This latter value was however inferred using VITA and, in view of the present comparison results, it is perhaps fortuitous that it is in agreement with the VISUAL result. Chen & Blackwelder themselves expressed surprise to the fact that their VITA crossing frequency was constant across the boundary layer (VITA was initially introduced by Kaplan & Laufer, 1968, to detect the turbulent /non-turbulent interface).

The inclination ϕ of the upstream interface is shown in Fig. 7, as a function of y/δ and for different R_m. ϕ is determined, for a particular value of y, from $\tan^{-1}(\Delta y/U_c\Delta t)$ where Δt is the delay between times of arrival of the interface at the two temperature sensors nearest the y location (Δy is the distance between these sensors) and U_c is the convection speed at a point midway between the sensors. ϕ varies little in the range $0.2 < y/\delta < 0.8$ and is essentially unaffected by R_m. The decrease in ϕ near the wall has been observed in other shear flows (e.g. Rajagopalan & Antonia, 1980a, in a turbulent duct flow and Phong-anant et al, 1980, in the atmospheric surface layer). The average value

---

[§]This value is equal to half the value given by Subramanian & Antonia (1979) who obtained $f\delta/U_1 \simeq 0.2$ but this value is thought to be incorrect as a result of the relatively short record durations and small number of wires (four only) used in that investigation.

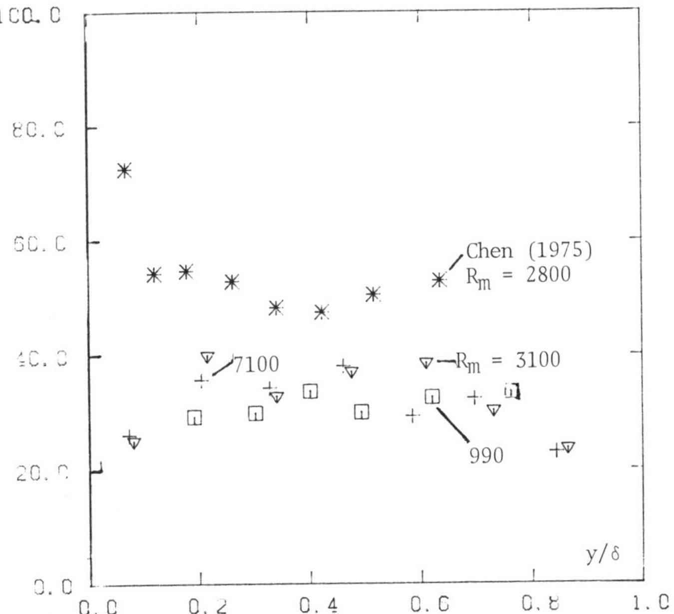

Fig. 7 Inclination of Upstream Interface.

of ϕ in the interval $0.2 < y/\delta < 0.8$ is about 35°, which is in good agreement with Falco's (1979) best estimate of the average inclination of the upstream boundary of the large eddies in his smoke photographs. Bandyopadhyay (private communication) has suggested that the large structure is made up of hairpin vortices, inclined at about 45° to the wall, which extend through the complete boundary layer thickness so that their tips lie on a line which makes an angle of about 20° to the wall. Chen & Blackwelder's (1978) inclination determined from cross-correlations between pulse trains (generated by VITA) at different y locations, is significantly larger (Fig. 7) than the present estimate of ϕ.

An approximate estimate of the contribution to the conventional Reynolds shear stress from the large scale structure has been obtained

$$\frac{\langle \tilde{uv} \rangle}{\overline{uv}} = \frac{1}{\overline{T}} \int_0^{|t_0|} \frac{\langle uv \rangle}{\overline{uv}} \, d|t| \qquad (t < 0)$$

where the integration limit $|t_0|$ is arbitrarily taken as the location where $\langle\theta\rangle$ first crosses zero along the negative time axis; $\overline{T}$ is the mean period between interfaces, here assumed given by $\overline{T}U_1/\delta = 2.5$ for all values of R_m. The product $|t_0|U_c$ can be loosely interpreted as a longitudinal length scale for the structure. As noted earlier, this scale has a tendency to increase as R_m decreases. Over the interval $0.1 < y/\delta < 0.5$, the ratio $\langle\tilde{uv}\rangle/\overline{uv}$ increases (Fig. 8) from about 20% at R_m = 990 to about 33% at R_m = 7100. The ratio $\langle\tilde{v\theta}\rangle/\overline{v\theta}$ (not shown here) exhibits a similar increase with R_m and is approximately equal to $\langle\tilde{uv}\rangle/\overline{uv}$.

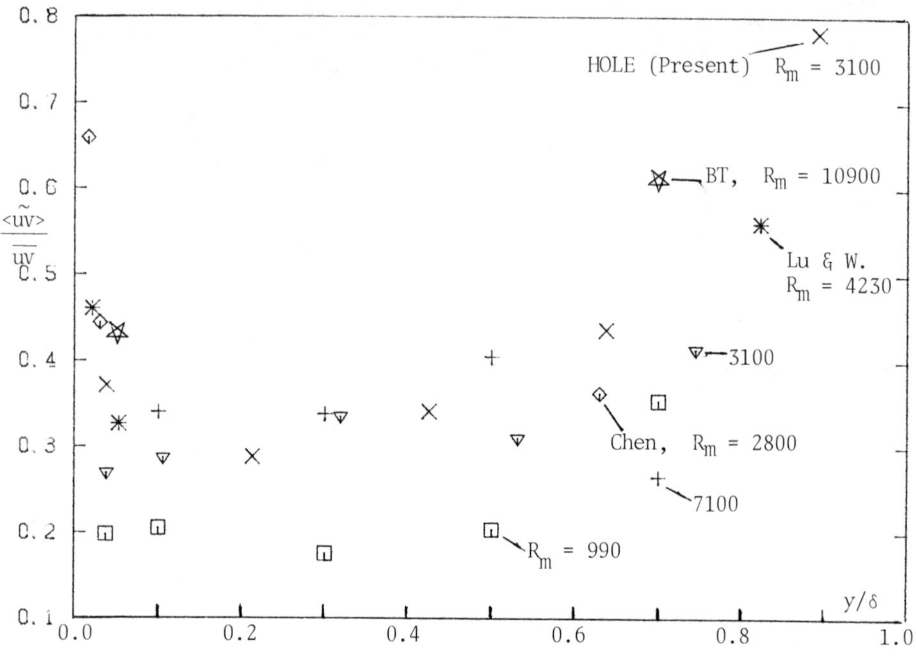

Fig. 8 Contributions to the Reynolds Shear Stress From the Large Scale Motion, as Estimated by Different Methods.

Contributions, at $R_m = 3100$, to $\overline{uv}$ from the region of the second quadrant ($u < 0$, $v > 0$) which corresponds to $|uv| > H\,u'v'$ have been computed for $H = 1.9$ (see Table 1), in the manner outlined by Lu & Willmarth (1973). Although the results (Fig. 8) compare favourably with those inferred from the VISUAL signatures, the agreement must be considered to be fortuitous since the contribution from the "hole" region of the second quadrant has been excluded from the HOLE estimates in Fig. 8. The relatively large amplitudes of the HOLE signatures of u and v in Fig. 1 suggest that HOLE would yield values of $\langle\tilde{uv}\rangle/\overline{uv}$, well in excess of those obtained with VISUAL. Also shown in Fig. 8 are the second quadrant contributions deduced from Lu & Willmarth's measurements at $R_m = 4230$ for $H = 1.9$. Estimates inferred from Thomas' (1977) signatures of uv are significantly higher (Fig. 8) than the present estimates. This is perhaps not surprising since the signatures of Thomas are enhanced by an iterative procedure used to sharpen the definition of the detection criterion.

Contributions to $\overline{uv}$ associated with the sharp change (near $t = 0$) in $\langle\theta\rangle$ have also been calculated. The integration was carried out over a time interval corresponding to the time delay $|t_m|$ between minimum and maximum values of $\langle\theta\rangle$. The resulting contributions are negligibly small (less than 1%) for $R_m = 990$ and represent about 2-3% at the other two values of R_m. The length scale $|t_m|U_c$ was associated by Chen & Blackwelder with the width of the internal shear layer. These authors found, with U_c assumed equal to U, that $|t_m|U_c/\delta$ increased from

0.05 to 0.1 across the layer. In the present experiment, $|t_m|U_c/\delta$ is constant, equal to about 0.042 across the layer. When $|t_m|U_c$ is normalised by (ν/U_τ), the resulting ratio is equal to about 40, which is considerably smaller than the length scale of about 200 for Falco's (1977) typical eddies. The normalised time derivative of the ensemble averaged temperature, as defined by $|(\delta/U_1\theta')\ (d\langle\theta\rangle/dt)|$ was found to be in the range 30 - 40 over the major part of the layer. No systematic variation with y/δ of this derivative was noticed. Both the magnitude and trend of this derivative are different from those obtained in Chen & Blackwelder. They found that the normalised derivative remained constant (= 22) across the layer.

ACKNOWLEDGEMENT

The support of the Australian Research Grants Commission is gratefully acknowledged.

REFERENCES

Antonia, R. A. 1980. The organized motion in a turbulent boundary layer. *Proc. 7th Australasian Conference on Hydraulics & Fluid Mechanics*, Brisbane, pp.155-162

Antonia, R. A. 1981. Conditional sampling in turbulent measurement. *Annual Review of Fluid Mechanics*, Vol. 13, pp.131-156.

Blackwelder, R. F. and Kaplan, R. E. 1976. On the wall structure of the turbulent boundary layer. *J. Fluid Mech.*, Vol. 76, pp.89-112.

Brown, G. L. and Thomas, A. S. W. 1977. Large structure in a turbulent boundary layer. *Phys. Fluids*, Vol. 20, pp.S243-S252.

Champagne, F. H. 1979. The temperature sensitivity of hot wires. *Proc. Dynamic Flow Conference - Dynamic Measurements in Unsteady Flows, 1978*, Marseille/Baltimore, pp.101-114.

Chen, C. H. P. 1975. The large scale motion in a turbulent boundary layer: a study using temperature contamination. Doctoral thesis, University of Southern California.

Chen, C. H. P. and Blackwelder, R. F. 1978. Large scale motion in a turbulent boundary layer: a study using temperature contamination. *J. Fluid Mech.*, Vol. 89, pp.1-31.

Coles, D. W. 1962. The turbulent boundary layer in a compressible fluid. *Rand Corp. Report R-403-PR, ARC 24478.*

Comte-Bellot, G., Sabot, J. and Saleh, I. 1979. Detection of intermittent events maintaining Reynolds stress. *Proc. Dynamic Flow Conference - Dynamic Measurements in Unsteady Flows, 1978*, Marseille/Baltimore, pp.213-229.

Falco, R. E. 1977. Coherent motions in the outer region of turbulent boundary layers. *Phys. Fluids*, Vol. 20, pp.S124-S132.

Falco, R. E. 1979. Comments on "large structure in a turbulent boundary layer". *Phys. Fluids*, Vol. 22, pp.2042-2043.

Head, M. R. and Bandyopadhyay, P. 1978. Combined flow visualization and hot-wire measurements in turbulent boundary layers. In *Coherent Structure of Turbulent Boundary Layers* (eds. C. R. Smith and D. E. Abbott), AFOSR/Lehigh University Workshop, p.98.

Huffman, G. D. and Bradshaw, P. 1972. A note on von Kármán's constant in low Reynolds number turbulent flows. *J. Fluid Mech.*, Vol. 53, pp.45-60.

Kaplan, R. E. 1973. Conditional sampling techniques. In *Turbulence in Liquids* (eds. G. K. Patterson and J. L. Zakin), University of Missouri-Rolla, pp.274-283.

Kaplan, R. E. and Laufer, J. 1968. The intermittently turbulent region of the boundary layer. *Report USCAE110*, University of Southern California.

Lu, S. S. and Willmarth, W. W. 1973. Measurements of the structure of the Reynolds stress in a turbulent boundary layer. *J. Fluid Mech.*, Vol. 60, pp.481-512.

Murlis, J. 1975. The structure of a turbulent boundary layer at low Reynolds number. Doctoral thesis, Imperial College, University of London.

Narahari Rao, K., Narasimha, R. and Badri Narayanan, M. A. 1971. The "bursting" phenomenon in a turbulent boundary layer. *J. Fluid Mech.*, Vol. 48, p.339.

Offen, G. R. and Kline, S. J. 1973. Experiments on the velocity characteristics of 'bursts' and on the interactions between the inner and outer regions of a turbulent boundary layer. *Report MD-31*, Department of Mechanical Engineering, Stanford University.

Rajagopalan, S. and Antonia, R. A. 1980a. Interaction between large and small scale motions in a two-dimensional turbulent duct flow. *Phys. Fluids*, Vol. 23, pp.1101-1110.

Rajagopalan, S. and Antonia, R. A. 1980b. Properties of the large structure in a slightly heated turbulent mixing layer of a plane jet. To appear in *J. Fluid Mech.*

Subramanian, C. S. and Antonia, R. A. 1979. Some properties of the large structure in a slightly heated turbulent boundary layer. *Proc. 2nd Int. Symp. on Turbulent Shear Flows*, London, pp.4.18-4.21.

Thomas, A. S. W. 1977. Organised structures in turbulent boundary layers. Doctoral thesis, University of Adelaide.

Van Atta, C. W. 1980. Conditional sampling techniques. To appear in *Handbook of Turbulence, Vol. 2*, Plenum Press.

Wallace, J. M., Brodkey, R. S. and Eckelmann, H. 1977. Pattern recognized structures in bounded turbulent shear flows. *J. Fluid Mech.*, Vol. 83, pp.673-693.

Wallace, J. M., Eckelmann, H. and Brodkey, R. S. 1972. The wall region in turbulent shear flow. *J. Fluid Mech.*, Vol. 54, pp.39-48.

Willmarth, W. W. 1975. Structure of turbulence in boundary layers. *Adv. Appl. Mech.*, Vol. 15, pp.159-254.

Willmarth, W. W. and Lu, S. S. 1972. Structure of the Reynolds stress near the wall. *J. Fluid Mech.*, Vol. 55, pp.65-92.

Wood, D. G. 1976. Studies in turbulent boundary layers. M.Eng.Sc. thesis, University of Sydney.

A Study of the Intermittent Phases in the Wall Region of a Turbulent Flow

N. VAN THINH
Laboratoire de Mécanique Expérimentale des Fluides
Université Pierre et Marie Curie
Orsay, France

ABSTRACT

Wall turbulence studies in the last years have led to a picture of the wall region which consists of a sequence of events including at least two intermittent phases: inrushes towards the wall and ejections outwards. Most of these studies were performed at relatively low Reynolds numbers and the aim of the present study is to provide more quantitative data about the evolution of the intermittent phases at higher Reynolds numbers. A conventional statistical analysis is first presented; the overall shape of the velocity probability density distributions, obtained by a recent and efficient estimation method, indicates the presence of intermittent phases which is attenuated at higher Reynolds numbers. In the conditional statistical analysis which is presented subsequently, two conditional sampling methods have been used to detect the intermittent phases. The importance of each phase has been clearly outlined for the investigated Reynolds numbers and the relative importance of the two inrush and ejection phases well evidenced for each wall layer.

1. INTRODUCTION

Recent studies of wall turbulence have brought evidence of coherent structures which contribute significantly to the turbulent transport processes. References to these studies can be made to the review articles by Laufer [1] and Willmarth [2]. These works, particularly the visual studies, have led to a picture of the wall region which consists of a sequence of events called the bursting phenomenon during which most of the turbulent production of energy was generated. It is well established that at least two phases are including in these events: an ejection phase of low-momentum fluid masses which lift up away from the wall vicinity and a sweep or inrush phase of high-momentum fluid masses towards the wall.

The application of high speed digital computing methods to turbulence research which was started by Frenkiel and Klebanoff [3] has contributed to a large part to the advances of wall turbulence studies by Zarić [4,5], Blackwelder and Kaplan [6], Wallace et al. [7], Blackwelder and Eckelmann [8]. In order to get a thick viscous sublayer, most of these studies were performed at relatively low Reynolds numbers and the important open question of how the observed wall structure will change with Reynolds numbers is not yet well known. The Reynolds number effect has been mentioned by Morrison et al. [9], Lu and Willmarth [10], Brodkey et al. [11] and also by Falco [12] fot the outer region of a boundary layer and by Sabot and Comte-Bellot [13] for the core region of a turbulent pipe flow. Due to the lack of systematic results, the presented opinions were in general of speculative nature. As examples, one may cite Morrison et al. [9] who stated in their summary that: "the streaky structure which is commonly regarded as being characteristic of the sublayer will probably not be detected at sufficiently high Reynolds numbers" or Lu and Willmarth [10] who, after comparing their results about the relative contribution of the two sweep and ejection phases to the Reynolds stress with those of Wallace et al. [14], concluded that the nature of the wall turbulent structure is similar above Reynolds numbers which correspond to their ones but at lower Reynolds numbers, significant changes occur. The aim of the present study is then to provide more quantitative data about the evolution of the intermittent phases at higher Reynolds numbers.

In the previously cited studies, the intermittent phases have been detected by different conditional sampling methods and as stated by Offen and Kline [15] and cited in the review article of Willmarth [2]: "there remain serious questions about what events are measured by each technique". Although none of the detection schemes seemed to be really adequate, we were convinced that, even with a relatively simple conditional technique, the expected evolution could be outlined with systematic results obtained at a wide range of Reynolds numbers and this has been the guiding philosophy of the present study. In a recent paper [16], Blackwelder discussed a method of pattern recognition called 'training' in which an algorithm was trained to recognize different events (e.g. sweep, ejection, etc.). The events were defined by associated characteristic parameters called 'features' which may include any combinations of amplitudes, derivatives, filtered signals, etc., from many different signals. An additional input to the algorithm, called training input, such as an electronic signal or visual observations, etc., determined which event is present. The technique appears then to be quite general in theory and may be a promising way in future conditional analysis but, in this study where a large amount of data must be processed, it cannot be applied due to the excessive cost. The Zarić's conditional sampling techniques [4,17], which have been developed for the analysis of a single probe signal and used successfully elsewhere, appeared to be adequate to our purposes and have then be adopted here.

The first part of this paper presents the conventional statistical analysis where is given a full description of the velocity probability density distributions obtained from a recent estimation method [18]. In the second part which concerns the conditional statistical analysis, results obtained from the complete Zarić's 'stochastic conditional sampling' technique [4] are first presented and subsequently, with a modified method based upon another Zarić's technique [17], we present results which confirm the general behaviour observed with the first method.

2. EXPERIMENTAL FACILITY AND TECHNIQUES

2.1. Experimental Facility

The experimental facility has been described elsewhere [19]. Measurements were made in air flow near a smooth wall of a rectangular channel. Table 1 summarizes the flow characteristics for three Reynolds numbers.

Table 1

U_a(m/s)	u_τ(m/s)	10^{-4} Re
1.93	0.10	4.32
4.89	0.24	10.39
7.68	0.35	17.71

Hot wire anemometry is employed. Probes of 5 μm diameter and 1.8 mm active length, tungsten wire are used with DISA anemometers and signal conditioners.

2.2 Signal Processing and Data Acquisition

The digital technique of data acquisition and analysis has been used for this study. An appropriate constant d.c. voltage was substracted from the signals which were amplified and sent over analog lines to a 3 channel data acquisition system where they were digitized in real time and stored on a digital tape. The sampling periods were $\tau = 480$ µs for the two first Reynolds numbers and $\tau = 240$ µs for the highest one. For each measuring point, a succession of 133620 instantaneous values were registered on the digital tape, these run times were sufficient to obtain accurate statistical estimates, particularly in the case of the spectral density estimates [20]. The analysis of the stored data was performed on an I.B.M. 370-168 digital computer and the results were plotted on a Benson digital incremental plotter. The various effects influencing the hot wire response have been taken into account [21].

2.3. Probability Density Estimates

The classical and commonly used probability density estimator, the histogram, is well known and is not reported here. It is only sufficient to mention the conflicting requirements on the mesh spacing h in this method [22]. A large value of h is needed to reduce the random error and a small value of h reduces the bias error. Practically, taking a large sample size N and $h \leq 0.2\,\sigma$ is satisfactory.

Quite recently, a new method to estimate the probability density, called the 'discretized maximum penalized likelihood estimator' or D.M.P.L.E., was developed by Scott [18]. The method optimizes a criterion function which combines the maximum likelihood principle and a penalty term. This estimator is shown to exhibit smaller errors than did most of the other ones. For a fixed sample, the D.M.P.L.E. solution approximates the continous solution as the mesh spacing h approaches zero. Then the D.M.P.L.E. is more robust in the choice of a small spacing than the histogram. For these reasons, this method is adopted in the present study and a value of $h = 0.05\,\sigma$ is used.

Velocity probability density distributions in the wall layers have been determined on the basis of $N_t = 133620$ instantaneous values for three Reynolds numbers and different non-dimensional distances from the wall ranging from $y^+ = 2$ to $y^+ = 75$. As in the spectral density estimation [20], the N_t values have been divided into $n = 262$ blocks of $N = 510$ values. For each block, an estimate $\hat{p}_i(u)$ of the probability density is obtained by the D.M.P.L.E. and the final p(u) is obtained as the ensemble mean:

$$p(u) = \frac{1}{n} \sum_{i=1}^{n} \hat{p}_i(u)$$

2.4. Conditional Sampling Techniques

As indicated in the introduction, two methods were used to detect the intermittent phases in the conditional analysis. The first one was the Zarić's principal conditional sampling technique called 'stochastic conditional sampling' or S.C.S. technique [4]. The detection scheme was built up to make the distinction of three different phases: the two intermittent inrush and ejection phases and the relatively undisturbed or calm phase; the detection function being in fact a combination of criteria based essentially on the instantaneous values of some velocity fluctuation and its time derivative. The second method was a modification of another Zarić's technique called 'cyclic conditional sampling' or C.C.S. technique [17] due to the character of the partial averages used in the sampling criterion. The two procedures will be discussed in detail in the next paragraph.

3. RESULTS AND DISCUSSIONS

3.1. Conventional Statistical Analysis

As pointed out by Zarić [5], velocity probability density distributions in the wall layers have been presented by a few authors only [23-29] with skewness and flatness factors presented only in [30], [31] and [32]. Where the digital method has been used to obtain the probability density estimates, it was only by the histogram method. In some of these studies, the Reynolds number influence has not been well evidenced due to the lack of accuracy in the probability density estimates. The most complete analysis was the work of Zarić [4] for a Reynolds number which is near our smallest one. The present study completes Zarić's work and gives a full description of the velocity probability density distributions for different Reynolds numbers. The results are presented in Fig. 1 for the probability density and in Fig. 2 for the skewness and flatness factors.

The general behaviour agrees with findings of other authors for the lowest Reynolds number. The presence of the intermittent phases can be observed in these distributions as has been stated by Zarić [5]: positive skewness observed close to the wall reflects the influence of high-momentum fluid inrushes while negative skewness observed further from the wall was the consequence of the influence of low-momentum fluid ejections. The distributions reported here indicate this presence at all Reynolds numbers. Some new features can be furthermore observed from these distributions:

(i) at about $y^+ = 5$, the skewness factor remains the same at different Reynolds numbers, it decreases with increasing Reynolds numbers in the viscous sublayer and tends to increase beyond $y^+ = 5$ and becomes positive for higher Reynolds numbers. These results seem to show an attenuation at higher Reynolds numbers of the relative importance of the two intermittent phases. This tendency needs however to be confirmed by the conditional analysis.

(ii) the change in the sign of the skewness factor observed by Kreplin [29] at $y^+ \simeq 13$, by Zarić at $y^+ \simeq 15$ and by Elena [31] at $y^+ \simeq 20$ disappears at the highest Reynolds number where the minimum in the distribution of the flatness factor is very attenuated. While the results from these studies were not accurate enough to assert a sufficient precision to these observed zero value locations in the skewness factor, there seems to be some evolution with Reynolds numbers and the present results permit an explanation of this tendency.

These distributions are well illustrated by the features observed in the plot of some representative records of the instantaneous normalized velocity fluctuations in Fig. 3.

3.2. Conditional Statistical Analysis

The Zarić's stochastic conditional sampling technique. In Zarić's S.C.S. method [4], an intermittency function is defined as:

$$I(t) = \begin{cases} 1 & Z > Z_s & \text{inrush phase} \\ 0 & Z_s \geq Z \geq -Z_e & \text{calm phase} \\ -1 & Z < -Z_e & \text{ejection phase} \end{cases}$$

where the criterion Z is a combination of different criteria and the values of the critical levels Z_s and Z_e differ from each other.

Zarić used five criteria, the most important one was $Z_1 = (u - U_m)(du/dt)$ for switching on the intermittency function (with U_m not necessarily the mean

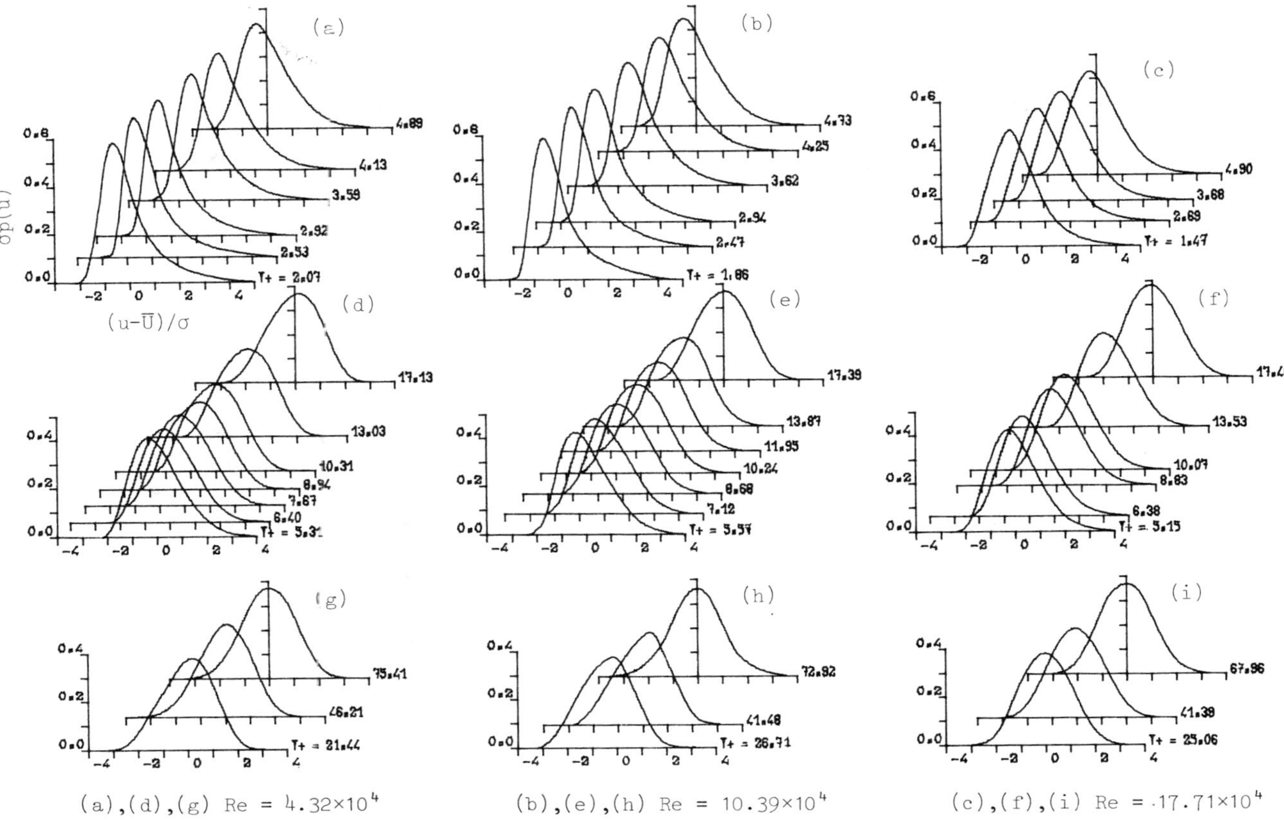

(a),(d),(g) Re = 4.32×10^4 (b),(e),(h) Re = 10.39×10^4 (c),(f),(i) Re = 17.71×10^4

FIG. 1. Velocity probability density distributions in the wall layers

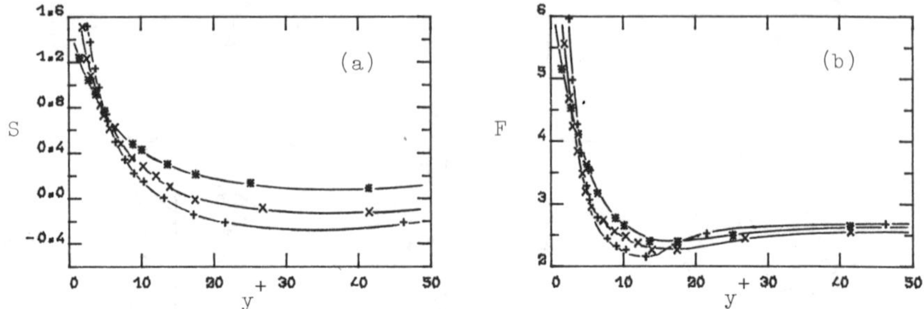

FIG. 2. Distribution of the skewness (a) and flatness factors (b) in the wall layers
+ Re = 4.32×10^4 ; × Re = 10.39×10^4 ; ✱ Re = 17.71×10^4

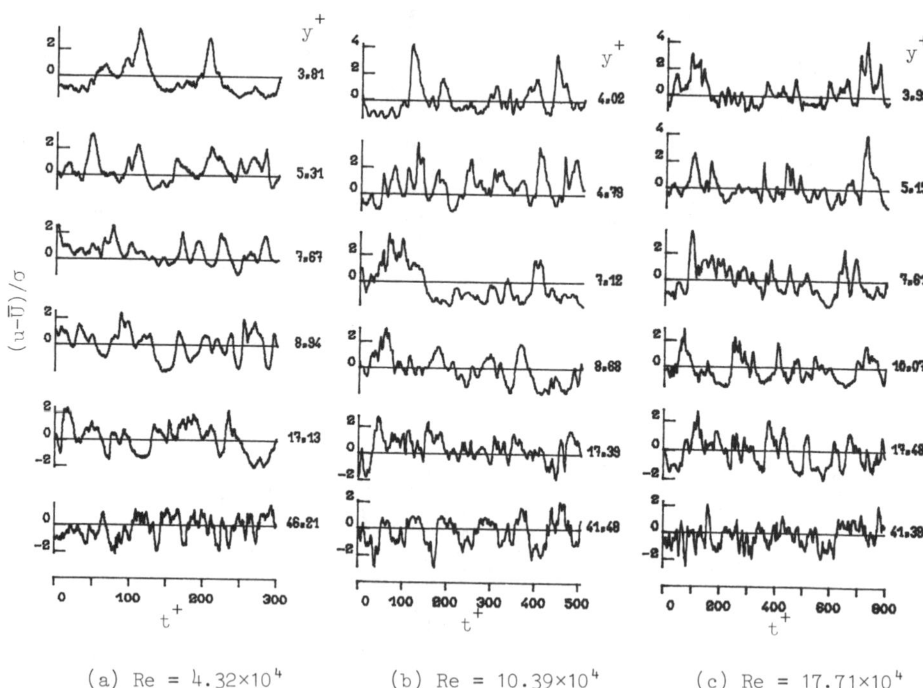

(a) Re = 4.32×10^4 (b) Re = 10.39×10^4 (c) Re = 17.71×10^4

FIG. 3. Plot of the instantaneous normalized velocity fluctuations

velocity but a value between the mean and the most probable value), the other ones are additional criteria. A set of the values of different critical levels (Z_{1s}, Z_{1e}, etc.) was established after a partial optimization and the different phases have been well separated by the procedure. The complete method is described in [4].

The method has been used in this study for the three Reynolds numbers, each one including six y^+ wall distances from the viscous sublayer to the logarithmic region. The critical levels used by Zarić have been taken for all Reynolds numbers. The great number of criteria increases the computational cost so the investigated points are limited, the major part of distances were lying in the buffer layer to show the evolution with y^+. Probability density distributions of different phases are given in Fig. 4. The intermittency factors of different phases:

$$\gamma_i = \overline{I(t)_i}$$

are shown in Fig. 5 and the corresponding contributions to the total variance of the signal:

$$e_i = \gamma_i(\sigma_i/\sigma)^2$$

(where $\sigma_i^2 = (\bar{U} - U_i)^2 + \overline{(u - U_i)^2}$ with U_i the corresponding mean values) are given in Fig. 6. The e_i values are indications of the relative contributions of the different phases to the turbulent energy.

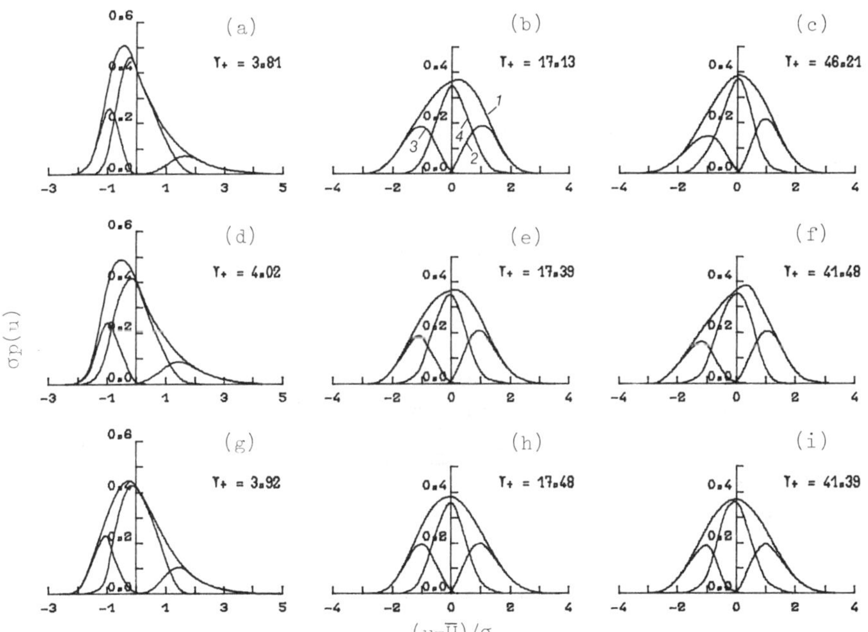

FIG. 4. Probability density distributions of the different phases
1 - total signal ; 2 - inrush phase ; 3 - ejection phase ; 4 - quiescent phase
(a),(b),(c) Re = 4.32×10^4; (d),(e),(f) Re = 10.39×10^4; (g),(h),(i) Re = 17.71×10^4

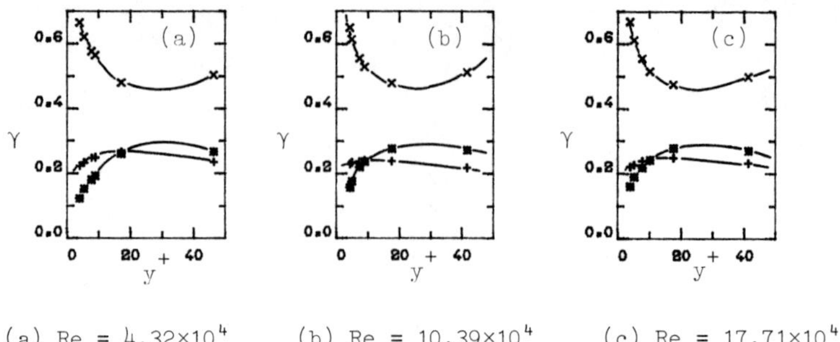

(a) Re = 4.32×10^4 (b) Re = 10.39×10^4 (c) Re = 17.71×10^4

FIG. 5. Intermittency factors of different phases in the wall layers
+ ejection phase ;× quiescent phase ;* inrush phase

The dual structure proposed by Zarić [4], i.e., the two intermittent phases of ejections and inrushes separated from the relatively undisturbed background flow, is observed at all Reynolds numbers. Results at the lowest Reynolds number agree well with Zarić's findings in the viscous sublayer where one can observe that for the three Reynolds numbers, the inrush phase contributes the most to the total variance. The differences with Zarić's results lie in the other layers: the relative importance of the two inrush and ejection phases is less pronounced, they tend to be of the same magnitude in the logarithmic region. This tendency is best accentuated at higher Reynolds numbers: the inrush contribution increasing and the ejection contribution decreasing to be nearly the same at the highest Reynolds number. The same behaviour can be observed in Fig. 7 where the maxima of the velocity probability density are given.

A modification of Zarić's cyclic conditional sampling technique. A preliminary conditional sampling technique was first proposed by Zarić [17] and called 'cyclic conditional sampling' or C.C.S. technique. The method consists of dividing the N_t total values into M groups each having $m = N_t/M$ digital values and forming the criterion:

$$Z = \frac{(\hat{u} - \overline{U})\,\chi^2}{\overline{\chi^2}\sqrt{\overline{(\hat{u} - \overline{U})^2}}}$$

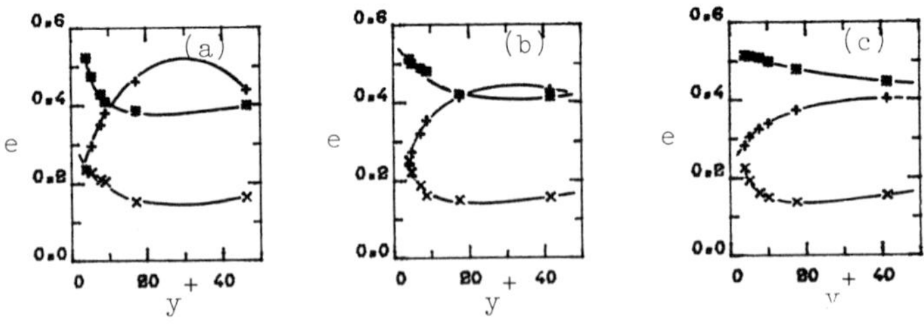

FIG. 6. Contribution of the different phases to the total variance
Notation as in Fig. 5

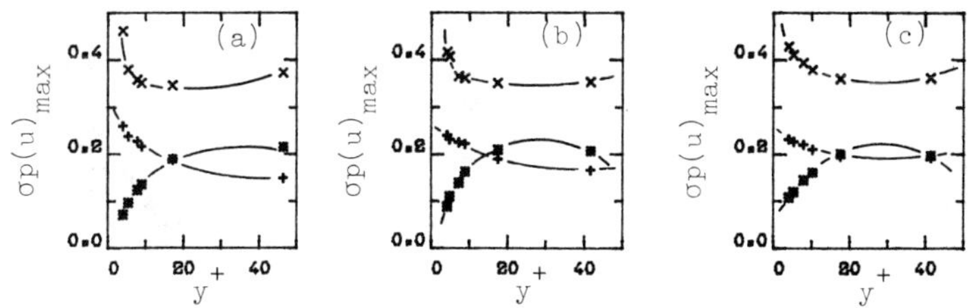

FIG. 7. Distribution of the maxima of the probability density of the different phases. Notation as in Fig. 5 and Fig. 6

where $\hat{u}$ and χ^2 are the mean and variance corresponding to the partial groups. The sampling criterion used then partial averages similar to short-time averages in [26] and [6]. Instead of acting on each instantaneous digital value as in the other method, the detection scheme works now on a group whose size m was fixed.

On the basis of the above criterion and taking $Z_s = Z_e = 5$, Zarić [17] performed preliminary conditional sampling at $y^+ = 2.5$ and $y^+ = 13.9$. The results of the analysis showed a certain degree of intermittent phase separation but this is not complete so the S.C.S. method has been prefered.

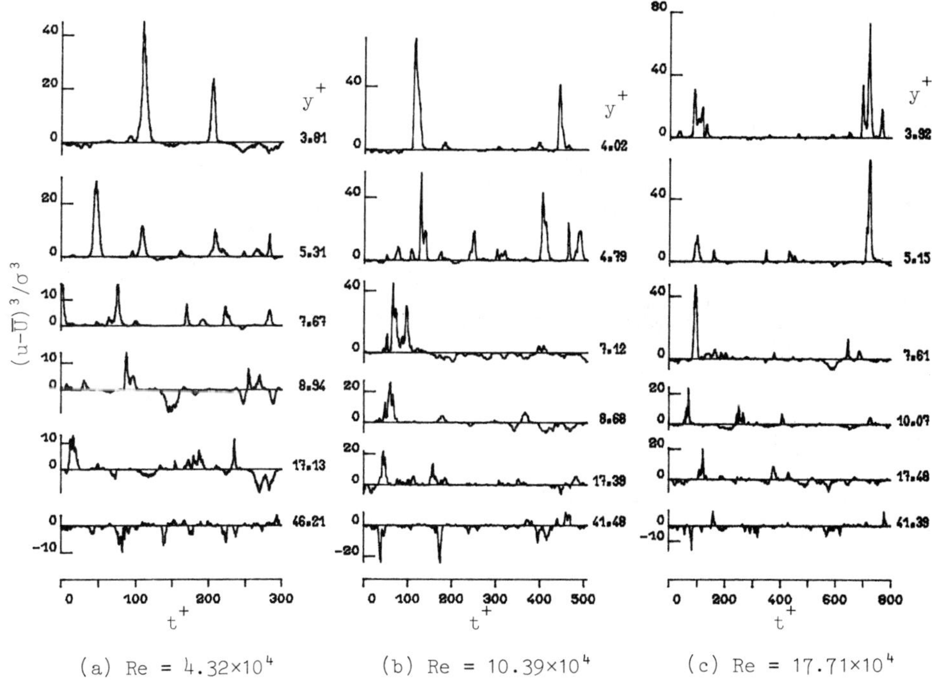

(a) Re = 4.32×10^4 (b) Re = 10.39×10^4 (c) Re = 17.71×10^4

FIG. 8. Plot of the triple products of the instantaneous normalized velocity fluctuations

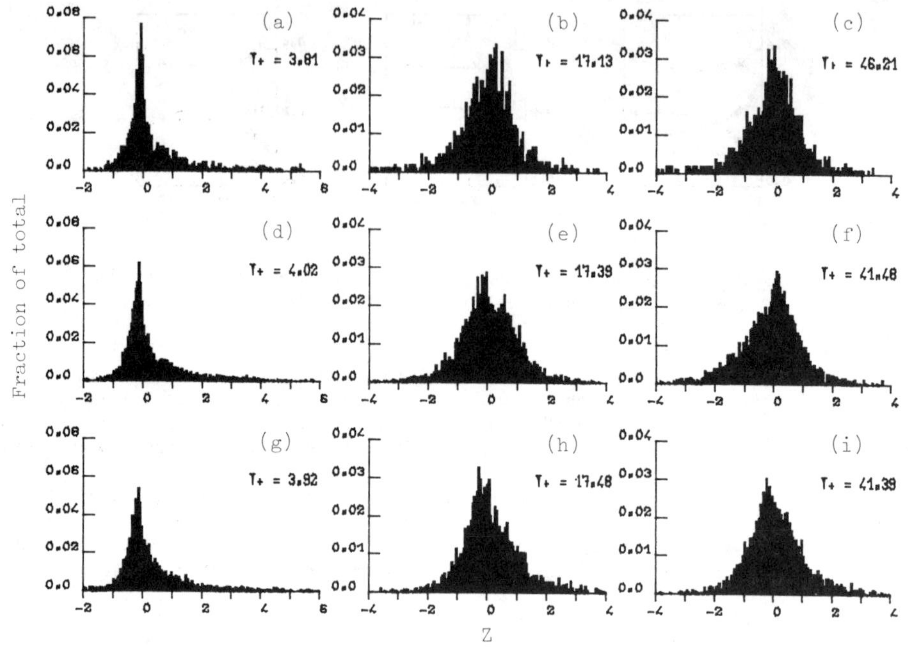

FIG. 9. Histograms of the criterion Z used in the C.C.S. technique (a),(b),(c) Re = 4.32×10^4; (d),(e),(f) Re = 10.39×10^4; (g),(h),(i) Re = 17.71×10^4

In this study, attention is given to find an appropriate choice of the size m of the partial blocks and the critical levels Z_e and Z_s. Following a procedure proposed by Elena [31], the triple products of the instantaneous normalized velocity fluctuations were performed and as examples, the same samples as in Fig. 3 are represented in Fig. 8. As can be seen in this figure, the triple products have the advantage of showing clearly periods of large amplitude and the sizes m are therefore taken to correspond to the mean durations of these periods. Fig. 9 represents the histograms of the criterion Z calculated with these sizes m. The same features found by Zarić [17] are observed for the three Reynolds numbers, namely, the maximum of the distributions which is located close to zero and the excursions of the criterion towards high positive values in the viscous sublayer. The following values are then choosen for this study: $Z_s = 0.9$ in the viscous sublayer, $Z_s = 0.1$ otherwhere and $Z_e = 0.05$ for all points.

Improvement of the method follows from the fact, inspired by the S.C.S. method, that the intermittent phases are well characterized by a variable duration rather than a constant one which was fixed by the size m. The additional criterion:

$$Z_3 = (u - U_m)$$

was then introduced with the critical levels $Z_{3s} = Z_{3e} = -0.5\,\sigma$ to switch off the intermittency function.

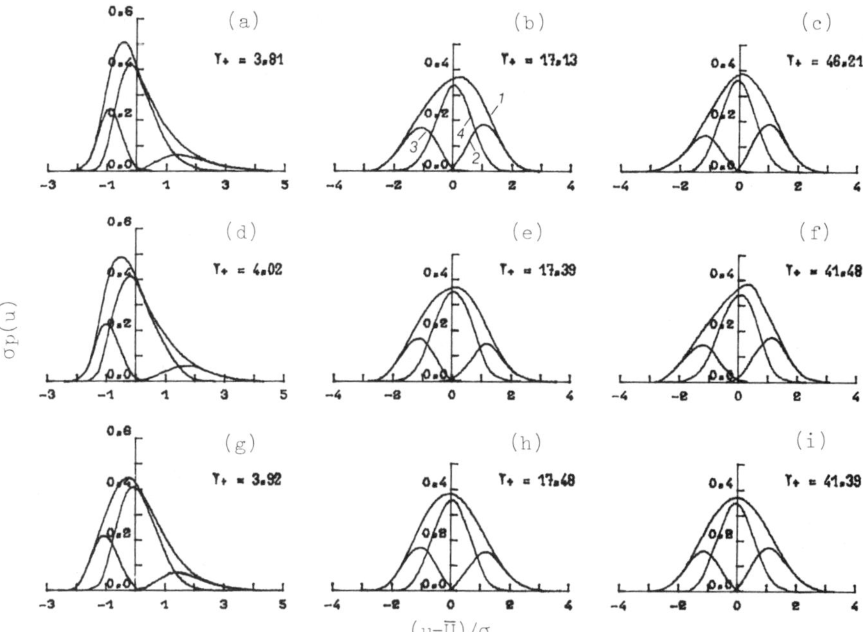

FIG. 10. Probability density distributions of the different phases obtained by the C.C.S. method. Notation as in Fig. 4

The results are given in Fig. 10. As in the S.C.S. case, the intermittent phases are well separated and the same evolution with wall distances is observed. The evolution with Reynolds numbers is less clear than in the S.C.S. case.

4. CONCLUSIONS

By the use of an efficient method which has not been used previously in turbulence studies, accurate estimates of the velocity probability density distributions in the wall layers have been obtained at different Reynolds numbers. The overall shape of these distributions indicates the presence of intermittent phases even at high Reynolds numbers. This presence is however seen to be attenuated at higher Reynolds numbers. The information on the wall turbulence structure given then by the conventional statistical approach needs to be confirmed by a conditional analysis.

The Zarić's 'stochastic conditional sampling' technique and a modified method based upon another Zarić's technique have been used successfully to detect the intermittent phases: both methods show the same behaviour: (i) at low Reynolds numbers, large contribution to the turbulent energy of the inrush phases in the viscous sublayer and of the ejection phases in the other layers; (ii) at higher Reynolds numbers, attenuation of the relative importance of the two intermittent phases.

Despite the relative great number of criteria needed in the method and although these criteria were not completely optimized, the Zarić's stochastic conditional sampling technique is found to be an useful tool in a conditional analysis. The modified cyclic conditional sampling method appears to have some degree of processing simplicity and is then shown to be an alternative to the analysis of the intermittent phases.

Finally, with the results accumulated over the last ten years, both visual and anemometry studies are found to be necessary to model turbulent flows more accurately. Evidence has also came out from these studies that a need of quantitative data over a wide range of Reynolds numbers is necessary to clarify the nature of the observed coherent structures. Anemometry studies, with appropriate conditional techniques, can bring out the characteristic features of events which exist in the wall layers and it is hoped that results in the present study provide some new interesting quantitative information and could eventually lead to a better understanding of turbulent processes in wall layers.

ACKNOWLEDGEMENTS

The author wishes to express his gratitude to Professor A. Fortier for his valuable suggestions and continued interest in this study and to Professor Z. Zarić for providing part of the programs used in the conditional analysis.

REFERENCES

1. J. Laufer, Ann. Rev. Fluid Mech., 7, 307-326 (1975)

2. W.W. Willmarth, Adv. Appl. Mech., 15, 159-254 (1975)

3. F.N. Frenkiel and P.S. Klebanoff, Phys. Fluids, 10, 507-520 (1967)

4. Z. Zarić, "Etude statistique de la turbulence pariétale", Boris Kidrič Institute, Belgrade (1974)

5. Z. Zarić, Int. J. Heat Mass Transfer, 18, 831-842 (1975)

6. R.F. Blackwelder and R.E. Kaplan, J. Fluid Mech., 76, 89-112 (1976)

7. J.M. Wallace, R.S. Brodkey and H. Eckelmann, J. Fluid Mech., 83, 673-693 (1977)

8. R.F. Blackwelder and H. Eckelmann, J. Fluid Mech., 94, 577-594 (1979)

9. W.R.B. Morrison, K.J. Bullock and R.E. Kronauer, J. Fluid Mech., 47, 639-656 (1971)

10. S.S. Lu and W.W. Willmarth, J. Fluid Mech., 60, 481-511 (1973)

11. R.S. Brodkey, J.M. Wallace and H. Eckelmann, J. Fluid Mech., 63, 209-224 (1974)

12. R.E. Falco, Phys. Fluids Suppl., 20, S 124-132 (1977)

13. J. Sabot and G. Comte-Bellot, J. Fluid Mech., 74, 767-796 (1976)

14. J.M. Wallace, H. Eckelmann and R.S. Brodkey, J. Fluid Mech., 54, 39-48 (1972)

15. G.R. Offen and S.J. Kline, Rep. No. MD-31, Dept. Mech. Eng., Stanford Univ. (1973)

16. R. Blackwelder, Phys. Fluids Suppl., 20, S 232-242 (1977)

17. Z. Zarić, "Wall turbulence statistical analysis", 4 th. All Union Heat Mass Transfer Conf., Minsk, S.S.S.R. May 15-19 (1972)

18. D.W. Scott, "Nonparametric probability density estimation by optimization theoretic techniques", Tech. Rep. No. 476-131-1, Rice Univ., Houston (1976)

19. N. Van Thinh, Disa Inf., 7, 13-18 (1969)

20. N. Van Thinh, Int. Seminar "Turbulent Flows with Heat and Mass Transfer Near a Wall", May 3-5,(1977), Orsay, France

21. N. Van Thinh, Letters in Heat and Mass Transfer, 6, 103-115 (1979)

22. J.S. Bendat and A.G. Piersol, "Random Data: Analysis and Measurement Procedures", Wiley-Interscience (1971)

23. P.S. Klebanoff, NACA Rep. 1247 (1955)

24. J. Marcillat, Thesis, Univ. d'Aix-Marseille (1964)

25. H. Eckelmann, "An experimental investigation of a turbulent channel flow with a thick viscous sublayer", Mitt. Max-Planck Instit., No. 48 (1970)

26. A.K. Gupta, Ph. D. Thesis, Univ. of Southern California (1970)

27. F.N. Frenkiel and P.S. Klebanoff, Phys. Fluids, 16, 725-737 (1973)

28. T.R. Heidrick, Ph. D. Thesis, Univ. of Manitoba (1974)

29. H.P. Kreplin, Mitt. MPI Strömungsforsch & AVA Göttingen, No. 63 (1976)

30. G. Comte-Bellot, "Ecoulement turbulent entre deux parois parallèles", P.S.T. Minist. de l'Air, No. 419, Paris (1965)

31. M. Elena, Int. J. Heat Mass Transfer, 20, 935-944 (1977)

32. H. Ueda and J.O. Hinze, J. Fluid Mech., 67, 125-143 (1975)

NOMENCLATURE

e_i	contribution of each intermittent phase to the total variance
F	flatness factor, = $\overline{u'^4}/\sigma^4$
h	mesh spacing in the probability density estimation
I	intermittency function
p(u)	probability density of u
Re	Reynolds number based on the hydraulic diameter and the average velocity
S	skewness factor, = $\overline{u'^3}/\sigma^3$
t	time

t^+	non-diemnsional time, = $t\, u_\tau^2/\nu$
u	instantaneous velocity, = $\overline{U} + u'$
$\overline{U}$	mean velocity
U_a	average velocity
U_i	mean velocity corresponding to each phase
U_m	velocity between the mean and the most probable value
$\hat{u}$	mean velocity corresponding to a conditional sampling group
u_τ	friction velocity
y	distance from the wall
y^+	non-dimensional distance from the wall, = $y\, u_\tau/\nu$
Z	conditional sampling criterion

Greek symbols

γ	intermittency factor, = $\overline{I(t)}$
ν	kinematic viscosity
σ	RMS value, = $\sqrt{\overline{u'^2}}$
τ	sampling period
χ^2	variance corresponding to a conditional sampling group

Subscripts

e	ejection phase
s	inrush phase

The Influence of Reynolds Number on Characteristics of Turbulent Boundary Layers

JANNIS ANDREOPOULOS, FRANZ DURST, and JOVAN JOVANOVIĆ
Sonderforschungsbereich 80
University of Karlsruhe
Karlsruhe, FRG

ZORAN ZARIĆ
Boris Kidrich Institute
for Nuclear Sciences
Belgrade, Yugoslavia

ABSTRACT

Hot-wire anemometer measurements are reported in flat plate turbulent boundary layers. Studies were carried out at various Reynolds numbers with particular attention being given to the near wall region. The measurements include mean velocity and higher order moments of streamline fluctuating components as well as their time derivatives. The results suggest a Reynolds number dependence of all properties upto Reynolds numbers of $Re_\delta \sim 10^5$.

INTRODUCTION

Experimental fluid mechanics has recently concentrated on the employment of more sophisticated detection schemes applied to time records of local velocities in order to detect so-called coherent structures that are claimed to make up the flow field. This research is based on the hypothesis that turbulence is not a completely random phenomena but yields time-dependent, repeatable, identifyable, deterministic flow patterns that appear superimposed on the small scale random flow field. Results of recent research also suggests that these deterministic structures play a major role in the process of maintaining turbulence by extracting energy from the mean flow; the coherent structures are claimed to be the centers of turbulence energy production. Many measurements made in different flow configurations, seem to support this idea and, at present, considerable effort is made to gain more information on these structures and their role in turbulent flows.

Up to now, studies of coherent structures in wall boundary layers have concentrated on low Reynolds number flows where various detection schemes have been applied to hot-wire signals to yield information on the coherent part of the flow. This has resulted in detailed information regarding the average signal expected from coherent structures at low Reynolds numbers. Similar information does not exist for high Reynolds number flows.

Strong Reynolds number dependences of the moments of the velocity probability density distributions are reported by Zaric

[7] that suggest a Reynolds number dependence of all flow phenomena and it appears that scaling with the wall variables is not applicable, at least, as far as the shape of the distributions is concerned. Since scaling with wall variables was reported in structural studies of Wallace et al.[9] at low Reynolds number flows, the question arises whether there is a difference between low and high Reynolds number flow phenomena. This is also suggested by cross-spectral measurements of Morrison [1], made in fully developed turbulent pipe flow. His measurements suggest drastic changes of flow structures close to the wall as the Reynolds number increases above $Re_D = 6 \times 10^4$. This finding and the results of Zaric [6], [7] initiated the present research activities of the authors and resulted in detailed investigations of the Reynolds number influence on statistical quantities of the flow field. Mean velocity values and higher order momentums of the streamwise, time-varying velocity were measured together with time-derivatives of these signals. The measurements were carried out in flat plate turbulent boundary layers. Reynolds numbers upto 1.5×10^5 were reached in a low turbulence wind tunnel.

Althought no direct structural measurements were carried out in the present work, the resultant moments might provide information on their existence. It is generally believed, e.g. see ref. [7], that deviations of the higher order momentums from Gaussian distributions are in some way related to a sequence of deterministic events as observed in the early visual studies of turbulent boundary layer flows, e.g. see Kline [8]. As a matter of fact, many of the detection schemes employed in conditional sampling utilize the deviations from Gaussian probability density distributions in order to detect coherent structures. For instance, the pattern recognition schemes suggested by Wallce et al. [9] utilizes the observed high values of skewness of the velocity derivative in order to detect the occurance of their coherent structures.

The lack of data on high Reynolds number boundary-layer flows initiated a collaborative effort to study the properties of high Reynolds number boundary layers. In a first attempt, velocity probability density measurements were carried out and are reported in this paper. The data suggest that Reynolds numbers, Re_δ, of the order of 100 000 are necessary in order to reduce the Reynolds number dependence observed with low Reynolds numbers. The low Reynolds number boundary layers still obey characteristics of transitional flows.

2. EXPERIMENTAL EQUIPMENT AND DATA ACQUISITION

The present measurements were performed in the large wind tunnel of the Institute of Hydromechanics at the University of Karlsruhe which is a close circuit wind tunnel with an entrance area contraction of 1 to 10 yielding a turbulence level of approximately .06 % at the maximum speed of 45 m/s. The cross-sectional shape of the test section is orthogonal with an area of approximately $2m^2$ The total length of the test section is approximately 7 m.

To carry out studies of boundary layer flows, a flat plate was installed inside the tunnel test section at a height of 50 cm from the bottom wall. The total length of the plate was 6 m and it was supported by strong bars over its entire length. In this way, plate deflections due to static pressure differences on top and below the plate were avoided. To avoid front plate separation, a trailing edge flap was incorporated and was adjusted to eliminate flow from the bottom of the plate to circulate around the leading edge. The material of the plate on which the boundary layers were studied was made of PVC (for Meica). This material was chosen to avoid hot-wire corrections close to the wall. The material was suggested in a theoretical and experimental study [13] which explains the actual mechanicsm of the increased heat transfer from hot-wires close to walls to be due to the strong heat conduction of the wall material. PVC has an extremely low heat conduction and is, therefore, well suited for near-wall velocity studies by means of hot-wire anemometry.

The traversing mechanism for the hot-wire probes had an automatic, externally driven stepping motor controlled by a driving unit. This unit permitted the precise positioning of the hot-wire in the three directions of a cartesian coordinate system with a step size of 25 μm. Special care was taken to position the prongs of the probe parallel to the wall surface in order to approach the wall as close as possible and to minimize flow disturbances by the wire prong s. An optimum between boundary layer disturbance and close-wall measurements was achieved by inclining the boundary layer wire probe approximately 5°.

Measurements were made with standard hot-wire equipment consisting of a DISA P05 gold-plated boundary layer probe operated with a DISA 55M01 constant temperature anemometer. The output of the anemometer was low-pass filtered at 10 kHz and processed by a Hewlett Packard 5455 Fast-Fourier Analyser. The signals were digitized at a rate of 20 kHz and at each point 22 sec. of real time data was recorded on a digital tape. The tape was thereafter analysed by means of soft-ware programs on the UNIVAC 1108 computer of the University of Karlsruhe. The tapes were also saved for later studies of the statistics of turbulent boundary layer flows.

3. EXPERIMENTAL RESULTS

The range of Reynolds numbers which has been covered in the present experiments is shown in Table 1. The lowest Reynolds number was chosen to be around 3×10^{4} for direct comparison with available data in the literature. At this low Reynolds number a fully-developed turbulent boundary layer is difficult to excist. Even with the use of a sand-paper trip close to the leading edge of the plate, the flow at the measuring position, 5 m from the leading edge, remains a developing turbulent boundary layer flow. Closer to the trip, features of fully-developed boundary layer flows might be found but the boundary layer shows features of a relaxing layer. This has also been observed by Wilmarth et al. [2] and Rao et al. [3] who suggested that completely fully developed turbulent boundary layer flows cannot be realized for Reynolds numbers smaller than 5×10^{4}.

The friction velocities u_{τ} given in Table 1 were computed from the slope of the measured velocity profiles only taking into account the data points in the viscous sublayer ($Re_{\delta} = 47 \times 10^{3}$). For the higher Reynolds numbers a Clausert chart was also employed to yield friction velocity values very close to those computed from the

U_o [m/s]	δ [mm]	u_τ [m/s]	Re_δ
7.0	104.80	0.196	47.10^3
22.61	88.19	0.812	128.10^3
28.79	81.28	0.999	150.10^3

Table 1:

velocity gradient. These were used to normalize the measured mean-velocity profiles for the aforementioned Reynolds numbers and the normalized velocity distributions are given in Figure 1. The difference between low and high Reynolds number flows, is clearly shown. At sufficiently high Reynolds number, only the wake region usually shows the Reynolds number dependence where it is claimed that the near wall region is Reynolds number independent. The present measurements suggest that untripped boundary layers and relaxed tripped boundary layers do not obey the log-law with the usual constants κ = .41 and b = 5.5. For small Reynolds numbers, these constants are also Reynolds number dependent. No physical explanation for this behaviour can be given at present but investigations are underway to have a map of various properties of flow Reynolds number boundary layers, tripped and untripped, in order to establish the physical reasons for the deviation from the log-law for higher Reynolds number flows.

Figure 2 presents the measured second-order moments of the local fluctuating longitudinal velocity, normalized by the corresponding friction velocities. The resultant distributions show a large difference between the low and high Reynolds number flows.

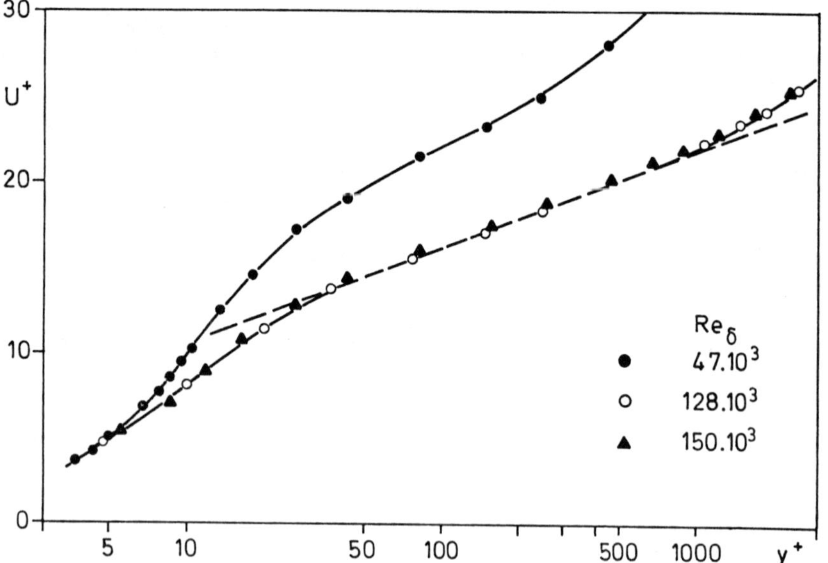

Figure 1: Normalized mean-velocity distributions

Over most of the boundary layer the rms-value of the velocity fluctuations is reduced with increasing Reynolds number as already indicated by Laufer [9]. For the two higher Reynolds numbers, the data for the second moments of velocity fluctuations are in close agreement as expected from von Karman's similarity hypothesis.

The third and fourth moments of the longitudinal velocity fluctuations are presented in Figures 3 and 4. These figures indicate that both quantities are considerably reduced in the wall-near region as the Reynolds number is increased. Both quantities gets closer to those expected for Gaussian distributions of velocity fluctuations.

The differences between low Reynolds number boundary layers and high Reynolds number boundary layers is indicated by the differences in the skewness distributions. For the high Reynolds numbers, the skewness is not chaning sign throughout the entire inner layer and the logarithmic region. This change of sign has often been taken as a support of the hypothesis (at low Reynolds numbers)that boundary layer flows contain coherent structures that are, in the average, located around the region of zero-skewness. Since at higher Reynolds numbers, skewness is not chaging sign close to the wall, the question is appropriate whether coherent structures are still present or whether they changed their properties such that the change in the measured skewness values results from these changes.

In order to explore the details of the Reynolds number influence on the flow field, statistical moments of the velocity time derivatives were computed using a formula suggested by Wallace et al. [9]:

$$\frac{dU(n)}{dt} = [2u(n-2)-u(n-1) + 2u(n+2) + u(n+1)]/10\Delta t$$

In Figure 5 rms-values of time-derivatives of the longitudinal velocity fluctuations are plotted normalized with the inner wall time-scale, ν, U_τ^2. This figure indicates a very strong Reynolds

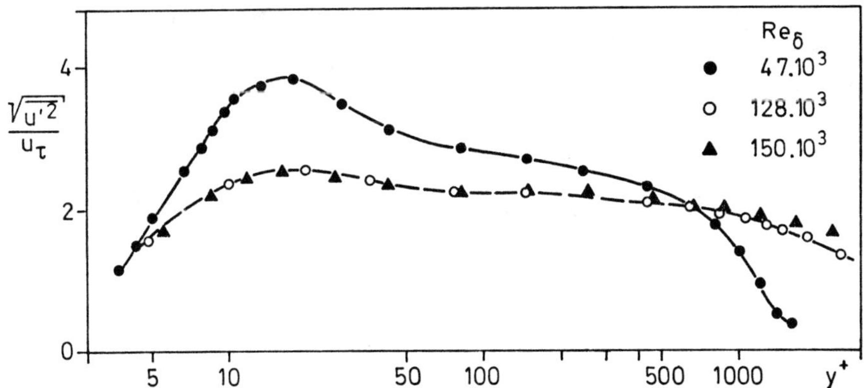

Figure 2: Normalized distributions of rms-values of longitudinal velocity fluctuations

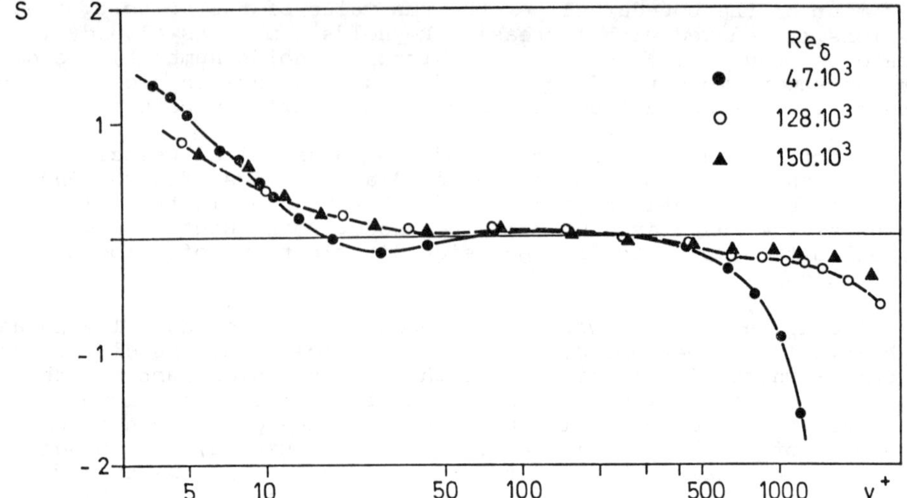

Figure 3: Distribution of skewness of longitudinal velocity fluctuations

number dependence of $(\partial U/\partial t)^2$ suggesting that the time derivative of the local instantaneous velocity fluctuations are even more sensitive to Reynolds number variations than the velocity fluctuations itself. It can readily be seen that the results in Figure 5 are consistent to those in Figure 1. Invoking Taylor's hypothesis, the time derivatives of the velocity fluctuations are related to the rate of dissipation of turbulent kinetic energy as indicated by the following formula:

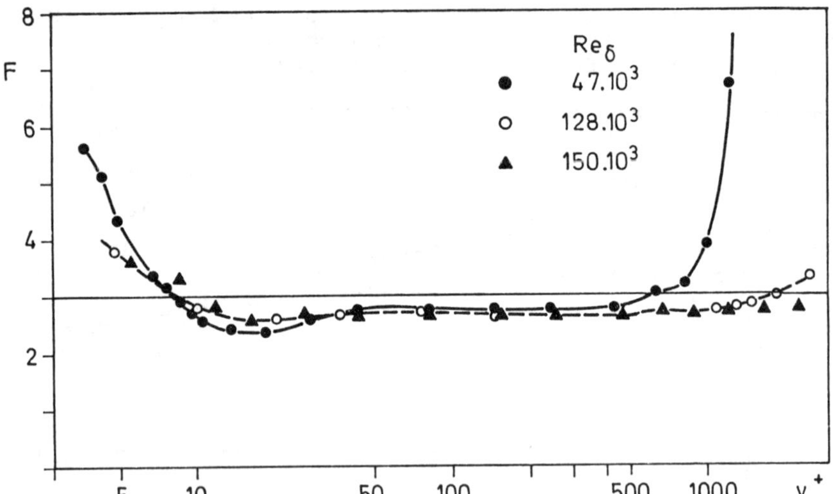

Figure 4: Distributions of flatness of longitudinal velocity fluctuations

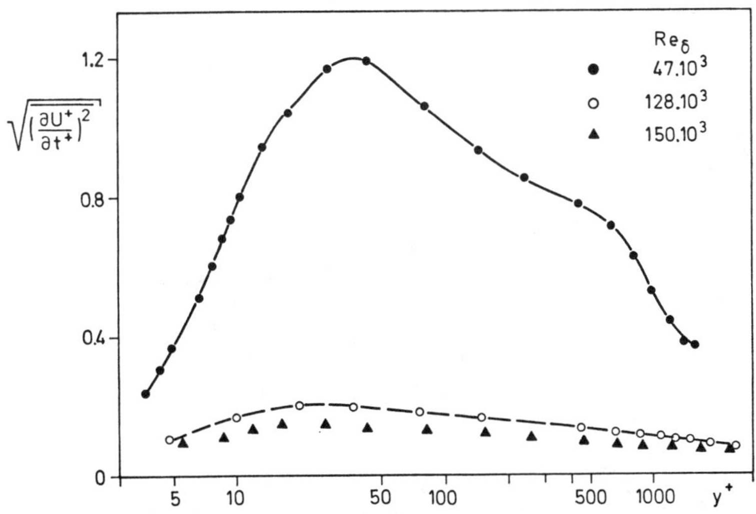

Figure 5: rms-value of time derivative of longitudinal velocity-fluctuations

$$W = 15 \frac{\nu}{U^2} \overline{(\frac{\partial u}{\partial t})^2}$$

Hence, at low Reynolds numbers higher dissipation rates near the wall were recorded. From Figure 1 it is clear that $(\partial u^+/\partial y^+)$ is higher at low Reynolds numbers. Higher values are also recorded for the rms-values of the longitudinal velocity fluctuations and it can therefore be inferred that close to the wall the quantity $\overline{uv}/u_\tau^2$ will reach higher values for the low Reynolds number case. This readily suggests that close to the wall higher turbulence production is also occuring at low Reynolds numbers. Hence, the higher dissipation rate suggested by Figure 5 are in agreement with the higher energy production suggested by Figures 1 and 2.

Time derivatives of the longitudinal velocity fluctuations were further processed to obtain the skewness and flatness factors for various Reynolds numbers. The skewness factor distribution of the time derivatives of the longitudinal velocity fluctuations are shown in Figure 6 and the corresponding flatness distributions in Figure 7. Both distributions show again the strong deviations of the flow Reynolds number signal from those at higher Reynolds numbers. These differences also show up on complete records of probability density distributions that were obtained for various Reynolds numbers and for various points in the flow field. Exampes of these measurements are shown in Figures 8a and 8b. Probability density measurements for the time derivative of the longitudinal velocity fluctuations were also recorded and examples are given in Figures 9a and 9b. From these distributions it can clearly be deduced that low Reynolds number flows exhibit features that are not present at high Reynolds numbers. The authors are aware that this conclusion is drawn for a limited number of measurements and further studies are underway at the University of Karlsruhe to support these conclusions. They are in agreement with findings by Willmarth and Lu [2] who pointed out

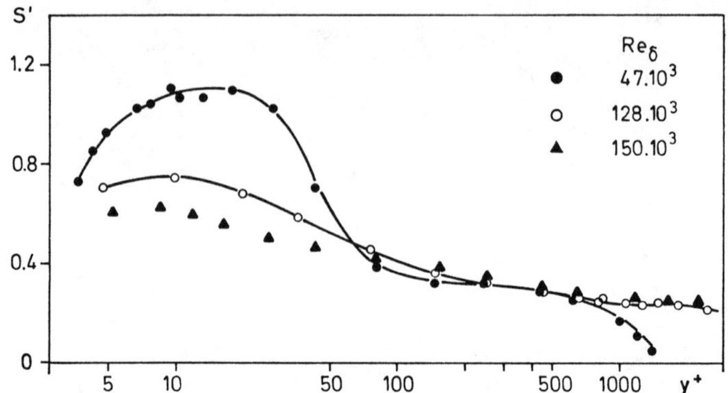

Figure 6: Distribution of skewness of time derivatives of longitudinal velocity fluctuations

that their investigations of coherent structures in boundary layers required mean velocities in excess of a minimum number below which the boundary layers showed behaviours of non-fully developed turbulent flows.

CONCLUSIONS AND FINAL REMARKS

The present study was performed to set up turbulent boundary layers for studies of flow structures at high Reynolds numbers. Strong differences between low Reynolds number turbulent boundary layers and high Reynolds number flows were observed and suggest strong differences between low Reynolds numbers coherent structures and those at high Reynolds numbers. The present study confirms the tendencies and

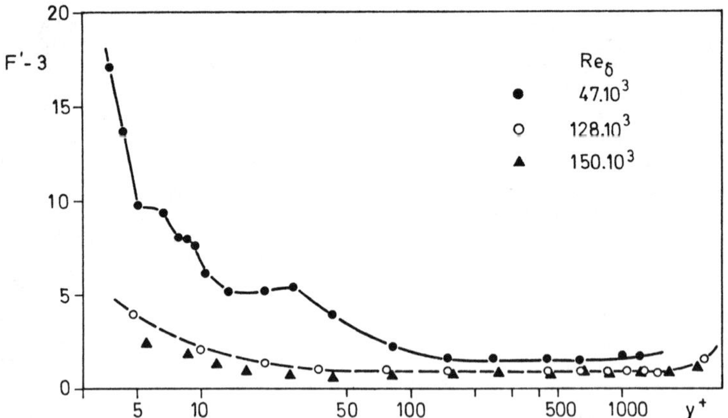

Figure 7: Distribution of flatness of time derivatives of longitudinal velocity fluctuations

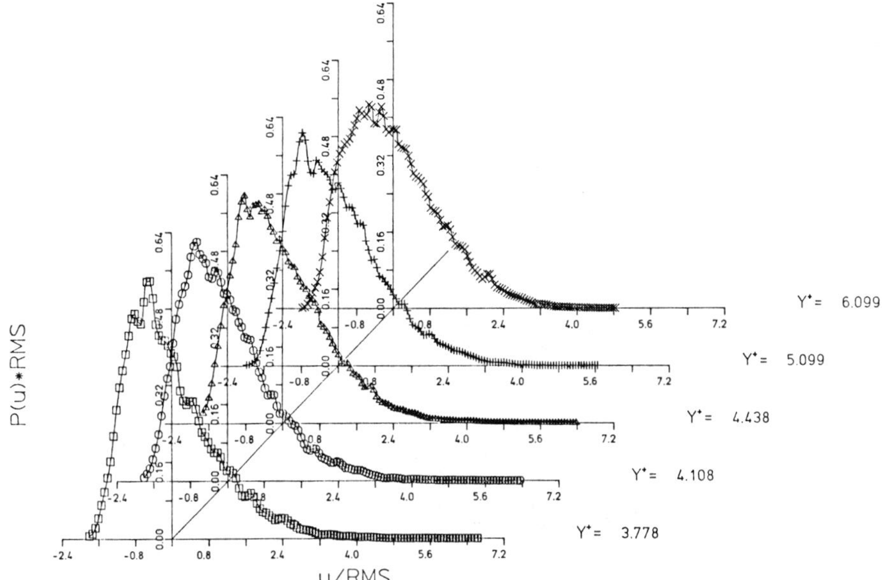

Figure 8a: Probability density distributions of longitudinal velocity fluctuations

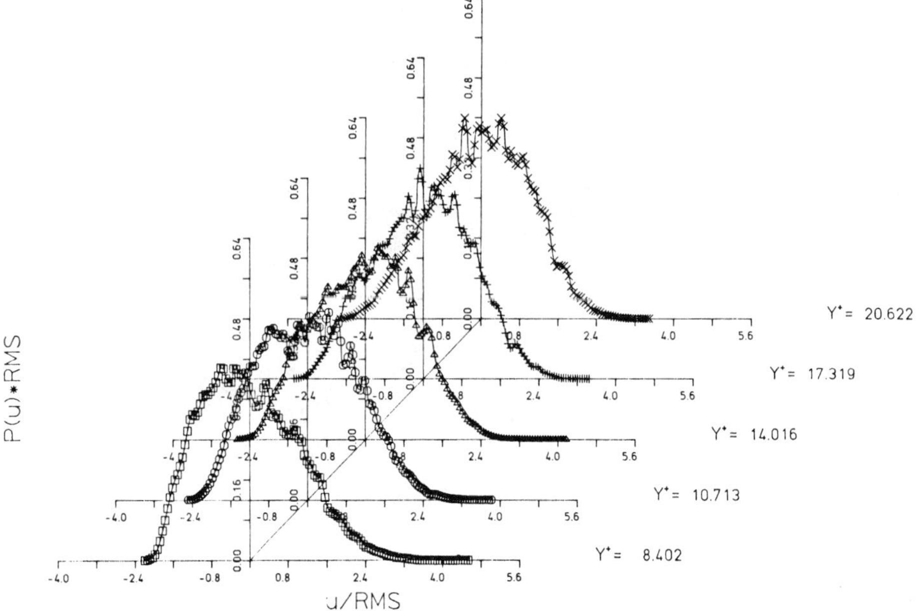

Figure 8b: Probability density distributions of longitudinal velocity fluctuations

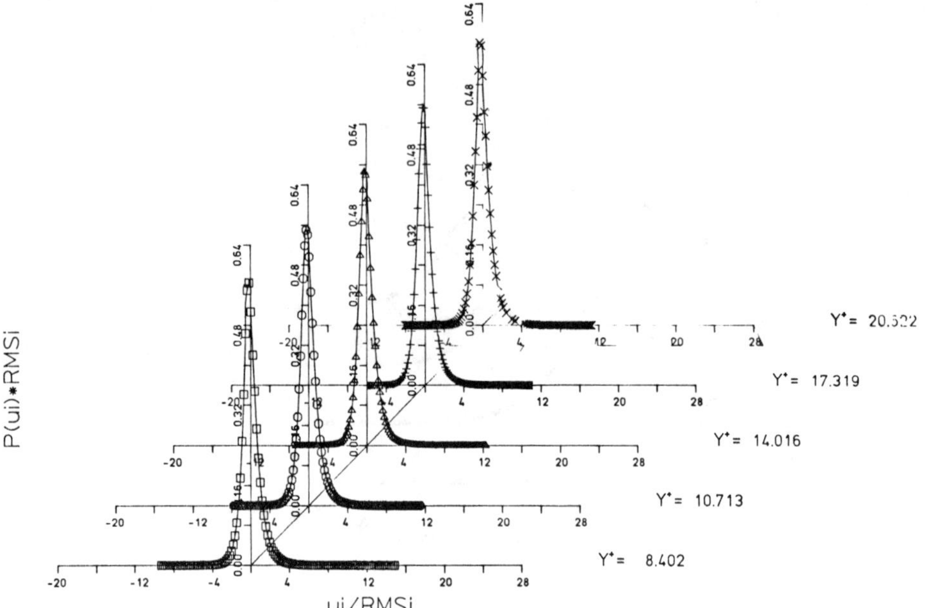

Figure 9a: Probability density distributions of time derivatives of longitudinal velocity fluctuations

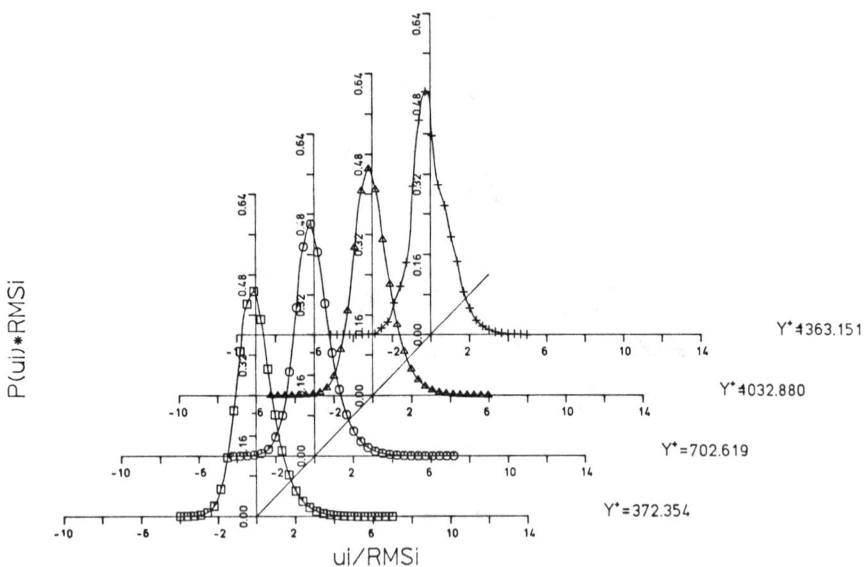

Figure 9b: Probability density distributions of time derivatives of longitudinal velocity fluctuations

quantifies these variations. Further studies are underway to support the conclusions drawn in this paper:

1. The moments of probability density distributions in a turbulent boundary layer appear to become Reynolds number independent for Re $\lesssim 10^5$.

2. Differences between low Reynolds number and high Reynolds number flows show up in all moments of the longitudinal velocity fluctuations. They are even more pronounced for the moments of the time derivatives of the longitudinal velocity fluctations.

Further studies are underway at the University of Karlsruhe to quantify differences between low Reynolds and high Reynolds number flows.

ACKNOWLEDGEMENTS

The present study is a collaborative effort between the Sonderforschungsbereich 80 at the University of Karlsruhe, W. Germany, and the Boris Kidrich Institute, Belgrade, Yugoslavia. The authors are grateful to acknowledge the financial support of their work by the Deutsche Forschungsgemeinschaft and the GermanJugoslavian Cooperation Programme administrated by the Internationales Büro, Kernforschungsanlage Jülich, W. Germany. The support of the Alexander von Humboldt Foundation through a scholarship to Jovan Jovanovic is also gratefully acknowledged.

The measuremenets were performed in the windtunnel of the Institute of Hydromechanics and computations carried out on the UNIVAC 1108 of the University of Karlsruhe.

References

[1] Morrison, W.R.B., Bullok, K.J. and Kronauer,R.E., Experimental evidence of waves in sublayer, J. of Fluid Mechanics, 47,4 pp. 639-656,1977.

[2] Willmarth, W.W. and Lu, S.S., Structure of Reynolds stress near the wall, J. of Fluid Mechanics, 56, pp. 65-92,1972.

[3] Rao, K.N., Narasinsha, R., and Narayaman, B., The bursting phenomenon in a turbulent boundary layer, J. Fluid Mechanics, 48, pp. 339-352,1971.

[4] Hinze.J.O., Turbulence, 2nd edition McGraw Hill Book Co., 1975.

[5] Laufer, J., New trends in experimental turbulence research, Annual Review of Fluid Mechanics, Vol. 7, 1975.

[6] Zaric, Z., Wall Turbulence Studies,Vol.8, 285/380 Academic Press, 1972.

[7] Zaric, Z., Statistical evidence of phenomena in wall layers of turbulent flow, Turbulent Forced Convection in Channels and Bundels, ed. by S. Kakac and D.B. Spalding, McGraw Hill Co, 1979, Vol. 1, pp. 377-403.

[8] Kline, S.J., Reynolds,W.C. Schrsub, F.A. and Rundstadler,P.W., The structure of turbulent boundary layers J. fluid Mechanics, 30, pp. 741-773, 1967.

[9] Wallace, J. Brodkey, R.B. and Eckelmann, H., Pattern-recognized structures in bounded turbulent shear flows, J. Fluid Mechanics, 83, pp. 673-693, 1977.

[10] Gupta, A.K., and Kaplan, R.E., Statistical characteristics of Reynolds stress in a turbulent boundary layer, Physics of Fluids, Vol. 15m No. 6, 1972.

[11] Laufer, J., Investigation of turbulent flow in two-dimensional channel, NACA Report, 1053, 1951.

[12] Blackwelder, R. and Kaplan, R.E. On the wall structure of the turbulent boundary layer, J Fluid Mechanics, Vol. 76, pp. 89-112, 1976.

[13] Bhatia, J.C., Durst, F. and Jovanovic, J., Corrections of hot-wire anemometer measurements near walls, submitted for publication to the Journal of Fluid Mechanics, 1980.

Measurement of Bursting Period and Test of Surface Renewal Model in a Turbulent Boundary Layer Disturbed by a Cylinder

KENJIRO SUZUKI and YASUO KAWAGUCHI
Department of Mechanical Engineering
Kyoto University
Kyoto 606, Japan

INTRODUCTION

Relaxing turbulent flows are frequently encountered in practices in the fields of fluid and thermal engineerings. For example, heat transfer augmentation is usually attained with the flows disturbed somehow and relaxing thereafter. Therefore, detailed studies on relaxing turbulent flows should be very important. The present paper deals with a two-dimensional turbulent boundary layer on a flat plate, disturbed by a cylinder located normal to the flow direction and parallel to and detached from the plate. The present flow is categorized into relaxing turbulent flow, and has a favourable feature for detailed study. Changes in cylinder position and in its size bring about the change in the length scale ratio of the turbulence generated in the wake of cylinder to the turbulence in approaching boundary layer flow, and the relaxation pattern of different restoring rate can easily be observed.

The boundary layer disturbed by a cylinder was studied first by Clauser [1], and recently by Marumo, Suzuki and Sato more in detail [2]. Further studies have been made on statistical turbulent quantities, augmented wall heat transfer in the downstream of cylinder, and tests of some mathematical turbulence models based on one-point correlation of turbulence quantities in the relaxing region of the flow after the cylinder [3-5]. One of the characteristic features found in the present boundary layer is that the restoration of the turbulence field lags behind that of the averaged flow field. Slow restoration of the turbulence field is more remarkable in the outer region of the boundary layer than in the near wall region. It may be interesting to study on coherent structure of the turbulent flow in such a non-equilibrium state. As a first step to the study, the bursting period is measured and is discussed first in the following.

Besides the mathematical models of turbulence based on one-point correlation, many works have been made also on the modelling of the bursting phenomenon. Combining the latter model with a numerical prediction technique employing one of the former models in the outer layer, it would enable in principle to replace the wall function method [6] often used for treating the low-Reynolds number turbulence near the wall. The modelling of the bursting phenomenon itself is attractive from the view point of heat and mass transfer from the wall. For example, an artificial generation of normal velocity to simulate the bursting phenomenon has succeeded to explain the dependency of the mass transfer from the wall on Reynolds number [7]. Among many such models, the simplest one by Einstein and Li [8] has recently been applied successfully to a transpired turbulent boundary layer [9]. Here is again paid attention to this simplest model, and its test in the present boundary layer is carried out in the second part of this article.

EXPERIMENTAL APPARATUS AND PROCEDURES

Measurements have been carried out in an open tunnel of 38cm × 38cm cross sectional area, used previously [2]. A cylinder of 8mm in diameter was inserted at a streamwise location of $Re_{xc} \simeq 1.2 \times 10^6$. as schematically shown in Figure 1. The cylinder was located at three positions with different height above the plate, i.e. H = 6, 15 and 33.5mm. Roughly speaking, these three positions correspond respectively to the edge of inner layer, the middle of outer layer, and the outer edge of the approaching boundary layer. The thickness of the boundary layer is about 27mm at test free stream velocity of 14m/s and at $Re_{xc} = 1.2 \times 10^6$ unless the cylinder was inserted. Each experiment at one position of cylinder is specified as the case S_1, S_2 and S_3, respectively, in the above order. An additional measurement is made for the case without cylinder, the undisturbed flat plate boundary layer, for comparison, and is specified as the case S_0.

The measurements are carried out at six streamwise locations, x = 37, 87, 187, 387, 637 and 837mm, after the cylinder along the median line of the flat plate with a constant-temperature hot wire anemometer. The hot-wire was positioned near the wall, within the range of the distance from the wall as $y^+ = 10 \sim 20$. Within this range, no effect of probe position on final results has been found. The bursting period was measured by applying the correlation function method [10]. Different from the usual statistical study of turbulence, the average of detected signal over long period of time is undesirable because it tends to smear out possible peaks in the correlation function, resulting from successive bursting events. Too shortly averaged correlation function is of irregular shape and is unsuitable also for the present experiment. The averaging period of time was fixed at 0.256 second in this study from the results of preliminary study, and the time interval between the first (at $\tau = 0$) and the second peaks was measured. When the third peak is remarkable, the time interval between the second and the third peaks was adopted as a datum of the bursting period. This inclusion of additional data did not alter any nature of the results obtained, and was efficient to reduce the labouring time. In Figures 2(a) and 2(b) are shown typical records of correlation function, one with long averaging period of time and other two with the averaging period of time adopted in this study. Oscillation of correlation curve found in Figure 2(a) is due to the existance of Kármán-like vortices and does not dissappear however the averaging time is increased.

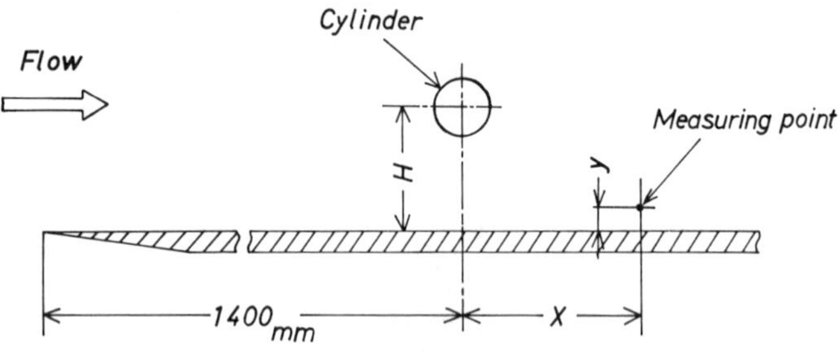

Figure 1. Flow configuration

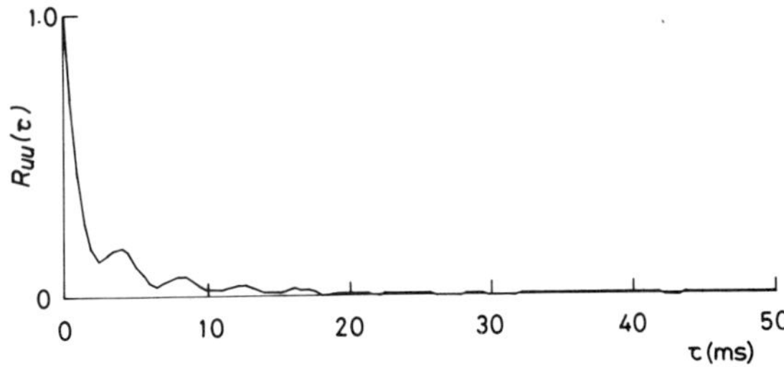

Figure 2(a). Correlation function, averaging period = 4.096 sec.

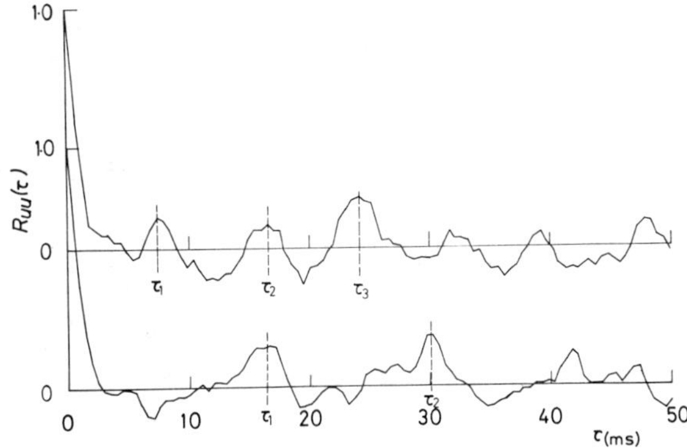

Figure 2(b). Correlation function, averaging period = 0.256 sec.

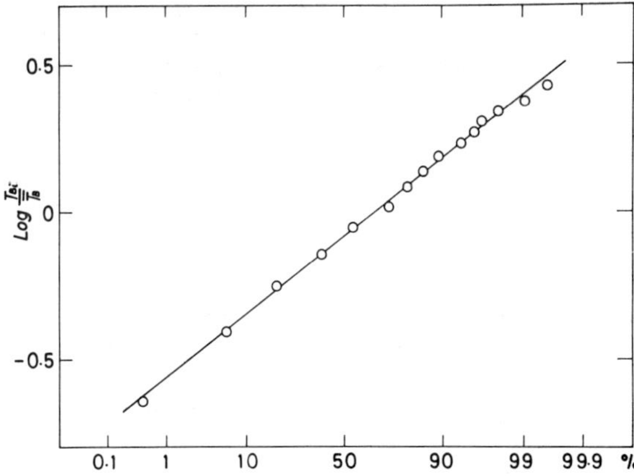

Figure 3. The distribution of the bursting period T_{Bi} plotted on probability paper.

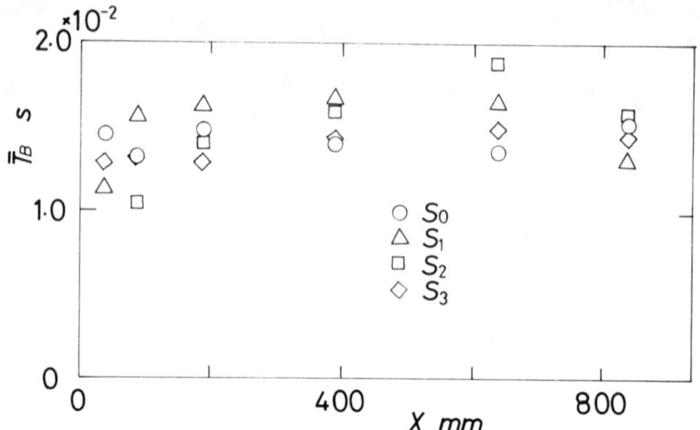

Figure 4. Mean bursting period measured in the disturbed and undisturbed boundary layer.

The bursting period T_{Bi} so measured scatters over a wide range of period. This is clearly seen from comparison of two typical records, obtained at the same position in the same case, with each other. The present distribution of T_{Bi} obeys a log-normal law, just as in the undisturbed flat plate boundary layer [11]. A typical example of the distribution is shown in Figure 3. A statistical mean of bursting period, $\overline{T_B}$, was evaluated by averaging the total of 200 data of T_{Bi} in every case. This number of data was confirmed in this experiment being sufficient for the evaluation of $\overline{T_B}$ within the accuracy of less than one percent.

EXPERIMENTAL RESULTS

In Figure 4 are plotted the data of the statistical mean period of four cases, S_0, S_1, S_2 and S_3, altogether. Somewhat surprizingly, the mean bursting period measured in the present disturbed boundary layer does not differ so much from the counterpart of the undisturbed boundary layer. It does not vary remarkably with the increasing distance from the cylinder, while the skin friction coefficient varies conspicuously with the distance, especially in the case S_1 [2]. The results of the case S_1 does not differ so much from those of other two cases S_2 and S_3, too.

Non-dimensional plots of the above results with the outer layer scale and with the inner layer scale are shown in Figures 5 and 6, respectively. In these Figures, U_e is the main stream velocity, U_τ the friction velocity, δ the boundary layer thickness, δ^* the displacement thickness, R_θ the Reynolds number based on the momentum thickness, and ν the kinematic viscosity. U_τ, δ, δ^* and R_θ are given in the reference of [2], or can be evaluated from the data shown there. The outer scale plot reveals a little influence of cylinder insertion, but all the data of the disturbed boundary layer still remain close to the values

$$\frac{\overline{T_B}\, U_e}{\delta^*} = 32, \qquad \frac{\overline{T_B}\, U_e}{\delta} = 5,$$

usually proposed for the flat plate boundary layer without pressure gradient,

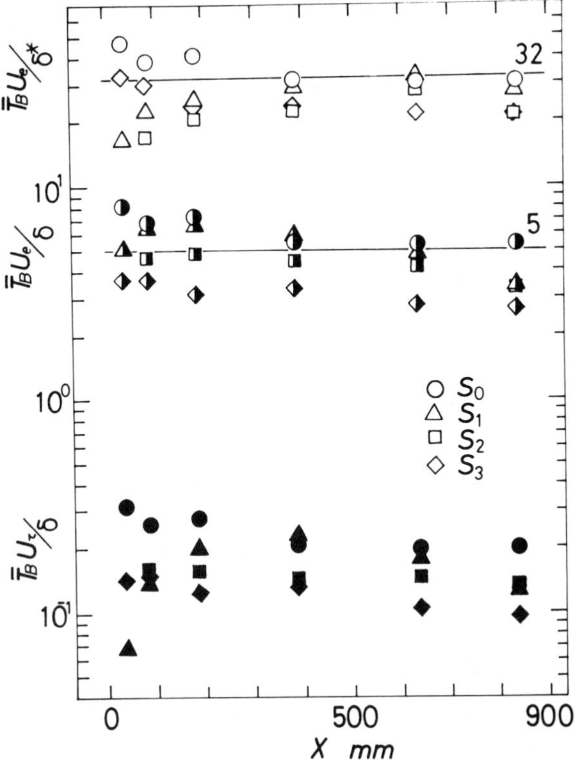

Figure 5. Non-dimensional plot of $\overline{T_B}$ with outer layer scale.

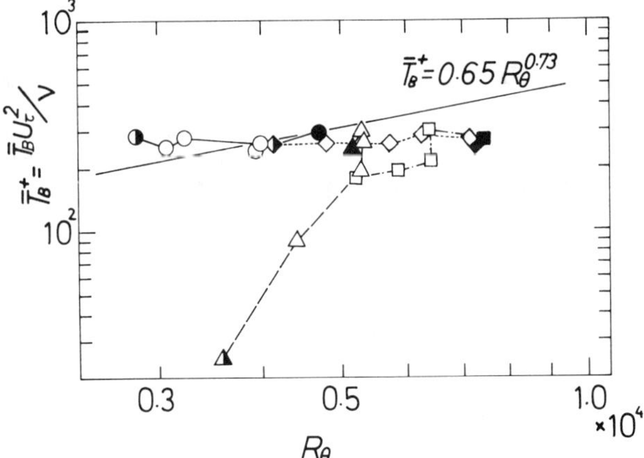

Figure 6. Non-dimensional plot of T_B with inner layer scale
(Half solid → Solid symbol : x increases)

and shown with the straight lines in Figure 5. The inner scale plot gives worse correlation of the data obtained, while the results tend to come closer to the counterpart of the flat plate boundary layer with the increasing distance, as seen in the cases S_1 and S_2 rather clearly. The solid line in Figure 6 shows the following relationship.

$$\overline{T_B}^{+} = \frac{\overline{T_B}\, U_\tau^{\,2}}{\nu} = 0.65 R_\theta^{\,0.73}$$

proposed for a flat plate [11]. The deviation from the equation of some results of the case S_1 at small distance from the cylinder x results from the small value of skin friction there [2] against the mean bursting period not so much different from that in the undisturbed case.

Unsuccess of inner scale plot may give a support to the concept that the bursting event is not the intrinsic instability near the wall such as being independent from the flow in the outer layer. Initially, the Kármán-like vortices shed from the cylinder were expected to affect the bursting period very much, but the results obtained proved not so. Incidentally, the shedding frequency of the vortices were 215, 250 and 325Hz respectively for the cases S_1, S_2 and S_3 [2], and far from the reciprocal of the mean bursting period. Further studies with various sizes of cylinders and at various values of Reynolds number are desired to comprehend the physics of unaffected bursting period.

TEST OF SURFACE RENEWAL MODEL

The surface renewal model by Einstein and Li [8] assumes the boundary layer approximation of equation of motion and neglects the contribution of convection terms in it. Thus, the governing equation of streamwise velocity u is as follows.

$$\rho \frac{\partial u}{\partial t} = \mu \frac{\partial^2 u}{\partial y^2} \qquad (1)$$

where t is the elapsing time after an occurrence of inrush-sweep motion of the fluid, y the distance from the wall, and ρ and μ are respectively the density and the viscosity of the fluid. The time interval between an ejection and its subsequent sweep events is ignored, so that the period of time of flow retardation process is taken being equal to a whole period of time of one sequence of ejection, sweep and retardation processes. If an initial condition for u at $t = 0$ is denoted by U_0, the solution of Eq. (1) is given as follows.

$$u = \frac{2U_0}{\sqrt{\pi}} \int_0^{\Xi} e^{-\xi^2}\, d\xi \; ; \qquad \Xi = \frac{y}{2\sqrt{\nu t}}$$

The shear stress acting on the wall at an instant t can be given by differentiating the above solution at $y = 0$ and multiplying the result by the viscosity μ. Averaging the instantaneous wall shear stress over the whole period of time of one sequence of events, the skin friction coefficient C_f can be evaluated.

The inrushing fluid velocity can be an initial condition for solving u, but it is unknown. Here are taken the averaged velocities at the following five

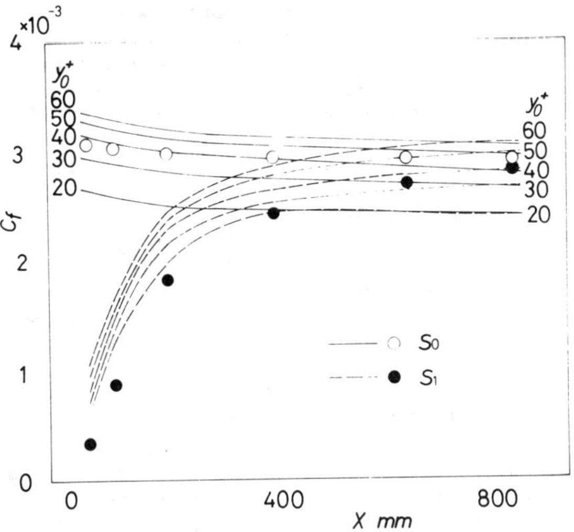

Figure 7. Skin friction coefficient (---- surface renewal model)

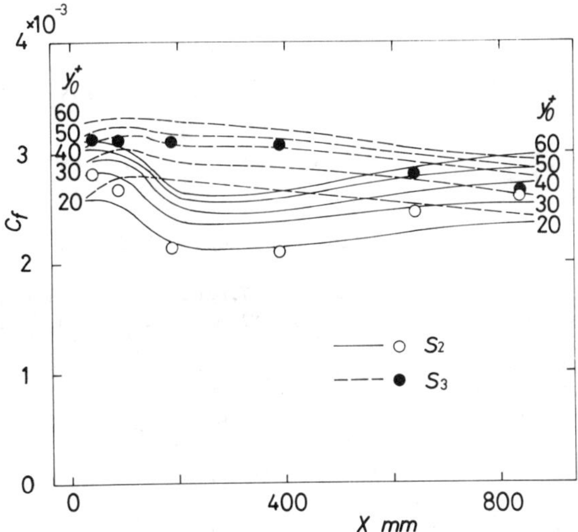

Figure 8. Skin friction coefficient (---- surface renewal model)

different positions,

$$y_0^+ = \frac{\rho y_0 U_\tau}{\mu} = 20,\ 30,\ 40,\ 50 \text{ and } 60.$$

and each of them is used as an assumed initial condition. Thus, the five results to be obtained are shown altogether in the following with y_0^+ being as a parameter. The averaged velocity at each position was read in the reference [2]. The whole period of one sequence of the events is assumed to be equal to the mean bursting period shown in the first part of this paper. Just for simplicity, the mean bursting period T_B is approximated by a constant value, i.e. 15ms, in all the calculation. As mentioned already, T_B does not vary remarkably either with the cylinder position or with the downward distance from the cylinder.

Figures 7 and 8 show the final results in the form of skin friction coefficient C_f and compare them with the experimental counterparts [2]. Both figures manifest rather good tendency of the calculated results, agreeing with the experiments. The surface renewal model may be concluded to have a potentiality of predicting the wall shear stress also in the present type of complicated wall turbulent flow. Similar works with more sophisticated model of bursting phenomenon may be interesting.

The same method was applied to heat transfer problem, adopting the averaged temperature at the above five positions as the initial condition for solving the heat conduction equation. The results obtained seem to give a good explanation of the augmented wall heat transfer in the downstream position of cylinder. The logarithmic law of the wall was used to evaluate the averaged temperature at each of the above five positions, but this is open to a question, especially, at the stations close to the cylinder. Further experiments on this and on temperature fluctuation are now in progress with a method devised previously [12].

ACKNOWLEDGEMENT

The authors express their thanks to Prof. T. Sato at Kyoto University for his continuous attention to this work.

REFERENCES

[1] Clauser, F.H. Advances in Applied Mechnics, Vol.IV, 2. New York: Academic Press. (1956)

[2] Marumo, E., Suzuki, K., and Sato, T., J.Fluid Mech. Vol.87, 121. (1978)

[3] Marumo, E., Suzuki, K., Sasaki, T., and Sato, T., Trans. J.S.M.E. Vol.46-407, 1211. (1980) (In Japanese)

[4] Marumo, E., Muramoto, R., Tanaka, J., and Sato, T., Preprint of J.S.M.E. Vol.774-10, 97. (1977) (In Japanese)

[5] Suzuki, K., Tanaka, J., Marumo, E., and Sato, T., Trans. J.S.M.E. Vol.46-408, 1447. (1980) (In Japanese)

[6] Launder, B.E. and Spalding, D.B., Computer Methods in Appl. Mech. and Engng., Vol.3, 269. (1974)

[7] Brodkey, R.S., McKelvey, K.N., Hershey, H.C. and Nychas, S.G., Int. J. Heat Mass Transfer. Vol.21, 593. (1978)

[8] Einstein, H.A., and Li, H., J. Engng. Mech. Div. Am. Soc. Civ. Engng. Vol.82 (EM2), 945. (1956)

[9] Senda, M., Suzuki, K., and Sato, T., Preprint of J.S.M.E. Vol.790-18, 96 (1979) (In Japanese)

[10] Kim, H.T., Kline, S.J., and Reynolds, W.C., J.Fluid Mech. Vol.50, 133. (1971)

[11] Rao, K.N., Narasimha, R., and Badri Narayanan, M.A. J.Fluid Mech. Vol.48, 339. (1971)

[12] Senda, M., Suzuki, K., and Sato, T., Proc. 2nd Int. Symp. on Turbulent Shear Flow. 9.17. (1979)

Anisotropy of Normal Reynolds Stresses in a Developing Boundary Layer of Non-Newtonian Fluid

ČEDO MAKSIMOVIĆ
Faculty of Civil Engineering
Bulevar Revolucije 73
Belgrade, Yugoslavia

ABSTRACT

In the paper presented are some results of an experimental study of the effect of drag reducing polymer solution injected into a developing boundary layer within a rectangular duct. Attention is paid to the anisotropy of normal Reynolds stresses. It is shown that the anisotropy is significantly increased in the polymer flow in comparison to water.

1. INTRODUCTION

In recent years there have appeared a lot of papers aiming to describe fundamentals and/or consequences of the so called Tom's effect. The major consequences of the effect such as changes in the mean flow pattern, drag reduction, gross flow properties are well documented by Hoyt [1] and Virk [2] although the understanding of fundamental causes of the phenomenon is far from being satisfactory.

The aim of this paper is to present some results which were obtained by a comprihensive experimental study of several quantities characterizing the structure of a non-zero pressure gradient flow within a square duct.

Several quantities have been measured and the emphasis was on the velocity field structure in the zone of the boundary layer development.

As the traditional devices for velocity measurements in liquid flow (such as Pitot-tube and hot film anemometer) have proved to be unsatisfactory, a Laser Doppler Anemometer (LDA) is used in the present study.

Interesting results on the effect of elongated polymer molecules on the turbulence structure are obtained.

2. EXPERIMENTAL APPARATUS

The experiments were carried out in a closed duct of square cross section with two working fluids i.e. water and nonuniform solution of polyethylene-oxide (WCR-301-Union Carbide) which was injected from the two slots (at the top and bottom of the duct).

The essenatial parts of the apparatus are presented in Fig.1.

The duct was made of plexiglas and was instaled into ordinary laboratory flume with glass walls.

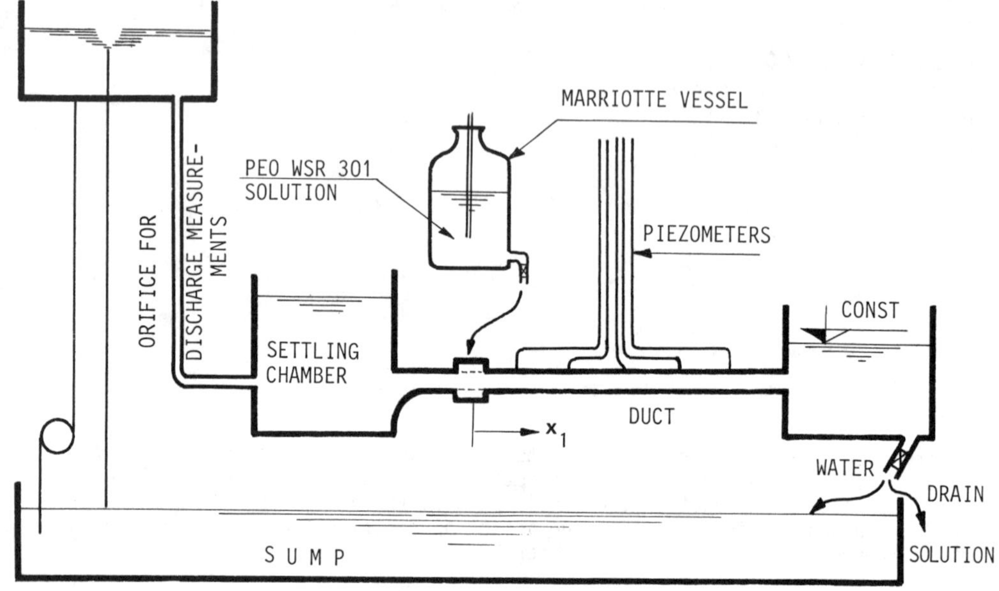

Fig.1. Experimental apparatus

Flow discharge was kept constant (Q_w = 3,26 dm^3/s) throughout all experiments and was controlled by an orifice plate flowmeter.

In order to avoid the polymer degradation effects, the "once through" system was applied.

The dosing rate of polymer solution was controlled by a Marriot vessel and a constant discharge of fresh solution of polymer was used in all experiments with injection.

The length of the duct downstream from the point of injection was 2,5 m and 2,9×10 cm in the cross section. The ratio of the length to hydraulic radius of the duct was two hundred.

The flow in such a cross section is known to be three-dimensional, due to the presence of strong secundary flows.

3. VELOCITY MEASUREMENT AND PROCESSING

Signifant errors have been reported in the velocity measurements of the polymer solution flow with Pitot-tube [3] and hot film anemometers, and Laser Doppler Anemometer (LDA) is proved to be superior for measurements in such flows.

The principle of LDA technique is well documented [4], and several approaches are possible. In the present study a forward scaterring configuration of the single component Laser Doppler Anemometer (of TSI) was used for velocity measurements.

Only one velocity component can be measured at one time with such a configuration, but after rotating the optical system for 90°, the other component can be measured. The same configuration could have been used also for Reynolds shear stresses measurement (see Logan [5]), but this was not done here.

The intensity of velocity component in the plane of laser beams crossing is related to the resulting output voltage from the tracker type processor by a very simple linear equation and therefore no special calibration was needed.

Two alternative processing procedures have been applied after the row voltage signal was obtained:

a) Analogue processing for obtaining the mean u_i and RMS $(\overline{u_i'^2})^{1/2}$ of the velocity component.
b) Digital processing by means of a mini computer,where the signal was fed after passing through a law pass filter, and converted to digital form by an A/D converter.

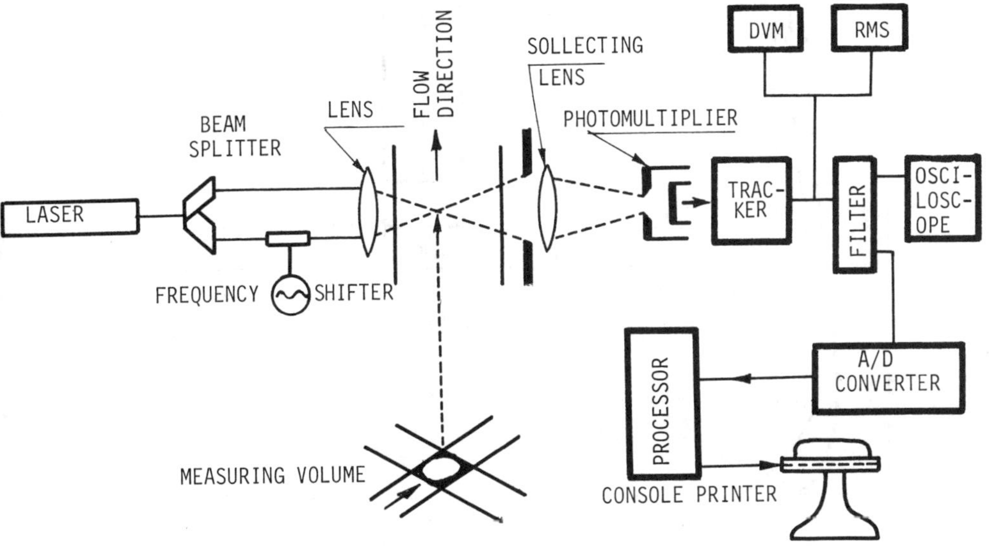

Fig.2. Analogue and digital processing of velocity data

In Fig.2. these two procedures are shown schematically.

The following turbulence characteristics of turbulence structure are obtained by digital processing for both horizontal (u_1) and vertical (u_2) velocity component:

- mean $$\bar{u}_i = \frac{1}{N} \sum_{j=1}^{N} u_i \quad (j) \tag{1}$$

- RMS $$= (\overline{u_i'^2})^{1/2} = \left[\frac{1}{N} \sum_{j=1}^{N} (u_i(j) - \bar{u}_i)^2\right]^{1/2} \tag{2}$$

- Power spectrum estimate

$$\hat{S}(f_k,u_i) = \frac{1}{N}\sum_{r=1}^{N} \frac{2\Delta t}{M\,\overline{u_i'^2}} \left\{ \sum_{j=1}^{M=1} u'_{ij}\, e^{-i\frac{2\pi jk}{M}} \right\}^2 \qquad k=1,2,\ldots,M/2 \tag{3}$$

where i is imaginary unit.

- Probability density estimate

$$P_i = P_{rob}\left[\, u_j < u_i < u_j + \Delta u)\right] \tag{4}$$

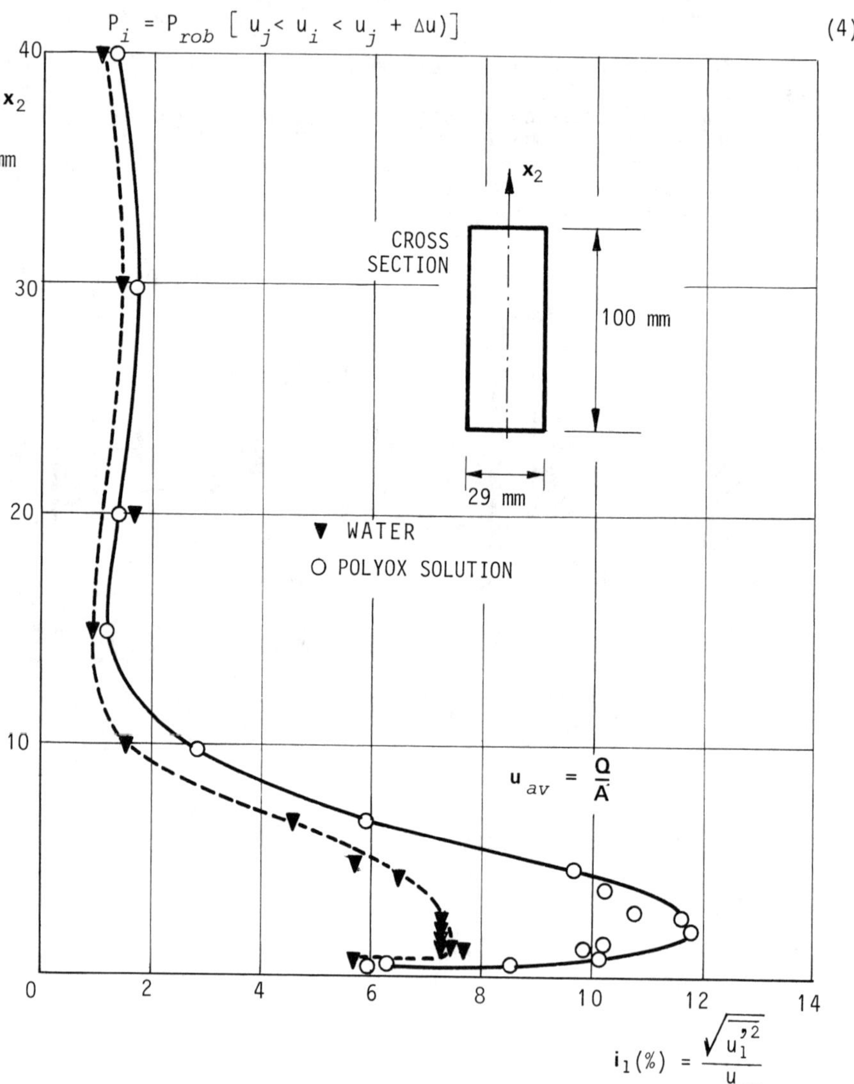

Fig. 3. Turbulence intensity of horizontal (u_1) velocity in vertical plane of simetry of the duct (x_1=240 mm)

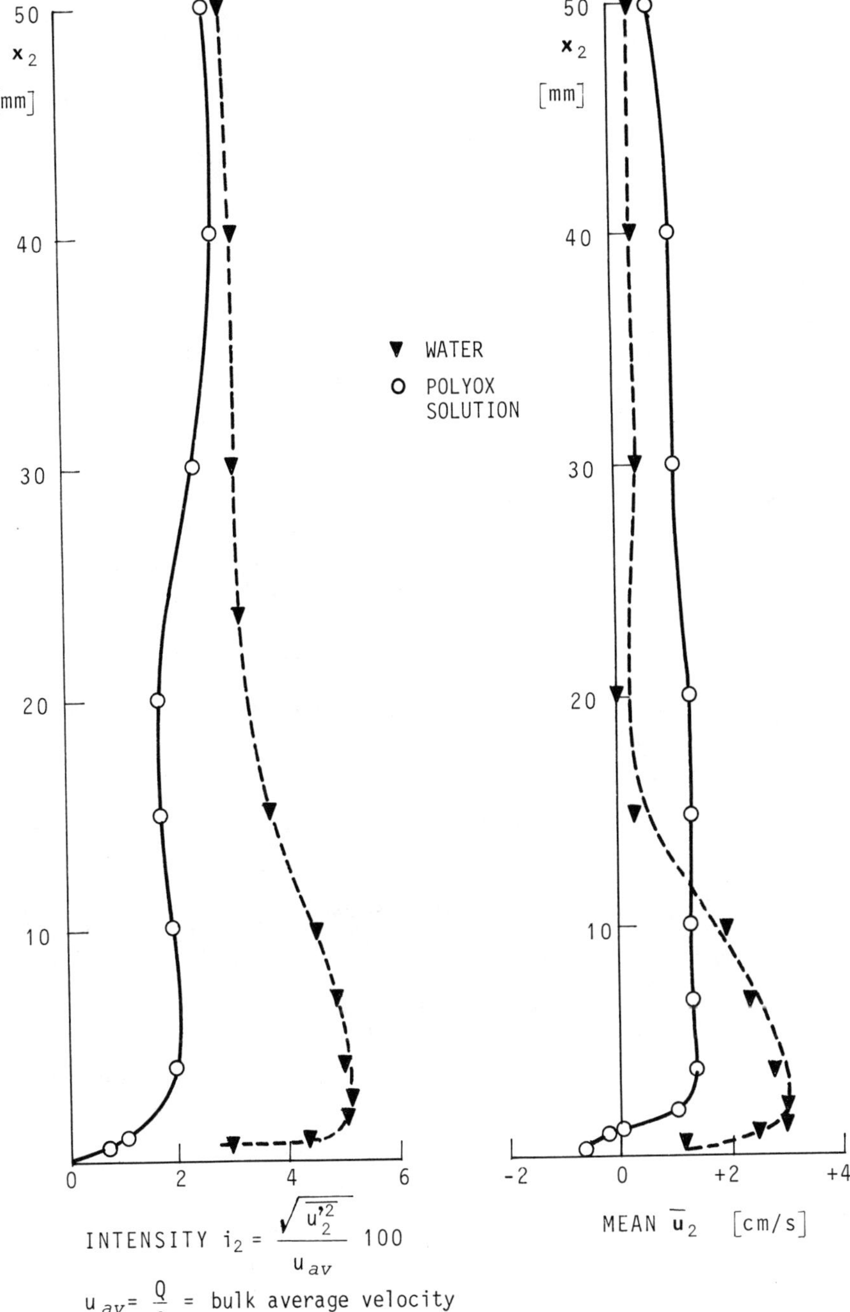

Fig. 4. Turbulence intensity and mean vertical velocity (x_1=240 mm)

Apart from this processing, the local averaged energy line slope and local everaged friction velocity was calculated from the piezometer readings.

More detailed results of the measurements are presented in Ref. [6].

For the sake of compleetness in Fig.3. and Fig.4. presented are the RMS values of horizontal and vertical component for both water and polymer flow, normalized by bulk mean velocity

$$u_{av} = \frac{1}{A} \int_A u_1 dA \tag{5}$$

When presented in this fashion turbulence intensity of the horizontal velocity component in the zone of maximum turbulent energy production is much higher for polymer flow than for water. On contrary, the intensity of vertical component issignificantly decreased. It is the fact to which the emphasis is given in this peper, and will be discussed in the preceding paragraphs.

4. MEASUREMENTS OF OTHER QUANTITIES

The preciding measurements were carried out in Hydraulic laboratory at the Department of Civil Engineering - Newcastle upon Tyne U.K. Apart from that, at the Hydraulic laboratory - Faculty of Civil Engineering the following additional measurements were done:

a) distribution of concentration of diffusing polymer solution in the nonuniform zone,
b) fluctuations of wall pressure beneath the developing boundary layer,
c) gross flow properties,
d) effect of injection of polymers into core region,
e) flow visualization studies.

All measurements are done for both fluids water and polymer solution respecttively.

Some preliminary results of the diffusion measurements are presented [7] and final results will be given elswhere. It is shown that, due to much lower diffusion rate in polymer solution flow, much higher drag reduction is obtained in comparison to the case if the polymers were uniformly distributed throughout the whole cross section. Even though in this flow geometry (square duct) the flow is three dimensional results are similar to those obtained in Fort Collins by several researches (see for example Ref. [8]) in circular pipe.

Results of pressure fluctuations shaw also unusual and unexpacted features and, as not yet beeing compleetely processed, will be presented elsewhere.

5. COMMENTS ON ISOTROPY

Reynolds stresses (σ_{ij}^R) are known to be obtainable by time averaging fluctuating parts of two velocity components and multiplying the resulting correlation by negative fluid density i.e.:

$$\sigma_{ij}^R = -\rho \overline{u_i' u_j'}$$

The simetric tensor of Reynolds stresses can thus be obtained:

$$\sigma_{ij}^{R} = \begin{Bmatrix} \boxed{\sigma_{11}^{R}} & \sigma_{12}^{R} & \sigma_{13}^{R} \\ \sigma_{21}^{R} & \boxed{\sigma_{22}^{R}} & \sigma_{23}^{R} \\ \sigma_{31}^{R} & \sigma_{32}^{R} & \sigma_{33}^{R} \end{Bmatrix} \tag{6}$$

It is also well known that the Reynolds stresses play the main role in the turbulent energy production i.e. the transfer of energy from the mean flow to turbulence.

Measurements of Logan [5] by LDA and other have shown that due to decrease of Reynolds stresses and modificantions in the mean velocity profile in the near wall zone the production of turbulence energy is significantly reduced. Results of Patterson at al [9] obtained by hot film anemometry are a good example for this,although some results have to be taken with scepticism because of the possible errors when the hot film is used in the polymer flow.

It is known that there exists a slight anisotropy of normal Reynolds stresses (marked by dashed lines in Eq. (6)), in Newtonian flow. The role of the anisotropy is discussed by Launder [10].

Owing to uncertain experimental measurements there has been some confusion in the recent past on the effect of anisotropy on the drag reduction.

Seyer and Metzner [11], for example erroneously conclude: "Assuming that the observed lowering of the axial intensity (relative to radial intensity) is real, a shift towards isotropy has occured. A lowering of the turbulent Reynolds stress is implied since for isotropic turbulence the Reynolds stress must be zero. This is of cause consistant with experimentally observed lowering of shearing stress of drag coefficient".

The results of the present study, as well as those of Patterson et al [11], and Logan [5] disagree with that conclusion.

Anisotropy is, in the present paper, represented by quantities

$$\delta_w = \frac{(-\rho\,\overline{u_1' u_1'}\,)_w^2}{(-\rho\,\overline{u_2' u_2'}\,)_w^2} \tag{7}$$

and

$$\delta_p = \frac{(-\rho\,\overline{u_1' u_1'}\,)_p^2}{(-\rho\,\overline{u_2' u_2'}\,)_p} \tag{8}$$

where subscripts w and p denote water and polymer respectively.

The powerful LDA enabled measurements as close as a quarter of a milimeter from the wall and in Figures 5,6 and 7. it is shown that anisotropy is significantly increased (almost for an order of magnitude) in the vicinity of the wall. Presented are the results for three downstream distances from the point of injection. The distances are indicated in Figures. Although the average near-wall concentration of polymer in the profile 5 (Fig.7) is much lower than in profile 3 (Fig.6) there is no apparent difference in the anisotropy pattern between these two profiles. The profile 1 (Fig.5) is very close to the point of injection and in accordance with classification of Poreh and Cermak [12] the profile is still in the intermediate zone of diffusion.

From the Figures it is evident that near the wall isotropy of normal Reynolds stresses is significant which, in accordance with Ref. 9 "opens the possibility

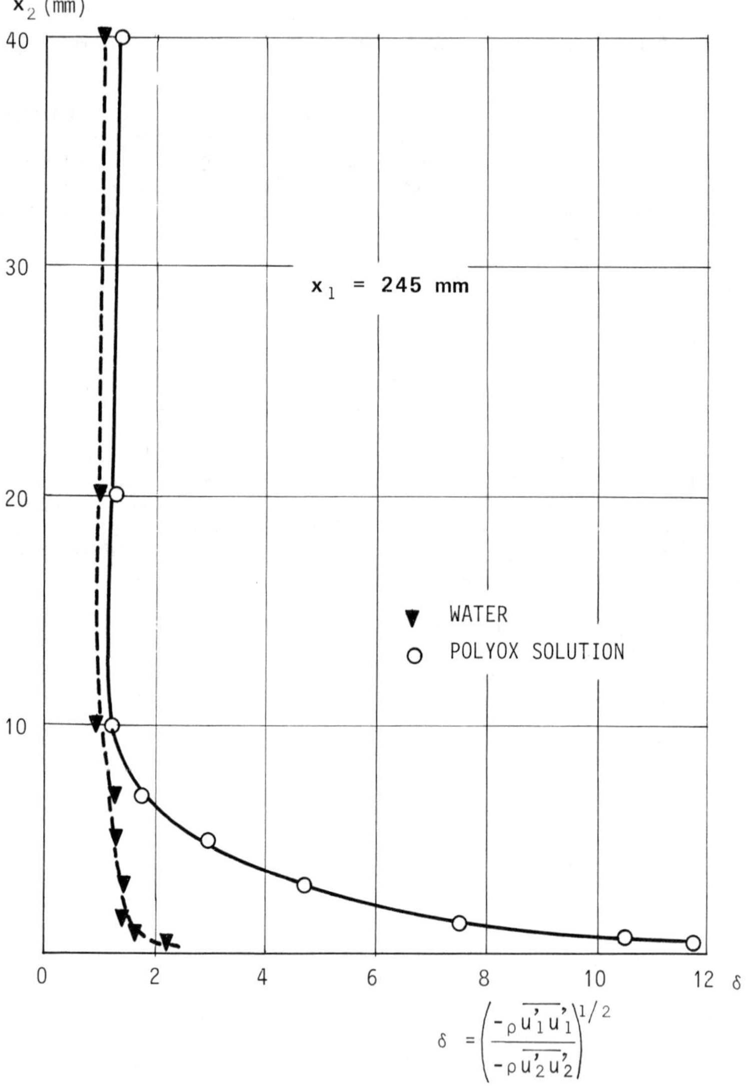

$$\delta = \left(\frac{-\rho \overline{u_1' u_1'}}{-\rho \overline{u_2' u_2'}} \right)^{1/2}$$

Fig.5. Anisotropy of normal Reynolds stresses in the midle of the duct x_1 = 245 mm. Profile 1.

of having very large longitudinal velocity fluctuations (or more likely shear stain fluctuations near the wall". Similar conclusions were drawn by Landahl [13], Khabakhpasheva [14] and others.

The increased anisotropy implies that the non-Newtonian drag, reducing fluid is more resistant to the transfer of energy from streamwise into spanwise components. Recent and not yet compleeted measurements of wall pressure fluctuations done by author underline the role of these fluctuations as they are significantly modified.

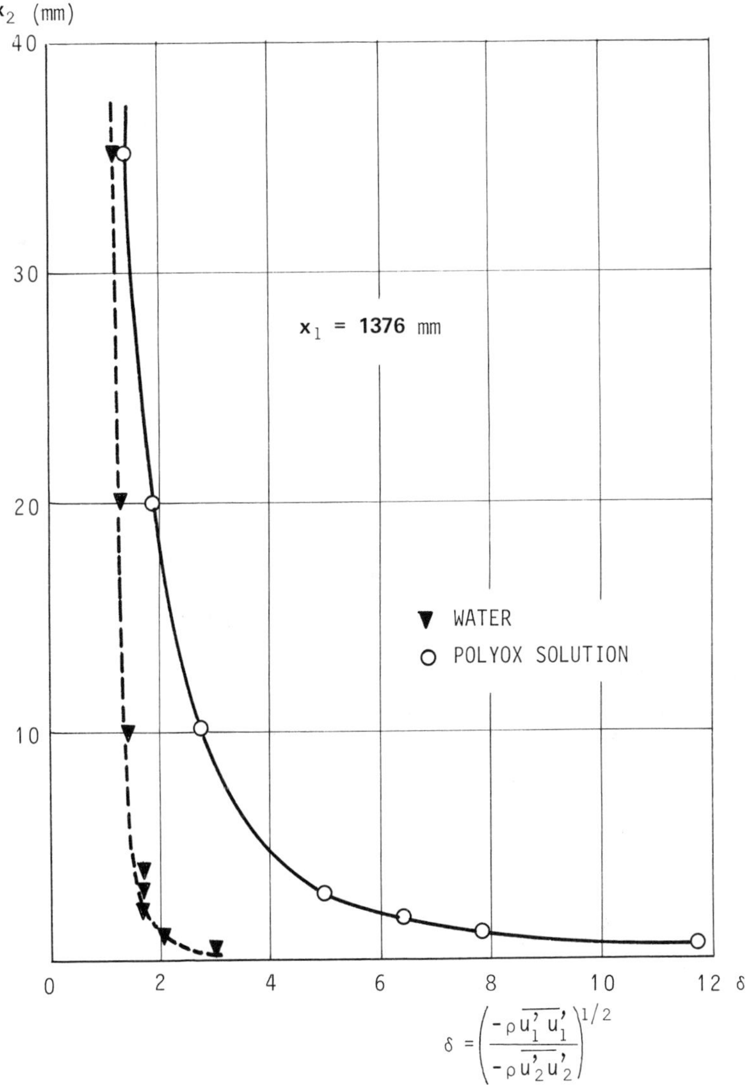

Fig.6. Anisotropy of normal Reynolds stresses in the midle of the duct x_1 = 1376 mm Profile 3.

6. CONCLUSION

1. The turbulence intensities in the streamwise direction are increased for polymer flow in the near wall region.
2. The turbulence intensity of spanwise component is significantly desreased in the same zone.
3. The results shaw that the anisotropy is significantly increased for polymer flow.
4. Both large eddies and small-scale ones are modified.

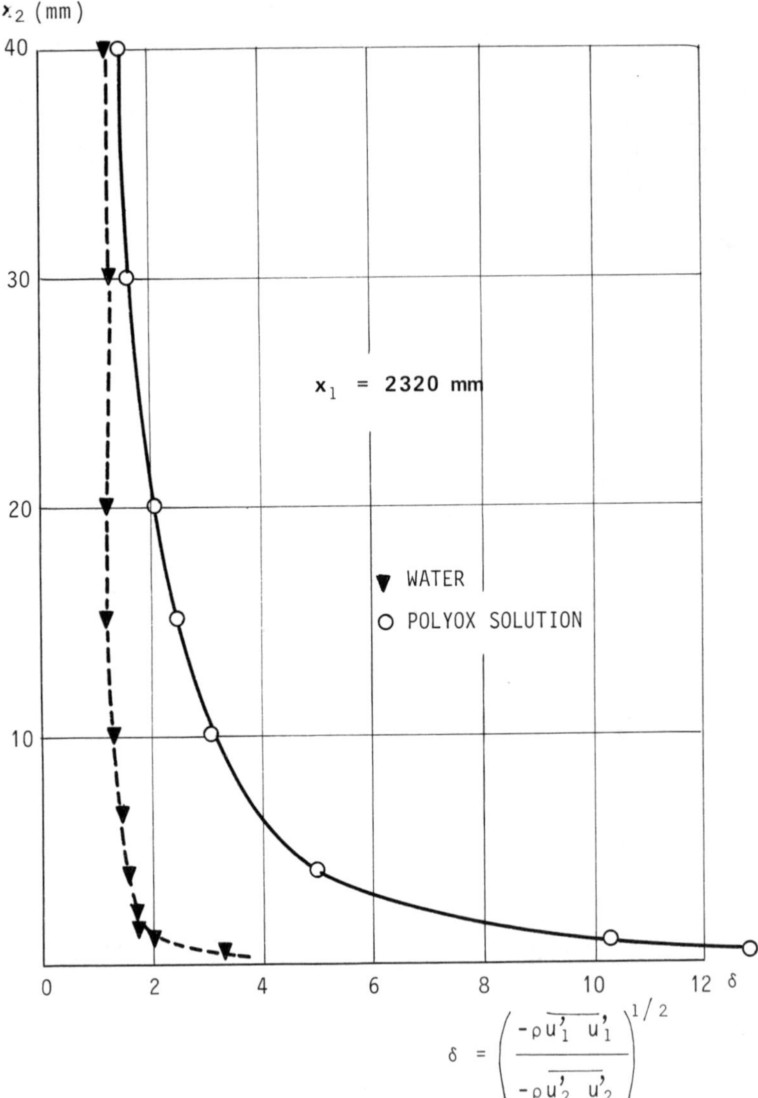

Fig.7. Anisotropy of normal Reynolds stresses near the end of the duct
x = 2320 mm   Profile 5.

7. ACKNOWLEDGEMENT

The author aknowledges the collaboration of Mr.M.Philipson and support of prof. P.Novak from the Department of Civil Engineering - Newcastle upon Tyne. The financial suppor of the Faculty of Civil Engineering Belgrade is also acknowledged.

REFERENCES

[1] Hoyt J.W. (1977). Polymer Drag Reduction - A Literature Review Proc. of Second Int.Conf. on Drag Reduction, BHRA, paper A1.

[2] Virk P.S. (1975). Drag Reduction Fundamentals. AICHE Journal Vol.21 (4) p.p. 625-655.

[3] Smith K.A.,Merrill E.W.,Mickley H.S.,Virk P.S. (1967). Anomalous Pitot Tube and Hot Film Measurements in Dilute Polymer Solutions. Chemical Engineering Science, Vol.22, p.p.619-626.

[4] Durst F.,Melling A.,Whitelaw J.(1976). Principles and Practice of Laser Doppler Anemometry. Acad.Pres.

[5] Logan S. (1972). Laser Velocity Measurement of Reynolds Stress and Turbulence in Dilute Polymer Solution. AIAA Journal Vol.10. 67,p.p.962-964.

[6] Maksimović Č., Philipson M. (1980). Stochastic Features of a Developing Boundary Layer with Diffusion of Drag Reducing Polymers. Third International Symposium on Stochastic Hydraulics, IAHR, Tokyo, p.p.667-676.

[7] Maksimović Č. (1980). Diffusion of Drag Reducing Additives in a Zone of Boundary Layer Development in Channel Flow. EUROMECH 130 - Diffusion and Dispersion in Open Channel Flow - Belgrade,June 1980.

[8] Ramu K.L.V., Tullis J.P. (1976). Drag Redustion and Velocity Distribution in Developing Pipe Flow. Journal of Hydronautics, Vol.10 Nº 2. p.p. 55-61

[9] Patterson G.K.,Chosnek J.,Zakin J.L.(1977). Turbulence Structure in Drag Reducing Polymer Solutions. Proc.IUTAM Symp. Structure of Turbulence and Drag Reduction. The Physics of Fluids, Vol.20, Nº 10,Part II,p.p.S89-S99.

[10] Launder B.E. (1980). Some Fundamental Problems of Modelling Turbulent Transport in Open Channel Flows - EUROMECH 130 - Belgrade, June 1980. Publ.Yugosl.Assoc.for Hydraulic Research.

[11] Seyer F.A.,Metzner A.B.(1969). Turbulence Phenomena in Drag Reducing Systems, AICHE Journal Vol.15, Nº 3, p.p.426-434.

[12] Poreh M.,Cermak J.E. (1964). Study of Diffusion from a Line Source in a Turbulent Boundary Layer. Int.Journal of Heat and Mass Transfer, Vol.7. p.p.1083-1095.

[13] Landahl M.L. (1977). Dynamics of Boundary Layer Turbulence. The Physics of Fluids, Vol.20, Nº 10, Part 2, S 55 - S 63.

[14] Khabakhpasheva E.M. (1976). Polymers and Turbulent Drag Reduction. Colloques Internationaux du C.N.R.S. Nº 233. Polymeres et Lubrification.

The Dominant Coherent Structure of the Circular Jet Organized by Controlled Perturbation

K. B. M. Q. ZAMAN and A. K. M. F. HUSSAIN
University of Houston, Texas 77004

ABSTRACT

The dominant large-scale coherent structure of the circular jet has been organized through controlled perturbation at the jet Strouhal number of 0.3, induced acoustically in two air jets via cavity resonance. Selected phases of the coherent structure have been investigated for three Reynolds numbers, and two limiting initial conditions: laminar and fully turbulent initial boundary layers. Data have been obtained by repeating the X-wire measurements over the spatial extent of the structure (thus without invoking the Taylor's hypothesis), the phase reference having been provided by the signal from a single-wire probe located on the jet centerline. The distributions of the phase-average structure properties, namely, longitudinal and lateral velocities, coherent vorticity and Reynolds stress, phase-random turbulence intensities and Reynolds stress, streamline and pseudo-streamfunction patterns have been educed.

The phase-random turbulence intensity is the largest at the structure center where the coherent vorticity is also the maximum. When the structure is located at $x/D \simeq 3$, the peak coherent vorticity is about $3U_e/D$ and is independent of the initial condition and the jet Reynolds number (Re_D). However, some properties of the structure show mild but systematic dependence on Re_D. With increasing Re_D, the structure cross-section, which is longitudinally elongated at lower Re_D, becomes more rounded and the structure becomes more energetic, and produces more coherent Reynolds stress. No distinct structure could be educed beyond about eight diameters from the jet exit even though the structures were organized initially through the controlled excitation.

1. INTRODUCTION

Spurred by the expectation that large-scale coherent structures are inherent to all turbulent shear flows and play a dominant role in transport phenomena in these flows, we have attempted a number of investigations of these structures in our laboratory. The objectives have been to: understand their characteristics; determine their significance in comparison with phase-random (small-scale) turbulence; provide detailed data for use in newer turbulence theories; and determine ways of controlling these structures and thus, presumably, transports of heat, mass and momemtum, and noise production.

Because of the large dispersion in the characteristics of the structures at any point in an unexcited shear flow, eduction of the naturally-occurring structure is not possible without sophisticated conditional sampling techniques. Because of this constraint, most of our studies have so far been limited to controlled excitation which provides easy phase reference for facilitating the eduction of the structure properties. The natural structures in a turbulent shear flow have to be classified on the basis of their characteristic parameters, e.g. shape, size, orientation, strength, convection velocity, etc., and ensemble averages will then have to be determined for narrow ranges of each parameter. However, if any particular structure with specific characteristics is dominant and is the one occurring most frequently, it may suffice to study that structure alone. For the axisymmetric jet, such a dominant structure does exist. This "preferred mode" of the axisymmetric jet scales on the jet diameter and occurs at the jet Strouhal number $St_D \simeq 0.30$. This has been confirmed by flow-visualization studies as well as by hot-wire studies under controlled excitation [1,2]. In the present study, the detailed properties of this preferred-mode structure have been explored by organizing it through controlled excitation and using the phase-averaging technique. The controlled excitation thus merely organized the most dominant natural structure and did not induce any artificial structure.

The objectives of the present investigation were to document and study the detailed characteristics of the preferred-mode coherent structure and its evolution, and to study the sensitivity of the structure properties to the initial condition and jet Reynolds number. The data are also expected to serve as calibrators of time-dependent and time-independent prediction codes for perturbed turbulent shear flows as well as find use in technological applications.

2. EXPERIMENTAL PROCEDURES

The experiments were carried out in a circular air jet facility, described in [2]. Most of the data presented are for a 7.62 cm diameter nozzle at the Reynolds number $Re_D = 55{,}000$. A thin sandpaper ring tripped the boundary layer at 4 cm upstream from the nozzle exit. The exit boundary layer was inferred to be fully-developed turbulent on the basis of the mean velocity and longitudinal fluctuation intensity profiles and $\tilde{u}$-spectra. The mean velocity profile consisting of distinct logarithmic and wake regions closely agreed with the flat plate equilibrium turbulent boundary layer in universal coordinates (u^+, y^+). The fluctuation intensity increased monotonically from the free stream value of about 0.25% to the peak values at $y^+ \simeq 13$. The untripped boundary layer profile agreed with the Blasius profile. The controlled excitation at the nozzle exit was induced by means of a loudspeaker attached to the first settling chamber of the two-chamber jet facility. This study was carried out at an excitation amplitude (u'_e/U_e) of 2%, measured at the jet exit centerline. This excitation did not produce any change in the exit mean velocity profile for both initial conditions. An analog (DISA) turbulence processor was used to obtain the instantaneous $\tilde{u}(t)$ and $\tilde{v}(t)$ signals from an X-wire, operated by DISA constant temperature anemometers. The data were acquired on a horizontal diametral plane on one side of the jet centerline, in the measurement region extending over $0 \gtrsim x/D \gtrsim 12$. At each spatial point, the phase-average quantities were computed online with a laboratory minicomputer (HP2100). A precision backlash-free traversing mechanism was used to traverse the measurement X-wire probe at different spatial points under computer control. The data acquisition procedure is schematically shown in Figure 1. The reference (single-wire) probe, placed in the flow, gave the triggering signal via a bandpass filter and a triggering device. The phase for data sampling could be selected through a potentiometer in the triggering device set to coincide with the smoke-laden structure stroboscopically frozen at the selected spatial location. Further details of the measurement procedures including phase-averaging, triggering technique, data acquisition and analysis have been discussed in [3].

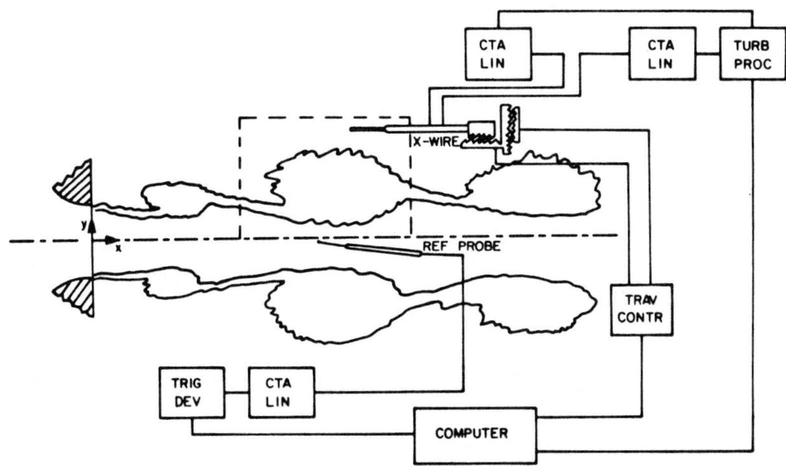

FIG. 1 Schematic of the flow and probe arrangement.

For the phase-average measurements, the triple decomposition,

$$\tilde{g} = G + \langle g_p \rangle + g_r; \quad \langle g \rangle = G + \langle g_p \rangle,$$

will be used in this paper. Here $\tilde{g}$ is the instantaneous value of a function, G its time average, $\langle g_p \rangle$ the periodic component of the phase-average and g_r the random component due to the background turbulence. The phase-average intensity of the last is defined as,

$$\langle g_r^2 \rangle^{\frac{1}{2}} = [\ \lim_{N\to\infty} \sum_{n=1}^{N} g_r^2\ (t + n\tau)]^{\frac{1}{2}}$$

The azimuthal (z) component of the vorticity is,

$$\Omega_z = \frac{\partial \langle v \rangle}{\partial x} - \frac{\partial \langle u \rangle}{\partial y},$$

where $\langle u \rangle$ and $\langle v \rangle$ are the longitudinal (x) and the transverse (y) components of the phase-average velocity. The phase-average of the velocity product $\tilde{u}\tilde{v}$ is,

$$\langle \tilde{u}\ \tilde{v} \rangle = UV + \langle u_p v_p \rangle + \langle u_r v_r \rangle.$$

$\langle u_p v_p \rangle$ is due to the periodic coherent structure and thus will be called the "coherent Reynolds stress." $\langle u_r v_r \rangle$ is the contribution of the background turbulence field.

The distribution of the velocity vector ($\langle u \rangle$, $\langle v \rangle$) can be used, via the method of isoclines, to obtain the phase-average streamlines (Ψ_p). Distribution of the single velocity component $\langle u \rangle$ can also be used to compute a pseudo-streamfunction ($\langle \Psi \rangle$), which should also give the streamline pattern. $\langle \Psi \rangle$ is defined as,

$$\langle\Psi\rangle = \int_0^y r(\langle u\rangle - U_{ref})\, dr/DU_e,$$

where U_{ref} is a chosen reference frame velocity. When U_{ref} is equal to the structure convection velocity, the educed streamline essentially coincides with the instantaneous streakline pattern.

3. RESULTS AND DISCUSSION

Data in Figures 2-7 pertain to the 7.62 cm diameter tripped jet at Re_D = 55,000 (the frequency f_p for the excitation cases was 44 Hz). Figures 8-15 include data for a 2.54 cm jet also. These data for different Reynolds numbers and initial conditions involve excitation at frequencies such that the St_D is 0.30 in all cases.

Contours of the phase-average voriticity Ω_z for a particular phase are shown in Figure 2 for the axial range $1.8 < x/D < 11.4$. For these data, the reference probe for the triggering signal was located at $x/D = 4.5$ on the jet centerline. The vorticity data are nondimensionalized by the excitation frequency f_p. It is interesting that the eduction process has been successful in bringing out, from the large-amplitude random velocity signals, not only the preferred-mode coherent structure but also its streamwise evolution. More impressive is the fact that these contours show the roll-up and subsequent evolution of an initially turbulent shear layer, into the jet column preferred mode structure. Note that the vertical scale in Figure 2 is magnified by a factor of 3, thus the apparently round vortex cross-section centered at $x/D \simeq 2.8$ is actually oval shaped, having its major axis parallel to the jet axis. Also note that the structure centered at $x/D \simeq 4.6$ is tilted, having its major axis at an angle with the jet axis. The structure tilt is a simple consequence of the shear across it.

In addition to the coherent structures at $x/D \cong 2.8$ and 4.6, there appears an apparent remnant of a vortical structure centered at $x/D \cong 6.6$. But still farther downstream, the phase-average measurements do not reveal any structure. A complete breakdown and randomization of the coherent vortices at this downstream location would explain the measured vorticity contours. However, instantaneously identifiable structures may still pass by this axial location at random phases, so that the phase-average measurements would reveal no structures. (Note that a periodic undulation in the $\Omega_z/f_p = 0.5$ contour, on the high-speed side, persist farther downstream indicating the possible remnants of the periodic coherent structure). From these data, one can at least infer that even though the coherent structures are initially organized by excitation at the preferred mode, the periodicity in their passage is practically lost beyond $x/D \cong 8$.

The contours of $\langle u\rangle$ corresponding to the data of Figure 2 are shown in Figure 3. Note that the computer-plotted data points (the small horizontal or vertical dashes) in this and many of the following figures are retained. The contours presented in all the figures are hand-drawn smoothed lines through these data points. The centers of the vortices, determined from the peak vorticity locations in Figure 2, are shown by the plus signs.

It is clear that the periodic component in the u-signal, measured on the jet centerline corresponding to the phase in Figures 2 and 3, have maxima at $x/D \cong 2.6$, 5.0, and 7.2. On this basis, the wavelength measured on the jet cneterline would be about 2.4D in the first 5 diameters and about 2.2D farther downstream. The wavelength value of 2.4D within the potential core agrees with that measured by Crow & Champagne [1] on the basis of comparision of

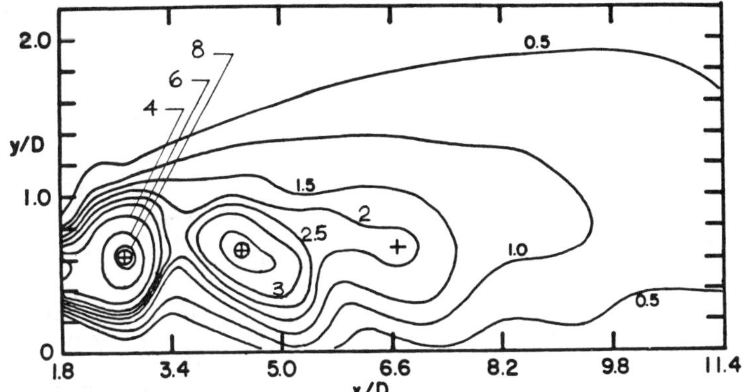

FIG. 2 Ω/fp

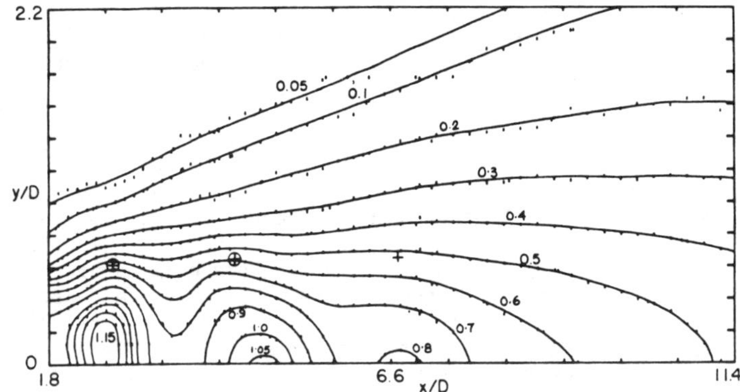

FIG. 3 $\langle u \rangle/U_e$

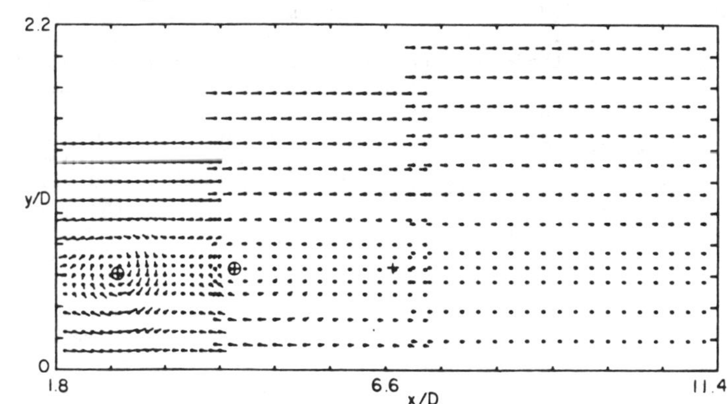

FIG. 4 Ψp

$\tilde{u}$-signals on the jet centerline. However, this value is far different from the actual center-to-center spacing of the vortices which is found to be 1.75D in Figure 2.

The center-to-center spacing between successive vortical structures has been checked by flow-visualization in the present flow to be 1.75D. On the other hand, the "wavelength" was measured independently on the jet centerline with the help of a (PAR) lock-in-amplifier, using the signal driving the loudspeaker as the reference. These measurements in the range $1 < x/D < 3$ yielded a wavelength of 2.25D, agreeing quite well with the corresponding value inferred from the educed vorticity contours in Figure 2.

Under the excitation, the fundamental frequency of the periodic waveform remains constant everywhere in the flow-field. This is apparent from the $\tilde{u}$-spectra anywhere on a (x,y) plane — the fundamental spectral spike does not shift in frequency. Thus, the above wavelengths should correspond to a longitudinal convection velocity of $0.72U_e$ on the jet centerline and $0.53U_e$ in the middle of the shear layer. These values agree excellently with those obtained by Hussain & Clark [4] for the most dominant structure in a natural axisymmetric shear layer by the method of double-Fourier transformation of the space-time correlation of $\tilde{u}(t)$. These data thus clearly show that the changes in the orientation (or tilting) of the individual structures, as they travel downstream, can satisfactorily explain the variation of convection velocity across a shear layer, also observed by other investigators [4, 5].

The spatial distributions of Ψ_p and $\langle\Psi\rangle$, corresponding to the data discussed above, are shown in Figures 4 and 5, respectively. For both these figures, U_{ref} used was $0.5U_e$. Figure 4 actually shows the distribution of the velocity vector ($\langle u\rangle$, $\langle v\rangle$), the dots indicating the origins of the vectors. Lines drawn tangential through these vectors would give the phase-average streamlines Ψ_p. While Ψ_p distribution in Figure 4 does not clearly reveal the structures except for the one centered at $x/D \simeq 2.8$, the $\langle\Psi\rangle$ contours clearly reveal the structures over a larger x-range and appear quite similar to the Ω_z contours. The failure of Ψ_p distribution to identify the structures, however, is due to the fact that the relative magnitudes of the velocity vectors in the downstream regions (Figure 4) become much smaller compared to those in the upstream region. If the streamlines Ψ_p were drawn through the vector patterns, using the method of iscoclines, a pattern, quite similar to the $\langle\Psi\rangle$ distribution in Figure 5, would result. Although the $\langle\Psi\rangle$ contours do not show the details of a structure (see later data and also [3]), the overall similarity of its distribution with that of Ω_z in the entire flow region is impressive, and shows that single-wire measurements are reasonably sufficient for the purpose of identification of structure shapes and boundaries. However, as structure identifiers, both Ψ_p and $\langle\Psi\rangle$ suffer from the fact that they are dependent on the choice of the reference frame velocity (see [3] for further discussion).

Contours of the perturbation velocity components $\langle u_p\rangle$ and $\langle v_p\rangle$ corresponding to the data in Figures 2-5 are shown in Figures 6 and 7, respectively. Note that the ordinate and the abscissa ranges of these two figures are trimmed in order to show the distributions clearly. Since V is relatively small, the contours of $\langle v\rangle$ are not radically different from those of $\langle v_p\rangle$ and are thus not shown. In the studies of the boundary layer spot, contours of $\langle u_p\rangle/U_\infty = \pm 0.02$ have been found to agree with the spot boundary and have been used as markers of the spot boundary [6, 8]. Comparison of contours in Figures 3, 6 and 7 with the vorticity contours in Figure 2 and the streamlines and pseudo-streamfunctions in Figures 4 and 5 show that none of the contours of

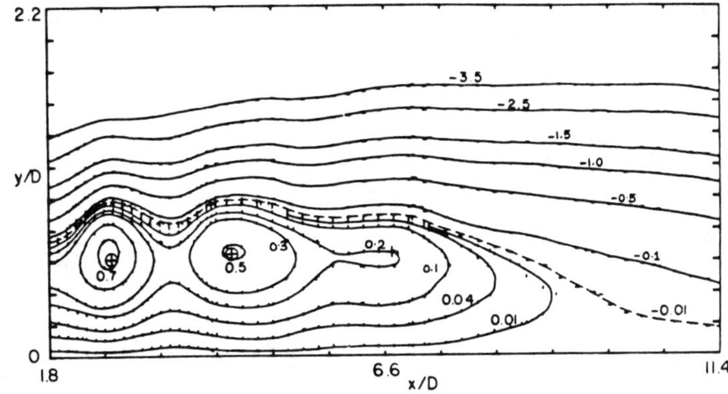

FIG. 5 <Ψ> (x10)

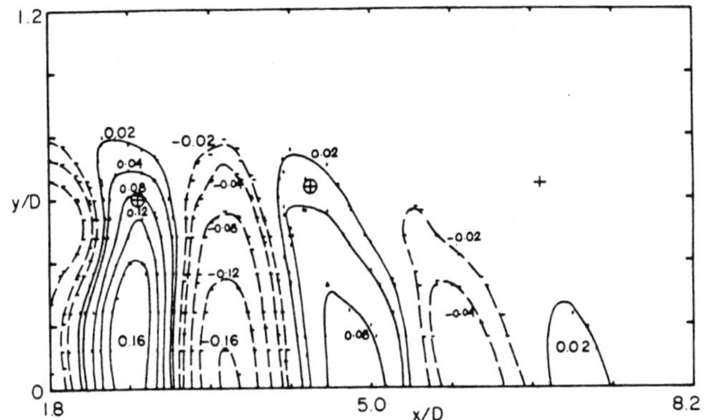

FIG. 6 $<u_p>/U_e$

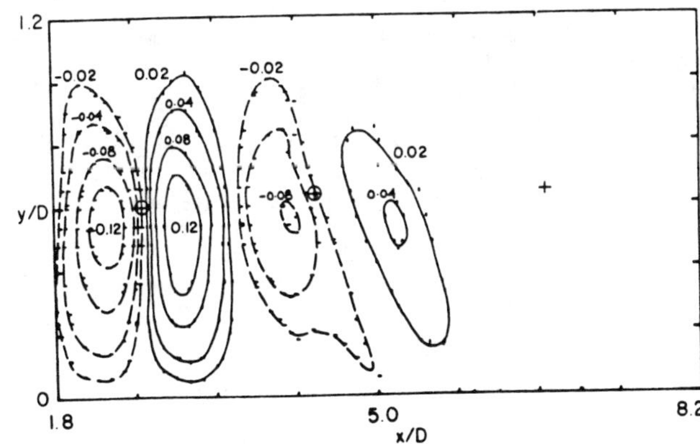

FIG. 7 $<v_p>/U_e$

$\langle u \rangle$, $\langle u_p \rangle$ and $\langle v_p \rangle$ can serve to identify the boundary of the coherent structure in the jet. Note that unlike the boundary layer spot case, $\langle u_p \rangle$ contours are not even closed. The $\langle v_p \rangle$ contours show radially outward transport of the jet core fluid at the front (downstream end) and radially inward transport of the ambient fluid at the back (upstream end) of the structure.

Spatial distributions of the phase-average measures were further explored in an attempt to determine the effects of the initial condition (i.e., fully turbulent vs. laminar initial boundary layer) and jet Reynolds number on the preferred-mode structure. Corresponding to the phase when the structure is centered at $x/D \simeq 3$, four different sets of vorticity contours are presented in Figures 8(a) - (d), all for excitation at $St_D = 0.30$. Figure 8(a) is for the same Re_D, f_p and u_e'/U_e as in Figure 2 (these data being retaken for a different phase). Figure 8(b) is for the same conditions as in 8(a) except that the exit boundary layer was laminar (inferred from the mean velocity and fluctuation intensity profile characteristics and $\tilde{u}$-spectrum). Data in figure 8(c) represents a turbulent exit boundary layer case in the 7.62 cm jet at a higher Re_D (=110,000). Figure 8(d) is for the 2.54 cm tripped jet at Re_D = 25,000. The characteristics of the initial boundary layers for both the cases in Figures 8(c) and (d) were checked to be fully developed turbulent, in accordance with the criteria discussed earlier.

Comparison of Figures 8(a) with (b) indicates that the structures in the region of measurement are essentially the same. The vortex center in Figure 8(b), however, is slightly downstream from that in Figure 8(a). This is due to the uncertainty in the trigger phase selection due to jitter. Nevertheless, it is evident that the structure characteristics viz., the aspect ratio, orientation and strength are essentially the same in the two flows. This is not unexpected in view of our observation based on the centerline fluctuation intensity data that the circular jet preferred mode occurs at $St_D \doteq 0.30$, irrespective of the initial boundary layer being laminar or turbulent [2]. The present data demonstrate for the first time that spatial features of the coherent structure for the jet-column preferred mode are independent of the initial boundary layer characteristics.

However, the structure in Figure 8(c), which is for excitation at St_D = 0.30 but for a higher Reynolds number (110,000) is somewhat different. Comparison with Figures 8(a), (b) shows that the structure aspect ratio and orientation are different but the normalized core vorticity levels are the same. Similarly, the differences in details, e.g. in the core size and aspect ratio of the structure in Figure 8(d) for the lower Re_D case as compared to the other cases should be apparent. Note, however, that the normalized peak vorticity in the cores in all the Figures 8(a) - (d) are the same. This is indeed striking since the excitation frequencies for the four cases are quite different; e.g. for Figure 8(d) it is about 4 times that in Figure 8(a). These data thus demonstrate that for the "jet-column mode" structures. the appropriate scaling factor for vorticity is the jet time scale D/U_e, irrespective of the jet exit speed, diameter or the initial condition.

Corresponding to the vorticity contours in Figures 8(a) - (d), the contours of $\langle u \rangle$ are shown in Figures 9(a) - (d), respectively. Similarly, the contours of $\langle u_p \rangle$ and $\langle v_p \rangle$ are shown in Figures 10(a) - (d) and 11(a) - (d), respectively. The $\langle u \rangle$ contours in Figure 9(a) - (d) exhibit only minor differences in details from one another. However, the perturbation velocity component $\langle u_p \rangle$, as well as $\langle v_p \rangle$, increase with increasing Reynolds number. These data, thus show that the preferred-mode coherent structures, as measured at $x/D \simeq 3$, become more energetic with increasing Reynolds number. As discus-

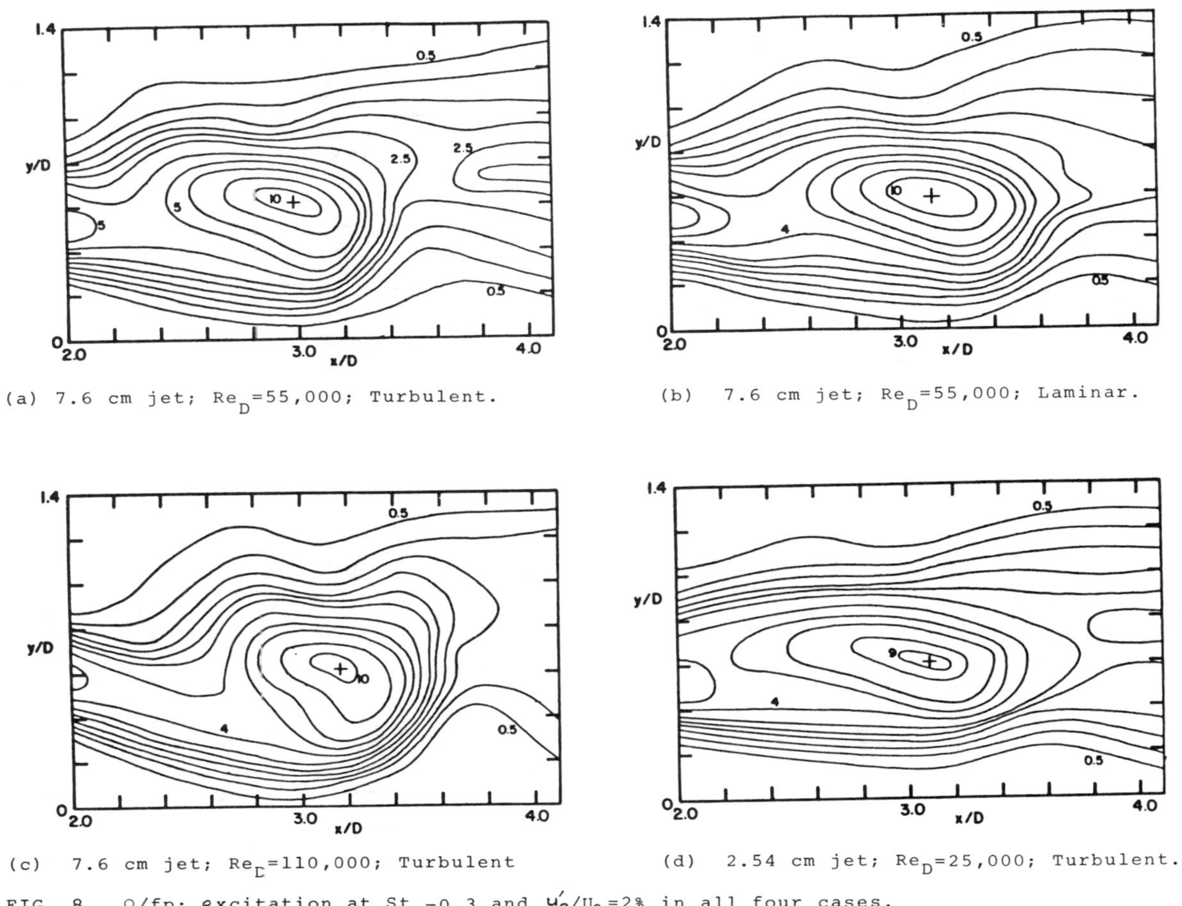

(a) 7.6 cm jet; Re_D=55,000; Turbulent.

(b) 7.6 cm jet; Re_D=55,000; Laminar.

(c) 7.6 cm jet; Re_D=110,000; Turbulent

(d) 2.54 cm jet; Re_D=25,000; Turbulent.

FIG. 8 Ω/fp; excitation at St_D=0 3 and u'_e/U_e=2% in all four cases.

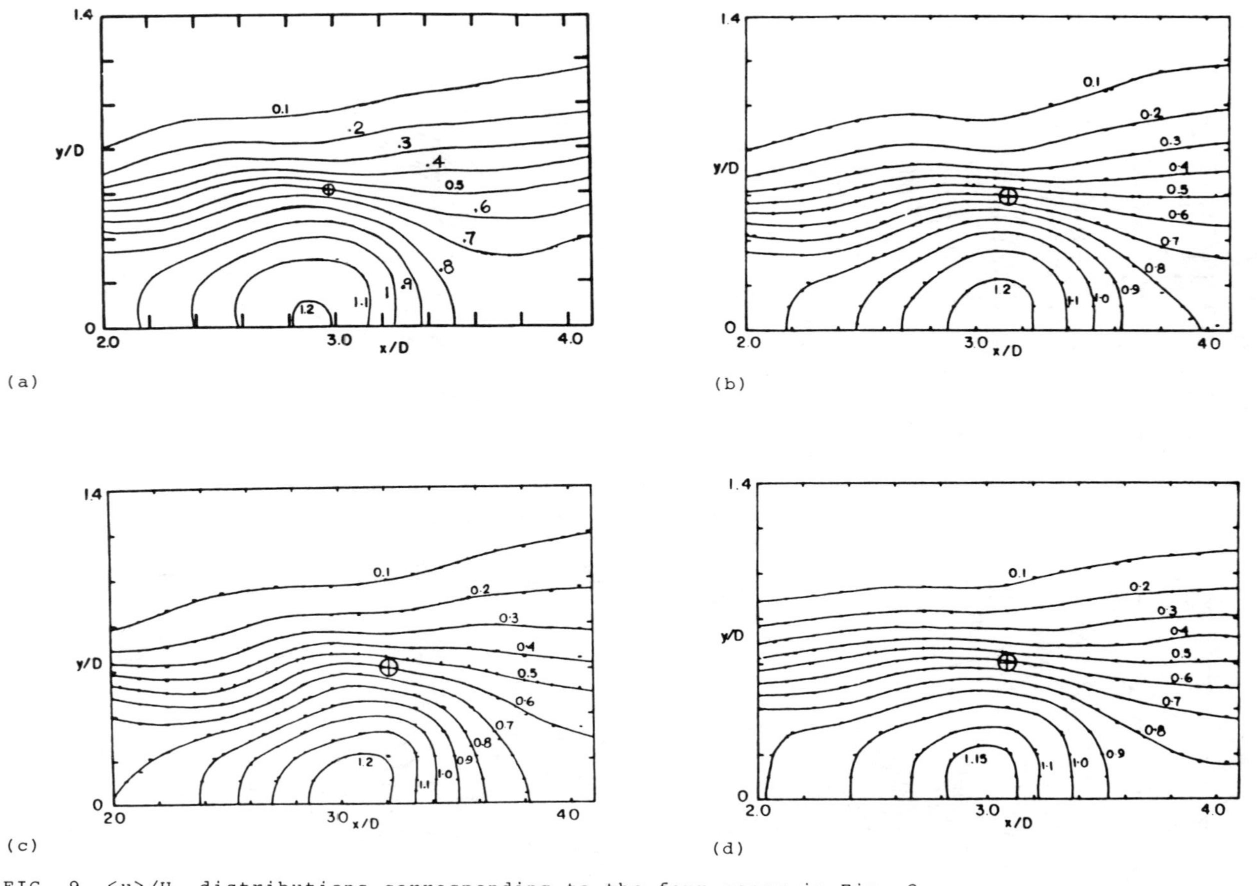

FIG. 9 $\langle u\rangle/U_e$ distributions corresponding to the four cases in Fig. 8.

sed in connection with Figures 3, 5, and 7, it is clear that none of the properties $\langle u \rangle$, $\langle u_p \rangle$ or $\langle v_p \rangle$ can be used for the identification of net structure shape and boundary.

$\langle u_r^2 \rangle^{1/2}$, $\langle v_r^2 \rangle^{1/2}$, Ψ_p and $\langle \Psi \rangle$ distributions corresponding to the four cases in Figures 8(a) - (d) are documented in Figures 12(a) - (d), 13(a) - (d), 14(a) - (d), and 15(a) - (d), respectively. Figures 12(a) - (d) show that the locations of the peak values of $\langle u_r^2 \rangle^{1/2}$ agree fairly with the structure centers identified by the peak coherent vorticity. Note that near the jet centerline, in all four cases, $\langle u_r^2 \rangle^{1/2}$ shows larger values downstream of the structure centers. This is to be expected because these are the regions of decelerating fluid motions in the jet core, corresponding to the phase of measurement. $\langle v_r^2 \rangle^{1/2}$, in all four cases, show peaks slightly upstream of the structure centers. These regions are locations of negative $\langle v_p \rangle$, as can be seen in Figures 11(a) - (d). Closer inspection of Figures 12 and 13 together with Figures 9, 10, and 11 show that in all four cases, peak $\langle u_r^2 \rangle^{1/2}$ occurs at locations of large $-\partial\langle u \rangle/\partial y$ (or $-\partial\langle u_p \rangle/\partial y$) gradient, and peak $\langle v_r^2 \rangle^{1/2}$ occurs at locations of large $\partial\langle v \rangle/\partial x$ (or $\partial\langle v_p \rangle/\partial x$) gradient.

The Ψ_p and $\langle \Psi \rangle$ distributions in Figures 14 and 15 show that these two properties can serve as identifiers of only the overall structure shape. Both of them fail to show the details of the structure shapes (e.g. compare Figures 14c and 15c with Figure 8c). The structure shapes revealed in Figure 14 by Ψ_p agree closely with that educed by Yule [9] in an unperturbed jet. This provides support to our contention that controlled perturbation does not induce an artificial structure but merely organizes and, presumably, enhances the natural structure.

4. SUMMARY

Previous investigations have established that the circular jet near field consists of large-scale coherent structures whose passage frequency sensed by a stationary probe correspond to the jet Strouhal number of $St_D \cong 0.3$. For initially laminar jets, this "terminal structure" is reached by about $x/D = 2$ through successive pairings of the initially rolled-up vortex rings. For jets starting with initially fully turbulent exit boundary layers, this structure is formed through roll-up of the turbulent mixing layer. Even though the jet near field is likely to have a variety of natural structures, the $St_D \cong 0.3$ mode appears to be most dominant and frequently occurring and thus has been termed as the "preferred mode."

Although the preferred mode — as identified from the maximum amplification of the centerline longitudinal fluctuation intensity — was found to occur at $St_D \simeq 0.3$, no previous study has attempted to characterize the associated coherent structure details, nor the sensitivity of the structure properties to the Reynolds number and the initial condition. In this study the dominant large-scale coherent structure — otherwise occurring randomly — has been organized through controlled excitation, and the spatial distributions of the structure properties have been educed through phase-locked hot-wire measurements. The induced structure is not an artificial one but a natural structure formed at controlled phase via the external excitation.

Both educed coherent structure properties and centerline $\tilde{u}$-spectrum show that even though the initial toridal vortical structures are organized through the controlled excitation, no structure could be educed beyond $x/D \simeq 8$. This may not necessarily mean that structures undergo complete breakdown before this location. It is also possible that distinct coherent structures might still exist except that these have large jitter farther downstream. It is also possible

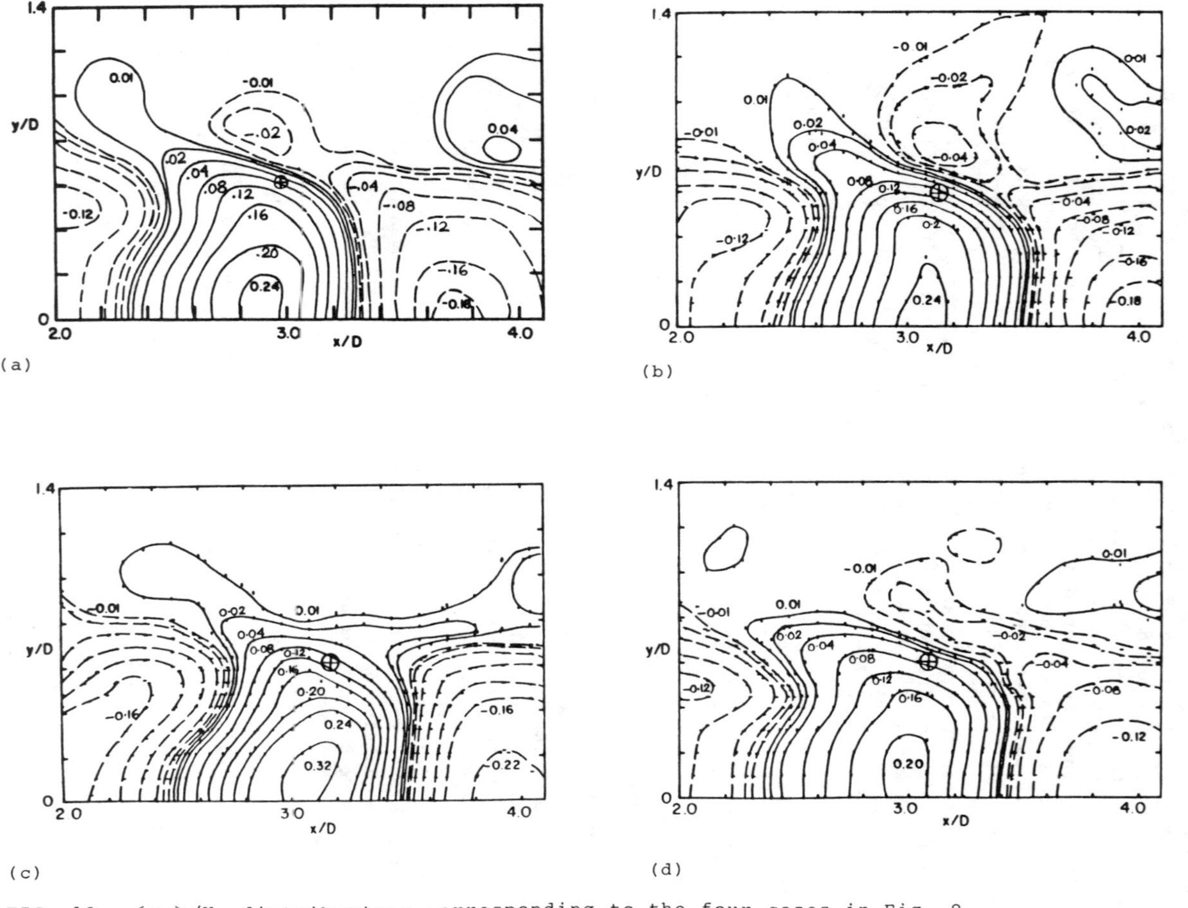

FIG. 10 $\langle u_p \rangle / U_e$ distributions corresponding to the four cases in Fig. 8.

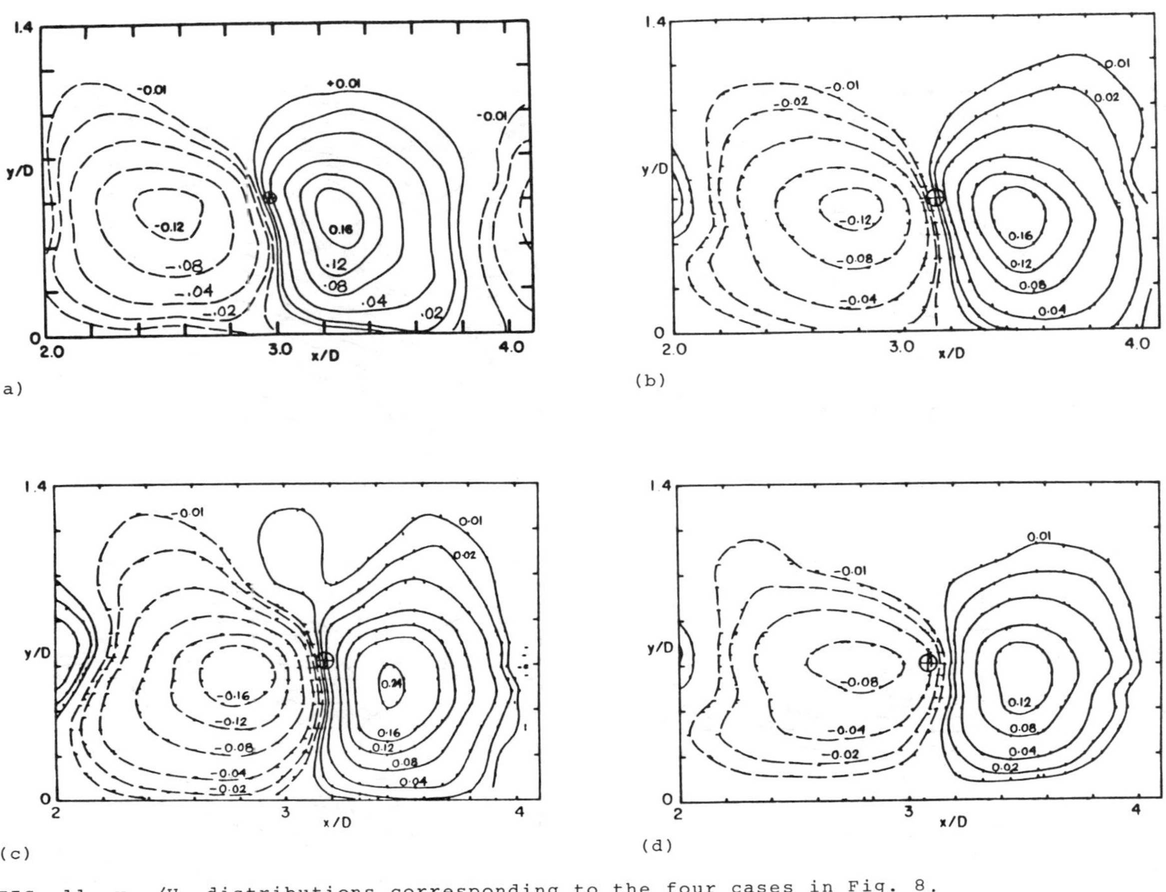

FIG. 11 $<v_p>/U_e$ distributions corresponding to the four cases in Fig. 8.

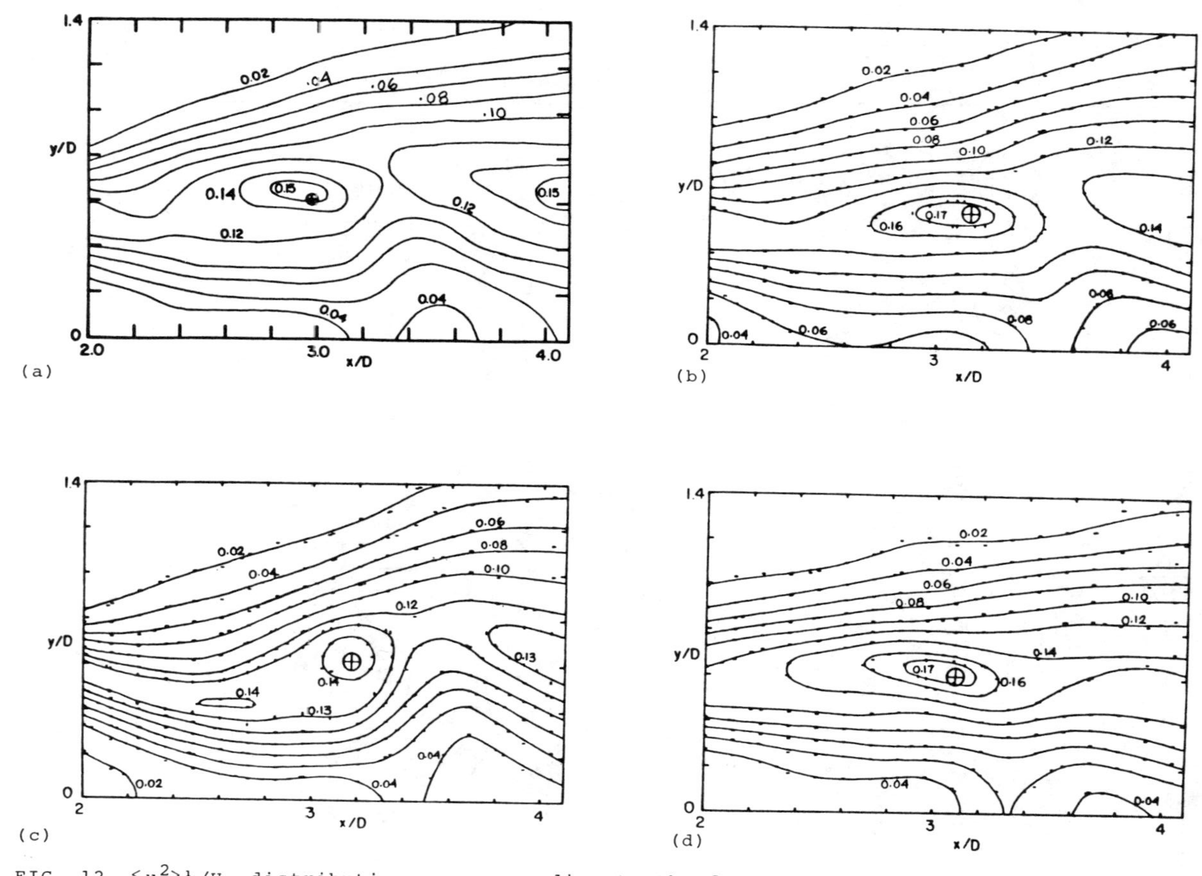

FIG. 12 $<u_r^2>^{1/2}/U_e$ distributions corresponding to the four cases in Fig. 8.

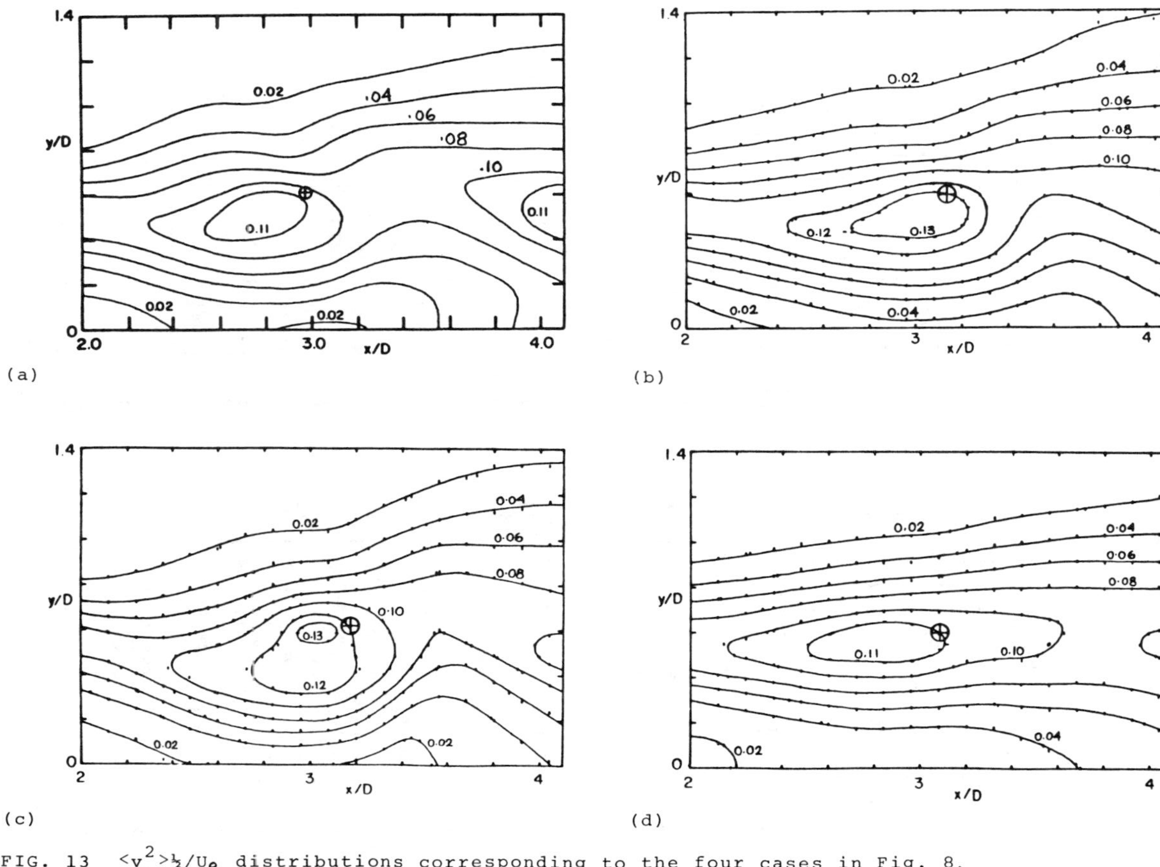

FIG. 13 $\langle v_r^2 \rangle^{1/2}/U_e$ distributions corresponding to the four cases in Fig. 8.

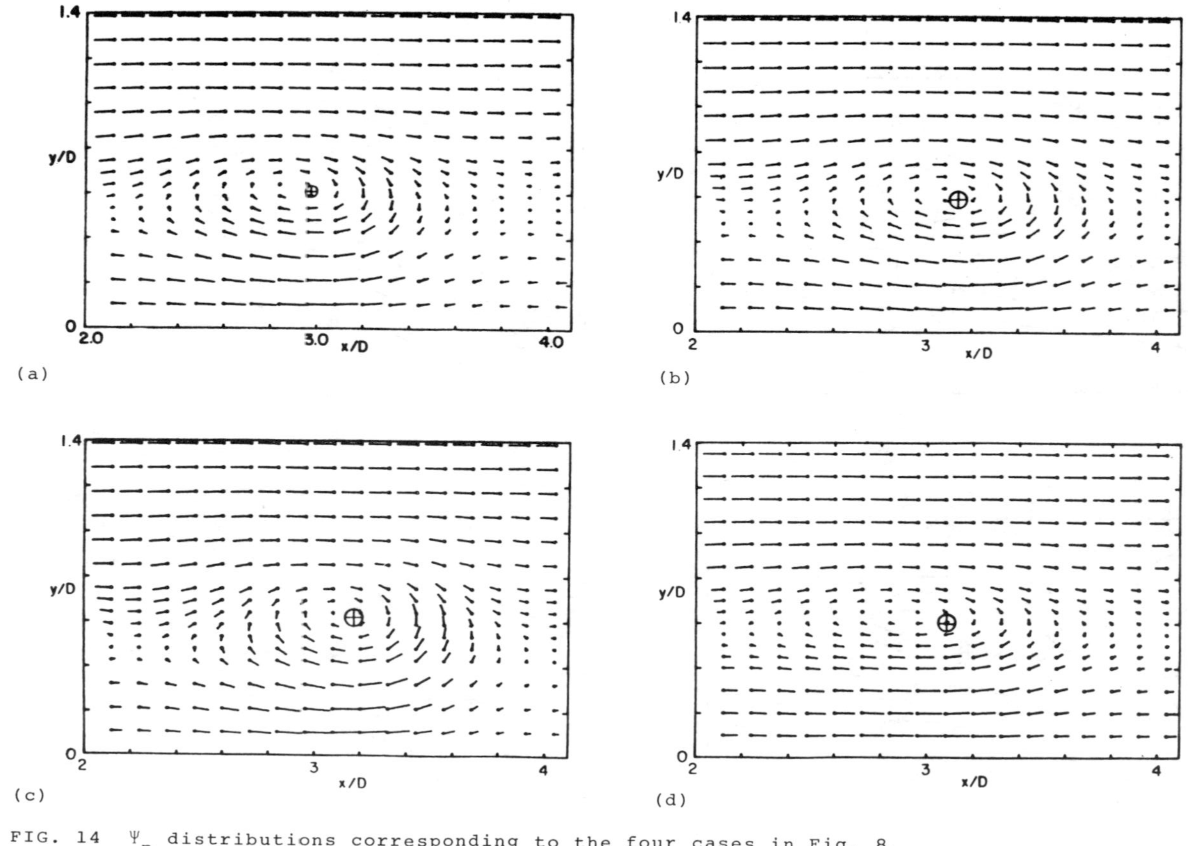

FIG. 14 Ψ_p distributions corresponding to the four cases in Fig. 8.

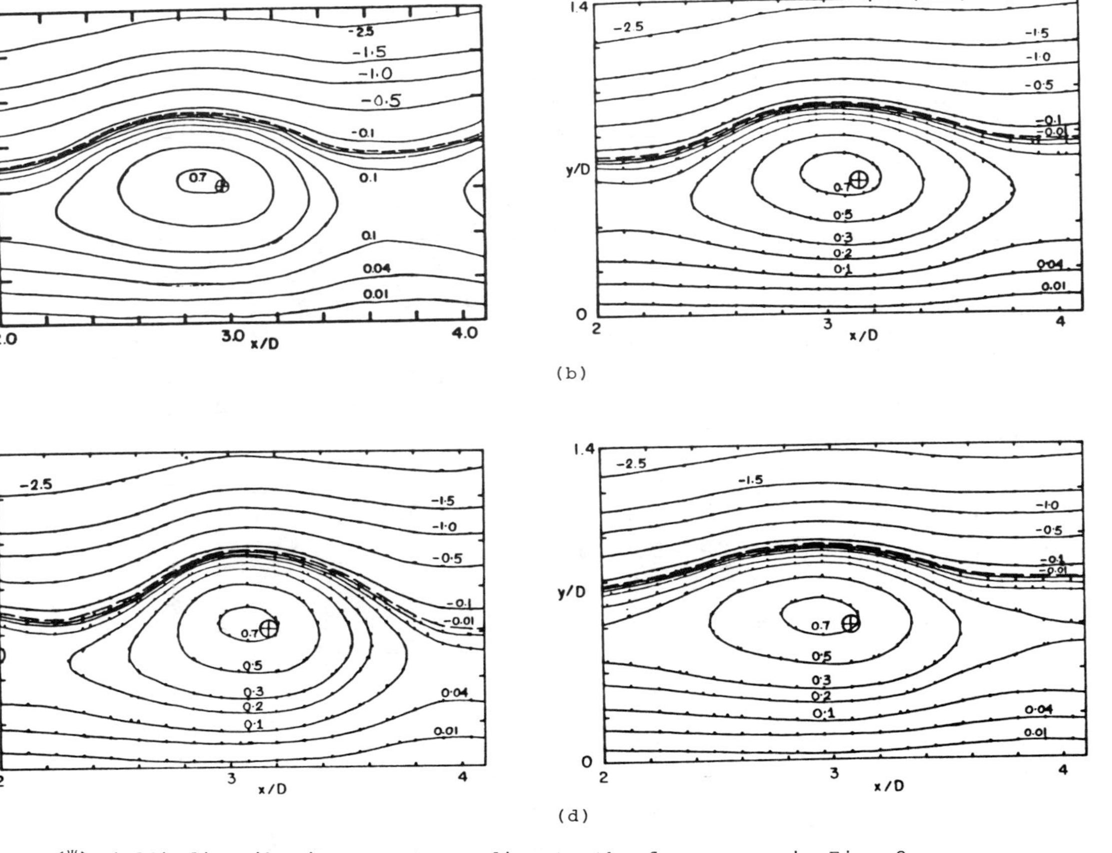

FIG. 15 $\langle\Psi\rangle$ (x10) distributions corresponding to the four cases in Fig. 8.

that the toroidal structures undergo division into lobed-substructures. Because of lack of aximuthal reference in the latter case, the eduction may produce no distinct structures. The reality is perhaps a combination of these three possibilities.

The contours of phase-average velocities $\langle u_p \rangle$, $\langle v_p \rangle$ are quite different from the contours of vorticity, streamlines or pseudo-streamfunctions. This indicates that, unlike the case of the boundary layer spot, contours of $\langle u_p \rangle$ or $\langle v_p \rangle$ cannot identify the structure boundaries in free turbulent shear flows. Contours of $\langle \Psi \rangle$ or Ψ_p are adequate for this purpose, except that these depend on the convection velocity of the reference frame with respect to which these are drawn. We have seen that these represent the structure boundaries well only when drawn in a frame convected downstream with the structure convection velocity.

The detailed structure properties show that the dominant coherent structure is essentially independent of the initial condition. At $x/D \simeq 3$, the structure peak vorticity is about $3U_e/D$. This peak is nearly coincident with the peaks of fluctuation intensities $\langle u_r{}^2 \rangle^{1/2}$, $\langle v_r{}^2 \rangle^{1/2}$ and is associated with the "saddle point" of the $\langle u_r v_r \rangle$ distribution.

Some features of the coherent structure show small but systematic dependence on the Reynolds number Re_D. With increasing Re_D, the peak positive and negative values of $\langle u_p \rangle$ and $\langle v_p \rangle$ increase monotonically and the structure cross-section tends to be more rounded. Consequently, the structure progressively becomes more rounded, becomes relatively more energetic, and produces relatively more coherent Reynolds stress. However, the time-average non-dimensional Rynolds stress does not change with Re_D, presumably because the dominant coherent structure occurs less frequently with increasing Reynolds number. Flow visualization studies in our laboratory tend to support this suggestion. Variation of structure properties with Re_D suggests that high Reynolds number jet behavior cannot be inferred from studies of jets at Re_D significantly below 10^5.

This research has been supported by the NASA Langley Research Center Grant NSG-1475 and the National Science Foundation Grant ENG-7822110.

REFERENCES

1. Crow, S. C. & Champagne, F. H. 1971 J. Fluid Mech. 48, 547.

2. Zaman, K.B.M.Q. & Hussain, A.K.M.F. 1980 J. Fluid Mech. (in press).

3. Hussain, A.K.M.F. & Zaman, K.B.M.Q. 1980 J. Fluid Mech. (in press).

4. Hussain, A.K.M.F. & Clark, A. R. 1980 J. Fluid Mech. (in press).

5. Lau, J. C. 1978 (submitted to J. Fluid Mech.).

6. Coles, D. & Barker, S. J. 1975 Turbulent Mixing in Nonreactive and Reactive Flows (ed. S. N. B. Murthy), Plenum Press, New York, 285.

7. Wygnanski, I., Sokolov, M. & Friedman, D. 1976 J. Fluid Mech. 78, 785

8. Zilberman, M., Wygnanski, I. & Kaplan, R. E. 1977 Phys. Fluids Suppl. 20. S258.

9. Yule, A. J. 1978 J. Fluid Mech. 89, 413.

Coherent Structures in Free Jets

A. ŽABICKI
Institute of Fluid Flow Machinery
Polish Academy of Sciences
Gdańsk, Poland

ABSTRACT

The description is given of an experimental setup and results of observations concerning the initial phases of eflux of a free jet of gas to environment. The jet was generated by means of a pressure-wave tube of the type of a diaphragm shock-tube. The resulting flow patterns, recorded with the use of two-colour flashes, provide information on the development of the loss of stability phenomenon.

1. INTRODUCTION

Investigation of coherent structures in two-dimensional or cylindrical free jets, apart from its importance for practical applications like sound generation and intensification of mixing processes, is significant in understanding of such phenomena as loss of stability , flow separation and the behaviour of large vortex structures in flows. In the research on two-dimensional and circular jets interest is focused on the spacing of tori appearing in the mixing zone,their velocity of propagation and possible effect of the coupling of tori and linear vortices.

The present paper is a complementation of the information contained in the works by Lau and Fisher [1] and Crow and Champagne [2] referring to the jet region within a distance between 2D and 4D from nozzle exit in a stationary jet with addiational acoustic stimulation. The region now examined is adjacent to the nozzle /below 2D/. The present results also confirm the phenomenon of coupling of the vortex tori analogous to the one observed for linear vortex by Browand and Winant [3] and also by Brown and Roshko [4] .

A pressure tube was used in the experiment, in an attempt to avoid the non-definable conditions due to possible large structures caused by boundary layers, ahead of the nozzle. Therefore the results given do not reflect directly the conditions met in practical applications,but they refer to a prefect shear discontinuity. This enables a more detailed examination of the initial phases of stability loss and also a relatively clear definition of the conditions of the experiment.

2. EXPERIMENTAL SET-UP

A diaphragm type shock tube of diameter 30 mm /constant throughout the length/ was used as the pressure tube for flow generation. A sharp-edged low-pressure section of the tube openning into atmosphere forms the flow generator proper. The tube dimensions are: 0.6 m in low-pressure and 5 m in elevated pressure section. This secures a steady efflux into atmosphere of about 35 ms duration. The length of low-pressure section is sufficient for the plane of separation between the driving and driven portion of air in the tube not to reach the nozzle during the experimental cycle.

A flake of mica 3 to 5 µm thick and initially strained at the circumference is used as a diaphragm. The diaphragm is broken by gradually increasing the strain, after filling the pressure section of the tube to a required level. The resulting pressure wave is characterized by a 30 µs rise time and a constant velocity at jet axis, of magnitude conforming with acoustic approximation.

The efflux from the tube starts at the instant of the pressure-wave arrival to the open tube end. This instant is monitored by a pressure sensor with an adequately tuned acoustic delaying line, which actuates a delaying system 6 and clock 8. A photodiode 10 switches of clock 8 and also clock 9 ; delaying system 6 starts a double flash-light 7 producing light pulses of about 2J energy and 10 µs duration. The flash-lights are equipped with colour filters improving the differantiation of the time sequence. Flow visualisation is effected by wetting the inside of the end section of the tube with trace amounts of titan tetrachloride applied in a narrow strip.

Thus a thin "cross-section" of resulting rings is revealed, recorded by camera 11 in natural scale. The diagramatic arrangement of the experimental set-up is given in Fig.1.

3. EXPERIMENTAL CONDITIONS

Since the subsequent phases of the development of the flow instability were examined in a separate test-run, a particular attention was attached to the repetitivity of experiments i.e. the operation of the synchronizing system and the flow observation itself.

The synchronizing system correctness has been checked from photographs of the very early stages of the motion, before the instability effects develop: the lengths of streaklines recorded in this period correspond to the registered duration of motion. Some conclusions on the repetitivity of the observations may be drawn from Figs. 2 and 3. Fig. 3 shows the motion of a starting torus, initiating the efflux. This torus is in contact with the jet at one side and with the motionless air at the other and it should move at a constant velocity if the flow was steady. In the present experiment it moved at constant velocity of about 50 % jet velocity after about 8 ms of motion. Fig. 3 shows the pitch of tori, as plotted from a number of photograps. The repetitivity of results, starting from about 10 ms of the motion is also visible in this figure.

All the data discussed in this paper refer to 500 mm WG initial pressure in the tube, which corresponds to 11.4 m/s jet

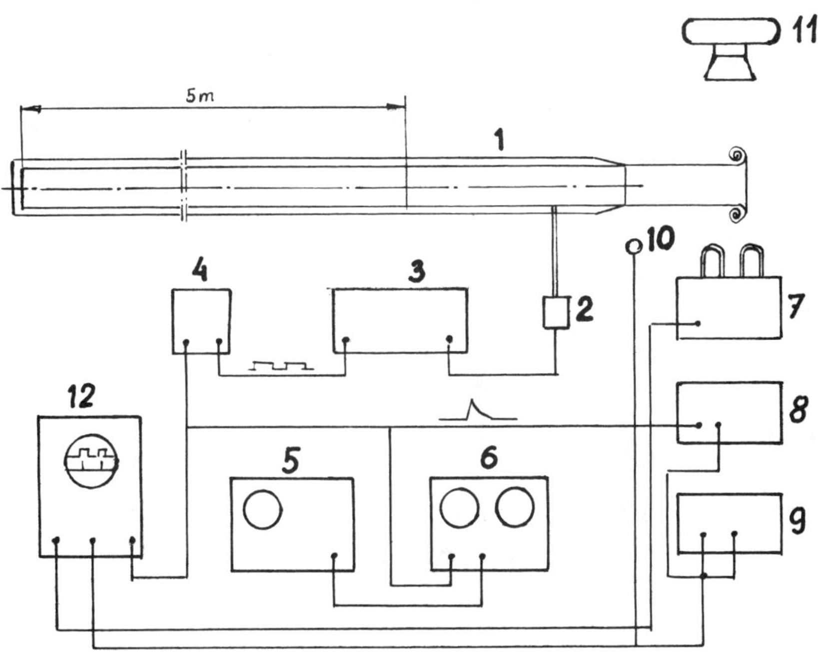

1 Pressure tube
2 Pressure sensor
3 Amplifier
4 Discriminator.
5 Square wave generator
6 Flash interval and delay control
7 Flash lights
8 Clock recording time delay τ_1
9 Clock recording interval between flashes τ_2
10 Photodiode
11 Photographic camera
12 Osciloscope

Fig. 1

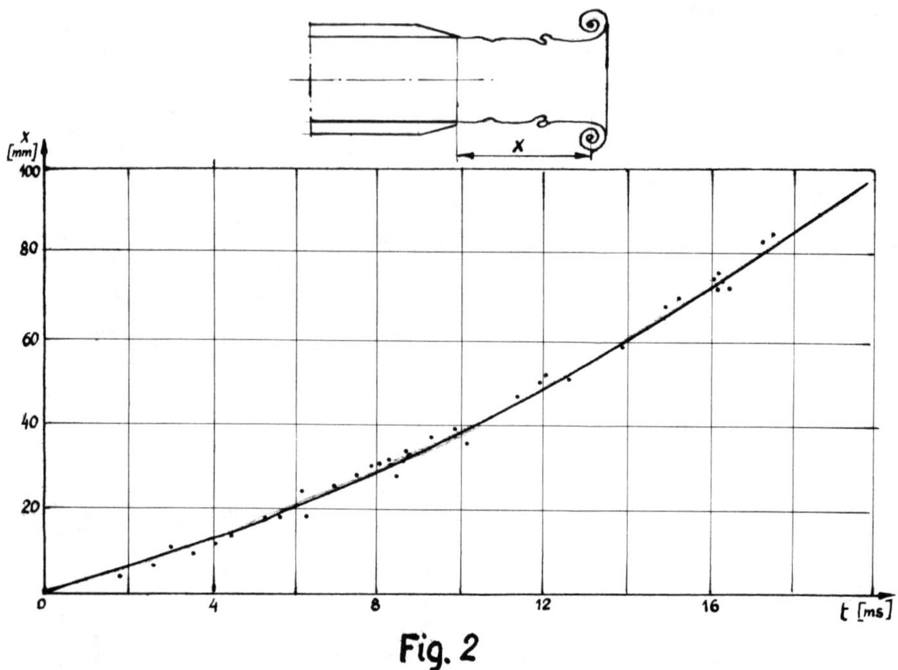
x
x
[mm]
100
80
60
40
20
0
4
8
12
16
t [ms]

Fig. 2

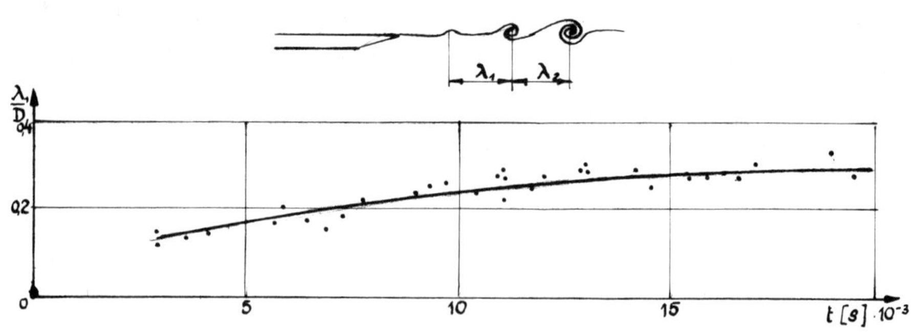
λ_1
λ_2
$\frac{\lambda_1}{D}$
0,4
0,2
0
5
10
15
t [s] · 10^{-3}

Fig. 3

velocity. This would give Re=22.4 x 10^4 in a fully developed stationary flow. Since the boundary layer thickness grows during 35 ms of motion only to 0.8 mm, the Reynold criterion has no significant meaning.

The onset and development of the flow instabilities was measured directly from the photographs, by means of a workshop microscope /as distance between the geometric centers of the vortex rings/.

4. RESULTS

The first symptoms of the loss of stability are regularly noticed at distance 0.4 D from the nozzle in form of a growing sinusoidal ondulation. In the initial period of 4 ms these disturbances do not develop into the form of tori, and they are included into the starting torus. Starting from 6 ms every second wave of the disturbance assumes the form of a vortex. The spacing /pitch/ of two vortices of shortest life is quite regular and equals about 0.2 D. The tori are brought closer to each other usually at a distance of 2 or 3 first pitches,after coupling the tori move together and no overtaking of an "older" by a "younger" one has been noticed. In the process of coupling the "older" torus travels distinctly a shorter way than the "younger". The fragments of vapour comprising two coupled tori do not manifest any regular internal structure and no conclusions as to their speed of rotation can be drawn. The tori after coupling, if they move away from each other again, manifest the some velocity as the uniformly spaced initial tori /0.6 to 0.7 of the jet velocity/.

The primary pitch of tori is not observed at distances beyond one D, the secondary pitch of vortices is very regularly twice the primary pitch.

No occurences of pairing of the coupled vortex rings have been observed: the photographic technique adopted is not well suited for that, due to possiible strong disturbances of the internal structure of the coupled rings.

As follows from fragmentary testing at velocities 2.5 times smaller and 2 times larger than the one used in regular tests, the primary spacing of vortices is inversely proportional to velocity. At small velocities no coupling of vortices has been observed probably due to a too short duration of steady motion and a stabilizing influence of the starting torus. At velocities 2 times higher, the pitches recorded were twice smaller and the vortices interacted vigorously.

The photographs in Figs. 4-10 show the flow patterns after 10 and 20 ms from the onset of motion, other conditions being specified in the figures. The blue streaklines have been recorded earlier than the red ones, as follows from the given time intervals between the blue and red flashes.

The effects of rings coming closer together and coupling are clearly visible in Figs. 7, 9, 10.Strong similarity between the recorded flow patterns is another evidence of the repetitivity of these experiments. It naturally does not mean that the same patterns would be repea ed in another experimental setup.

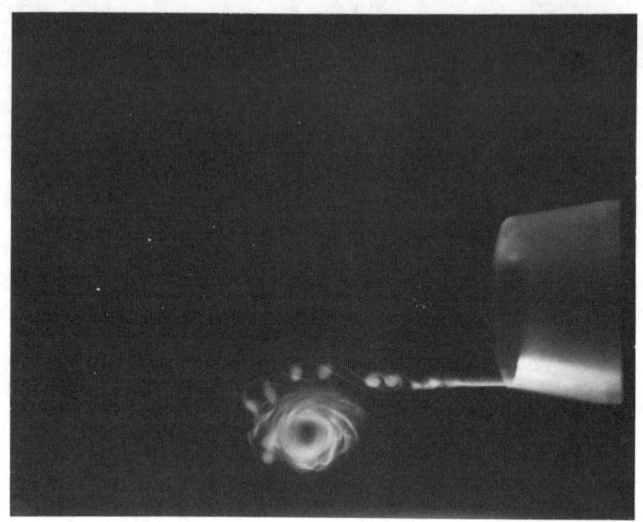

Fig. 4 - Flow pattern after 9,801 ms from the start of motion /blue streaklines/, red streaklines recorded 0,425 ms later

Fig.5 - Blue pattern 10,194 ms, red - 0,652 ms later

Fig. 6 - Blue pattern 9,985 ms, red - 0,966 ms later

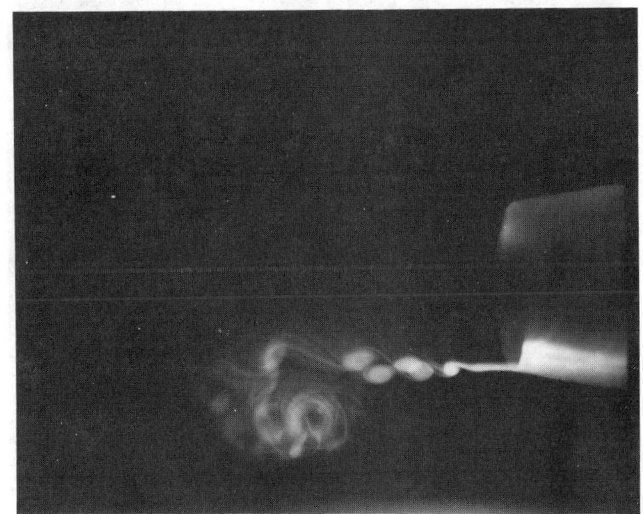

Fig. 7 - Blue pattern 10,049 ms, red - 1,322 ms later

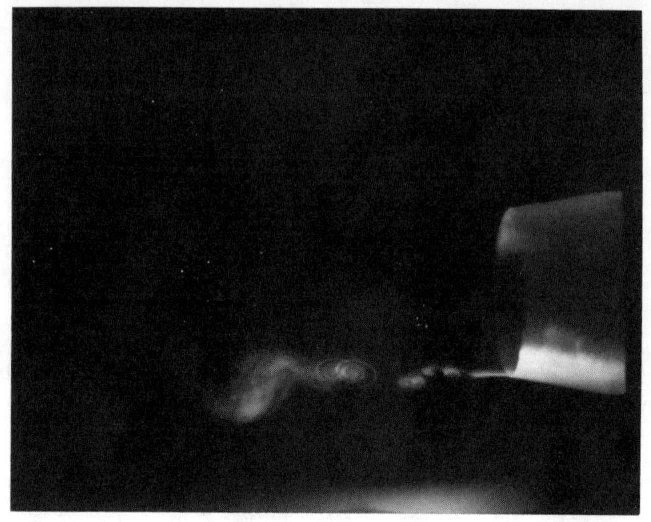

Fig. 8 - Blue pattern 20,527 ms, red - 0,506 ms later

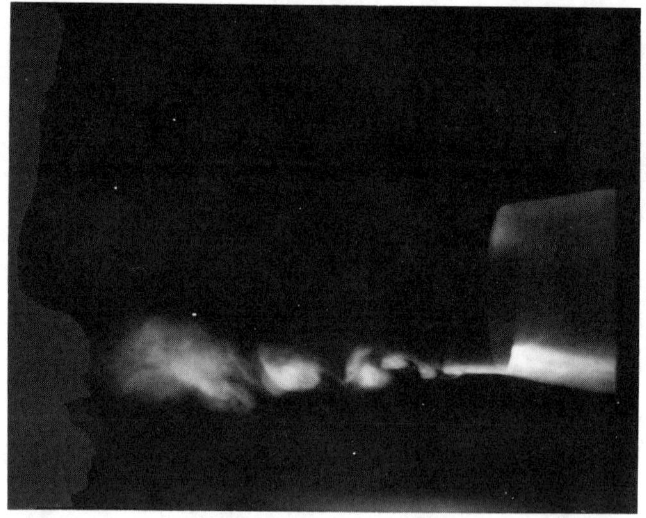

Fig. 9 - Blue pattern 21,417 ms, red - 0,736 ms later

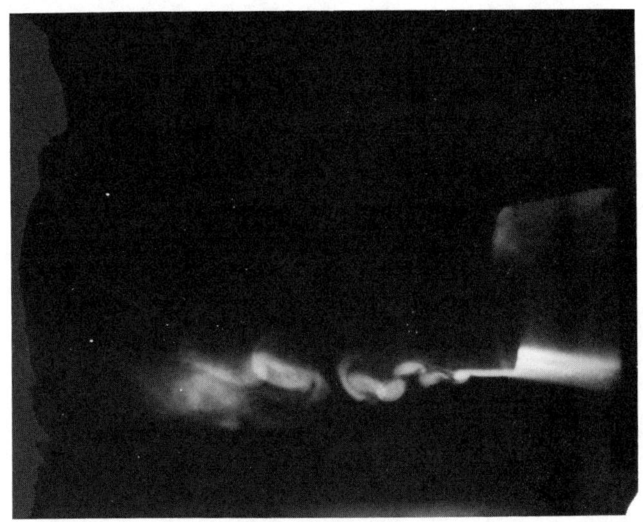

Fig. 10 - Blue pattern 21,364 ms, red - 1,114 ms later

5. CONCLUSIONS

The experimental conditions chosen in the present work differ from the experiments made in conditions of steady flow for reasons discussed earlier. Therefore it is difficult to compare directly some detail quantitative results like e.g. the measured distances between vortex rings. The presented material contains the experimental confirmation of the course of vortex rings coupling as thay move away from nozzle exit.

Strouhal number based on the spacing of rings, frequency and jet velocity is constant and amounts to about 0.3 for primary and paired vortices. This value is lower than those recorded in other experiments.

The photographed flow patterns imply a considerable regularity of the development of motion in spite of basically stochastic character of the loss of stability.

REFERENCES

1. Lau, J.C. and Fisher,M.J., 1975, J.F.M. Vol. 67

2. Crow, S.C. and Champagne,F.H., 1971 , J.F.M. Vol. 48

3. Winant, C. D. and Browand, F. K., 1974, J.F.M., Vol. 63

4. Brown, G. and Roshko,A., 1974, J.F.M. Vol. 64

Characteristics of Coherent Structures in Complex Turbulent Shear Flows

J. F. KEFFER
Department of Mechanical Engineering
University of Toronto
Toronto, Canada

J. G. KAWALL and F. GIRALT
Facultat de Quimica
University of Barcelona
Tarragona, Spain

C. BÉGUIER
Institut de Mécanique Statistique de la Turbulence
Marseille, France

ABSTRACT

A slightly heated, asymmetric, plane turbulent jet, a flow in which there exist regions of negative production of both turbulence kinetic energy and its thermal equivalent, $\overline{\theta^2}$-stuff, has been investigated by means of hot-wire anemometry and digital signal processing techniques. The results that are presented here include one-point coherence functions, phase spectra and cross-correlations of lateral velocity and temperature signals, which were measured simultaneously at various lateral locations within the flow, and one-point cross-correlations of signals derived from the measured velocity and temperature fluctuations by the conditioning of the latter with respect to their signs. The phase spectra reveal that there are definite (i.e., non-random) phase differences between the Fourier components of the velocity and temperature signals for certain frequency ranges within which the corresponding coherence functions display distinct broad-band peaks. Thus, these spectral results serve to show that large-scale, coherent structures exist within the flow. As well, they lend support to the view that the occurrence of a zone of negative production within a turbulent shear flow is linked to the presence in that flow of turbulent motions that are basically bimodal in nature. For lateral locations within and very close to the region where negative production of $\overline{\theta^2}$-stuff occurs, the cross-correlations of the unconditioned signals display extrema at non-zero lag times, whereas, for locations far away from this region, they peak at zero lag times. These traits are compatible with those of the corresponding phase spectra, in that the latter display phase angle differences of about 90° for locations inside and in the immediate vicinity of the negative production zone and roughly 180° for locations well beyond the limits of this zone. The cross-correlations of the sign-conditioned signals enable dominant correlation modes to be established. The dominant mode pertaining to the negative production region is found to be consistent with the mathematical model developed by Béguier, Fulachier and Keffer [1] to describe the lateral turbulent transport of heat within such a region. On the other hand, the dominant modes for points outside the negative production region are found to be in accordance with the conventional, gradient-type diffusion model.

1. INTRODUCTION

There now exists definitive empirical evidence that certain isothermal and heated turbulent shear flows with asymmetric lateral profiles of mean quantities contain localized regions within which the positions where the lateral gradients of the mean quantities are zero do not coincide with those where the associated, lateral turbulent transport terms are zero (see, e.g., [1] to [6]). The result is that the net production of turbulence kinetic energy and/or its

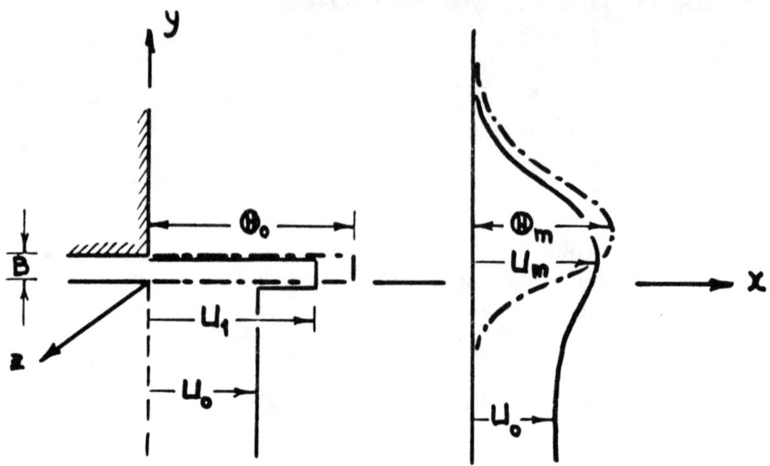

Fig.1. Definition sketch. B = slot width; U_1, U_0 = reference velocities; Θ_0 = reference temperature difference; U_m = local maximum mean velocity; Θ_m = local maximum mean temperature difference; x, y, z = Cartesian co-ordinates.

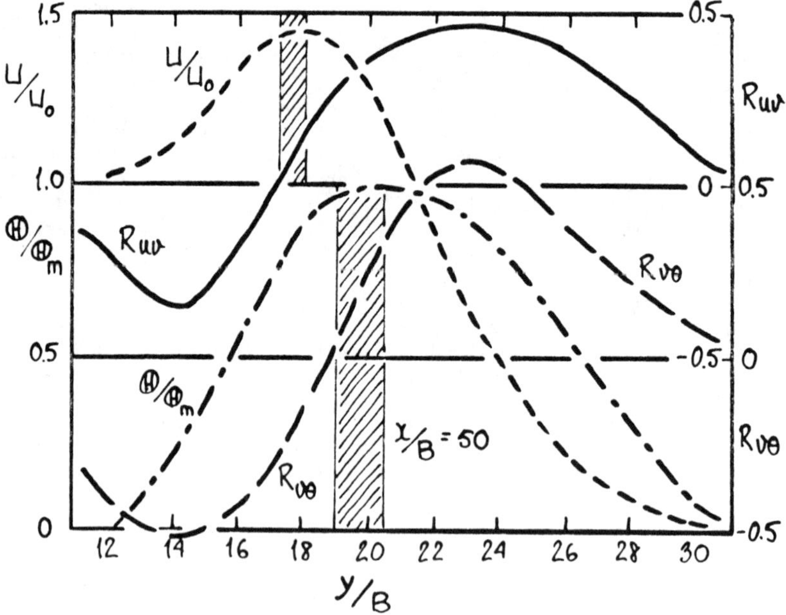

Fig.2. Lateral profiles of mean quantities and lateral turbulent transport terms.

thermal equivalent, $\overline{\theta^2}$-stuff, is negative inside these regions.

Béguier et al [1], who examined a turbulent mixing layer with an asymmetrical distribution of temperature, found that a zone of negative production of $\overline{\theta^2}$-stuff occurred in the flow. These researchers have established that this anomalous "energy" production stemmed from the dominance of a large-eddy, counter-gradient bulk transport over small-eddy gradient diffusion, and they have developed a double-eddy model that characterizes the lateral turbulent heat transport within the negative production zone. This model is analogous to the one that was introduced independently by a number of researchers (e.g., Béguier [4] and Hinze [7]) to describe lateral turbulent momentum transport within flow regions where negative production of turbulence kinetic takes place.

Recently, Keffer, Kawall, Giralt and Béguier [6], using coherence functions and co-spectra, analysed both the above-mentioned mixing layer and a slightly heated, asymmetric, plane turbulent jet — which is a more complex flow than the mixing layer, inasmuch as it contains a region of negative production of turbulence kinetic energy as well as a region of negative production of $\overline{\theta^2}$-stuff. Their results are consistent with the bimodal eddy models, and demonstrate that the Fourier components which control the momentum and heat transport processes within the negative production zones are, in fact, large-scale coherent structures.

In the current work, we review some of the results for the heated, asymmetric jet that were obtained by Keffer et al [6], and we present additional statistical properties of this flow, which pertain to the negative production of $\overline{\theta^2}$-stuff. These statistics complement those reported previously in [1] and [6], providing further information on the characteristics of the coherent structures that are imbedded in complex, asymmetric shear flows.

2. EXPERIMENTAL DETAILS

The heated, asymmetric jet was generated by a blower-type wind tunnel, equipped with a heating unit. This tunnel, which is located in the Institut de Mécanique Statistique de la Turbulence (IMST) is described fully by Béguier [8]. A definition sketch of the flow is presented in Fig. 1. Velocity and temperature signals were measured simultaneously by means of DISA constant-temperature anemometer systems, in conjunction with a constant-current anemometer, special probes and electronic circuitry for processing of the anemometer signals, which were designed and constructed at IMST. The details of the measurement technique are described by Béguier et al [1]. Measurements were taken in the vertical centre-plane of the flow (i.e., the plane given by $z = 0$) at $x/B = 50$ (see Fig. 1).

The (processed) analogue signals were low-pass filtered at 5000 Hz and recorded on an AMPEX FM tape recorder at IMST. The signals were subsequently digitized by means of a PDP-11 computer at the University of Toronto Computer Centre (UTCC) and the resulting digital data, processed on the UTCC IBM 370-165 computer, with the aid of fast Fourier transform techniques developed in the Department of Mechanical Engineering, University of Toronto. The digitization rate used was 10,000 points per second, so that the problem of aliasing was effectively eliminated. Approximately 12.3 seconds of every signal were analysed.

3. PROPERTIES OF THE FLOW

The statistical properties of the flow, that are examined here are as follows:

(i) The mean velocity, U, and the mean temperature, Θ.

(ii) The non-dimensional, lateral, turbulent momentum and heat transport terms, R_{uv} and $R_{v\theta}$.

(iii) The one-point cross-correlation function for the lateral velocity and temperature signals, $v(t)$ and $\theta(t)$, which is defined as

$$\rho_{v\theta}(\tau) = \frac{\overline{v(t)\ \theta(t+\tau)}}{v'\theta'}$$

$$= \left[\lim_{T\to\infty} \frac{1}{T} \int_0^T v(t)\ \theta(t+\tau)dt\right] / v'\ \theta' ,$$

where τ is a time lag and v' and θ' are the root-mean-square values of the signals.

(iv) The autocorrelation of $\theta(t)$, $\rho_{\theta\theta}(\tau)$, and the autocorrelation of $v(t)$, $\rho_{vv}(\tau)$.

(v) The coherence function (or spectrum) for $v(t)$ and $\theta(t)$, which describes the coherence or degree of correlation between $v(t)$ and $\theta(t)$ at frequency n. This function is given by

$$\Gamma_{v\theta}(n) = \frac{C^2_{v\theta}(n) + Q^2_{v\theta}(n)}{E_{vv}(n)\ E_{\theta\theta}(n)} .$$

Here, $C_{v\theta}(n)$ is the co-spectrum and represents the correlation between the in-phase components of $v(t)$ and $\theta(t)$ at frequency n; $Q_{v\theta}(n)$ is the quadrature spectrum and represents the correlation between 90° out-of-phase components at frequency n: $C_{v\theta}(n)$ and $Q_{v\theta}(n)$ are the real and imaginary parts respectively, of the Fourier transform of $\rho_{v\theta}(\tau)$, i.e.,

$$C_{v\theta}(n) - i\ Q_{v\theta}(n) = \int_{-\infty}^{\infty} \rho_{v\theta}(\tau)\ \exp(-i2\pi n\tau)d\tau .$$

$E_{vv}(n)$ and $E_{\theta\theta}(n)$ are the normalized autospectra of $v(t)$ and $\theta(t)$ respectively, which are defined such that

$$\rho_{vv}(0) = \rho_{\theta\theta}(0) = \int_0^{\infty} E_{vv}(n)dn = \int_0^{\infty} E_{\theta\theta}(n)dn = 1.0 .$$

(vi) The phase spectrum, which describes the phase relationship between the Fourier components of $v(t)$ and $\theta(t)$. This spectrum is given by

$$\Phi_{v\theta}(n) = \tan^{-1}\left[\frac{Q_{v\theta}(n)}{C_{v\theta}(n)}\right].$$

(vii) The cross-correlation functions for the sign-conditioned lateral velocity and temperature signals. These functions are given by

$$\rho_{v\theta}^{++}(\tau) = \frac{\overline{I_v^+(t)\, v(t)\, I_\theta^+(t+\tau)\, \theta\,(t+\tau)}}{v'\theta'} ,$$

$$\rho_{v\theta}^{+-}(\tau) = \frac{\overline{I_v^+(t)\, v(t)\, I_\theta^-(t+\tau)\, \theta\,(t+\tau)}}{v'\theta'} ,$$

$$\rho_{v\theta}^{-+}(\tau) = \frac{\overline{I_v^-(t)\, v(t)\, I_\theta^+(t+\tau)\, \theta\,(t+\tau)}}{v'\theta'} ,$$

and

$$\rho_{v\theta}^{--}(\tau) = \frac{\overline{I_v^-(t)\, v(t)\, I_\theta^-(t+\tau)\, \theta\,(t+\tau)}}{v'\theta'} ,$$

where

$$I_v^\pm(t) = \begin{cases} 1 & \text{if } v(t) \gtrless 0 , \\ 0 & \text{otherwise} , \end{cases}$$

and

$$I_\theta^\pm(t) = \begin{cases} 1 & \text{if } \theta(t) \gtrless 0 , \\ 0 & \text{otherwise} . \end{cases}$$

It should be noted that

$$\rho_{v\theta}(\tau) \equiv \rho_{v\theta}^{++}(\tau) + \rho_{v\theta}^{+-}(\tau) + \rho_{v\theta}^{-+}(\tau) + \rho_{v\theta}^{--}(\tau) .$$

4. RESULTS AND DISCUSSION

Lateral distributions of U , Θ , R_{uv} and $R_{v\theta}$ are shown in Fig. 2, where it is seen that both the U profile and the Θ profile are distinctly asymmetric, and that the positions where the lateral mean gradients, $\partial U/\partial y$ and $\partial\Theta/\partial y$, are zero do not coincide with the positions where the associated, lateral turbulent transport terms, $\overline{uv}$ and $\overline{v\theta}$, are zero. The zeros of $\partial U/\partial y$ and $\overline{uv}$ demarcate the zone in which the net production of turbulence kinetic energy is negative, and the zeros of $\partial\Theta/\partial y$ and $\overline{v\theta}$, the zone in which the net production of $\overline{\theta^2}$-stuff is negative. We note that these zones, which are indicated by the "shaded" portions of Fig. 2, are quite significant in extent and do not overlap.

Representative one-point $v\theta$ coherence functions are presented in Fig.3. These functions exhibit unmistakable, broad-band peaks with magnitudes in excess of 0.25, centred at about 25 Hz. In view of the findings of Budny, Kawall and Keffer [9], pertaining to the cylinder-generated wake, we interpret the presence of these regions of frequency-centred-activity in the coherence curves as signifying that large-scale, coherent structures exist within the

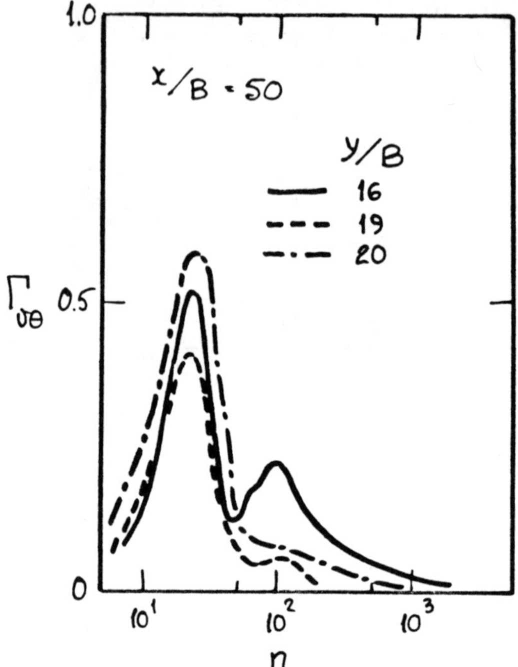

Fig.3. One-point coherence functions of v and θ signals.

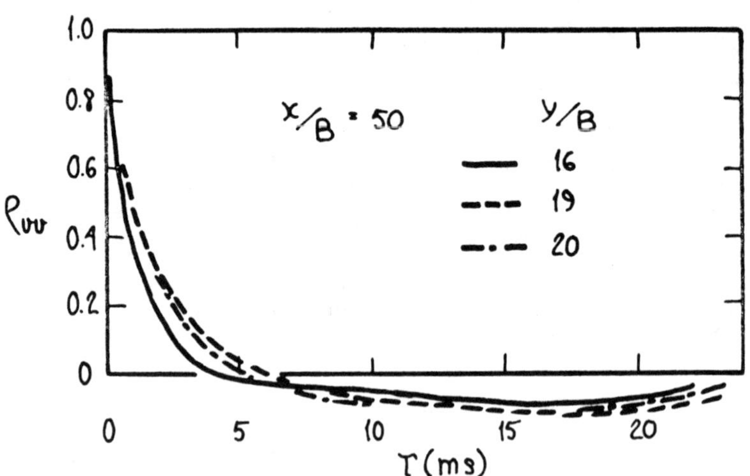

Fig.4. Autocorrelations of v signals.

heated, asymmetric jet. A marked gap separating a "low frequency, high coherence" region from a "high frequency, low coherence" region can be seen in the $v\theta$ coherence curve for the location, y/B = 16, which is outside the zone of negative production of $\overline{\theta^2}$-stuff (see Fig. 2). Such a gap is also apparent in $v\theta$ coherence functions for other locations outside the negative production zone; and a spectral gap occurs in the $v\theta$ co-spectra for locations both within and outside this zone (see [6]). The implication of these coherence and co-spectral results is that the turbulent motions that are present in the heated, asymmetric jet are basically bimodal in nature. In other words, there are, in essence, two distinct types of motions within the flow, viz., low frequency motions generated by the large-scale, coherent structures, and high frequency motions generated by small-scale, relatively diffuse eddies. Moreover, it can be inferred that the low frequency motions produce counter-gradient, bulk convection, while the high frequency motions effect gradient diffusion. Since, as Fig. 3 indicates, for locations inside the negative production zone (e.g., y/B = 20, which is roughly in the centre of the zone), frequency-centred-activity in the $v\theta$ coherence spectra is negligibly small at high frequencies, the conclusion follows that dominance of the large-scale, counter-gradient transport process over the small-scale, gradient process is responsible for the occurrence of the zone of negative θ^2-stuff production within the heated, asymmetric jet [6].

Figures 4 and 5 show typical v and θ autocorrelations. The absence of oscillations in these curves (and of peaks in the various auto-spectra) clearly establishes that there were no anomalous periodic components in the signals, which might possibly have given rise to the regions of frequency-centred-activity in the coherence functions.

Representative, one-point $v\theta$ phase spectra are depicted in Fig. 6. A comparison between these curves and the corresponding coherence functions shown in Fig. 3 indicates that there are definite phase differences between the Fourier components of the v and θ signals for the same frequency ranges within which there is significant coherence. Notice that, for frequencies beyond which the coherence is effectively zero, the phase becomes random, as expected; as well, the phase variability is smaller within the low frequency region,where the coherence is high,than it is within the high frequency range,where the coherence is relatively low. Clearly, these phase results (like the coherence data) signify that large-scale, coherent structures occur in the flow, and that the turbulent motions are basically bimodal in nature.

Figure 7 illustrates typical $v\theta$ cross-correlations. These results serve to demonstrate that, for locations in the immediate vicinity of and within the negative production zone (e.g., y/B = 19 and y/B = 20), peak, positive and/or negative, correlations between the v and θ signals are generated at non-zero lag times, whereas, for locations far away from the zone (e.g., y/B = 16), peak correlations are generated at zero lag times. These traits are compatible with those of the phase spectra, in that the latter display phase differences of about 90° in the case of locations inside and very close to the negative production zone (see Fig. 6) — for which the peak correlations occur at non-zero lag times — and roughly 180° in the case of locations well beyond the zone limits (see Fig. 6) — for which the peak correlations occur at zero lag times.

The three $v\theta$ cross-correlation curves shown in Fig. 7 are plotted along with the corresponding cross-correlations of the sign-conditioned v and θ signals in Figs. 8, 9 and 10. As indicated in Section 3, four different sign-conditioned signals were generated, and, hence, there are four different correlation mechanisms or "modes", viz., a $v^+\theta^+$ "mode", a $v^+\theta^-$ "mode", a $v^-\theta^+$ "mode" and a $v^-\theta^-$ "mode". By combining these "modes" linearly, we obtain

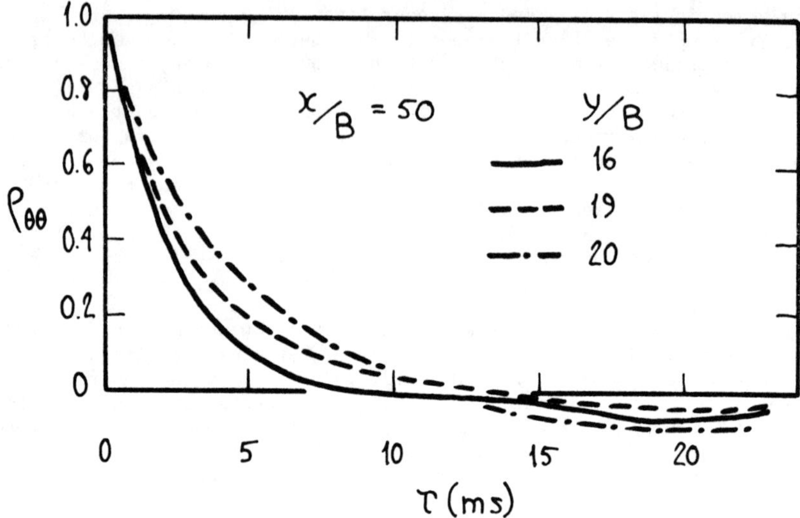

Fig.5. Autocorrelations of θ signals.

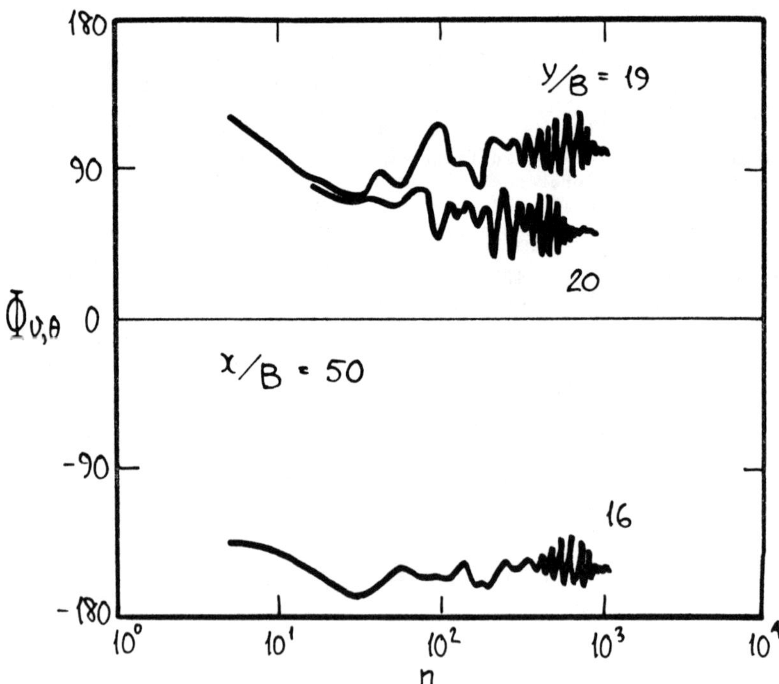

Fig.6. One-point phase spectra of v and θ signals.

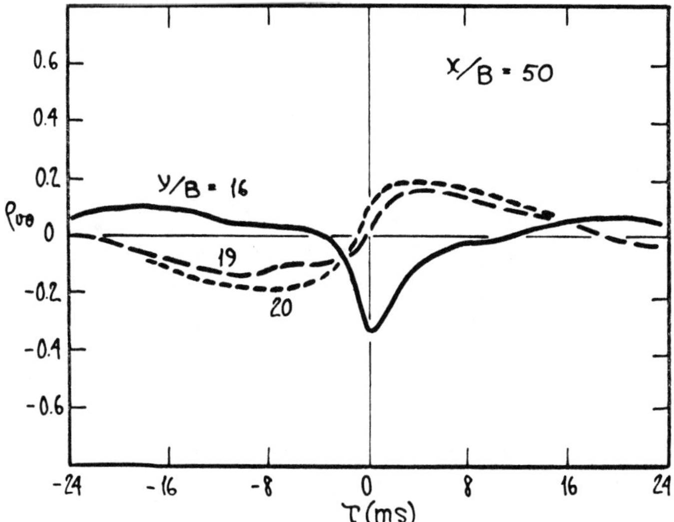

Fig.7. One-point cross-correlations of v and θ signals.

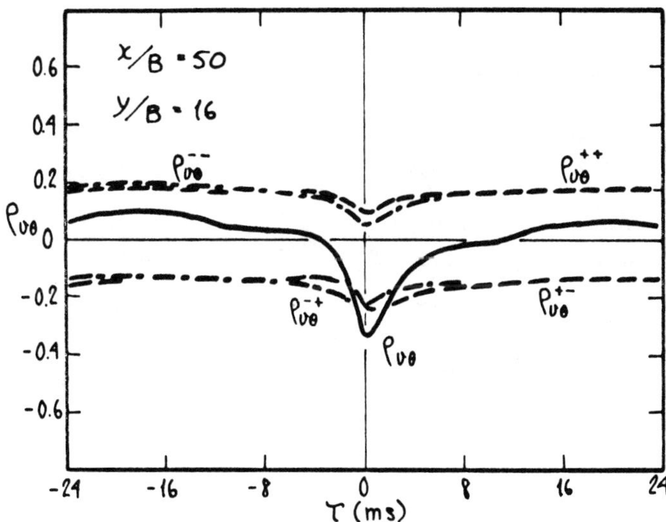

Fig.8. One-point cross-correlations of unconditioned and sign-conditioned v and θ signals.

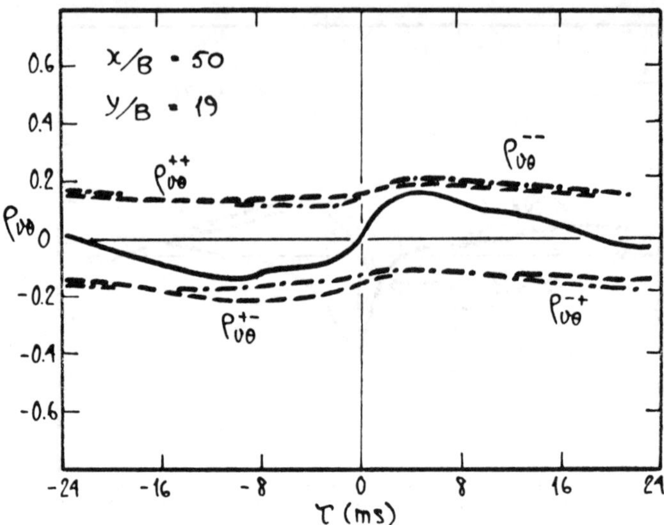

Fig. 9. One-point cross-correlations of unconditioned and sign-conditioned v and θ signals.

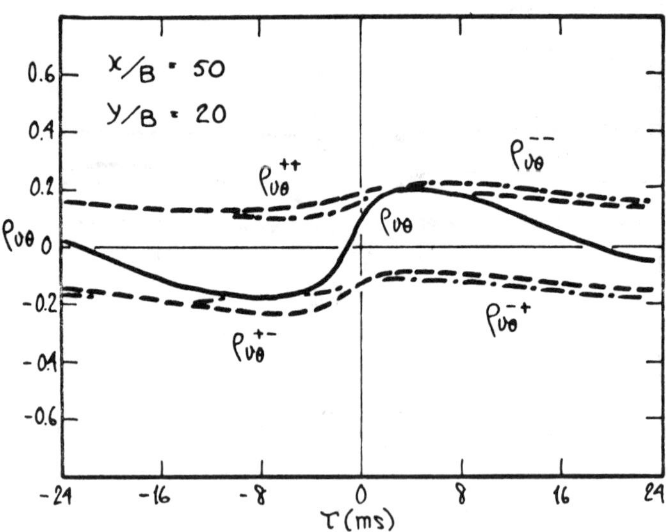

Fig.10. One-point cross-correlations of unconditioned and sign-conditioned v and θ signals.

the conventional cross-correlation of the (unconditioned) v and θ signals (see Section 3). It can be seen that the various sign-conditioned signals remain strongly correlated when the conventional cross-correlation functions are effectively zero. This is a consequence of the fact that the mean values of the various sign-conditioned signals are not zero. We note that the trends of the curves shown in Figs. 8, 9 and 10 are similar to those of the cross-correlations of unconditioned and sign-conditioned u and v signals, obtained by Fraunié [10], for an asymmetric wake.

The average values of normalized, unconditioned and sign-conditioned "vθ product" signals for the various locations at which measurements were taken are reported in Table 1. Plots of segments of the "vθ product" signals pertaining to a location within the negative production zone are shown in Fig.11. It is apparent from Table 1 that the dominant correlation mode within the negative production region of the flow is the $v^+\theta^+$ "mode", which is consistent with the model developed by Béguier et al [1] to describe the lateral turbulent transport of heat within such a flow region. In contrast, the dominant correlation modes outside the negative production region are the $v^+\theta^-$ and $v^-\theta^+$ modes. This is consistent with the conventional, gradient-type diffusion model.

In brief, we have presented a number of results that pertain to the structural features of a heated, asymmetrical, plane turbulent jet in which regions of negative production of turbulence kinetic energy and of $\overline{\theta^2}$-stuff exist; and we have identified, in certain of these results, signatures of both large-scale, coherent structures and small-scale, relatively diffuse eddies. It is anticipated that a proposed examination of simultaneous traces of u, v, θ, uv and vθ signals will provide deeper insight into the physics of the flow.

Table 1. Unconditioned and sign-conditioned vθ correlation coefficients

y/D	$\rho_{v\theta}(0)$	$\rho^{++}_{v\theta}(0)$	$\rho^{--}_{v\theta}(0)$	$\rho^{+-}_{v\theta}(0)$	$\rho^{-+}_{v\theta}(0)$
12.0	-0.2111	+0.1849	+0.0620	-0.0812	-0.3768
13.0	-0.2364	+0.1649	+0.0613	-0.1019	-0.3607
14.0	-0.3119	+0.1239	+0.0513	-0.1472	-0.3399
15.0	-0.3425	+0.0972	+0.0409	-0.1910	-0.2896
16.0	-0.3334	+0.0849	+0.0494	-0.2342	-0.2335
17.0	-0.2582	+0.0964	+0.0819	-0.2498	-0.1867
18.0	-0.1042	+0.1194	+0.1300	-0.2074	-0.1461
19.0	+0.0081	+0.1489	+0.1492	-0.1598	-0.1302
20.0	+0.0886	+0.1792	+0.1572	-0.1247	-0.1232
21.0	+0.0936	+0.1958	+0.1391	-0.1164	-0.1249

ACKNOWLEDGEMENTS

This research was supported by Natural Sciences and Engineering Research Council Canada, Grant No. A-2746 and the CNRS of France.

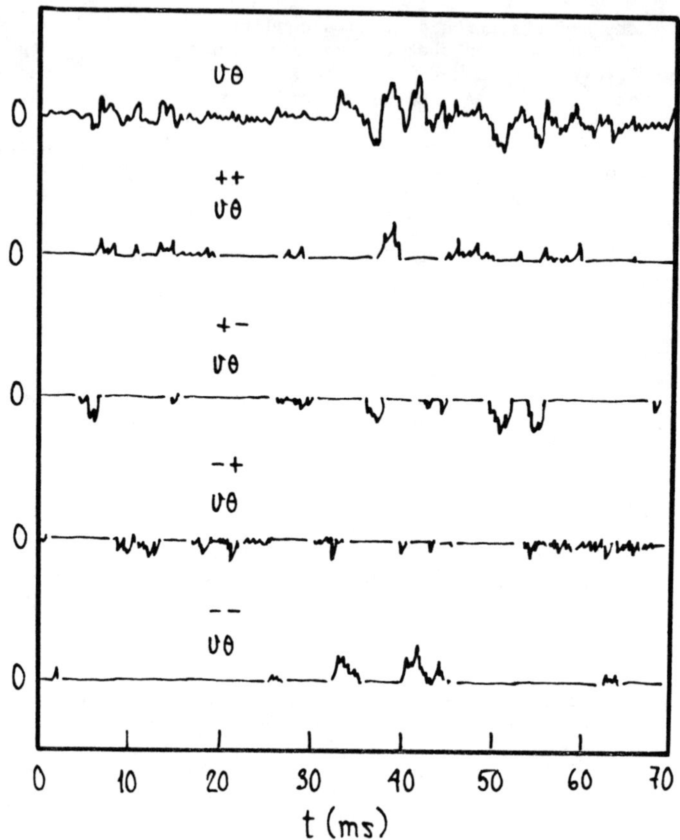

Fig.11. One-point "vθ product" signals.

REFERENCES

1. Béguier, C., Fulachier, L. and Keffer, J.F. 1978. The turbulent mixing layer with an asymmetrical distribution of temperature. Journal of Fluid Mechanics, Vol. 89, pp. 561-587.

2. Béguier, C., Giralt, F., Fulachier, L. and Keffer, J.F. 1977. Negative production in turbulent shear flows. Lecture Notes in Physics, Vol. 76, pp. 22-35.

3. Eskinazi, S. and Erian, F.F. 1969. Energy reversal in turbulent flows. The Physics of Fluids, Vol. 12, pp. 1988-1999.

4. Béguier, C. 1969. Écoulements plans dissymétriques en régime turbulent incompressible. C.R. Acad. Sci. Paris, A268, pp. 69-72.

5. Palmer, M.D. and Keffer, J.F. 1972. An experimental investigation of an asymmetrical turbulent wake. Journal of Fluid Mechanics, Vol. 53, pp. 593-610.

6. Keffer, J.F., Kawall, J.G., Giralt, F. and Béguier, C. 1979. Analysis of turbulent structures in complex shear flows. 2nd Symposium on Turbulent Shear Flows, Imperial College, London.

7. Hinze, J.O. 1970. Turbulent flow regions with shear stress and mean velocity gradient of opposite sign. Dutch J. Appl. Sci. Res., Vol. 22, pp. 163-175.

8. Béguier, C. 1971. Etude du jet plan dissymétrique en régime turbulent incompressible. Université de Provence, Marseille, Thèse de Doctorate ès Sciences.

9. Budny, R.S., Kawall, J.G. and Keffer, J.F. 1979. Vortex street evolution in the wake of a circular cylinder. 2nd Symposium on Turbulent Shear Flows, Imperial College, London.

10. Fraunié, P. 1979. Contribution à l'etude des champs dynamique et thermique dans le sillage de deux cylindres de section circulaire. L'université d'Aix-Marseille II, Marseille, Thèse de Docteur-Ingenieur.

HEAT AND MASS TRANSFER IN TURBULENT FLOWS

Suction Effects on the Structure of Turbulent Boundary Layer on a Heated Porous Wall

LOUIS FULACHIER, MAX ELÉNA, EDMOND VEROLLET, and RÉGIS DUMAS
Institut de Mécanique Statistique de la Turbulence
Laboratoire Associé au C.N.R.S. No. 130
12 Avenue de Général Leclerc - 13003 Marseille

ABSTRACT

The purpose of this paper is to analyse the suction effects on the mechanism of turbulence generation and on the structure of the flow. The experiments are carried out in the turbulent boundary layer of a flat plate. The boundary layer was already fully developed at the beginning of the porous wall. The results relating to the viscous sublayer and the buffer layer have been obtained in a cylindrical pipe flow. The wall was slightly heated and heat is generally used as a passive contaminant in order to describe the turbulent field structure. The mean temperature and velocity profiles and the turbulent quantities are measured. The anisotropy of turbulent field is studied through a proposed criterion. Some typical statistical balances are presented which emphasize the influence of suction.

1. INTRODUCTION

Among the studies of boundary layer turbulent flows with mass transfer at the wall, few of them concern the suction. These researches have a practical interest, especially in the field of exchangers and gaseous diffusion. They present an interest of a fundamental nature owing to the modifications of the flow structure under the effect of the suction. At first, researches relate to the global characteristics modifications of wall flows, as mean velocity or temperature distributions, and wall exchange coefficients [1]. Analytical expressions of mean velocity profiles [2, 3, 4, 5, 6] and mean temperature profiles [7, 8, 9] have been proposed. Works of Weissberg [10], to our mind, have brought first results relative to the longitudinal component of velocity fluctuation. Next studies, first at the I.M.S.T., have pointed out the reduction, by suction, of Reynolds stresses, turbulence production [11, 12, 13], diffusion and dissipation [7,8]. More recently, a similar study has been carried out in duct flow [14].

In other respects, suction effect on turbulent terms connected to temperature fluctuations has been brought to light from direct measurements [15, 7, 8, 16, 17]. Let us quote also the works relative to blowing influence on a turbulent boundary layer with rough wall [18].

Experimental results which are presented later, allow to analyse suction influence on the structure and the mechanisms of a fully developed turbulent boundary layer with heated flat plate. In internal region our purpose is, especially, to point out the part which can be played by mass transfer on the turbulence generation mechanism. In particular, this analysis allows to have an idea of suction influence on coherent structures for which schemes have been proposed and discussed in recent articles [19, 20, 21], without mass transfer. Moreover the suction effect on various transfer processes can be analysed by means of local statistical balances which allow also to complete the knowledge of the mutual influence of various boundary layer zones [7]. This study is made from kinetic energy balances of macroscopic and turbulent motion, enthalpy and temperature variance balances.

2. EXPERIMENTAL CONDITIONS AND MEASUREMENTS METHODS

Measurements have been carried out in two types of flows. Zones close to the wall, especially the viscous sub-layer, have been analysed from measurements made in a cylindrical pipe with a circular pipe [16]. Beyond $Y^+ = y\, u^*/\nu \simeq 20$ (y = distance from the wall, $u^* = (\tau_w/\rho_w)^{1/2}$ = skin-friction velocity where τ_w and ρ_w are respectively the skin-friction stress and the volumic mass at the wall, $\nu = \mu/\rho$ = kinematic viscosity) the analysis has been made, mainly, from experimental results obtained in a flat plate boundary layer.

In these various experimental investigations, the walls could be slightly heated (maximum temperature differences $\theta_w - \theta_e \simeq 20$ K). With maximum velocity which is adopted ($\bar{u} \sim 10$ m s^{-1}), even near by the wall, it can be considered that heat is behaving as a passive contaminant, concerning at least the fluctuations [8]. In the analysis of suction influence on coherent structures, heat has been essentially used to mark the fluid.

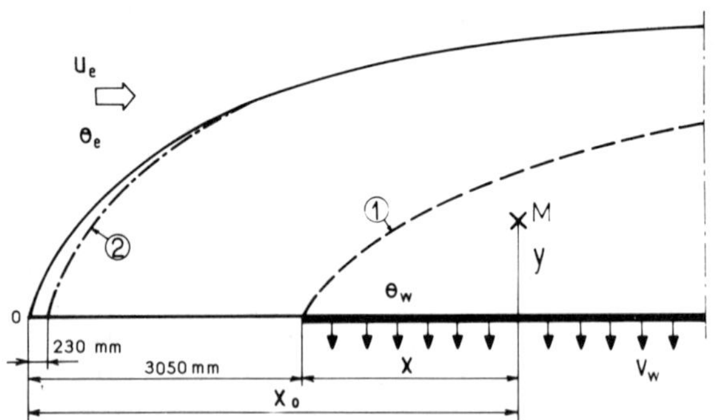

Fig. 1. Flat plate boundary layer

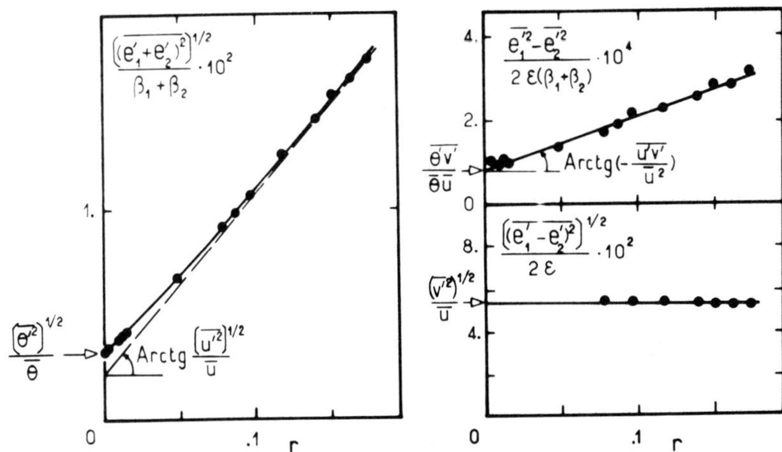

Fig. 2. Fluctuations diagrams for X wires

Indeed, its diffusion by turbulence allows to describe turbulent field, and more especially large-scale structures.

Concerning suction rates $A = -\rho_w v_w/\rho_e u_e$ they are moderate so that the boundary layer does not resorb itself. The value of the maximum rate used, A = .0030, is of the same order as the wall skin-friction coefficient without suction.

2.1 Experimental Conditions

Cylindrical pipe. It concerns a 1116 mm long pipe, D = 76.6mm in diameter. Its porous wall can be heated. It is preceded by a 52 D long non-heated tight pipe. Measurements are made in a section located at 12,8 D from the beginning of the porous part where thermal transfers affect the section completely. Without suction the flow characteristics are : maximum velocity u_e = 8.3 m/s, Reynolds number $R_e = u_e \delta^{**}/\nu$ = 1400 (δ^{**} momentum thickness), u_o^* = .37 m/s, $\theta_w - \theta_e$ = 25 K (θ_e isothermal flow temperature) and temperature scale $\theta_o^* = (q_w)_o/C_p \rho_w u_o^*$ = 1.32 K defined from the heat transfer at the wall q_w. Suction reduced rate which is used, $A^+ = -v_w/u^*$, is .049.

Flat plate. Concerning thermal conditions, two experimental cases have been considered, as it is shown on figure 1. At the section where measurements are performed ($X \simeq 10\ \delta_o$), boundary layer characteristics are (see, e.g. [17]) : boundary layer thickness at $u = u_e$, δ_o = 62 mm, u_e = 12 m/s, u_o^* = .47 m/s ; $\theta_w - \theta_e$ = 21 K, θ_o^* = 1.27 K, non isothermal layer thickness $\delta_\theta \simeq$ 32 mm, in the case 1 ; $\theta_w - \theta_e$ = 22 K, θ_o^* = 1.07 K, $\delta_\theta \simeq \delta_o$ in the case 2.

2.2 Measurements Methods

Measurements methods have been described with details in previous papers [7, 8, 16]. However, it is interesting to recall that many measurements have been performed with compensated constant current anemometers by "fluctuations-diagram" method [22], which is enlarged to X wires [23]. Figure 2 shows an example of so-obtained diagrams. $r = (\alpha_1 + \alpha_2)/(\beta_1 + \beta_2)$ is the ratio between the longitudinal velocity fluctuations sensibility $(\alpha_1 + \alpha_2)$ and the temperature fluctuations sensibility $(\beta_1 + \beta_2)$. ε is the transverse velocity fluctuations sensibility. It is shown [23] that :

$$\frac{\overline{(e'_1 + e'_2)^2}}{(\beta_1 + \beta_2)^2} = \frac{\overline{u'^2}}{\bar{u}^2} r^2 - 2\frac{\overline{\theta' u'}}{\bar{\theta}\,\bar{u}} r + \frac{\overline{\theta'^2}}{\bar{\theta}^2} \qquad (1)$$

$$\frac{\overline{e'^2_1} + \overline{e'^2_2}}{2\varepsilon(\beta_1+\beta_2)} = -\frac{\overline{u'v'}}{\bar{u}^2} r + \frac{\overline{\theta' v'}}{\bar{\theta}\,\bar{u}} \qquad (2)$$

$$\frac{\overline{(e'_1 - e'_2)^2}}{4\,\varepsilon^2} = \frac{\overline{v'^2}}{\bar{u}^2} \qquad (3)$$

A similar method has been adopted in order to determine some spectral densities of temperature F_θ, of velocity F_u and the cospectra relating to θ', u' and v' [24, 8].

Moreover the temperature fluctuations measurements close to the wall have been carried out with a compensated constant current anemometer and platinum cold wires, 1 μ m in diameter ; the heating intensity is .15mA ; that corresponds to an overheat coefficient of .0003. With these experimental conditions the relative maximum error owing to u' contamination on $\overline{\theta'^2}$ is about .001. These cold wires have an aspect ratio of about 600 so that the end effects play a minor part.

3. SUCTION EFFECT ON THE STRUCTURE OF THE INTERNAL AND EXTERNAL ZONES OF THE BOUNDARY LAYER

3.1 Internal Zone

Simplifying hypothesis. The main hypothesis, which is made, is to suppose in internal zone the variations with regard to x are negligible. That involves, from the equations (14), (15), (18) of chapter 4.

$$\bar{\rho}\,\tilde{v} = \rho_w\, v_w \qquad (4)$$

$$\tau - \rho_w\, v_w\, \bar{u} = \tau_w \qquad (5)$$

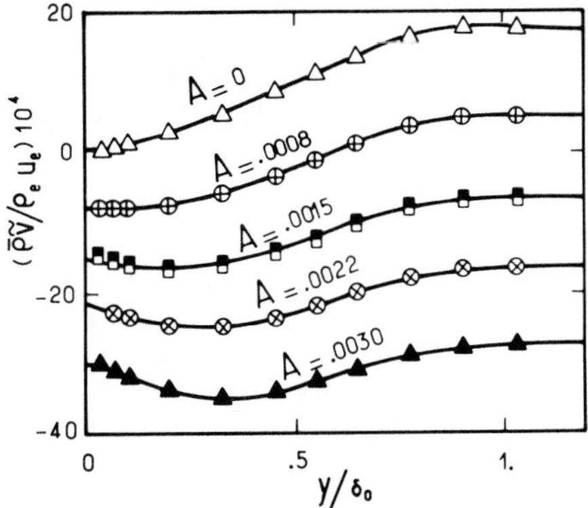

Fig. 3. Mass flux density perpendicular to the wall

$$q - \rho_w v_w C_p (\theta_w - \bar{\theta}) = q_w \tag{6}$$

The velocity $\tilde{v} = \overline{\rho v}/\bar{\rho}$ is the mass weighted mean velocity [25]. τ is the total shear stresses :

$$\tau = \bar{\mu} \frac{\partial \bar{u}}{\partial y} - \bar{\rho}\, \overline{u'v'} \tag{7}$$

q is the total heat flux :

$$q = - \bar{k} \frac{\partial \bar{\theta}}{\partial y} + \bar{\rho}\, C_p\, \overline{\theta' v'} \tag{8}$$

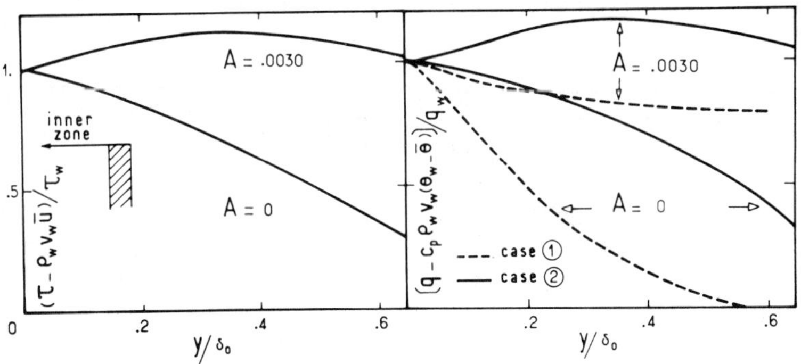

Fig. 4. Total shear stresses and heat fluxes

Experiments (Fig. 3 and 4) show that used suction rates lead to a better verification of the hypothesis expressed by the relations (4), (5), (6). Let us notice that it is necessary to take into account deviated momentum and heat flux when suction is applied.

Temperature and velocity profiles. Using (4) and (5), and a mixing-length hypothesis independent of suction rate, one obtains different expressions for the mean velocities [7]. Among these, the expression proposed by Stevenson [4] expresses the best experimental results obtained on the flat plate [Fig. 5] :

$$\frac{2}{A^+}\,[1 - (1-A^+u^+)^{1/2}] = \frac{1}{K}\,\mathrm{Log}\,Y^+ + C_o \qquad (9)$$

$u^+ = \bar{u}/u^*$; The Karman constant K, and C_o are supposed independent of suction. Taking into account the Boussinesq hypothesis, and assuming constant the turbulent Prandtl number, the relations (5), (6), (9) lead to the mean temperature profile expression [7, 8, 9] :

$$\frac{2}{A^+}\,[1 - (1-A^+\theta^+)^{1/(2Pr_t)}] = \frac{1}{K}\,\mathrm{Log}\,Y^+ + F_o \qquad (10)$$

where $\theta^+ = (\theta_W - \bar{\theta})/\theta^*$; F_o is independent of A^+.

So, as indicated on the figure 5, one can deduce, in the fully turbulent zone, the temperature profiles with suction from the velocity profile without suction. In the case 1, the experimental values do not fit only one curve ; it seems to be linked to the fact that thermal boundary layer is not completely developed.

Viscous sublayer and buffer layer. In order to point out the various sequences very close to the wall, u' and θ' fluctuations recordings are indicated on the figure 6. In fact, to emphasize the greatest amplitude fluctuations, the cubes u'^3 and θ'^3 have been given.

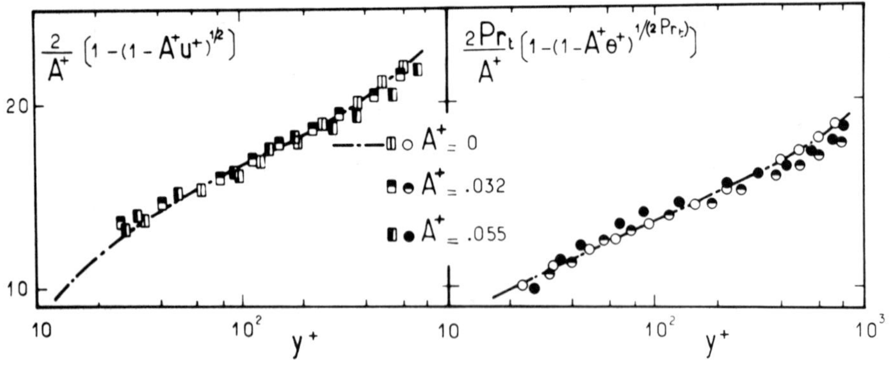

Fig. 5. Mean temperature and velocity profiles (case n°2)

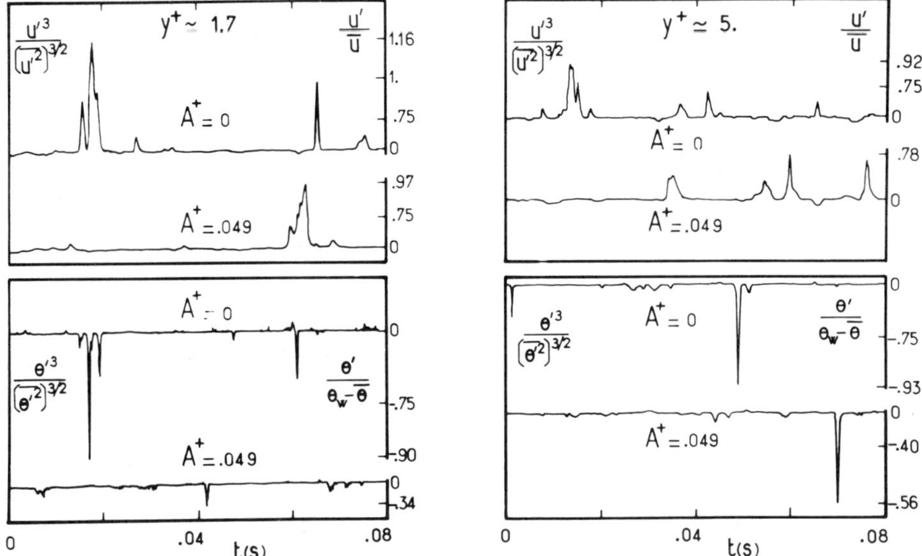

Fig. 6. Recordings of longitudinal velocity and temperature fluctuations

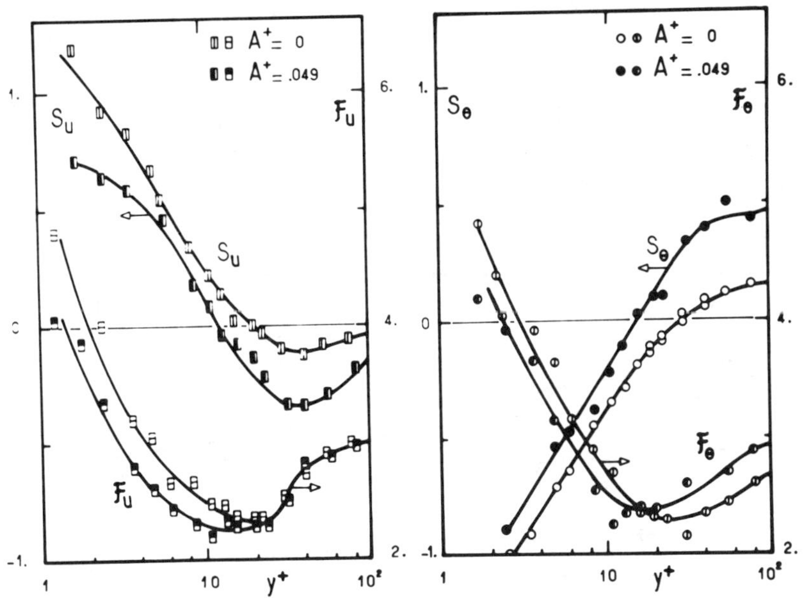

Fig. 7. Skewness and flatness factors

In the viscous sublayer, only colder inflows are detected. They originate from zones a little more distant from the wall. That expresses the high values of the skewness factors S_u and S_θ (Fig. 7).

These sequences correspond to the sequences that Corino and Brodkey [26] have called sweep events. That confirms the intermittent character of the sublayer which involves the high values of the flatness factors $\mathcal{F}_u$ and $\mathcal{F}_\theta$ (Fig. 7).

Concerning the u' component, suction does not appreciably reduce the number of very high rate perturbations $(>10\ (\overline{u'^2})^{1/2}$

Moreover, u' lower levels are notably reduced (25 %). Besides, u' high levels are practically not reduced while these of θ' undergo a very important diminution. That is more prominent when approaching the wall. Then θ' depending on three components u', v', w' of fluctuating velocity vector [8], we can assume [16] that suction mainly reduces w', because v' is of lower magnitude. So, in the sublayer, since, without suction, the turbulent flow can be considered twodimensional, it would become quasi-lamellar with suction. That is in good agreement with the increase, by suction, of large u' fluctuations duration (Fig. 6), expressed by the longitudinal stretching of structures. Analogous conclusions can be drawn from the anisotropy criterion defined in 3.3 (Fig. 20). Concerning the number of turbulent perturbations, it can be thought [27], that it is linked to the frequency N_L which, with the Taylor hypothesis, is relating to the integral length L.

The value of $N_L^* = N_L\ \nu/u^{*2}$ at the wall -equal to about .006 without suction (Fig. 8)- is to be brought nearer to the value of the inflow number $N_a^* = .010$, directly obtained in the same conditions [16]. So, it appears, from this figure, that suction reduces the inflow number which is directly linked to the process of generation of turbulence close to the wall.

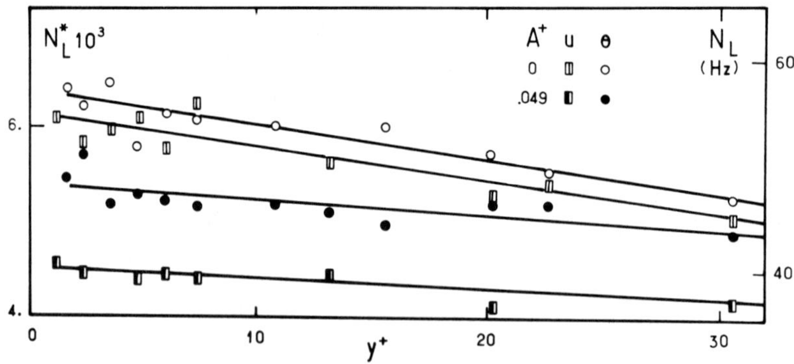

Fig. 8. Integral frequency

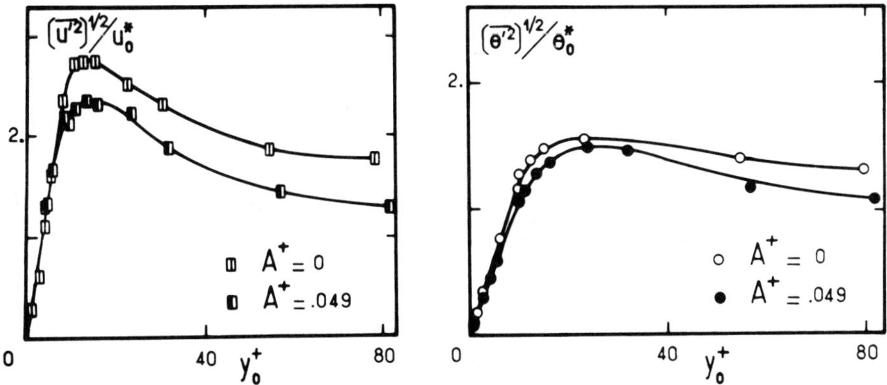

Fig. 9. Standard deviations of longitudinal velocity and temperature fluctuations

That reduction is perhaps due to the decrease of the transverse turbulent motion, in a plan parallel to the wall, during the inrush phase. That would lead to a larger stability of the flow if the scheme proposed by Laufer [28] is assumed. In this scheme, the coming of large speed flow triggers off a three-dimensional instability in the viscous sublayer region. If we consider the three dimensional scheme indicated by Blackwelder [29] -where the turbulence bursts originate from the internal zone by the interaction of outflows and inflows- suction would have the effect to reduce the three-dimensional character. These results can give an explanation of the reduction by suction, of the turbulent production pointed out by Favre, Dumas and Verollet in 1960 [11].

Figure 9 shows suction effect on the standard deviations $(\overline{u'^2})^{1/2}$ and $(\overline{\theta'^2})^{1/2}$. Let us notice that, without suction, we find again [30] in the sublayer :

$$\frac{(\overline{u'^2})^{1/2}}{u^*} = .3\ y^+$$

Besides, we have [16] :

$$\frac{(\overline{\theta'^2})^{1/2}}{\theta^*} = .15\ y^+$$

The decrease phenomenon of turbulence production can be directly estimated, in the zone which stretches from the wall to $y^+ \simeq 30$, with the statistical equation of the kinetic energy

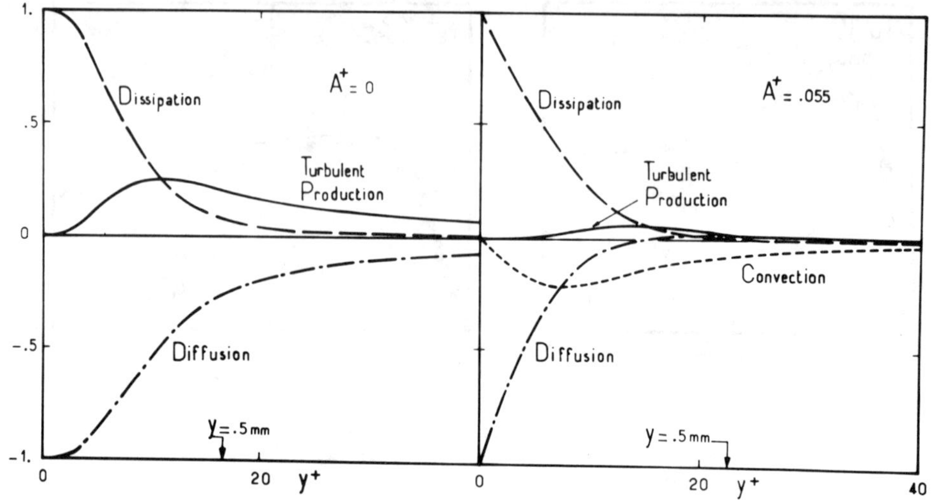

Fig. 10. Balances of kinetic energy equation of mean motion close to the wall (estimate)

balance of the macroscopic motion (see(16)). Indeed, neglecting the derivatives with regard to x [7], this equation is :

$$A^+u^+ \frac{du^+}{dy^+} - \frac{d}{dy^+}(\tau^+u^+) + \left(\frac{du^+}{dy^+}\right)^2 - \frac{\overline{u'v'}}{u*^2}\frac{du^+}{dy^+} = 0 \qquad (11)$$

convection — diffusion — viscous dissipation — turbulent production

Figure 10 shows the results obtained by taking into account relations (4) and (5), and by using the Van Driest profile as velocity distribution, for A = 0, and that deduced from previous measurements [12], for A = .0030.

Let us notice that the maximum of production is approximately located at the same distance of the wall ($y^+ \simeq 11$ for $A^+ = 0$, $y^+ \simeq 15$ for $A^+ = .055$). The results presented on this figure will be commented on chapter 4 with more details.

Fully turbulent region. Experimental results under-mentioned concern the suction effect in the fully turbulent region analysed through turbulent quantities which characterize transfer processes.

Figure 11 displays the total shear stresses τ : suction increases strongly the friction at the wall while suction decreases it anywhere else. The bump in the curve (A = .0030) is due to upstream boundary conditions : a close to the wall region, where the decrease of turbulent production is strong, is adjoining the

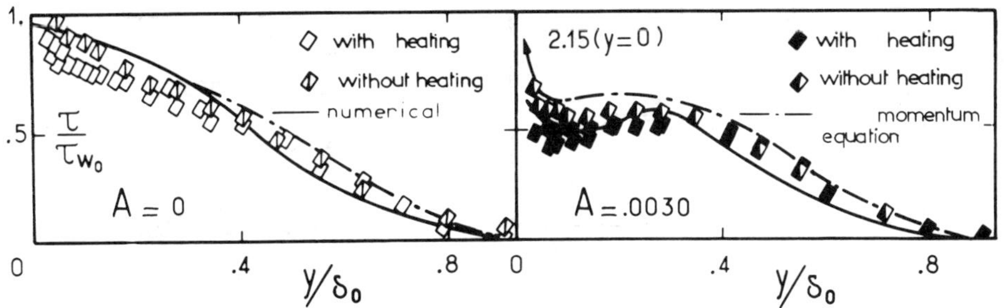

Fig. 11. Shear stresses

external region which is not very altered by suction (Cf 3.2 and 4). This behaviour is the same for the heat fluxes q in the case n° 2 but does not exist in the case n° 1 since the beginning of heating coincides with the beginning of the porous plate (Fig. 12).

On the other hand, suction has not noticeable effect on the correlation coefficients (Fig. 13).

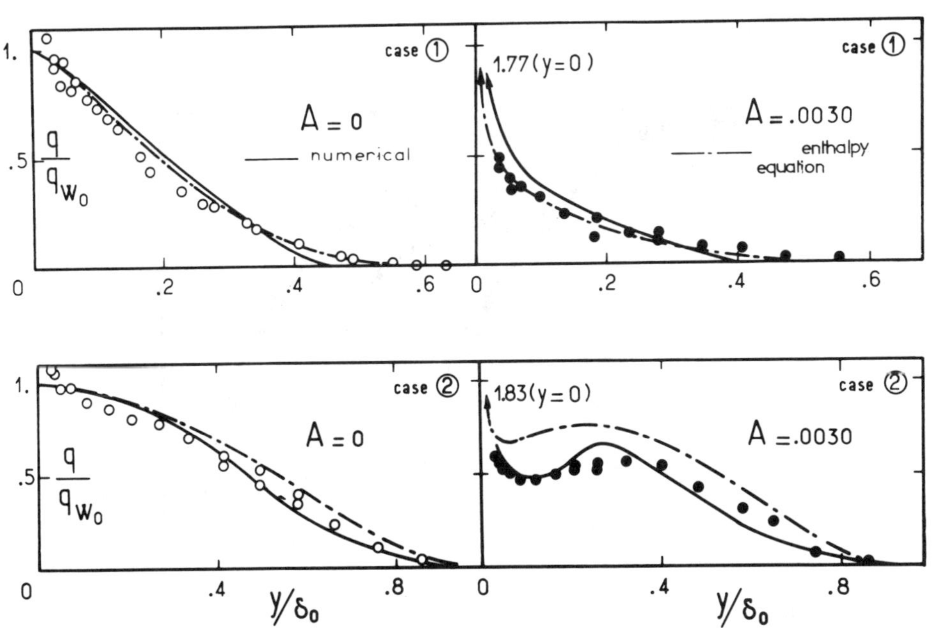

Fig. 12. Heat fluxes

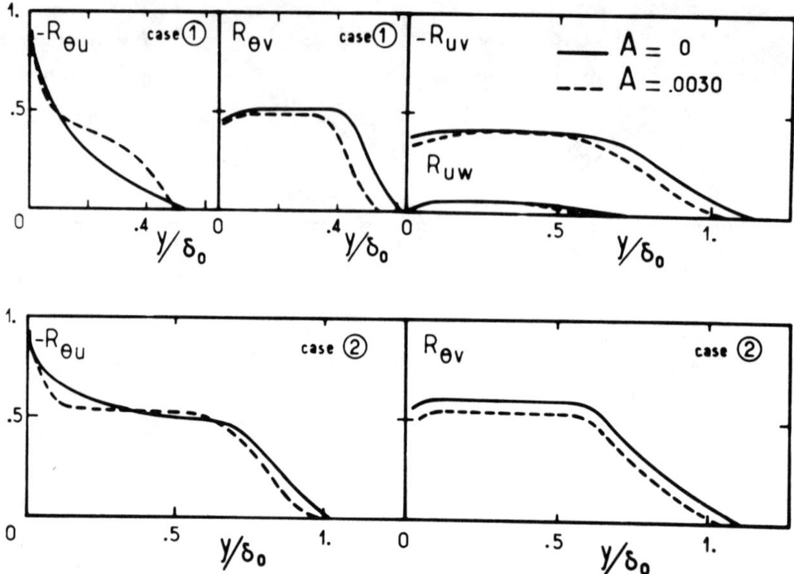

Fig. 13. Correlation coefficients

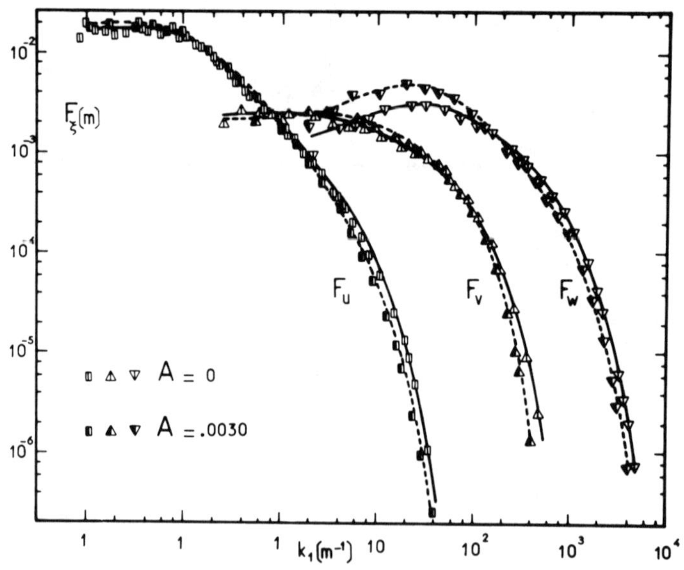

Fig. 14. Velocity spectra (Y/δ_o = .065)

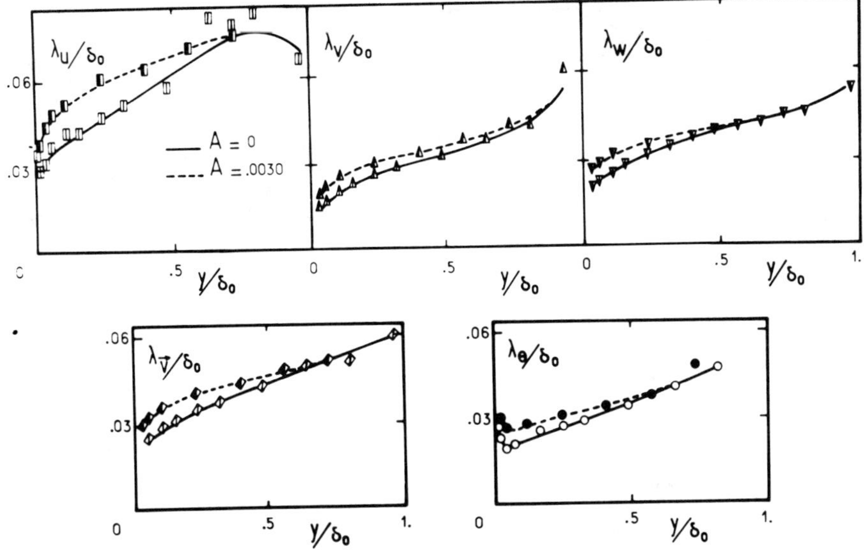

Fig. 15. Taylor microscales

Concerning the spectral repartitions, figure 14 shows a typical example of Taylor spectra F_u, F_v, F_w, normalized to unity, corresponding respectively to u', v', w' fluctuations. They are plotted as a function of the wave number $k_1 = 2\pi n/\bar{u}$, where n is the frequency. Suction decreases the contribution of high wave numbers so that the Taylor microscales increase (Fig. 15). The lengths λ_θ and $\lambda_{\vec{v}}$ are respectively relating to temperature fluctuations θ' and velocity vector fluctuations $\vec{v}'$. The length $\lambda_{\vec{v}}$ is defined by

$$\frac{\overline{q'^2}}{\lambda^2_{\vec{v}}} = \frac{\overline{u'^2}}{\lambda^2_u} + \frac{\overline{v'^2}}{\lambda^2_v} + \frac{\overline{w'^2}}{\lambda^2_w} \tag{12}$$

with $\overline{q'^2} = \overline{u'^2} + \overline{v'^2} + \overline{w'^2}$

Besides, the experiments have established rather conveniently the θ' and $\vec{v}'$ spectra were similar for the wall flows [8, 31], at least concerning the lower frequency range that includes about 80 % of the variance (Fig.16). The $\vec{v}'$spectrum $F_{\vec{v}}$ is defined as :

$$\overline{q'^2}\, F_{\vec{v}} = \overline{u'^2}\, F_u + \overline{v'^2}\, F_v + \overline{w'^2}\, F_w \tag{13}$$

Without suction, the flat part in the spectrum F_θ, for

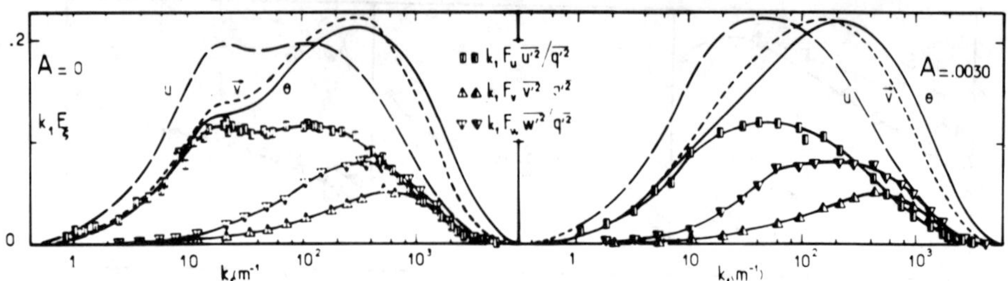

Fig. 16. Spectral analogy

$15\ m^{-1} < k_1 < 30\ m^{-1}$, is explained by the strong correlation between θ' and u' in this spectral range corresponding to the flat portion of the spectrum F_u (Tchen's law : $F_u \propto k_1^{-1}$). It is the same for $F_{\vec{v}}$, by analogy. Since suction decreases production, the spectrum Fu does not verify the k^{-1} law and consequently F_θ has no more a flat part.

An example of suction effect on longitudinal macroscale L_u is shown on figure 17. With or without suction, L_u is about half boundary layer thickness in the central zone, which squares with the slight decrease pointed out. Nearer the wall L_u is increasing with suction, which agrees with the structures stretch already disclosed in the viscous sublayer.

3.2 External Zone

From the results above mentioned it appears that suction has only a little effect on the external zone -see for example shear

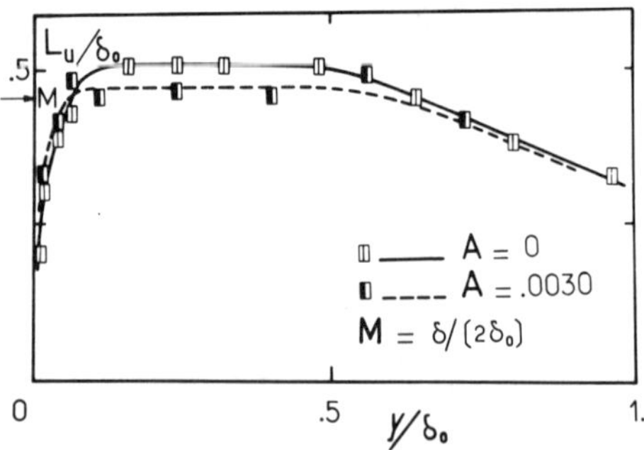

Fig. 17. Longitudinal macroscale

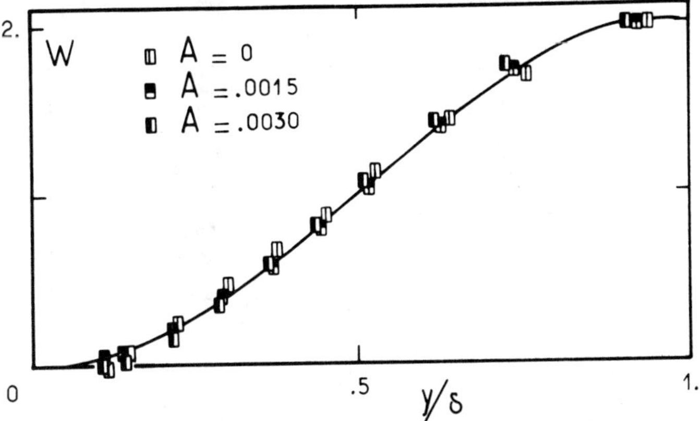

Fig. 18. Wake function

stresses (Fig. 11), heat fluxes (Fig. 12, case n° 2). Suction just reduces slightly the boundary layer thickness (from 62 to 57 mm at usual measurement station, for A = .0030).

Concerning the mean velocity $\bar{u}$, it is generally written in this region of the flows :

$$\bar{u} = \bar{u}_c + \bar{u}_s$$

where $\bar{u}_c$ is the constraint velocity deduced from the law of the wall and $\bar{u}_s$ is the wake velocity. From Coles [32], one obtains :

$$\frac{\bar{u}_s}{u^*} = \frac{\pi(x)}{K} \, W(\frac{y}{\delta})$$

W is the wake function which can be considered as universal and $\pi(x)$ a parameter which depends on pressure gradient. Black and Sarnecki [3] have extended this hypothesis to suction or injection wall flows : they suppose the W function is unchanged and that suction has an effect only on the parameter π. This is proved by the experimental results given here (Fig. 18) in which the values of π are .58, .51 and .41 respectively for A = 0, A = .0015 and A = .0030. For a pressure gradient equal to zero Coles finds π = .55.

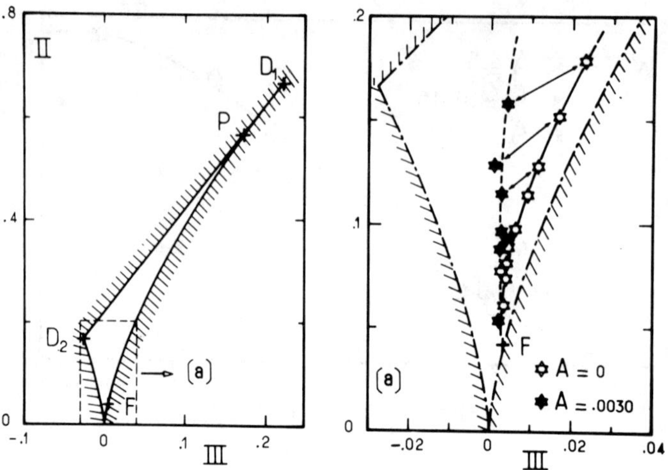

Fig. 19. Invariants

3.3 Miscellaneous Results

Suction effect on the anisotropy of turbulent field. Since Reynolds stresses $\overline{u'_i \, u'_j}$ are decreasing when suction is applied [17] (see for example Fig. 11), it was attractive to determine if this decrease involves or not a tendency to isotropy of turbulent field [33, 34]. In particular, we have considered the invariants

$$I = b_{ii} = 0$$

$$II = b_{ik} \, b_{ki} = \frac{1 + (\sigma_2/\sigma_1)^2 + (\sigma_3/\sigma_1)^2}{(1 + (\sigma_2/\sigma_1) + (\sigma_3/\sigma_1))^2} - \frac{1}{3}$$

$$III = b_{ik} \, b_{kl} \, b_{li} = \frac{1 + (\sigma_2/\sigma_1)^3 + (\sigma_3/\sigma_1)^3}{(1 + (\sigma_2/\sigma_1) + (\sigma_3/\sigma_1))^3} - II - \frac{1}{9}$$

of deviator tensor

$$b_{ij} = \frac{\overline{u'_i \, u'_j}}{\overline{q'^2}} - \frac{1}{3} \, \delta_{ij}$$

σ_i are the principal components of Reynolds stresses tensor ($\sigma_1 > \sigma_2$ and σ_3).

Figure 19 shows II as a function of III, according to diagram proposed by Lumley and Newman [35]. From previous equations, the point O corresponds to an isotropic turbulent flow, D_1 (2/9, 2/3)

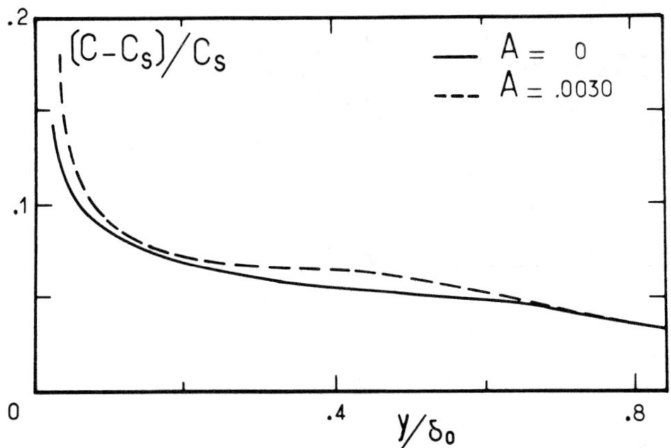

Fig. 20. Anisotropy criterion

corresponds to one-dimensional turbulence, D_2 (-1/36, 1/6) corresponds to isotropic two-dimensional turbulence. Along the diagonal straight line (D_1D_2: 18 III - 9 II + 2 = 0) the turbulence is two-dimensional. The right and left curved boundaries (D_1OD_2 : $II^3 = 6\,III^2$) correspond to axisymmetric turbulence. Besides all homogeneous turbulence must be found within the hachured area. Moreover, we have established [34] the point P (.171, .564) corresponds to the state of the turbulence for the distance from the wall tending to zero and the point F (1/288, 1/24) corresponds to induced turbulence.

The experimental values (Fig. 19 a) point out a large effect of suction in this representation. However, concerning tendency to isotropy, interpretation of these results appears to be impossible if the choice of a physically convenient metric is not taken into account. Therefore a global criterion of anisotropy degree has been selected [34] from geometrical considerations. Since, for an isotropic turbulent field, the ellipsoïd of Reynolds stresses tensor must be a sphere, the proposed criterion is based on

$$C = S/V^{2/3}$$

(S = surface, V = volume of ellipsoïd) which is minimum for a sphere ($C_s = (36\pi)^{1/3}$). The degree of anisotropy is given by $(C - C_s)/C_s$. This ratio is increased by suction (Fig. 20). This result is corroborated by spectral analysis (Cf 3.1, Fig. 14) : suction increases the contribution of large structures which characterize the flow anisotropy. This increase of anisotropy is in good agreement with a previously mentioned result (Cf. 3.1) : in viscous sublayer, flow would become quasi-lamellar owing to suction.

Suction effect on a, a_θ, Pr_t, B. The parameters

$$a = -\overline{u'v'}/\overline{q'^2} \ , \ a_\theta = \overline{\theta'v'}/((-\overline{u'v'})\overline{\theta'^2})^{1/2}$$

used in prediction methods, as Bradshaw and Ferriss method [39], are

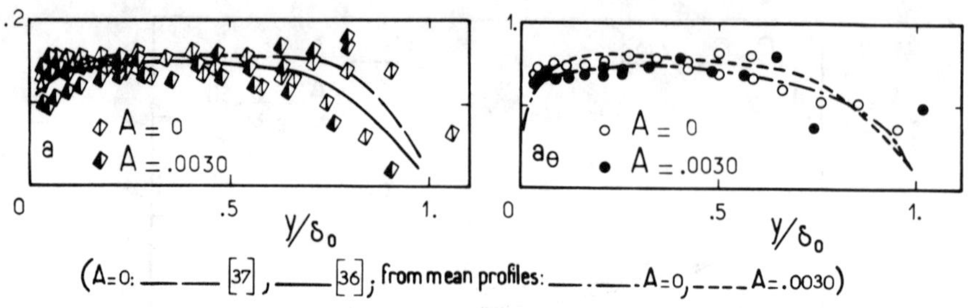

Fig. 21. a and a_θ distributions

little modified by suction (Fig. 21). It is the same for the turbulent Prandtl number Pr_t (Fig. 22). These quantities are not constant through the boundary layer. But the parameter B proposed [8]

$$B = \left(\frac{\overline{q'^2}}{\overline{\theta'^2}}\right)^{1/2} \frac{|\partial\bar{\theta}/\partial y|}{\partial\bar{u}/\partial y}$$

does not depend on suction and on distance to the wall in the major part of the boundary layer. B is related to the previous parameters by :

$$B = Pr_t \; a_\theta / a^{1/2}$$

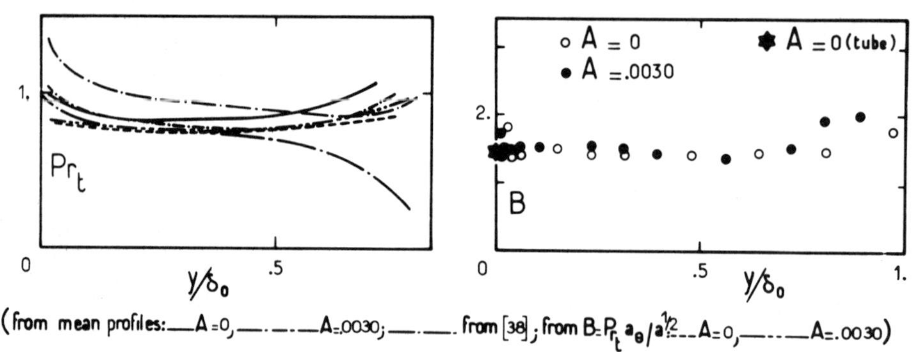

Fig. 22. Pr_t and B distributions

4. ANALYSIS OF SUCTION EFFECT ON TURBULENCE MECHANISM FROM LOCAL STATISTICAL BALANCES

In order to analyse suction effect on the various transfer processes, on the different zones of the boundary layer and on their interaction, statistical balances have been carried out [7]. Some typical examples are shown here.

The statistical momentum equations used have been obtained [7] from Favre general equations [25]. The velocity $\tilde{v}$ can be deduced from continuity equation :

$$\frac{\partial \bar{\rho}\bar{u}}{\partial x} + \frac{\partial \bar{\rho}\tilde{v}}{\partial y} = 0 \qquad (14)$$

by means of mean temperature and longitudinal velocity profiles. For $\bar{v}$ determination it is necessary to take $\overline{\rho' v'}/\bar{\rho}$ into account :

$$\tilde{v} = \bar{v} + \overline{\rho' v'}/\bar{\rho}$$

This term is very large with respect to $\bar{v}$, this at the very least close to the wall, and with no suction (Fig. 23). The X momentum equation is only reported here :

$$\bar{\rho}\bar{u}\,\frac{\partial \bar{u}}{\partial x} + \bar{\rho}\tilde{v}\,\frac{\partial \bar{u}}{\partial y} + \frac{dp_e}{dx} - \frac{\partial \tau}{\partial y} + \frac{\partial}{\partial x}\,\bar{\rho}(\overline{u'^2} - \overline{v'^2}) = 0 \qquad (15)$$

4.1. Balances of Kinetic Energy Equation of Mean Motion

$$\underbrace{\frac{1}{2}\left(\bar{\rho}\bar{u}\,\frac{\partial \bar{u}^2}{\partial x} + \bar{\rho}\tilde{v}\,\frac{\partial \bar{u}^2}{\partial y}\right)}_{\text{Convection}} + \underbrace{\bar{u}\,\frac{dp_e}{dx}}_{\text{pressure work}} - \underbrace{\frac{\partial}{\partial y}\left(\bar{\mu}\bar{u}\,\frac{\partial \bar{u}}{\partial y}\right)}_{\text{viscous diffusion}} + \underbrace{\bar{\mu}\left(\frac{\partial \bar{u}}{\partial y}\right)^2}_{\text{viscous dissipation}}$$

$$+ \underbrace{\frac{\partial}{\partial y}(\bar{\rho}\,\bar{u}\,\overline{u'v'})}_{\text{turbulent diffusion}} - \underbrace{\bar{\rho}\;\overline{u'v'}\,\frac{\partial \bar{u}}{\partial y}}_{\text{turbulent production}} = 0 \qquad (16)$$

The values of these various terms, normalized by $[(\tau_w)_o^3 / \rho_w (\delta_2)_o^2]^{1/2}$ -where $(\delta_2)_o$ is the momentum thickness without suction- are shown on figure 24 as a function on the $y/(\delta_2)_o$.

The main difference between the two cases, A = 0 and A = .0030, lies in a strong modification of convection and diffusion curves in the fully turbulent zone, with a very strong increase of viscous diffusion and dissipation very close to the wall. The turbulent production decreases but the curve trends are not modified.

Without suction, a little part of the energy, which is taken out of the external zone, is transformed, where it is, into turbulent kinetic energy (Fig. 24). Nearly all the kinetic energy of mean motion is diffused in the inner zone. In this region, this energy is transformed into heat by dissipation for the benefit of

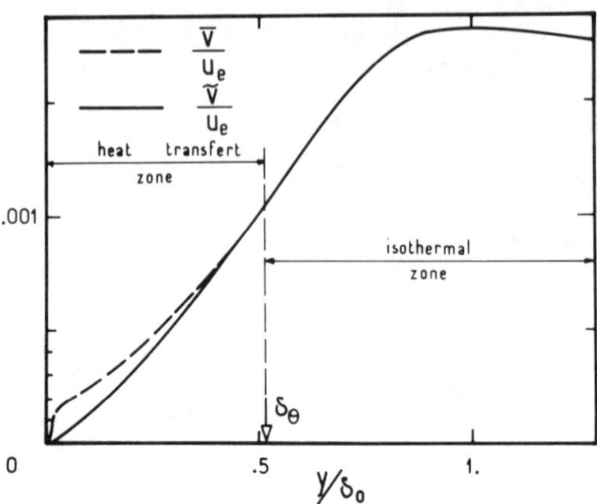

Fig. 23. Distributions of $\tilde{v}$ and $\bar{v}$ without suction (case n°1)

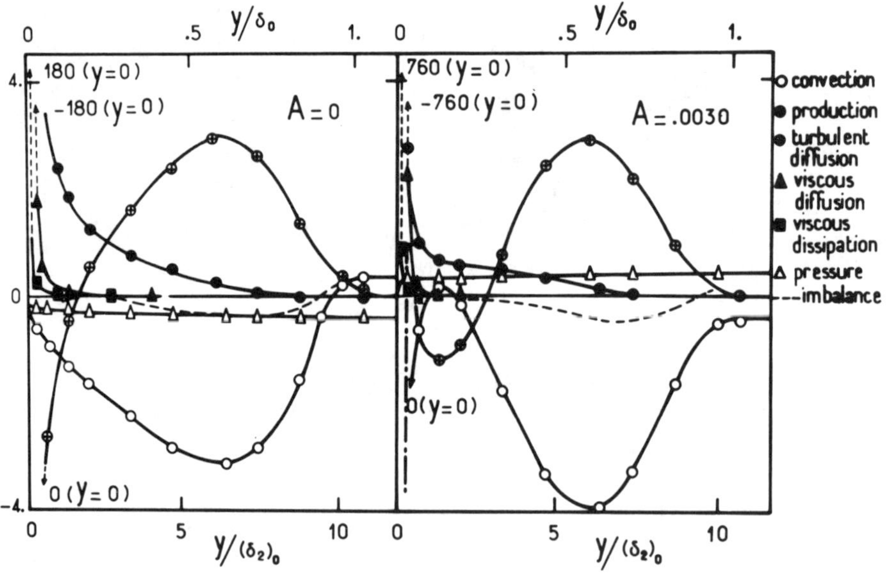

Fig. 24. Balances of kinetic energy equation of mean motion

turbulence. This viscous dissipation is strong, very close to the wall (Fig. 10) and negligible in the fully turbulent zone. The maximum turbulent production is located at about $y^+ = 15$ where it is practically equal to the dissipation. At this distance (Fig. 7), the skewness factors S_u and S_θ are equal about to zero and the flatness factors $\mathcal{F}_u$ are minimal ($\mathcal{F}_u \simeq \mathcal{F}_\theta \simeq 2.3$). This can be related to the harmonic waves characterized by S = 0 and F = 1.5. Then, it can be thought that this is an hydrodynamic instability zone as it has been assumed [40].

With suction, the phenomenon is more complex. The energy is diffused in the inner zone but in a very thin layer very close to the wall (Fig. 10). In this layer, the viscous effects play a leading part so that the energy is here transformed into heat by dissipation to the detriment of the turbulent production. The decrease of the production explains the reduction of various turbulent quantities, as for example the Reynolds stresses (Fig. 11). This result is also in good agreement with the diminution of the perturbations number (Fig. 8). Besides, it can be noted that the maximum production is located practically at the same distance to the wall as in the case A = 0 ($y \simeq .3$ mm).

The complexity of results concerning turbulent diffusion and convection is owing to the interaction between boundary layer internal zone and external zone. This one, which depends strongly on upstream conditions, is practically not influenced by suction (Fig. 24) since the boundary layer is fully developed at the beginning of the suction zone. On the other hand the internal zone depends essentially on wall conditions, i.e. on suction.

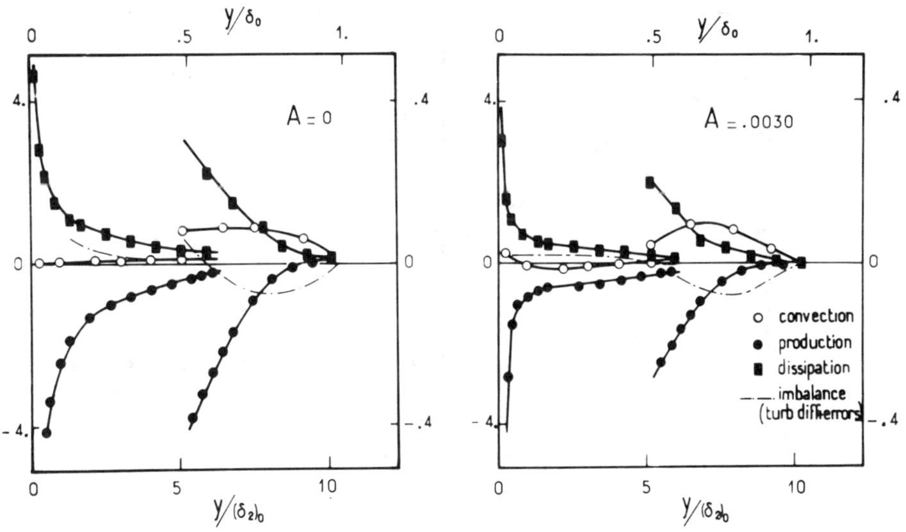

Fig. 25. Balances of kinetic energy equation of turbulent motion

4.2 Balances of Kinetic Energy Equation of Turbulent Motion

$$\frac{1}{2}\left(\bar{\rho}\bar{u}\,\frac{\partial \overline{q'^2}}{\partial x} + \bar{\rho}\tilde{v}\,\frac{\partial \overline{q'^2}}{\partial y}\right) + \bar{\rho}\,\overline{u'v'}\,\frac{\partial \bar{u}}{\partial y} + \frac{\partial}{\partial y}\left(\frac{1}{2}\,\bar{\rho}\,\overline{q'^2 v'} + \overline{p'v'}\right)$$

Convection turbulent production turbulent diffusion

$$- \frac{1}{2}\frac{\partial}{\partial y}\left(\bar{\mu}\,\frac{\partial \overline{q'^2}}{\partial y}\right) + \bar{\mu}\,\overline{\frac{\partial v'_\alpha}{\partial x_\beta}\frac{\partial v'_\alpha}{\partial x_\beta}} = 0 \qquad (17)$$

viscous diffusion dissipation

The viscous diffusion is negligible with respect to dissipation, except close to the wall.

Experimental values of these various terms, normalized by $[(\tau_w)_o^3 / \rho_w (\delta_2)_o^2]^{1/2}$, are shown on figure 25, except the turbulent diffusion term which is included in the imbalance.

The convection of turbulent motion kinetic energy is small in all the boundary layer, so that the major part of energy generated is transformed into heat by dissipation. The dissipation is reduced by suction as the turbulent production but on a smaller scale beyond the fully turbulent zone. The largest values of dissipation seem to be located nearer the wall than it is without suction.

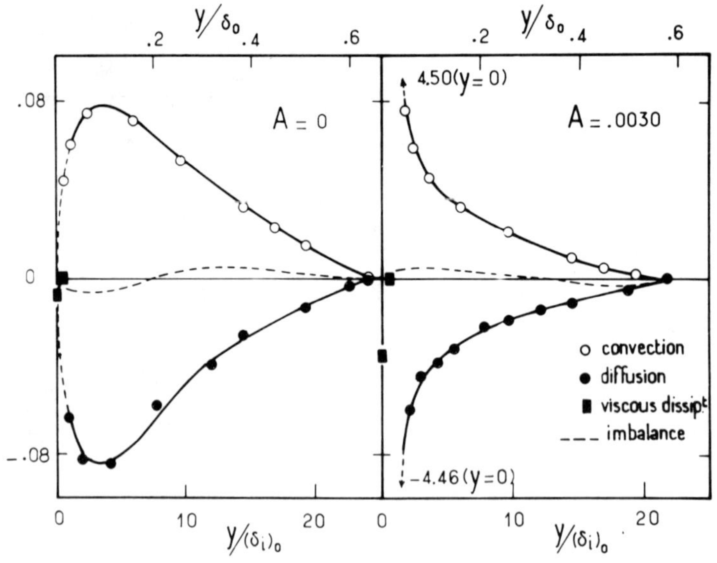

Fig. 26. Balances of enthalpy equation (case n° 1)

4.3 Balances of Enthalpy Equation

$$C_p(\bar{\rho}\bar{u}\frac{\partial\bar{\theta}}{\partial x} + \bar{\rho}\tilde{v}\frac{\partial\bar{\theta}}{\partial y}) - \bar{\mu}(\frac{\partial\bar{u}}{\partial y})^2 - \frac{\partial}{\partial y}(\bar{k}\frac{\partial\bar{\theta}}{\partial y})$$

convection — viscous dissipation — molecular diffusion

$$+ \frac{\partial}{\partial y}(\bar{\rho}\, C_p\, \overline{\theta' v'}) = 0 \qquad (18)$$

turbulent diffusion

The experimental values of these terms, normalized by $(q_w)_o/(\delta_i)_o$ -where $(\delta_i)_o$ is the enthalpy thickness without suction- are plotted on figure 26 for the case n° 1.

The viscous dissipation of mean motion kinetic energy, which is transformed into heat, is not negligible only into a thin layer very close to the wall ($y^+ \lesssim 15$). Consequently the convection, i.e. the enthalpy variation following the macroscopic motion, is mainly due to the molecular and turbulent diffusion.

Without suction, enthalpy variation is maximum in a zone close to the wall ($y \simeq 6$ mm) and remains large enough up to distances far from the wall. This is owing to the proximity of heated part beginning, i.e. to thermal boundary layer which is not developed.

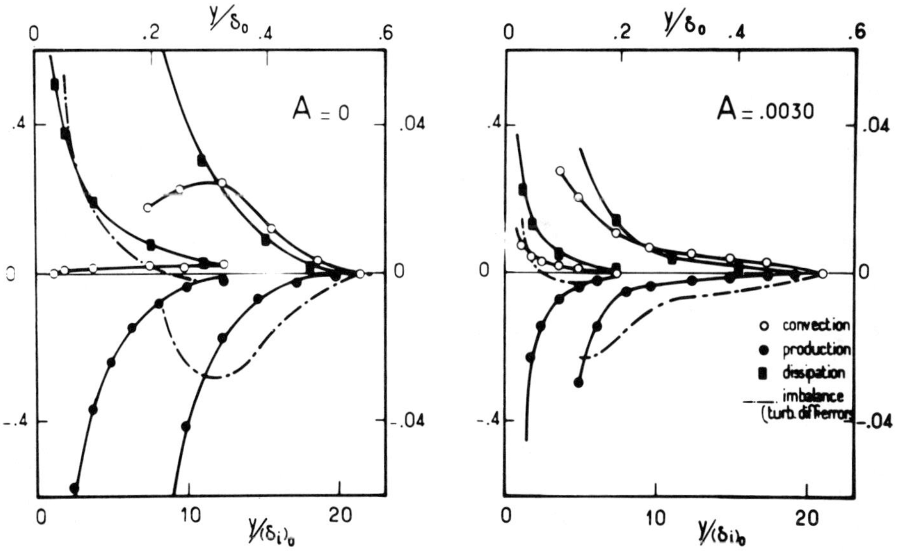

Fig. 27. Balances of temperature variance equation

Suction delays the thermal expansion, an important part of heat being used to increase the deviated fluid temperature. We have shown [7] that the enthalpy variation corresponding to the motion parallel to the wall ($C_p \bar{\rho} \bar{u} \frac{\partial \bar{\theta}}{\partial x}$) is reduced by suction, while the variation corresponding to the motion normal to the wall ($C_p \bar{\rho} \tilde{v} \frac{\partial \bar{\theta}}{\partial y}$) -practically null without suction- is very large. So, the strong heat exchanges are concentrated close to the wall.

4.4 Balances of Temperature Variance Equation

$$\underbrace{\bar{\rho}\bar{u} \frac{\partial \overline{\theta'^2}}{\partial x} + \bar{\rho}\tilde{v} \frac{\partial \overline{\theta'^2}}{\partial y}}_{\text{convection}} + \underbrace{2\bar{\rho}\, \overline{\theta' v'} \frac{\partial \bar{\theta}}{\partial y}}_{\text{turbulent production}} + \underbrace{\frac{\partial}{\partial y}(\bar{\rho}\, \overline{\theta'^2 v'})}_{\text{turbulent diffusion}}$$

$$\underbrace{- \frac{1}{C_p} \frac{\partial}{\partial y} \left(\bar{k} \frac{\partial \overline{\theta'^2}}{\partial y}\right)}_{\text{molecular diffusion}} + \underbrace{2 \frac{\bar{k}}{C_p} \overline{\frac{\partial \theta'}{\partial x_\alpha} \frac{\partial \theta'}{\partial x_\alpha}}}_{\text{"dissipation"}} = 0 \qquad (19)$$

Figure 27 displays the experimental values of these terms, normalized by $\rho_w u_o^* \theta_o^{*2} / (\delta_i)_o$, in the case n° 1. The turbulent diffusion term is included in the imbalance.

Without suction, the results obtained before [41] or after [42] are similar.

The curves trends are analogous to those concerning turbulent kinetic energy (Fig. 25). However, suction effect is more noticeable. It is due to the boundary conditions (case n° 1).

5. CONCLUSION

The suction at the wall introduces a fundamental modification on the classical assumptions concerning the inner layer, i.e. the inertial terms are no more negligible in the momentum equation. The law of the wall $\tau = \tau_w$ (or $q = q_w$) becomes $\tau = \tau_w - \rho v_w \bar{u}$ (or $q = q_w - \bar{\rho} C_p v_w(\theta_w - \bar{\theta})$) taking into account the deviated momentum (or deviated thermal flux). So, the mean velocity profiles are directly affected by suction even at a low rate (e.g. suction rate much lower than the skin friction coefficient), at least in the sublayer and the inertial zone. Let us remark that the suction effects are more complex when suction is applied on a boundary layer, which has been previously developed and as it is the case in these experiments. This complexity is pointed out in the balance of kinetic energy of mean motion. But the external zone is practically not modified. Suction involves a diffusion of kinetic energy in a layer very close to the wall, which is thinner than with no suction. Suction tends to reduce the development of thermal boundary layer when the beginning of heating is coincident with that of suction.

Strong heat exchanges are concentrated very close to the wall.

Measurements show that the production of turbulence is decreasing when the suction increases. Besides, the rate of the bursting process is also reduced by suction. This result and other considerations lead us to assume that suction reduces the instability of the viscous sublayer. Indeed, the structure of the flow would tend to become less three-dimensional.

ACKNOWLEDGEMENTS

The authors write to express their gratitude to M. ASTIER who has carried up all the experimental facilities. They also acknowledge Dr. I. Dekeyser for his scientific collaboration. These studies were supported in particular by Commissariat à l'Energie Atomique contracts.

REFERENCES

1. Mickley, H.S. and Ross, R.C. Squyers, A.L. and Steward, W.E. 1954. Heat mass and momentum transfer for flow over a flat plate with blowing or suction. NACA TN 3208.

2. Kay, J.M. 1948. Boundary layer flow along a flat plate with uniform suction. London, Aero. Res. Council, R and M, n° 2628.

3. Black, T.J. and Sarnecki, A.J. 1958. The turbulent boundary layer with suction or injection. Aero. Res. Council, report 20.501.

4. Stevenson, T.N. 1963. A law of the wall for turbulent boundary layers with suction or injection. College of Aero. Cranfield, rep. aero. n° 166.

5. Tennekes, H. 1964. Similarity laws for turbulent boundary layers with suction or injection. Delft, Doct. Thesis.

6. Simpson, R.L., Kays, W.M. and Moffat, R.J. 1967. The turbulent boundary layer on a porous plate : an experimental study of the fluid dynamics with injection or suction. Stanford Univ., Thermosciences Div., Dpt. Mech. Eng., report H.M.T. 2.

7. Verollet, E. 1972. Etude d'une couche limite turbulente avec aspiration et chauffage à la paroi. Univ. de Provence, I.M.S.T., Marseille, Doct. Thesis. (1977. C.E.N. Saclay, Rap. CEA-R-4872).

8. Fulachier, L. 1972. Contribution à l'étude des analogies des champs dynamique et thermique dans une couche limite turbulente. Effet de l'aspiration. Univ. de Provence, I.M.S.T., Marseille, Doct. Thesis.

9. Verollet, E., Fulachier, L. and Dekeyser, I. 1977. Etude phénoménologique d'une couche limite turbulente avec aspiration et chauffage à la paroi. International J. of Heat and Mass Transfer Vol. 20, n° 2, pp. 107-112.

10. Weissberg, H.L. 1956. Velocity profiles and friction factors for turbulent pipe flow with uniform wall suction. Oak Ridge, Tennessee, Physics report, K-1264.

11. Favre, A., Dumas, R. and Verollet, E. 1960. Couche limite sur paroi plane poreuse avec aspiration. Xème Congrès Int. de Méc. Appl. Stresa. (1961. Paris, Publ. Sc. Techn. Min. Air n° 377).

12. Favre, A. and Dumas, R. Verollet, E. and Coantic, M. 1965. Couche limite turbulente sur paroi poreuse avec aspiration. Naples, Meeting AGARD. (1966 J. de Mécanique, vol. 5, n° 1, pp. 3-28).

13. Andersen, P.S., Kays, W.M. and Moffat, R.J. 1975. Experimental results for the transpired turbulent boundary layer in an adverse pressure gradient. J. of Fluid Mechanics, vol. 69, part 2, pp. 353-375.

14. Schildknecht, M., Miller, J.A. and Meier, G.E.A. 1979. The influence of suction on the structure of turbulence in fully developed pipe flow. J. of Fluid Mechanics, vol. 90, part I, pp. 67-107.

15. Verollet, E. and Fulachier, L. 1969. Mesures de densités de flux de chaleur et de tensions de Reynolds dans une couche limite turbulente avec aspiration à la paroi. C.R. Acad. Sc., T. 268, Série A, pp. 1577-1580.

16. Elena, M. 1975. Etude des champs dynamique et thermique d'un écoulement turbulent en conduite avec aspiration à la paroi. Univ. de Provence, I.M.S.T., Marseille, Doct. Thesis.

17. Fulachier, L., Verollet, E. and Dekeyser, I. 1977. Résultats expérimentaux concernant une couche limite turbulente avec aspiration et chauffage à la paroi. International J. of Heat and Mass Transfer, Vol. 20, n° 7, pp. 731-739.

18. Pimenta, M.M., Moffat, R.J. and Kays, W.M. 1978. The structure of a boundary layer on a rough wall with blowing and heat transfer. New-York, ASME, 78-HT-3.

19. Praturi, A.K. and Brodkey, R.S. 1978. A stereoscopic visual study of coherent structures in turbulent shear flow. J. of Fluid Mechanics, vol. 89, p. 251.

20. Blackwelder, R.F. 1978. The bursting process in turbulent boundary layer. Lehigh Univ. Workshop on coherent structures in turbulent boundary layers.

21. Elena, M., Fulachier, L. and Dumas, R. 1979. Etude experimentale des apports et des éjections de fluide dans une couche limite turbulente. AGARD Conf. Proc. n° 271 - pp. 2-1, 2-21.

22. Kovasznay, L.S.G. 1953. Development of turbulence measuring equipment. Washington. NACA T.N. 2939.

23. Verollet, E. 1962. Contribution aux méthodes de mesure de turbulence de vitesse et de température par l'anémomètre à fils chauds. Fac. Sc., I.M.S.T., Marseille, Doct. 3ème cycle thesis. (1969. Paris. Publ. Sc. Techn. Min. Air n° 449).

24. Fulachier, L. 1971. Répartitions spectrales des fluctuations de flux de chaleur dans une couche limite turbulente. Paris. C. R. Acad. Sc. T. 272 série A, pp. 1022-25.

25. Favre, A. 1965. Equations des gaz turbulents compressibles. J. de Mécanique, vol. 4, pp. 361-421.

26. Corino, E.R. and Brodkey, R.S. 1969. J. of Fluid Mechanics, vol. 37, part I, pp. 1-30.

27. Elena, M. and Dumas, R. 1978. Champs dynamique et thermique d'un écoulement turbulent en conduite avec aspiration à la paroi. Conf. Proc. VI° International Congress on Heat Transfer, FC(b)-11, pp. 239-244, Toronto.

28. Laufer, J. 1972. Recent developments in turbulent boundary layer research. London. Symp. math., vol. 9, Ac. Press.

29. Blackwelder, R.F. 1979. Coherent structures in turbulent boundary layers. AGARD Conf. Proc. n° 271, pp. 24-1, 24-7.

30. Laufer, J. 1953. The structure of turbulence in fully developed pipe flow. NACA T.N. 2954, report 1174.

31. Fulachier, L. and Dumas, R. 1976. Spectral analogy between temperature and velocity fluctuations in a turbulent boundary layer. J. of Fluid Mechanics, vol. 77, part II, pp. 257-277.

32. Coles, D. 1956. The law of the wake in turbulent boundary layer. J. of Fluid Mechanics, vol. 1, part II, pp. 191-226.

33. Fulachier, L., Dekeyser, I. and Verollet, E. 1977. Effet de l'aspiration sur l'anisotropie du champ turbulent d'une couche limite. Grenoble, IIIème Congrès Français de Mécanique.

34. Dekeyser, I., Fulachier, L. and Verollet, E. 1977. Anisotropie du champ turbulent d'une couche limite. Influence de l'aspiration à la paroi. J. de Mécanique, vol. 16, n° 4.

35. Lumley, J.L. and Newman, G.R. 1977. The return to isotropy of homogeneous turbulence. J. of Fluid Mechanics, vol. 82, part I, pp. 161-178.

36. Gaviglio, J. 1958. Sur quelques problèmes de mesures de turbulence effectuées à l'aide de l'anémomètre à fils chauds parcourus par un courant d'intensité constante. I.M.S.T., Marseille, Doct. Thesis.

37. Klebanoff, P.S. 1955. Characteristics in a boundary layer with zero pressure gradient. Washington, NACA, report 1247.

38. Simpson, R.L., Whitten, D.G. and Moffat, R.J. 1970. An experimental study of the turbulent Prandtl number of air with injection and suction. International J. Heat and Mass Transfer, vol. 13, n° 1, pp. 125-143.

39. Bradshaw, P. and Ferriss, D.H. 1968. Calculation of boundary layer development using the turbulent energy equation : "Heat transfer with small temperature differences." N.P.L. Aero. report 1271.

40. Blackwelder, R.F. and Kaplan, R.E. 1971. Intermittent structures in turbulent boundary layers. London, AGARD Conf. Proc. n° 93, pp. 5.1-5.7.

41. Johnson, D.S. 1955. Turbulent heat transfer in a boundary layer with discontinuous wall temperature. Baltimore Depart. of Aero., the Johns Hopkins Univ.

42. Antonia, R.A., Danh, H.Q and Prabhu, A. 1977. Response of a turbulent boundary layer to a step change in surface heat flux. J. of Fluid Mechanics, vol. 80, pp. 153-177.

Transport Processes in Turbulent Boundary Layer under High-Level Free Stream Turbulence

B. P. MIRONOV, V. N. VASECHKIN, V. N. MAMONOV, and N. I. YARYGINA
Institute of Thermophysics
Siberian Branch of the USSR Academy of Sciences, Novosibirsk

ABSTRACT

Experimental investigation of effect of free-stream turbulence on the processes of heat transfer and friction in turbulent boundary layer on solid and porous surface also downstream of porous cooling region are reported. It has been shown that at large blowing the free-stream turbulence substantially intensifies transport processes.

1. INTRODUCTION

Many practical problems associated with production of power plants, increasing the chemical engineering process intensity, etc. require systematic studies of complex influence of various factors on the turbulent transport processes. Joint effect of compressibility, nonisothermality, substance crossflow, longitudinal pressure gradient, etc. on the turbulent heat and mass transfer and friction is considered in [1,2]. One of the most common perturbing factors affecting the turbulent transfer processes practically in all tangible designs is the high turbulence level of the mainstream. A relatively small increase of skin friction, heat and mass transfer at the free stream turbulence level of several percent encouraged the systematic studies in the field of the free stream turbulence effect on the laminar-to-turbulent transition where this effect was more perceptible.

However, under the joint influence of various factors depending on their variation range there can occur such situationswhen the transport process intensity under the effect of one of the factors, changes by an order and more. Examples are found in the intensity variation of heat and mass transfer processes on a penetrable surface at large injections and different temperature factors. In this case at the expense of an appreciable change of magnitudes of the critical injection parameters the heat output changes essentially [1,2].

Intensifying effect of the high level mainstream turbulence upon total heat transfer and friction of streamline bodies has long been known [3] and many research works are devoted to this problem. In the transverse flow around bodies a very strong effect appears in the neighbourhood of the frontal point. It is associated with penetration of turbulent vortices into internal regions of the laminar boundary layer. Theoretical studies reveals a dependence of penetrating power of vortices on the ration of their scales and boundary layer thickness [4,5,6]. A similar picture is observed in a laminar flow on a flat plate. Besides the heat transfer intensification in a laminar boundary layer in a high-level turbulence flow there occurs a displacement of the transition point upstream and a contraction (decrease) of the transition zone. In early works [7,8,9] the influence of the free stream turbulence on heat transfer in the turbulent boundary layer was not revealed. At the same time dynamic characteristics of the boundary layer essentially depend on external turbulent perturbations. The boundary layer thickness increases, velocity profiles become more filled, intensity of longitudinal and transverse velocity pulsations grows, the longitudinal and transverse turbulence scales change, skin friction factor increases. In [10] it is shown that for the recent different experiments, beyond the square-meshed grids the following expression is valid:

$$C_f / C_{fo} = 1 + C\,Tu_o \qquad (1)$$

which takes account of the effect of the free stream turbulence on the skin friction factor C_f. The constant value C is equal to 3.2. Some experimental data confirm that the value (C_f / C_{fo}) becomes nearly constant at $Tu_o > 0.08$ 10 . Relation (1) was adequately confirmed in [11] and 12 , but in the latter the value of the constant $C = 2$. In [12] it is pointed out that relations like (1) are tentative ones before obtaining additional experimental data.

In fact, an analysis of the experiment made in 13 showed that at fixed Tu_o the relation C_f / C_{fo} quickly diminishes with growing L / δ, where L is dissipation scale of external flow turbulence, and δ is thickness of dynamic boundary layer. This conclusion was confirmed in [12] though estimation of L / δ there was made using the equation of turbulent energy of decaying uniform turbulence, the equation being transformed so that for this estimate it was sufficient to measure only the longitudinal component of pulsation velocity. In the same paper and in [11] it was shown that in a flow with high level turbulence, the Reynolds analogy parameter grows since turbulization of the external flow results in more intense rise of heat transfer as against friction. In this case, the friction and heat transfer growth is asymptotic. The absence of the free stream turbulence effect on heat transfer in the turbulent boundary layer in early works [7,8,9] the authors of 12 explain by low Reynolds numbers and high turbulence scales. From the above consideration it is seen that the determining factors of the mechanism of the free stream turbulence effect on heat transfer processes are the scale and intensity of the main flow turbulence. However, an important role is played by various additional conditions of the external flow past body. In [14] it is shown that at $Pr \gg 1$, when the thermal boundary layer is submerged in the dynamic layer, heat transfer is affected by flow turbulization only slightly. A similar picture of interaction of thermal and dynamic layers is observed in the presence of the adiabatic starting length.

In the literature there are only few works known that deal with the studies of protection properties of a gas screen at high level of the external flow turbulence [15,16,17]. The former two are devoted to the study of the effectiveness of slot tangential and the latter to the porous cooling effectiveness in a strongly turbulized boundary layer. It was found that protection properties of both cooling types decreased. In [17] working substance was water ($Pr \approx 5.4$).

The present paper describes different cases of the effect of the external flow turbulence level on friction and heat transfer processes under combined action of two perturbing factors. In real designs such turbulence may arise due to the combustion chamber operation, specific character of gas flow in the gas turbine blading, etc. Investigations were performed on an impenetrable surface at stepwise change of wall boundary conditions (thermal stage), on a penetrable surface and beyond the porous injection region.

2. HEAT TRANSFER FROM SOLID SURFACE

Study of heat transfer at zero injection was performed on the surface of a cylinder subjected to a longitudinal air flow in a wind tunnel with a working channel 205 X 205 X 600 mm^3. The test section is described in detail in [18]. Cylinder diameter is 38.9 mm, length of heat transfer region is 150 mm. 12 thermocouples were built into the heat transfer region; the first one at a distance of 5 mm from the beginning, next eight thermocouples spaced 10 mm apart, and the last three spaced 20 mm apart. Before the heat transfer section a cone 180 mm in length was fixed. Heat transfer was measured for two positions of the model with respect to the grid: at a distance $(x_g - x_o) = 290$ mm and 470 mm at the velocity 50 to 51 m/s.

The turbulence in the mainstream was generated by a biplanar grid with 40 mm square meshes, the grid was welded from steel cylinderical rods 12 mm in diameter. In the grid knots and along its perimeter two-petale elastic flags were fixed, this provided a high turbulence intensity. (Fig.1).

Turbulence measurements lengthwise at a height 60 mm from

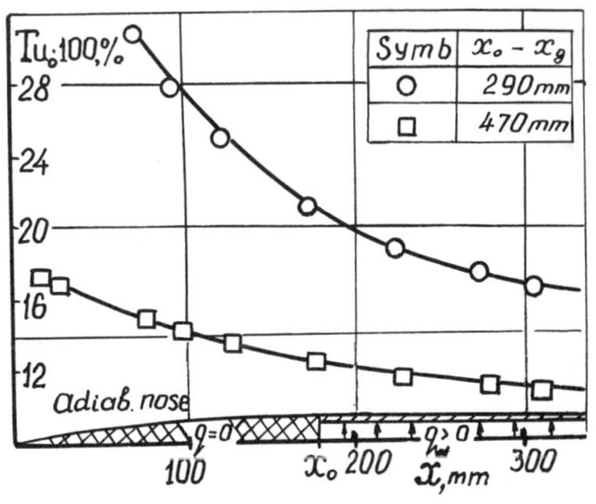

Fig. 1.
Variation of turbulence level of external flow along test section. Turbulence generator is grid with flexible flags 100 mm in length, grid mesh M = 40 mm, rod diameter 12 mm.

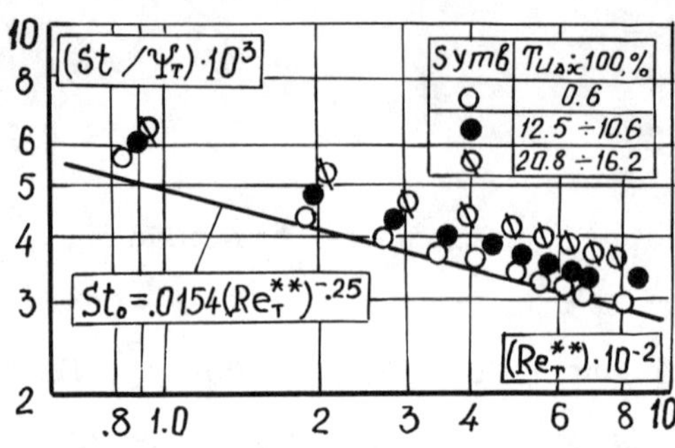

Fig. 2.
Heat transfer from a cylinder with longitudinal flow around it under different conditions of external flow turbulence. Starting length is $x_o = 180$ mm.

the channel axis and in several sections across the channel above the heat transfer region were performed using measuring equipment of DISA Co. production. A sensitive element 8μm in diameter and 4-5 mm in length was placed normal to the direction of the main air flow. The probe registered the pulsation component of the velocity vector. At turbulence level lower than 0.2 its difference from the pulsation component of the longitudinal velocity does not exceed 3-4 %. Hence, further the turbulence level implies the intensity of pulsation component of the longitudinal velocity $\sqrt{\overline{u'^2}}/u_o$.

The average velocity profiles in the boundary layer were measured by an impact tube with 0.5 mm high inlet hole. Measurement of the average temperature was made using Bredfield thermoprobe [19] consisting of nichrome-constantan thermocouple ° 0.03mm

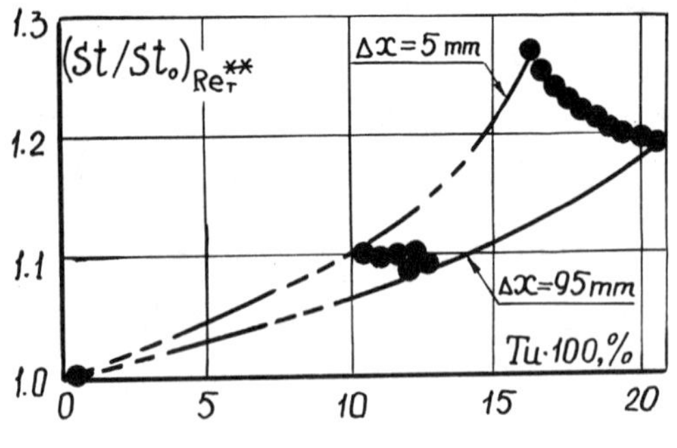

Fig. 3.
Dependence of local relative coefficient of turbulent heat transfer on external flow turbulence level.

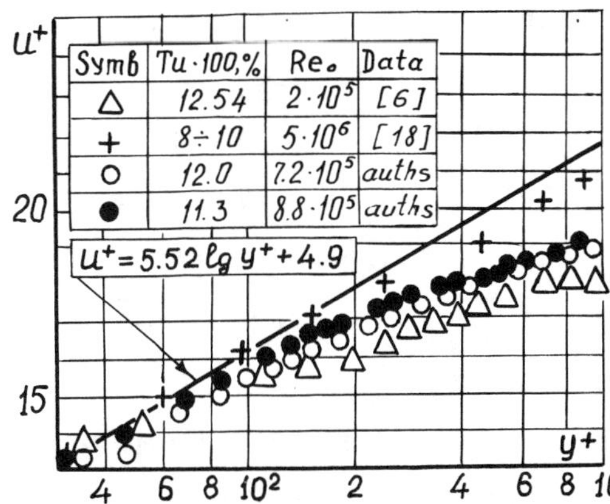

Fig. 4

Comparison of average velocity profiles in turbulent boundary layer obtained by different authors at same turbulence level of external flow. Universal coordinates u^+ and y^+ are calculated from C_f.

in diameter, freely elongated in a groove on the front end of the wedge-shaped wooden substrate. The size of the probe working surface is 0.2 X 1.25 mm^2.

The turbulence level over heat transfer region Tu_o

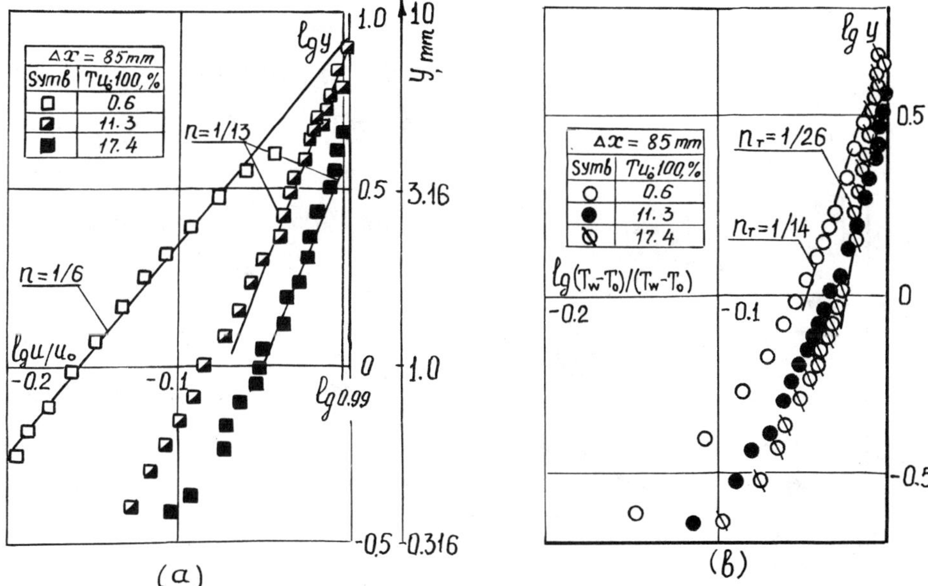

Fig.5.Distribution of average velocity and temperature in boundary layer at three levels of free stream turbulence (u_o= 50 ± ± 2 m/s, x_o= 180 mm). Comparison of external parts with 1/n exponent law.

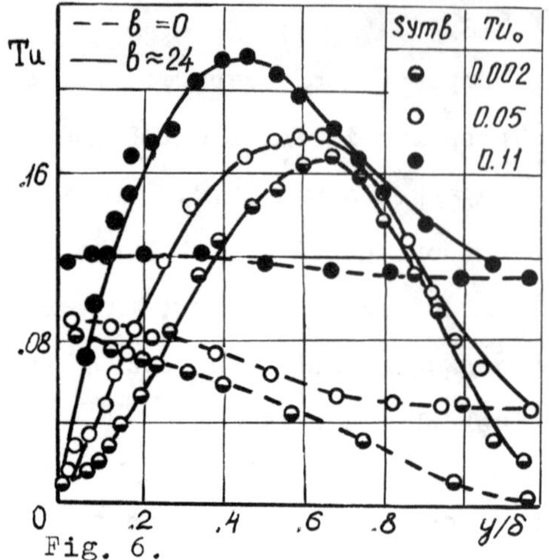

Fig. 6.
Distribution of intensity of turbulent velocity pulsations across turbulent boundary layer with and without injection.

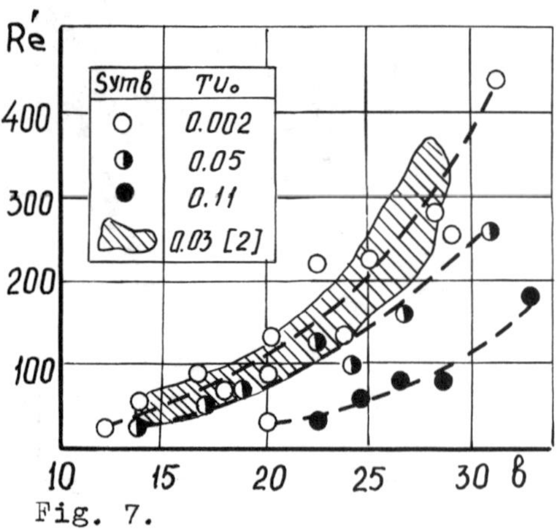

Fig. 7.
Free stream turbulence effect on thickness of dynamic blow-off layer.

in the first position of the model decreased from 0.21 to 0.16%; in the second position - from 0.12 to 0.11. The dynamic boundary layer thickness at the end of the heat transfer region did not exceed 8 mm.

Figures 2 and 3 represents experimental results illustrating the dependence of heat transfer on the mainstream turbulence level. As seen from the diagrams, heat transfer increases by 10 % at $Tu_{\Delta x}$=0.12 - 0.11, and by 20 - 25 % at $Tu_{\Delta x}$=0.21 - 0.16, and slightly increases downstream. According to [10], the last-mentioned circumstance may be associated with the decrease of the turbulence scale downstream.

In our experiments the effect of the free stream turbulence on transfer is weaker than on friction. The estimate of the relative friction factor C_f/C_{f_0} was obtained from experimental profiles of the average velocity represented in universal coordinates $\{u^+, y^+\}$. Figure 4 shows

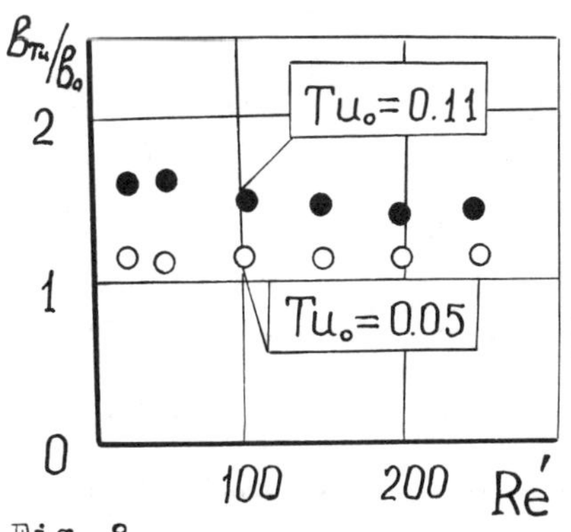

Fig. 8.
Ratio of injection parameter in turbulized external flow to injection parameter at low level of external flow turbulence for same Reynolds numbers (Re' =idem).

that profiles in the near-wall part are described by the universal logarithmic law if the dynamic velocity is calculated by the value of the skin friction factor which is equal to $1.3\,C_f$. In most experimental works this value was obtained as asymptotic: at $Tu > 0.08 - 0.10$ the skin friction factor does not exceed it.

Results of [6, 20] shown in the same Figure are in good agreement with our data.

The correlation of the average velocity and temperature profiles in coordinates $u/u_0 = f_1(y/\delta)$ and $(T_w - T)/(T_w - T_0) = f_2(y/\delta_h)$ (Fig.5) indicates a considerable non-similarity of them even in the presence of a relatively small adiabatic starting length that

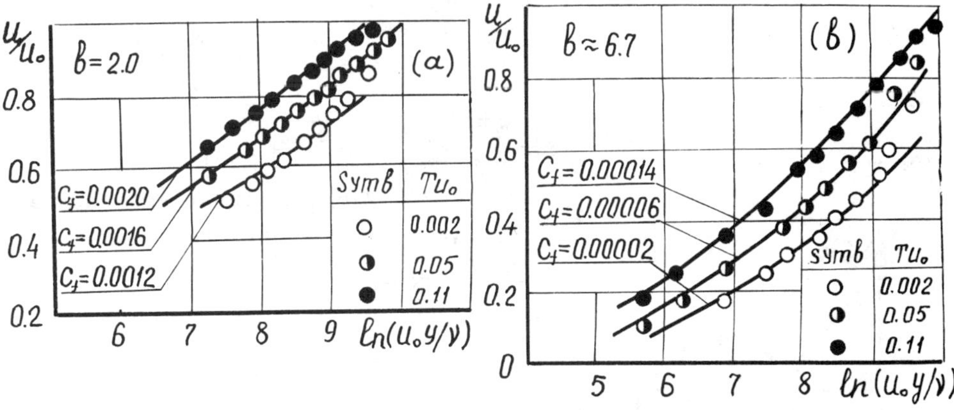

Fig. 9.
Determination of friction factor on penetrable plate according to Clauser method.

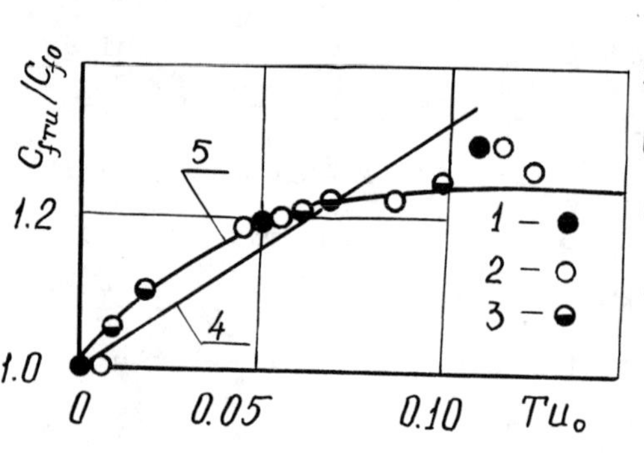

Fig. 10.

Effect of external flow turbulization on skin friction in turbulent boundary layer on impenetrable surface. 1 – authors data, 2 – [6], 3 – [25], 4 – Eq.(1), 5 – Eq.(4).

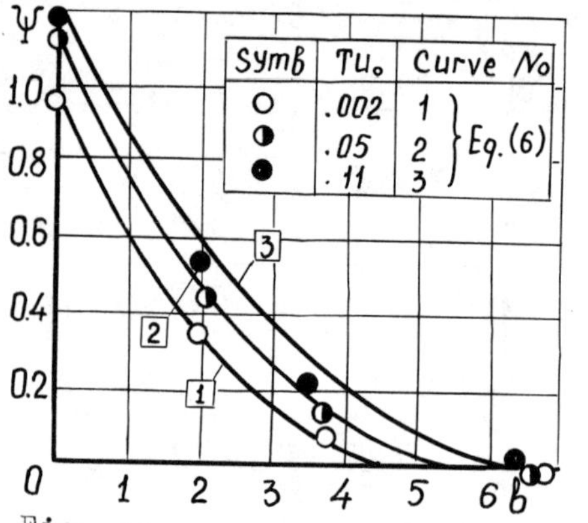

Symb	Tu_o	Curve No
○	.002	1
◑	.05	2 } Eq.(6)
●	.11	3

Fig. 11.

Skin friction in turbulent boundary layer with injection in low level and turbulized external flow.

is a fairing 180 mm in length. Non-similarity remains at a higher level of the free stream turbulence though increase of mixing in the boundary layer seems to result in decrease of the starting length effect. Such a fact indicates the dominant role of boundary conditions on the streamline surface in the development of

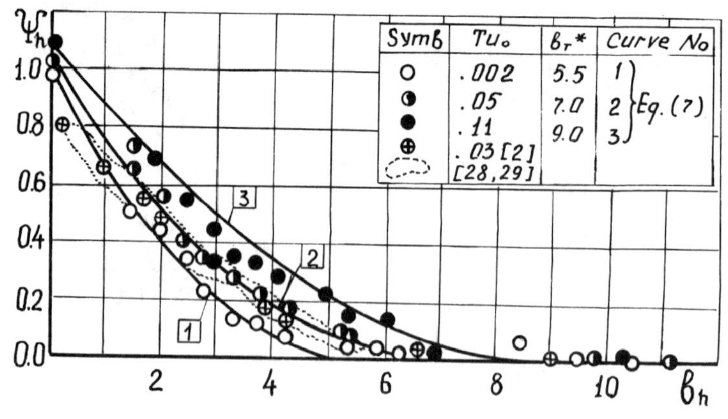

Fig. 12.

Heat transfer in turbulent boundary layer with injection in low level and turbulized external flow.

wall turbulence processes.

This is also evidenced by the pulsation intensity distribution of the velocity vector over the thickness of the boundary layer at different values Tu_o (Fig.6). It is seen that the largest variations of the distribution under the effect of Tu take place in outer part of the boundary layer.

3. HEAT TRANSFER AND FRICTION ON PENETRABLE SURFACE

Experiments were performed in the test installation described in detail in [21]. A working model was a hollow Plexiglass plate 1052 X 152 X mm. Three porous sections were cemented in the upper wall of the model. These sections formed a porous surface 840 mm in length on which a boundary layer developed. Each section was 278 X 120 X 8 mm made of Plexiglass powder. The model was installed in the working chamber of the wind tunnel. The working chamber was 1400 mm in length, its cross section was 200 X 200 mm. Experiments were performed during injection of the air into the air under isothermal conditions at the temperature 20°C. The flexible upper wall of the model provides a gradientless flow past the model. The mainstream turbulence level was measured by grids [21]. Variation of the turbulence intensity over the model length is similar to that in Fig.1.

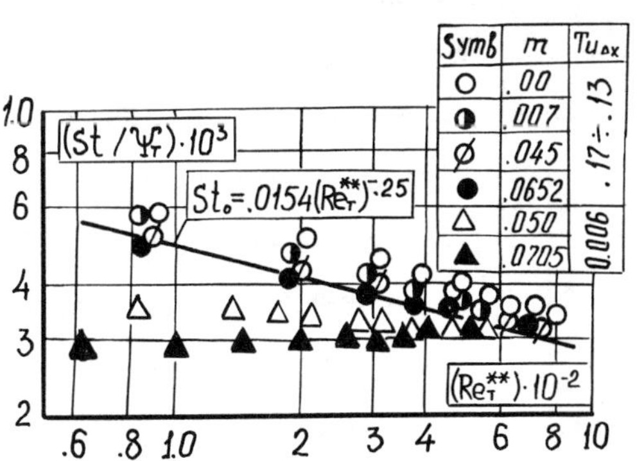

Fig.13.Turbulent heat transfer intensity in uniform porous film cooling zone.

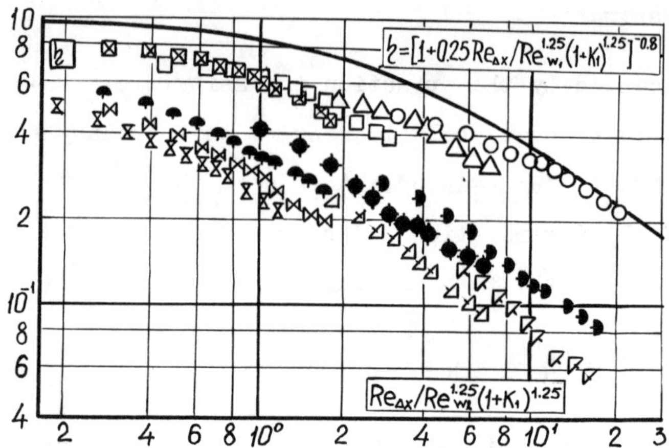

Fig. 14. Effectiveness of porous film cooling at different levels of external flow turbulence.

Measurements were carried out in the section $x = 430$ mm (x is the distance from the model beginning) at the mainstream velocity $u_o = 10$ m/s. In experiments without grid the parameter Tu_o assumed the value 0.002, with grid 1 the value was about 0.05, with grid 2 it was more than 0.11. From energy spectra of the velocity pulsations the estimate of the longitudinal integral scale of turbulence was also obtained: 8-9 mm at $Tu_o = 0.002$, 15-20 mm at $Tu_o = 0.05$ and 50-60 mm at $Tu_o = 0.11$. The thickness of the boundary layer in this section depending on the injection intensity varied from 20 to 90 mm. Measurement technique of the average velocity profiles and turbulence intensity is similar to that described in previous section.

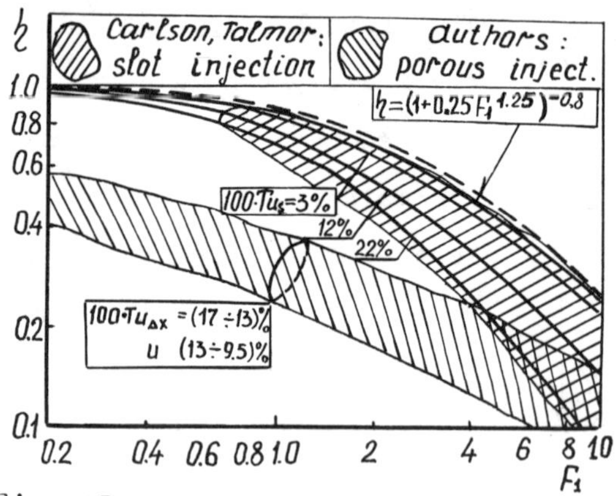

Fig. 15. Comparison of effectivenesses of uniform porous and tangential (inclination angle 20°) film cooling.

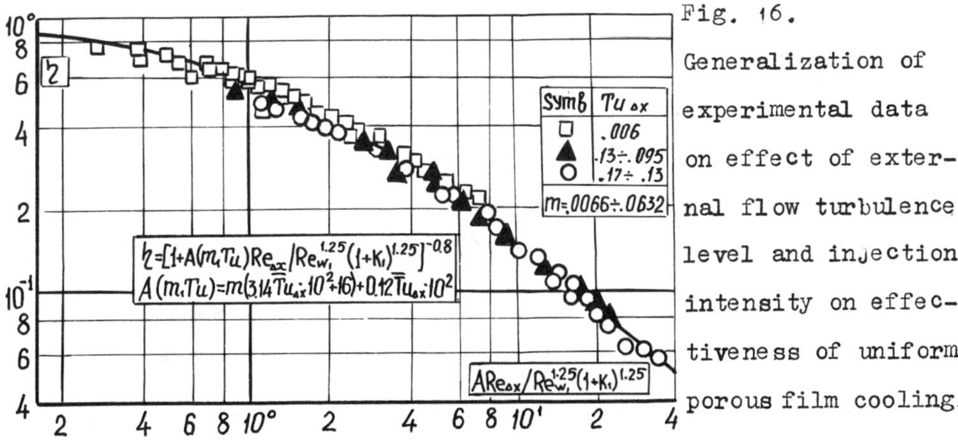

Fig. 16. Generalization of experimental data on effect of external flow turbulence level and injection intensity on effectiveness of uniform porous film cooling,

Variations of the intensity of turbulent pulsations of the velocity vector over the boundary layer thickness at injection parameter $b = 24$ and different values of Tu_0 are shown in Fig.6. It is seen that in case of strong injection the free stream turbulence effect proves to be much stronger and manifests itself over the whole length of the boundary layer. As was mentioned in [22] increased Tu_0 values led to the change of the pulsation level in the immediate vicinity to the wall. At the same time, there remains a zone near the wall (blow-off layer) where these pulsations are invariable [22,23,24]. Dependences of the demensionless blow-off thicknesses on the injection parameters at different Tu_0 were obtained using the technique from [2,24] (Fig.7.). Practically constant values b_{Tu}/b_0 provided Re'= idem determined from Fig.7. and shown in Fig.8 enable us to evaluate relative change of the critical injection parameter versus the level of the free stream turbulence. Such a method proved to be succesful in experimental determination of the influence of non-isothermality, foreign gas injection, pressure gradient on the critical injection parameters [2]. According to Fig. 8

$$b^*_{Tu} / b^*_0 \approx b_{Tu} / b_0 = 1 + 3.8\, Tu_0 \; . \tag{2}$$

Taking into account expression (2) and proceeding from relative change of friction and heat transfer on the penetrable surface [1]

$$\Psi = (1 - b/b^*)^2 \tag{3}$$

an appreciable change of turbulent transport processes at injections close to critical ones may by expected. The value of the friction factor in a wide range of variation of the injection parameters «b» and Tu_0 were determined in the same manner as is described in the previous section of the present paper, i.e. by Clauser method. Figure 9 (a,b) illustrates some examples of agreement between the C_f values of calculated grid and experimental results. Figure 10 represents the results of estimate of relative change of the friction factor at $b=0$ (impenetrable surface), here are also plotted experimental data of [6,25], calculation by formula (1) and by formula

$$\Psi_{Tu} = C_{fTu} / C_{f0} = 1 + 0.25\, th\,(20\, Tu_0) \tag{4}$$

according to [25].

Experimental and estimated data on (2), (3) and (4) are given in Fig. 11. It is seen that at low b the effect of parameter Tu_0 is relatively small. At near-critical injections this effect is significant. A similar picture was observed also in heat transfer studies in [21] (Fig.12). On an impenetrable surface a relative change of the heat transfer coefficient versus parameter Tu_0 (in the range $0 \leq Tu_0 \leq 0.11$) according to Fig.3 can be approximated by a straight line:

$$\Psi_{STu} = (St_{Tu}/St_o)_{Re_T^{**}} = 1 + 0.85\,Tu_o \qquad (5)$$

Taking (2), (3), (4), (5) into account the experimental data on friction and heat transfer can be well generalized by the relations:

$$\Psi = \Psi_{Tu}\,(1 - b/b^*_{Tu})^2 \quad , \qquad (6)$$

$$\Psi_h = \Psi_{STu}\,(1 - b_h/b^*_{hTu})^2 \quad . \qquad (7)$$

4. HEAT TRANSFER DOWNSTREAM OF INJECTION REGION. POROUS FILM COOLING.

The investigation was carried out using the test equipment described in section 2. A uniform porous injection before the heat transfer region was adjusted so that the air goes into the porous section through a thermally insulated pipe located inside a long cylindrical housing. The heat transfer region was separated from the porous wall by a heat-insulating ring 15 mm in thickness. In a low level turbulent flow ($Tu_o \approx 0.006$) the porous injection was not only as a film cooling but served to increase the effective length of the starting adiabatic portion. The turbulizing grid was so mounted that the housing passes through its central hole, and the injection zone practically starts immediately after the ends of the flexible petals. Violation of the geometry of the square-meshed grid resulted in a situation that with good transverse uniformity of the average velocity field, some increase in the turbulence level in direction from the working region surface towards the channel wall by five absolute percent in the starting section of the heat transfer region, and by two absolute percent in its end section. In the analysis of the experimental data, the values of the turbulence level at the external boundary of the dynamic boundary layer were used as characteristic.

In a high-level turbulent flow the studies of thermal processes in the porous film cooling zone were performed for two turbulence levels. In the first case the turbulence level on the external boundary of the boundary layer varied from 0.13 to 0.095, in the second case it varied from 0.17 to 0.13.

The results of the heat transfer studies are shown in Fig.13. As it is known at relatively small injections and low level of the free stream turbulence the most convinient relation for the calculation of heat transfer of an impenetrable surface is [1]:

$$St_o = 0.0128\, Re_T^{**-0.25}\, Pr^{-0.75} \tag{8}$$

(in determining the temperature drop from the wall adiabatic temperature $/ T_w - T_w^* /$). When the free stream turbulence level grows the same situation remains.

At low Tu_o and strong injections ($m \geq 0.03$) heat transfer beyond the injection region is appreciably lower (about two times) than that from (8). This may be caused by the generation of a specific vortex region: due to strong injections the gas injected into the boundary layer initiates a flow of a flow-past a back flow type. As a result the gas injected is entrained into a stable vortex motion, longer resides at the wall, and the film cooling effectiveness thereby increases in the cooling region. As a consequence, heat transfer coefficients decrease.

Because of mixing intensification caused by a high level of the free stream turbulence, the vortex region seem to destroy and heat transfer intensity increases (Fig.13). Thus at higher turbulence level of the external flow the applicability of relation (8) covers the range of substantially larger injections.

In section 3 it was shown, that the increase of the free stream turbulence level causes enhancement of heat transfer from the porous surface, i.e. deterioration of protection properties of porous cooling. The same tendency is observed for the film cooling zone when analysing its effectiveness (Fig. 14).

The deviation of experimental points for a low-level turbulent flow from the Kutateladze-Leontyev standard relation [1] is, first, due to the fact that the average velocity profile in a long cylindrical portion (X_o = 800 mm) is filled more than that on a flate plate [26] and, second, is due to a great length of the injection region. The effect of this factor was revealed in [27].

The above results indicate a substantial (by 2 - 4 times) loss of the porous film cooling effectiveness at higher level of the free stream turbulence. These results and the data from [15] on the study of the slot tangential effectiveness in a gradientless flow are shown in Fig. 16. In these experiments turbulizing grids were installed somewhat removed upstream from the injection point. They provided the level of turbulence Tu_o = 0.03; 0.12 and 0.22 in the slot location.

In Fig. 15

$$F_1 = Re_{\Delta x}^{0.8} / Re_{w_1} (1 + K_1) \quad \text{for porous film cooling}$$

and

$$F_1 = (Re_{\Delta x}^{0.8} / Re_s)(\nu_o / \nu_s)^{0.8} \quad \text{for slot cooling ;}$$

$$K_1 = (T_{w_1} - T')/(T_o - T_{w_1}), \quad Re_s = s\, u_s / \nu_s$$

It is seen from Fig. 16 that the nature of the free stream turbulence effect on the effectivenes of both cooling types is qualitatively the same but the turbulence affects the porous film cooling stronger than the slot cooling. This result of interest requires further experimental tests.

Result of an empirical generalization of the experimental data of Fig. 14 are shown in Fig. 16. It should be emphasized that in this direction research should involve a comprehensive investigation into the flow structure.

6. CONCLUSION

The results reported in this paper not only are of a proper value inasmuch as they broaden the conception of peculiarities of the turbulent transfer processes under various conditions. They allow to state that increasing the free stream turbulence depending on the variation range of other perturbing factors can substantially influence the transport process intensity. If this fact is ignored this may result in substantial erros in concrete engineering designs. It is sufficient to say that friction and heat transfer at large injections are examples of that.

NOMENCLATURE

x, y - coordinates;

δ, δ_h - thicknesses of dynamic and thermal boundary layer;

δ' - thickness of dynamic blow-off layer;

u, v - longitudinal and transverse components of average velocity;

u' - pulsation component of u;

$Tu = \sqrt{\overline{u'^2}}/U_o$ - flow turbulence level;

T - temperature;

Re^{**}, Re_T^{**}, Re' - Reynolds number over pulse loss thickness, energy loss thickness and blow-off layer thic thickness;

c_f - friction factor;

St - Stanton number;

c_{f0}, St_0 - friction factor and Stanton number on impenetrable plate at low level of free stream turbulence;

$b = \frac{\rho_w U_w}{\rho_0 U_0} \frac{2}{c_{f0}}$; $b_h = \frac{\rho_w U_w}{\rho_0 U_0} \frac{1}{St_0}$ - dynamic and thermal parameters of injection;

$\Psi = c_f / c_{f0}$; $\Psi_h = St/St_0$ - relative friction and heat transfer coefficient;

b^*, b_h^* - critical dynamic and thermal parameters of injection;

$\Psi_T = 4/\left(1 + \sqrt{\frac{T_w}{T_0}}\right)^2$ - function of heat transfer non-isothermality;

x_0 - heat transfer region beginning; $\Delta x = x - x_0$;

$F_1 = Re_{\Delta x}^{0.8} / Re_{w1}(1+K_1)$, $F_1 = (Re_{\Delta x}^{0.8} / Re_s)(\nu_0/\nu_s)^{0.8}$ - determining function for porous and slot screens;

$K_1 = \frac{T_{w1} - T'}{T_0 - T_{w1}}$ - thermal parameter of porous film cooling;

$m = \frac{(\rho U)_w}{\rho_0 U_0}$; $m = U_s/U_0$ - relative coolant flow rate; for porous and slot cooling;

$\eta = \frac{(T_w - T_{w1})}{(T_{w1} - T_0)}$ - film cooling effectiveness.

Subscripts:

o - parameters at external boundary layer;

w - wall parameter;

1 - parameters at injection zone end;

Tu - parameters in turbulized external flow;

REFERENCES

I. Кутателадзе С.С., Леонтьев А.И. I972. Тепломассообмен и трение в турбулентном пограничном слое. М., Энергия.

2. Кутателадзе С.С., Миронов Б.П., Накоряков В.Е., Хабахпашева Е.М. I975. Экспериментальное исследование пристенных турбулентных течений. Новосибирск, "Наука", Сибирское отделение.

3. Хинце И.О. I963. М., Физматгиз.

4. Sutera S.P., Maeder P.F., Kestin J. 1963. On the sensitivity of heat transfer in the stagnation-point boundary layer to free-stream vorticity. J. Fluid Mech., Vol. 16, N 4, pp 491-520.

5. Sunden B. 1979. A theoretical investigation of the effect of free-stream turbulence on skin friction and heat transfer for a bluff body. Int. J. Heat and Mass Transfer, Vol.22, N 7, pp. 1125-1136.

6. Дыбан Е.П., Эпик Э.Я. I975. Микроструктура пограничных слоев и процессы переноса в них повышенной турбулентности внешнего потока. В кн. : "Пристенное турбулентное течение". Труды ХУШ Сибирского теплофизического семинара, ч.П, Новосибирск, ИТФ СО АН СССР, с. 30-42.

7. Edwards A., Furber B.N. 1956. The influence of free-stream turbulence on heat transfer by convection from an isolated region

on a plane surface in parallel air flow. Proc. Inst. Mech. Engrs., London, Vol. 1970, p. 941.

8. Джунхан, Серови. 1967. Влияние турбулентности и градиента давления набегающего потока на профили скоростей в пограничном слое на плоской пластине и на теплопередачу. - Теплопередача, № 2, с. 58.

9. Kestin J., Maeder P.F., Wang H.E. 1961. Influence of turbulence on the transfer of heat from plates with and without pressure gradient. Int. Developm. in Heat Transfer, Int. Heat Transfer Conf. ASME, p. 432.

10. Брэдшоу. 1975. Сложные турбулентные течения (обзор). - Теоретические проблемы инженерных расчетов, № 2, с. 101-111.

11. Пядишюс А.А., Кажимекас П.-В.А., Шланчяускас А.А. Теплопередача пластины потоку воздуха повышенной турбулентности. Труды АН Лит.ССР, сер.Б, т.5 (108), с.91-99.

12. Симонич, Брэдшоу. 1978. Влияние турбулентности внешнего потока на теплообмен в турбулентном пограничном слое. - Теплопередача, т.100, № 4, с. 122-129.

I3. Hancock P.E., 1978. Effect of free-stream turbulence on turbulent boundary layers. Ph. D. Thesis, Imperial College, Univ. of London.

I4. Кажимекас П.-В.А. 1978. Влияние турбулентности внешнего потока жидкости на теплоотдачу пластины.- Труды АН Лит.ССР, сер.Б, т.6 (I09).

I5. Carlson L.W., Talmor E. 1968. Gaseous film cooling at various degree of not-gas acceleration and turbulence levels. Int. J. Heat and Mass Trans., Vol. 11, N 11, pp. 1695-1713.

I6. Brown A., Minty A.G. The effect of meanstream turbulence intensity and pressure gradient on film cooling effectiveness for cold air injection slits of various aspect-ratios. ASME Paper 75-WA/HT-17.

I7. Миронов Б.П., Луговской П.П., Тарарин В.Н. 1977. Влияние степени турбулентности основного потока на массообмен за участком пористого вдува. - В кн. : "Турбулентный пограничный слой при сложных граничных условиях", Новосибирск, ИТФ СО АН СССР, с.5-I4.

I8. Васечкин В.Н., Миронов Б.П., Ярыгина Н.И. 1977. Влияние предыстории потока на процессы турбулентного переноса.- В кн.:"Турбулентный пограничный слой при сложных граничных условиях", Новосибирск. ИТФ СО АН СССР, с.60.

I9. Методы и средства измерения температуры газовых потоков. Обзоры ОНТИ ЦАГИ, I97I, № 359, с.53.

20. Pichal V. 1972. Die Turbulente Grenzschicht bei hochturbulenter Außenströmung. ZAMM, Bd. 52, N 10, s. 407-416.

2I. Мамонов В.Н. 1977. Теплообмен на проницаемой пластине при повышенной степени турбулентности набегающего потока. - В кн. "Турбулентный пограничный слой при сложных граничных условиях". Новосибирск, ИТФ СО АН СССР, с.60-72.

22. Миронов Б.П., Алимпиев А.И., Зеленгур А.А., Луговской П.П., Мамонов В.Н. 1975. Экспериментальные характеристики турбулентного пограничного слоя на проницаемой поверхности.- В кн.: "Пристенное турбулентное течение". Труды ХУШ Сибирского теплофизического семинара: ч.П, Новосибирск, ИТФ СО АН СССР, с. 7-29.

Free-Stream Turbulence Effects on Heat and Mass Transfer from Spheres

KAHORU TORII and MITSUTOSHI YOSHIDA
Department of Mechanical Engineering
Yokohama National University
Yokohama, Japan

ABSTRACT

The effects of free-stream turbulence on the heat and mass transfer from spheres were investigated theoretically as well as experimentally. Both the average and the local mass transfer coefficients around the sphere supported with a slender rod at the rear were measured by means of naphthalene sublimation technique. The flow conditions covered the Reynolds numbers between 3500 and 140000 and the intensities of free-stream turbulence Tu, from below 0.3% to 8.5%, generated with isotropic turbulence grids. The appearance of the peak of the local Sherwood number distribution at the rear hemisphere with the increase of the Reynolds numbers and turbulence intensities was found to be attributed to the intermittent reattachment of the separated free shear flow, from flow-visualization and the measured velocity-fluctuations and wall-pressure distributions around the spheres. A two-equation model of turbulence (Jones & Launder-1973) was tried to be applied to predict the experimental results.

1. INTRODUCTION

The effects of free-stream turbulence on heat or mass transfer from a solid body in a flow have been found to be remarkable on the surface with pressure gradient. As for the sphere, however, the results of previous researches are quite discrepant each other, qualitatively as well as quantitatively [1]-[13], while the agreements are good in case of cylinders. These discrepancy and uncertainty may be mainly due to much more difficulty in performing exact experiments excluding the other effects such as support, surface-imperfection and heat-loss, which is caused by the three-dimensional shape of sphere. In the preliminary experiment the average mass transfer rates from a sphere, obtained by using a crossflow support in a flow without free-stream turbulence were found to be several percent higher than those obtained by using a rear support, even if its diameter was 0.3 mm. The reliable heat transfer experiments on sphere so far, were performed by Yuge [5], and by Raithby and Eckert [8], but both of them did not measure the local heat transfer. The local heat transfer data of the spheres have been provided by Cary [1] and Wadworth [4] and covered the wide range of the Reynolds number from 20000 to 1300000. However, they may not be reliable since they employed the cross-flow support which was found to affect the flow tremendously as suggested by Raithby and Eckert [8], [14].

The present research aims at obtaining the true knowledge of the free-stream turbulence effects on heat and mass transfer from spheres, by overcoming those experimental difficulties. In the present experiments, a special device for measuring the local mass transfer coefficient was employed to permit to hold the sphere from the rear. In order to investigate the flow around the sphere, especially the separated flow on the rear hemisphere, flow visualization techniques such as oil-film and smoke-tracer methods were employed and the wall-pressure distributions around the spheres were also measured. The correlation between the fluid dynamic characteristics and the local mass transfer distributions are discussed. The overall mass transfer were also measured and compared with the heat transfer results obtained with an electrically heated aluminium-sphere in air-flow to investigate the free-stream turbulence effects on the analogy between heat and mass transfer.

The flow conditions covered Reynolds numbers between 3500 and 140000 and the intensities of free-stream turbulence from below 0.3% to 8.5%, generated with isotropic turbulence-grids. The experiments were conducted particularly in the critical flow regime, where the transport phenomena are so sensitive to the thermal and fluid-dynamic conditions or disturbances that reliable results may not be existent yet in the literature. The mass transfer experiments enable us to eliminate the heat-losses by conduction through the support, free convection and radiation and to avoid the measurement-error or non-uniformity of the surface temperature, which may be inevitable in heat transfer experiments.

As for theoretical analyses, many have been performed about the turbulent flows and heat transfer with employing mathematical model of turbulence. Most of them were concerned with internal flows or free jets and few were done with external boundary layers around bodies. Especially in case of the existence of free-stream turbilence Priddin's Report [15] on heat transfer of gas-turbine blade was the only one, so far as the present authors knew. Here, a two-equation model of turbulence (k-ε model, Jones and Launder [16]) incorporated into the boundary layer program of Patankar and Spalding [17] was tried to be applied to predict the boundary layer development and the heat or mass transfer rates.

2. EXPERIMENTAL APPARATUS

A more detailed description of the experiments will be given elsewhere [18].

2.1 Mass Transfer Experiments

Fig. 1 shows a schematic drawing of the experimental apparatus. The six naphthalene spheres of 10, 20, 30, 40, 50 and 76 mm in diameter d were used and placed in the center of a 300x300mm^2 square duct connected to a low-speed, low-turbulence wind tunnel. The sphere was held at the rear by a slender support less than 0.1 d with two tightened crossed-wires of 0.3 mm diameter to prevent it from vibrating. The turbulence in the test section was generated by inserting a turbulence-grid into a slot at its upstream end. The desired turbulence intensity in free-stream Tu could be set by changing the distance between the turbulence grid and the front stagnation point of the sphere, x. Tu

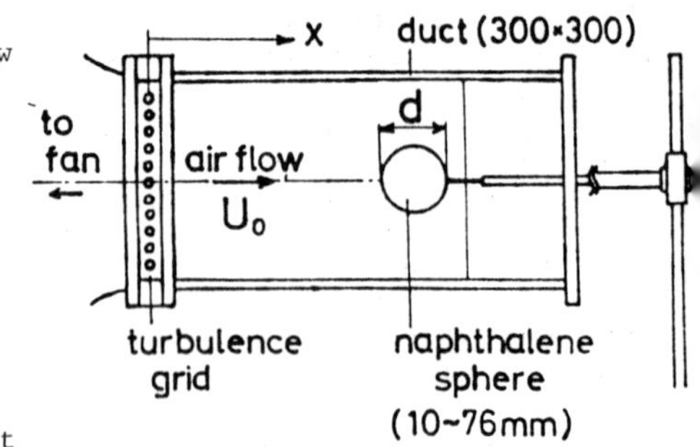

Fig.1. Experimental apparatus.

was evaluated at x as the intensity in the test duct without the sphere in place. The grid having a 16.5 mm square mesh made of 6 mm-dia rod was used since the structure of turbulence generated by it had been fully investigated by Sato and Yamamoto [19]. The longitudinal integral scales of turbulence ranged from 0.3 to 1.1 M in the present experiments. The effects of the scale of turbulence were disregarded in the present research because they had been found to be little by Raithby and Eckert [8]. The effects of tunnel blockage were considered to be well corrected with the equation recommended by Leppert et al [20], since the blockage ratio was as small as 0.05 at the maximum. Local mass transfer rates were obtained from measurements of the surface contour of the naphthalene sphere both before and after a data-run, using a special device. The accuracy of the data reduction is supported by the closure of the overall mass balance which was evaluated by comparing the overall mass transfer obtained by direct weighing with that from a numerical integration of the local mass transfer distribution.

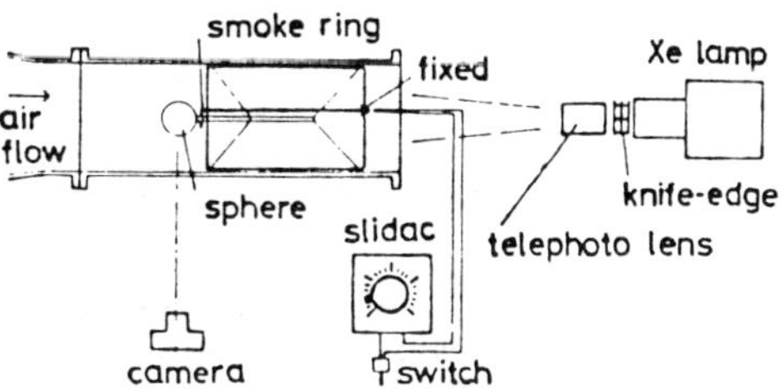

Fig.2. Flow visualization setup.

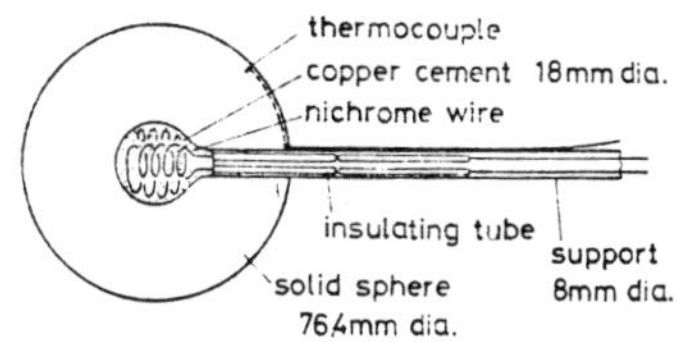

Fig.3. Heat transfer model.

2.2 Pressure Distributions

The wall pressure distributions around the sphere were measured with a hollow sphere of 100 mm diameter which was made of duralumin, had 24 pressure-taps of 0.8 mm on the equator and was supported from the rear as in the heat and mass transfer experiments.

2.3 Flow Visualization Experiments

The flow was visualized by means of oil-film and smoke-tracer. The oil-film technique can visualize not only the distribution of the skin friction around the sphere qualitatively but also the separation line quantitatively. The oil-film is the mixture of kerosene, titanium oxide white powder, a little linseed oil and oleic acid, the ratio of which is changed according to the flow velocity. A plastic, black, hollow sphere of 101.5 mm diameter was held with a 8 mm rear support suspended in a frame of rods by 0.3 mm steel wires as shown in Fig.2. The frame was inserted in the test duct for a few minutes until the flow patterns of the oil-film were fixed. The smoke-tracer technique was employed to visualize the separated, recirculating region at the rear hemisphere. The experimental setup is also shown in Fig.2. A smoke-ring of 29 mm diameter made of a 0.32 mm nichrome wire was placed downstream of the sphere, was coated with liquid paraffin and

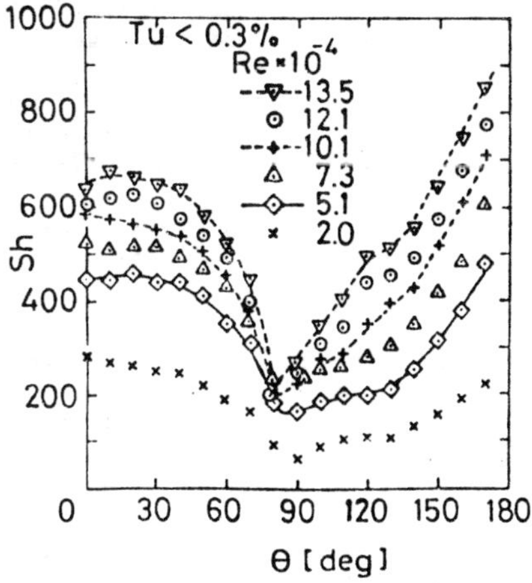

Fig.4. Local Sherwood number without free-stream turbulence.

was directly heated by electric current to generate the smoke without disturbing the flow.

2.4 Heat Transfer Experiments

The overall heat transfer rates were obtained with an aluminium aphere shown in Fig.3. The heater was made small to facilitate a uniform wall-temperature condition.

3. EXPERIMENTAL RESULTS

Further details will be given elsewhere [18].

3.1 Local Mass Transfer Coefficients

$R_e \times 10^4$	T_u [%]	
● 13.4	<0.3	—— Merk's Method
◑ 10.1	<0.3	(S_c=2.56)
○ 7.3	<0.3	······ Present Formula
◇ 10.1	4.8	Eqs. (4), (5)
△ 10.1	8.4	

Fig.5. Local mass transfer rates around spheres

For near zero turbulence intensity (Tu<0.3%), the distributions of the local mass transfer coefficients around the spheres vary with Reynolds number as shown in Fig.4. The profiles seem to be similar for all Reynolds numbers, as they decrease gradually, take the minimum values at 80° to 90° from the front stagnation points and increase towards the rear stagnation points. The present results of the laminar boundary layer on the front hemisphere show good agreement with the theoretical predictions in Fig.5, which proves the high accuracy of the present measurements for mass transfer. The theoretical prediction was done by Merk's method [21] which can economically solve the laminar boundary layer equations with Meksyn's wedge variable. Flössling's formula (Re=157200) was used as the velocity at the edge of the boundary layer around the sphere. As for the rear hemisphere, the dependence of the local sherwood number on the Reynolds number was found to be well correlated with $Sh/Re^{0.719}$ as shown in Fig.5. The details of the circumferential variation of the exponent of the Reynolds number are shown in Fig. 6.

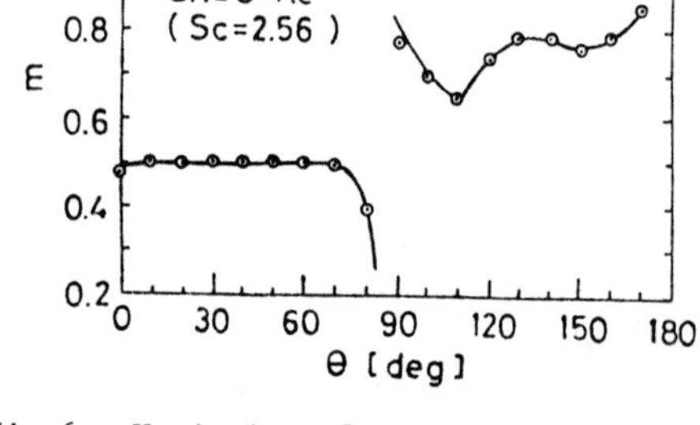

Fig.6. Variation of exponent m of Re.

According to these results, the following correlating equations for the local Sherwood numbers are obtained for two regions before and after separation-points as

$$Tu<0.3\%,\ 0°\leq\theta\leq80°:\ Sh/\sqrt{Re} = 1.843 - 0.267\,(\theta/90)^2 - 1.391\,(\theta/90)^4 \qquad (1)$$

$$Tu<0.3\%,\ 90°\leq\theta\leq170°:\ Sh/\sqrt{Re} = 0.0525 + 0.0100\,(\theta/180)^2 + 0.141(\theta/180)^4 \qquad (2)$$

which are compared with the present data in Fig.5.

When the turbulence is introduced in the free-stream flow, the remarkable changes appear in the distributions of the local Sherwood numbers around the spheres as shown in Fig.5 in comparison with the results for zero turbulence intensity. The typical results illustrating the effects of the free-stream turbulence intensity on the local Sherwood number distributions are given for

Re=72000 and 134000 in Fig.7. At the region from the front stagnation point to about 80° or 100°, the local Sherwood numbers increase more mildly with the turbulence intensity than at the rear hemisphere, and have similar profiles to those of zero turbulence intensity. Their augmentation increases remarkably with turbulence intensity above 5%. On the other hand, their change is drastic at the rear hemisphere and their peak appears at the angle of 110° or 120°. The value of the peak increases with the turbulence intensity and becomes, at Tu=8.3% and Re=134000, 2.5 times higher as at Tu<0.3%. Fig. 8 illustrates the dependence of the local Sherwood number distributions on Reynolds number for the turbulence intensity of 4.7% or 8.2%, as an example. It may be seen that the effect of the free-stream turbulence becomes more remarkable at the higher Reynolds number. Therefore, the peak appearing near 120° seems to increase both with the turbulence intensity and with the Reynolds number, which exert their multiplication-effect.

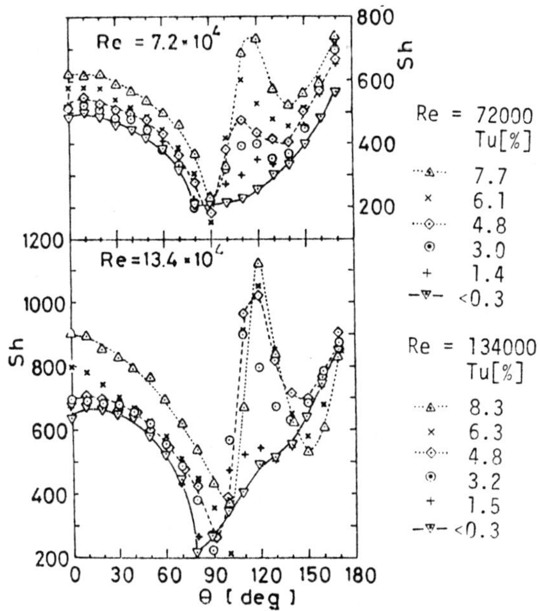

Fig.7. Effects of free-stream turbulence on local Sherwood number distributions at certain Reynolds numbers.

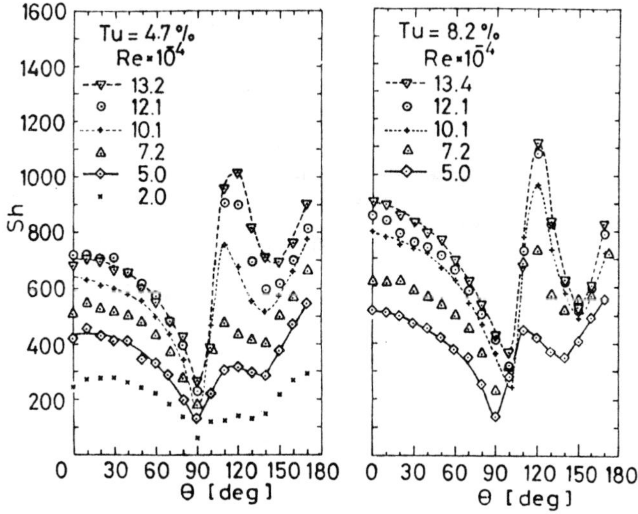

Fig.8. Effects of Reynolds number on local Sherwood number distributions at certain turbulence intensity.

Paying attention to such a remarkable effect of the free-stream turbulence at the rear hemisphere, the minimum or maximum values of the local Sherwood numbers and their angular positions θ_1, θ_2, θ_3, were correlated with the turbulence intensities and Reynolds numbers. The changes of their angular positions with the turbulence intensity and Reynolds number are mapped in Fig.9. The number in the figure denotes the tenth digit of the angle. For zero turbulence intensity, the first minimum point θ_1 moves upstream from 100° to 80° as the Reynolds number increases. The effects of the turbulence intensity do not appear at the Reynolds numbers below 5×10^4, above which θ_1 moves downstream with the increase of the turbulence intensity. The change becomes more remarkable at the higher Reynolds numbers. Since the maximum point θ_2 may be inevitably influenced by the movement of θ_1, it shows somewhat complex dependence on the turbulence intensity and Reynolds number. The second minimum point θ_3 moves downstream from 130° to 150° as the turbulence intensity and Reynolds number increase. The dependences of the maximum or minimum Sherwood numbers at the stagnation point, θ_1, θ_2 and θ_3 on the Reynolds numbers at various turbulence intensities were found to be well correlated with equations (3) to (6) respectively as

$$\theta=0: \quad Sh=121.3\ (Re/4000)^m, \quad m=0.481 + 4.32\ Tu^{1.6} \tag{3}$$

$$\theta=\theta_1: \quad Sh=0.0267\ Re^m, \quad m=0.781 \tag{4}$$

$$\theta=\theta_2: \quad Sh=62.8\ (Re/10000)^m, \quad m=0.739 + 3.48\ Tu^{0.8} \tag{5}$$

$$\theta=\theta_3: \quad Sh=35.6\ (Re/5000)^m, \quad m=0.792 + 2.22\ Tu. \tag{6}$$

3.2 Flow Visualization

The oil-film technique was found to be very useful to observe the flow near the surface of the sphere. In a flow without free-stream turbulence, the oil-film of the front hemisphere was washed away by the skin-friction to the separation point, at which it flowed down and left a single line (LS). The measured angle of the separation point was 79.5° which agreed with 80° measured by Taneda [22]. The oil-film of the rear hemisphere was left unchanged. In the flow with free-stream turbulence, the oil-flow pattern changed as the turbulence intensity and the Reynolds number increased. The pattern for Re=134000 and Tu=8.2% shows the four lines appearing at θ=104°, 114°, 130° and 158°, which are expressed by LS, RA, TS and A respectively. The dependence of the angular position of LS, RA and A lines on the Reynolds number for various turbulence intensities are shown in Fig.10. They move downstream with the increase of the turbulence intensity, which is more remarkable at high Reynolds numbers. It seems that the effect of free-stream turbulence decreases to disappear as the Reynolds number decreases. The mixture-ratio of the oil-film had to be so delicately adjusted to visualize the line TS that it often failed to appear and was difficult to be shown in the figure.

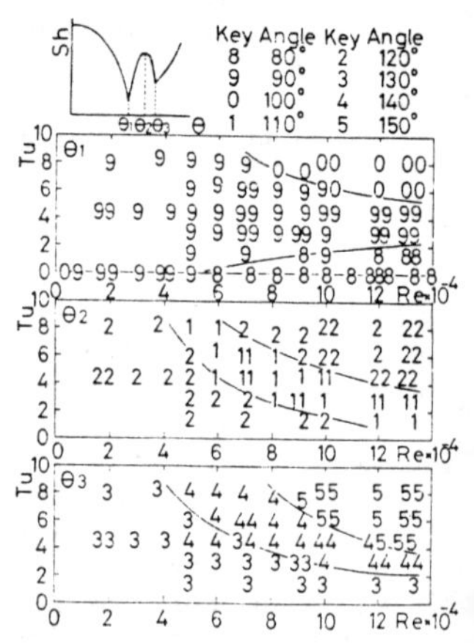

Fig.9. Maps of θ_1, θ_2, θ_3.

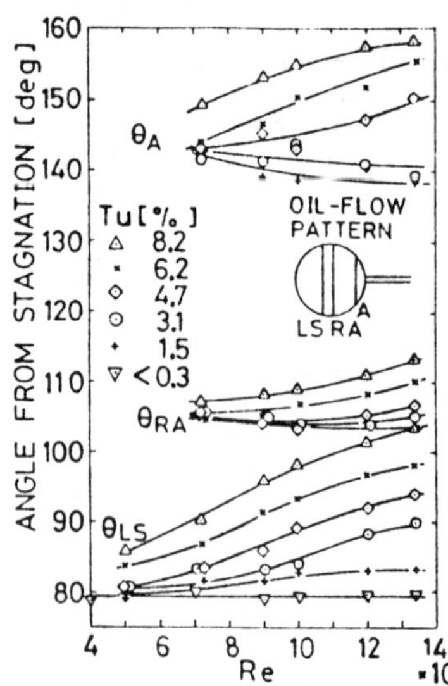

Fig.10. Angular change of lines appearing in oil-flow patterns.

The separation-point or the upper limit of the reverse flow-region behind the sphere was successfully visualized by means of a smoking ring of an electrically heated, fine wire placed 10 mm downstream from the rear stagnation point. It showed for Tu=0 and low values of Re how the reverse flow coming back along the support, impinged to the rear surface of the sphere, crept upstream along the surface up to the front hemisphere and then flowed away downstream with the main stream. In the flow with free-stream turbulence, the reverse flow coming back from the rear is confined in a narrower region as the turbulence intensity or the Reynolds number increases. The visualized flows are consistent with the oil-flow patterns. By comparing each other, the four lines in oil-flow patterns may be identified as the first LS is the separation line of the laminar boundary layer, the second line RA is the line appeared rather upstream of the reattachment-line of the separated free-shear layer, the third line TS is the separation-line of the turbulent boundary layer and the fourth line A is the attachment-line of the reverse flow coming from the rear.

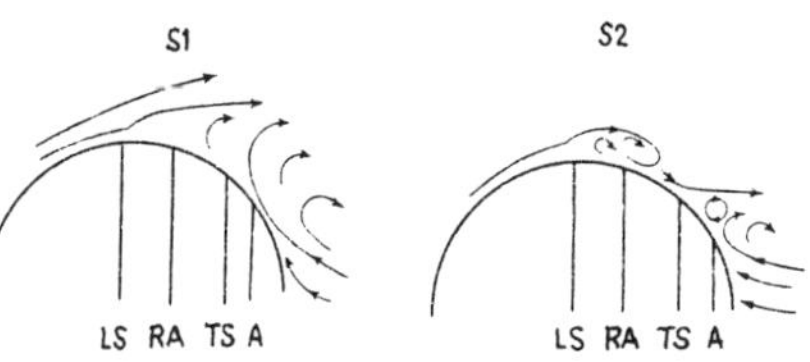

Fig.11. Schematic drawing of flow around a sphere.

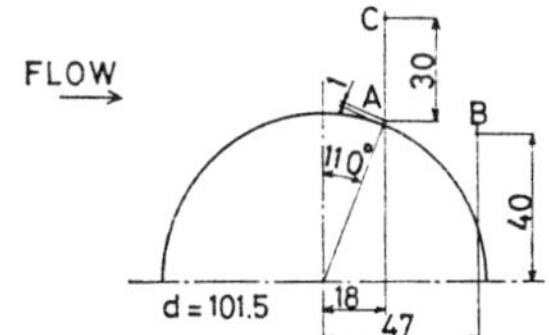

Fig.12. Locations of hot-wire probes.

According to these flow visualization, the flow around the sphere may be schematically illustrated in Fig. 11. In the sub-critical range where the flow has the Reynolds number below the critical value and no free-stream turbulence, the flow around the sphere was found to develop the laminar boundary layer to its separation as the flow-pattern of S1, while, in super-critical flow range where both Reynolds number and turbulence intensity exceed the critical values, it was found to develop the laminar boundary layer to its separation, the separated turbulent free-shear layer to its reattachment, the turbulent boundary layer to its separation and the impingement of the reverse flow as the flow-pattern of S2. The flow in the critical flow range was farther investigated for Re=72000 and Tu=6.0%, by analysing the instanteneous velocity obtained with hot-wire anemometer probes located at the points A, B and C in Fig.12. From these investigations, it may be concluded in the critical flow range that the flow takes intermittently two flow-patterns of S1 and S2 illustrated in Fig.11. This is the reason why the second and third lines appear in the oil-flow pattern for Re=72000 and Tu=4.5%, though the smoke creeps up forward of the third line TS. In the local Sherwood number distribution, the rise of the peak near 110° or 120° with the increase of the turbulence intensity in the critical flow range may be due to the increase of the period taking the flow pattern of S2. Figure 11 should not be considered to well describe the complex structure of the real flow but to be drawn roughly for explaining convenience. According to the present flow visualization by smoke, it really seems that the reverse flow might move in spiral motions, which is left for a future study. Taneda [22] tried to draw the flow around the sphere in the supercritical flow range without free-stream turbulence, streaming in complex spiral motions.

3.3 Wall-Pressure Distribution

For Re=120000 and Tu=4.7% as a typical example, Fig.13 shows how well the pressure distribution, the local Sherwood numbers distribution and the oil-flow pattern correspond to one another. The first minimum point θ_1 of the local Sherwood number distribution shows good agreement with the laminar-separation point LS appearing both in the pressure distribution and in the oil-flow pattern. The local Sherwood number decreases at the front hemisphere as the laminar

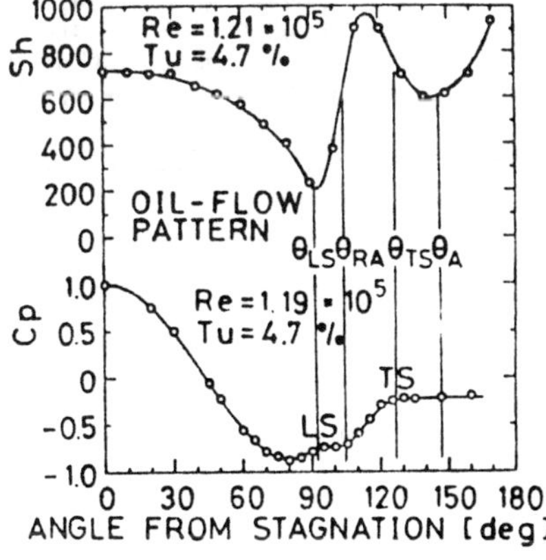

Fig.13. Correspondence among local Sherwood number, wall-pressure and oil-flow pattern.

boundary layer develops to separate. The separated free-shear layer making transition to the turbulent one reattaches intermittently rather downstream of the second RA line of 105° and develop the turbulent boundary layer. These cause the peak of the local Sherwood number at 115° and give the relatively high value until the turbulent separation point TS near 130° appearing both in the pressure distribution and in the oil-flow pattern. The reverse flow impinges on the rear stagnation point, giving the peak of the Sherwood number. The line A in the oil-flow pattern is found to correspond to the angle of the second minimum Sherwood number. The drag coefficients were calculated from the pressure distributions. The critical Reynolds number Re_{cr} conveniently defined as the one at which C_d is 0.3, was found to be well correlated with the correlation proposed by Sucker and Brauer [23]:

$$Re_{cr} = 378500/\exp(6\sqrt{Tu}). \tag{7}$$

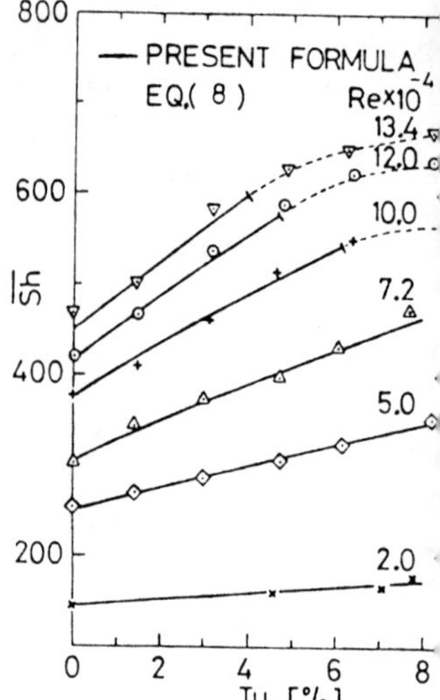

Fig.14. Dependence of overall mass transfer on turbulence intensity at various Reynolds numbers.

3.4 Average Mass Transfer Coefficients

The dependence of the average Sherwood number on the Turbulence intensity for various Reynolds numbers is shown in Fig.14. It shows that the Sherwood number was increased by approximately 150% as a result of increasing the turbulence intensity from near zero to 8%. For Reynolds number less than 72000, the Sherwood number increases linearly with the free-stream turbulence intensity, but for Reynolds number larger than 100000, its increasing rate starts to decrease at a certain high value of turbulence intensity. The average Sherwood numbers were well correlated with

$$\overline{Sh} = 2 + 83.5\ (Re/8000)^m,\ m = 0.59 + 1.41\ Tu^{0.8}: \tag{8}$$

$$1.6\times10^4 < Re < 1.4\times10^5,\ Re\exp(6\sqrt{Tu}) < 4.4\times10^5,\ Sc=2.56.$$

The second condition of the above equation was deduced from equation (7), considering that Reynolds number of C_d-minimum is a usual definition of critical Reynolds number which is 440000 for Tu=0. In other words, the flow may be for convenience regarded as the supercritical flow for $Re\exp(6\sqrt{Tu}) - 440000$. It also implies that Reynolds number of $C_d = 0.26$ may be defined as the critical Reynolds number. In Fig.14 the solid lines represent equation (8), and the broken lines denote the supercritical flow range (which is, strictly speaking, the critical flow range) judged by this criterion. These results are consistent with what has been obtained in the previous sections.

3.5 Heat Transfer Results

It was found that the dependences of the average Sherwood and Nusselt numbers on the Reynolds numbers at the same values of the turbulence intensity were well correlated with

$$\overline{Sh} - 2 = A\ Re^m Sc^n;\ \overline{Nu} - 2 = B\ Re^m Pr^n. \tag{9}$$

The exponent n at various turbulence intensities was obtained by using equation (9) from the heat and mass transfer results. In conclusion, the overall mass transfer from the sphere was found to be well correlated with

$$\overline{Sh} = 2 + (57.3 - 64.3\,Tu)(Re/8000)^m Sc^n,$$
$$m = 0.590 + 1.41\,Tu^{0.8},\quad n = 0.4 + 1.25Tu: \quad (10)$$
$$1.6\times10^4 \leq Re \leq 1.4\times10^5,\ Re\ \exp(6\sqrt{Tu}) \leq 4.4\times10^5,$$
$$0 \leq Tu \leq 0.08,\ 0.7 \leq Sc \leq 2.6.$$

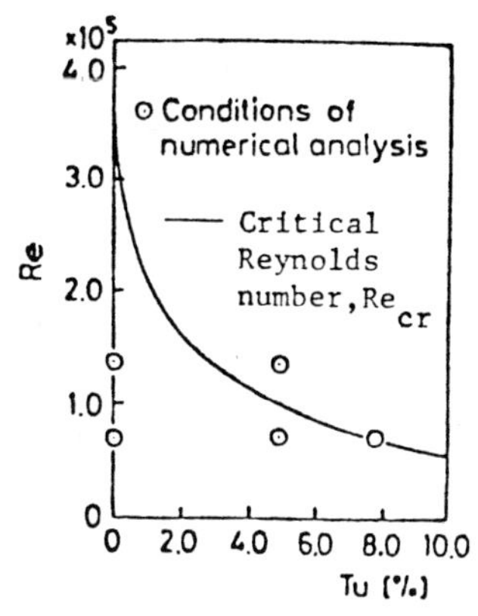

Fig.15. Variation of critical Reynolds number with turbulence intensity and analytical points.

4. THEORETICAL ANALYSIS

Numerical analyses were performed in five typical cases, of which conditions are shown in Fig. 15. The curve in the figure is equation (7) and represents the critical flow condition; the region below the curve indicates the subcritical flow regime.

Preceding the analysis, the distribution of free-stream turbulence intensity around the sphere was experimentally investigated. The decay of turbulence downstream from the isotropic turbulence grid was confirmed to be well predicted by the present model.

4.1 Governing Equations

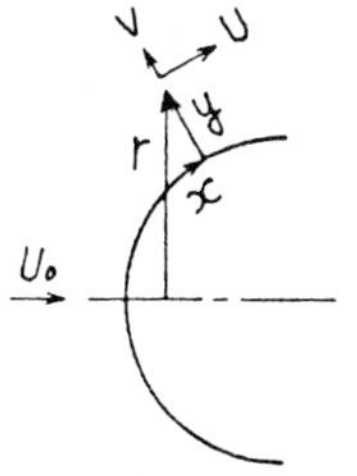

Fig.16. Coordinate system.

Figure 1 shows the coordinate system employed for the present analysis. Introducing the turbulent viscosity μ_t defined by

$$\mu_t = \tau_t/(\partial U/\partial y) = -\rho\overline{uv}/(\partial U/\partial y), \quad (11)$$

the continuity and mean-motion equations for the boundary layer are:

$$\frac{\partial(r\rho U)}{\partial x} + \frac{\partial(re V)}{\partial y} = 0, \quad (12)$$

$$\rho U\frac{\partial U}{\partial x} + \rho V\frac{\partial U}{\partial y} = \frac{1}{r}\frac{\partial}{\partial y}\left[r(\mu+\mu_t)\frac{\partial U}{\partial y}\right] - \frac{\partial p}{\partial x}. \quad (13)$$

The boundary conditions are at the wall

$$y=0\ :\ U = V = 0, \quad (14)$$

and at the edge of free boundary

$$y=\infty:\ U_\infty(x)\frac{dU_\infty(x)}{dx} = -\frac{1}{\rho}\frac{dp}{dx}. \quad (15)$$

The conservation-equation for a scalor quantity C is

$$\rho U\frac{\partial C}{\partial x} + \rho V\frac{\partial C}{\partial y} = \frac{1}{r}\frac{\partial}{\partial y}\left[r\left(\frac{\mu}{\sigma} + \frac{\mu_t}{\sigma_t}\right)\frac{\partial C}{\partial y}\right], \quad (16)$$

where C and σ stand for the mass fraction of the species (naphthalene vapor) or the enthalpy and the Schmidt or Prandtl number respectively for mass or heat transfer processes. The boundary conditions for equation (16) are:

$$C = C_w \text{ at } y = 0\ ;\ C = C_\infty \text{ at } y = \infty. \quad (17)$$

4.2 The k-ε Turbulence Model

The local turbulent viscosity in equation (11) was determined from the two-equation model of turbulence (k-ε model) proposed by Jones and Launder [16] as

Table 1. Constants of k-ε model[24]. $(R_t = k^2/\varepsilon\nu)$

σ_k	σ_ε	C_1	C_2	C_3	C_μ
1.0	1.3	1.44	$1.92[1.0-0.3\exp(-R_t^2)]$	2.0	$0.09\exp[-3.4/(1+R_t/50)^2]$

$$\rho U \frac{\partial k}{\partial x} + V\frac{\partial k}{\partial y} = \frac{1}{r}\frac{\partial}{\partial y}\left[r\left(\mu+\frac{\mu_t}{\sigma_k}\right)\frac{\partial k}{\partial y}\right] + \mu_t\left(\frac{\partial U}{\partial y}\right)^2 - \rho\varepsilon - 2\mu\left(\frac{\partial\sqrt{k}}{\partial y}\right)^2 \tag{18}$$

$$\rho U\frac{\partial \varepsilon}{\partial x} + \rho V\frac{\partial \varepsilon}{\partial y} = \frac{1}{r}\frac{\partial}{\partial y}\left[r\left(\mu+\frac{\mu_t}{\sigma_\varepsilon}\right)\frac{\partial \varepsilon}{\partial y}\right] + c_1\frac{\varepsilon}{k}\mu_t\left(\frac{\partial U}{\partial y}\right)^2 - c_2\rho\frac{\varepsilon^2}{k} + c_3\frac{\mu\mu_t}{\rho}\left(\frac{\partial^2 U}{\partial y^2}\right)^2 \tag{19}$$

$$\mu_t = c_\mu \rho k^2/\varepsilon. \tag{20}$$

The boundary conditions:

$$y = 0 : k = \varepsilon = 0 \tag{21}$$

$$y = \infty : U_\infty(x)\frac{dk_\infty}{dx} = -\varepsilon_\infty;\ U_\infty(x)\frac{d\varepsilon_\infty}{dx} = -c_2\frac{\varepsilon_\infty^2}{k_\infty} \tag{22}$$

In the above equation set, the c's and σ's retain the values assigned to them in Table 1, which was recommended for a turbulent free shear layer by Launder, Priddin and Sharma [24].

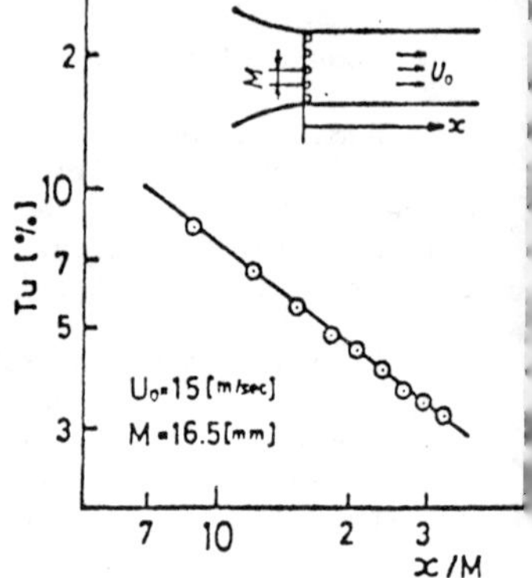

Fig.17. Decay of turbulence downstream of a grid.

4.3 The Turbulence at the Free Boundary

The delay of grid-turbulence Figure 17 shows the experimental results of the decay of the turbulence intensity downstream of a grid having a 16.5 mm square mesh made of 6 mm-dia rod in the test duct. The results were found to be almost independent of the velocity of the flow. The complete isotropy of the present grid turbulence had been found in the region of $9 \leq x/M < 30$ by Sato and Yamamoto [19]. For the isotropic turbulence, the turbulence kinetic energy k can be calculated from the measured turbulence intensity Tu as

Table 2. Constants in Eq.(25)

U_0 [m/sec]	$9 \leq x/M \leq 27$	
	A	n
15	1.34×10^{-3}	1.54

$$k = 1.5\,(U_o\cdot Tu)^2; \tag{23}$$

and the decay of the kinetic energy of the isotropic turbulence can be expressed by

$$\frac{d\overline{u^2}}{dt} = -\frac{2}{3}\varepsilon,\ \text{where } t = \frac{x}{Uo}\ . \tag{24}$$

From Fig. 17, the following correlation was obtained:

$$\overline{u^2} = At^{-n} \tag{25}$$

where A and n are constants and their values are shown in Table 2. With the insertion of equation (25) into (24) there results the rate of the dissipation of the turbulence kinetic energy ε as

$$\varepsilon = 1.5\ n\ At^{-(n+1)} \tag{26}$$

Thus, the values of k and ε in Fig.18 were determined from the experimental data with equations (23) and (26). For the decay of the isotropic grid turbulence, equations (18) and (19) reduce to

$$U_0 \frac{dk}{dx} = -\varepsilon \qquad (27)$$

$$U_0 \frac{d\varepsilon}{dx} = -c_2 \frac{\varepsilon^2}{k} \qquad (28)$$

For the initial values of k and ε, required to start the integrations of the above equations, the experimental values at nine meshes downstream of the grid (x/M=9) where the turbulence produced by the grid had been checked to become isotropic and homogeneous, were utilized. The predictions of the decay of turbulence energy and its dissipation rate, downstream of the grid, show good agreements with the measurements in Fig.18. It also confirms that the turbulence produced by the present grid is isotropic at $x/M \geq 9$ since the present model was constructed by the originators so that it, when applied to the calculation of the decay of isotropic grid-turbulence, accorded with experiment.

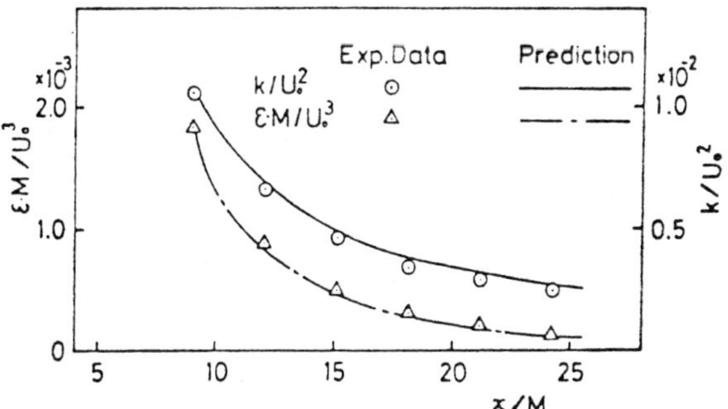

Fig.18. Comparison between the measured and predicted decays of turbulence from a grid.

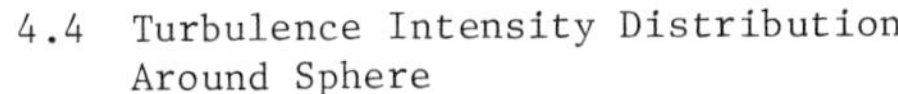

4.4 Turbulence Intensity Distribution Around Sphere

The mean and fluctuating components of the velocity at the edge of the boundary layer around the sphere were measured with hot-wire anemometer. Figure 19 shows the measured variations of the turbulence intensity at the edge of the boundary layer around the front hemisphere Tu_∞, which is defined by

$$Tu_\infty = \sqrt{\overline{u_\infty^2(x)}}/U_\infty(x) \qquad (29)$$

In Fig. 20, they are replotted by utilizing another definition of turbulence intensity Tu' defined by

$$Tu' = \sqrt{\overline{u_\infty^2(x)}}/U_0 \qquad (30)$$

which may represent the dimensionless turbulence energy. Now it should recall that the turbulence intensity Tu is evaluated at the front stagnation point of the sphere as the intensity in the test duct without the sphere in place. The comparison between Figures 19 and 20 revealed the remarkable difference between the variations of Tu_∞ and Tu'. Judging only from Tu_∞ in Fig.19, it seems as if the turbulence were strongest near

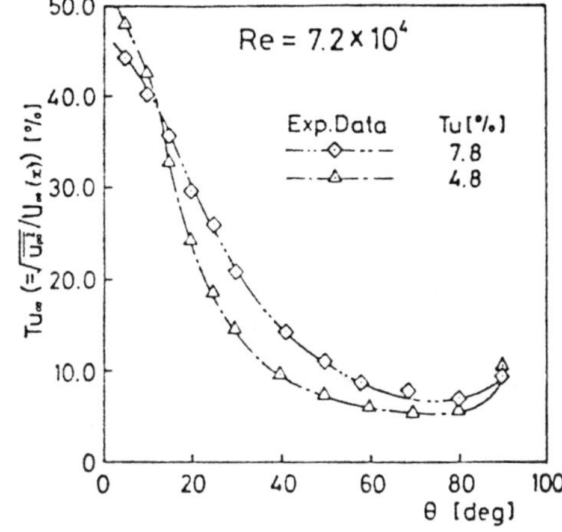

Fig.19. Variation of turbulence intensity Tu_∞ at the edge of boundary layer around a sphere.

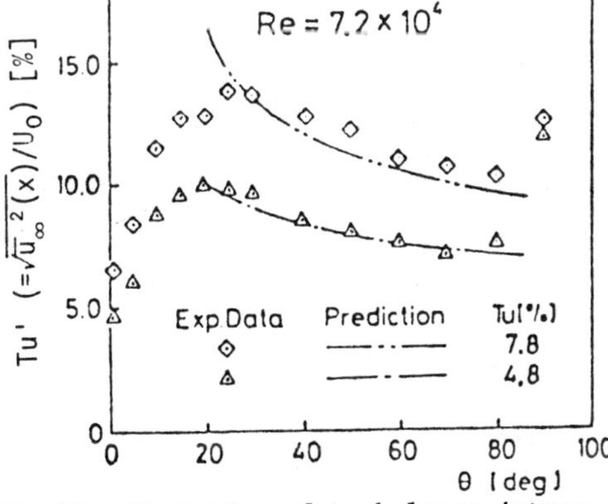

Fig.20. Variation of turbulence intensity Tu' at the edge of boundary layer a ound a sphere.

the stagnation point. According to Fig.20, however, the turbulence energy itself (Tu') was found to be rather weakest and nearly equal to the turbulence energy of the approaching flow (Tu). Though Tu_∞ decreases monotonously with θ up to θ=80°, $\tilde{T}u'$ increases for 0° < θ<20° and begins to decrease at θ = 20°. It was found that the value of Tu' at θ = 0° was nearly equal to the value Tu and that the maximum value of Tu' near θ = 20" was approximately two times as much as the value of Tu' at θ = 0°. The predicted curves in Fig.20 will be discussed later.

5. COMPARISON BETWEEN PREDICTION AND EXPERIMENT

5.1 Without Free-Stream Turbulence

The local mass transfer coefficients around the sphere without free-stream turbulence were obtained by solving equations (12), (13) and (16) for Tu = 0%, k = ε = 0 and μ_t=0 by means of GENMIX computor program [17]. The variation of the velocity at the edge of the boundary layer in equation (15) was approximated by a polynomial expression:

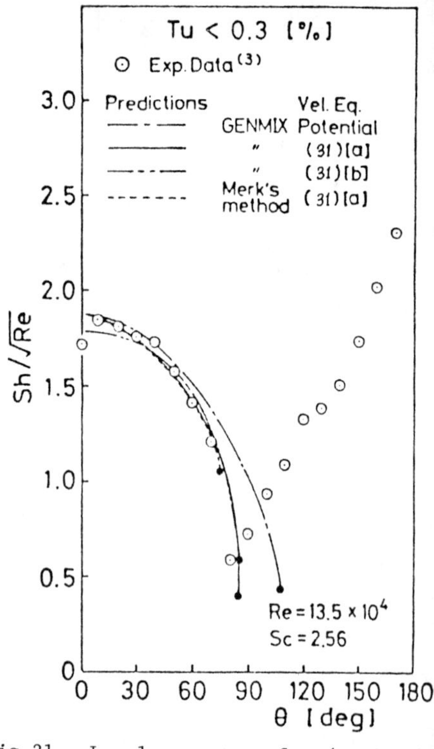

Fig.21. Local mass transfer (Tu<0.3%).

$$\frac{U_\infty(x)}{U_0} = A_1\left(\frac{x}{d}\right)^1 + A_3\left(\frac{x}{d}\right)^3 + A_5\left(\frac{x}{d}\right)^5 + A_7\left(\frac{x}{d}\right)^7 \quad (31)$$

where the A's retain the values assigned to them in Table 3. The values of line [a] of Table 3 was assigned by Froessling from Fage's data [25]. Twenty cross-stream intervals were employed in the finite-difference procedure. Assuming that the region in which the second term of the right-hand side of equation (31) is less than 0.1 percent of the first term must be the region of the stagnation flow, the computation started there at x_0, avoiding the stagnation point:

$$x_0 < \sqrt{0.001 \times |A_1/A_3|} \times d. \quad (32)$$

At the starting point x_0, the linear profiles of velocity and mass concentration was utilised for the initial profiles so that the governing set of equations were solved to get the profiles of the one step downstream. The similar-profiles to them were assigned to the new ones at point x_0. This iteration continued until it converged and then the computation proceeded to the next step. The computed profile of velocity at point x_0 agreed with the exact solution for axially symmetrical stagnation-flow [26] within the maximum error of 0.9%. The predictions show good agreement with the experimental data of the local Sherwood number for the laminar boundary layer around the sphere, except the prediction utilizing the pressure distribution of potential flow, as in Fig.21. Though Merk's approximate solution [21] predicted the separation point (denoted by the closed circle in the

Table 3. Values of coefficients in Eq.(31).

	Tu	Rex10^-4	A_1	A_3	A_5	A_7	Investigator
a	—	15.7	3.000	-3.497	4.739	-5.148	Frossling
b	<0.3	13.5	2.728	-1.612	0.039	-1.214	Author

figure; defined as the point with zero velocity-gradient at the wall) as 0-75°, the present solution utilizing GENMIX predicted the separation point as θ=84° which was closer to the experimental result. The total computation-times of the former and the latter were 18.5 s and 5.9 s respectively with HITAC 8700/8800.

5.2 Free-Stream Turbulence Effects

The local mass transfer rates around the sphere with free-stream turbulence were predicted and compared with the present experimental results in Fig.22. The prediction for Tu<0.3% was done by means of the procedure described in the previous section. The predictions for Tu=4.8 and 7.8% was not made from θ=0° but from θ=20° by employing the k-ε model. The reasons are as follows. Equation (22), to which the k-equation (18) reduces at the free boundary, leads to

$$\frac{dk_\infty}{dx} < 0, \qquad (33)$$

since ε never gets negative and U_∞ also never at least between the front-stagnation and separation points. It means that the present model should not be applied to the region where k_∞ increases with x. The measurements show in Fig.20 that the turbulence energy increases with x for 0°<θ<20°. This is the reason why the computation started at θ=20°

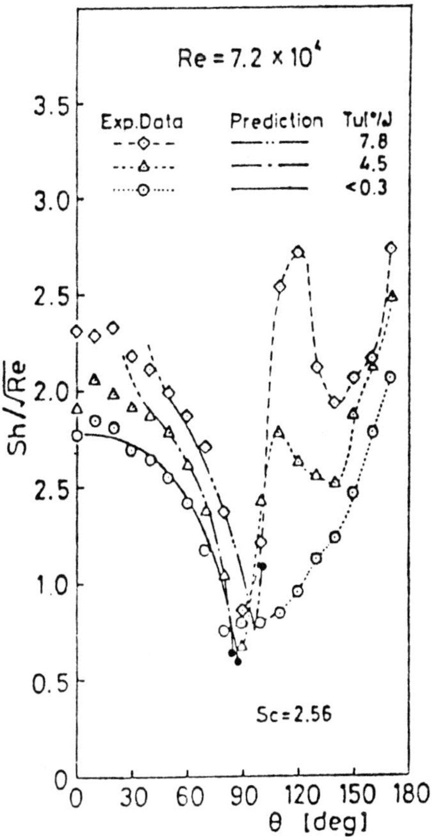

Fig.22. Comparison between the predicted and the measured local Sherwood numbers around spheres.

For the value of k_∞ at θ=20°,

$$Tu' = 2.1 \times Tu \qquad (34)$$

was assumed from the experimental results in Fig.20. Assuming that the turbulence may be isotropic at the edge of the boundary layer, equations (23), (30) and (34) reduce to

$$k_\infty = 1.5\ (2.1 \times U_0 \times Tu)^2, \text{ at } \theta = 20°. \qquad (35)$$

The value of ε_∞ at θ=20° was calculated from equation (24) with equation (35). The variations of the velocity at the edge of the boundary layer were approximated by the polynomial expressions of equation (31) from the experimental data of pressure distributions. The values of

Table 4. Values of coefficients in Eq.(31).

Tu	Rex10^{-4}	A1	A3	A5	A7	Appli. range
7.8	7.2	2.8456	-1.7067	-0.4253	0.3256	0 ≤ θ ≤ 95
4.8	"	2.8036	-1.9177	0.9834	-1.9871	0 ≤ θ ≤ 85
<0.3	"	2.6738	-0.4134	-4.6656	3.7247	0 ≤ θ ≤ 85
4.8	13.5	2.8237	-1.4451	-0.6891	0.5211	0 ≤ θ ≤ 100

Table 5. Values of coefficients in Eq.(36).

Tu	Rex10^{-4}	B0	B1	B2	——	Appli. range
7.8	7.2	-1.6907	7.1772	-4.3006	——	95 ≤ θ ≤ 115
4.8	13.5	-5.0475	14.3637	-8.0542	——	100 ≤ θ ≤ 120

A's are given in Table 4. As the computation was not able to reach the separation point within the applicable range of equation (31) in case of supercritical flow-condition, the following approximation was utilized there:

$$\frac{U_\infty(x)}{U_0} = B_0 + B_1\left(\frac{x}{d}\right)^1 + B_2\left(\frac{x}{d}\right)^2 \tag{36}$$

The assigned values of B's are given in Table 5. In this way the predictions of the local Sherwood number were made, which also provided those of Tu' shown in Fig.20.

Figure 23 shows predictions of the local Nusselt number around the sphere for Tw=70°C, T_∞=20°C and Pr=0.71.

The turbulence energy distributions and the turbulence energy balance in the boundary layers around the spheres were predicted as shown in Fig.24. The k-distributions are shown in dimensionless forms as k/u_*^2 vs y^+. The k-equation (18) may be transformed into

$$0 = -\left(U\frac{\partial k}{\partial x} + V\frac{\partial k}{\partial y}\right) + \frac{1}{r\rho}\frac{\partial}{\partial y}\left[r\left(\mu + \frac{\mu_t}{\sigma_k}\right)\frac{\partial k}{\partial y}\right]$$

$$+ \frac{\mu_t}{\rho}\left(\frac{\partial U}{\partial y}\right)^2 - \varepsilon - 2\frac{\mu}{\rho}\left(\frac{\partial\sqrt{k}}{\partial y}\right)^2 . \tag{37}$$

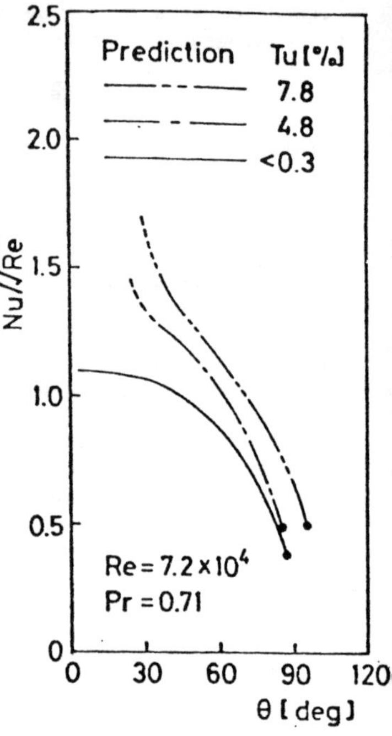

Fig.23. Predicted local Nusselt numbers around a sphere.

The terms of the right-hand side of the above equation denote in order (i) convection (ii) diffusion (iii) generation (iv) dissipation and (v) dissipation due to molecular viscosity. They were multiplied by ν/u_*^4 to be non-dimensionalized in Fig.24.

Figure 25 shows the predicted distributions of total shear stress, turbulent shear stress and turbulent viscosity in the same three cases as in Fig.24. In the boundary layer at θ=100° for Tu=7.8%, the turbulent motion was found to be much stronger than in other two cases. It may be supposed that the boundary layer has already made transition to turbulent one at θ=100° for Tu=7.8%.

In the finite-difference procedure, fifty cross-stream intervals were employed of which thirty were distributed within the region where y^+ was less than 50. The total computational time was within half a minute.

Subcritical flow range The flow of Tu=4.8% in Figures 22 to 25 may be subcritical, judging from Fig.15. The predictions of local Sherwood number by means of the k-ε model show good agreement with the measurements at θ>40°, though the agreement is poor at 20°<θ< 40° as in Fig.22. It is also well predicted that the flow develops the laminar boundary layer and then separates as it may be seen in Figures 24 and 25. It is confirmed that even this kind of an external boundary layer may also be well predicted by the k-ε model.

Critical flow range The flow of Tu=7.8% in Figures 22 to 25 is just at the critical Reynolds number. The agreement between the prediction and the experiment is as good as in the subcritical flow range, as shown in Fig.22. It is different from the subcritical flow that the predicted local Sherwood number begins to increase rapidly at θ=95°. This may be due to the transition from

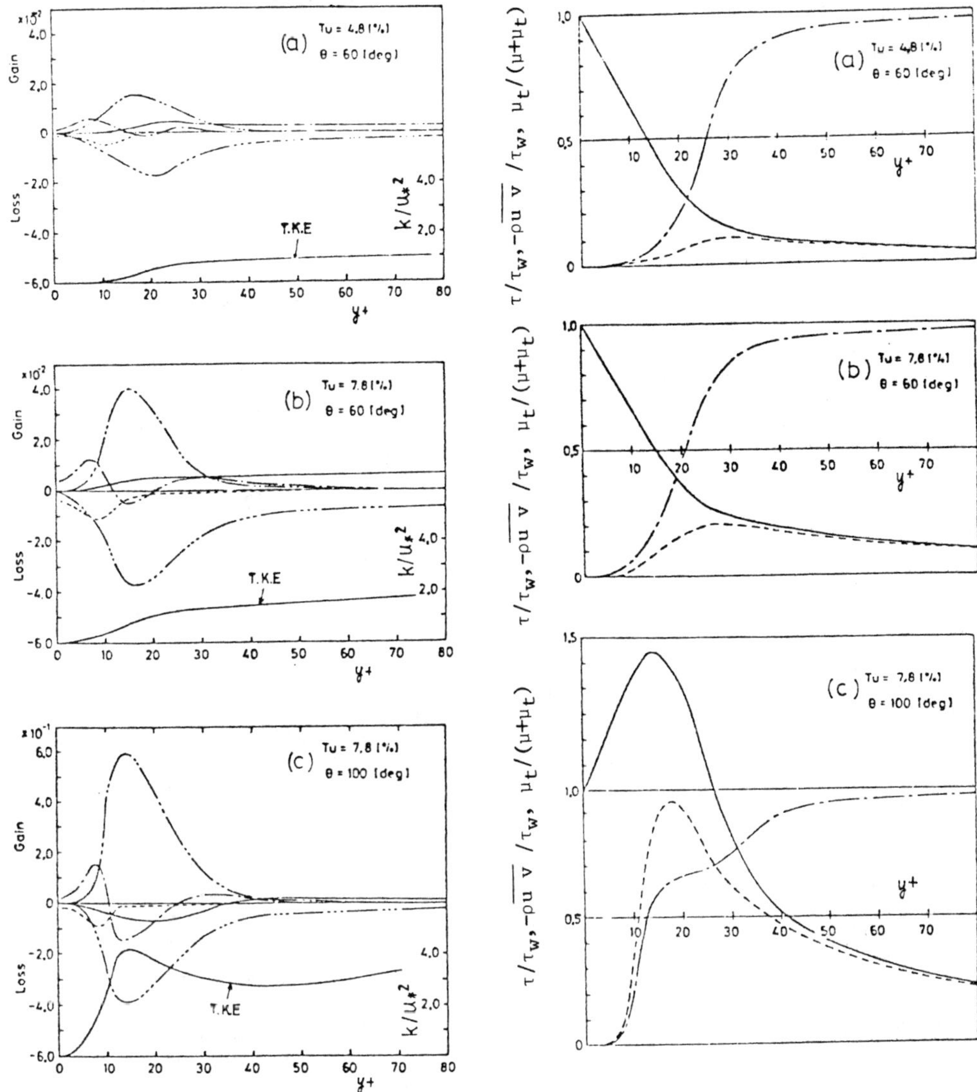

Fig.24. Prediction of turbulence kinetic energy distributions and energy balances in the boundary layers on spheres. ——:Convection, — · — :Diffusion, — ·· — :Generation, — ··· — :Dissipation(1), ---- :Dissipation(2).

Fig.25. Prediction of total and turbulent shear-stress and turbulent viscosity distributions in the boundary layer. —— : τ/τ_w , ---- : $-\rho uv/\tau_w$, — · — : $\mu_t/(\mu+\mu_t)$.

laminar to turbulent boundary layer. In Fig.23, the prediction of local Nusselt number, however, does not show such increase. This may be because the temperature of the surface was given 50°C higher than the free stream; because the viscosity near the wall is so high that the flow may separate without making transition to turbulent boundary layer. Thus, it appears in the critical flow range that the pridiction is so sensitive to the conditions that it may change the flow pattern from turbulent to laminar separations in an unstable manner with some slight variation of the conditions. On the other hand, the intermittent reattachment of the turbulent boundary layer downstream of the laminar separation was observed in the experiments. This discrepancy may be inevitable because the present model is composed of parabolic partial differential equations and cannot predict the reattachment but cannot help choosing either of the two flow-patterns.

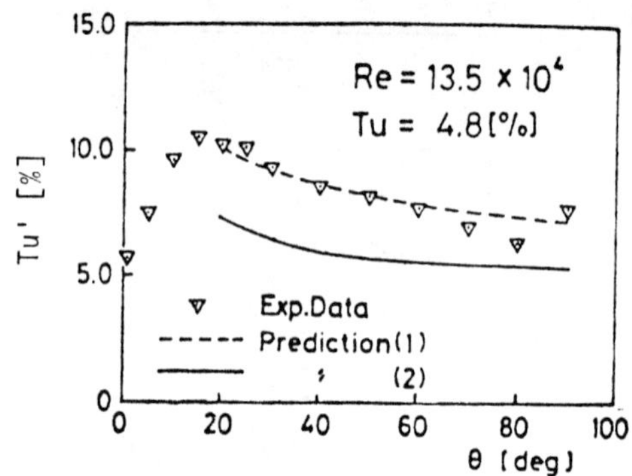

Fig.26. Variation of turbulence intensity Tu' at the edge of the boundary layer around a sphere.

<u>Supercritical flow range</u> Experimental results of the variation of Tu' around the sphere with Tu=4.8% and Re=135000 which is higher than the critical Reynolds number, are shown in Fig.26. It was found that the profil in Fig.26 was almost identical with that for Tu=4.8% and Re=72000 (subcritical range) in Fig.20, depending only on the free-stream turbulence intensity. Thus, by assuming $Tu'=2.1\times Tu$ at $\theta=20°$, the prediction of the local Sherwood number was made, which was shown in Fig.27 with broken line as prediction 1. This prediction, however, gave a rather higher value even in the front hemisphere than the experiment. The solid line denotes Prediction 2 which assumed $Tu'=1.5\times Tu$ at $\theta=20°$. It can make the agreement much better at the cost of the much disagreement of Tu' as shown in Figures 26 and 27. At least in supercritical flow condition, the k-ε model and the present analysis must be more improved, especially near the free boundary, relating to the entrainment rate.

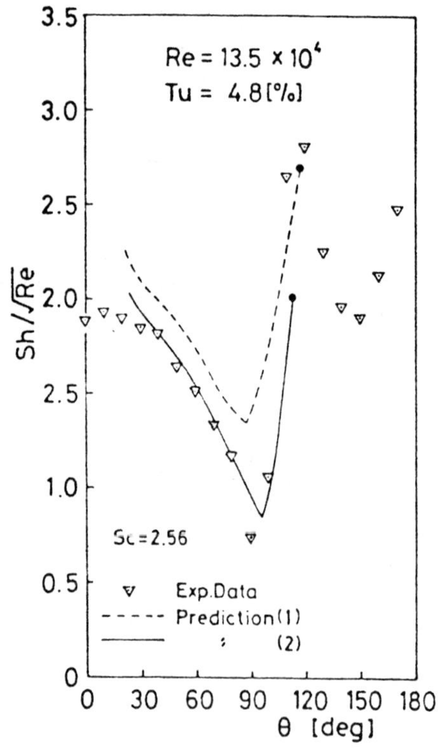

Fig.27. Comparison between the predicted and the measured local Sherwood numbers in supercritical flow range.

6. CONCLUSION

From the above investigations, the followings are to be concluded.

(1) The augmentation of the average Sherwood number with the increase of the free-stream turbulence intensity is mainly due to the peak of the local Sherwood number appearing at 110° or 120° from the front stagnation point and due to the increase at the rear hemisphere. In the laminar boundary layer on the front hemisphere, the local Sherwood number does not increase remarkably with the turbulence intensity until it becomes 4 or 5%.

(2) The flow around a sphere develops the laminar boundary layer to its separation in the subcritical flow range, while, in supercritical flow range, the laminar boundary layer to its separation, the separated, turbulent free-shear layer to its reattachment, the turbulent boundary layer to its separation and the impingement of the reverse flow. In the critical flow range, the reattachment occurs intermittently. The height of the peak of the local Sherwood number may be closely related to the intermittency factor of the reattachment.

(3) The laminar separation point moves downstream with the increase of the free-stream turbulence intensity. This movement is more remarkable as the Reynolds number is higher. The first minimum point of the local Sherwood number is found to agree with the laminar separation point.

(4) The empirical formula of the overall heat and mass transfer was obtained as equation (10).

(5) The turbulence energy at the edge of the boundary layer around the sphere (Tu') was found to be rather the weakest at the front stagnation-point and nearly equal to the turbulence energy of the approaching flow (Tu). The measured Tu' took a maximum value of 2.1 Tu at $\theta=20°$ and its variation was independent of Reynolds number.

(6) The excellent precision of the experiments was confirmed by the agreement between the experimental results and the theoretical predictions of the local mass transfer rates of the laminar boundary layer around the sphere in the uniform flow without free-stream turbulence.

(7) The predictions showed good agreement with the experiments in the subcritical or critical flow-condition but were rather poor in the super-critical flow-condition.

(8) The present model of turbulence is not applicable to the boundary layer in which the turbulence energy increases downstream at its free boundary.

(9) Further improvement of the model is to be needed, especially near the free boundary.

ACKNOWLEDGEMENTS

We are grateful to many staffs and students of our Heat Transfer Laboratory for their various contributions in performing the experiments.

NOMENCLATURE

- C, may stand for mass fraction or enthalpy;
- C_d, drag coefficient of sphere;
- C_p, pressure coefficient of sphere;
- d, sphere diameter;
- k, turbulence kinetic energy;
- m, exponent of Reynolds number;
- M, grid mesh dimension;
- n, exponent of Prandtl or Schmidt number, Eq.(9);
- Nu, local Nusselt number;
- $\overline{Nu}$, average Nusselt number;
- p, pressure;
- Pr, Prandtl number;
- r, radial coordinate, Fig.16;
- Re, Reynolds number;
- Re_{cr}, critical Reynolds number;
- Sc, Schmidt number;
- Sh, local Sherwood number;
- $\overline{Sh}$, average Sherwood number;
- t, time scale, x/U_0;
- T, temperature;
- Tu, free-stream turbulence intensity, $\sqrt{\overline{u^2}}/U$;
- Tu', turbulence intensity, Eq. (30);
- Tu_∞, turbulence intensity at free boundary, Eq. (29);
- u, fluctuating component of streamwise velocity;
- u_*, friction velocity, $\sqrt{\tau w/\rho}$;
- U, mean velocity in the x direction;
- U_0, velocity in empty duct or approaching velocity, Fig.1;
- V, mean velocity in the y direction;
- x, streamwise coordinate, Fig.1 or 16;
- y, cross-stream coordinate, Fig.16;
- y^+, dimensionless value of y, $\rho u_* y/\mu$;
- ε, turbulence kinetic energy dissipation rate
- θ, angle subtended at center of sphere, measured from front stagnation point;
- μ, dynamic viscosity;
- ρ, density;
- σ, molecular Schmidt or Prandtl numbers;
- τ, shear stress;

Subscripts

- t, denotes equivalent turbulent value;
- w, denotes wall value;
- ∞, denotes value at the edge of free boundary of the layer;
- overbars, signify space or time averages as appropriate.

REFERENCES

1. Cary, J.R. 1953. The determination of local forced-convection coefficients for spheres. Trans. ASME, Vol. 75, pp. 483-487.

2. Garner, F.H. and Grafton, R.W. 1954. Mass transfer in fluid flow from a solid sphere. Proc. Roy. Soc. (London), A224, pp. 64-82.

3. Sato, K. and Sage B.H. 1958. Thermal transfer in turbulent gas streams - Effect of turbulence on macroscopic transport from spheres. Trans. ASME, Vol. 80, pp. 1380-1388.

4. Wadsworth, J. 1958. The experimental examination of the local heat transfer on the surface of a sphere when subjected to forced convective cooling. NRC of Canada report MT. 39.

5. Yuge, T. 1960. Experiments on heat transfer from spheres including combined and forced convection. J. Heat Transfer, Vol. 82, pp. 214-220.

6. Short, W.W., Brown R.A.S. and Sage B.H. 1960. Thermal transfer in turbulent gas streams - Effect of turbulence on local transport from spheres. Trans. ASME, Ser. E, Vol. 27, pp. 393-402.

7. Brown, W.S., Pitts, C.C. and Leppert, G. 1962. Forced convection heat transfer from a uniformly heated sphere. J. Heat Transfer, Vol. 84, pp. 133-140.

8. Raithby, G.D. and Eckert, E.R.G. 1968. The effect of turbulence parameters and support position on the heat transfer from Spheres. Int. J. Heat Mass Transfer, Vol. 11, pp. 1233-1252.

9. Galloway, T.R. and Sage B.H. 1968. Thermal and material transfer from spheres - Prediction of local transport. ibid., Vol. 11, pp. 539-549.

10. Pei, D.C.T. 1969. Effect of tunnel blockage and support on the heat transfer from spheres. ibid., Vol. 12, pp. 1707-1709.

11. Gostkowski, V.J. and Costello, F.A. 1970. The effect of free stream turbulence on the heat transfer from the stagnation point of a sphere. ibid., Vol. 13, pp. 1382-1386.

12. Newmann, L.B., Sparrow, E.M. and Eckert, E.R.G. 1972. Free stream turbulence effects on local heat transfer from a sphere. J. Heat Transfer, Vol. 94, pp. 7-16.

13. Hayward, G.L. and Pei, D.C.T. 1978. Local heat transfer from a single sphere to a turbulent air stream. Int. J. Heat Mass Transfer Vol. 21, pp. 35-41.

14. Raithby, G.D. and Eckert, E.R.G. 1968. The effect of support position and turbulence intensity on the flow near the surface of a sphere. Waerme-und Stoffuebertragung, Vol. 1, pp. 87-94.

15. Priddin, C.H. 1975. The behavior of the turbulent boundary layer on a curved, porous walls. Ph. D Thesis, University of London.

16. Jones, W.P. and Launder, B.E. 1973. The calculation of low Reynolds number phenomena with a 2-equation model of turbulence. Int. J. Heat Mass Transfer, Vol. 16, pp. 1119-1130.

17. Spalding, D.B. 1977. A general computer program for two-dimensional parabolic phenomena. Williams Clowes & Sons, Ltd.

18. Torii, K., Isobe, M., Miura, N. and Horikoshi, T. 1981. Free-stream turbulence effects on heat and mass transfer. 1st. and 2nd. reports. Bull. JSME, Vol. 24-187.

19. Sato, Y. and Yamamoto, K. 1974. The local structure of isotropic turbulence. Bull. of the Faculty of Engineering, Yokohama National University, Vol. 23, pp. 91-120.

20. Vliet, G.C. and Leppert, G. 1961. Forced convestion heat transfer from an isothermal sphere to water. J. Heat Transfer, Vol. 83, pp. 163-175.

21. Merk, H.J. 1959. Rapid calculations for boundary-layer transfer using wedge solutions and asymptotic expansions. J. Fluid Mech., Vol. 5, pp. 460-480.

22. Taneda, S. 1978. Visual observations of the flow past a sphere at Reynolds numbers between 10^4 and 10^6. J. Fluid Mech., Vol. 85-1, pp. 187-192.

23. Sucker, D. und Brauer, H. 1975. Fluid dynamik bei quer angestroemten Zylindern. Waerme-und Stoffuebertragung, Vol. 8, pp. 149-156.

24. Launder, B.E., Priddin, C.H. and Sharma, B.I. 1977. The calculation of turbulent boundary layers on spinning and curved surfaces. J. Fluid Engng., pp231.

25. Fage, A. 1936. Experiments on a sphere at critical Reynolds numbers. A.R.C. R. & M. no. 1766.

26. Schlichting, H. 1968. Boundary layer theory. 6th ed., pp 87, McGraw Hill.

Some Characteristics of Turbulent Flows at Elevated Temperatures

I. GÖKALP, P. GOUGAT, A. LASEK, and F. MARTIN
Laboratoire d'Aérothermique de C.N.R.S.
4ter, route des Gardes
92190 Meudon, France

ABSTRACT

Turbulent dynamic field at the exit section of a vertical tube of 50 mm is investigated for Reynolds numbers ranging from 5 000 to 20 000 and for mean flow temperatures ranging from the ambient to 510°C. Laser Doppler Anemometry is used. Mean axial velocity and its fluctuations' profiles are not modified by the temperature increase. On the axis of the exit section, turbulence intensities show a dependence on the mean flow temperature for constant Reynolds numbers. One-dimensional power spectra are modified for elevated temperature flows: their microscalar fractions become less energetic. Longitudinal integral scales and dissipation microscales increase with increasing mean temperature.

1. INTRODUCTION

Dynamic and thermal turbulent field studies are generally effectuated in the validity domain of the passive contaminant hypothesis for the temperature. Following this hypothesis the mean flow temperature and its fluctuations are low enough so that the density and the transport coefficients' variations do not modify the dynamic field.

This hypothesis seems justified when the objective is to study the dynamic field and/or the thermal field convected by the former in weakly heated turbulent flows. But, in several practical applications this hypothesis is not valid. A pertinent example is the case of turbulent combustion where the mean temperature and also the temperature fluctuations are high.

In conventional turbulence studies, the influence of mean temperature is taken into consideration through the bias of kinematic viscosity which depends on temperature. As a matter of fact, the following statements are generally assumed :

a) the Reynolds number involving the kinematic viscosity is the universal parameter determining the behaviour of the velocity fluctuations in the absence of buoyancy forces ;

b) when considering two dynamic turbulent fields having respectively viscosities ν_1 and ν_2, the microscalar fractions of their power spectrum are similar in a normalized representation involving the Kolmogoroff microscale and the kinematic viscosity [1] ;

c) when a thermal field is convected by a dynamic field and when the Prandtl number is equal unity, the microscalar fractions of the dynamic and thermal spectra are analogous [2].

From the point of view of the turbulent combustion, these three statements seem to be questionable at least regarding the influence of the mean temperature. In particular, the following questions may be raised in turbulent flows at elevated mean temperature :

a) is the incorporation of the mean temperature effect in the Reynolds number through the viscosity sufficient to determine the behaviour of the velocity fluctuations ?

b) is, the similarity between the microscalars of the power spectra conserved in turbulent flows at different elevated mean temperatures ?

c) is the analogy between the dynamic and thermal fields is conserved in turbulent flows at elevated mean temperature with Prandtl number being always unity ?

Starting from a similar point of view, Lin and Lin [3] tested the validity domain of the passive contaminant hypothesis for the temperature. Their results obtained for mean temperature increases less than 100°, show no difference in the spectral behaviour of the temperature fluctuations.

In a previous study [4], we investigated by resistance thermometers (cold wires), the behaviour of turbulent thermal fields with increasing mean temperature in a pipe. This work has shown appreciable spectral differences for a temperature increase of 400°.

Present work is an extension of this investigation to turbulent dynamic fields with the use of Laser Doppler Anemometry and where the exit section of a tube is chosen as the test section for the commodity of the optical measurements. This choice of the test section brings a new dimension to the investigation. Indeed, whilst the main objective is not the study of turbulent jets, we compared our spectral measurements to those performed in jet studies. These comparisons are however limited because spectral measurements in jets are generally concerned with sections at least 2 diameter downstream of the exit section.

2. EXPERIMENTAL SET UP AND SIGNAL PROCESSING

An axisymmetric vertical turbulent jet is established at the exit of a brass tube of 700 mm length and 50 mm in diameter. The tube is placed downstream of a centrifugal fan. The flow is heated with an electrical heater of 10kW power, located between the fan and the tube. This heater is a cylindrical block of keramic of 200 mm length and 90 mm in diameter. The block is traversed by 60 channels of 5,5 mm diameter, containing the electrical resistances and through which the air flows before entering the tube.

The mean velocity of the cold flow may be regulated between 1 m/s and 20 m/s. The mean temperature of the flow may be regulated between the ambient and 600°C and it is measured by a chromel-alumel thermocouple probe.

In dynamic field measurements, the laser Doppler velocimeter is used in forward scatter. An optical base, on both sides of the issuing jet, permits a simultaneous displacement of the receiving and transmitting optics with an accuracy of 1/10 mm in the horizontal direction and 1/100 mm in the vertical one. The Argon laser of 2 Watts remains fixed. The flow is seeded at the entrance of the fan with zirconium oxyde.

The Doppler signal obtained through the photodetector is processed in a counter, giving the mean axial velocity $\overline{U}$ and its fluctuations $\sqrt{\overline{u^2}}$, Figure 1 shows the block diagram of the measuring equipment. The analogic signal supplied by the counter is tape-recorded and used to obtain the power spectral density of the velocity fluctuations by a real time spectrum analyser [5]. The analyses are performed on frequency ranges from 0 to 5 000 Hz. The counter gives a data rate of 20 000 per second which is greater than three times the maximal frequency of the 0-5 000 Hz range [6].

3. MEAN VELOCITY AND VELOCITY FLUCTUATIONS MEASUREMENTS

Figure 2 shows, for the exit section (z/R=0), the radial profiles of the mean axial velocity $\overline{U}$ normalized by the centerline velocity U_c, and of the intensity of the axial velocity fluctuation $u/\overline{U}$, for mean temperature increases ranging from $\Delta T = 0°$ to $\Delta T = 490°$. These profiles are similar to those in fully developed pipe flows and present no modification with increasing mean temperature. Mean temperature profiles presented on figure 3 are not modified by the temperature increase.

One point measurements are performed on the center of the exit section for $\overline{U}$, and u. Figures 4 and 5 present the behaviour of these quantities versus the temperature increase for different Reynolds numbers based on the tube diameter. It is observed that not only $\overline{U}$ but also u increases with increasing mean temperature.

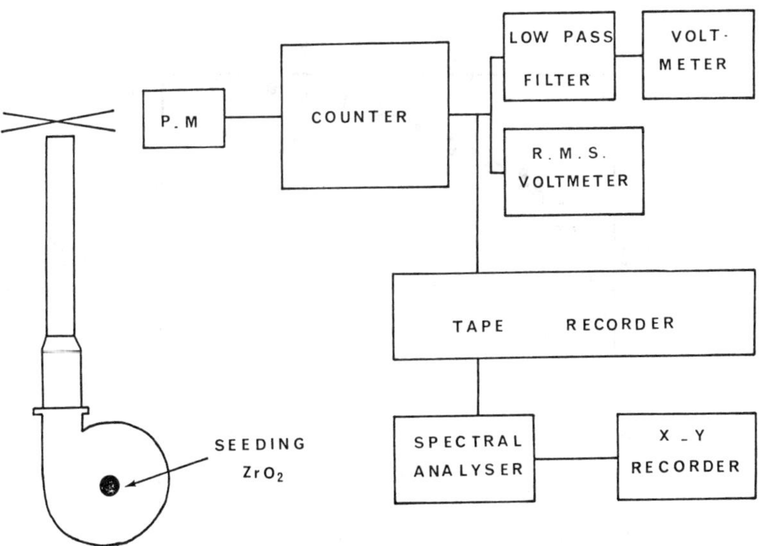

Fig. 1. Experimental set up and signal processing.

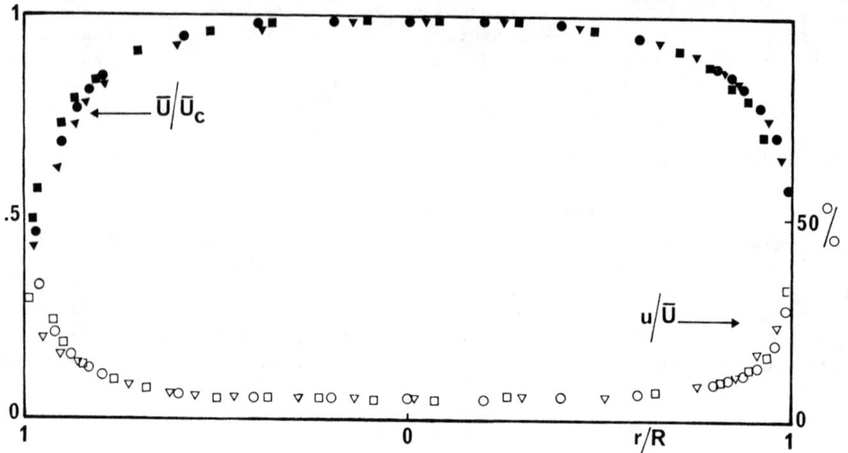

Fig. 2. Mean axial velocity and its fluctuations' profiles
●, O : $\overline{U}_c$ = 7,6 ms^{-1} ; ΔT = 0°
▼, ∇ : $\overline{U}_c$ = 16,6 ms^{-1} ; ΔT = 374°
■, □ : $\overline{U}_c$ = 19 ms^{-1} ; ΔT = 490°

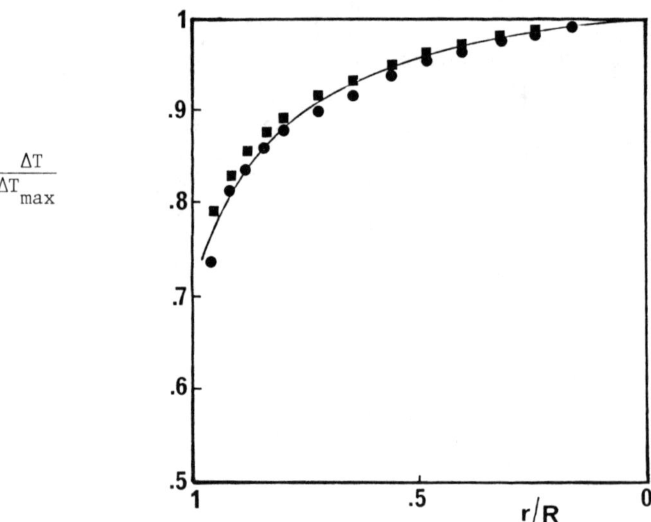

Fig. 3. Mean temperature increase, $\frac{\Delta T}{\Delta T_{max}}$ profiles

● ΔT = 258 ; ■ ΔT = 490

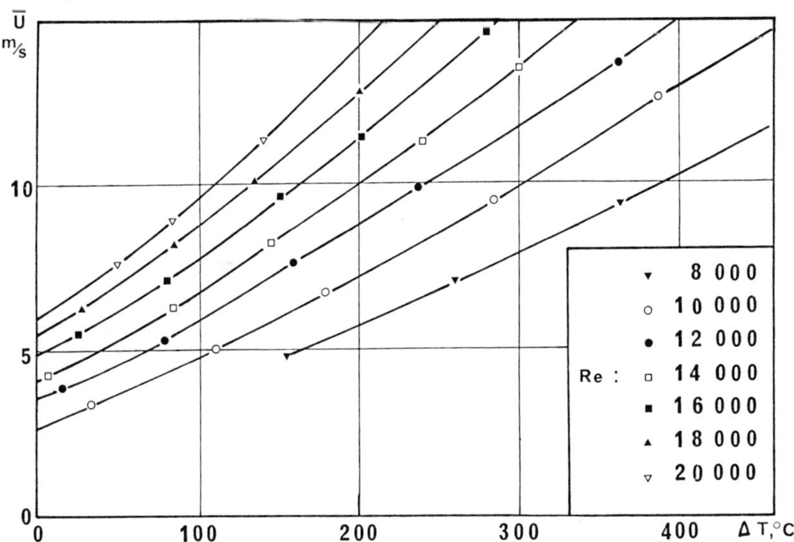

Fig. 4. Axial mean velocity $\overline{U}$ versus ΔT for different Reynolds numbers, at z/R=0, r/R=0.

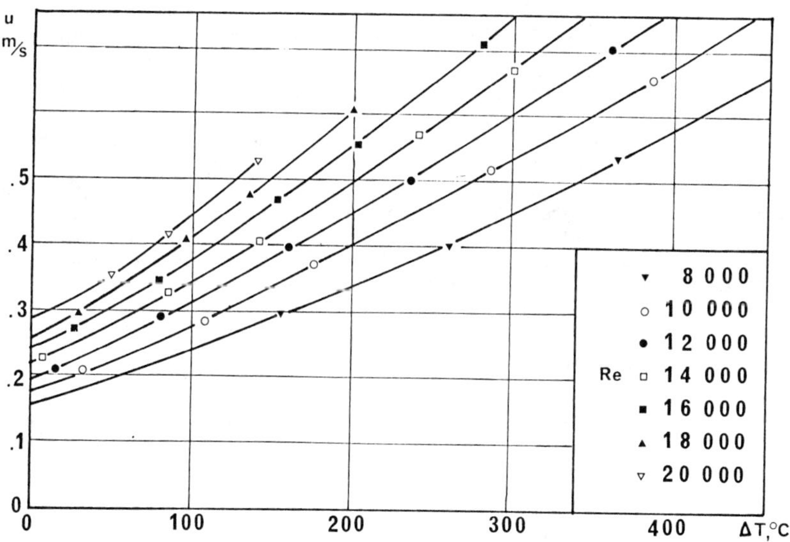

Fig. 5. Axial velocity fluctuations u versus ΔT for different Reynolds numbers, at z/R=0, r/R=0.

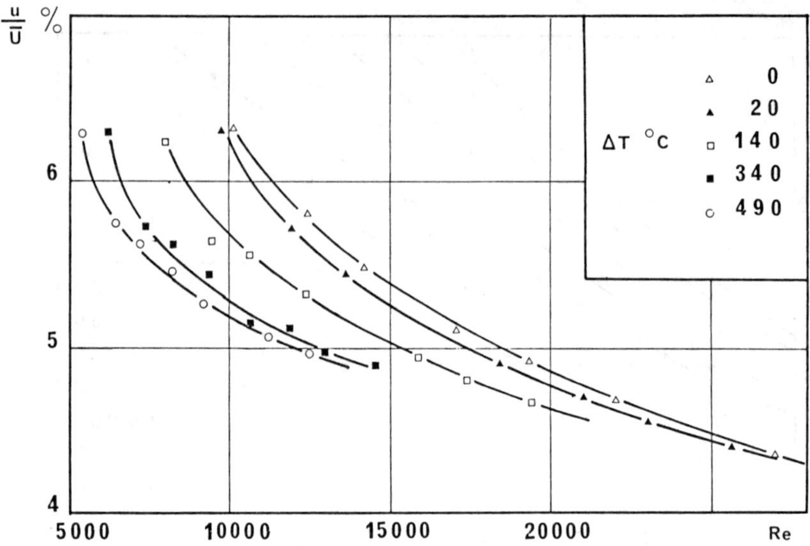

Fig. 6. Axial turbulence intensity versus the Reynolds number for different mean temperature increases, at z/R=0, r/R=0.

Figure 6 shows the behaviour of the turbulence intensity $u/\overline{U}$ versus the Reynolds number for different mean temperature increases. Two remarks follow from these curves :

- one usual, concerns the decrease of the turbulence intensity with the Reynolds number,
- the second remark concerns an additional dependence of the turbulence intensity on the mean temperature increase. It is noteworthy that this last dependence occurs whilst the influence of the temperature is already involved in the Reynolds number through the kinematic viscosity.

It is also observed that these curves seem to tend to a unique one for higher Reynolds numbers. This tendency may correspond to the results obtained in fully developed pipe flows [7].

4. SPECTRAL MEASUREMENTS

Normalized one dimensional spectra, $E_1(k_1)/\overline{u^2}$ for $\Delta T = 0°$ at (z/R=0, r/R=0) ; (z/R=-1, r/R=0) and (z/R=0, r/R=0.9) are presented on figure 7 in logarithmic coordinates. These spectra correspond to a mean exit velocity of 4.5 m/s giving a Reynolds number of 14 000.

As observed in many other measurements, as well with hot wire as with Laser Doppler Anemometry [8,9,10], the velocity spectra have pronounced peaks in the potential core of the jet. Present measurements (fig. 7) show this peak at the center of the exit section (z/R=0). It is interesting to note that the spectrum at z/R=-1, that is inside the tube, presents also the same peak. In

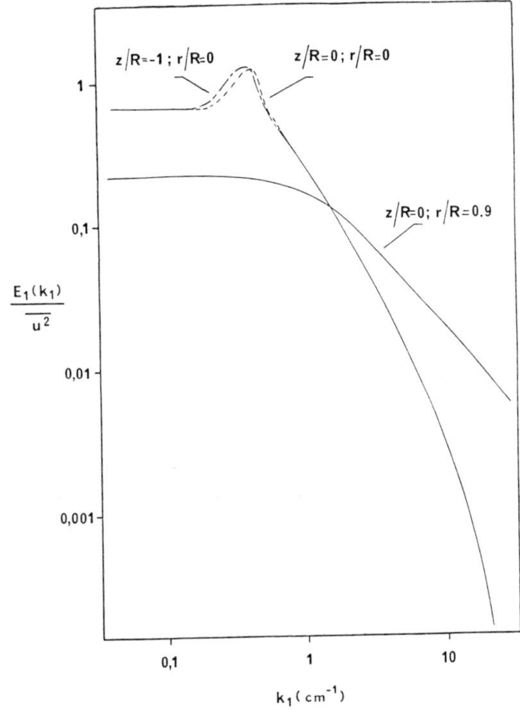

Fig. 7. Normalized one-dimensional spectra for Re = 14 000 at ambient temperature.

fact, spectra at z/R=0 and z/R=-1 are superposed. This indicates that even at z/R=-1, the laser signal contains informations about the local turbulence fluctuating velocity, and also about the hydrodynamic pressure field associated with the vortices formed at the jet exit. In the mixing region (z/R=0, r/R=0.9) the spectrum peak disappears and the spectrum falls monotically with increasing frequency, due to the masking of the pressure by the local turbulence.

On the other hand, as shown on figure 8, the spectral peak decreases and disappears with increasing exit velocity. This result is consistent with the observations of Ko and Davies [9] for a constant exit velocity of 72 m/s. Indeed, these authors relate a complete absence of any peak, except for the 50 Hz hum, within the first quarter diameter downstream of the nozzle exit, indicating that no near field pressure is present in the signal ; however, they observe this peak for this high velocity at section farther downstream.

The observed spectral peak can be expressed in terms of the Strouhal number, $f_p D/U$, where f_p, D and U are respectively the peak frequency, jet diameter and exit velocity. Present results yields a Strouhal number of 0.34, which agrees with previous hot wire results at comparable exit velocities.

On figure 7, spectra corresponding to the axial positions present a flat and constant spectral level at low wave number side up to the spectral peak. An inertial subrange with a universal slope of -5/3 is observed in the wave number range between k_1 = 0.5 cm^{-1} and 10 cm^{-1}. For higher wave numbers, the slope increases to -3.7.

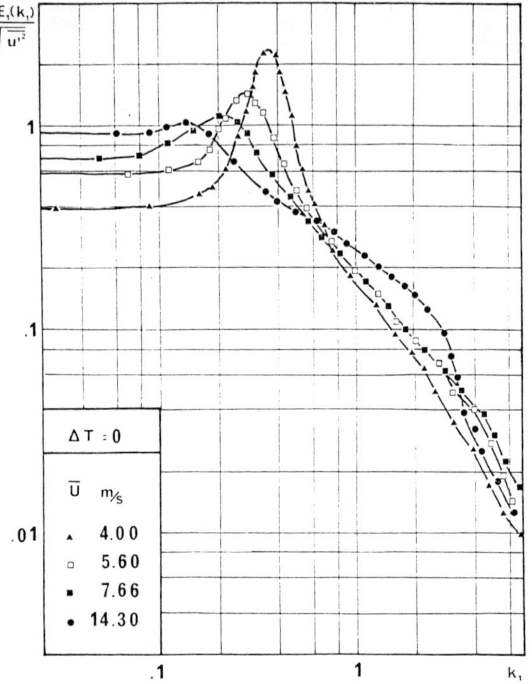

Fig. 8. Normalized one-dimensional spectra at ambient temperature for different exit velocities.

The longitudinal integral scale, determined from the one-dimensional spectrum by

$$L_u = \frac{\pi}{2} \lim_{k_1 \to 0} \frac{E_1(k_1)}{\overline{U^2}}$$

is 11.1 mm at z/R=0 and 10.2 mm at z/R=-1. The longitudinal dissipation microscale, obtained from the integration of the dissipation curve as

$$\frac{1}{\lambda_u^2} = \int_0^\infty k_1^2 \frac{E_1(k_1)}{\overline{u^2}} dk_1$$

is 3,8 mm at z/R=0 and 3.65 mm at z/R=-1. The corresponding turbulence Reynolds number is 40.

The u-component spectrum corresponding to the mixing region (z/R=0, r/R=0.9) is also presented on figure 7. Compared to the two others, this spectrum is less energetic in its macroscalar fraction and more energetic in its microscalar fraction. In a wave number range between $k_1 = 2\ cm^{-1}$ and $30\ cm^{-1}$ it presents a reduced slope of -1.1 ; the slope is -2.6 for higher wave numbers. This spectrum gives an integral scale of 3.45 mm and a dissipation microscale of 0.5 mm.

Figure 9 presents u-component spectra at the same positions as previously but for an elevated temperature flow with ΔT = 490°. With a mean exit velocity of 18 m/s, the Reynolds number is kept constant at 14 000. The spectral peak is also observed in this hot case at axial locations ; it disappears in the mixing region. The corresponding Strouhal number is decreased compared to the cold flow

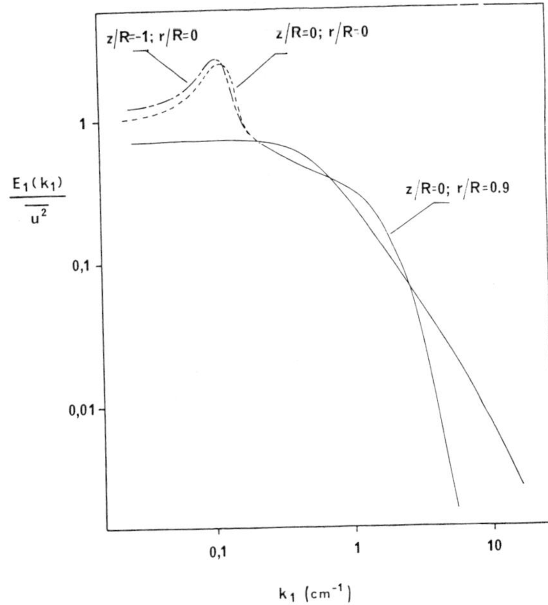

Fig. 9. Normalized one-dimensional spectra for Re = 14 000 at ΔT = 490

and is equal to 0.2. In these measurements at constant Re, both velocity and temperature vary between cold and hot cases. Other measurements at constant velocity and for different temperatures show that there is no temperature effect on the Strouhal number (fig. 10).

For axial spectra, the peak is followed by a reduced slope region (-1.1) up to $k_1 = 1.2\ cm^{-1}$. The slope of -5/3 is observed in a narrower range comprised between $k_1 = 1.4\ cm^{-1}$ and $2.5\ cm^{-1}$. Higher wave numbers present a steeper slope of -4.7. The integral scale is increased in the hot case and reaches 18.5 mm at z/R=0 and 15.9 mm at z/R=-1. The dissipation microscale is also increased : it reaches 8 mm at z/R=0 and 7.8 mm at z/R=-1. The turbulence Reynolds number becomes 65. The spectrum corresponding to the mixing region presents approximately the same slopes as in the cold flow. Its integral scale increases to 11.5 mm and its dissipation microscale to 2.2 mm.

The comparison of the u-component spectral behaviour between cold and hot flows, at the same Reynolds number, is given on figure 11, where the spectra $k_1E_1(k_1)/u^2$ versus k_1, corresponding to the axis of the exit section, are shown in a semi logarithmic scale.

In the cold flow case, the spectral peak due to the near field pressure and the one corresponding to the most energetic eddies are superposed and a single peak appears centered around $k_1=0.4cm^{-1}$. In the hot case, the near field pressure peak is shifted towards lower wave numbers, which become more energetic. On the contrary, the high wave numbers of the hot flow spectrum are less energetic than the cold ones. This result agrees with the increase of the dissipation microscale with increasing mean temperature.

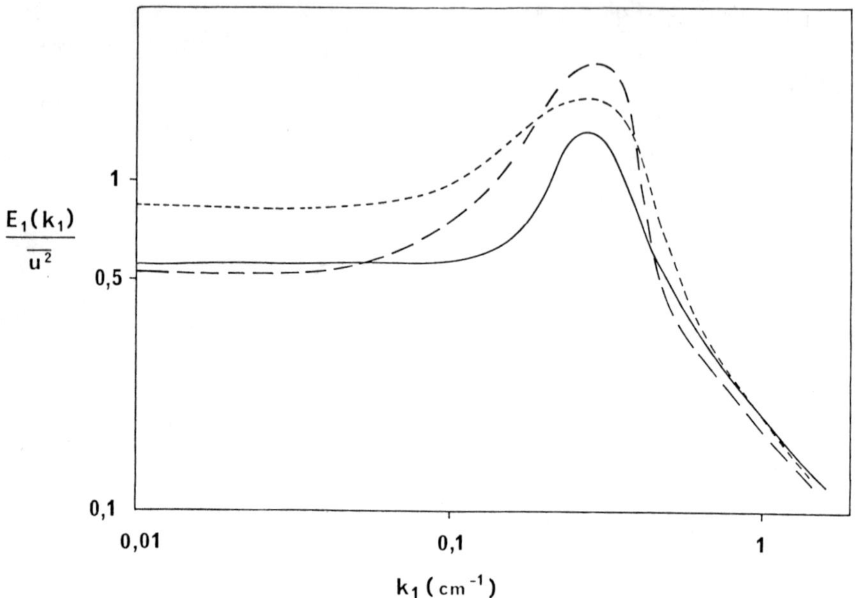

Fig. 10. Normalized one-dimensional spectra at a constant exit velocity $\overline{U}$ = 5,6 ms^{-1} for different mean temperatures, at z/R=0, r/R=0. — ΔT = 0° ; – – ΔT = 250° ; --- ΔT = 550°.

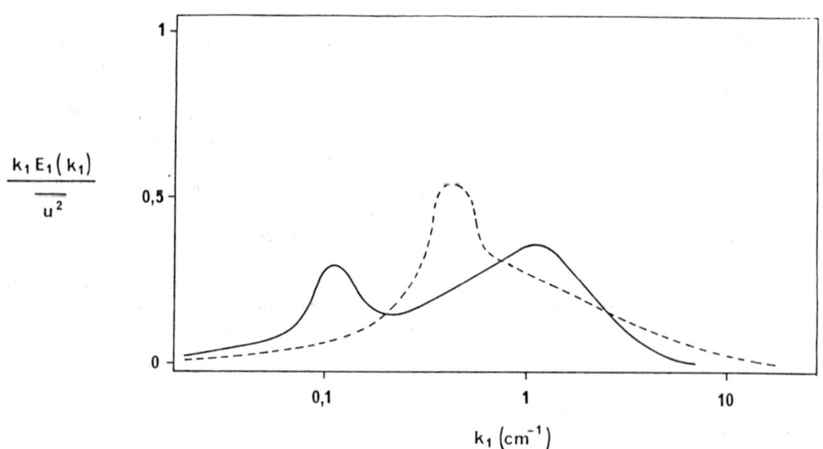

Fig. 11. Normalized one-dimensional spectra for Re = 14 000, at z/R=0, r/R=0. --- ΔT = 0° ; — ΔT = 490°.

In order to check spectral comparisons, it is necessary to discuss the Laser Doppler Anemometry accuracy in non-isothermal flows. Indeed, Mizushina et al. [11] pointed out that in non-isothermal turbulent flows, the laser beams encounter a succession of hot and cold eddies of various temperatures, which modify the phase and direction of laser beams and result in a measured apparent velocity. In such cases, corrections are necessary for the turbulent intensities and the power spectral density functions. In particular, the high frequency range of the velocity spectra is over-estimated because of the contribution of a fictitious velocity to the apparent velocity. In the present measurements the eventuality of such an error is not taken into account. However, the following remark can be made. The objective of the presented spectral measurements is to compare them between the cold and the hot turbulent flow cases. As mentioned above, these comparisons show that the high frequencies are less energetic in the hot case. On the other hand, absolute temperature fluctuations increase with increasing mean temperature for a given cold flow dynamic turbulence [4]. We can them assert that the spectral error due to the interaction of the laser beams with turbulent temperature fluctuations, which must increase with increasing temperature, does not alter our spectral comparisons.

5. CONCLUSION

The principal conclusions of this study are the following :

1) The normalized mean axial velocity and mean temperature radial profiles, at the exit section of a vertical tube, are not modified by the temperature increase. This is also true for the radial profiles of the axial velocity fluctuations.

2) The behaviour of the turbulence intensity $u/\overline{U}$ versus the Reynolds number is less conventional. Indeed, whilst the Reynolds number stays constant, $u/\overline{U}$ is still affected by the temperature increase in the Reynolds number ranges considered.

3) The spectral behaviour of the axial velocity fluctuations is modified by the mean temperature increases for flows at constant Reynolds numbers. The microscale fractions of the spectra at elevated temperatures become less energetic in normalized scales. The cold flow u-spectrum presents a well-defined inertial subrange. The corresponding hot flow spectrum presents the slope of -5/3 in a narrow wave number band, in spite of its increased turbulence Reynolds number. These dynamic field spectral observations agree with our previous results concerning thermal fields in turbulent flows at elevated temperatures [4].

4) Longitudinal integral and dissipation scales increase with increasing mean temperature at constant Reynolds numbers.

ACKNOWLEDGMENTS

Part of this work is financially supported by the A.T.P. "Combustion Turbulence". Contract C.N.R.S.

The authors would like to thank Mrs. Fleury for the careful typing of the manuscript.

REFERENCES

1. Rey, C. and Mathieu, J. 1978. Etude spectrale des mécanismes d'échange entre champ cinématique et champ scolaire. International J. of Heat and Mass Transfer, vol.21, n° 8, pp. 1009-1018.

2. Fulachier, L. and Dumas, R. 1978. Spectral analogy between temperature and velocity fluctuations in a turbulent boundary layer. J. Fluid Mechanics, vol.77, pp. 257-278.

3. Lin, S.C. and Lin, S.C. 1970. Study of strong temperature mixing in subsonic grid turbulence. Physics Fluids, vol 16, pp. 1587-1598.

4. Gökalp, I. and Lasek, A. 1979. Fluctuations de température dans un écoulement turbulent à température élevée. International J. of Heat and Mass Transfer, vol 22, n° 9, pp. 1309-1317.

5. Gougat, P. and Martin, F. 1976. Analyse de Fourier et corrélation de vitesse en aérodynamique instationnaire. Agard Report n° 210.

6. Max, J. 1972. Méthodes et techniques de traitement du signal. Masson

7. Laufer, J. 1954. The structure of turbulence in fully developed pipe flow. NACA TN 2954

8. Bradshaw, P. Ferris, D.H. and Johnson, R.F. 1964. Turbulence in the noise-producing region of a circular jet. J. Fluid Mechanics, vol 19, pp.591-624.

9. Ko, N.W. and Davies, P.J.A.L. 1971. The near field within the potential core of subsonic jets. J. Fluid Mechanics, vol.50, pp. 49-78.

10. Lau, J.C. Morris, P.J. and Fisher, M.J. 1979. Measurements in subsonic and supersonic free jets using a laser velocimeter. J. Fluid Mechanics, vol.93, pp. 1-28.

11. Mizushina, T. Ogino, F. Ueda, H. and Komori, S. 1979. Application of Laser Doppler velocimetry to turbulence measurements in non-isothermal flow. Proceedings Royal Society London, A. 366, pp. 63-79.

Momentum and Temperature Fluxes in a Shock Wave—Turbulence Interaction

JEAN-FRANÇOIS DEBIEVE, HENRI GOUIN, and JEAN GAVIGLIO
Institut de Mécanique Statistique de la Turbulence
Laboratoire Associé au Centre National de la Recherche
Scientifique No. 130
12, Avenue Général Leclerc, 13003-Marseille, France
and Office National d'Etudes et de Recherches Aérospatiales
29, Avenue Division Leclerc, 92320-Châtillon sous Bagneux, France

ABSTRACT

An analysis is proposed, which allows to relate the mean flow and the Reynolds stress tensor evolution in a shock wave - turbulence interaction. From another classical analysis which is extended here to the case of a strong pressure gradient, the corresponding thermal field evolution is obtained.

1. INTRODUCTION

The shock wave - turbulence interaction is encountered in various configurations, for example in turbomachinery. This states the problem of understanding the behaviour of the turbulent field in such a non equilibrium flow, and more particularly the problem of predicting the Reynolds stresses and the heat fluxes evolutions. In order to try to explain experimental results dealing with both the mechanical and the thermal aspects of the phenomena, an analysis is made. Besides, for an equilibrium turbulent flow, when the Reynolds stresses are known, it is possible to obtain information on the correlative heat fluxes. This approach uses the classical analogies based on the hypothesis that the thermal turbulent field is diffused by the vorticity field. The aim of this paper is to answer the question: is it possible to extend such a classical way to a shock wave - turbulence interaction?

As a consequence, at first the Reynolds stress behavior is analyzed through a transformation of its evolution equation. This does not lead to a complete calculation method, but it provides an algebraic expression relating the value of the Reynolds stress on both sides of the shock wave to the mean flow properties.

Then, the validity of classical relations between the velocity field and the thermal field is examined, mainly the " Strong Reynolds Analogy " relation which is commonly used in supersonic flows. The value of the analysis was checked against experiments performed in the case of a turbulent boundary layer interacting with a shock wave on a corner, without separation (fig. 1). It must be noted that the analysis does not concern specific problems of the whole layer, but only the part of the turbulent flow in which the apparent thickness of the shock wave is negligible when compared to the tickness δ of the layer.

2. EXPERIMENTAL DATA

2.1. Apparatus Instrumentation

The experiments are made in a turbulent boundary layer on an adiabatic wall, submitted to a compression produced by a corner (6° incidence) without separation. They are performed at I.M.S.T. in the S. 8 nozzle of the continuous running supersonic wind tunnel; the dimensions at the sonic neck are 52 × 170 (mm × mm) . The nominal Mach number is $M_\infty = 2{,}32$; the nominal stagnation pressure is $P_{T_{ref}} \simeq .506 \cdot 10^5$ N/ m^2; the stagnation temperature is $\theta_o \simeq 293$ ° K.; the nominal Reynolds number based on the momentum thickness is $R_\theta \simeq 4{,}2 \cdot 10^3$ and the boundary layer thickness is $\delta \simeq 9{,}6$ mm. The experiments consist in classical mean flow measurements (schlieren observations; pressure data obtained by using probes; piezo resistant captors; and hot wire anemometry) ; and in turbulent field measurements performed using a constant current hot-wire anemometer. The bandwidth extends to 300 KHz (- 3 dB). Compensation of the thermal lag is adjusted in situ and corrections for compensation circuit imperfections are made [1 , 2]. The material of the wire is platinum plated tungsten. The diameter is ϕ = 3,8 microns as for mean stagnation temperature measurements, and 2,5 microns with a length of .7 mm as for turbulence measurements.

2.2. Mean Flow Data

Stagnation temperature distributions (fig. 5) were obtained by the hot - wire method given by Laufer and Mc Clellan [3]. It appears that the interaction induces only slight changes in the initial distribution. The wall pressure data are given on fig. 2. p_o increases through the interaction , over a distance at the wall which is approximately equal to the boundary layer thickness δ, $\Delta p_o / p_{o_{ups}} \simeq .4$. Static and total pressure are measured in the whole field.

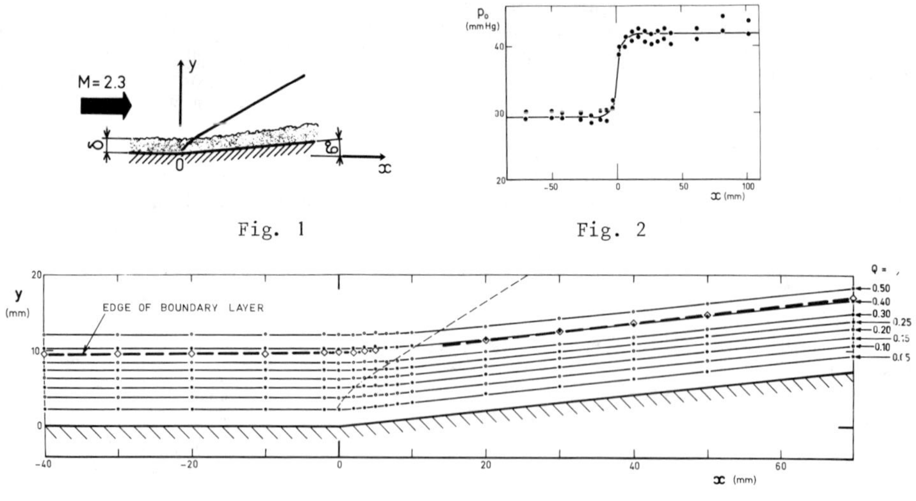

Fig. 1

Fig. 2

Fig. 3

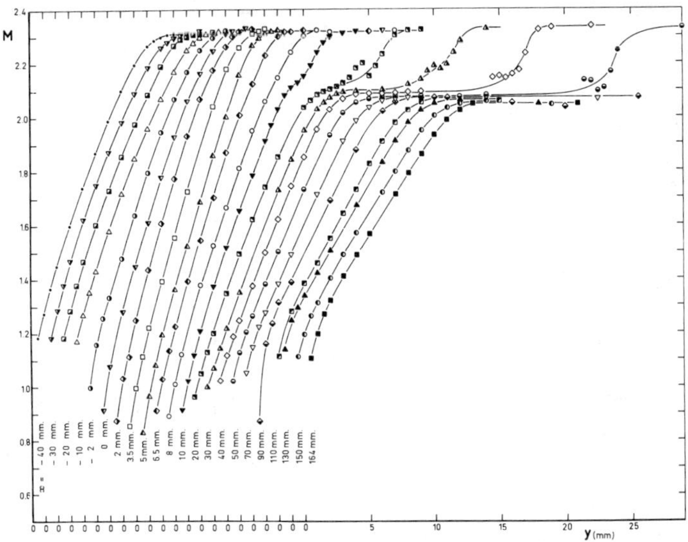

Fig. 4

These measurements allow to determine : the Mach number profiles, fig. 4 ; the mean streamlines, on which the mean mass flux $Q = \int_0^y \bar{\rho}\tilde{U}\,dy$ is constant, (fig. 3). $\bar{\rho}$ is the mean specific mass, $\tilde{U}$ is the mean mass weighted velocity [4].

Integral parameters have been determined : displacement thickness δ^*, momentum thickness θ and the shape factor H, fig. 6. Excepted in the shock wave region $x \simeq 0$, these parameters are not much altered by the interaction effect. The skin friction coefficient c_f is computed by using a two parameter formula proposed by Chew [5] for such flows. This formula is an extension to compressible boundary-layers of a well known relation due to Ludwieg and Tillman, fig. 7. The interaction does not change much the friction coefficient c_f. The downstream values of c_f are slightly lower than the upstream values, as shown by Adamson and Messiter [6], in the case of turbulent supersonic boundary layers.

2.3. Turbulent Flow Data

Velocity and temperature variances of the fluctuations, $\overline{u'^2}$, $\overline{\theta'^2}$ and the correlation coefficients $R_{u\theta}$ were obtained by hot-wire anemometry ; the classical Morkovin's and Kovasnay's fluctuation diagram techniques were used. The interpretation of the hot - wire data is based on the hypothesis that, out of the shock, like in a quasi equilibrium flow (Laufer, [7]), " within the layer, the fluctuations may be described primarily in terms of vorticity and entropy modes", the sound mode being negligible.

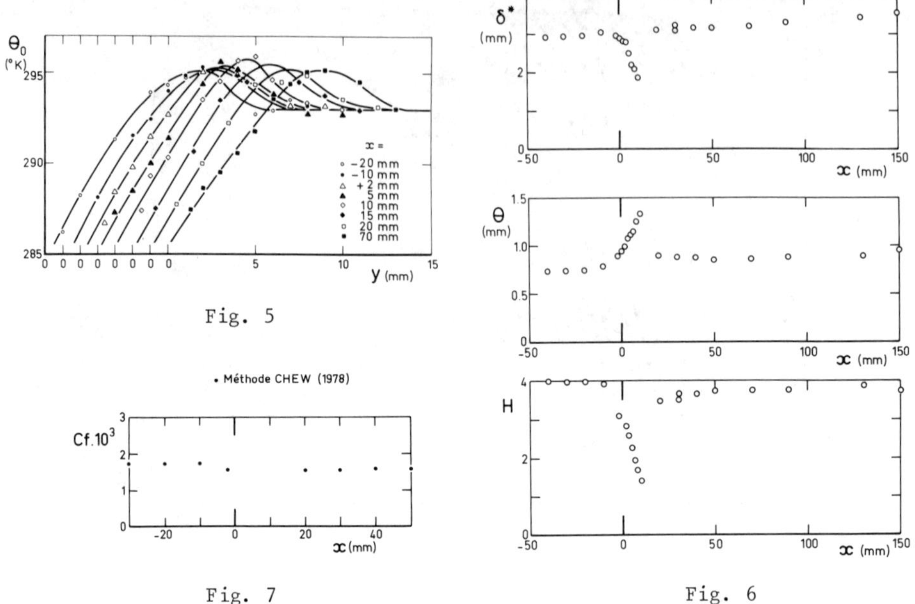

Fig. 5

Fig. 7

Fig. 6

At first, the evolution of $\overline{u'^2}$, $\overline{\theta'^2}$ and $R_{u\theta}$ have been carried out through the shock, along a mean streamline, corresponding to the mass flux Q = .1 ($y/\delta \simeq .4$ upst.). In this case fig. 8 shows an increase (or a "step") about 35 %, through the shock, of longitudinal velocity and temperature variances $\overline{u'^2}$ and $\overline{\theta'^2}$. The correlation coefficient between longitudinal velocity and thermal fluctuation $R_{u\theta}$ keeps a value about - 0.85 which is not significantly changed by the interaction.

Then, the evolution of the velocity and turbulence intensities and their correlation coefficients were carried out through the shock, in two sections of the flow, upstream (x = - 15 mm, $x/\delta \simeq -1,6$) and downstream (x = + 20 mm, $x/\delta \simeq 2,1$) . Fig. 9.

Whatever the distance to the wall , there is an increase of the levels through the shock. Anew, $R_{u\theta}$ remains practically unchanged.

As for the total temperature fluctuations, their level $\sqrt{\overline{\theta_o'^2}}/\overline{\theta}$ remains practically unchanged through the shock and is lower than the level of the temperature fluctuations θ'; but it is not negligible.

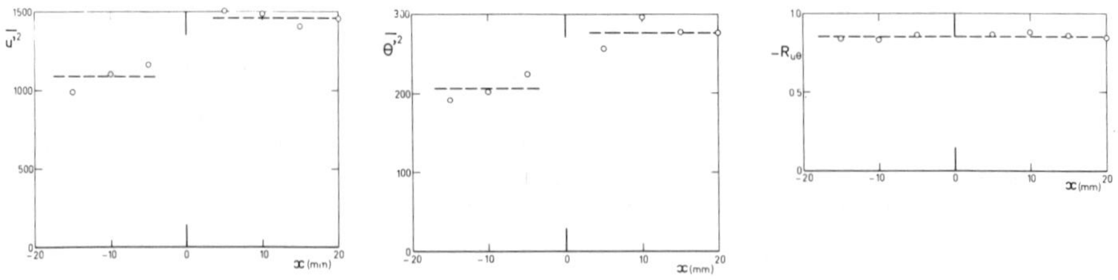

Fig. 8

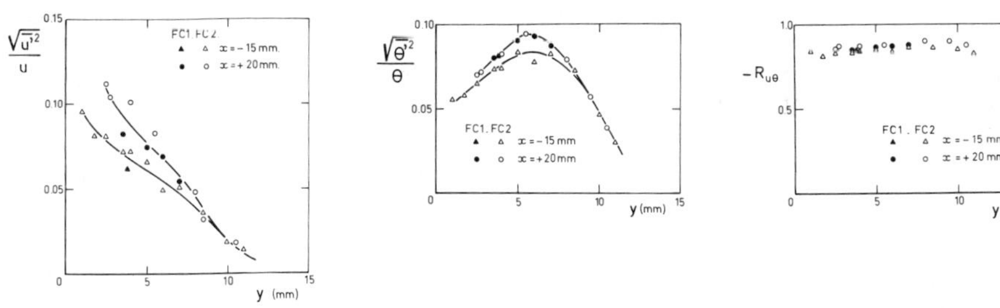

Fig. 9

3. THE VELOCITY TURBULENT FIELD

3.1. Analysis of the Reynolds Stress Evolution in a Shock Wave - Turbulence Interaction

Because of the presence of unknown and very strong gradients in the shock region, classical approaches based on the Reynolds stress equation are difficult to set up. So, the Reynolds stress evolution is here examined using a Lagrangian type transformation of the Reynolds stress transport equation.
This enables to separate the geometrical effects due to the strong mean deformation, from the specific effects corresponding to turbulent source terms.
A discontinuity model will provide an algebraic expression relating the Reynolds stress values on both sides of the interaction expressed as a function of the mean flow and of sources located on the shock.

Transformation of the Reynolds stress equation. In order to apply the Reynolds stress equation to discontinuous flows, let us transform this equation. With classical notations, the latter is written, in a continuous case:

$$(\partial_t \overline{\rho u_i' u_j'} + \partial_k \overline{\rho u_i' u_j'}\, \tilde{U}_k + \overline{\rho u_i' u_k'}\, \partial_k \tilde{U}_j + \overline{\rho u_j' u_k'}\, \partial_k \tilde{U}_i)/\overline{\rho} =$$

$$(\overline{u_i' \partial_k p_{jk}} + \overline{u_i' \partial_k p_{ik}} - \partial_k \overline{\rho u_i' u_j' u_k'})/\overline{\rho} \quad (1)$$

p_{ij} is a component of the viscous and pressure stresses.
Let $(\ldots)^*$ be the matricial transposition, T the Reynolds stress tensor ($T = \overline{\rho u' u'^*}/\overline{\rho}$), and S the right side of equ. 1. S will be called here turbulent sources, they exclude transport and kinetic production terms.
The matricial form is obtained as:

$$D_t T + \{ (\partial \tilde{U}/\partial x) T + T (\partial \tilde{U}/\partial x)^* \} = S \quad (2)$$

D_t is the mean material derivative and $D_t T$ is the transport term.
$\{\ldots\}$ is the kinetic production.
S represents the other turbulent sources.
In order to interprete the Reynolds stress evolution, it is efficient to write this equation in a Lagrangian type form. Let us consider a particular or reference position of the medium, representative of the mean motion Lagrangian variable (see appendix 1). Let A be the linear tangential transformation associated to the mean flow (see appendix 1) and T_o the tensor given by the linear tangential transformation,

$$T_o = A^* T A \tag{2}$$

Two cases must be examined : firstly, the mean motion is irrotational; secondly, the mean motion is rotational. In order to clarify the analysis, the later case will be examined in appendix 2.

If rot $\tilde{U} = 0$, the Reynolds stress equation writes in the very simple form :

$$D_t T_o = A^* S A \tag{3}$$

The problem is to study:

i) the strong deformation effect due to the mean flow (geometrical approach, equ. 2),

ii) the effect of the turbulent sources S, equ. 3.

At once it will be sought for the nature of T_o, by studying the limiting case where the S sources are neglected. This exhibits the effect of the mean flow geometry. In Lagrangian variables, T_o is an invariant for the motion (equ. 4) ; in Eulerian variables, it corresponds to a zero value of the T_o convective derivation (equ. 5) :

$$D_t T_o = 0 \tag{4}$$

$$D_t T + (\partial\tilde{U}/\partial x)\, T + T\,(\partial\tilde{U}/\partial x)^* = 0 \tag{5}$$

This convective derivation which corresponds to transport and kinetic production terms, has a simple geometrical interpretation. As we already showed [8], relation (5) implies that, for all vector $\mathbf{M}$ moving with the mean motion such as $D_t M_i = (\partial_i \tilde{U}_k)\, M_k$, we obtain :

$$D_t\, \overline{\rho (u' . \mathbf{M})^2} / \bar{\rho} = 0$$

This means that following the mean stream, the turbulent kinetic energy in the m direction is proportional to $\mathbf{M}^{-2}$:

$$\overline{\rho\, u'^2_m} / \bar{\rho} = \text{cte} / \mathbf{M}^2$$

The kinetic energy $\overline{\rho\, u'^2_m} / \bar{\rho}$ increases in the contracted direction, and conversely. In the same way, when $D_t T_o = 0$ relation (2) indicates that the Reynolds tensor at the time t is obtained applying to its initial value T_o the distorsion (A^{-1}). The Reynolds stress tensor can be interpreted as the image of a T_o tensor the nature of which is similar to that of the Piola - Kirchhoff tensor. When source terms are not set equal to zero, relation (2) indicates that T_o can change along the mean flow, due to the source accumulative effects. This will be taken into account hereafter.

Formulation of the problem in the case of a shock wave - turbulence interaction. The evolution of the Reynolds stresses through the shock wave is examined by substituting to the strong gradient region a discontinuity on a fixed surface. The use of the Reynolds stress equation in the case of a shock wave raises many difficulties due to terms involving mean gradients. On the contrary, relations (2) and (3) can be used easily. Indeed, the Reynolds stress discontinuity, termed [T] , is expressed using equ. 2 and taking into account the mean motion discontinuity [A] and the T_o step associated with the turbulent sources S :

$$T_2 = \{ I - N [\widetilde{U}]^*/\widetilde{U}_{N_2} \} \; T_1 \; \{ I - [\widetilde{U}] N^*/\widetilde{U}_{N_2} \} + A_2^{-1*} [T_o] A_2^{-1} \qquad (6)$$

[ ] indicates a discontinuity, see appendix 1.
N is the direction locally normal to the fixed surface equivalent to the shock. I is the unit matrix. 1 and 2 means upstream and downstream. More details are given in appendix 1.

Dealing with the last term $A_2^{-1*}[T_o] A_2^{-1}$ of relation (6), in order to express the contribution to the T_o evolution of the sources, equ. 3 is formally extended to a discontinuity model. In order to do this, let h and δ be the Heaviside and Dirac symbols ; $f(x)$ the distance to the shock, such as $f = 0$ on the shock, $f < 0$ upstream, $f > 0$ downstream.

T_o and $A^* S A$ can be expressed in the form [9] :

$$T_o = h(-f) \, T_{o1} + h(f) \, T_{o2}$$

Let G_1, G_2, G' defined by :

$$A^* S A = h(-f) \, G_1 + h(f) \, G_2 + G' \, \delta(f)$$

From these relations and from equ. 3, the T_o discontinuities form can be expressed as :

$$[T_o] = \frac{1}{\widetilde{U}_{N_1}} \{ G' + T_{o2} (\widetilde{U}_{N_1} - \widetilde{U}_{N_2}) \}$$

and finally equ. 6 becomes :

$$T_2 = \widetilde{U}_{N_1} / \widetilde{U}_{N_2} \{ I - N [\widetilde{U}]^*/\widetilde{U}_{N_2} \} \; T_1 \; \{ I - [\widetilde{U}] N^*/\widetilde{U}_{N_2} \} + A_2^{-1*} G' A_2^{-1} / \widetilde{U}_{N_2}$$

So, the step of the Reynolds stresses is expressed by taking into account the characteristics of the mean flow and the only part of the sources which corresponds to Dirac functions G' located on the shock surface ; one must notice that the source steps disappear. Hence, at this point of the calculation, hypothesis will only be made on the form of the source evolution, but not on the absolute source values. The contribution to G' of each source term is not examined. For example, in the following application, the phenomenological assumption $G' = 0$ brings a great simplification and seems to agree with experiments. This corresponds to two possible cases :

- The sources S are continuous through the shock ;
- Their discontinuities are only in a step form (but not with a greater degree).

Then, the step of the Reynolds stress is directly expressed in terms of the mean flow characteristics :

$$T_2 = \widetilde{U}_{N_1} / \widetilde{U}_{N_2} \{ I - N [\widetilde{U}]^*/\widetilde{U}_{N_2} \} \; T_1 \; \{ I - [\widetilde{U}] N^*/\widetilde{U}_{N_2} \} \qquad (7)$$

This assumption $G' = 0$ cannot be directly justified, so that experiments are needed to check its validity. However, the hypothesis is suggested on one hand by the fact that the part of inertial terms (transport, production) which are in a Dirac form have been already taken into account. On the other hand, this assumption, which is equivalent to the existence of continuous source terms or step discontinuity source terms, is suggested by rapid distorsion theories [10] or such considerations [11]. These rapid distorsion theories or considerations assume that there exist characteristic relaxation times for some source terms. For example, dissipation is assumed to be constant when the condition $q' L / \Lambda \tilde{U} \ll 1$ is fulfilled. The criterion compares the time scale corresponding to the times $L / \tilde{U}$ during which a fluid particle is submitted to the distorsion, to the turbulent characteristic time scale Λ/q'. q' is the r.m.s. of the velocity fluctuations; L is the mean distorsion spatial scale; Λ is a characteristic length scale of turbulence.

Before comparing data and results deduced from relation (7), we shall give a geometrical illustration of this relation giving the Reynolds stress evolution. On fig. 10, the Reynolds stresses are represented by closed curves (or by closed surface in a three dimensional representation) as shown below.

Point m on this curve corresponds to the variance $Om = \overline{\rho u'^2_m} / \overline{\rho}$ of velocity fluctuation in this direction. This can be also written:

$$\overline{\rho u'^2_m} / \overline{\rho} = \overline{\rho (u' \cdot m)^2} / \overline{\rho} = T_{ij} m_i m_j$$

where m_1, m_2, m_3 are the components of the unit vector in this direction. Relation (7) reads :

$$m_i m_j (\overline{\rho u'_i u'_j} / \overline{\rho})_2 = (\tilde{U}_{N_2} / \tilde{U}_{N_1}) K_{il} (\overline{\rho u'_l u'_k} / \overline{\rho})_1 m_i m_j K_{jk}$$

with

$$K_{pq} = \delta_{pq} - \frac{1}{\tilde{U}_{N_2}} N_p (\tilde{U}_{N_2} - \tilde{U}_{N_1})$$

Fig. 10 corresponds to the experimental case : upstream $x = -15$, $y/\delta = .4$, downstream $x = +20$, along the same streamline.

3.2. Comparison Between Computed and Experimental Data

The comparison between computed and experimental data is made only for the Reynolds stress component $\overline{\rho u'^2_s} / \overline{\rho} \simeq \overline{u'^2_s}$, where s is the direction of the mean local velocity, fig.11. At present time, only this component is measured , but it is necessary to know all the upstream components of the Reynolds stress. Only $\overline{u'^2_s}$ was measured in the initial boundary layer, and the other initial Reynolds stress components were deduced from $\overline{u'^2_s}$. The assumption was that the anisotropy coefficients are like in Klebanoff's measurements [12] in an incompressible equilibrium boundary layer. Each component of the downstream Reynolds stress is computed from relation (7) and from mean flow measurements. Morever the following simplifications are used :

a) the mean velocity discontinuity $[\tilde{U}]$ is normal to the shock wave. This is justified when the total resulting stress (pressure, viscosity, Reynolds stress) is diagonal in the shock frame (see appendix 2) .

b) The normal N is defined by classical shock wave relations, and by the measured Mach number ratio which verifies these relations.

The velocity fluctuation variance $\overline{u'^2_s}$ in the direction s of the mean flow is deduced from the downstream Reynolds stress through the relation $\overline{u'^2_s} = T_{ij} s_i s_j$ and is compared to the measured quantities.

The comparison shows that most of the evolution is taken into account by the proposed analysis.

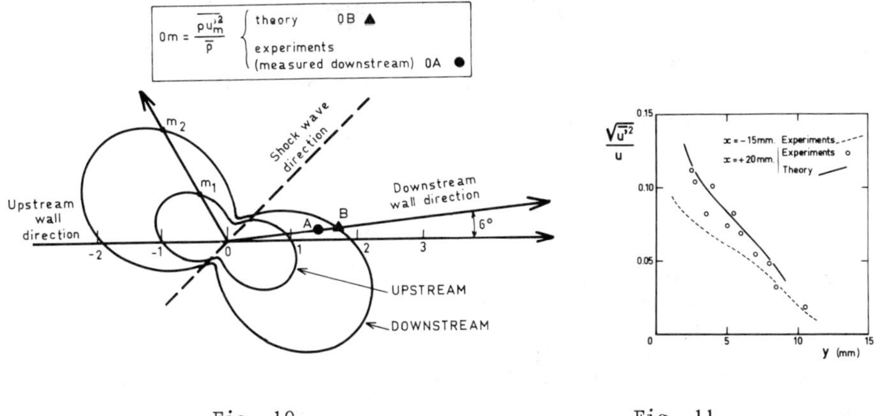

Fig. 10 Fig. 11

4. THE THERMAL TURBULENT FIELD

In turbulent supersonic flows, even without heat sources, turbulent heat fluxes exist. In order to get information on the evolution of these fluxes, it is convenient to use classical analogies based on the hypothesis that the behaviour of the thermal turbulent field depends on the velocity field. It has been seen (§ 2) that velocity and temperature intensities increase through the shock. The aim is here to answer the question: how the classical relations between turbulent thermal field and velocity field become in such a strong pressure gradient ?

The evolution of the velocity-temperature correlation coefficient and the validity of the so called "Strong Reynolds Analogy" are hereafter examined. The classical Reynolds Analogy which introduces a turbulent Prandtl number is often used as a phenomenological representation of the thermal fluxes in low speed flows. It is also valid at supersonic speeds (Meier and Rotta [13]). The Strong Reynolds Analogy (S.R.A., Young, [14]; Morkovin, [15]) especially concerns supersonic flows. It gives a relation between temperature and longitudinal velocity fluctuations, but it is used in calculation methods involving transversal turbulent heat fluxes (Rotta, [16], Wilcox and Alber, [17], Bradshaw, [18]). Although the way to prove it is somewhat paradoxical, it is often in good agreement with the experiments.

From similitude considerations between momentum and stagnation temperature equations, Young deduces the following instantaneous relations:

$$\theta'_o = 0 \quad \Rightarrow \quad -(\gamma - 1) M^2 u'/\overline{U} = \theta'/\overline{\theta} \qquad (8)$$

(γ ratio of specific heats, c_p and c_v ; M, Mach number).

This is a consequence of the stagnation temperature definition $\theta_o = \theta + \overline{U}^2/2c_p$ which leads to the more general relation:

$$\frac{\theta'_o}{\overline{\theta}} = \frac{\theta'}{\overline{\theta}} + (\gamma - 1) M^2 \frac{u'}{\overline{U}} \qquad \text{for } \frac{u'}{\overline{U}} \ll 1 \qquad (9)$$

Relations (8) are given to be valid in the case of a boundary layer developing on an adiabatic flat plate, without pressure gradient. They correspond to very restrictive conditions.

Results. Experimentally, (8) cannot be directly checked. But it is possible to measure the relative fluctuations levels $\sqrt{\overline{u'^2}}/\overline{U}$ and $\sqrt{\overline{\theta'^2}}/\overline{\theta}$, and the correlation coefficient $R_{u\theta}$. The instantaneous relation (8) corresponds to the two statistical relations :

$$\theta'_o = 0 \iff \begin{cases} (\gamma - 1)M^2\sqrt{\overline{u'^2}}/\overline{U} = \sqrt{\overline{\theta'^2}}/\overline{\theta} & (10) \\ R_{u\theta} = -1 & (11) \end{cases}$$

Experiments were performed in order to check the validity of the three last conditions. It has been previously found [19] that, even in equilibrium flows, only relation (10) is verified, and that this result also accurs in non equilibrium flows, (expansions, mixing flows, compression region . . .), despite of the pressure gradient. Morever, θ'_o is not null, while $R_{u\theta}$ does not differ much from the usual values found in many self preserving flows [19]. But this had to be examined in the case of a shock wave - boundary layer interaction.

- Fig. 12 shows that, even in this case, the ratio $\dfrac{(\gamma - 1)M^2\sqrt{\overline{u'^2}}/\overline{U}}{\sqrt{\overline{\theta'^2}}/\overline{\theta}}$ is not much different from 1.

- Dealing with the correlation coefficient (fig. 13), $R_{u\theta}$ everywhere keeps a value $\simeq$ -.85, as in an equilibrium turbulent flow, and not the value - 1 given by the relation (11).

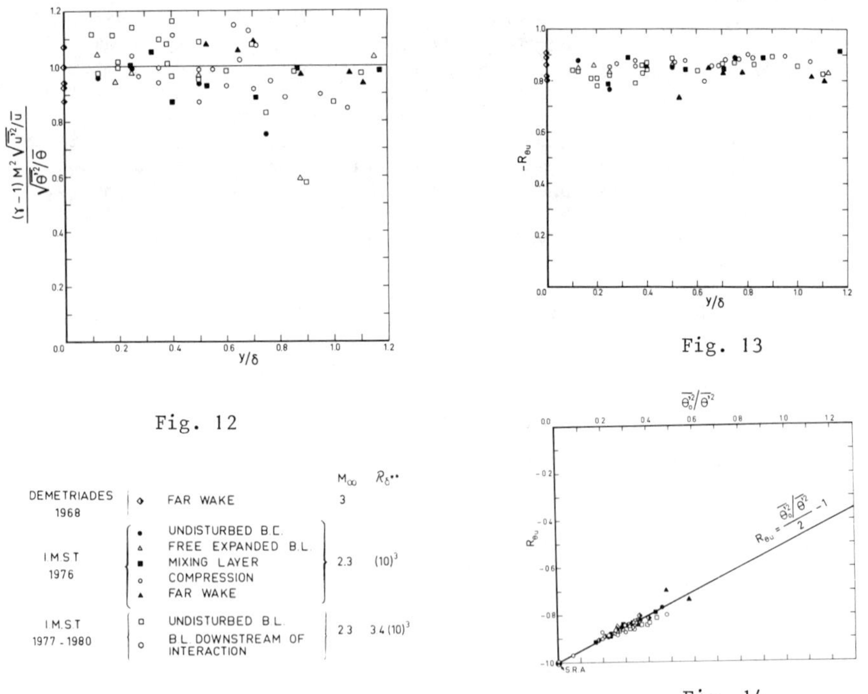

Fig. 12

Fig. 13

Fig. 14

- Finally, as for θ'_o, the hypothesis $\overline{\theta'^2_o} = 0$ is not justified. Indeed, fig. 14 which is explained below shows that the ratio $\overline{\theta'^2_o} / \overline{\theta'^2}$ is not zero. As a consequence of relations (9),(10), and with the condition $\theta'_o \neq 0$, the relation $R_{u\theta} = \frac{(\overline{\theta'^2_o} / \overline{\theta'^2})}{2} - 1$ holds. Only the point $R_{u\theta} = -1$, $\overline{\theta'^2_o} / \overline{\theta'^2} = 0$ corresponds to the limiting case of the S.R.A. represented by (8).

These properties are grouped on fig. 15. The diagram is built on the remark that a covariance has scalar product properties, so that the r.m.s. value of the sum of two random variables corresponds to a geometrical sum. This remark is applied to relation (9) in which each term is divided by $\sqrt{\overline{\theta'^2}}/\overline{\theta}$ in order to have a convenient scale. Hence, on fig. 15, MO = 1 corresponds to $\sqrt{\overline{\theta'^2}}$, OP corresponds to $\sqrt{\overline{u'^2}}$, and MP corresponds to $\sqrt{\overline{\theta'^2_o}}$.

Morever, the correlation coefficients read : $\cos(\widehat{MO,OP}) = R_{u\theta}$, $\cos(\widehat{MO,MP}) = R_{\theta\theta_o}$ and $\cos(\widehat{OP,MP}) = R_{\theta_o u}$. Everywhere in the flow, the latter is small, for example $R_{u\theta_o} \simeq .3$.

The experimental data shows that, although $\overline{\theta'^2_o} \neq 0$ and $R_{u\theta} \neq -1$, the S.R.A. relation (10) is valid, even downstream of the shock wave.

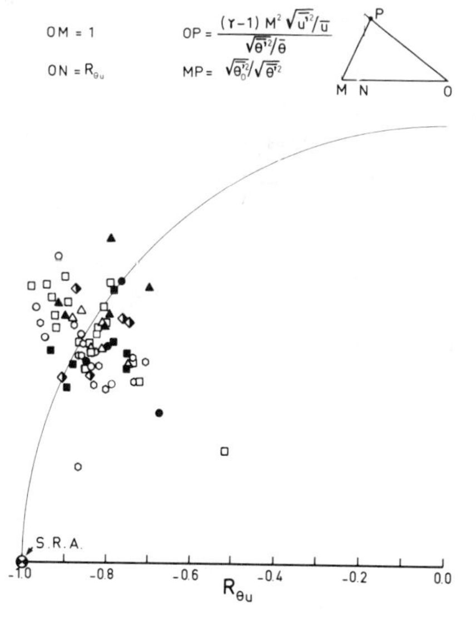

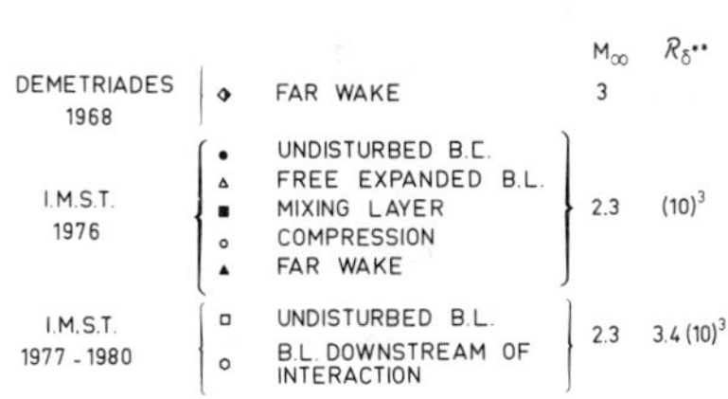

Fig. 15

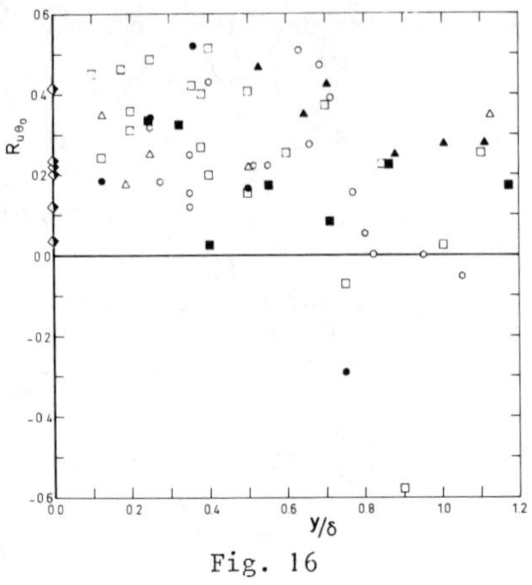

Fig. 16

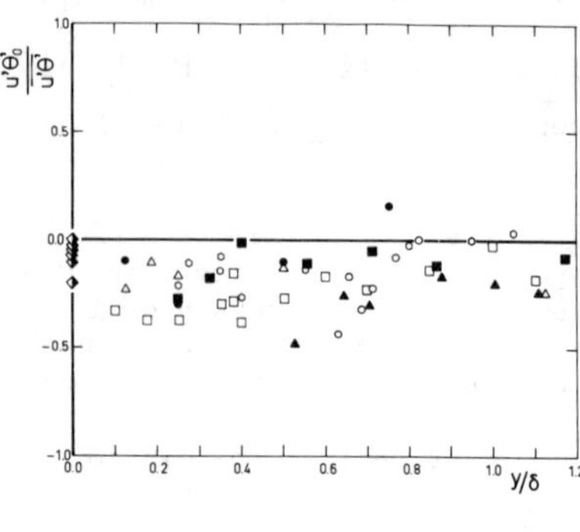

Fig. 17

In order to avoid the difficulties which arise to establish the fact that $\dfrac{(\gamma-1)M^2\sqrt{\overline{u'^2}}/\overline{U}}{\sqrt{\overline{\theta'^2}}/\overline{\theta}}$ does not differs much from 1, another hypothesis can be made : not only the stagnation temperature intensity $\sqrt{\overline{\theta_o'^2}}$ is smaller than $\sqrt{\overline{\theta'^2}}$, but also $R_{u_i\theta_o}$ is smaller than $R_{u_i\theta}$. This hypothesis is consistent with the experiments, fig. 16. This leads to generalize an hypothesis on the fluxes due to Morkovin [15], who assumes that $\overline{v'\theta'_o}$ is very small as compared to $\overline{v'\theta'}$.

As for the longitudinal fluxes, fig. 17 shows that the ratio between the total and the static temperature fluxes is low, for example - 0,5 to 0.

The exact relation $\overline{u'\theta'_o} = \overline{u'\theta'} + \dfrac{\overline{U}}{c_p}\overline{u'^2}$ leads to

$$\frac{(\gamma-1)M^2\sqrt{\overline{u'^2}}/\overline{U}}{\sqrt{\overline{\theta'^2}}/\overline{\theta}} = -R_{u\theta}\left(1-\frac{\overline{u'\theta'_o}}{\overline{u'\theta'}}\right) \tag{12}$$

When (12) is applied, equation (10) is verified with a good approximation : this is due to the fact that $-R_{u\theta}$ is not far from unity, and is multiplied by a factor slightly greater than 1. So the afore mentionned hypothesis on the fluxes is midler than the S.R.A.

As for the transversal fluxes, there is a lack of experimental data. The exact relation $\overline{v'\theta'_o} = \overline{v'\theta'} + \dfrac{\overline{U}}{c_p}\overline{u'v'}$ leads to :

$$\frac{(\gamma-1)M^2\sqrt{\overline{u'^2}}/\overline{U}}{\sqrt{\overline{\theta'^2}}/\overline{\theta}} = -\frac{R_{v\theta}}{R_{uv}}\left(1-\frac{\overline{v'\theta'_o}}{\overline{v'\theta'}}\right) \tag{13}$$

Should the ratio of the correlation coefficients have the same signs and values as in subsonic heated flows [15], (13) would indicate that $\overline{v'\theta'_o}$ is weak but not negligible as compared to $\overline{v'\theta'}$. Both these terms are positive, at least in an equilibrium flow. This conclusion is consistent with some new measurements recently performed [20]. Morever, it can be remarked that when the total enthalpy flux can be neglected, a relation between the Reynolds stress $\overline{\rho u'v'}$ and the heat flux $c_p \overline{\rho\theta'v'}$ can be obtained. This result is sometimes deduced from the instantaneous relation (9). In addition, when the mean total stagnation temperature is constant, (13) is consistent with a Prandlt number equal to unity. This corresponds to the first Reynolds Analogy. Now, a reason for which the midler forms (12) and (13) of the S.R.A. are not altered by the strong pressure gradient, appears. Indeed, from a gradient model, based on the constant mean stagnation temperature gradient $\partial\theta_o/\partial y$ through the shock, it is inferred that the corresponding stagnation temperature fluxes by turbulence are small, and in addition that they are not much altered by the interaction.

5. APPENDIX 1. PROOF OF THE RELATION (6)

Expression (6) relates the velocity, the normal to the shock surface and the linear tangent application associated to the mean stream.
Demonstrating this tensorial relation can be done using a physical space represented by a quadri-dimensional time-space; the spatial part admits a canonical Euclidean metrix.

5.1. Motion of a Turbulent Fluid

In order to describe the transformation analytically, let us introduce a fixed rectangular coordinate system.
We refer to the coordinate triple $\begin{vmatrix} x^1 \\ x^2 \\ x^3 \end{vmatrix}$ of a particle in the "mean stream", and denote it x.

$z = \begin{vmatrix} t \\ x^1 \\ x^2 \\ x^3 \end{vmatrix}$ is the column vector of " Eulerian variables ".

At time t, the fluid occupies the Euclidean differentiable manifold $\mathcal{D}_t$ called configuration,[21 , 22]. At the initial time $t = 0$, it occupies the reference position $\mathcal{D}_o$, [23].

$X = \begin{vmatrix} X^1 \\ X^2 \\ X^3 \end{vmatrix}$ denotes Lagrangian coordinates of the particle. x is determined by a function of X and t, and the motion can be represented by the transformation $x = \varphi(t,X)$.
(In coordinate applications $x^i = \varphi^i(t,X^1,X^2,X^3)$, $i \in \{1,2,3\}$).
Let us denote $\mathcal{W}$ the time-space set described by $\mathcal{D}_t$ between the time t_1 and t_2; $\mathcal{W}_o$ the cartesian product $(t_1,t_2) \times \mathcal{D}_o$.

Denoting $Z = \begin{vmatrix} t \\ X^1 \\ X^2 \\ X^3 \end{vmatrix}$, the fluid motion is described by the application from $\mathcal{W}_o$ into $\mathcal{W}$:

$$z = \Phi(Z)$$

(In coordinate applications $\begin{cases} t = t \\ x^i = \varphi^i(t,X^1,X^2,X^3) , \quad i \in \{1,2,3\} \end{cases}$).
This application is bijective.

5.2. Kinematic of Shock Waves

The "mean shock wave" is schematized in $\mathcal{D}_t$ by the moving surface (C_t) described by the equation $f(t,x^1,x^2,x^3) = 0$. (f is supposed to be differentiable). This equation represents a three - manifold of $\mathcal{W}$. Its image by the application Φ is the three - manifold (C_o) of $\mathcal{W}_o$ given by the equation :

$$f\{t,\varphi^1(t,X^1,X^2,X^3),\varphi^2(t,X^1,X^2,X^3),\varphi^3(t,X^1,X^2,X^3) = 0$$

(We note $f_o(t,X^1,X^2,X^3) = 0$).
In the fixed system, the one - form $\frac{\partial f}{\partial z}$, represented by the row vector $\left(\frac{\partial f}{\partial t} , \frac{\partial f}{\partial x^1} , \frac{\partial f}{\partial x^2} , \frac{\partial f}{\partial x^3} \right)$, is the annulator of tangent vectors at (C) [24].

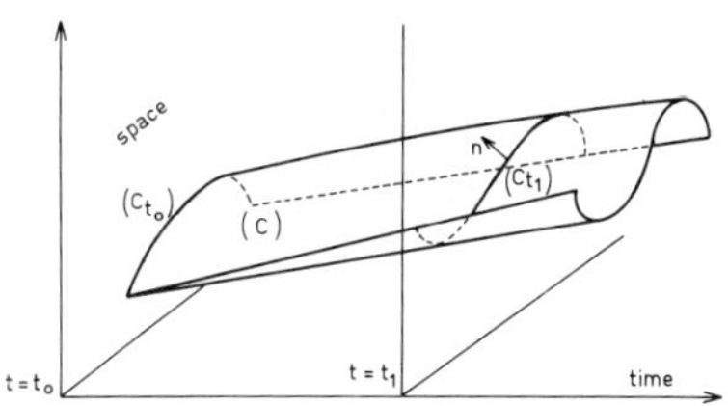

Fig.18. Representation of the shock surface in a quadri-dimensional time-space.

(C_t) is supposed to be orientable; $(\;)^*$denotes the transposition ; then there exist a unique scalar μ such that $n = \mu\,(\frac{\partial f}{\partial x})^*$ is the unit normal oriented vector at (C_t).

$M(t,x)$ is a point of (C_t), V its velocity. The relation:

$$\frac{\partial f}{\partial t}\,dt + \frac{\partial f}{\partial x^1}\,dx^1 + \frac{\partial f}{\partial x^2}\,dx^2 + \frac{\partial f}{\partial x^3}\,dx^3 = 0$$

implies
$$\frac{\partial f}{\partial t} + \frac{\partial f}{\partial x^1}\,V^1 + \frac{\partial f}{\partial x^2}\,V^2 + \frac{\partial f}{\partial x^3}\,V^3 = 0$$

where V^1, V^2, V^3 denote the components of V in the fixed system.

We deduce :
$$n^*\,V = -\mu\,\frac{\partial f}{\partial t}\,.$$

It is the velocity of (C_t) at M with respect to the fixed system. It is denoted by g. The velocity of (C_t) at M with respect to the fluid is $c = g - n^*\tilde{U}$.

$u = n^*\tilde{U} - g$ represents the mean velocity of the fluid at M with respect of (C_t).

Let N^* be $(-g\,,n^*)$, it follows : $N^*\begin{vmatrix}1\\ \tilde{U}\end{vmatrix} = u.$

In the same way, the one - form represented by the row vector

$\frac{\partial f_o}{\partial Z} = (\frac{\partial f_o}{\partial t}, \frac{\partial f_o}{\partial X^1}, \frac{\partial f_o}{\partial X^2}, \frac{\partial f_o}{\partial X^3})$ is the annulator of tangent vector at (C_o).

We denote N_o^* the linear form associated to $\frac{\partial f_o}{\partial Z}$ which defines the flow of a vector field through (C_o). By analogy, we write $N_o^* = (-g\,,n_o^*)$.

<u>Consequences.</u> The volumic mass and the velocity of a particle being discontinued during the shock crossing, we suppose that the application Φ is bijective, bicontinuous from $\mathcal{W}_o$ into $\mathcal{W}$ but differentiable in the complementary of (C_o)

Out of (C_o), we can define the linear tangent application A of Φ.

In the fixed system, it is represented by the matrix :

$$A = \begin{vmatrix} 1 & 0 & 0 & 0 \\ \tilde{U}^1 & & & \\ \tilde{U}^2 & & (\frac{\partial x^i}{\partial X^j}) & \\ \tilde{U}^3 & & & \end{vmatrix} \qquad i,j \in \{1,2,3\}$$

Here $\tilde{U}^1$, $\tilde{U}^2$, $\tilde{U}^3$ are the components of $\tilde{U}$ and $dx = \tilde{U}\,dt + A\,dX$, where A is represented by the matrix of generic elements $(\frac{\partial x^i}{\partial X^j})$, $i,j \in \{1,2,3\}$.

Properties. Let [a] denote the step $a_2 - a_1$, [25], where a_2 and a_1 are respectively the values of a upstream and downstream the shock. The symbol . designating any "product law", it follows, $[a.b] = [a]\, b_1 + a_2 [b]$. The discontinuity of $\tilde{U}$ implies the discontinuity of the linear tangent application.

Firstly, Φ restricted at (C_o) is differentiable :

$$N_o^* \, dZ = 0 \quad \Rightarrow \quad [A]\, dZ = 0$$

Then, it exist a Lagrange multiplier, represented by the four-vector L, such as:

$$[A] = L \, N_o^*$$

$$\begin{vmatrix} o & 0 \\ [\tilde{U}] & [A] \end{vmatrix} = L\,(-g_o, n_o^*), \text{ where } \quad L = \begin{vmatrix} o \\ L \end{vmatrix} \quad \text{and} \quad L = \begin{vmatrix} L^1 \\ L^2 \\ L^3 \end{vmatrix}.$$

Then, $[\tilde{U}] = -L\, g_o$ and $[A] = L\, n_o^*$.

Denote $n'_o = -\frac{n_o}{g_o}$, $[A] = [\tilde{U}]\, n_o'^*$. (14)

If the tangential component of the velocity is countinuous during the shock crossing, $[A] = [u]\, n\, n_o'^*$.

Secondly, Φ transforms any tangent vector at (C_o) into tangent vector at (C):

$$N_o^* \, dZ = 0 \quad \Rightarrow \quad N^* dz = 0.$$

The restriction of Φ at (C_o) being differentiable,

$$N^* dz = N^* A \; dZ \, .$$

It follows: $N_o^* \, dZ = 0 \Rightarrow N^* A\, dZ = 0$ and there exists a scalar Lagrange multiplier such as : $N^* A = \chi\, N_o^*$.

From $N^* A = (n^*\tilde{U} - g \,,\, n^*A)$, we deduce : $(u \,,\, n^* A) = \chi\,(-g_o, n_o^*)$.

Let be $u\,(1\,,\,\frac{n^*A}{u}) = \mu\,(1\,,\,n_o'^*)$, where $\mu = -\chi\, g$, the relation follows:

$$n_o'^* = n^* \frac{A_1}{u_1} = n^* \frac{A_2}{u_2} \tag{15}$$

and

$$n^* [\frac{A}{u}] = 0$$

$$(8) + (9) \Rightarrow \quad [A] = [\tilde{U}]\, n^* \frac{A_k}{u_k} \qquad \text{(where 1, upstream, 2, downstream)} \tag{16}$$

5.3. Discontinuity of a Two Covariant Tensor

The image of a tensor of $\mathcal{D}_t$ in $\mathcal{D}_o$ is obtained with the linear tangent application A, and the discontinuity of such a tensor during the shock crossing follows from the discontinuity of its image in $\mathcal{D}_o$ and from the discontinuity of the linear tangent application.

Let M be a two-covariant tensor and M_o its image in $\mathcal{D}_o$, J_α and J_β being two vectors of $\mathcal{D}_t$ with respective images $J_{\alpha,o}$ and $J_{\beta,o}$.

By application of M to the two vectors, we deduce : $M(J_\alpha, J_\beta) = M_o(J_{\alpha,o}, J_{\beta,o})$.
Now, $J_\alpha = A J_{\alpha,o}$ and $J_\beta = A J_{\beta,o}$.
$\mathcal{D}_t$ having an Euclidean metrix, using matrix, we can write :

$$\forall \quad J_{\alpha,o}, J_{\beta,o}, \qquad J_\beta^* M J_\alpha = J_{\beta,o}^* M_o J_{\alpha,o}$$

$$J_{\beta,o}^* A^*MA J_{\alpha,o} = J_{\beta,o}^* M_o J_{\alpha,o}$$

$$M = A^{-1*} M_o A^{-1} \tag{17}$$

Consequently, $[M] = [A^{-1*} M_o A^{-1}] = A_2^{-1*}[M_o A^{-1}] + [A^{-1*}] M_{o,1} A_1^{-1}$

$$= A_2^{-1*} M_{o,2} [A^{-1}] + [A^{-1*}] M_{o,1} A_1^{-1} + A_2^{-1*}[M_o] A_1^{-1}$$

Now $\qquad [A^{-1}] = - A_2^{-1} [A] A_1^{-1} = - A_1^{-1} [A] A_2^{-1}$.

It follows, $[M] = - A_2^{-1*} M_{o,2} A_2^{-1} [A] A_1^{-1} - A_2^{-1*} [A^*] A_1^{-1} M_{o,1} A_1^{-1} + A_2^{-1*} [M_o] A_1^{-1}$.

$$[M] = - M_2 [A] A_1^{-1} - A_2^{-1*} [A^*] M_1 + A_2^{-1} [M_o] A_1^{-1}$$

Relation (17) implies:

$$A_2 A_1^{-1} - I = \frac{[\tilde{U}]\ n^*}{u_1} \quad \Rightarrow \quad \begin{cases} (A_1 A_2^{-1})^* - I = - \dfrac{n\ [\tilde{U}]^*}{u_2} \\ (A_1 A_2^{-1}) - I = - \dfrac{[\tilde{U}]\ n^*}{u_2} \end{cases}$$

$$M_2 - M_1 = - M_2 (A_2 A_1^{-1} - I) - (I - (A_1 A_2^{-1})^*) M_1 + A_2^{-1*} [M_o] A_1^{-1}$$

$$M_2 A_2 A_1^{-1} = (A_1 A_2^{-1})^* M_1 + A_2^{-1*} [M_o] A_1^{-1}$$

$$M_2 = (A_1 A_2^{-1})^* M_1 A_1 A_2^{-1} + A_2^{-1*} [M_o] A_2^{-1}$$

It follows:

$$M_2 = (I - \frac{n\ [\tilde{U}]}{u_2}) M_1 (I - \frac{[\tilde{U}]\ n^*}{u_2}) + A_2^{-1*} [M_o] A_2^{-1} \tag{18}$$

This relation expresses the value behind the shock of a tensorial quantity using its value above the shock, the discontinuity of its image M_o in $\mathcal{D}_o$, and the geometry of the shock.
The left hand of the Reynolds equation (1') is written in form of transport equation for a covariant tensor, [26], so we take the Reynolds stress tensor with the structure of a quadratic form. It is a two-covariant tensor on which formula (18) can be applied. The relation (6) follows.

6. APPENDIX 2. CASE OF A NON IRROTATIONAL MEAN FLOW

When rot $\tilde{U} \neq 0$, the Reynolds stress equation (1') can be written in the form:

$$D_t T + (\partial\tilde{U} / \partial x)^* T + T (\partial\tilde{U} / \partial x) = S + S' \tag{1''}$$

where S' is an additional term $S' = T\Omega - \Omega T$, and $\Omega = (\partial\tilde{U} / \partial x - \partial\tilde{U} / \partial x^*)$.

The later term corresponds to the antisymmetrical matrix associated with the curl.
Equation (3) can be formulated as :

$$D_t T_o = A^* S A + A^* S' A \quad \text{where} \quad T_o = A^* T A \ .$$

Term S' is zero in the case where the Reynolds stress tensor is isotropic. In the same way, the form of the turbulent kinetic energy equation does not depends on the fact that $\text{rot}\, \tilde{U} = 0$ or $\text{rot}\, \tilde{U} \neq 0$. Indeed, this equation is obtained by contracting (1"), since $\text{Tr.}\ S' = 0$.

Proof of the Relation (6) in the Case of a Rotational Mean Motion.

It can be remarked that the condition $\text{rot}\, \tilde{U} = 0$ is not a restriction to the shock wave application :
As for the Reynolds stress tensor, one can write $\tilde{U} = h(-f)\, \tilde{U}_1 + h(f)\, \tilde{U}_2$.
It follows that the distribution derivative writes :

$$\partial\tilde{U} / \partial x = \partial h/\partial f \ (- \tilde{U}_1 N^* + \tilde{U}_2 N^*) + h(-f)(\partial\tilde{U}_1 / \partial x) + h(f)(\partial\tilde{U}_2 / \partial x)$$

and $$\Omega = \delta(f) \{ (\tilde{U}_2 - \tilde{U}_1) N^* - N (\tilde{U}_2^* - \tilde{U}_1^*) \} + h(-f)\, \Omega_1 + h(f)\, \Omega_2$$

Now $(\tilde{U}_2 - \tilde{U}_1)\, N^* - N\, (\tilde{U}_2^* - \tilde{U}_1^*)$ is the antisymmetrical matrix associated with the vector product $N \times (\tilde{U}_2 - \tilde{U}_1)$.

At this point, one can remark that, according to experiments, this last term is negligible. In fact, mean mass and momentum equations show that if the stress tensor is diagonal on the shock frame, the discontinuity $[\tilde{U}]$ is normal to the shock surface, and $N \times [\tilde{U}] = 0$.
The total stress is the sum of : a scalar tensor, the pressure which is predominant, and an additional Reynolds stress which is almost diagonal in the shock frame, (see fig. 10). It follows : $\Omega = h(-f)\, \Omega_1 + h(f)\, \Omega_2$ and in the sense of distributions,

$$S' = T\, \Omega - \Omega\, T = T_{o,1}\, \Omega_1\, h(-f) + T_{o,2}\, \Omega_2\, h(f) \ .$$

The source S' has Heaviside terms only and is not modified in the discontinuity calculus.

REFERENCES

1. Gaviglio, J. and Dussauge, J.P. 1977. Proc. Symp. Flow in Open Channels and Closed Circuits. Gaithersburg, Md. U.S.A., N.B.S. Publ. 484, T.P. O.N.E.R.A. 1977 - 168.

2. Gaviglio, J. 1978. Sur les méthodes de l'anémométrie par fil chaud des écoulements turbulents compressibles de gaz. J. de Méca. Appl. , Paris, vol. 2, n° 4, pp. 449 - 498.

3. Laufer, J. and Mc Clellan R. 1957. Measurements of Heat Transfer from Fine Wires in Supersonic Flows. Jet Propulsion Laboratory, California Inst. of Tech., Report n° 20 - 101.

4. Favre, A. Kovasznay, L.S.G. Dumas, R. Gaviglio, J. and Coantic, M. 1976. La turbulence en Mécanique des Fluides, Gauthier - Villars ed., Paris.

5. Chew, Y.T. 1978. Two - Parameter Skin - Friction Formula for Adiabatic Compressible Flow. A.I.A.A. Journal, t. 16, n° 2, pp. 186 - 188.

6. Adamson, T.C. and Messiter, A.F. 1980. Analysis of Two - Dimensional Interactions Between Shock Waves and Boundary Layer. Annual Review of Fluid Mechanics, t. 12, pp. 103 - 138.

7. Laufer, J. 1962. Sound Radiation from a Turbulent Boundary Layer. Colloques Internationaux du C.N.R.S., n° 108. Ed. C.N.R.S., Paris, pp. 381 - 393.

8. Debiève, J.F. 1978. Methode d'analyse de la production des tensions de Reynolds en écoulement turbulent compressible. Comptes Rendus Acad. Sc., Paris, t. 286, série A, pp. 705 - 708.

9. Tavéra, G. and Burnage, H. 1977. Sur les ondes de choc dans les écoulements instationnaires. Comptes Rendus Acad. Sc., Paris, t. 284, série A, pp. 571 - 573.

10. Hunt, J.C.R. 1973. A Theory of Turbulent Flow Round Two - Dimensional Bluff Bodies. J. of Fluid Mech., t. 61, n° 4, pp. 625 - 706.

11. Dussauge, J.P. and Debiève, J.F. 1978. Influence d'un fort gradient de pression sur les perturbations instationnaires de vitesse dans le cas d'un écoulement turbulent supersonique. 15 ème Coll. d'Aérodynamique Appl., Marseille.

12 Klebanoff, P.S. 1954. Characteristics of Turbulence in a Boundary Layer with Zero Pressure Gradient. N.A.C.A. Tech. Note 3178.

13. Meier, H.U. and Rotta, J.C. 1971. Temperature Distributions in Supersonic Turbulent Boundary Layers. A.I.A.A. Journal, t. 9, n° 11, pp. 2149 - 2156.

14. Young, A.D. 1951. The equations of Motion and Energy and the Velocity Profile of a Turbulent Boundary Layer in a Compressible Fluid. Cranfield, The College of Aeronautics, rep. n° 42.

15. Morkovin, M.V. 1962. Effects of Compressibility on Turbulent Flows. Colloques Internationaux du C.N.R.S., n°108. Ed. C.N.R.S., Paris, pp. 367 - 380.

16. Rotta, J. 1967. Effect of Streamwise Wall Curvature on Compressible Turbulent Boundary Layers. Phys. of. Fluids, pp. S174 - S180.

17. Wilcox, D.C. and Alber, I.E. 1972. A Turbulence Model for High Speed Flows. Proc. Heat Transfer and Fluid Mechanics. 1972, pp. 231 - 252.

18. Bradshaw, P. 1977. Compressible Turbulent Shear Layers. Annual Review Fluid Mech., t. 9, pp. 33 - 54.

19. Gaviglio, J. Dussauge, J.P. Debiève, J.F. and Favre, A. 1977. Behavior of a Turbulent Flow, Strongly Out of Equilibrium, at Supersonic Speeds. The Physics of Fluids, vol. 20, n° 10, pp. S179 - S192.

20. Anguillet, J.P. 1980. Sur les mesures de flux de quantités de mouvement et de chaleur au moyen de l'anémomètre à fil chaud en écoulement supersonique. Th. de 3 ème cycle, Univ. Aix - Marseille II.

21. Truesdell, C. and Toupin, R.A. 1960. The Classical Field Theories. Handbuch der Physik. Vol. III/1. Springer - Verlag.

22. Serrin, J. 1959. Mathematical Principles of Classical Mechanics. Handbuch

der Physik. Fluid Dynamics 1, vol. VIII/1. Springer-Verlag.

23. Casal, P. 1969. Cours de Mécanique des Milieux Continus, Polycop. Univ. Aix-Marseille 1.

24. Gouin, H. 1978, Contribution à une étude géométrique et variationnelle des milieux continus. Th. de Doctorat d'Etat, Univ. Aix-Marseille 1.

25. Hadamard, J. 1949. Leçons sur la propagation des ondes et les équations de l'hydrodynamique. Chelsea Publishing Company.

26. Brillouin, L. 1960. Les tenseurs en Mécanique et en Elasticité. Masson Ed., Paris.

ACKNOWLEDGEMENTS

The authors wish to thank doctor Dussauge for his pertinent remarks; misters Borel, Cayrac, Morand and Soscia for the precious help they brought to the preparation of the experiments and of the manuscript.

Measurement of Turbulence Correlations in a Two-dimensional Supersonic Wake

J. P. BONNET and T. ALZIARY DE ROQUEFORT
Centre d'Etudes Aerodynamiques & Thermiques
43 Rue de l'Aerodrome
86000 Poitiers, France

ABSTRACT

Study of turbulence in a supersonic plane wake at high Reynolds number is performed by means of hot-wire anemometry. Classical analog devices are used to measure unfiltered space-time correlation allowing to give scales and convection velocities, in agreement with Demetriades's results. However the behaviour of transverse correlation is not in agreement with the "double-roller eddy" model proposed by Grant. The use of a tape recorder, sampling of the signal and Fast Fourier Transform techniques leads to coherence function and phase-velocity, giving some additional informations about the convection velocity of eddies. Intermittency function inside the wake is also given. Isocorrelation plots show elongated, quasi-ellipsoïdal forms with angles of inclination about 30 degrees, result confirmed by high speed schlieren visualizations.

1. INTRODUCTION

Measurements giving informations about the structure of turbulence in plane wakes are still few. The important experimental work done by Grant [1] has been extensively used by Lumley and Payne [2] and Townsend [3] giving a conceptual model of "double-roller eddy" for the structure of the wake. The measurement of Grant [1] is an extensive direct study of the nine components of the double velocity correlation tensor, without any conditional sampling or frequency filtering. The work done by Thomas [4] investigates the structure of turbulence in a plane wake by means of conditional sampling and gives essentially intermittency, conditional probability densities and statistics of the turbulent bursts. For high speed wakes, specially supersonic ones, the lack of data is quite striking. A.Demetriades [5, 6] has studied the axisymmetric compressible wake (M=3) by measuring intermittency, scales, and space-time correlations coefficients in two directions. Some other measurements were made concerning space correlation coefficients in ballistic range by Fox [7, 8] and general informations on hypersonic wakes by Finson [9]. Extensive study of turbulent quantities in an axisymmetric wake strongly out of equilibrium at supersonic speed has been made by Gaviglio et alii [10], putting emphasis on auto-correlations measurements. The recent study of Townsend [11] using a rake of several hot-wires seems to confirm the double-roller eddy model. It appears that in the field of wakes, and particularly plane ones and for high speed flows at large Reynolds numbers, there is a real dearth of data permitting to construct a conceptual model of turbulence structure. In a supersonic wake, it appears many experimental difficulties, essentially the large frequency bandwidth of the signals due to the high speed and the coexistence of different modes due to compressibility.

2. EXPERIMENTAL DESIGN AND MEASUREMENT TECHNIQUES

2.1 Wind Tunnel

The wind tunnel used in the present study is a blowdown wind tunnel with a maximum running time of about 1 minute. The dimensions of the test section are 15×15 cm. The supersonic flow is created by a two-dimensional nozzle giving a freestream Mach number $M_\infty = 1.975$. The total pressure is 6 bars and total temperature about 250 K decreasing about 0.3 deg per second. With these conditions the Reynolds number R_e is equal to $0.96.10^6$ cm^{-1}.

The model used for the wake generation is a flat plate with the leading edge located upstream from the sonic throat. It is a 15 cm wide steel plate with a length in the supersonic flow equal to 0.8 meter ; the thickness is 0.3 cm and the trailing edge is bewelled with 3 degrees double wedge angle. The boundary layers at the trailing edge are fully-developped turbulent ones, and have a thickness of about 1 cm.

The x direction is the streamwise one, z is parallel to the trailing edge (the mean motion is homogeneous in z direction) and y is normal to the plane of the wake (the maximum of mean gradients are along the y axis).

2.2 Measurement Apparatus

We use C.T.A. because it is more easy to use than C.C.A., specially in blowdown wind tunnels with short running time and unsteady total temperature. The frequency response in phase and amplitude was tested by electric tests and by heating of the wire using a modulated laser-beam [12]. Very satisfactory response is obtained up to 500 KHz, using DISA 55 M 12 symmetrical bridge anemometer, at an overheat ratio of 0.8. Earlier spectral data show that the useful energetic range did not extend much beyond 300-400 KHz.

The wires are Pt. Plated Tungsten, $\phi = 5$ μm, $\ell = 1$ mm long. Each wire and H.W.A. is tested, before use, in a specially designed, small-dimensions, calibration wind tunnel in order to eliminate wires giving too high strain-gage effects and to make electronic adjustments.

As shown by Demetriades, the interference between the shock waves shed by the upstream probe and the downstream one is very localized and doesn't affect the greater part of space correlations.

The data were collected during a run either by keeping a stationnary hot-wire at a point (X,Y,Z) in the wake and traversing a second wire (X+x', Y+y', Z+z') (changing y' step by step), or moving simultaneously the two wires located at (X,Y,Z) and (X,Y,Z+z') (changing Y step by step). The two wires are connected to identical C.T.A.

During the run, the two A.C. output signals of the C.T.A. are recorded, using a magnetic tape recorder Bell & Howell type M 14 G, using two F.M.II channels with 0-500 KHz bandwidth; the DC levels and displacement synchronisation signal are recorded on FMI channels (0-80 KHz bandwith) at a tape speed of 120 inches/sec.

The usual reproduce speed used is 7,5 ips giving 16:1 ratio for time-dilatation.

2.3 Analog Measurements

The measurement of unfiltered space and space-time correlations are made by analog way using classical devices.

The two signals are first used during the run to give direct space correlations then replayed after the run to give space and space-time correlations.

Two RMS voltmeters (DISA 55 D 35) are used in addition to one turbulence processor (52 B 25 DISA) associated with the time-delay unit (55D75). These devices gives $\sqrt{\overline{e_1'^2}}$, $\sqrt{\overline{e_2'^2}}$, $\overline{e_1'(t)e_2'(t+\tau)}$ where e_i is the A.C. output signal of the C.T.A., τ is the time-delay imposed to signal 2.

The correlation coefficient is then obtained by calculating

$$R \simeq R_{11}(x',y',z';\tau) = \frac{\overline{\rho u'(X,Y,Z,t_o)\ \rho u'(X+x',Y+y',Z+z';t_o+\tau)}}{\sqrt{\overline{\rho u'^2(X,Y,Z,t_o)}}\ \sqrt{\overline{\rho u'^2(X+x',Y+y',Z+z',t_o)}}}$$

As shown by Demitriades [6], the complexity of mode-correlation resolution is considerable ; comparisons between restored and unrestored correlations show that the differences are rather minor and "an approximate notion of the most complex feature of the wake turbulence can be obtained by looking at the "raw" signals" [6].

2.3 Digital Processing

For the study of intermittency or filtered correlations the recorded signals are digitized and stored on a mini computer for later use. In the case of correlation studies the sampling frequency taking into account the speed ratio of the magnetic tape recorder was 320 KHz per channel. The interval between the sampling of each channel introduces an inter-channel time delay of $1.56\ 10^{-6}$ s which was taken into account in the numerical procedure. The output signals from the recorder are low-pass filtered below the Shannon frequency with two identical Schlumberger Active Filters FAB 24 (24 dB/octave).

3. BASIC FLOW AND INTERMITTENCY

The free-stream turbulence of the wind-tunnel is essentially due to sound-mode disturbances generated upstream from the sonic throat. In spite of the installation of acoustic mufflers and several screens in the settling chamber the turbulence level in the external region out of the wake is about $4\ 10^{-3}$ for $(\rho u)'/\overline{\rho u}$.

Measurements show that the mean value of total temperature is nearly constant inside the wake. This result can be explained by the fact that the wall temperature of the flat plate is nearly equal to the stagnation temperature in the settling chamber. The Mach number profiles are computed from static and total pressure measurements and, together with the total temperature profiles, they allow to obtain the mean velocity profile given on figure 1.

All the correlation measurements are made at the same x station located 140 mm downstream from the trailing edge, $x = x^*/\delta_o = 14$, where δ_o is the thickness of half defect velocity profile, $\delta_o = 10$ mm. All the lengths are non-dimensional ones (reference length is δ_o).

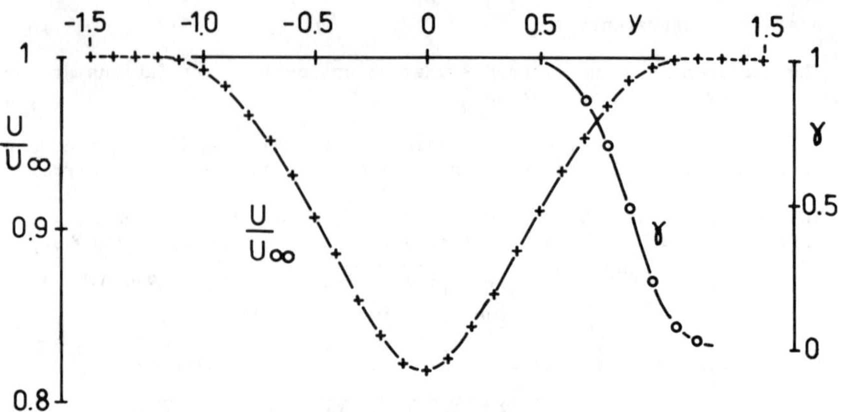

Fig.1 - Mean velocity and intermittency factor in the supersonic wake at x = 14

The Reynolds number based upon the free-stream conditions and δ_o at this abscissa is equal to $0.96\ 10^6$. The two-dimensionality of the flow was checked and found satisfactory over 80% of the wake span.

The intermittency factor γ was determined for several Y positions following a digital procedure similar to those used by several authors (see for example [13] and [14]).

- First, digitalization at a sampling frequency of 530 KHz of 12 independant samples e(t) of 2048 points each.
- Second, numerical time-derivation : $\left|\frac{\partial e}{\partial t}\right| = \left|\frac{e(t+\Delta t)-e(t)}{\Delta t}\right|$
- Third, low-pass numerical filtering at 52 KHz of the serie $\left|\frac{\partial e}{\partial t}\right|$.
  The digital filtering method consists in a Fast Fourier Transform analysis of the serie followed by a FFT synthesis. This filter is an ideal and very versatile one.
- Fourth, an absolute threshold level is imposed, equal to 0.2 of the maximum value of $\left|\frac{\partial e}{\partial t}\right|$ filtered obtained at the point $\gamma = 0.5$.

The Taylor microscale λ_s can be estimated from the microscale time λ_T by $\lambda_s = \bar{u}\lambda_T$ and λ_T can be computed from the samples by

$\lambda_T = \sqrt{\frac{2\ \overline{e'^2}}{\overline{(\frac{\partial e'}{\partial t})^2}}}$ [15] . In the present case, $\lambda_T = 2.6\ 10^{-6}$ sec giving a value of $\lambda_s = 1.2$ mm. The sampling time interval is $1.9\ 10^{-6}$ sec. Following Hedley and Keffer [13] the sampling time interval should be much smaller than the microscale time because, according to the high value of the Reynolds number, the ratio of the Taylor microscale to the Kolmogorov scale is very high (typically greater than $2\ 10^2$). This condition was not satisfied in the present study which is to be considered as preliminary and better results should be obtained in the near future by using a sampling frequency higher than 1 MHz.

The profile of intermittency factor γ, plotted fig.1, agrees well with the measurement of Thomas [4] in a plane turbulent wake at subsonic speed.

4. SPACE AND SPACE-TIME CORRELATIONS

4.1 Space-Time Correlations in x Direction

The correlation coefficient $R_{11}(x',0,0;\tau)$ is measured for two positions i.e on the wake axis Y = 0 and near the point of maximum shear Y = .6. For the first Y position, mechanical arrangement does not allow to put the two wires at exactly the same Y location and in fact the quantity $R_{11}(x',.02,0,\tau)$ is measured instead of $R_{11}(x',0,0,\tau)$. The results are plotted for time delays up to 10 10^{-6} sec on the figure 2.

The two sets of curves are rather similar and for τ = 0 one can notice significant values of the correlation for x>.5. This suggest the existence of structures with a typical scale $O(\delta_0)$

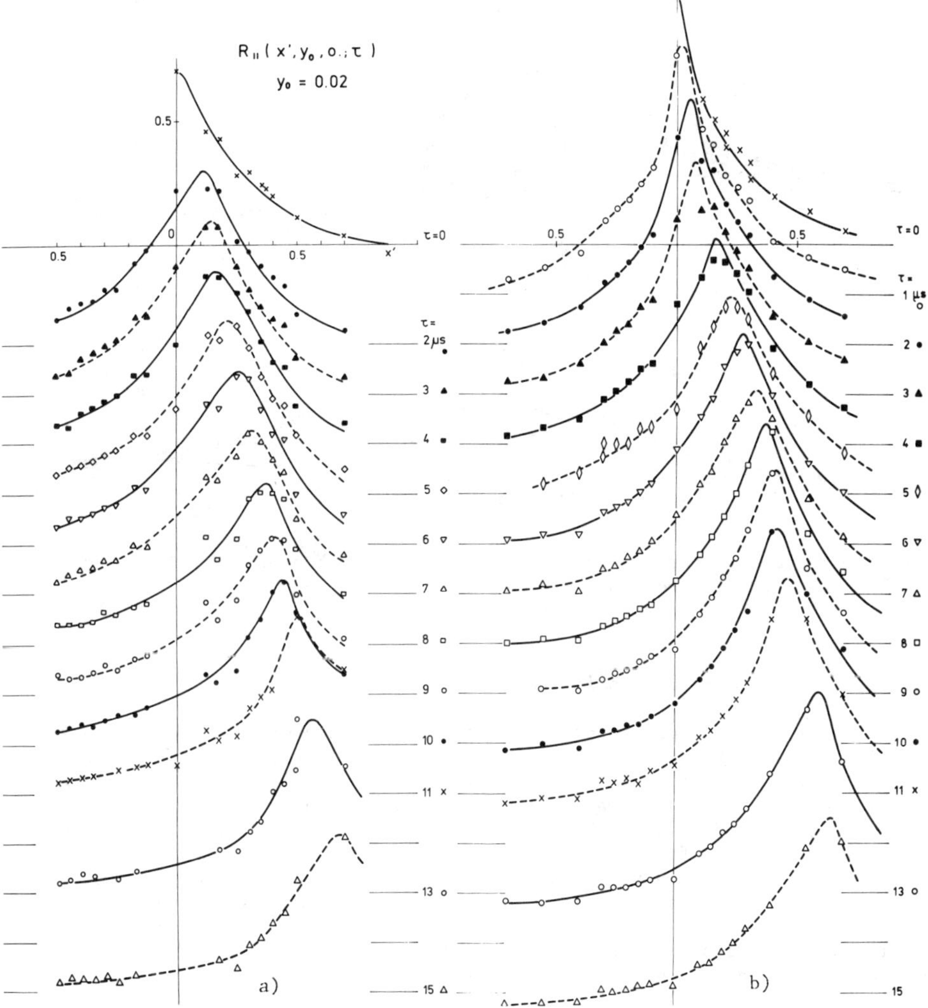

Fig.2 - Space-time correlation $R_{11}(x',0.,0.;\tau)$ at two Y locations a) Y = 0 and b) Y = 0.6

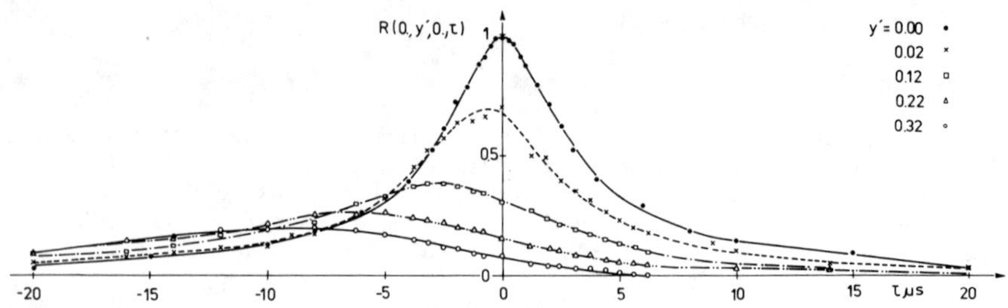

Fig.3 - Space-time correlation $R_{11}(0.,y',0.;\tau)$ at Y = 0

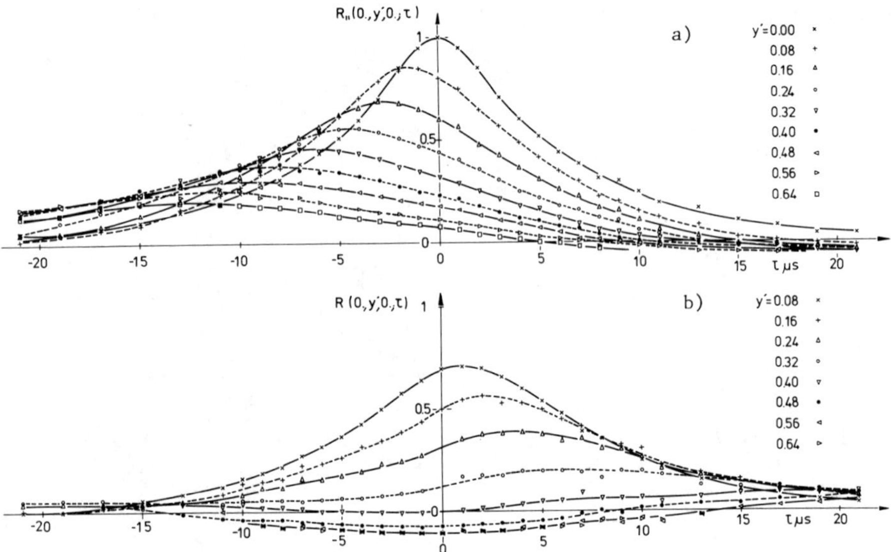

Fig.4 - Space-time correlation $R_{11}(0.,y',0.;\tau)$ at Y = 0.3
a) y'>0 towards the external flow
b) y'<0 towards the axis

4.2 Space-Time Correlations in Y Direction

The correlation coefficient $R_{11}(0,y',0;\tau)$ is measured for 3 positions in the wake : on the axis Y = 0, at the maximum shear point Y = .6 and between these two stations Y = .3. When the stationnary probe is not on the wake axis the second probe is moved away from the stationnary one successively towards the axis (y'< 0) and towards the external stream (y'>0). We can notice on figures 3-4-5 the striking difference between the two sets Y'>0 and Y'<0 for both Y=.3 and Y=.6. The wake axis appears as a particular location, and significant negative correlations are obtained for y'<0. The minimum correlation is obtained for τ = 0 and the space correlation $R_{11}(0,y',0;0)$ is plotted on the figure 6. This behaviour is not in agreement with the earlier measurements of Grant [1] in a subsonic plane turbulent wake.

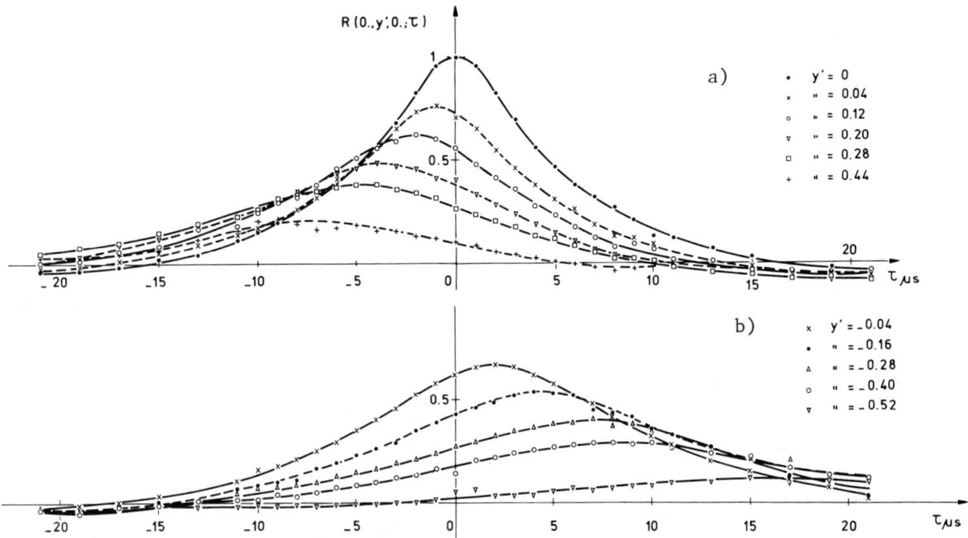

Fig.5 - Space-time correlation $R_{11}(0.,y',0.;\tau)$ at Y = 0.6
a) y'>0 towards the external flow
b) y'<0 towards the axis

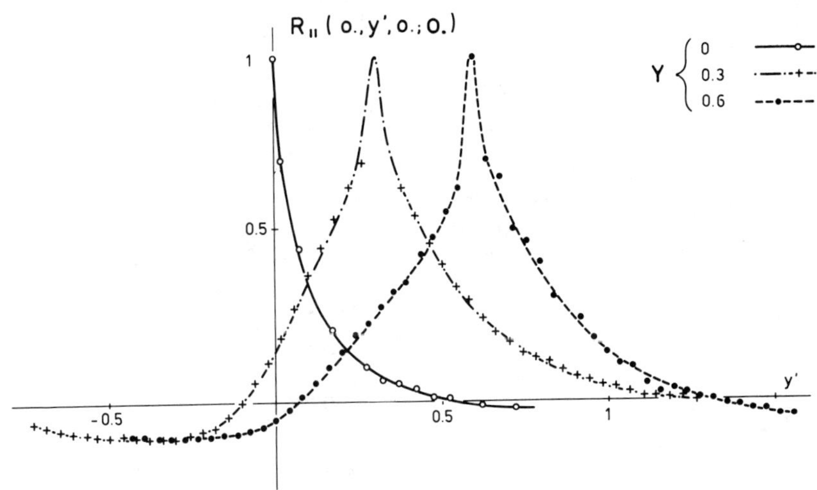

Fig.6 - Transverse correlation $R_{11}(0.,y',0.;0.)$ at three locations Y = 0 ; 0.3 ; 0.6

4.3 Space-Time Correlations in Z direction

The correlation coefficient $R_{11}(0.,0.,z';0.)$ is plotted figure 7 for Y = 0 and Y = 0.6, and the evolution following Y is plotted on figure 8. The influence of Y location of the two probes seems to be not very important and the coefficient R_{11} does not reach significatively negative values at Y = 0.6 ;

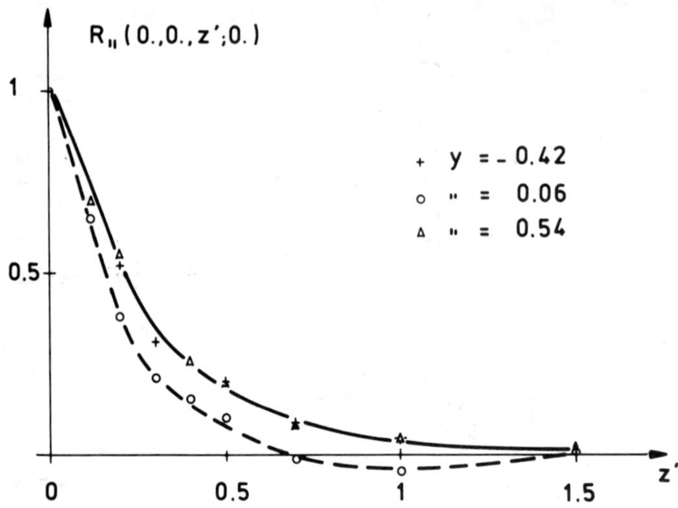

Fig.7 - Space correlations $R_{11}(0.,0.,z;0.)$

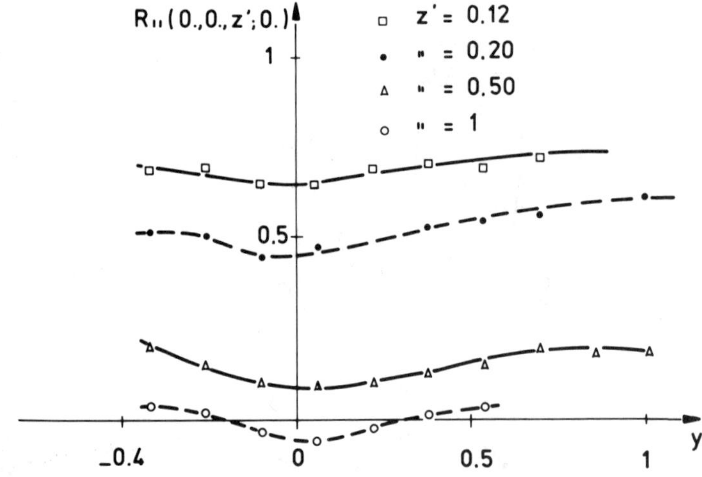

Fig.8 - Evolution, following Y location, of the space correlation $R_{11}(0.,0.,z;0.)$

this behaviour is quite different from the results of Grant [1] (this author's results show relatively large negative values for R_{11} when z' increases). The correlation tends towards zero for large z separation and this seems to preclude the possibility of flapping of the wake as a whole which should give some positive correlation. Space-time correlations $R_{11}(0.,0.,z;\tau)$, plotted on figure 9, show a small dissymetry between positive and negative values of τ.

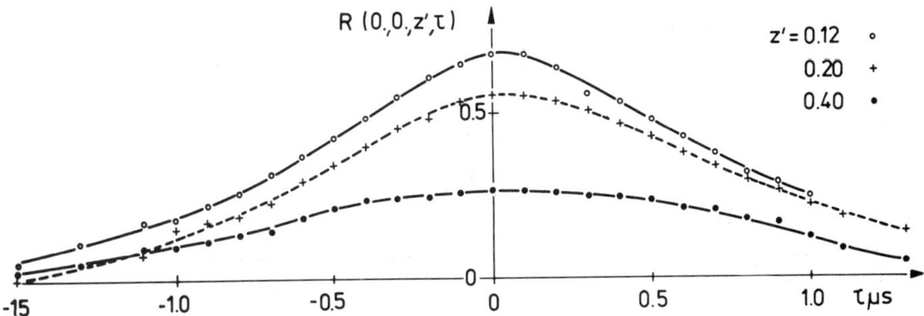

Fig.9 - Space-time correlations $R_{11}(0.,0.,z';\tau)$ at Y = 0.6

4.4 Space Correlation in X-Y Direction

For two Y position (Y = 0 and Y = 0.6) we plot on figure 10 the values of $R_{11}(x',y',0.;0.)$. One can see important differences between the two stations. It shall be noticed that for these measures, at Y = 0.6, no measures were made for y'<0 (towards the axis).

4.5 Space Correlations in Y-Z Direction

With the stationnary probe at Y = 0.6 we made measurements for several y' and z' separations, giving the coefficient $R_{11}(0.,y',z';0.)$ plotted on figure 11. Here again the wake axis plays a particular part and negative correlations are obtained when the probes are located on both sides of the symmetry plane. As expected the correlation decrease towards zero when Z increases precluding the possibility of flapping.

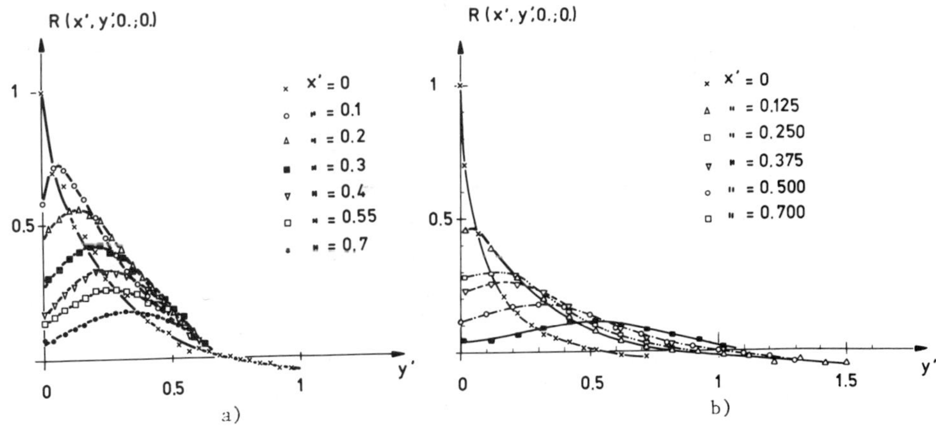

Fig.10 - Space correlation $R_{11}(x',y',0.;0.)$
a) Y = 0.6 b) Y = 0.0

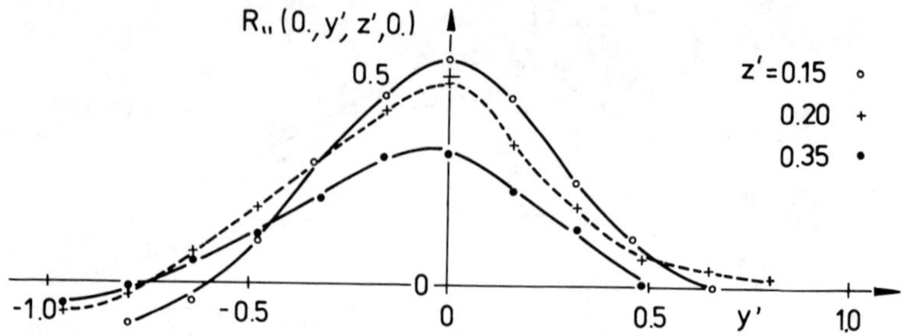

Fig.11 - Space correlation $R_{11}(0.,y',z';0.)$ at Y = 0.6

4.6 Integral Scales

The integral scales are defined following the expressions :

$$\begin{cases} \Lambda_{x_Y} = \int_0^{\infty} |R_{11}(x',0,0,0)|\, dx' \quad \text{at Y location} \\ \Lambda_{y^+_Y} = \int_0^{+\infty} |R_{11}(0,y',0,0)|\, dy' \\ \Lambda_{y^-_Y} = \int_{-\infty}^{0} |R_{11}(0,y',0,0)|\, dy' \\ \Lambda_{z_Y} = \int_0^{+\infty} |R_{11}(0,0,z',0)|\, dz' \\ \Lambda_{T_Y} = \int_0^{\infty} |R_{11}(0,0,0,\tau)|\, d\tau \end{cases}$$

They are given on table 1.

- TABLE 1 -

	Y = 0	Y = .3	Y = .6
Λ_x	.22		.22
Λ_{y+}	.11	.122	.175
Λ_{y-}		.195	.23
Λ_z	.17		.295
Λ_T	5.2 μs	6.7 μs	5.7 μs
Λ_x/Λ_{y+}	2	1.8*	1.25
Λ_x/Λ_{y-}		1.13*	.96
Λ_x/Λ_z	1.3		.74
$\Lambda_x/u_e\Lambda_T$	.94		.86

* computed with Λ_x = .22.

The value 2 obtained for the ratio Λ_x/Λ_y on the axis is higher than the result of Grant [1] (1.66) and similar to the results obtained by Gutmark [16] in a plane jet (2.19), Comte Bellot et Corrsin [17] in grid turbulence (2.) and Champagne and alii [18] for nearly homogeneous shear flow (2.05).

5. CONVECTION VELOCITIES AND TAYLOR HYPOTHESIS

5.1 Convection Velocities

Following Favre et alii [19], the convection velocity can be obtained by plotting iso-correlation curves in X and τu_e coordinates. Several convection velocities can be deduced : with τ=cste : $(u_c)_\tau$, with x'=cste$(u_c)_x$ or intermediate solution $(u_c)_{\tau x}$.

The three definitions give quite similar results (figure 12), the iso-correlation curves being very elongated ellipses, in a similar way to the case of a frozen-pattern.

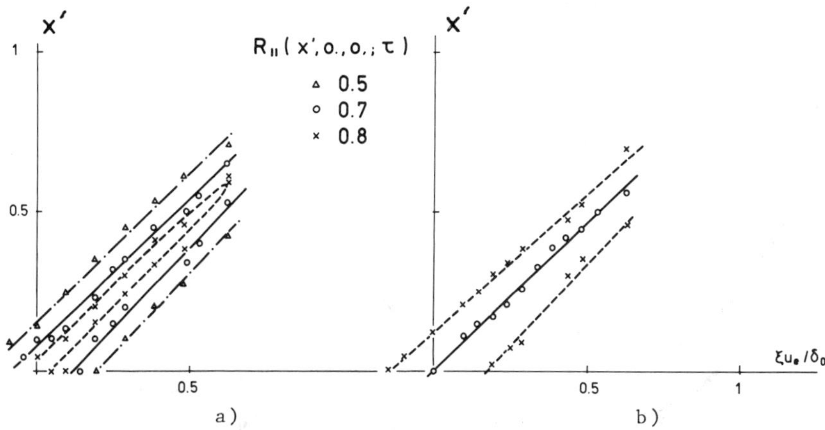

Fig.12 - Contour plots for $R_{11}(x',0.,0.;0.)$
a) Y = 0.6 b) Y = 0.0

The convection velocity reaches a value very near the external velocity u_e : $u_c/u_e \simeq 0.94$ for two position inside the wake.

This result can be confirmed by plotting τ_{opt}, the optimum time delay obtained for $R_{11}(x',0.,0.;z)$, versus x' separation (figure 13) as done by Demitriades [6]. The value of u_c/u_e obtained, nearly equal to 1 and independant of Y location, confirms the measurement of Demetriades [6].

5.2 Taylor Hypothesis

In order to test the validity of the Taylor hypothesis we compare (figure 14) the values of $R_{11}(x',y',0.;0.)$ and of $R_{11}(0.,y',0.;\tau)$ at the same locations Y = 0 and Y = 0.6.

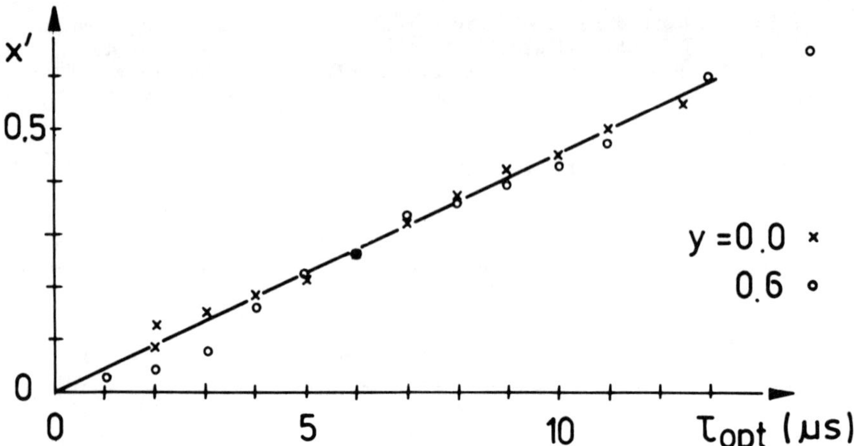

Fig.13 - Optimum time delay, from $R_{11}(x',0.,0.;\tau)$
for Y = 0.6 and Y = 0.0

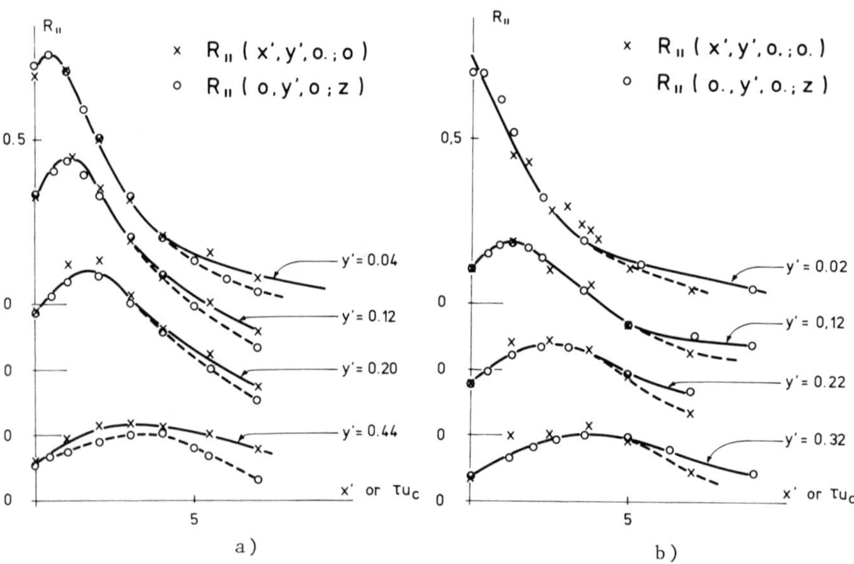

Fig.14 - Space and space-time correlation comparison -
a) Y = 0.6 b) Y = 0

As already shown by Demetriades, some discrepancies appear between auto-correlation and space-correlation when the y separation is too high, i.e. near the wake edge.

5.3 Evolution of Optimum Space Correlation

Following Dumas and alii [20], we have plot on figure 15 the evolution of optimum space-correlation coefficient $R_{11}(x',0.,0.;\tau_{opt})$ versus the

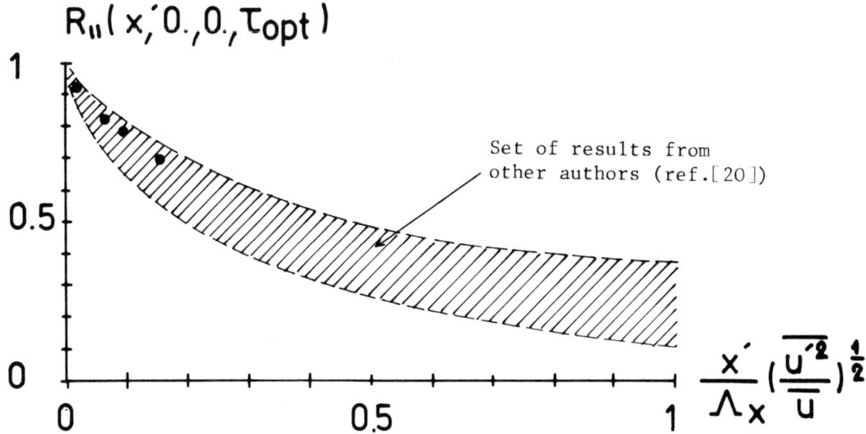

Fig.15 - Evolution of optimum space-time correlation at Y = 0.6

similarity parameter $x'/\Lambda_x.(\overline{u'^2}/\bar{u})^{1/2}$ at Y = 0.6. Although the range of x' separation covered in the present measurements is too small to draw definitive conclusions, the agreement with results obtained by various authors seems rather satisfactory.

5.4 Filtered correlations

The definition of the convection speed through unfiltered correlation measurements gives an overall result nearly independent of the scale of the turbulent structures. In order to look at some possible dependence of the convection speed upon the structure scale we applied a numerical procedure instead of doing filtered correlation measurements by analog means. The method consists in computing the cross-spectra of the two signals from which one can deduce the coherence function and, using the phase angle, the convection speed versus the frequency [19].

The following technique is adopted :

- First collect by digitization of the recorded signals 56 samples of 2048 values from each hot-wire output. In order to avoid aliasing errors the two signals are passed through paired active low-pass filters. The cutoff frequency is 8KHz giving an equivalent true value of 128 KHz.
- Second, taper each sample over 1/10 of its length in order to avoid leakage due to finite length and, using the FFT algorithm, compute the complex Fourier transform which gives the real and imaginary part of the cross-spectrum for 1024 values of the frequency.
- Third, ensemble average the 56 cross-spectra, then frequency smooth over 16 neighbouring values of the frequency, following the procedure described in ref.[21]. The result is then an average of 896 complex values of the cross-spectra for 64 center frequencies. The sampling frequency being 320 KHz on each channel, the resulting frequency wideband is 2.5 KHz. The estimates are chi-square variables with 1792 degrees of freedom and the normalized standard error is 3%.
- Fourth, calculate the coherence function $\gamma^2(f) = |G_{12}(f)|^2/G_1(f).G_2(f)$ and

the phase angle, then correct the phase angle $\psi(f)$ by the amount due to the inter-channel time interval and finally, after smoothing the phase angle in order to avoid difficulties at very low frequencies, calculate the convection velocity $u_c(f) = 2\pi f x'/\psi(f)$.

The figures 16 and 17 show, for two Y locations, the square root of coherence function (γ), the unrestored phase angle and the convection velocity plotted adimensionnaly : $u_c^*(f) = u_c(f)/u_c(f_o)$ where f_o is the first central frequency, here 2.5 KHz.

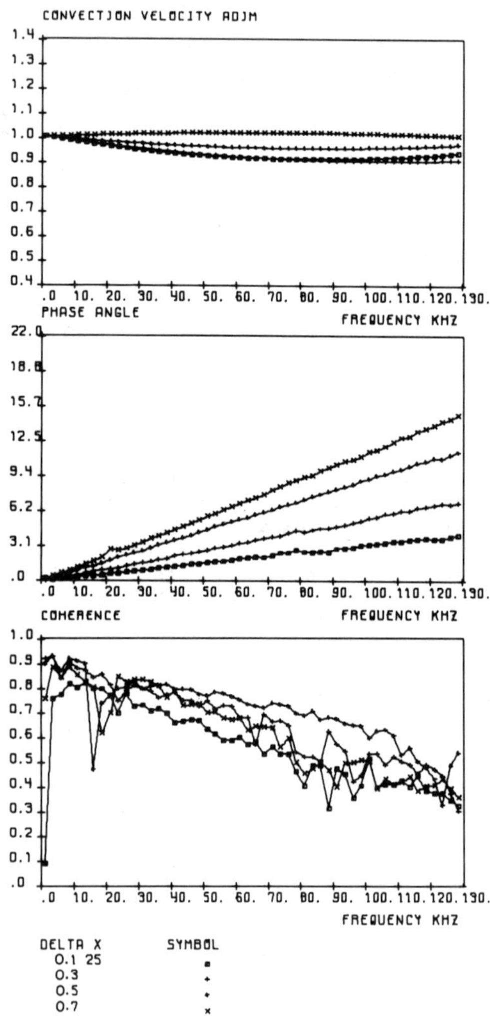

Fig.16 - Coherence, phase angle and adimensional velocity at Y = 0

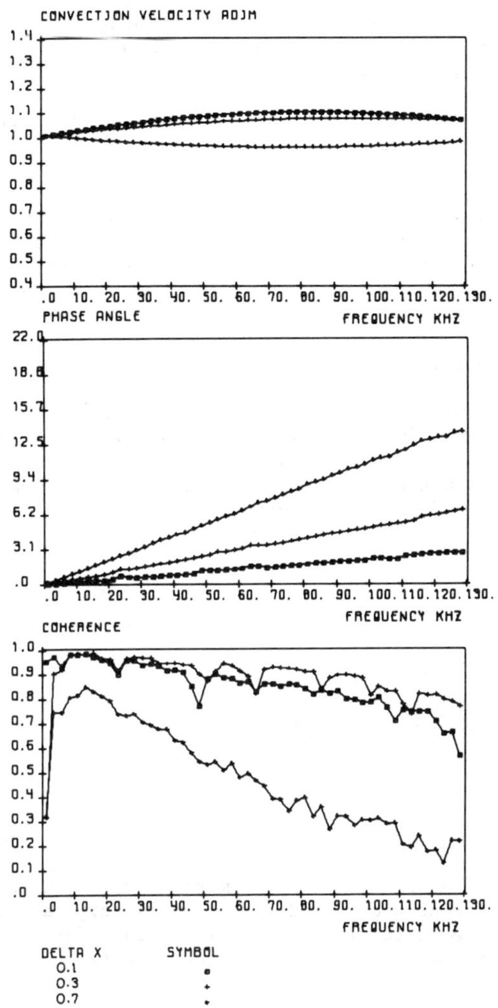

Fig.17 - Coherence, phase angle and adimensional velocity at Y = 0.6

It appears that the convection speed depends upon the x' length and the frequency but the observed variations are rather small. The absolute value of the convection speed for a frequency about 30 KHz, corresponding to the energy containing range, agrees very well with the value nearly equal to u_e determined in § 5.2 from correlation measurements.

Comparison of the figure 16 and 17 shows a significant difference : on the wake axis Y = 0 the convection speed decreases when the frequency increases and the reverse is true for Y = .6. This result on the figure 16 could be expected to some extent as, presumably, the velocity should tend towards the local mean value $\bar{u} = .8u_e$ when the fluctuation scale decreases. This result will be even more pronounced if, as suggested by Favre and alii [22], the velocities are determined on distances x′ consistent with the frequency i.e. small separation between the two probes for high frequency.

Analysis of the results for the coherence function is delicate because, even for 1792 degrees of freedom, the statistical scattering is rather high especially for low values of the coherence. Apart the effect of finite bandwidth the coherence function can be viewed as the filtered correlation for optimum time delay. One should then expect a decrease of the coherence as the frequency or the x' length increases. This behaviour can be found on the figure 17 between the positions x' = .3 and x' = .7 but the apparent independance upon x' which appears on the figure 16 is not yet explained.

6. ISOCORRELATION AND VISUALIZATIONS

From the measurements of space-correlations R(x',y',0.;0.) one can plot the iso-correlation curves in x' and y' axis. The figure 18 shows elongated, quasi ellipsoïdal forms inclined with an angle about 30 degrees relative to the wake axis. This result can be compared with the more complete map of isocorrelations abtained from space and space-time correlation measurement given in figure 19.

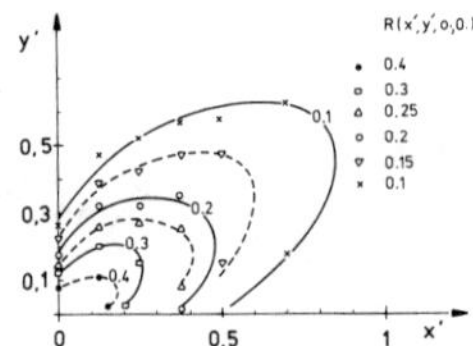

Fig.18 - Contour plot for space-correlation R_{11}(x',y',0.;0.)

The first conclusion is the good agreement between figure 18 and figure 19a), i.e. between the shapes obtained either by direct space correlation or by use of Taylor hypothesis to give x' = $\tau u_c/\delta_0$; therefore it can be concluded that the mutual aerodynamic interference between the two probes is not very important.

Second, the map of figure 19 gives an interesting description of the shapes encountered :

- in the plane X-Y the shapes are quite affected near the axis, especially for the Y = 0.3 location where the isocorrelations curves are tight to the wake axis
- in the planes Y-Z and Z-X the shapes present a peculiar behaviour along the Y or X axis.

The 3-Dimensionnal representation of the isocorrelation surfaces looks like ellipsoïdal volume inclined on the axis with a kind of depression on the leeward side near the symmetry plane of the wake. This description requires a more extensive study in order to give definitive conclusions and to be explained in terms of specific structures.

In order to confirm those results, a high speed film has been made using schlieren visualizations. The camera used permits to take a set of 25 pictures

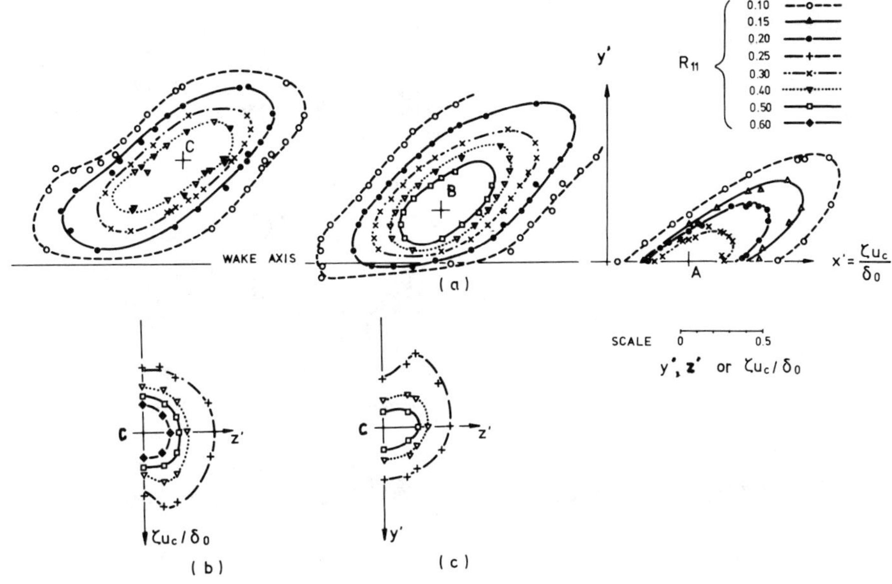

Fig.19 - Contour plots for space and space-time correlations
Fixed wire locations A : Y=0. ; B : Y=0.3 ; C : Y=0.6
a) $R_{11}(0.,y';0.;\tau)$; b) $R_{11}(0.,0.,z';\tau)$; c) $R_{11}(0.,y',z';0.)$

at a speed of 200.000 frames per second ; the figure 20 shows a typical result with 4 successive frames with a time-separation of 5 µs followed by 10 frames, with a time-separation of 15 µs. The arrow indicates the x-evolution of a given structure during the set of visualizations. The shapes encountered seem to agree well with the results of figure 19a), and the convection velocity calculated from the time-evolution of a typical shape observed on the pictures is in quite good agreement with the convection velocity calculated from space-time correlations.

7. CONCLUSION

Space and space-time correlation measurements in a supersonic plane wake at high Reynolds number put in prominent position elongated structures tilted in the xy plane with an angle of about 30 degrees relative to the x axis. This result is confirmed by high speed motion pictures. However the correlation distributions does not agree very well with the double roller eddy model of Grant.

ACKNOLEDGMENTS

This study has been performed under contract D.R.M.E. 78-573

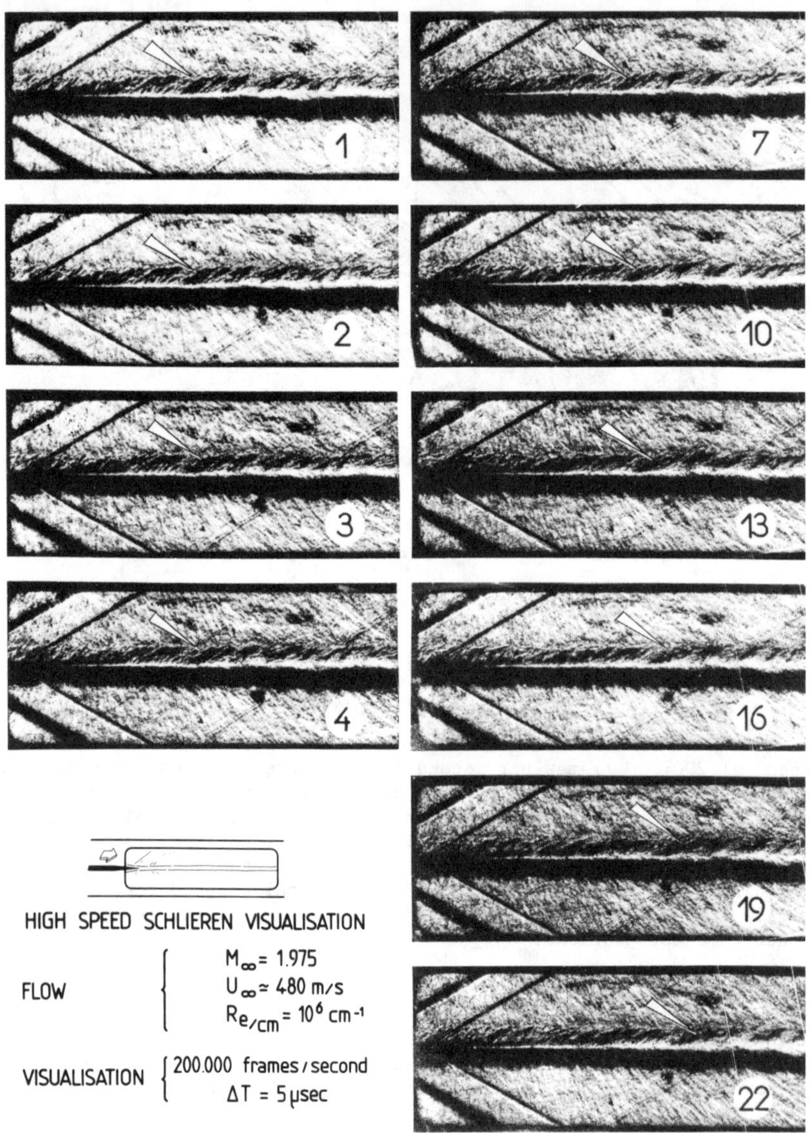
1
2
3
4
7
10
13
16
19
22
HIGH SPEED SCHLIEREN VISUALISATION
FLOW
$M_\infty = 1.975$
$U_\infty \simeq 480$ m/s
$R_{e/cm} = 10^6$ cm^{-1}
VISUALISATION
200.000 frames / second
$\Delta T = 5\,\mu$sec

Fig. 20

REFERENCES

1. GRANT, M.L. The large eddies of turbulent motion. J. Fluid Mech. (1958) p.4, pp.149-190.

2. PAYNE, F.M. and LUMLEY J.L. Large eddy structure of the turbulent wake behind a circular cylinder. Physics of Fluids Sup., (1967) S194-S196.

3. TOWNSEND, A.A. The structure of turbulent shear flows. Cambridge University Press, 1976.

4. THOMAS, R.M. Conditional sampling and other measurements in a plane turbulent wake. J. Fluid Mech. (1973), Vol. 57, p.3, pp. 549-582.

5. DEMETRIADES A. Turbulent front structure of an axisymmetric compressible wake. J. Fluid Mech. (1968) Vol. 34, p.3, pp. 465-480.

6. DEMETRIADES A. Turbulence correlations in a compressible wake. J. Fluid Mech. (1976), Vol. 74, part 2, pp. 251-267.

7. FOX, J., WEBB, W.M., JONES, B.G. and HAMITT, A.G. Hot-wire measurements of wake turbulence in a ballistic range. AIAA J., Vol. 5 N° 1, Jan. 1967, pp. 99-102.

8. FOX, J. Space correlation measurements in the fluctuating turbulent wakes behind projectiles. AIAA J., Vol. 6, N° 2, Feb. 1968, pp. 233-238.

9. FINSON, M.L. Hypersonic wake aerodynamics at high Reynolds numbers. AIAA J., Vol. 11, N° 8, Aug. 1973, pp. 1137-1145.

10. GAVIGLIO, J., DUSSAUGE, J.P., DEBIEVE, J.P. and FAVRE, A. Behaviour of a turbulent flow, strongly out of equilibrium, at supersonic speeds. The Physics of Fluids, Vol. 20 N° 10, p.II, Oct. 1977, p. S179-S192.

11. TOWNSEND, A. The structure of turbulent shear flow. 2nd Edition, Cambridge University Press, 1976.

12. BONNET, J.P. and ALZIARY de ROQUEFORT, T. Determination and optimization of frequency response of constant temperature hot-wire anemometers in supersonic flows. Rev. Sci. Instr. 51(2), Feb. 1980, p. 234-239.

13. HEDLEY, T.B. and KEFFER, J.F. Turbulent/non-turbulent decisions in an intermittent flow. JFM (1974) Vol. 64, p. 4, pp. 625-644.

14. ARNAL, D. and JUILLEN, J.C. Techniques d'analyse conditionnelle d'un signal fil chaud pour l'étude de l'intermittence de transition. La Recherche Aérospatiale, 1979, Vol. 1, p. 23-31.

15. TENNEKES, M. and LUMLEY, J.L. A first course in turbulence. MIT Press 1972.

16. GUTMARK, E. and WYGNANSKI, I. The planar turbulent jet. J. of Fluid Mech., Vol. 73, part 3, p. 465-495, 1976.

17. COMTE-BELLOT, G. and CORRSIN S. Simple Eulerian time correlation of full and narrow-band velocity signals in grid generated "isotropic" turbulence. J. of Fluid Mech., Vol. 48, part 2, p. 273-337, 1971.

18. CHAMPAGNE, F.H., HARRIS, V.G. and CORRSIN, S. Experiments on nearly homogeneous turbulent shear flow. J. of fluid Mech., Vol. 41, part 1, p. 81-139, 1970.

19. FAVRE, A., KOVASZNAY, L.S.G., DUMAS, R., GAVIGLIO, J. and COANTIC, M. La turbulence en mécanique des fluides. Gauthiers-Villars, 1976.

20. DUMAS, R., FULACHIER, L., ARZOUMANIAN, E. and FAVRE, A. Etude de la structure de la turbulence dans une couche limite par les corrélations spatio-temporelles. Journal de Physique, T. 37, 1976, p. C1-182.

21. BENDAT, S. and PIERSOL, A.G. Random Data : Analysis and measurement procedures, Wiley-Interscience, 1971.

22. FAVRE, A., GAVIGLIO, J. and DUMAS, R. Structure of velocity space-time correlations in a boundary layer. The Physics of Fluids Sup., Vol. 10, p. S138-145, Sept. 1967.

Heat Transfer and Associated Turbulent Processes in Pulsed Turbulent Flows under Conditions of Acoustical Resonance

P. ANDRÉ and R. CREFF
Laboratoire d'Energétique et de Mécanique
Université d'Orléans
45046 Orleans Cedex, France

I - INTRODUCTION

Results concerning turbulent heat transfer processes between the heated wall of a cylindrical duct and a pulsed air flow are presented. They show important local increases in heat transfer rates for particular conditions of the pulsation ; this suggests a modification of the convective turbulent transfer mechanisms.

Many previous works have looked for changes in heat transfer rates involved by flow pulsations ; these pulsations can be generated by a periodical change in pressure gradient or by superimposition of an acoustic vibration field or also by a forced vibration of the wall limiting the flow. A careful insight of the available results about heat transfer modifications induced by pulsations or vibrations reveal great differences : noticeable increases, till 150 %, and more for certain circumstances, decreases for other cases or negligible effects [1,2,3] . This diversification can be explained by the large variety of the used experimental sets and their boundary conditions.

. Experimental parameters of the present study : heat transfer processes presented here, have been undertaken in a cylindrical circular duct of large diameter (100 mm) and length L = 9.50 m. The air flow is generated by depression and its rate is modulated by a sonic nozzle with a time periodic variable section. The range of the mean Reynolds number extends from 10^5 to 2.10^5 (mean velocity $\sim$ 25 m/s). The flow rate modulation frequencies are included between 0 and 40 Hz and the modulation amplitude rate has been fixed to 36 %. The choosen modulation system gives a nearly zero harmonic component rate to the modulation definition. Then, instantaneous velocities and pressures can be written in the form :

$$u(x,r,t) = \text{Re}\{ U_0(x,r) + U_M(x,r)\, e^{j\omega t + \varphi(x)} \} + u'$$

$$p(x,t) = \text{Re}\{ p_0(x) + p_M(x)\, e^{j\omega t + \psi(x)} \} + p'$$

With regard to previous works oftenly refering to small diameter ducts, the present tube size and the modulation generation system confer two new characteristics to the study. The former allows investigations of the dynamic and thermal fluid fields by means of probes inside the flow.

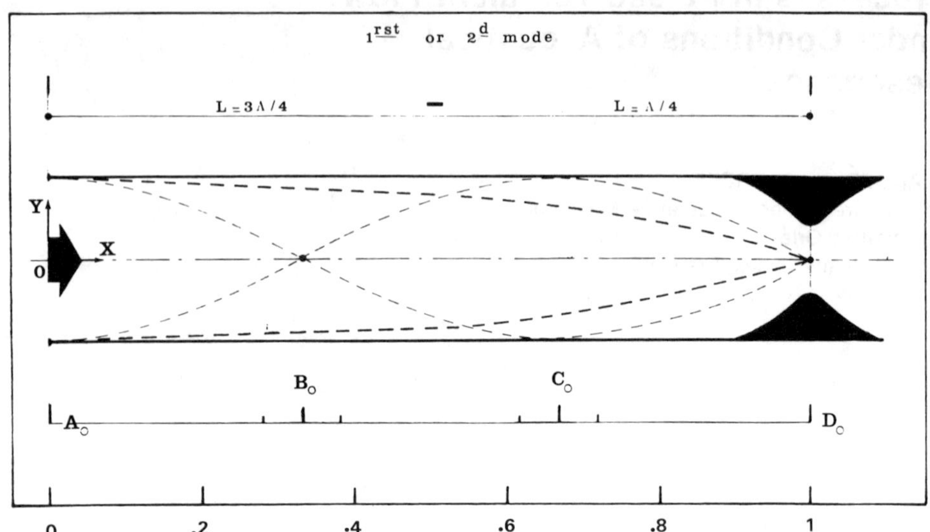

Fig. 1. Scheme of the velocity modulation amplitude for the 1st and 2nd resonance modes (L = $\Lambda/4$ for f_{01} , L = $3\Lambda/4$ for f_{02})

. Aims : in the range of the used frequencies, there are two resonance frequencies associated to the coupled system : flow and duct, f_{01} = 9.2 Hz (first mode) and f_{02} = 27.7 Hz (second mode). Scheme of the velocity modulation amplitude corresponding to the two modes is presented in fig. 1. Moreover, the static pressure modulation amplitude at the wall presents staggered extremums (i.e. : B_o : velocity modulation node for f_{02} and pressure modulation antinode). A non resonant pulsation induces no noticeable modification on the local heat transfer between wall and fluid, whereas for the resonance, and in the velocity antinode sections it appears an improvement about 150 %. We tried to cheek if these increases of the heat transfer coefficients could be related to a modification of the temperature and velocity fluctuation levels.

2 - STUDY OF THE THERMAL FLUID FIELD

As a first stage, fluid temperatures and velocities have been investigated in the particular sections corresponding to the velocity antinode and node of the 2nd resonance mode (C_o, B_o respectively). In the same way of our previous work [4], the heat transfer between the wall and the fluid field has been localized on a short length of the wall tube (600 mm) which allows a constant wall temperature equal to 80° C. A measuring system, (emitting fluxmeter), permits to establish the heat balance in the tube section and then to give the convective transfer coefficients $h(x_i)$. The choice of such a longitudinal localization of heat transfers give us the possibility to consider a nearly isentropic development of the flow field and then a better understanding of the influence of the various dynamic parameters on the heat transfer. Mean steady and unsteady temperatures have been detected along the radius R by means of a microthermocouple (fast response time : 3 ms). The use of a such thermocouple is of great interest in pulsed flows by comparison to the constant current anemometry ; as well known, the microthermocouple is not sensitive to the local direction of the flow as it is for a hot-wire.

The good agreement between the local flux behaviour at the wall with the mean temperature profiles has been checked in sections C_o and B_o for the f_{02} frequency (fig. 2). These profiles are compared to the non pulsed flow profile with $\Theta_m = (T_r - T_c)/(T_w - T_c)$. The flux increase in section C_o agrees with the strong temperature gradient. On the other hand, in section B_o the transfer rate for the pulsed flow is closely related to those of the mean non pulsed flow. Moreover, if we study the factor $g = (T_1 - T_\infty)/(T_2 - T_\infty)$ where T_1 and T_2 are the wall temperatures at the boundaries of the heated section, g illustrates a wall temperature gradient [5] . This quantity tends to 1 under resonance condition and especially at the velocity antinode showing homogeneity in wall temperatures. Then, we may suppose that, in the conduction boundary layer, flux lines are directed normally to the wall in mean time terms. As confirmed by the mean temperature profiles the major result would be a preferential heat transport from the wall to the center axis during the resonance regime.

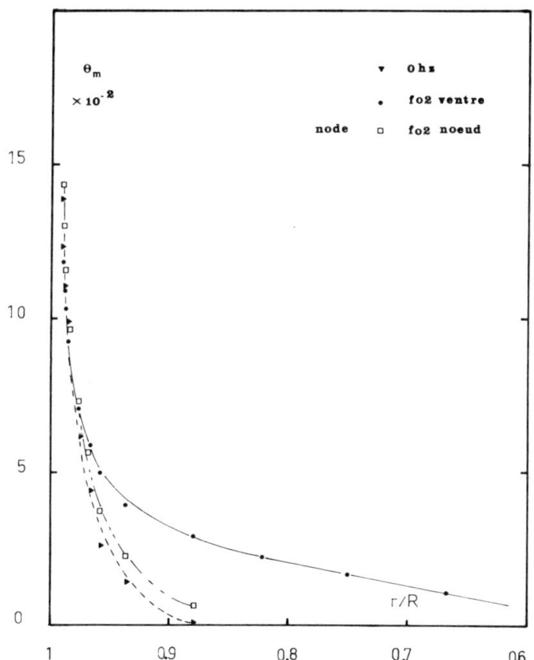

Fig. 2. Mean temperature profiles in the node and antinode section (2nd mode)

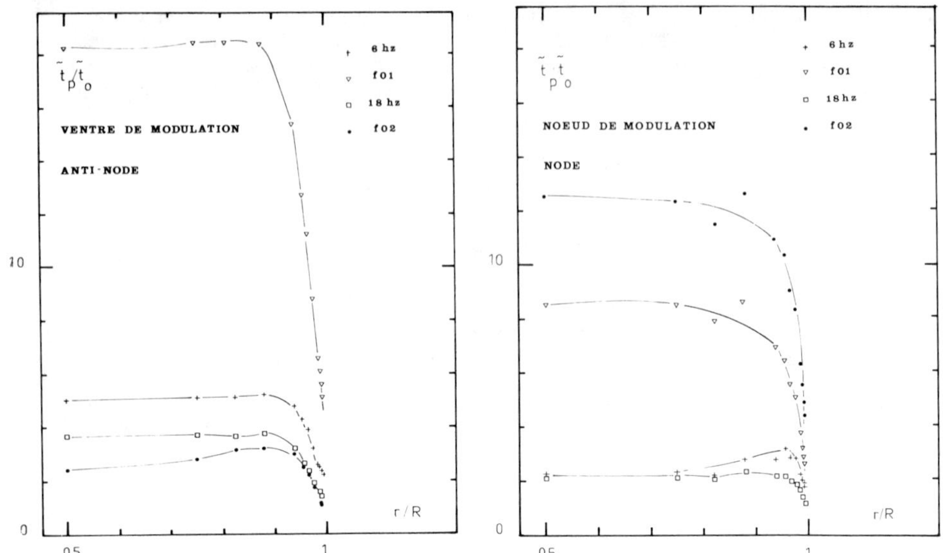

Fig. 3 and 4 : Temperature fluctuation rates along the radius in the antinode and node sections (2nd mode)

The quantity $\tilde{t}_p/\tilde{t}_o$ characterizes the local intensity of the root mean squared temperature fluctuations in pulsed flows with respect to those of the corresponding non pulsed flow. Along the reduced radial coordinate r/R, $\tilde{t}_p/\tilde{t}_o$ varies in the following manners:

i) in section C_o : (fig. 3) : there is a small change in the local fluctuations from the wall to the axis for the 2nd resonance mode whereas they grow quickly along a small radial distance for the first mode. This distance is nearly constant whatever the frequency is.

ii) in section B_o : (fig. 4) : for the two regimes far from resonance modes (6 and 18 Hz) local intensities do not differ greatly from those of the steady flow. For the two resonance modes, f_{01} , f_{02} , the ratios $\tilde{t}_p/\tilde{t}_o$ are of an order of magnitude of 12 an 8 respectively in the potential core.

This reduced quantity is plotted in fig. 5 and 6 versus the pulsation frequency for different radial points. In a general manner, fluctuation rates decrease as approaching the wall and for any frequency.

The strong fluctuation intensities found in section C_o for the f_{01} mode (fig. 5) and in section B_o for f_{01} and f_{02} (fig. 6) are in good agreement with those concerning relative static pressure fluctuations at the wall $\tilde{p}/\tilde{p}_o$ and which have been already stated. Indeed, for the 2nd resonance mode, the section C_o refers to a velocity antinode and a static pressure node ; inversely with the section B_o ; for the 1st mode the section C_o is in the vicinity of a pressure antinode (D_o , fig. 1) and the section B_o a pressure node (A_o).

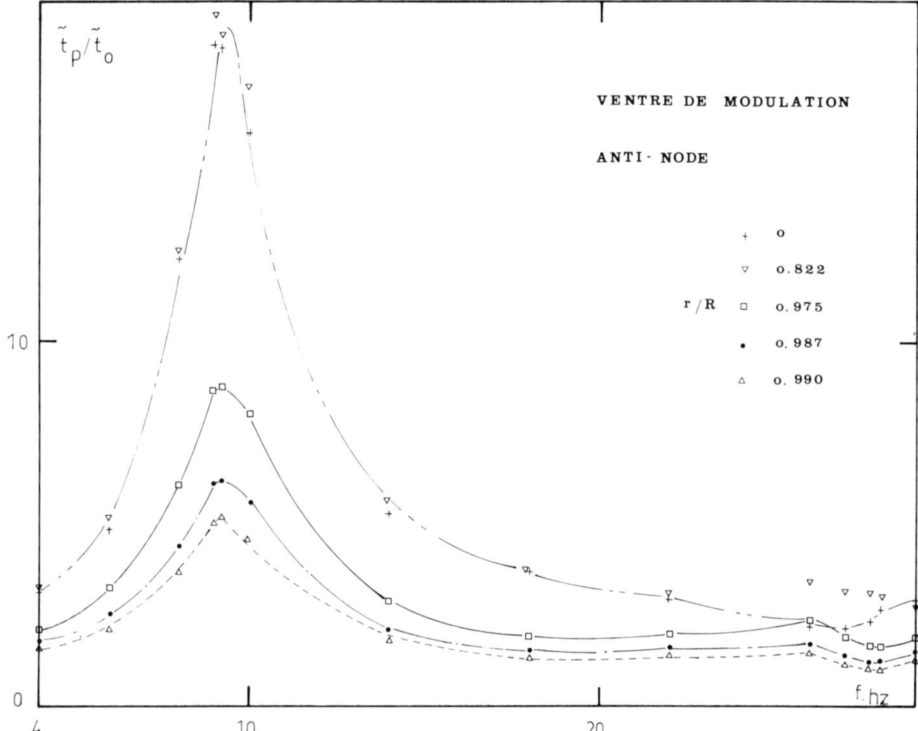

Fig. 5. Temperature fluctuation rates vs pulsation frequency in the antinode section (2nd mode)

3 - STUDY OF THE DYNAMIC FLOW FIELD

In the previous study [4] , the quasi steady state approximation has shown that local heat transfer modifications could not be only related to the macroscopic dynamic effects such as the local mean velocity modulation. Then we tried to look through a possible coupling between pulsation and turbulence and if this could be related to heat transfer rate variations. The study of such a coupling is of interest particularly for the resonance regimes.

To illustrate the velocity modulation evolution, the axis component U_r of the mean local velocity refered to the center axis value U_c is plotted on fig. 7 for the f_{02} resonance case. The antinode and node profiles are compared to those of the steady flow. This velocity has been measured by hot wire anemometry. At the node B_o, the unsteady flow profile is very close to the steady one, while, in section C_o, the over-modulation is traduced by a strong wall gradient.

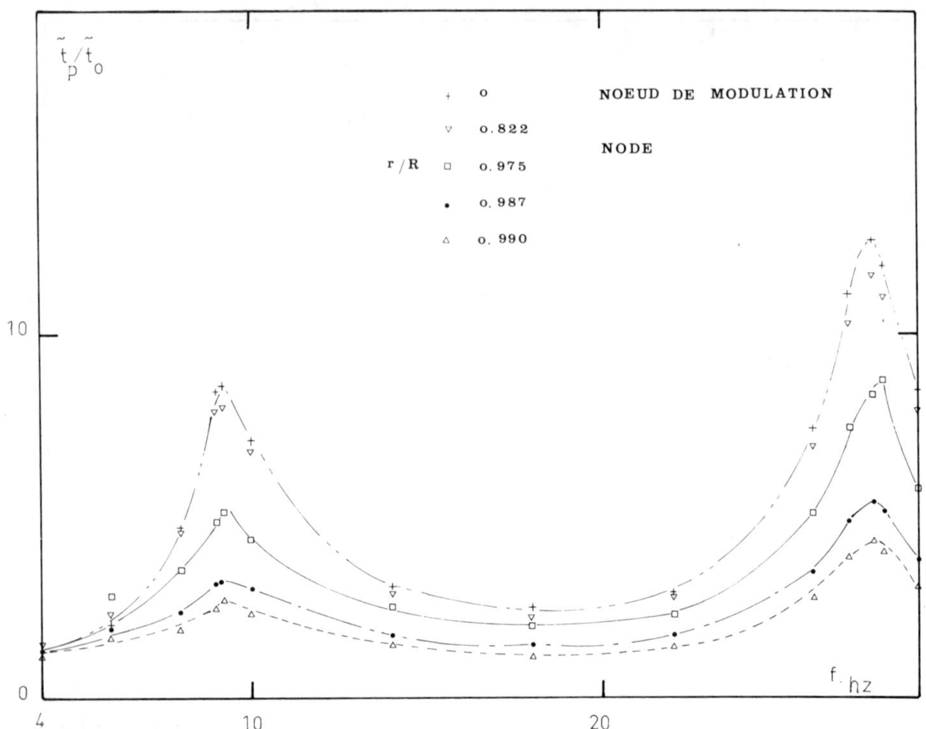

Fig. 6 : Temperature fluctuation rates vs pulsation frequency in the node section. (2nd mode)

By examining oscillogramms of the instantaneous velocity, as a first analysis, changes in the fluid field can be easily characterized in the two situations of a non resonant or resonant pulsation. The oscillogramm shapes shown in fig. 8 concern the 6 Hz regime (8 b) and the 2nd resonance mode (8 a, c, d) in the section C_o. In terms of comparison, the fundamental pulsed component is also viewed in (8 b). From a series of such oscillogramms taken in 14 different radial positions from the axis to the wall it appears, in the 2nd resonance mode, that :

i) by the fact of a local modulation rate : $(U_M - U_m)/2\,U_o$ greater than 100 %, the hot wire detects a reverse velocity with respect to the flow direction at each period of the pulsation ; then, a reverse flow is created during a fraction of the fundamental period. The highest amplitude of the negative pulsed component increases from the axis to the wall showing a progressive variation of the local modulation rate along the radius.

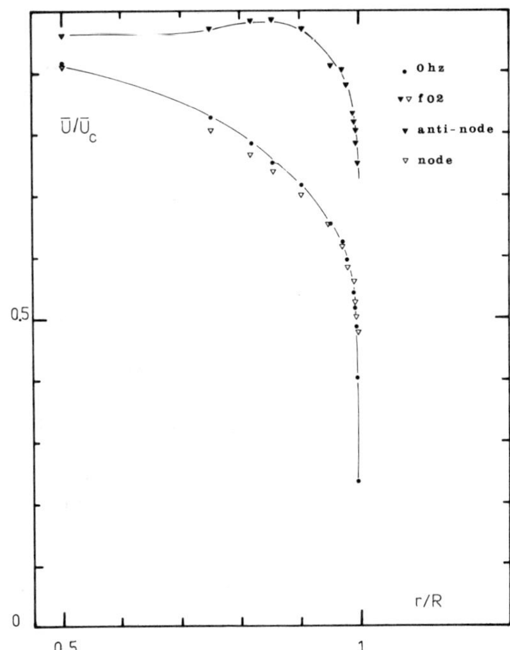

Fig. 7 Non dimensionalized mean velocity along the radius in the node and antinode sections (2nd mode)

ii) near the wall (8c), the strong intensity fluctuations superimposed on the fondamental signal are essentially detected at the moment of each extremum of the period ; this phenomenon remains when removing from the wall whereas the intensity rate decreases. A comparable result has been obtained by COUSTEIX et al. [6] in the study of a developing turbulent boundary layer along the flat plate in presence of a pulsed flow.

The dependence of the fluctuations of turbulent origin with the pulsation (forced fluctuations) is looked through the local intensity $i_k = \tilde{u}_p/U_r$ along the radius. A bandwidth frequency study is then undertaken in the following manner :

i_1 : corresponds to fluctuation frequencies from 0 to 40 Hz ; this represents more precisely the local intensity of forced fluctuations resulting from the pulsation ; then i_2 : 0 to 100 Hz, i_3 : 100 - 200 Hz, i_4 : 200 - 400 Hz, i_5 : 400 - 630 Hz.

As previously, it is of interest to compare the behaviour of those different intensities in the two sections, C_o and B_o, for the 2nd resonance mode and for a non resonant regime, i.e. : 18 Hz. For the latter, sections C_o and B_o do not state as particular sections as it is for the f_{02} mode. Thus, an homogeneous radial evolution for the different i_k appears in those sections for the 18 Hz frequency . Concerning f_{02} , several facts can be pointed out :

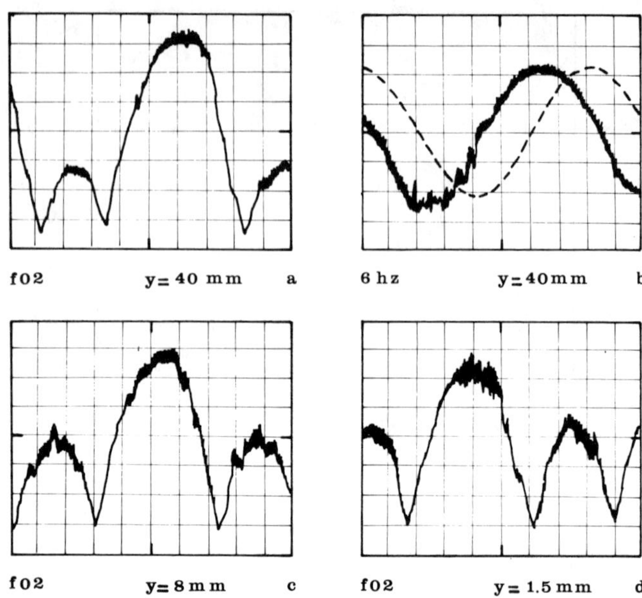

Fig. 8. Instantaneous Velocity for different cases.

i) in the bandwidth (0 - 40 Hz), (fig. 9) i_1 varies along two inverted profiles in the antinode and node sections though the value at the wall remains nearly constant in the two cases. Obviously, i_2 follows a same way.

ii) for bandwidthes over 100 Hz (fig. 10) : the decrease from the wall to the axis changes in similar manners whatever the section and the frequency range are. The intensity values in the two sections become progressively the same, as the fluctuation frequency increases. At the antinode and for frequencies included between 100 and 200 Hz, the fluctuation generation rate becomes 150 to 200 % of those obtained at the node.

By relating these two characteristic pulsed regimes, it can be stated that the difference between fluctuation intensities in the first frequency ranges ($<$100Hz) comes from the only local modulation effect (fig. 9 et 11). On the other way, the flow pulsation at the 2nd resonance mode induces noticeable intensities greater than those encountered in non resonant regimes and for fluctuation frequencies over 100 Hz (fig 10 et 12). This conclusion is extended to other pulsation frequency cases in the section C_o and B_o. The study of the longitudinal component velocity can check an existing connexion between the resonant pulsation and dynamic fluctuations over 100 Hz, which is not established outside the 2nd resonance mode.

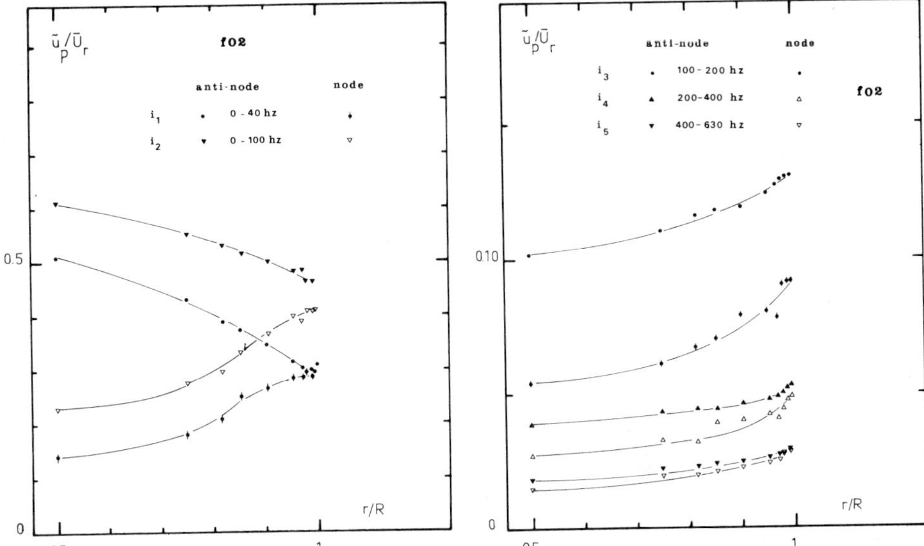

Fig. 9 and 10 : velocity fluctuation intensities along the radius for the second mode in the node and antinode sections.

4 - DISCUSSION - CONCLUSION

From a macroscopic view, the resonance phenomenon induces variations of the temperature, velocity and pressure fonctions of the flow field which are in agreement with the localized heat transfer increases ; thus, the mean temperature profiles agree with the local increase or the non modification of transfered heat fluxes ; there is conformity between the temperature fluctuation and the pressure fluctuation evolutions in terms of the pulsation frequency. Moreover, the study of the wall temperature gradient shows a preferential heat transport to the flow axis, under resonance condition and in a velocity antinode section. We could suppose the existence of a radial component of the velocity allowing us to justify this particular direction between the wall and the flow center line for the heat transfer.

A more precise analysis shows that the resonance condition promotes a noticeable increase of the fluctuation levels in the frequency range over 100 Hz and particularly at each extremums of the pulsation period.

Whereas many data are concerned with the influence of pulsations on heat transfer in a general manner, few of them are relevant about the resonance case [7-8-9-10]. This particular condition always leads to an increase in the rate of the heat exchange at the velocity modulation antinode. The GALITSEISKII's interpretation [10] has to be noted : by studying the oscillating boundary layer thickness δ_{os} and for a frequency range from 70 Hz to 1000 Hz, it has been shown that for a value of the ratio δ_{os} δ_o less than 3, the effect of the pulsations on the heat transfer is maximum. Beyond this value the transfer decreases. The ratio δ_{os}/δ_o can be written in the form:

$$\frac{\delta_{os}}{\delta_o} = 0.1\sqrt{\frac{2\tau_w}{\rho_o}\cdot\frac{1}{\omega\nu}}$$

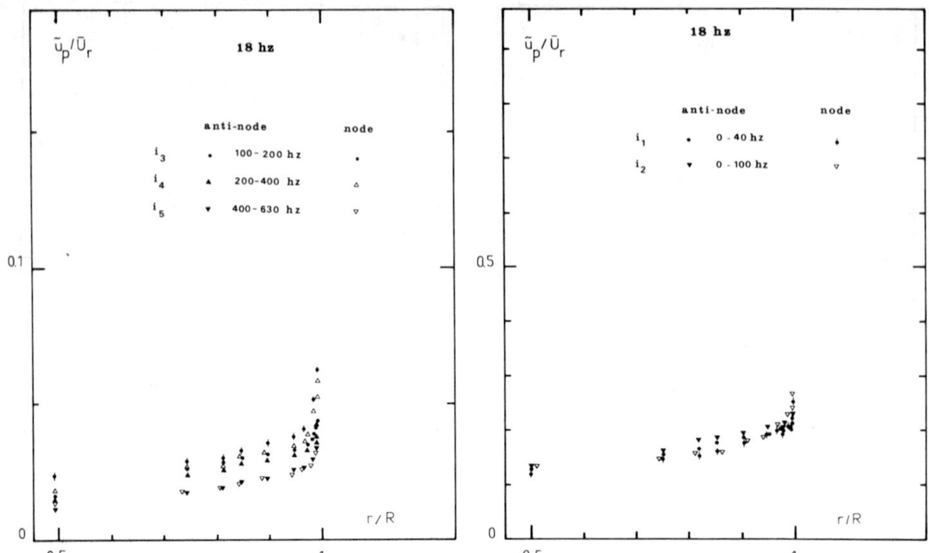

Fig. 11 and 12. Velocity fluctuation intensities along the radius for the 18 Hz regime in the node and antinode sections of the 2nd mode.

for a mean Reynolds number between 10^4 and 10^7. In our case, for the second resonance mode (f_{02} = 27.7 Hz) this quantity equals 2.97 ; there is a good accordance with this criterion. However, for the first resonance frequency (f_{01} =9.2 Hz), where the local heat transfer increase has about the same value as for the second mode, then the ratio δ_{os}/δ is equal to 5.

Considering presently the known results about heat transfer and pulsation influence and especially with the resonance condition, it seems that the observed heat transfer rate increases come from the following conjugate effects :

- the local over-modulation at the velocity antinodes (or pressure nodes)
- the likely existence of a non negligible radial component of the velocity
- a periodical change of the thermal and dynamic boundary layer thicknesses
- an intensity increase of the superimposed fluctuations to the pulsation illustrating a coupling between pulsation and turbulence.

Yet, a study of the instantaneous functions may allow a better understanding of the respective influences leading to a local flux increase.

AKNOWLEDGEMENTS :

We wish to thank the DGRST - Direction Générale de la Recherche Scientifique et Technique - for the support of this work by the contract n° 79.7.0094

NOMENCLATURE

f_{01}, f_{02} : first and second associated resonance frequencies

g : dimensionless wall temperature gradient

$h(x_i)$: local heat transfer coefficient

i_k : $\tilde{u}_p / U_r$: local intensity of the velocity fluctuation for different frequency bandwidthes

$\tilde{p}$: root mean squared pressure fluctuations at the wall in pulsed flow

$\tilde{p}_o$: root mean squared pressure fluctuations at the wall in steady flow

r : radial coordinate

$\tilde{t}_p$: local root mean squared temperature fluctuations in pulsed flow

$\tilde{t}_o$: local root mean squared temperature fluctuations in steady flow

$\tilde{u}_p$: local root mean squared velocity fluctuations

x : longitudinal coordinate

δ_{os} : oscillating boundary layer thickness :

δ_o : stationnary boundary layer thickness

ν : kinematic viscosity

ρ_o : fluid specific mass

τ_w : wall shear stress

ω : pulsation :

Λ : wave length for resonance cases

L : duct length

Θ_m : dimensionless mean fluid temperature

R : duct radius

T_r : local fluid temperature at radius r

T_c : local fluid temperature on the center axis

T_1, T_2 : wall temperature at the boundaries of the heated section

T_w : wall temperature

T_∞ : ambiant fluid temperature inlet

U_M , U_m : maximum and minimum amplitude of the pulsed velocity component

$U_\circ$: mean velocity

U_r : local mean velocity at radius r

REFERENCES

1. LEMLICH R. (1961) Vibration and pulsation boost heat transfer Chem. Enging. 15, pp 171-176

2. GOSSE J. MARTIN M. (1970). Etude aérothermique des écoulements pulsés. Rev. Gen. de Thermique n° 103-104, pp 887-893

3. ERDAL KARAMERCAN O., GAINER J.L. (1979). The effect of pulsations on heat transfer. Ind. Eng. Chem. Fundam. Vol 18,n°1, pp 11-15

4. CREFF F., ANDRE P., PLAN M. (1980) Détermination des coefficients de transferts thermiques locaux entre la paroi d'une conduite cylindrique et un écoulement d'air turbulent pulsé. Trans. C.S.M.E. Vol. 6n° 1.

5. ANDRE P., CREFF R., CRABOL J. Effets de la condition particulière de résonance acoustique sur les transferts thermiques en écoulements turbulents pulsés et en canalisation cylindrique. (In submission to the Int. J. of Heat and Mass Transfer)

6. COUSTEIX J., HOUDEVILLE R., DESOPPER A. (1978) Résultats expérimentaux et méthodes de calcul relatifs aux couches limites turbulentes en écoulement instationnaire. Unsteady Aerodynamics. Agard Conference Proceedings n° 227. pp 17.1, 17.16

7. JACKSON T.W., PURDY K.R. (1965) Resonant pulsating flow and convective heat transfer I. of Heat Transfer ; nov. pp. 507-512

8. PAWLOWSKI M. (1971) Wymiana ciepxa przy pulsacji rezonansowej. Inzynieria Chemiczna. Vol. 1, 4, pp. 471-488

9 FEDOROV B.N. (1977) Experimental study of the effect of acoustic oscillations on heat transfer in a gas flow. Inzhenerno-Fizicheskii Zhurnal ; vol. 32, n° 1, pp. 61-67 (Translation Heat Transfer Soviet Research UDC 536-244)

10. GALITSEISKII B.M., NOZDRIN A.A., RYZHOV Y.A., YAKVSH E.V. (1976) Investigation of local heat transfer coefficients under conditions of the resonance oscillations of a gas in channels. I.F.Z. Vol 31, n°2 pp. 197-201 (Translation UDC 536.242)

Some Recent Developments in the Theory of Premixed Turbulent Flames

PAUL A. LIBBY
Department of Applied Mechanics and Engineering Sciences
University of California, San Diego
La Jolla, California 92093 USA

ABSTRACT

We review the application to the description of premixed turbulent flames of simple geometry of a recently developed theory based on second order closure and avoiding gradient transport assumptions throughout. It is shown that the gradient normal to the flame of the mean pressure acts differentially on parcels of heavy reactant and light product to cause significant effects absent in constant density turbulence; countergradient diffusion and the production of turbulent kinetic energy. A similar phenomena is predicted for oblique flames with undeflected mean streamlines; in this case the gradient tangent to the flame of the Reynolds stress acts differentially so as to cause parcels of reactant and product to follow mean streamlines which differ from one another and from the mean streamline. The implications of these results for experimental combustion research and for the development or predictive methods for turbulent combustion are emphasized.

1. INTRODUCTION

A planar turbulent flame involving premixed reactants provides a relatively simple flow suitable for the study of the fundamental processes arising in such flames and more generally in variable density, nonreacting turbulence. The problems of describing those processes can be conveniently categorized into those dealing with chemical effects and those involving fluid mechanical effects.[1] The challenge of turbulent reacting flows is that there is a two-way interaction between these effects: the turbulence alters the chemical behaviour while the heat release accompanying the chemical reaction in many cases of practical interest alters the turbulence. The Bray-Moss theory[2] of premixed turbulent combustion is highly effective in handling chemical effects in flows involving fast chemical reactions and intense turbulence and thus in permitting attention to be focused on fluid mechanical effects. It is the purpose of the present paper to review some recent theoretical developments[3-5] which appear to have significant implications relative not only to the phenomenology of turbulent reacting flows but to all turbulent flows involving variable densities due either to high speeds or heterogeneous composition.

In such a review it is inappropriate to discuss all of the details of the theory involved. The indicated references should be consulted in this regard. Rather we outline the main features of the theory and then discuss the physical implications of the results, the comparison of predictions with experiment and the significance of the results relative to the phenomenology of turbulent flows.

It is convenient for purposes of exposition to consider two flame configurations, each two-dimensional in the mean, as shown in Figs. 1a-b. In one the mean velocity normal to the flame is increased by heat release while the mean velocity tangent to the mean flame surface is either zero or unaltered within the flame. This case is termed a normal flame even though the mean velocity tangent to the flame surface need not be zero. The important characteristics of such a flame are that there are no turbulent shear stresses within the reaction zone and that the mean streamline is unconstrained. The normal flame represents an idealization of turbulent unconfined flames which can be established in the laboratory.[6] The second flame configuration may be considered to be an opposite limiting case; the mean streamline is undeflected within the reaction zone and as a consequence turbulent shear stresses arise. We term this case that of an oblique flame; it is an idealization of the situation which arises when turbulent flames are established within a duct.[7]

1.1. Elements of the Theory - The Thermochemistry

The thermochemical situation we consider is that described by Bray and Moss.[2] A variable $c(\underset{\sim}{x}, t)$ which may be termed either a progress variable characterizing the state of an arbitrarily complex chemical system[8] or as we prefer for simplicity the mass fraction of product normalized to the value of unity in fully-burned mixture plays a central role in determining the thermochemical state. With frequently employed assumptions the density and temperature are determined by c and by a heat release parameter τ according to the equations

$$\rho = \rho_o/(1 + \tau c), \quad T = T_o\ (1 + \tau c)$$

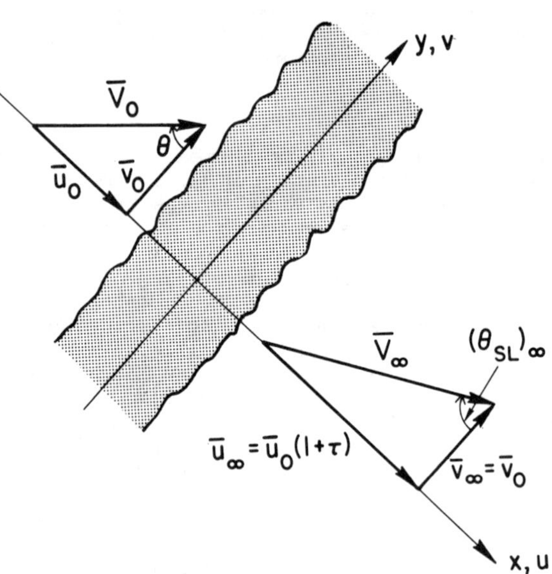

Fig. 1a. Idealized flame configurations. An oblique flame with an unconstrained mean streamline, termed a normal flame.

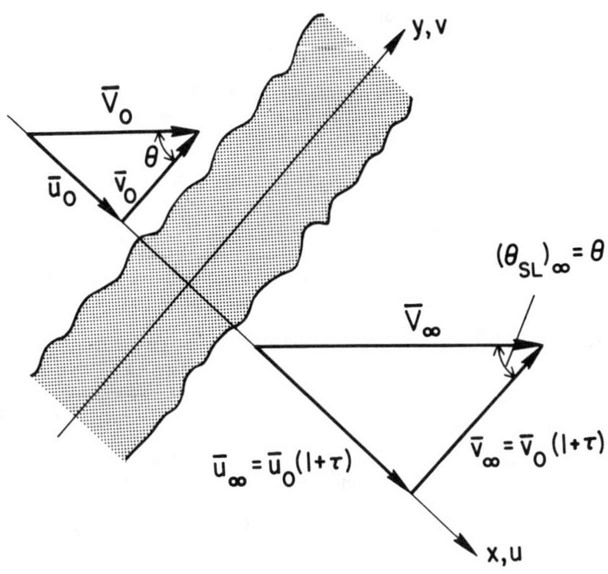

Fig. 1b. Idealized flame configurations. An oblique flame with an undeflected mean streamline, termed an oblique flame.

If applied to fully burned mixture denoted by the subscript ∞, we have

$$\rho_\infty = \rho_0/(1 + \tau), \quad T_\infty = T_0\ (1 + \tau)$$

which indicate that the parameter τ is readily determined by the temperature ratio (T_∞/T_0).

The important role of the variable $c(\underset{\sim}{x}, t)$ suggests consideration of the probability density function (pdf) $P(c; \underset{\sim}{x})$ as shown schematically in Fig.2a. Such a pdf for a point within the reaction zone has entries at the ends of the 0-1 range corresponding to reactants and to fully burned products respectively; we idealize such entries as delta functions of strength $\alpha(\underset{\sim}{x})$ and $\beta(\underset{\sim}{x})$ as shown.* Interior entries correspond to the passage of the point in question of gas undergoing chemical reaction. It is convenient to normalize the interior distribution given by $f(c; \underset{\sim}{x})$ so that $\alpha(\underset{\sim}{x}) + \beta(\underset{\sim}{x}) + \gamma(\underset{\sim}{x}) = 1$.

*It is interesting to note that in calculations it is sometimes necessary to perform the analogy of "gating" used in the conditioned sampling technique of experimental turbulence in order to remove singularities as $c \rightarrow 0,1$ and to determine the strengths of the delta functions (cf. Libby et al[9]).

It is central to the Bray-Moss model that the kinetics of the reactions are such $\gamma \ll 1$, i.e., that the fraction of time that reacting mixture is present at the point in question is small. Thus we deal with a nearly bimodal distribution of the product concentration $c(\underset{\sim}{x}, t)$. In this case the details of the interior distribution are largely inessential; when $f(c; \underset{\sim}{x})$ is required, rough estimates based, e.g., on the structure of an unstrained laminar flame, are adequate. Moreover, the strengths of the delta functions are simply related to the mean values of c.

A conceptually and practically important consequence of bimodality is the the variable $c(\underset{\sim}{x}, t)$ serves as a conditioning function analogous to the well known intermittency function which discriminates between irrotational and turbulent fluid.[10-12] We find that such a function can guide the modeling required to achieve closure; to amplify this point we consider Fig.2b which shows the joint pdf $P(u, c; \underset{\sim}{x})$ where $u(\underset{\sim}{x}, t)$ is the velocity component normal to the flame. Again entries at the ends of the range 0-1 are dominant so that the conditioned pdf's $P(u, 0; \underset{\sim}{x})$ and $P(u, 1; \underset{\sim}{x})$ enter the analysis. It is found that the conditioned statistics of the velocities, e.g., of the u-component given by these conditioned pdf's, facilitate the modeling and lead

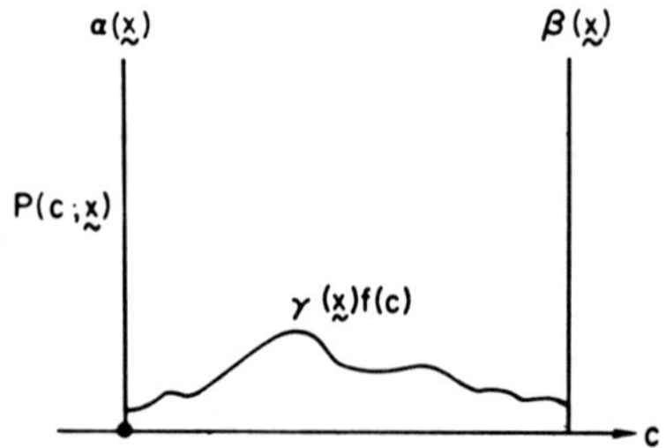

Fig. 2a. The probability density functions (From reference 3). The single variable pdf $P(c; \underset{\sim}{x})$.

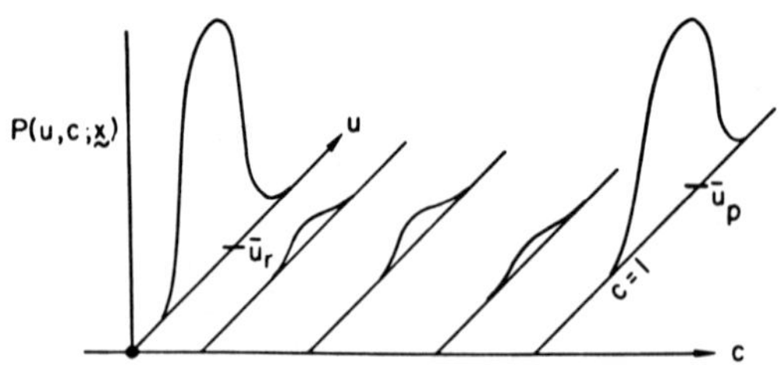

Fig. 2b. The probability density functions (From reference 3). The joint pdf $P(u, c; \underset{\sim}{x})$.

to simple, physically appealing representations of various terms as required for closure. Thus we are interested in $\overline{u_r}(\underset{\sim}{x})$ and $\overline{u_p}(\underset{\sim}{x})$, the conditioned velocities within reactant and product respectively, and in $(\overline{u'^2})_r$ and $(\overline{u'^2})_p$, the conditioned intensities within reactant and product respectively. These quantities are closely related to the so-called zone averages obtained by conditioned sampling in experimental turbulence (cf., e.g., Kovasznay et al[10]). Moss[13] has recently provided conditioned data from a premixed open flame by a combination of light scattering from seeded particles providing $\rho(t;\, \underset{\sim}{x})$ and laser doppler velocimetry providing $V(t;\, \underset{\sim}{x})$ where V denotes a particular velocity component determined by the arrangement of the instrumentation. We discuss later the results of Moss in more detail in the context of comparison between prediction and experiment; here it is sufficient to note that the techniques of conditioned sampling are being introduced into experimental combustion research and that theory is effectively utilizing the concept of conditioned variables, a refreshing distinction from the disregard of intermittency in current turbulence phenomenology.

1.2. The Fluid Mechanics of Turbulent Reacting Flows

Having clarified the thermochemical description of the flames we deal with, we next turn to fluid mechanical considerations. In this regard it is worth noting that in a recent review Libby and Williams[1] point out that the fluid mechanics of turbulent reacting flows with degrees of heat release of practical interest is subject to considerable uncertainty and that increased research attention devoted to the various problems connected thereto is required. In the earlier theories[14-16] of turbulent flames such as those shown in Figs. 1a-b the modeling of the fluid mechanical features follows general practice and carries over with little alteration the various models used with more or less success in predictive methods for constant density turbulence. Our emphasis more recently is on a careful assessment of the various models required in the theory of premixed turbulent combustion; results to date clearly discourage the general practice. The main processes examined in this review are turbulence production and turbulent transport which is almost universally described in terms gradient transport. In discussing a theory which avoids such transport we note other points of departure from current practice and our reasons for doing so.

There are a variety of reasons for questioning the applicability of gradient transport approximations in turbulent flames. Briefly, such flames are relatively thin, involve large changes in mean velocity and thus large rates of mean strain. Large density inhomogeneities are also involved. In addition there are small gradients in mean pressure across flames, gradients which might be expected to act differentially on parcels of heavy reactant and light products. Bilger[17] has pointed out the possible significance of such pressure effects in nonpremixed turbulent flames. Given this situation it seems appropriate to develop a theory which carefully avoids gradient transport assumptions and which can therefore provides a theoretical assessment of the validity of such assumptions.

2. NON-GRADIENT THEORY OF NORMAL PREMIXED FLAMES

We find it convenient to employ Favre averaging for the description of turbulent reacting flows with significant density variations due to heat release. This method of averaging[18] was developed specifically for high speed turbulent boundary layers but is now widely used for the analysis of any turbulent flow with variations in density.* With this averaging it is appropriate to modify the usual notation; thus, for example, the velocity component $u_i(\underset{\sim}{x}, t) = \tilde{u}(\underset{\sim}{x}) + u''(\underset{\sim}{x}, t)$ where $\tilde{u} = \overline{\rho u}/\bar{\rho}$.

The equations needed for the development of a nongradient theory of normal flames are thus:

$$\frac{d}{dx}(\bar{\rho}\,\tilde{u}) = 0 \tag{1}$$

$$\frac{d}{dx}(\bar{\rho}\,\tilde{u}^2 + \overline{\rho u''^2}) = -\frac{d\bar{p}}{dx} \tag{2}$$

$$\frac{d}{dx}(\bar{\rho}\,\tilde{u}\,\tilde{c} + \overline{\rho u''c''}) = \bar{w} \tag{3}$$

$$\frac{d}{dx}\left(\bar{\rho}\,\tilde{u}\,\frac{\overline{\rho u''^2}}{\bar{\rho}} + \overline{\rho u''^3}\right) + 2\,\overline{\rho u''^2}\,\frac{d\tilde{u}}{dx} = -2\,\overline{u''}\,\frac{d\bar{p}}{dx} - \bar{\chi}_u \tag{4}$$

$$\frac{d}{dx}\left(\bar{\rho}\,\tilde{u}\,\frac{\overline{\rho v''^2}}{\bar{\rho}} + \overline{\rho u''v''^2}\right) = 0 \tag{5}$$

$$\frac{d}{dx}\left(\bar{\rho}\,\tilde{u}\,\frac{\overline{\rho u''c''}}{\bar{\rho}} + \overline{\rho u''^2c''}\right) + \overline{\rho u''c''}\,\frac{d\tilde{u}}{dx} + \overline{\rho u''^2}\,\frac{d\tilde{c}}{dx} = -\,\overline{c''}\,\frac{d\bar{p}}{dx} + \overline{u''w} - \bar{\chi}_{uc} \tag{6}$$

In these equations pressure fluctuations are neglected on the grounds that in the thin reaction zones under consideration their influence is expected to be significantly less than that of the mean pressure gradient. In addition molecular transport is neglected compared with turbulent transport, a commonly made assumption for large turbulence Reynolds numbers.

---

* Bilger[19] provides an extensive discussion of the arguments in favor of Favre-averaging for the description of turbulent reacting flows and against the casual treatment of density variations usually resorted to when conventional averaging is used.

The two pressure gradient terms, those in Eqs. (4) and (6), are special in the following sense: If Eqs. (1)-(6) are compared with their counterparts for constant density flows, it is found that a termwise identification of the two sets can be made with the important exception of the pressure gradient terms which are absent in constant density flows. The implication is that these terms represent an interaction between pressure gradients and density inhomogeneities. It is to be expected on physical grounds that a mean pressure gradient acts differentially on heavy and light parcels of gas so as to accelerate to a greater extent the latter in the direction of decreasing pressure. Our results show that such differential action leads not only to countergradient diffusion but to the production of turbulent kinetic energy. Note that for the thermochemical system considered here the multipliers of the pressure gradient terms are to lowest order in γ given by

$$\overline{c''} = \tau\, \tilde{c}(1 - \tilde{c})/(1 + \tau \tilde{c})$$

$$\overline{u''} = \tau\, \overline{\rho u''c''}/\bar{\rho}(1 + \tau \tilde{c})$$

To close these equations models are needed for the third order correlations, $\overline{\rho u''^3}$, $\overline{\rho u''v''^2}$ and $\overline{\rho u''^2 c''}$, for the velocity-chemistry correlation $\overline{u''w}$ and for the dissipation terms $\bar{\chi}_u$ and $\bar{\chi}_{uc}$. Libby and Bray[20] show that the usual models for dissipation are inoperative in thin reaction zones, presumably because they describe molecular processes requiring considerably greater flow lengths to be effective than are available. However, they also show that consideration of the conditioned velocities and of the structure of laminar flamelets across which the changes from one conditioned velocity to the other occur leads to alternative models reflecting flamelet dissipation. The resultant models are used here; thus we have

$$\bar{\chi}_{uc} = \kappa_1 \bar{w}\, \overline{\rho u''c''}/\bar{\rho}\, \tilde{c}(1 - \tilde{c}) \tag{7}$$

$$\bar{\chi}_u = \kappa_2 \bar{w} \left(\overline{\rho u''c''}/\bar{\rho}\, \tilde{c}(1 - \tilde{c})\right)^2$$

where κ_1 and κ_2 are empirical constants, believed to be close to one another and of order unity.

The conditioned statistics of the u-velocity component are effective in guiding the modeling of the third order correlations and of $\overline{u''w}$. To obtain some indication of this role consider the following equations which are accurate to lowest order in γ:

$$\overline{\rho u''c''} = \bar{\rho}\, \tilde{c}(1 - \tilde{c})\, (\bar{u}_p - \bar{u}_r) \tag{8}$$

and

$$\overline{\rho u''^2}/\bar{\rho} = \tilde{c}(1 - \tilde{c})\, (\bar{u}_p - \bar{u}_r)^2 + (1 - \tilde{c})\, \overline{u'^2_r} + \tilde{c}\, \overline{u'^2_p}\,. \tag{9}$$

From Eq. (8) we see that the mean flux of product is proportional to the difference in the conditioned mean velocities. According to gradient transport in a flame $\overline{\rho u''c''} < 0$ implying that $\bar{u}_r > \bar{u}_p$. If it is recalled that the velocity downstream of the flame where only product exists can be considerably greater than the upstream velocity where only reactants exist, this implication cannot be considered physically appealing. Equation (8) and the relation between the unconditioned velocity $\tilde{u}$ and the two conditioned velocities permit the latter to be eliminated in terms of prime variables, $\tilde{u}$ and $\overline{\rho u''c''}$.

Equation (9) provides one equation involving the conditioned velocity intensities but we need a second empirical equation in order to express in terms of prime variables those intensities which appear in the third order correlations. Libby and Bray[3] use a model for $\overline{u'^2_r}$ based on rapid straining of the reactants but the experimental results of Moss[13] indicate that the conditioned intensity of reactants experiences a modest increase and not the rapid decrease such a model predicts. Accordingly, in Bray et al[4] it is assumed that

$$(\overline{u'^2_p} - \overline{u'^2_r}) = \kappa_3 (\bar{u}_p - \bar{u}_r)^2 \tag{10}$$

where κ_3 is an additional empirical constant. The model given by Eq. (10) makes the difference in the conditioned intensities proportional to the square of the difference in the conditioned velocities, a relation with some physical appeal. The use of Eq. (10) effectively completes the modeling called for if only the lowest order approximations in γ are made.

The closed set of equations is most conveniently solved if the x-coordinate is replaced as the independent variable by $\tilde{c}$, the mean product concentration. As a consequence the equations are solved in the 0-1 range with singular points requiring special treatment at the ends. An advantage of this change in variables is that no model for the chemical source term is called for unless spatial distributions are desired. Accordingly, attention focuses on Eq. (4) for the intensity of the velocity fluctuations and on Eq. (6) for the mean flux of product. Equation (2) provides an expression for the mean pressure gradient while Eq. (3) permits elimination of $\bar{w}$ required inter alia in Eqs. (7). Equation (5) implies that fluctuations of the two velocity components parallel to the mean flame surface are unaltered in passage through the flame. It is thus clear that initially isotropic turbulence becomes anisotropic within turbulent flames.

With the modeling based on the pdf's indicated in Figs. 2a-b there result relatively few empirical constants available for adjusting predictions to experimental results. In addition the values of the three obvious constants, the κ_i's, are constrained somewhat by the nature of the singularities at the ends of the 0-1 range. In this same regard we note that those singularities result in the final formulation involving a single eigenvalue which we take to be I_∞, a quantity proportional to the intensity of the velocity fluctuations downstream of the flame. Thus the change in the turbulence through the flame is determined as part of the solution. In addition to the values of κ_i,

$i = 1\text{-}3$, we must select the value of a parameter $I_o = \overline{u'^2}/\overline{u}_o^2$ where $\overline{u'^2}_o$ is the intensity in the oncoming reactants and $\overline{u}_o$ is the turbulent flame speed. A wide variety of data subject to considerable scatter but from various investigations indicates $I_o = 0.23$.

3. COMPARISON WITH EXPERIMENT AND IMPLICATIONS OF RESULTS

We now discuss the comparison of the theory for normal flames outlined here with experimental results. Bray et al[4] consider the theory in the light of the experimental results of Moss[13] obtained in an open premixed turbulent flame. The values of the heat release parameter τ and of I_o are given directly by the experiment as 6.5 and 0.16 respectively. The nature of the flow configuration and of the data collected is such that significant interpretation is called for in order to carry out the comparison with a theory for normal flames. Nevertheless by selecting on the basis of casual numerical experimentation in contrast to the use of optimization techniques the empirical constants, namely $\kappa_1 = 0.85$, $\kappa_2 = 0.25$ and $\kappa_3 = 0.1$, they obtain the encouraging results shown in Figs.3a-d. The comparison is carried out in terms of the distributions with $\tilde{c}$ of the normalized intensity $\overline{\rho u''^2}/\overline{\rho}\overline{u'^2}_o$; the nondimensional mean flux of product $\overline{\rho u''c''}/\rho_o\overline{u}_o$; the conditioned velocities in the form $(\overline{u}_r/\overline{u}_o)$ and $(\overline{u}_p/\overline{u}_o)$; and the conditioned inten-

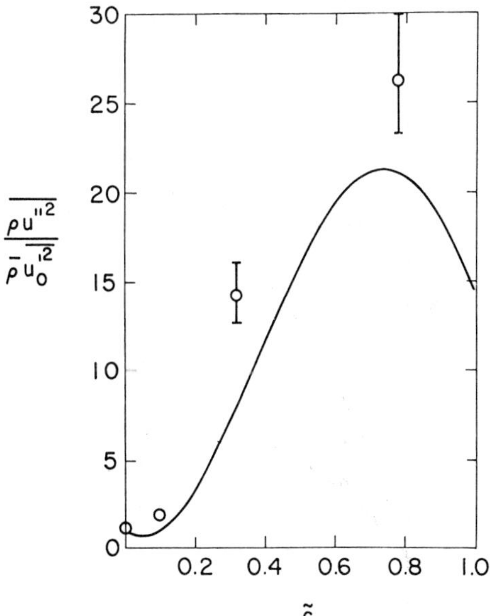

Fig. 3a. Comparison between theory and prediction for normal flames (From reference 4). The normalized intensity of the velocity fluctuations.

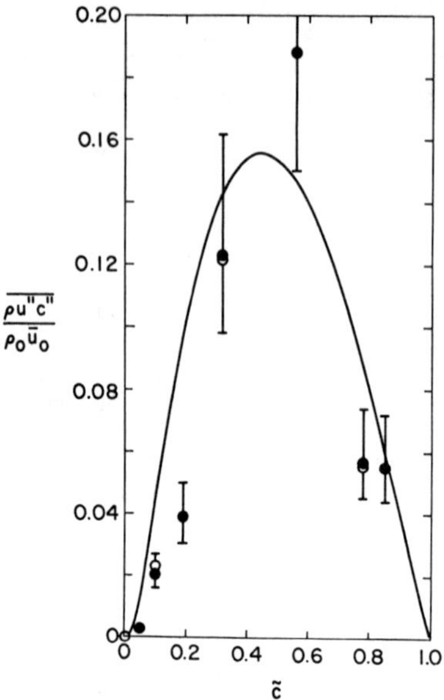

Fig. 3b. Comparison between theory and prediction for normal flames (From reference 4). The non-dimensionalized mean flux of product.

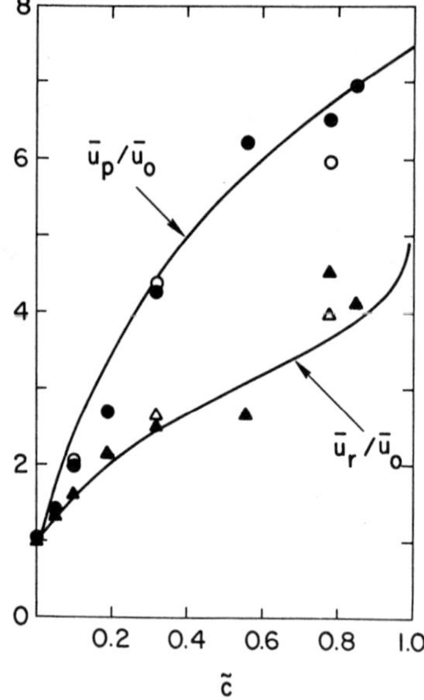

Fig. 3c. Comparison between theory and prediction for normal flames (From reference 4). The non-dimensionalized conditioned velocities.

sities $(\overline{u'^2_r}/\bar{u}_o^2)$ and $(\overline{u'^2_p}/\bar{u}_o^2)$. The error bars relate to the uncertainties associated with interpretation of the data in the form permitting comparison with a theory of normal flames.

The agreement with respect to the first three quantities shown in Figs. 3a-c is considered satisfactory given the the uncertainties in the experimental data; the absence of evidence indicating a decline in the normalized intensity on the hot side of the flame is attributed to the open flame configuration since careful consideration of the individual terms in Eq. (4) indicates that a well-defined peak in intensity should exist within the reaction zone for values of the heat release parameter greater than two or three.

The comparison of predicted and measured values of the conditioned velocity intensity within products shown in Fig. 3d is relatively poor; if additional experiments confirm this discrepancy, a revision of the model used for the conditioned intensities is indicated. It should be understood that such a revision will not alter significantly the agreement shown in the other figures.

3.1. Some Implications from the Results

We now turn to a discussion of the implications of the results from the theory of normal premixed flames which in the light of the agreement shown in Figs. 3a-d must be considered to be reasonably supported by experiment. For

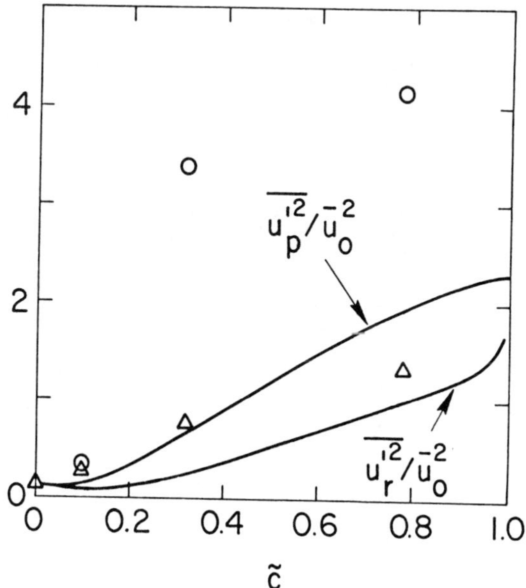

Fig. 3d. Comparison between theory and prediction for normal flames (From reference 4). The non-dimensionalized conditioned intensities of the velocity fluctuations.

low degrees of heat release, i.e., for values of $\tau < 1$, the theory predicts distributions of mean flux of product in qualitative agreement with gradient transport and in rough quantitative agreement with earlier theories based on such transport. In addition in these cases the intensity of the velocity fluctuations and the turbulent kinetic energy are reduced by dilatation due to heat release. Thus the new theory is in accord with gradient transport approximations for small degrees of heat release. In view of the long history of successful calculations for constant density flows we consider such accord an important characteristic of the new theory.

Since the value of the heat release parameter τ used in the calculations shown in Figs. 3a-d, namely 6.5, is within the range of values of practical interest, we consider the implications of the predicted results aside from the experimental data. From Fig. 3b we see that the mean flux of product is positive throughout most of the reaction zone. There exists a small region near the cold side with negative flux;* Libby and Bray[3] interpret this as a flame-stabilizing region. The physical explanation of the predicted countergradients resides in the differential effect discussed earlier of the mean pressure gradient on parcels of heavy reactant and light product. Mathematically this effect is due principally to the $\overline{c''}(d\overline{p}/dx)$ term in the flux equation, Eq. (6).

We see from Fig. 3a that the intensity of the velocity fluctuations initially decreases from its upstream value, then increases to a peak and finally declines to a value considerably above its initial value. This result is in contrast to the prediction of earlier theories of normal flames[14-16] which do not account for the effect of pressure gradient, particularly of the $\overline{u''}\,(d\overline{p}/dx)$ term in the intensity equation, Eq. (4). For values of τ greater than four the effect of the pressure gradient overcomes the effect of dilatation and leads to an overall increase in the intensity of the u-component fluctuations. If this result is combined with the constancy of the intensities of the other two velocity components, we see that the turbulent kinetic energy can increase even for a normal flame free of Reynolds stresses. As noted by Bray et al[4] this finding removes a significant apparent discrepancy between experimental results from a variety of turbulent flames and earlier theoretical predictions, apparent because the experiments seldom permit direct and careful comparison with predictions for normal flames.

In concluding this discussion we emphasize several points: The pressure effects we find significant are thermochemically unimportant and are self-induced. A flame embedded in a flow field may be subjected to a pressure gradient which either enhances or reverses the self-induced pressure gradients due to heat release. As a consequence turbulent transport and turbulence production are altered along the lines suggested by our calculations. Starner and Bilger[21] provide data showing the effects of pressure gradients imposed on a premixed turbulent flame. The anistropy of the turbulence due to passage through a flame implies that predictive methods accounting only for turbulent kinetic energy may not adequately account for significant processes.

---

*Models for the conditioned intensities of the general form of Eq. (10) assure that $(dF/d\tilde{c})\,(\tilde{c} = 0) < 0$ and thus that this region always exists.

4. EXTENSION TO OBLIQUE FLAMES

Consider now turbulent flames such as those shown by Fig. 1b, i.e., infinite planar flames which are oblique to the oncoming reactants and which have mean streamlines undeflected within the reaction zone. A consequence of the assumption of an undeflected streamline is that

$$\tilde{v} = \tilde{u}/\tan\theta \tag{11}$$

Since some premixed flames are highly oblique (cf., e.g., Wright and Zukoski[7] where $\theta = 6\text{-}8^{\circ}$), we see from Eq. (11) that the mean tangential velocities can be relatively large.

The equation of mean momentum conservation in the y-direction leads directly to a relation for the Reynolds stress, namely to[14]

$$\overline{\rho u''v''} = -\tau\,\tilde{c}\,\rho_o\bar{u}_o^2/\tan\theta \tag{12}$$

so this important quantity is obtained without further approximation.

4.1. The Describing Equations

Now it is found that Eqs. (1)-(4) and (6), those used to describe normal flames, apply without alteration for oblique flames. Moreover, if in Eq. (5) $\overline{\rho v''^2}/\bar{\rho}$ is replaced by $\overline{\rho w''^2}/\bar{\rho}$, our earlier considerations apply so that this latter intensity is unaltered within the reaction zone. However, Eq. (5) becomes with Eq. (11) taken into account

$$\frac{d}{dx}\left(\bar{\rho}\,\tilde{u}\,\frac{\overline{\rho v''^2}}{\bar{\rho}} + \overline{\rho u''v''^2}\right) + 2\,\frac{\overline{\rho u''v''}}{\tan\theta}\,\frac{d\tilde{u}}{dx} = -\bar{\chi}_{vv} \tag{13}$$

and we see that the intensity $\overline{\rho v''^2}/\bar{\rho}$ is changed by obliquity. The mean flux of product in the y-direction is described by

$$\frac{d}{dx}\left(\bar{\rho}\,\tilde{u}\,\frac{\overline{\rho v''c''}}{\bar{\rho}} + \overline{\rho u''v''c''}\right) + \frac{\overline{\rho u''c''}}{\tan\theta}\,\frac{d\tilde{u}}{dx} + \overline{\rho u''v''}\,\frac{d\tilde{c}}{dx} = \overline{v''w} - \bar{\chi}_{vc}\,. \tag{14}$$

Equations (13) and (14) depend on the solution for a normal flame via the production term in Eq. (14).

4.2. Modeling

Some additional modeling is called for to close Eqs. (13) and (14) but generally we follow the methodology and perspectives used for normal flames.

In particular we imagine a generalization of the joint pdf's shown in Fig. 2b to $P(u, v, c; \underset{\sim}{x})$, a pdf dominated for the thermochemical system considered here by entries corresponding to c = 0,1 and thus by the conditioned pdf's $P(u, v, 0; \underset{\sim}{x})$ and $P(u, v, 1; \underset{\sim}{x})$. Conditioned correlations of the u- and v-components in reactants and products naturally arise.* For example, the following are found to be important equations:

$$\overline{\rho v'' c''}/\bar{\rho} = \tilde{c}(1 - \tilde{c})(\bar{v}_p - \bar{v}_r) + O(\gamma) \; . \tag{15}$$

$$\overline{\rho u'' v''}/\bar{\rho} = (1 - \tilde{c})\,\overline{u'v'}_r + \tilde{c}\,\overline{u'v'}_p$$

$$+ \tilde{c}(1 - \tilde{c})(\bar{u}_p - \bar{u}_r)(\bar{v}_p - \bar{v}_r) + O(\gamma) \tag{16}$$

Equation (15) is the counterpart of Eq. (8) for the mean flux in the x-direction.

Several comments in connection with Eq. (16) are appropriate; we see that the correlation involves two separate contributions: Those due to correlations within reactants and products may be considered true turbulence while the second contribution is associated with bimodality. When either c = 0 or c = 1, Eq. (16) degenerates properly so that unconditioned and conditioned correlations are identical. Finally, in the theory of flame structure discussed here the second contribution dominates so that the modeling of the conditioned correlations is not crucial to the results unless comparison with experimental data on those correlations is to be carried out.

In using Eq. (16) to achieve closure of Eqs. (13) and (14) we assume $\overline{u'v'}_r = 0$; such an assumption appears to be physically defensible and has as a consequence that Eqs. (12) and (16) yield an explicit relation for $\overline{u'v'}_p$ and thus that the triple correlation $\overline{\rho u'' v'' c''}$ in Eq. (14) is expressible in terms of prime variables.

4.3. Further Modeling and Results

An interesting consequence of the modeling developed along these lines is that Eq. (14) can be solved independently of Eq. (13) for the mean flux of product in the y-direction in the form $(\overline{\rho v'' c''}/\rho_o \bar{u}_o) \tan\theta$ independent of the flame orientation determined by θ. The distributions of the two mean fluxes of product obtained from such a solution are shown in Fig. 4 for $\tau = 6.5$,

---

*It appears that the description of multidimensional flows involving premixed turbulent combustion described by the Bray-Moss model will be facilitated by exploitation of the ideas of conditioned statistics of the velocity components and by incorporation of conditioned data from closely coordinated experimental efforts.

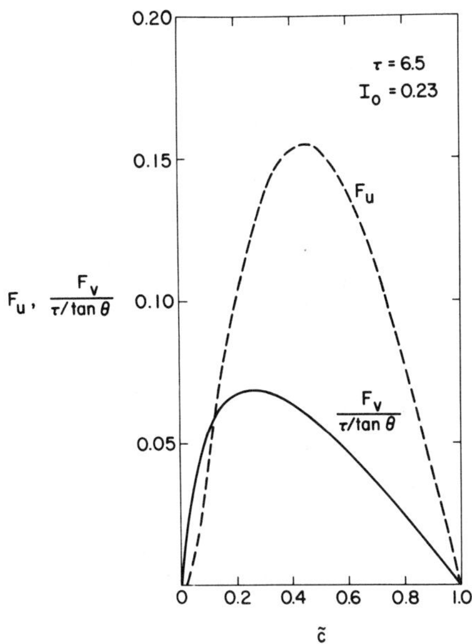

Fig. 4. The distributions of mean flux of product: $F_u = (\overline{\rho u''c''}/\rho_o u_o)$, $F_v = (\overline{\rho v''c''}/\rho_o u_o)$.

$I_o = 0.23$. As is to be expected on physical grounds the mean flux in the y-direction is zero for a normal flame ($\theta = 90^o$) and very large for highly oblique flames.

Equation (13) calls for additional modeling distinct to the consideration of obliquity and to the higher dimensionality of the pdf's required thereby. We assume, for example,

$$\frac{(\overline{u'v'^2})_p}{(\overline{u'^2v'})_p} = \left(\frac{(\overline{v'^2})_p}{(\overline{u'^2})_p}\right)^{1/2} \tag{17}$$

The measurement of the conditioned statistics entering this equation is possible by extension of the techniques used by Moss[13] and would be invaluable in the development of theory. In the absence of appropriate data we use results from turbulent shear flows with constant fluid properties to establish some plausibility for Eq. (17) but it must be considered largely a conjecture. The remaining models required to close Eq. (13) follow the lines indicated earlier.

The angular dependence of the intensity $\overline{\rho v''^2}/\bar{\rho}$ is not as simple as that prevailing for the flux $\overline{\rho v''c''}/\bar{\rho}$. However, it is found numerically that for

$\theta < 60^{\circ}$, $(\overline{\rho v''^2}/\bar{\rho}_o\bar{u}_o^2)\tan^2\theta$ is nearly independent of θ. The distributions across the reaction zone of the three intensities in appropriate nondimensional form are shown in Fig. 5. The implication from these results is that for oblique flames which are highly swept highly anisotropic turbulence arises with only $\overline{\rho w''^2}/\bar{\rho}$ remaining unaltered within the flame.

4.4. The Various Streamlines

One of the interesting findings from our consideration of oblique flames relates to the mean streamlines followed by parcels of reactant and product. Of course the streamlines of parcels of reactant randomly terminate where such parcels are destroyed by chemical reaction and the streamlines of parcels of product randomly originate at such terminal points. However, we can consider the streamlines of reactant parcels which survive to the hot side of the reaction zone and streamlines of product parcels which are created at the cold side of the flame.

The assumption of an undeflected mean streamline may off-hand be taken to imply that the conditioned velocity components are likewise constrained so that the parcels of reactant and product follow the mean streamline. However, if it is recalled that there is a mean force field in the y-direction associated with the variability of the Reynolds stress given by Eq. (12) and that such a force field acts differentially on parcels of heavy reactant and light

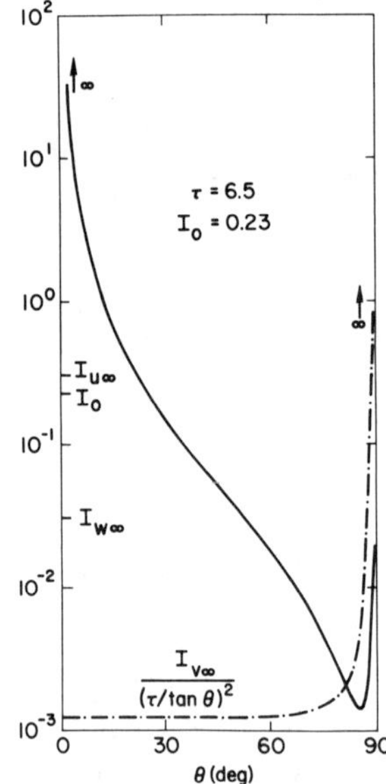

Fig. 5. The distributions of the intensities of the velocity fluctuations: $I_u = (\overline{\rho u''^2}/\bar{\rho}_o\bar{u}_o^2)$, $I_v = (\overline{\rho v''^2}/\bar{\rho}_o\bar{u}_o^2)$.

product, we are lead to the conclusion that the streamlines of the two sets of parcels are not colinear with the mean streamline. Since we express the two sets of conditioned velocities in terms of prime dependent variables, we can determine the distributions of the mean streamlines through the reaction zone. We have for either set of parcels the relation

$$\tan\,(\theta_{SL})_{r,p} = (\overline{u}/\overline{v})_{r,p} \quad . \tag{18}$$

The results of calculations for a typical oblique flame are shown schematically in Fig. 6a. For such a flame the parcels of reactant behave as though unconstrained, i.e., $(\theta_{SL})_r > \theta$. The y-wise force associated with the Reynolds stress is relatively ineffective in accelerating reactant parcels in the y-direction. On the contrary that force acts so effectively on product parcels that $(\theta_{SL})_p < \theta$.

In Fig. 6b we show the comparable situation for the case of an oblique flame with an unconstrained mean streamline and thus with no y-wise force field. Because of the absence of such a field $\overline{v}_r = \overline{v}_p = \overline{v}_o = u_o/\tan\,\theta$. However, the conditioned u-velocity components are unequal with the consequence that again there are three distinct streamlines; reactant parcels turn more than the mean, product parcels less.

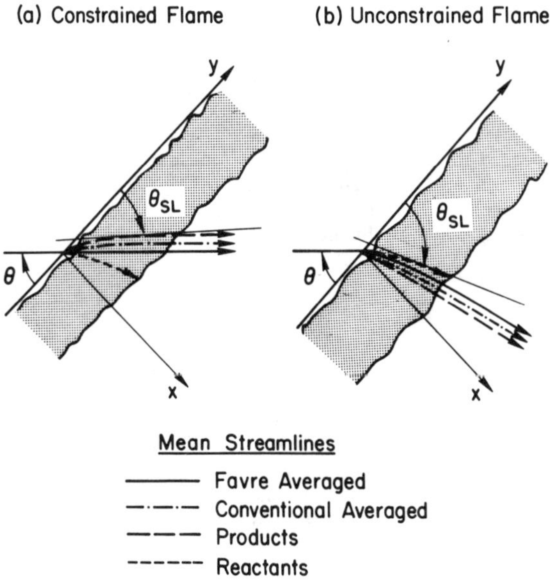

Fig. 6a. Schematic representation of the various mean streamlines. An oblique flame with an undeflected mean streamline.

Fig. 6b. Schematic representation of the various mean streamlines. An oblique flame with an unconstrained mean streamline.

The differences in the pairs of conditioned velocity components are predicted to be sufficiently large as to be measurable in the laboratory. It would be of considerable interest to have data permitting assessment of these theoretical predictions.

5. CONCLUDING REMARKS

In this review of some recent theoretical work related to premixed turbulent flames we have demonstrated the important dynamic role of mean pressure gradients acting differentially on heavy reactants and light products. A consequence of this action is that the usual gradient transport mechanism which describes turbulent transport in constant density flows is inapplicable; in particular the flux has a magnitude and sign not determined by the gradient in the associated direction of the associated mean quantity but rather by the differential action of the mean pressure gradient. In addition this differential action can lead to an additional mechanism for the production of turbulent kinetic energy. Comparison with the limited experimental data which are suitably detailed support the theoretical predictions. Throughout this review we have pointed out the need for, and the possibilities to obtain, additional experimental data.

An oblique flame with an undeflected streamline is predicted to involve large degrees of anisotropy and to have significantly different mean streamlines for parcels of reactant and product. The differences in the conditioned velocities in the two orthogonal directions are predicted to be sufficiently large as to be measurable in the laboratory. An assessment of the theoretical predictions by comparison with experiment would be of interest.

Although these results are developed in connection with premixed turbulent flames with simple geometries, they would seem to have significance for all turbulent flows with density inhomogeneities whether due to high speed or heterogeneous composition due to mixing or combustion. The extent of this significance must of course be investigated but it seems clear that a variety of interesting phenomena can arise in variable density turbulence and that our perspective relative to such turbulence must be altered.

ACKNOWLEDGEMENTS

The research reported here is largely supported by the Office of Naval Research of the United States Navy as part of Project SQUID under Contract N00014-79-C-0355. It is part of a continuing international collaboration with Professor K. N. C. Bray and Dr. J. B. Moss of the University of Southampton, a collaboration facilitated by a NATO Research grant. The author is pleased to acknowledge the essential contributions to this research by Professor Bray and Dr. Moss and by Mr. Goro Masuya of the National Aerospace Laboratory, Kakuda Branch, Japan who is spending a year at the University of California San Diego.

REFERENCES

1. Libby, P.A. and Williams, F.A., "Some Implications of Recent Theoretical Studies in Turbulent Combustion," AIAA J. (To appear).

2. Bray, K.N.C. and Moss, J.B., "A Unified Statistical Model of the Premixed Turbulent Flame," Acta Astronautica, 4, March-April 1977, 291-319.

3. Libby, P.A. and Bray, K.N.C., "Countergradient Diffusion in Premixed Turbulent Flames," AIAA J. (To appear).

4. Bray, K.N.C., Moss, J.B. and Libby, P.A., "Turbulence Production in Premixed Turbulent Flames," Combust. Sci. Tech. (Submitted).

5. Masuya, G. and Libby, P.A., "Theory of Oblique Turbulent Flames with Premixed Reactants," AIAA J. (Submitted).

6. Palm-Leis, A. and Strehlow, R.A., "On the Propagation of Turbulent Flames," Combust. Flame 13, 1969, 111-129.

7. Wright, F.H. and Zukoski, E.E., "Flame Spreading from Bluff-Body Flame-holders," Eighth Symposium (International) on Combustion, Williams and Wilkens, Baltimore, 1962, 933-978.

8. Bray, K.N.C., "The Interaction Between Turbulence and Combustion," Seventeenth Symposium (International) on Combustion, Combustion Institute, Pittsburgh, 1979, 223-233.

9. Libby, P.A., Bray, K.N.C. and Moss, J.B., "Effects of Finite Reaction Rate and Molecular Transport in Premixed Turbulent Combustion," Project SQUID Technical Report UCSD-10-PU, May 1978.

10. Kovasznay, L.S.G., Kibens, V. and Blackwelder, R.F., "Large Scale Motion in the Intermittent Region of a Turbulent Boundary Layer," J.Fluid Mech. 41, 1970, 283-325.

11. Libby, P.A., "Prediction of the Intermittent Turbulent Wake of a Heated Cylinder," Phys. Fluids 19, 1976, 494-501.

12. Dopazo, C., "On Conditioned Averages for Intermittent Turbulent Flows," J.Fluid Mech. 81, 1979, 433-438.

13. Moss, J.B., "Simultaneous Measurements of Concentration and Velocity in an Open Premixed Turbulent Flame," Comb. Sci. Tech. 22, 1980, 115-129.

14. Bray, K.N.C. and Libby, P.A., "Interaction Effects in Premixed Turbulent Flames," Phys. Fluids 19, 1976, 1687-1701.

15. Libby, P.A. and Bray, K.N.C., "Variable Density Effects in Premixed Turbulent Flames," AIAA J. 15, 1977, 1186-1193.

16. Libby, P.A., Bray, K.N.C. and Moss, J.B., "Effects of Finite Reaction Rate and Molecular Transport in Premixed Turbulent Combustion," Combust. Flame 34, 1979, 285-301.

17. Bilger, R.W., "Turbulent Jet Diffusion Flames," Prog. Energy Comb. Sci. 1, 1976, 87-109.

18. Favre, A., "Statistical Equations of Turbulent Gases," in Problems of Hydrodynamics and Continuum Mechanics, SIAM, Philadelphia, 1969, 231-266.

19. Bilger, R.W., "Turbulent Flows with Nonpremixed Reactants," in Turbulent Reacting Flows, ed. P.A.Libby and F.A.Williams, Springer-Verlag, Berlin-Heidelberg-New York (To appear in 1980).

20. Libby, P.A. and Bray, K.N.C., "Implications of the Laminar Flamelet Model in Premixed Turbulent Combustion," Combust. Flame (To appear).

21. Starner, S.H. and Bilger, R.W., "Measurements of Velocity and Concentration in Turbulent Diffusion Flames with Pressure Gradients," AIAA-80-0205, 1980.

Investigation of Intermittent Phenomena in Two Phase Boundary Layer

NAIM H. AFGAN
Nuclear Sciences Institute Boris Kidrič
Beograd, Yugoslavia

1. INTRODUCTION

Essential characteristic of any two phase system is its intermittency in density in time and space coordinate[1]. This is a reason that most of our efforts in dealing with two phase system have been focused to the interaction processes between the subsystems with different density. As our present knowledge in continuous mechanics as well as in thermodynamics allows us to deal only with homogenous systems, our mean task in the analysis of two phase system is either to make it homogenous or to find corresponding relation which describes the process at the boundary of separated subsystems. Either of these approaches leads us to the substantial simplification which brings us to the limitation of its physical meaning. It is a reason why our present effort is devoted to the analysis of two phase system as a collective behaviour of the dispersed phase in continuous phase. This approach enables us to construct physical models of two phase system taking into account its internal structure.

In order to be more specific we will take liquid-vapour two phase system for our further consideration in describing the essential idea of this approach. In order to take into consideration heat transfer process between the phases special attention has been paid to the boiling two phase system. In this case the simplest case is behaviour of the single bubble at the heating surface. There are many different approaches to problem of the mathematical description of the physical processes in the vicinity of liquid-vapour interface, which takes into account all relevant parameters which effect behaviour of the single bubble in the surrounding liquid /2/. Experimental verification of this processes has proved (Fig. 1) that mathematical description of the processes is sufficiently accurate so that behaviour of the single bubble in the surrounding superheat liquid can be predicted within the prescribed accuracy. If the number of single bubbles in the space increases there will be mutual interaction between them which will cause generation of multilateral interaction between them. In order to describe this interaction and its influence on the total behaviour of the two phase system it is needed to have adequate tool for the description of the

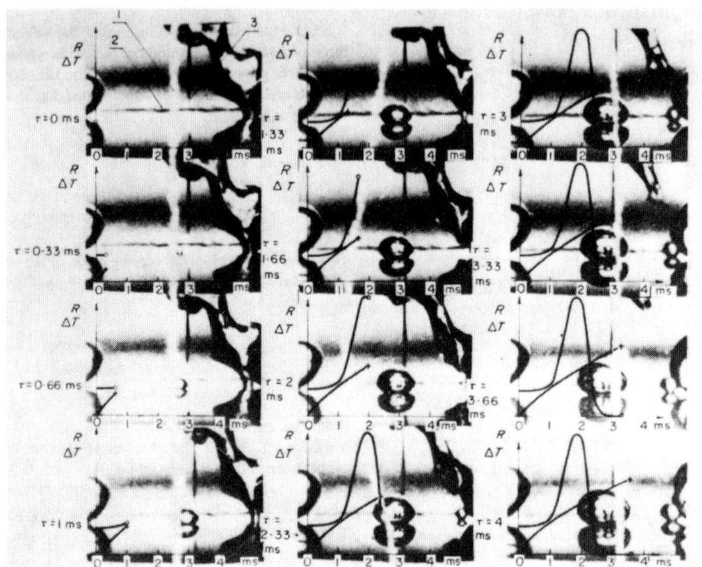
1
2
3
R
ΔT
τ=0 ms
τ=0·33 ms
τ=0·66 ms
τ=1 ms
τ=1·33 ms
τ=1·66 ms
τ=2 ms
τ=2·33 ms
τ=3 ms
τ=3·33 ms
τ=3·66 ms
τ=4 ms
0 1 2 3 4 ms

Fig. 1

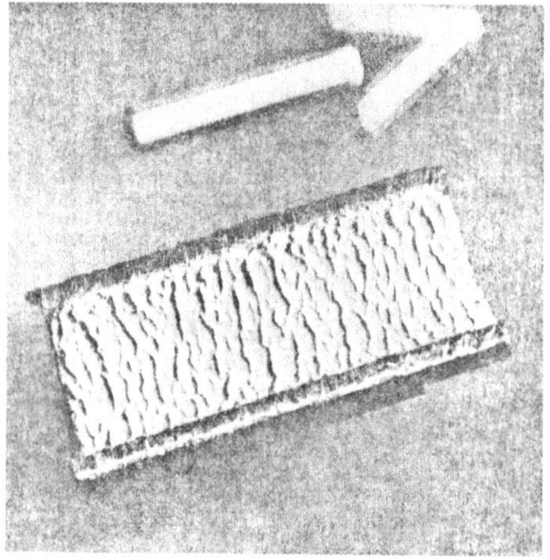

Fig. 1a

structure of two phase system. It is obvious that these interactions will generate corresponding motion of continuous phase. This fluid motion deserves special attention because its regulative might help us in understanding some of the picture which is connected with the interaction of the dispersed phase with the continuous phase. That is a main insentive for our present interest in the collective behaviour of the dispersed phase in the two phase system.

It should be emphasized that this mass transport process which is a result of the secondary motion of continuous phase of two phase system is very often cause of the adverse behaviour of two phase system. It is recognized that corrosion problems in modern boiler tubes are determined by the hydrodynamic conditions in the tube /3/. It was observed that some deposits are formed like sand ripples on a beach. In a boiler tube the scale of these ripples is very much smaller but they still have a characteristic possessed by certain other "ordered" roughness on a surface; they offer a resistance to the flow of fluid over them which is much greater than the resistance of an irregular surface profile. The occurances of burst of turbulence in flowing layer of liquid on a smooth surface in now will be recognized. It is intriguing to find that the distances associated with this phenomena are very close to the wavelenghts of ripple deposits in boiler tube. This is also one of the reasons that our attention is focused to analyse two phase flowing system in order to investigate the secondary motion of the fluid in continuous phase. Fig 1a

2. VOID FRACTION RATE

Void fraction is one of the essential parameters of two phase flow. It is a function of space and time. If we consider any point of two phase system in the instant time coordinate it will be noticed that void fraction will change from 0 to 1, forming intermittency function in the time. Second option for the consideration of two phase system is that the controlled time interval would be sufficiently long that the time average value of void fraction could be obtained. In the first case we are faced with either phase in any instant time. In the second option we obtain average value of the void fraction giving a possibility to form homogeneous two phase system with the parameters formed on the basis of homogenous mixing of two phases. It is obvious that two phase system exists in certain space the time interval which is neither instant and very often nor sufficiently long to allow average value to be obtained. So that we need to define void fraction accordingly which will satisfy our requirement that controlled time for void to be defined is related to the system in consideration.

$$\alpha = \lim_{\Delta\tau \to \Delta\tau_0} \frac{\Sigma \Delta\tau_i}{2\Delta\tau}$$

where $\Delta\tau_i$ - time fixed point has spent in vapour phase

$\Delta\tau$ - controlled time interval

It follows that void fraction rate change is:

$$\frac{d\alpha}{dt} = \frac{\Delta\alpha}{\Delta t} = \frac{1}{\Delta t} \lim \sum \frac{\Delta\tau_i}{2\Delta\tau}$$

From the other side it can be shown that for any entity in the controlled time a following conservation law could be proved /4,5,6/.

$$\frac{\partial}{\partial\tau} \sum_{\Delta\tau_p} \Theta \, d\tau = \sum_{\tau} \frac{\partial\Theta}{\partial\tau} d\tau + \sum_{disc \in 2\Delta\tau} \frac{\vec{\Theta} \cdot (\vec{\upsilon}_p \cdot \vec{n}_p)}{(\upsilon_p \cdot \vec{n}_p)}$$

where τ_p = total time interval a fixed point has spent in the vapour phase within the controlled time interval $2\Delta\tau$

With the assumption that Θ represents the temperature of the system and its value at the interface between the liquid and vapour phase is Θ^*, it can be writen /5/:

$$\frac{\partial}{\partial\tau} \sum_{i \in 2\Delta\tau} \Theta^* \Delta\tau_{pi} = \sum_{disc\ 2\Delta\tau} \Theta^* \frac{(\vec{\upsilon}_p \cdot \vec{n}_p)}{(\vec{\upsilon}_p \cdot \vec{n}_p)}$$

If this equation is divided with controlled time interval it could be obtained that

$$\Theta^* \frac{\partial\alpha}{\partial\tau} = \frac{1}{2\Delta\tau} \sum_{disc\ 2\Delta\tau} \Theta^* \frac{(\vec{\upsilon}_p \cdot \vec{n}_p)}{|\vec{\upsilon}_p \cdot \vec{n}_p|}$$

It means that void fraction rate multiplied with the temperature at the liquid-vapour interphase represents volumetric flux of vapour phase within the controlled time interval.

If the conservation equation is applied to the liquid phase, it will be obtained

$$\frac{\partial}{\partial\tau} \sum_{(2\Delta\tau - \Delta\tau_p)} \Theta d\tau = \sum_{(2\Delta\tau - \Delta\tau_p)} \left[\frac{\partial\Theta}{\partial t} d\tau \right] + \Theta^* \frac{\partial\alpha}{\partial\tau}$$

This means that

$$\Theta^* \frac{\partial\alpha}{\partial\tau} = \frac{\partial}{\partial\tau} \sum_{\Delta\tau_L} \Theta \, d\tau - \sum_{\Delta\tau_L} \frac{\partial\Theta}{\partial\tau} d\tau$$

This analysis shows that the determination of the void fraction rate represents one of the essential parameters of two phase system. Its analysis will give us a chance for the better understanding of the processes in the liquid phase and its behaviour. It is obvious that the statistical characteristic of the void fraction rate and its regularity will enable us to learn more about physical processes involved.

3. TEMPERATURE FLUCTUATION IN TWO PHASE SYSTEM

In order to prove that the presented analysis could be applied to a real two phase system a boiling two phase system in the vicinity of heated wall with pool boiling on horizontal surface was selected. The temperature fluctuation has been measured at same fixed point in two phase boundary layer. The obtained temperature signal represents continuous function in time which at the boundary between the phases has sudden change of temperature for a temperature difference between the two phases. The time which fixed point has spent in each phase can be determined by finding the detector which represents a selection criterion for its determination. For this purpose adequate methodology was developed /7,8/ which enables us to form an intermittency function corresponding to the selected criterion for the detector function. On the basis of our present knowledge of the vapour bubble behaviour and its growth in the two phase boundary layer detector function has been selected. In order to obtain nobiased results there were introduced addition criterions which came out from the particularity of temperature fluctuation in boiling system. With the application of these conditions to the temperature signal obtained at the fixed point in two phase boundary layer of boiling liquid the following three regions can be obtained: superheated liquid layer, surrounding liquid and vapour phase.

For our further analysis we are particularly interested in vapour phase so that the intermittency function which defines the time fixed point spent in vapour phase is determined with

$$I\ (x,\ \tau) = \begin{cases} 1 & M(x) \quad P \\ 0 & M(x) \quad P \end{cases}$$

Using definition (2.1) it means that

$$\alpha = \sum_{i \in 2\Delta\tau} I_i(\tau)$$

The local void rate at fixed point can be obtained from the following expression

$$\frac{\partial \alpha}{\partial \tau} = \frac{\partial}{\partial \tau} \sum_{i \in 2\Delta\tau} I_i(\tau)$$

Where $I_i(\tau)$ is intermittency function which defines vapour phase.

Using second intermittency function /9/ it was possible to define two different regions in the liquid phase. One corresponding to the liquid superheat layer and the second to the surrounding liquid. On the basis of these data processing it was possible to determine temperature fluctuation in any of these two regions. Fig. 2 shows typical signal to be analysed and also criterions for the detectors function together with the obtained intermittency functions for the selection of corresponding subregions. Using digital computer analysis of continuous signal it was possible to determine the probability density function for the temperature fluctuation in the individual region namely liquid superheat layer and surrounding liquid. Fig. 3 shows typical example of the probability distribution of temperature fluctuation in liquid superheat layer and surrounding liquid for the signal obtained at the distance δ = 1 mm from heated wall and heat flux q = 19 W/cm^2. It is obvious that the temperature fluctuation in the liquid superheat layer is deterministic and reperesents temperature distribution in the vicinity of liquid--vapour interface. Probability distribution of the temperature fluctuation in the surrounding liquid is non gausian with very pronounced effect of third moment due to presence of organized motion of the liquid in the vicinity of heated wall of boiling surface. With the increase of the distance from the heated wall this effect is diminished leaving the probability distribution of temperature fluctuation to be close to gausian one. Fig. 4 a,b shows the dispersion of the probability density distribution. As it was pointed out earlier it can be seen that the value of $\Theta^{2\ P}$ is very small in comparison with the corresponding value of $\Theta^{2\ L}$. Also $\Theta^{2\ P}$ is constant at all distances from the wall while $\Theta^{2\ L}$ changes with the distance. This also confirms that the intensity of turbulence changes with the distances as would be expected.

CONCLUSION

These two examples of the investigation of the intermittent phenomena in two phase system have shown that the turbulence structure in this system has important role in understanding heat transfer phenomena in the vicinity of interface between the phases. It is also noticed that this problem requires further investigation besides the methodological approach to the problem as well as the experimental technique for the measurement corresponding parameters which will enable us to do adequate analysis of the two phase system.

NOMENCLATURE

α - void fraction
τ - time
$\Delta\tau$ - time interval
Θ - temperature
Θ^* - interface temperature
Θ^P - vapour mean temperature
Θ^L - liquid mean temperature

υ_p - interface velocity
n_p - normal vector
I - intermittency function

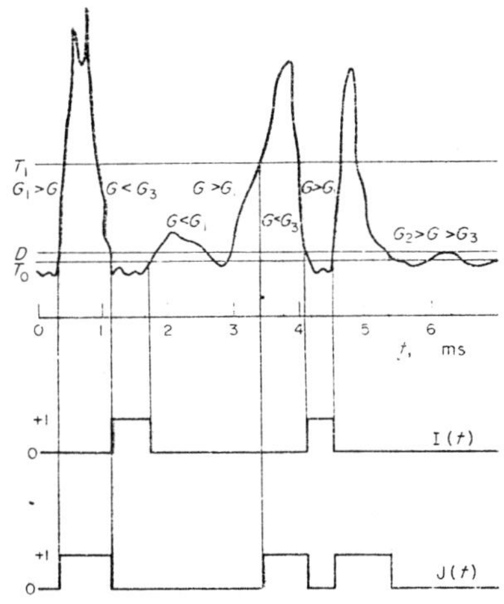

Fig. 2

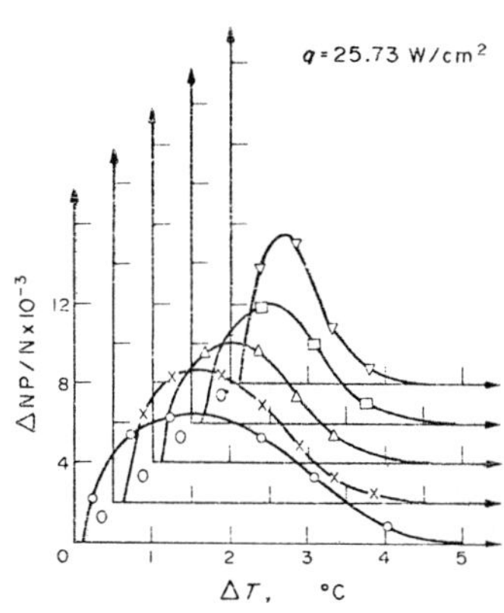

Fig. 3b

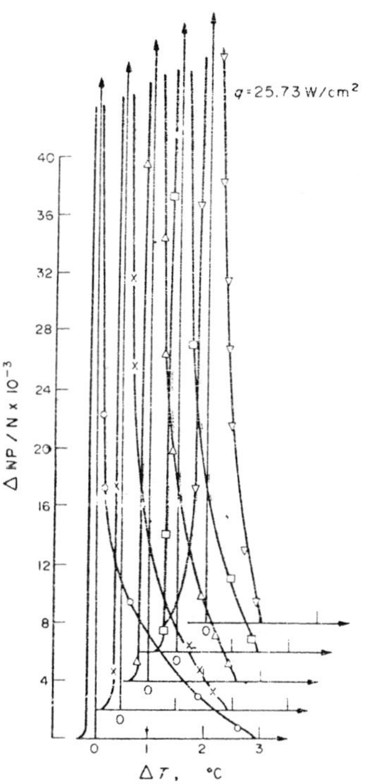

Fig. 3a

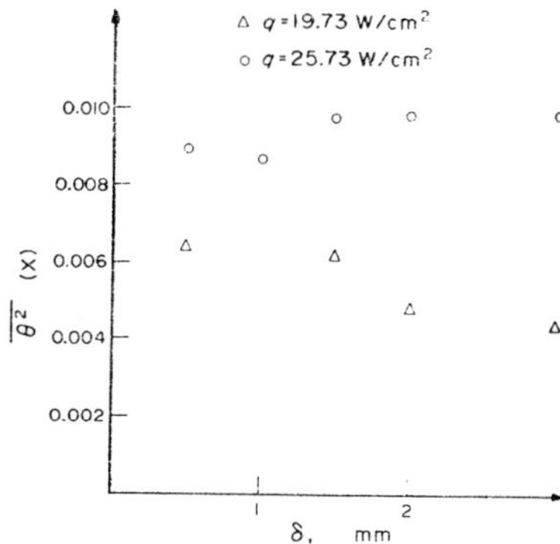

Fig. 4a

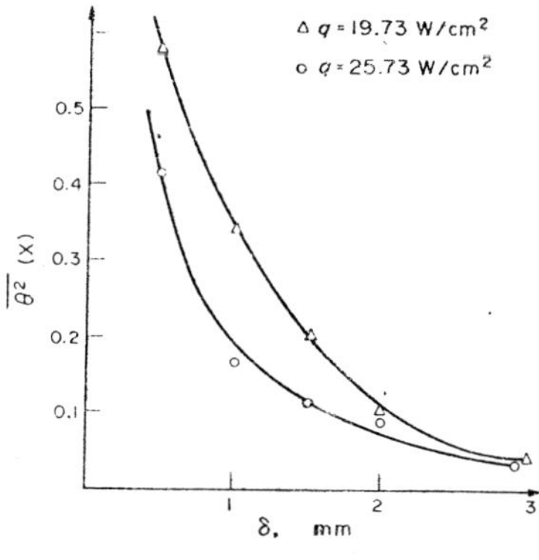

Fig. 4b

REFERENCES

1. N.H. Afgan, Advances in Heat Transfer, Vol. 11, pp. 1-60, Academic Press, (1975)

2. E. Ganić, N. Afgan, International Journal of Heat and Mass Transfer, Vol. 18, pp. 301-309, (1975)

3. E.C. Firman, Combustion, pp 21-28, July 1975

4. R.I. Nigmatulin, Two Phase, Momentum Heat and Mass Transfer in Chemical Process and Energy Engineering Systems, Vol. 1, (1978), Hemisphere Publishing Corporation, Washington

5. N. H. Afgan, The Method for Local Void Fraction Rate Determination in Two Phase Flow (to be published)

6. Ph. Vernier, J.M. Dehaye, Energie Primaire, Vol. 4, No 1-2, (1968)

7. L.S.G. Kovasznay, V.Kibens and R.F. Blackwelder, Journal of Fluid Mechanics, Vol. 41, pp. 283-325, (1970)

8. N. Afgan and L. Jović, International Journal of Heat and Mass Transfer, Vol. 21, pp. 427-434, (1978)

9. N.H. Afgan, Proceeding of Sixth International Heat Transfer Conference, Vol. 6, pp. 407-417, (1978)

General Discussion on Heat and Mass Transfer in Turbulent Flows

LOUIS FULACHIER
Institut de Mécanique Statistique de la Turbulence
Laboratoire Associé au C.N.C.S. No. 130
12 Avenue du Général Leclerc-13003 Marseille

The topics of the two sessions 4 and 5, Heat and Mass Transfer in Turbulent Flows, are of a crucial interest for this Symposium. Indeed, the principal purpose of the Symposium is to contribute to a better understanding of heat and/or mass transfer mechanisms in turbulent flows. Besides, as it has been shown in the first topic (physics of Isothermal Flows) a large emphasis was placed on large-scale coherent structures in various turbulent flows. Indeed, numerous experiments have demonstrated the persistence of these structures in turbulent shear flows and consequently have cast doubt on the usual local transport relations and even on usefulness of Reynolds averaging, used in pratically all modelling approaches. A pure statistical approach appears to be not sufficient to describe, for example, the bursting process in a turbulent boundary layer.

During the three first previous sessions some results were relating to slightly heated turbulent flows, without mass transfer, but heat was considered as a passive contaminant used mainly to analyse the structure of turbulent flows itself.

In general, most of results concerning the stucture of turbulent flows with heat and/or mass transfer are often related only to global quantities. But only few results provide a detailed study of turbulence structure, as :

- The influence of heat and/or mass transfer on coherent structures in free flows or on bursting process in wall flows.

Or as :

- The role played by these coherent structures in the heat and/or mass transfer.

Let us quote, for example, the Symposium on Turbulence which has been held in Berlin in August 1977 and which concerns for a large part the coherent structures. Only one paper is relating to the influence of heat on the coherent structures and it is relating to buoyancy effects! "Buoyancy effects on the large scale structure of free turbulent shear flows" by E. J. Hopfinger (Structure and Mechanisms of Turbulence - Proc. Symp. on Turbulence, Berlin 1977. Springer-Verlag, Berlin Heidelberg New-York 1978). However, it can be noticed that it exits some experiments concerning the instantaneous values of turbulent stresses, and longitudinal and transverse heat fluxes which have been performed in various slightly heated shear flows. In these experiments conditional sampling are used. It is then possible to have some informations on the influence of large structure -more or less coherent- on heat transfer.

It is difficult to do a synthesis of the eight papers which constitute the two sessions 4 and 5. Indeed they concern very different topics : suction effect on a turbulent boundary layer as well as turbulent flows at high temperature with or without chemical reaction ; two phases boundary layer as well as supersonic flows ; heat transfer in pulsed turbulent ducted flows as well as free stream turbulence effects on heat and mass transfer from solid wall. Most of these results do not concern the coherent structures. I think the two main reasons are firstly the difficulties to carry out measurements as sophisticated as measurements performed in isothermal flows without mass transport ; secondly, as Professor Libby has already underlined it, conditional measurements, in particular, has not given, for this moment, many implications in the modelling of isothermal flows.

Before the beginning of discussion, I want to recall a suggestion of Leslie S. G. Kovasznay concerning the impact of large scale structures on the turbulent transport of momentum, of heat or chemical concentration. The most important progress would be to offer a rational alternative to the use of the gradient diffusion assumption. Besides this Boussinesq hypothesis is very wrong in asymmetrical flows. As one possible alternative the suggestion is offered to express the turbulent transport in an integral form instead of the usual gradient diffusion term. Thus the spread of a transportable quantity at a point x, y, z, t, would be calculated from a quadruple integral, with three spaces and one time variables, taking into account the form and the statistical occurence of the large scale structures through the kermel k :

$$\varphi(x,y,z,t,) = \iiiint \varphi(\xi,\eta,\zeta,\tau)k(\xi,\eta,\zeta,\tau,x,y,z,t)d\xi d\eta d\zeta dt$$

I want also to point out that, if we consider a statistical view point, it is necessary to use the mass weigted averaging, so called Favre averaging, when, for example the temperature difference is not very weak.

If you agree, the discussion could be orientated around the following ideas :

- How structures play a role in heat and/or mass transfers ?

- What is the influence of heat and/or mass transfer on these structures ?

- What is happening when Prandtl or Schmidt number is different from unity ?

- What experiments could be suggested ?

I would like to have the opinion of the audience on these problems.

Dr. C. Rey

One means of incorporating variable density effect in the description of turbulent reacting or buoyant flows is to employ Favre averaging ; in this case density variations and fluctuations does not appear explicitly but their effects are contained implicitly in the various correlations or pressure terms. Now, my feeling is that a formal simplification of the equations is not always the better way when we want to understand a detailed mechanism of the turbulence. For example when buoyant flows are concerned, Favre averaging leads to relatively simple equations without approximations. But it is not possible to know exactly the weight of dilatation effect on inertia (usually neglected when the Boussinesq approximations are used) and to analyse separately the buoyancy, because these terms do not appear explicitly in the equations. If one wants to analyse the dilatation effects for different Reynolds number flows, Reynolds number effect on the balance between dilatation phenomena and the other's can be only studied if the different term varying with density are distinctly written. Then for these kinds of application I think that Favre averaging gives a global look of the mechanisms and we are not actually used to this reasoning.

Dr. L. Fulachier

This is an interesting idea about which Dr. Rey will speak more extensively in his communication on the simulation of a buoyant stratified atmosphere in session 6. I think we'll have then more precise details about the bounds of Boussinesq approximations. For myself, I would like that the influence of Richardson number to be clearly specified, in particular on the calculation of the kinetic energy production terms.

Dr. A. Dinkelacker

The chairman has asked what sort of future investigations might be of interest. I think it might be interesting to investigate flows along walls which have a fine longitudinal structure in their temperature and/or in their heat transfer qualities. This for the following reason : in many investigations on turbulent flows longitudinal vortices have been observed which have a typical lateral distance in wall quantities of about $\ell^+ = 100$. If one introduces artificially longitudinal structures at the wall - say by longitudianl lines of different wall temperature - which can influence the formation of these longitudinal vortices it might be possible to influence in this way the turbulence close to the wall, and hence to influence both heat transfer as well as skin friction. In particular I think it might be possible that wall configurations which have $\ell^+ < 100$, say $\ell^+ = 30$, might lead to a reduction in skin friction.

Dr. L. Fulachier

In order to have significant effect in the experiment you suggest, the heat transfer must be strong enough. It would be besides necessary to obtain directly the velocity field in order to determine ℓ^+. In this case the heat could no more be considered as a passive contaminant enabling an analysis of turbulent structures. Therefore it seems that Laser-Doppler measurements could yield conveniently these informations.

Prof. Mizushina and Ueda

In order to investigate further detail of the turbulence structure in the flows with heat and mass transfer, measuring technique of fluctuating velocity components should be established urgently. Hot wire anemometry with a temperature compensation circuit have been used. However, it has some disadvantages in measuring velocity components perpendicular to the mean direction and in measuring turbulent fluctuation in very slow motion, e.g. in thermally stratified flows.

Laser Doppler anemometry is useful. However, the following effects should be noted in its application in the non-isothermal flows ; the one is the bending of the laser beams due to the mean temperature gradients (mean refractivity gradient) in the flow field. The point of intersection of the two beams, i.e. measuring point should be determined directly by some instrument or estimated from the information on the mean temperature field. The another important effect is the interaction of the laser beams with turbulent temperature fluctuations along the beam paths. In the non-isothermal turbulent flows, the laser beams encounter a succession of hot and cold pockets formed by turbulent eddies of various sizes and various temperatures, which modify the phase and direction of the laser beams, and result in a measured apparent velocity. The correction for such interaction effect needs to be determined experimentally, owing to the lack of detailed knowledge of turbulence along the beam paths. The correction method may be referred to Mizushina, T., Ogino, F., Ueda, H. and Komori, S. (1979, Proc. R. Soc. London, A366, 63-79) and Hong, N.S., Jones, A.R. and Weinberg, F.J. (1977, Proc. R. Soc. London, A353, 77-85).

Dr. L. Fulachier

I thank you for these references and your comment concerning measuring techniques. Obviously, in turbulent flows with heat (and mass) transfer, the temperature-velocity vector correlations are of a great interest for the understanding of transfer mechanisms. If L.D.A. techniques are adopted, important difficulties are encountered, in particular velocity components must be determined with L.D.A. and temperature with cold wire. And this must be done simultaneously. In gases, problems linked with particles increases this difficulty. In contrast it is possible, with hot and cold wires techniques to obtain these instantaneous turbulent heat fluxes, in particular in turbulent flows with moderate heating. However the complexity of this problem is increasing when the temperature is more and more elevated.

Prof. R.A. Antonia

I believe there is considerable scope for reliable measurements of temperature fluctuations in turbulent shear flows. It is somewhat surprising that the transport equation for $\overline{\theta^2}$ (θ is the temperature fluctuation) has not received more attention by the experimentalist since all the terms in the equation can be obtained. In particular, all three contributions to the destruction term of $\overline{\theta^2}$ can be mesured without resort to local isotropy.

I should also mention one particular flow which has not been successfully "modelled". I am referring to the calculation of the boundary layer immediately downstream of a sudden change in surface heat flux or temperature. Experiments have been carried out by Blom and Fulachier, in the case of a sudden increase in surface temperature, and by Antonia, Danh and Prabhu for a sudden increase in surface heat flux. The calculation of the momentum boundary layer presents no great difficulties since the flow is approximately self-preserving at the change and the magnitude of the temperature or heat flux change is such that temperature is a passive marker of the layer. The assumption of a constant turbulent Prandtl number Pr_t does not lead to agreement between the calculated mean temperature and heat flux profile and the measured distribution near the change. Perhaps surprisingly, calculations that include a transport equation for the heat flux, do not seem to calculate the flow any more accurately than those which assume $Pr_t \simeq$ constant. The boundary layer downstream of a sudden decrease in surface temperature has been measured by Charnay et al at Lyon. Bert Browne (University of Newcastle) and I are currently trying to calculate this flow but our attempts to date indicate that this flow may be less amenable to calculation than that corresponding to a sudden increase in surface temperature. The relaxation distance appears to be significantly greater in the case of a sudden decrease than for a sudden increase in surface temperature.

Dr. L. Fulachier

I agree with you. The experiments of I.M.S.T. have been carried out in turbulent boundary layer immediately downstream of a sudden change in surface temperature (cf. e.g. our paper session 4). Dr. I. Dekeyser has applied the Patankar - Spalding and Bradshaw-Ferris programs in order to calculate mean temperature profiles. He finds effectively near the beginning of these sudden change in surface temperature a rather large discrepancy between measurements and calculations. However, for away, the Bradshaw's program gives even a turbulent heat flux quite closely the experiments (I. Dekeyser, thesis, 1972 I.M.S.T. - L. Fulachier, E. Verollet, I. Dekeyser, Int. J. Heat Mass Transfer 20, p. 731, 1977).

Prof. E.J. Hopfinger

Prandtl (Schmidt) number effects are possibly more important in wall shear flows than in free turbulent flows. A quantity which is the most sensitive indicator of molecular effects is the amount of mixing which occurs. The measurements of reaction product in a

shear layer between two streams of chemically reacting fluids with $Sc \simeq 600$ made by Breidenthal (Ph. D. Thesis, Caltech 1978) show that the amount of product is independent of Reynolds number when the large-scale Reynolds number is greater than about 5×10^3. And, more importantly, the numerical value of the measured product is only by about a factor of two lower than in a shear layer with $Sc \simeq 0.7$. Below the transition Reynolds number (which corresponds approximatly to the appearance of small scales) mixing is strongly Schmidt number dependent. The difference in product above the transition Reynolds number may be ascribed to a difference in the physical processes which limit the mixing event (small scale diffusion limited or large scale entrainment limited). Experimental error is another factor : while the technique used in the case of $Sc \simeq 0.7$ tends to overestimate mixing, the optical technique used for underestimate it . In any event, it is of interest to note that above the transition Reynolds number a difference in Schmidt number by three orders of magnitude has only a small effect on mixing.
Concerning the active scalar effects it is well known that a scalar quantity may rapidly become dynamically active and alter substantially the large scale dynamics. There is however a great lack of quantitative information on the nature in which the turbulent processes are modified. This is particularly so in wall shear flows. I just want to mention in this connection that the dynamics of the large scale structures may be affected by gradients as well as by fluxes in scalar quantity. Usually a flux generates a gradient or vice versa but the respective importance depends on the flow situation.

Dr. L. Fulachier

I thank Prof. Hopfinger to have called our attention to the molecular Prandtl or Schmidt numbers effects. I would add to his remarks that the turbulent Prandtl or Schmidt numbers depend also on the molecular ones, what gives a further complexity to this problem. Concerning the effects of active scalar it is necessary to get experiments as those described in sessions 1, 2 and 3 using for example the L.D.A. taking into account in particular the corrections previously recalled by Prof. Mizushina and Ueda.

Dr. C. Rey

Comparison between turbulent velocity and temperature fields require to separate between two situations.

- The first is the boundary layer one, when the initial conditions generating the two turbulent fields are not predominant downstream in comparison with the turbulent production from the wall. Then if the wall conditions are relatively the same along a sufficiently part of the plate upstream, comparison of heat and mass transfer can be made and the analogy between the temperature energy spectrum and the velocity vector energy spectrum is well suported by the data in the lower frequency range (Fulachier 1972) which includes about 80 % of the total energy.

- The second case concerns the free flows. Here a large part of the turbulent phenomena can be related to the initial condition and the memory of the turbulence plays an important role. Then if the initial conditions generating the two fields are not the same, although the temperature field is transported by the velocity one, the properties of the two fields taken at the same point, are different. In the case of the experimental work of Morel (1979) considering a turbulent wake behind an asymmetrically heated plate, it is shown that the kinematic and thermal fields grow with some degrees of independance, then it is possible to represent the turburlent properties of the two fields with respect to their own axes ans scales. Then the non dimensional representation of the mean and turbulent and sampled quantities of the temperature fields, using the temperature axe and scales, gives an artificial self preserving temperature field, within the wake (Morel and Rey, to be published) ; Fulachier and Dumas J. Fluid. Mech. (1976), vol. 77, part 2, pp. 257-277 ; Morel, Rey, Awad, Mathieu and Schon Phys. Fluids, 22 (4) April 1979.

Dr. L. Fulachier

The analogy we have proposed between the temperature and the velocity vector spectra can be applied principally in wall flows : for example, in the case of a simple boundary layer on a heated flat plate but also in the case of atmospheric surface layer. In these different situations, the turbulence is continously renewed at the wall. This is no more the case of previously heated free-flows such as wakes. So, it is not surprising that this spectral analogy is not so well working although the physical idea must not be rejected, at first sight.

Dr. A. Rosengard

Mass transfer, specially related with turbulence structure, may occurs in different ways, and these ways interfact in the relation between momentum and mass transfer. For example :

- The fluid is a gaseous mixture of two isotopes, or a mixture of two gases of different masses, with very percent of them : mass transfer does not influence momentum transfer (but of course is influenced by momentum transfer).

- The fluid is a mixture of two different gases, so that composition variations induce specific mass variation ; mass tranfer induce local specific mass variations, which yield the velocity field and local turbulence variations.

- With multiphase flows, boundary conditions are modified by mass transfer, and the structure of turbulence becomes eventually different.

We must mention the particular case of blowing or suction of a gas flow along a porous wall. Mass transfer has here two consequences on the flows :

- a global effect by addition or removal fluid
- a local effect by modification of the velocity field near the wall.

These two consequences induce modifications on the turbulence structure ; we must depart the specific influence of each of them.

Dr. L. Fulachier

You have reason to bring to our mind these precisions. I add to what you said that, in addition of the interest of this kind of mass transfers themselves, blowing and suction through a wall could be means to modify the behaviour of the structures close to the wall. In particular, as I already underlined in my communication (session 4), this can be a procedure to act upon the bursting process what could yield valuable informations about its mechanism.

Prof. K. Suzuki

I would like to comment on two points related to the direction of discussion suggested by the chairman of this general discussion.

In combustion problems, often occurs such a situation that coherent structure affects strongly the nature of the phenomenon. For example, in a diffusion flame suspended in a wake of a thin wall separating fuel and oxidizer flows, the flame is formed in the region where coherent eddy motion is remarkable. The coherent motion governs the mixing pattern of two streams that encountered each other at the downstream tip of the separating wall. It seems from some experiments also that the change in fluid density due to the occurrence of combustion, and accompanied heat and mass transfer can alter some features of the coherent motion there. These have been one of the main concerns of combustion scientists.

Favre average referred to by Dr. Fulachier has also been paid attention by some people working on combustion problems. It could give a means to attack on the problems in which density fluctuation occurs severely.

Dr. L. Fulachier

Concerning Favre average, Prof. P.A. Libby has already pointed out here its importance in particular in combustion problems. But even in the case of density variations relatively moderate it is suitable to use yet this averaging procedure.

At this point of these discussions it appears that few of the different intervenors have approached the two first points related to the mechanism of the dependence between coherent structures and heat or mass transfer. It seems therefore of a great interest to perform new experiments similar to those exposed in the three first sessions concerning the structures in turbulent flows without heat and mass transfers.

Now I thank you very much for your contributions which have brought some lights upon various points and which have animated this general discussion.

ENVIRONMENTAL FLOWS

Turbulent Transport in the Mixed Convection over a Heated Horizontal Plane

T. MIZUSHINA
Kyoto University
606 Kyoto, Japan

H. UEDA, S. MITSUMOTO, and S. KOMORI
National Institute for Environmental Studies
305 Ibaraki, Japan

K. FUKUI
Himeji Institute of Technology
671-22 Himeji, Japan

ABSTRACT

Turbulent motions and transport processes in the boundary layer over a heated horizontal plane were investigated experimentally, in which the flow changed from forced convection, via mixed convection to natural convection with increasing the heat flux at the plane.

In the mixed convection, buoyancy forces enhance the turbulent transport, and the eddy diffusivities of momentum and heat, ε_M and ε_H, and the ratio $\varepsilon_H/\varepsilon_M$ increase. However, stability dependence of the $\varepsilon_H/\varepsilon_M$ is significantly different between the wall and outer regions; in the wall region the $\varepsilon_H/\varepsilon_M$ increases slightly with increasing instability, while it changes drastically in the outer layer. Taking into account this fact, friction factor, Nusselt number and profiles of mean velocity and temperature may be well predicted. So far as the heat transfer is concerned, transition from natural to forced convection occurs rapidly and so the mixed convection regime is rather narrow.

1. INTRODUCTION

Thermal or density stratification causes substantial changes in the motions and turbulent transport processes in the atmospheric environment. In the inland area, the lower atmosphere is heated from below by the earth surface which has diurnally varying temperature, forming mixed convection there. The mixed convection is also motivated by the abrupt change of the surface temperature between the sea and the land, when the sea breeze comes into the coastal region in the daytime. These mixed convection layers are often accompanied by an inversion cap and they contain pollutant within them. Thus, their behavior affects the air pollution seriously.

These mixed convection layers are under unstably stratified conditions and the upward and downward motions of the turbulent eddies are predominant, so that the vertical eddy diffusivities of momentum, heat and mass are enhanced. The influence of buoyancy on the turbulence structure and the turbulent transport processes has been investigated in detail in the previous reports [1, 2]. The stability dependence of the turbulence quantities is universally correlated with a local parameter, i.e., gradient Richardson number, $Ri=-g\,\partial\bar{\rho}/\partial z/[\bar{\rho}(\partial\bar{u}/\partial z)^2]$, where g is the acceleration due to gravity, $\bar{\rho}$ the time-mean density, $\bar{u}$ the time-mean velocity and z the distance in the vertical direction.

On the other hand the field observation was carried out to examine the stability dependence of the eddy diffusivity of heat in the lower atmosphere, and confirmed that the laboratory results cited above can be applied to the atmosphere above the surface layer [3] and that the stability dependence of the turbulent transport processes in that layer is different from that in the surface layer. The difference of these stability dependences is attributed to the effect of the presence of the ground on the fluctuating pressure field as indicated by Gibson & Launder [4].

In accordance with these facts, the present work was undertaken as an attempt to understand the turbulent transport processes in the whole region of mixed convection. A laboratory experiment was performed on the mixed convection over a heated plane installed horizontally in a wind tunnel. It is similar to the experiment on the duct flow done by Townsend [5]. Here, particular attention was paid to the behavior in the wall region.

2. EXPERIMENT

The experiments were carried out in a wind tunnel with a 1.5m wide, 1.0m high and 19m long test section, shown schematically in Fig. 1. A controlled flow of air was obtained from a centrifugal blower driven by a direct current motor. The air temperature was regulated to within 0.2K and the uniformity of the velocity in the inlet cross section was within ±1.0% of an air velocity of 1m/s.

Heat transfer equipment (Fig. 2) was installed horizontally in the free stream of the wind tunnel. The heated plane was a copper plate of 0.9m wide, 2.58m long and 3mm thick. The surface was electrically heated by a 0.5mm thick stainless steel sheet, with a teflon sheet between this and the copper. A uniform heat flux condition may be assumed at the measuring stations. To minimize the backward heat loss a subheater of a stainless steel sheet was attached to the back surface of a 24mm thick wooden plate, insulated by a 25mm thick rock-wool. To prevent heat flux toward the leading and side edges the surrounding surface around the heated plate was covered with plate-heaters, of which temperatures were controlled. In order to promote transition to turbulent boundary layer flow, sandpaper was put on the initial 240mm section (Fig.2.).

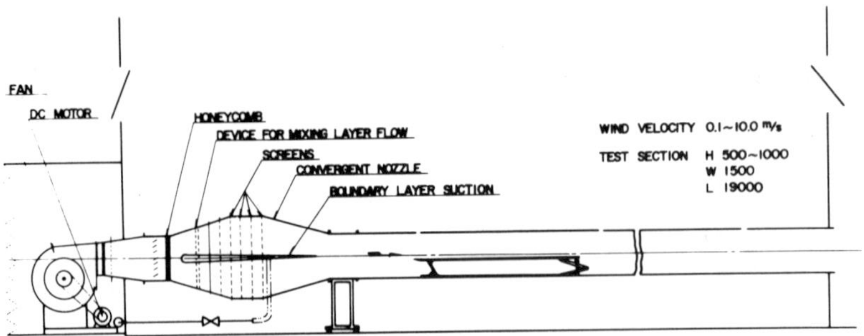

Fig.1. Schematic diagram of the wind tunnel

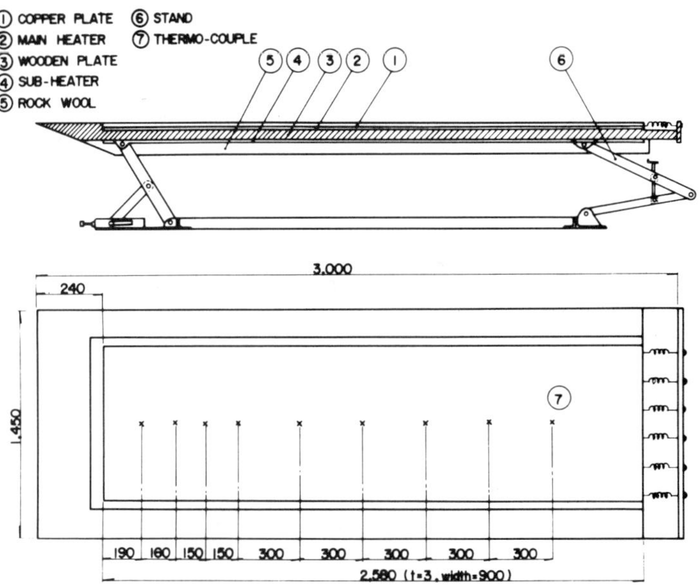

Fig.2. Heat transfer equipment

Measurement of the temperature and its fluctuation was made respectively by Philips Ac025 thermocouples and by a resistance thermometer of Wollaston wire, 1.0μm in diameter and 0.4mm long, operated by a DISA 55M20 constant current temperature bridge. A DISA 55X00 two-color laser Doppler velocimeter with a frequency shifter [6] and a total pressure tube with a rectangular opening, 1.2mm wide and 0.6mm high, were used for measuring the time mean and fluctuating velocity components.

The voltage outputs from the instruments were digitized and recorded on a magnetic tape. Those signals were processed with the HITAC 8450 computer system in the Data Processing Center of the National Institute for Environmental Studies. The sampling interval and the sampling size were 0.01 second and about 30000, respectively.

The experiments were performed with the free stream velocity ranging from 0.2 to 5m/s and with the surface temperature ranging from 5 to 66K above the ambient temperature.

3. RESULTS AND DISCUSSION

3.1. Heat Transfer Coefficient

Convective heat flux was determined by the heat balance, i.e., heat input

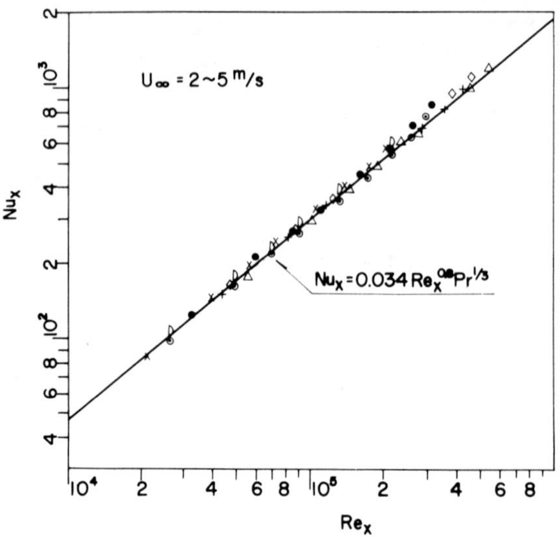

Fig.3. Correlation of local heat transfer coefficient for forced convection

minus radiation and conduction contribution. This contribution was at most 9%, and its estimated error was improved by comparing it with the heat flux determined from the measured temperature gradient near the wall.

With low heating rate and high free stream velocity, the flow may be assumed to be forced convection. As seen in Fig. 3, the local heat transfer coefficient can be correlated well by

$$Nu_x = 0.034\ Re_x^{0.8}\ Pr^{1/3} \tag{1}$$

This agrees well with the previous investigation [7] for flat plate boundary layer flow.

If the free stream velocity is decreased while the heat flux is kept constant and high, buoyancy forces become dominant and the flow should resemble natural convection and the heat transfer rate becomes independent on the free stream velocity and correlated by Rayleigh number Ra_x only. Experimental results are shown in Fig. 4 and can be expressed in the form

$$Nu_x = \alpha\ Ra_x^{1/3} \tag{2}$$

where α=0.155.

Turbulent natural convection over a heated horizontal plane in stagnant surrounding has been investigated by Vliet [8], Fujii & Imura [9], Miller & Gebhart [10] and others. They correlated heat transfer results in the form of Eq.(2) but values of the coefficient α are scattered. This might be due to the facts that flow separation takes place in the center portion of the heated plate and that the definition of ambient fluid temperature is arbitrary since the ambient fluid is stratified gradually in the container. The present experiment has an advantage because of the existence of weak free stream.

It was found that the value of α was equal to 0.155 (Table 1) and this agrees with α=0.16 obtained for $Ra_L < 2x10^8$ by Fujii & Imura and α=0.14 recommended by McAdams [11].

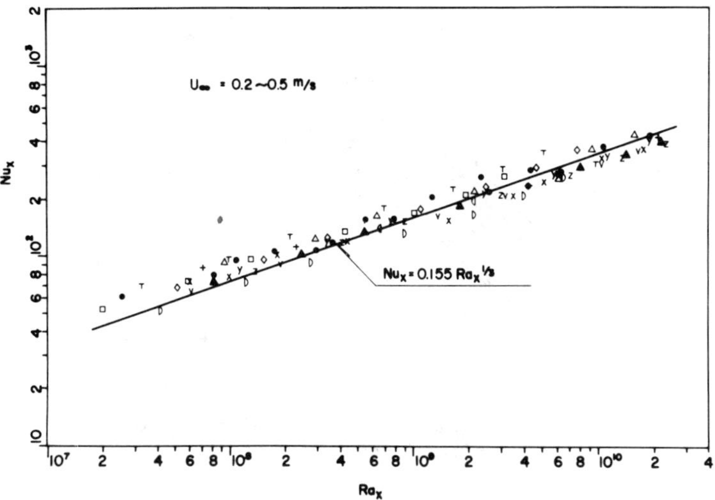

Fig.4. Correlation of local heat transfer coefficient for natural convection

Table 1. Comparison of the coefficient α

α	Investigators	Ra
0.155	Present results	$2 \times 10^7 \sim 2 \times 10^{10}$
0.119	Miller & Gebhart	$3 \times 10^5 \sim 3 \times 10^7$
0.13	Fujii & Imura	$> 5 \times 10^8$
0.16	Fujii & Imura	$< 2 \times 10^8$
0.14	McAdams	$2 \times 10^7 \sim 3 \times 10^{10}$

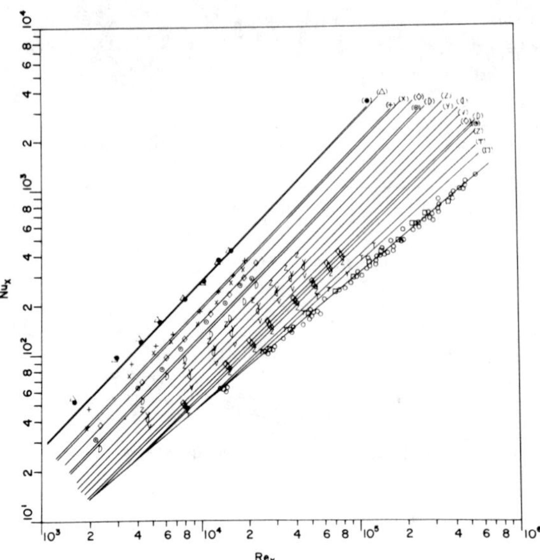

Fig.5. Variation of Nu_x with Re_x for mixed convection

For the mixed convection together with forced and natural convection the present results are shown in Figs. 5,6 and 7. Figures 5 and 6 show the dependence of the heat transfer rates, Nu_x and St_x, on Re_x with the temperature difference $\Delta\bar{\theta}$ as a parameter. In the figures symbol o represents the values for forced convection. Heat transfer is enhanced with increasing $\Delta\bar{\theta}$, and for the strongest heating, i.e., $\Delta\bar{\theta}$ = 66K (symbols Δ, • in those figures) the heat transfer rate becomes about 8 times larger than that for forced convection. The same experimental results were replotted in the form of St_x vs. Re_x. As the flow condition approaches natural convection, St_x does not change with X so much and begins to depend only on $\Delta\bar{\theta}$.

The present results were plotted as a function of Ra_x in Fig. 7 in which the lowest line represents the correlation (2) for natural convection. With increasing the free-stream velocity U_∞, Nu_x increases remarkably. Such a deviation from equation (1) and (2) is significant. For convenience this behavior in the mixed convection is assumed to be represented in the form

$$Nu_x = [(Nu_{x,\ forced})^n + (Nu_{x,\ natural})^n]^{1/n} \tag{3}$$

The Nusselt number Nu_x was calculated for various values of n and was compared with the experimental results. As seen in Table 2, the best-fit value of n was 3.94 and the calculated results from equation (3) were shown in Figs. 5,6 and 7. From the fact that the value n is relatively large, it may be concluded that the transition from natural to forced convection occurs rapidly so far as the heat transfer coefficient is concerned.

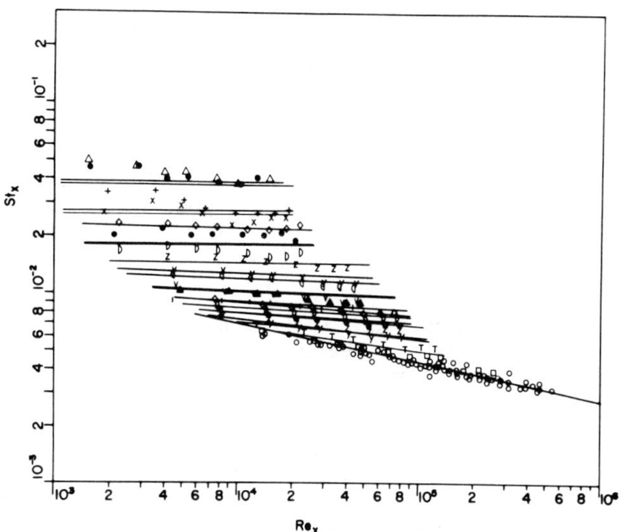

Fig.6. Variation of St_x with Re_x for mixed convection

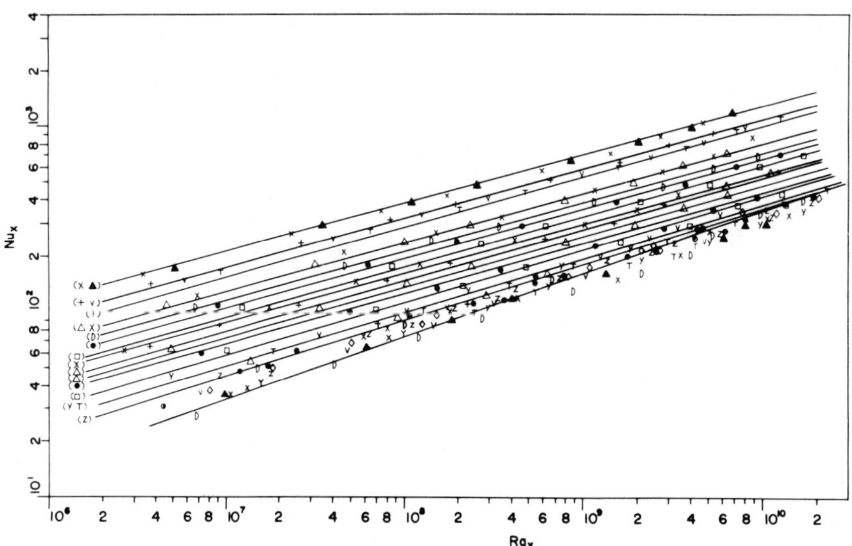

Fig.7. Variation of Nu_x with Ra_x for mixed convection

Table 2. Deviation of the calculated results for various values of n

n	2	3	3.94	4	5
Error %	15.5	9.5	8.26	8.5	8.6

3.2. Eddy Diffusivities of Momentum and Heat

For forced convection measurement shows that a van Driest-type formula [12] may be applied for the eddy diffusivity of momentum ε_{MO} or the mixing length ℓ_0, that is,

$$\ell_0^+ = \kappa z^+[1-\exp(-z^+/26)] \qquad (4)$$

where κ is von Kármán's constant and suffix 0 denotes the neutral condition and the superscript denotes the values normalized by wall parameters. The ratio of eddy diffusivity of heat ε_{HO} to that of momentum ε_{MO}, i.e., the reciprocal turbulent Prandtl number $(\varepsilon_H/\varepsilon_M)_0$ may be assumed to be equal to 1.2.

With increasing thermal stratification unstably, however, buoyancy forces enhance the turbulent transport and the eddy diffusivities of momentum and heat increase. They, as well as other turbulence quantities, were found to be well correlated with the single local stability parameter Ri[2]. The stability dependence of ε_M and ε_H are expressed as

$$\varepsilon_M/\varepsilon_{MO} = (1-25Ri)^{1/3} \qquad (5)$$

$$\varepsilon_H/\varepsilon_M = (\varepsilon_H/\varepsilon_M)_0(1-Rf/Rf_{crit})/(1-Rf)^2 \qquad (6)$$

where Rf is the flux Richardson number and Rf_{crit} is the so-called critical Richardson number and is determined to be 0.1. These relations obtained in a laboratory experiment are valid in the atmosphere above the surface layer. That is confirmed from a field observation made by Ueda et al [3]. However, from an examination of those results together with other field observations made in the ground level, it was found that those relationships are not applicable in the atmospheric surface layer. Recently, Gibson and Launder [4] proposed a second-order turbulence model to take into account the modification of the fluctuating pressure field by the presence of the ground, and predicted the different buoyancy effects observed in the surface layer and stratified free shear flow with good accuracy. In the present case mixed convection in the outer layer of the boundary layer which is referred to as 'similarity region' by Townsend, is substantially similar to the stratified free shear flow and the stability dependence of the eddy diffusivities is supposed to be correlated well by equations (5) and (6), although its determination from the experiment is difficult because of very small gradients of velocity and temperature.

In the wall region where the production rate of turbulence has its maximum, Townsend [5] and Arya [13] made measurements of mean flow and temperature but did not present the results on the stability dependence explicitly. In the present investigation it was determined from the measured profiles in the region of $10 < z^+ < 30$. For the eddy diffusivity of momentum it was found that the stability dependence did not change with height, z, within the experimental error. However, the stability dependence of the ratio $\varepsilon_H/\varepsilon_M$, i.e., the reciprocal turbulent Prandtl number in the wall region is significantly different from that in the outer region; as seen in Fig.8, the $\varepsilon_H/\varepsilon_M$ increases slightly with increasing instability in the wall region, while it changes drastically in the outer region. Reasons for this behavior may lie in the nature of mixed convection. In the wall region the turbulent transport is influenced by both the shear stress and buoyancy. The effect of shear stress is damped rapidly with height and so in the outer layer the convection is determined only by the buoyancy and by the presence of the surface as a constraint on the motion. In Fig.8 predicted result from Gibson and Launder's theory is also shown. Agreement with the present results is very good. On the other hand, Townsend determined the ratio of $\varepsilon_H/\varepsilon_M$ to be 1.4± 0.2 from the value of the surface stress. This value may be regarded as an averaged value for heights in the wall region for mixed convection, and is in agreement with the present results.

In Fig.9 another comparison is made of the so-called shear functions ϕ_M and ϕ_H which are defined as

$$\phi_M = (\kappa z/u^*)/\partial\overline{u}/\partial z \tag{7}$$

$$\phi_H = \kappa z u^*(\partial\overline{\theta}/\partial z)/(-\overline{w\theta}) \tag{8}$$

Both ϕ_M and ϕ_H obtained in the present experiment are in accordance with field observations, especially with the results of Pruitt et al.[14].

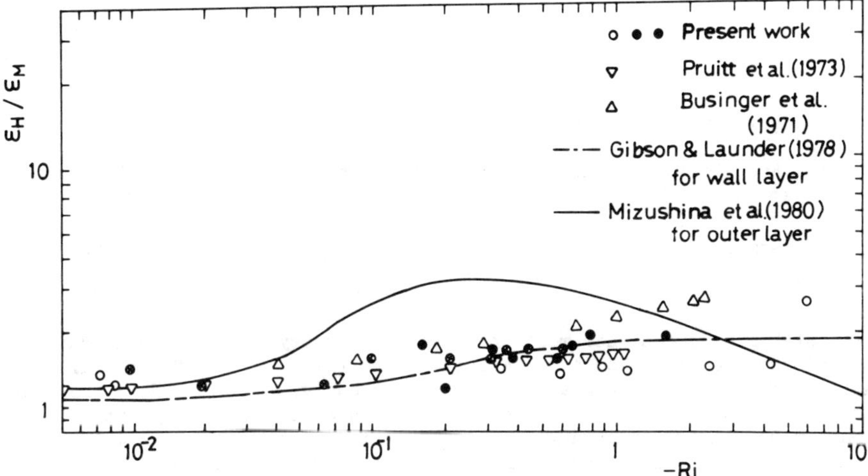

Fig.8. Variation of $\varepsilon_H/\varepsilon_M$ with Ri in unstably stratified flows

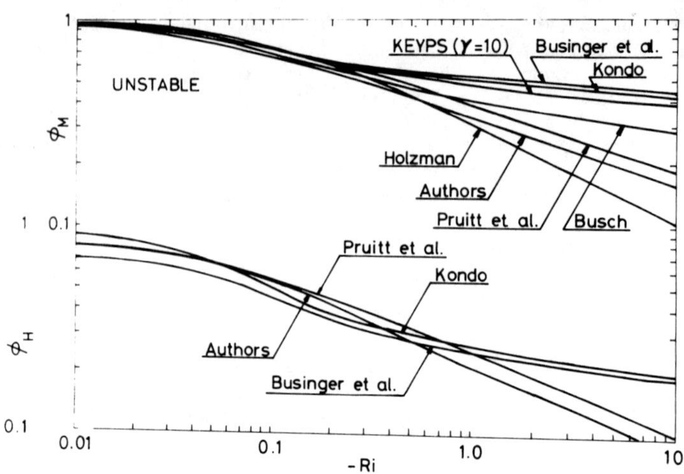

Fig.9. Comparison of the present results with those of other field observations in the atmospheric surface layer

3.3. Boundary Layer Calculation

Using Escudier's eddy diffusivity formulation for neutral conditions [15], i.e.,

$$\ell^+ = \kappa z^+[1-\exp(-z^+/26)] \qquad \text{for } z^+ < \lambda^+ \tag{9}$$

$$\ell^+ = \kappa\lambda^+ \qquad \text{for } \lambda^+ < z^+ < \delta^+ \tag{10}$$

and the stability dependence mentioned above, boundary layer calculation was performed numerically. The basic equations are the equation of continuity, the equations of motion with buoyancy term and the energy equation. Here the static pressure gradient in the flow direction arises due to the development of a thermal or densimetric boundary layer, and varies with height z as well as the distance from the leading edge. That was estimated by an assumption of hydrostatic equilibrium. Those equations were transformed into the difference form proposed by Patankar and Spalding's method [16], and were integrated numerically.

In Fig.10 a comparison between the predicted and experimental results is shown. For forced convection the predicted relations of St_x and Cf_x vs. Re_x are in accordance with the experimental results and with the previous investigations [17,18]. Here, the experimental values of the wall shear stress τ_w were determined from the velocity gradient near the wall, which might have a reasonable accuracy because of thick viscous sublayer.

As an example of mixed convection a calculation was made for strongly unstable flows, i.e., flows with $\Delta\theta$ = 38.14K and 66.25K, respectively, and with weak free-stream velocity U_∞ = 0.5m/s. As pointed out by Townsend [5], the flow near the surface in the initial region from the leading edge is accel-

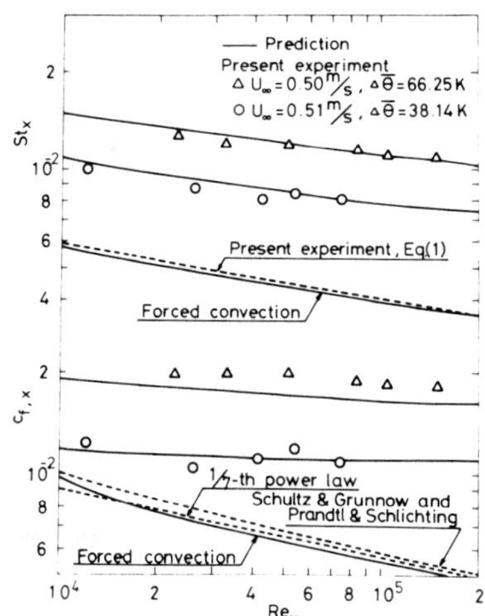

Fig.10. Comparison between the prediction and experiment

erated so that the velocity profile has a maximum similar to a wall jet. This is due to large horizontal gradients of temperature and so static pressure near the wall, which are produced by heating the cold air. Far from the leading edge, the profiles of velocity and temperature are nearly identical and the flow is nearly fully-developed. The changes of Cf_x and St_x with Re_x are shown in Fig.10. They are almost constant along the flow direction but significantly larger than those for forced convection. The predicted results agree fairly well with the experimental ones.

3.4. Profiles of Mean Velocity and Temperature and Large Eddy Structure

Profiles of mean velocity and temperature are shown in Fig. 11-(a) and -(b) for forced and mixed convections, respectively. Difference between these convections is obvious, and predicted very well. The profiles for mixed convection have large gradients in the limited region $z^+ < 40$. They are almost uniform in the outer region and there exist large fluctuations of velocity and temperature which is caused by the buoyancy-driven motion, i.e., upward movement of hot eddies and downward movement of cold eddies.

Flow visualization by a smoke method reveals some insights into the fluid motions in the outer region (Plate 1). Because of the presence of strong wind shear in the inner layer, long eddies with axes aligned with the flow direction are formed, and plumes ejected from the viscous sublayer are nearly two-dimensional in form. That motion seems to be substantially similar to the bursting phenomena observed in the wall region of neutral turbulent flow by Kline et al. [19] and others, but the difference in the mixed convection is stronger ejection of the plumes which retards the mean flow strongly. This causes intermittent and strong positive-spikes of temperature, θ, which are accompanied by the upward motion, w, and results in a stronger correlation between w- and θ-fluctuation.

Motion in the outer layer is seen to be composed of large structure of longitudinal vortex-roll of which axis oscillates in its direction. This is very similar to that in natural convection, in particular oscillatory convec-

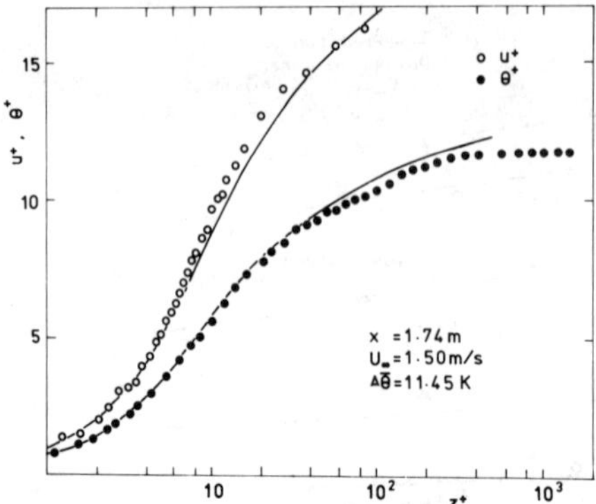

Fig.11-(a). Profiles of velocity and temperature for forced convection

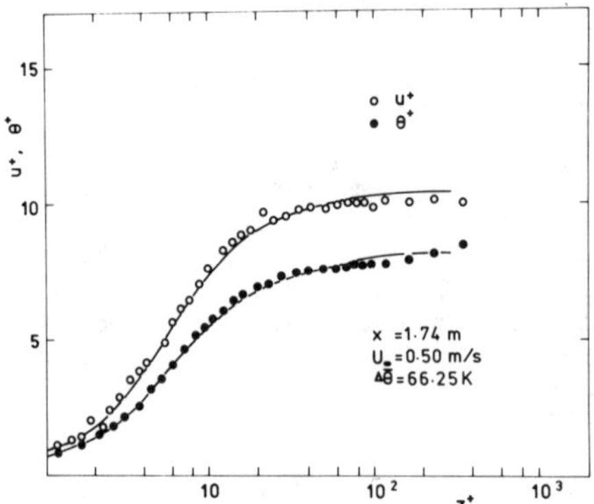

Fig.11-(b). Profiles of velocity and temperature for mixed convection

tion found by Clever and Busse [20]. As a result of the secondary motion like vortex-roll eddy diffusivities are enhanced in the vertical direction but decreased in the spanwise direction rather than in the neutral flow case.

CONCLUSION

(1) In mixed convection buoyancy forces enhance turbulent transport. The eddy diffusivities of momentum and heat, ε_H and ε_M, and the ratio $\varepsilon_H/\varepsilon_M$ increase. However, the stability dependence of $\varepsilon_H/\varepsilon_M$ is different between the wall and outer regions.

(2) The friction factor and Nusselt number for mixed convection are significantly larger than those for forced convection and do not change so much along the flow direction. They, as well as velocity and temperature profiles, are predicted well, taking into account the difference of the stability dependence between the inner and outer regions.

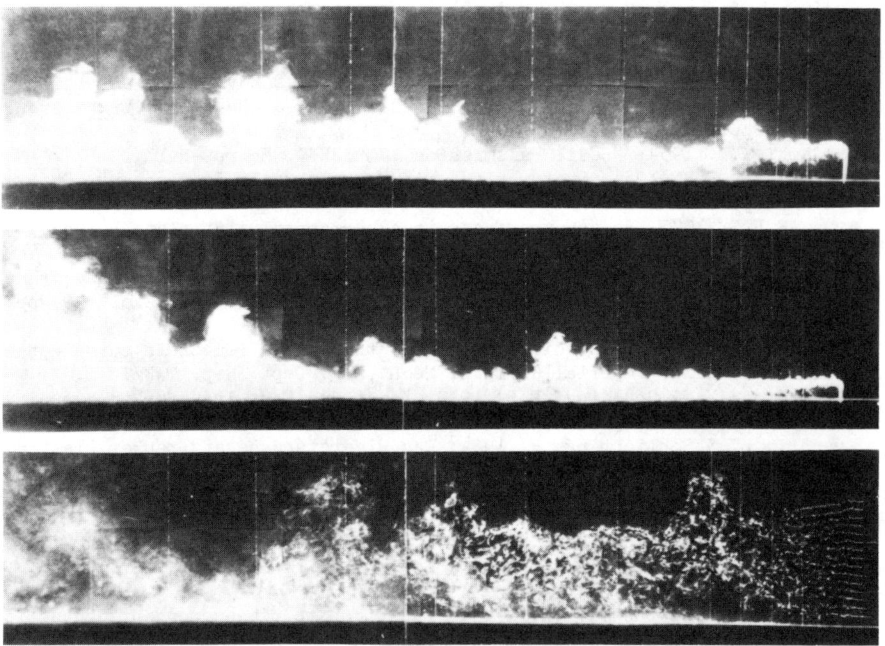

Plate 1. Flow visualization by a smoke method

REFERENCES

1. Mizushina, T., Ogino, F., Ueda, H. & Komori, S. 1980. Turbulence in unstably stratified flow, Submitted to J. Fluid Mech.
2. Mizushina, T., Ogino, F., Ueda, H. & Komori, S. 1978. Buoyancy effect on eddy diffusivities in thermally stratified flow in an open channel, Heat Transfer 1978, Vol. 1, pp.91-96.
3. Ueda, H., Mitsumoto, S., Komori, S. & Mizushina, T. 1980. Buoyancy effects on the turbulent transport processes in the lower atmosphere, submitted to Quart. J. Roy. Meteor. Soc.
4. Gibson, M.M. & Launder, B.E. 1978. Ground effects on pressure fluctuations in the atmospheric boundary layer, Int. J. Heat Mass Transfer, Vol. 86, pp. 491-511.
5. Townsend, A.A. 1972. Mixed convection over a heated horizontal plane. J. Fluid Mech. Vol. 55, pp.209-227.
6. Mizushina, T., Ogino, F., Ueda, H. & Komori, S. 1979. Application of laser Doppler velocimetry to turbulence measurement in non-isothermal flow. Proc. Roy. Soc. Lond. A366, pp.63-79.
7. Seban, R.A. & Roughty, D.L. 1956. Heat transfer to turbulent boundary layers with variable free stream velocity. Trans. ASME., Vol. 78, pp. 217-223.
8. Vliet, G.C. 1969. Natural convection local heat transfer on constant-heat-flux inclined surfaces. Trans. ASME., J. Heat Transfer, Vol. 91, pp.511-516.

9. Fujii, T. & Imura, H. 1972. Natural-convection heat transfer from a plate with arbitrary inclination. Int. J. Heat Mass Transfer, Vol. 15, pp.755-767.
10. Miller, R.M. & Gebhart, B. 1978. An experimental study of the natural convection flow over a heated ridge in air. Int. J. Heat Mass Transfer, Vol. 21, pp.1229-1239.
11. McAdams, W.H. 1954. "Heat Transmission", 3rd Ed., McGraw-Hill.
12. van Driest, E.R. 1956. On turbulent flow near a wall. J. Aeronaut. Sci. Vol. 23, pp.1007-1011.
13. Arya, S.P.S. 1975. Buoyancy effects in a horizontal flat-plate boundary layer. J. Fluid Mech. Vol. 68, pp.321-343.
14. Pruitt, W.O., Morgan, D.L. & Lourence, F.J. 1973. Momentum and mass transfers in the surface boundary layer, Quat. J. Roy. Met. Soc., Vol. 99, pp. 370-386.
15. Escudier, M.P. 1965. The distribution of the mixing length in turbulent flows near walls. Imperial College, Mech. Eng. Dept. Rep. TWF/TN/1.
16. Patankar, S.V. & Spalding, D.B. 1970. "Heat and Mass Transfer in Boundary Layers" Intertext Books, London.
17. Prandtl, L. & Schlichting, H. 1934. Das Widerstandsgesetz rauker Platten. Werft, Reederei, Hafen 1-4.
18. Schultz-Grunow, F. 1940. Pulsierender Durchfluss durch Rohre, Forschg. Ing. -Wes. Vol. 11, pp.170-187.
19. Kline, S.J., Reynolds, W.C., Schraub, F.A. and Runstadler, P.W. 1967. The structure of turbulent boundary layers. J. Fluid Mech. Vol. 30, pp. 741-773.
20. Clever, R.M. and Busse, F.H. 1974. Transition to time-dependent convection. J. Fluid Mech. Vol. 65, pp.625-645.

A Comparison of Spectral and Cospectral Characteristics of Dynamic and Thermal Turbulent Fields in Weakly Stratified Boundary Layers

PATRICE MESTAYER and MICHEL COANTIC
Institut de Mécanique Statistique de la Turbulence
12, avenue du Général Leclerc
13003 Marseille, France

JEAN-PIERRE PAGES and JOSEPH SAISSAC
Observatoire du Pic du Midi
65200 Bagneres de Bigorre, France

ABSTRACT

The turbulent fluctuations of temperature and two components of velocity have been measured with hot and cold wires in the I.M.S.T. Air-Sea Interaction Simulation Tunnel and at the Grau-de-Vendres micro-meteorological site in near neutral conditions. Direct and indirect estimates of the fluxes of momentum and sensible heat are compared. The non-dimensional production and dissipation functions for kinetic energy and temperature variance in Monin-Obukhov theory are estimated. The spectral and cospectral macrostructures are compared with the Kansas data. Fulachier & Dumas's extension of the Reynolds analogy to the spectral structures of the temperature and velocity fields is verified.

1. INTRODUCTION

The similarity theory of Monin and Obukhov (1953-1954) (see Monin and Yaglom, 1965-1971) postulates that, in the atmospheric surface layer where the vertical fluxes can be considered as constant, the structure of the turbulent field is dependent on only four parameters besides the elevation z over the ground : the air density, ρ, the surface shear stress, τ_0, and vertical heat flux, H_o, and a buoyancy parameter, g/θ_p, where g is the gravity and θ_p the absolute potential temperature. These five parameters can be combined to form the dimensionless variable z/L where L is Monin-Obukhov length scale

$$L = \frac{u_*^{\,3}}{\kappa(g/\theta_p).H_o/(\rho C_p)} \tag{1}$$

and κ is Von Karman's constant, introduced by Obukhov (1946). So, any variable representative of a property of turbulence at the elevation z, normalized by a combination of the above four parameters, or rather of the more practical parameters ρ, $u_* = (\tau_0/\rho)^{1/2}$, $H_o/(\rho C_p)$ and g/θ_p, can be expressed as a function of z/L that is supposed to be universal. This, in particular, leads to the "flux-profile relations" expressing the vertical mean gradients,

$$\partial\overline{U}/\partial z = \frac{u_*}{\kappa z} \quad \phi_m(z/L) \tag{2}$$

$$\partial\overline{\theta}_p/\partial z = \frac{\theta_*}{\kappa z} \quad \phi_h(z/L) \tag{3}$$

where $\theta_* = -\overline{w'\theta'}/u_* = -H_o/(\rho C_p u_*)$. Similarly, the dissipation rate ε of turbulent kinetic energy, and the destruction rate χ_θ of temperature variance $\overline{\theta'^2}$

can be expressed by means of two "universal" functions, after

$$\varepsilon = \frac{u_*^{\,3}}{\kappa z}\,\phi_\varepsilon\,(z/L)\ ,\ \chi_\theta = \frac{2\,u_*\,\theta_*^{\,2}}{\kappa z}\,\phi_\chi\ (z/L) \tag{4}$$

Theoretical considerations and a number of experimental investigations have led to several propositions for these four functions (see a survey of these works in Yaglom, 1977). The results of the 1968 Air Force Cambridge Research Laboratories experiments in Southwest Kansas have been extensively studied to determine the flux-profile relationships (Businger et al., 1971), the parameterisation of the terms in the turbulent kinetic energy and temperature variance budget equations (Wyngaard & Coté, 1971) and the variations with stability of the spectra and cospectra (Kaimal et al., 1972 ; Wyngaard & Coté, 1972). Although these results are often considered as standards, they present some uncertainties, mainly due to instrumental problems (Haugen et al., 1971 ; Lynch & Bradley, 1974), in particular about their estimation of Von Karman's constant: 0,35 instead of 0.40, its commonly accepted value (Wieringa, 1980).

Indeed, measurements in near neutral conditions are not easy in the atmosphere : on one hand the very low levels of temperature fluctuations imply the use of probes having very high signal-to-noise ratios ; on the other hand, these conditions are often associated with unstationarities which raise problems during the signal processings to define integration times. The I.M.S.T. Air-Sea Interaction Tunnel offers interesting possibilities to simulate the near neutral surface layer in the 0.75 meter thick fully developed turbulent boundary layer over the water surface (Coantic et al., 1969). Indeed, the measurements that we present here were made at a wind velocity of 7.5 m/s with a ± 10°C difference between the air and water surface temperatures and fluctuations of 0.5°C r.m.s., while the stability parameter z/L had values of only +0.006 and -0.007 at the measurement level $z = 0.25$m.

We present here the results of turbulence measurements realized in the tunnel with a multi-hot-wire probe and on the Grau-de-Vendres micro-meteorological site with the same probe and a sonic anemometer, in near neutral conditions. The estimates of the ϕ's are presented, along with estimates of the momentum and heat fluxes obtained by direct and indirect methods. The spectra and cospectra of u', w' and θ' are compared, in non-dimensional variables, with the curves proposed by Kaimal et al. (1972) for the neutral stratification. We also present a comparison between thermal and dynamic spectral characteristics in the frame of the spectral extension of the Reynolds analogy by Fulachier and Dumas (1976).

2. EXPERIMENTAL PROCEDURES

2.1. The I.M.S.T. Tunnel

The Air-Sea Interaction Simulation Tunnel has been described (Coantic et al., 1969) as a micro-meteorological wind tunnel combined with a wind-wave tank. The floor of the air test-section is the water surface in the tank, so that turbulent boundary layers 40m long are generated over the air-water interface and under the solid wall of the test-section ceiling. The experiments took place in a section 34.20 m downstream of the initial section. The values of the main parameters of the flow in the boundary layer are given in Table 1, under the names ME.MO.MI. IV.1 for the weakly stable experiment and ME.MO.MI. IV.2 for the weakly unstable one (ME.MO.MI. stands for Mesures de Moments Mixtes). The mean temperatures of air and water and the air humidity content were

Run	z m	$\overline{U}$ m/s	$\overline{\theta}$ °C	$\partial\overline{U}/\partial z$ s^{-1}	$\partial\overline{\theta}/\partial z$ °C/m	z/δ	U_∞ m/s	$\theta_{surf.}$ °C
ME.MO.MI. IV.1	.25	7.5	23.8	4.0	4.5	.333	8.86	15.0
ME.MO.MI. IV.2	.25	7.4	17.1	3.9	-3.7	.333	8.74	24.5
ME.MO.MI. II.1	7.3	3.9	11.8	.11	.0058			
ME.MO.MI. II.2	7.3	3.9	11.6	.084	.0089			

Run	$\overline{u'^2}^{1/2}$ m/s	$\overline{w'^2}^{1/2}$ m/s	$\overline{u'^2}^{1/2}/\overline{U}$	ε m^2s^{-3}	χ_θ $°C^2s^{-1}$	η mm	R_λ
IV.1	.70	.54	.09	.653	.476	.27	616
IV.2	.67	.52	.09	.634	.398	.27	561
II.1	.62	.38	.18	$7.55\ 10^{-3}$	$.81\ 10^{-4}$	.78	5670
II.2	.66	.39	.15	$8.04\ 10^{-3}$	$1.85\ 10^{-4}$	.78	4700

Table 1. Mean and turbulent parameters.

controlled by the tunnel thermal regulations and accurately measured in the experiment section with thin thermocouples and an E.G. & G. 137C3 dew-point meter. The mean velocity profiles were measured with Pitot tubes and hot-wire anemometers, the mean temperature profiles with a thin thermocouple and a cold-wire thermometer (Figure 1). They are in fairly good agreement with the data of Selva (1979).

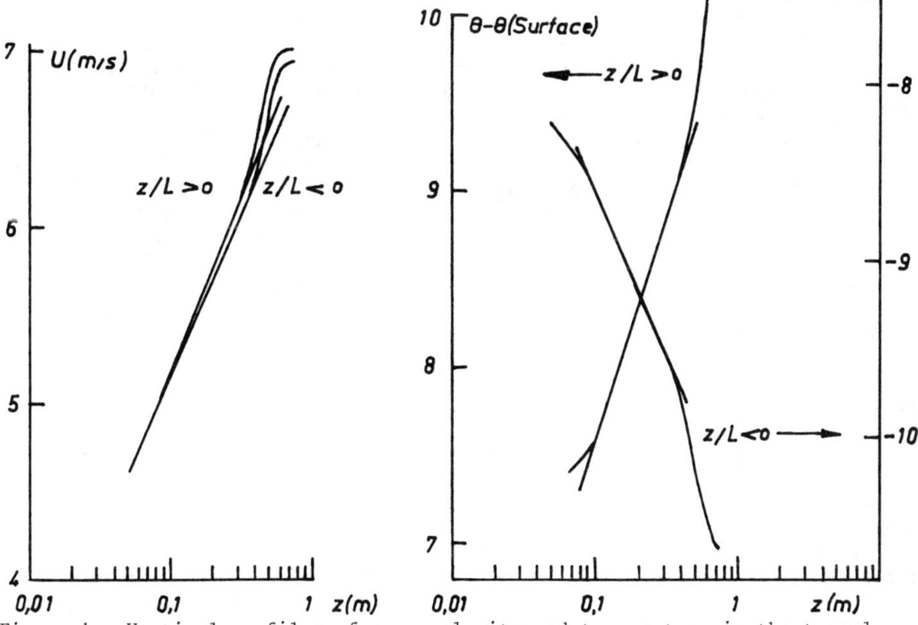

Figure 1. Vertical profiles of mean velocity and temperature in the tunnel.

2.2. The Micro-Meteorological Site

The experimental site of the Observatoire du Pic du Midi group is located at Grau-de-Vendres, 30 km south of Béziers (Hérault), 1 km north of the river Aude mouth in the Mediterranée Sea. The wind was blowing from W - NW over a dried marsh. The ground is flat, scarcely sprinkled with dwarf bushes, for a fetch of at least 25 km in this direction. The site is equipped with a 50 m high tower, a 12 m telescopic orientable mast and automatic recording systems. The tower instrumentation includes, at five levels, Wheather Measure propeller aerovanes and thermocouples. Their signals are electronically corrected for their non-linear response, processed through damping circuits with 2 minutes time constants, sampled and digitally recorded. The telescopic mast is located at 50 m from the tower, in the north direction. At the top of the mast a bar supported the turbulence measurement instrumentation : a 2-way E.G.&G. sonic anemometer, a fast response bi-vane and the I.M.S.T. turbulence array including the multi-wire probe, a thin thermocouple connected to an electronic ice-point unit and a thin Pitot tube connected to an A.C.B. electric anemometer. The mast head orientation is remotely actuated so that, by looking at the bi-vane output records, it is possible at anytime to set the instrumentation in the average wind direction. During our measurements the instruments were raised at the height of 7.3 m and the vertical positioning was carefully checked. The sonic anemometer and the propeller aerovanes had been calibrated in the large tunnel of the Laboratoire de Banlève at the E.N.S.E.E.I.H.T. of Toulouse. The 2-way sonic anemometer was set as a vertical X and the signals were electronically decoded and linearized to obtain the values of the longitudinal and vertical components of the instantaneous velocity as well as the fluctuating temperature. These last signals were input in an automatic covariance computation device ; they were also recorded for the further digital processing.

2.3. The Multi-Wire Probe

The probe is composed of three individual hot wire sensors : a single hot-wire, a temperature cold wire and a double X-wire sensor. The three of them are mounted on sliding guides allowing to set them in an optimal configuration ; the measuring volume is 2 mm wide and 1.5 mm high and deep. The cold wire is slightly ahead of the hot-wires. The hot-wires were DISA 55 P 01 and 55 P 61 probes (1.25 mm in length, 5 μ in diameter). The Wollaston cold wire was 0.6 μ in diameter and 0.3 mm in length ; it was operated in a constant current bridge with a current of .6 mA. The performances of these probes and the choices of their operating parameters are described in Mestayer & Chambaud (1979) : with the help of a digital processing to simultaneously linearize the signals and separate the variables, $u'(t)$, $w'(t)$ and $\theta'(t)$ can be measured with no contamination up to about the Kolmogorov frequency $\overline{U}/2\pi\eta$. The effective spatial resolution (Mestayer, 1980) allowed measurements of scales as small as the Kolmogorov micro-scale η (see Table 1).

2.4. The Signal Processing

The turbulent signals from the hot wire anemometers and cold wire thermometer were, after adequate low-pass filtering, digitized on line during the laboratory experiments. For each of the two experiments, three successive recordings were made after sampling at the rates of 312.5, 2500 and 10000 samples per second, with a 15 bit Preston GMAD-3 analog-to-digital converter. Each digital record contained 299 520 samples of each variable, their durations being respectively 958, 120 and 30 seconds. The spectra and cospectra were computed by means of complex Fast Fourier Transform routines with an equivalent number of degrees of freedom of 585. On site, after proper analog conditioning (low pass filtering and amplification) the multi-wire probe signals were simultaneously recorded with the signals from the sonic anemometer and the slow response probes (bi-vane, thermocouple and Pitot tube) on two analog FM recorders : a 14 track Ampex FR 1300 and a 7-track Schlumberger MP 5522 at 76 cm/s with a 51 db signal-to-noise ratio. Two 14 minutes reels have been selected, corresponding to runs ME.MO.MI. II.1 and 2 in Table 1. These tapes have been played back in the I.M.S.T. laboratory and the preconditioned signals were successively digitized at 2.5, 10, 80, 625 and 5000 samples per second. In order to analyse only a reasonable number of data, the recordings were not continuously sampled at the highest two rates. Only segments of records were digitized. The segments had lengths of 4.92 seconds at 625 Hz and 0.92 seconds at 5000 Hz and were equally spaced over the 14 minutes of the recordings. The equivalent numbers of degrees of freedom of the spectra and cospectra computed in the five frequency bands are respectively 4, 17, 132, 585 and 585. With that procedure there was strictly no need for any artificial adjustement between the estimates in the various frequency bands. The spectra and cospectra, computed from the three or five digital recordings, were only "cleaned" of the noises and the filters roll-offs (Mestayer, 1980). For the tunnel spectra, eleventh order polynomial fits were computed by least squares regression of the spectral density estimate logarithms. For all other curves, the smoothings were obtained by moving averages on logarithmically increasing numbers of estimates.

3. PROFILES, FLUXES AND DISSIPATION RATES

3.1. Vertical Profiles

The vertical profiles of mean velocity and temperature measured in the tunnel boundary layer exhibit the typical logarithmic shape (Figure 1). The Figure 2 shows the vertical repartitions of wind velocity and potential temperature measured at the five levels of the site tower and averaged over the 14 minutes of the two ME.MO.MI. II runs. They do not present the regular logarithmic shapes. One must note that the potential temperature differences are very small. The vertical repartitions of the gradient Richardson number seem to indicate that the measurements were made during the beginning of the night fall transition : a thermal inversion was developing at the lowest levels, with a weak stability growing with the wind abating. This tendency increased afterwards leading to the disparition of the dynamic stationnarity, which was still acceptable during the two runs. The time traces of the signals from the anemometers and bi-vane indicate that the wind direction was well centered, normal to the probes, with maximum excursions of ± 10 degrees. Actually only the measurements from the lower two levels of the tower were used to compute the mean gradients $\partial \bar{U}/\partial z$ and $\partial \bar{\theta}_p/\partial z$.

3.2. Friction Velocity and Scaling Temperature

In the constant stress layer, the vertical fluxes of momentum and heat follow the relations

$$\begin{aligned} \tau &= -\rho\,\overline{u'w'} = \rho\, u_*^{\,2} = \tau_o \\ H &= \rho\, C_p\, \overline{w'\theta'} = -\rho\, C_p\, u_*\, \theta_* = H_o \end{aligned} \qquad (5)$$

which define the friction velocity u_* and the scaling temperature θ_*. The covariances $\overline{u'w'}$ and $\overline{w'\theta'}$ can be obtained by averaging the instantaneous products of the centered velocity and temperature signals. Yet, this method necessitates a careful preliminary study of the integral scales. Mestayer et al. (1978) have

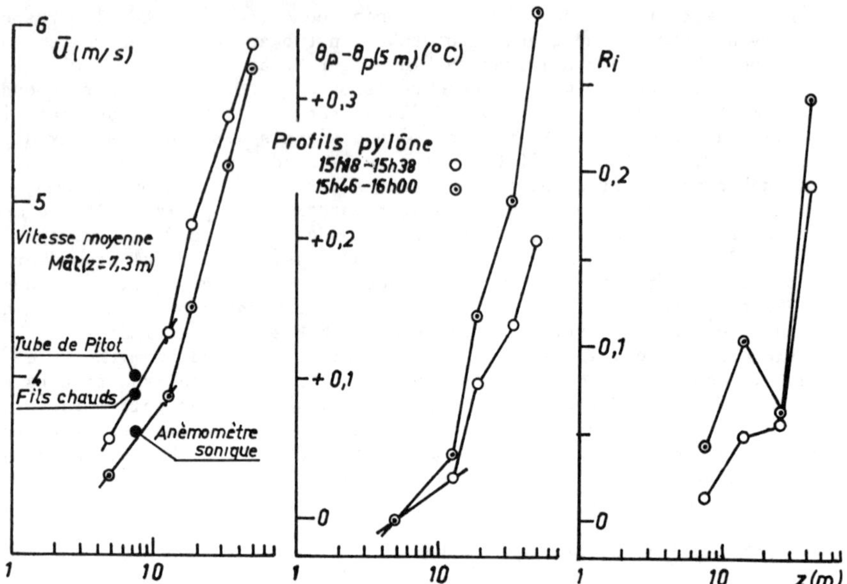

Figure 2. Vertical repartitions of mean velocity, temperature and Richardson number in the atmosphere

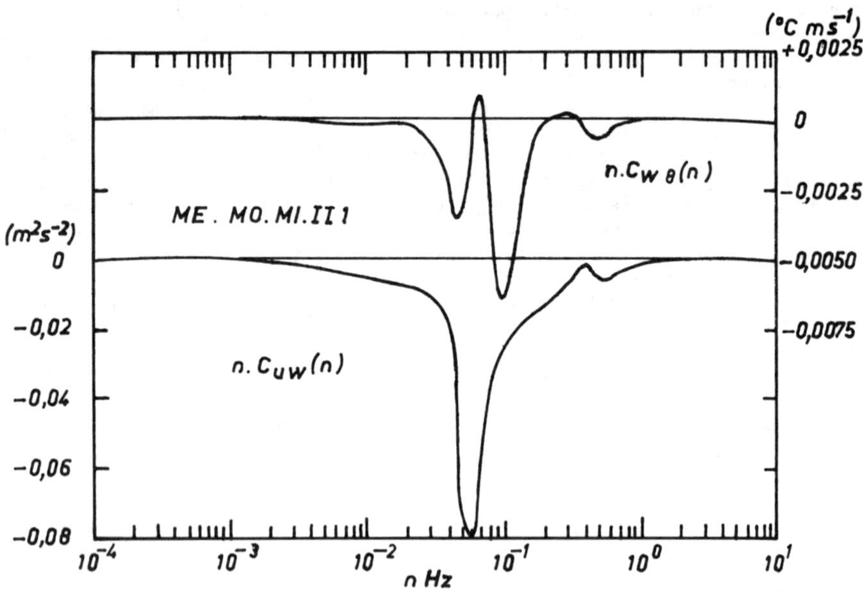

Figure 3. u,w and w,θ cospectra in the atmosphere

shown that an *a priori* choice of these scales could lead to large errors in the flux determination : this will also be illustrated in § 4.3. So in both laboratory and in situ experiments, the covariances have been estimated by means of the cospectra of the un-centered signals

$$\overline{u'w'} = \int_0^\infty C_{u,w}(n)\, dn \quad , \quad \overline{w'\theta'} = \int_0^\infty C_{w,\theta}(n)\, dn \tag{6}$$

computed from the frequencies of 0.61 Hz in the tunnel and 0.005 Hz in the atmosphere and averaged over the whole records. This method allows to determine the lowest frequencies to take into account, to eliminate the noise effects and to detect eventual contributions of opposite signs. In the tunnel, the cospectra have regular shapes between 0.6 and 1000 Hz, as will be shown in § 4.4. In the atmosphere, the smoothed cospectra of Figure 3 (first run) show that some scales, close to the maximum of the momentum cospectrum, contribute to a positive "counter-gradient" heat flux. These contributions could be remnants of the pre-inversion turbulent mechanisms, thus indicating that the transition was not a continuous process but rather a phenomenon beginning close to the surface and growing simultaneously in scale and altitude. In the flux computation, we nevertheless took into account all the lobes of the w'θ'cospectra, the positive ones being quite small.

3.3. Dissipation Rates

The dissipation rate of kinetic energy $\varepsilon = \nu\, \overline{\frac{\partial u_i}{\partial x_j}\left(\frac{\partial u_i}{\partial x_j} + \frac{\partial u_j}{\partial x_i}\right)}$ and the destruction rate of temperature variance $\chi_\theta = \mathcal{D}_\theta\, \overline{\frac{\partial \theta}{\partial x_i} \cdot \frac{\partial \theta}{\partial x_i}}$ (ν is the kinematic viscosity of air and $\mathcal{D}_\theta$ the molecular diffusivity for temperature) have been estimated in the tunnel from the derivative integrals

$$\varepsilon = \frac{15\,\nu}{\overline{U}^2} \int_0^\infty E_{\partial u/\partial t}(n)\, dn \quad , \quad \chi_\theta = \frac{6\,\mathcal{D}_\theta}{\overline{U}^2} \int_0^\infty E_{\partial\theta/\partial t}(n)\, dn \tag{7}$$

which assumes that local isotropy is verified and Taylor's hypothesis is valid. There are however converging indications that the local isotropy assumption is not valid for temperature (Mestayer et al., 1976 ; Mestayer, 1980). As for Taylor's hypothesis, the works of Lumley (1965) and Wyngaard & Clifford (1977) would indicate a 6% over-estimation of ε and a 2% over-estimation of χ_θ. The temperature cold vire was able to measure even the smallest scales of the flow without attenuation, its length being close to the Kolmogorov scale η and its time constant being around 10 µs. The straight hot wire, being operated in a constant temperature bridge had a sufficient frequency response (Mestayer & Chambaud, 1979). Its spatial integration ($1/\eta \simeq 4$) results in an attenuation of about 8% in the estimation of ε (Wyngaard, 1968) which nearly compensates the overestimation due to Taylor's hypothesis. As a consequence no correction has been applied on the results obtained from equations (7) for the tunnel measurements.

In the atmosphere, the Kolmogorov scale is much larger (see Table 1) so that the spatial integration was negligible. As the turbulence intensity was rather large (15 to 18%) the values of ε and χ_θ have been corrected for Taylor hypothesis overestimations (8 and 5 %, according to Wyngaard & Clifford, 1977). Actually, since the high frequency spectrum of $\partial\theta/\partial t$ presented a large amount of noise picked-up by the wire itself from a nearby radar emission, the value of χ_θ has been estimated in the inertial sub-range of the temperature spectrum from the relation

$$E_\theta(n) = (2\pi/\overline{U})\ \beta_\theta\ \varepsilon^{-1/3}\ \chi_\theta\ (2\pi n/\overline{U})^{-5/3} \tag{8}$$

with $\beta_\theta = 0.41$ (Champagne et al., 1977).

4. RESULTS AND DISCUSSION

4.1. Flux estimations

The values of the vertical fluxes of momentum and heat estimated by three methods in the tunnel and four methods in the atmosphere are compared in Table 2. As already stated the cospectral method is but a properly controlled direct method. The cospectra have been computed from the linearized and decontaminated temperature signal from the cold-wire in the multi-wire probe and from the u' and w' decontaminated signals from the X-wires in that probe, in the tunnel. In the atmospheric experiment, we had doubts about the determination of the X-wire sensitivities (Mestayer, 1980) so that we combined the cold-wire temperature signal to the u' and w' sonic anemometer signals.

The automatic covariances were computed from the u', w' and θ' sonic anemometer signals electronically multiplied and centered about 2 minutes averages, then averaged over the 14 minutes of the runs

$$\overline{u'w'} = \overline{\overline{uw}^{\,2\,\text{min.}} - \overline{u}^{\,2\,\text{min.}} \cdot \overline{w}^{\,2\,\text{min.}}}^{\,14\,\text{min.}} \tag{9}$$

This system is routinely used for the long duration surveys of the fluxes.

The profile method derives the fluxes from the relations

$$\begin{aligned} -\overline{u'w'} &= u_*^{\,2} = (\kappa z)^2 \left(\frac{\partial \overline{U}}{\partial z}\right)^2 \\ \overline{w'\theta'} &= -\,u_*\,\theta_* = \frac{(\kappa z)^2}{0.74}\,\frac{\partial \overline{U}}{\partial z}\cdot\frac{\partial \theta_p}{\partial z} \end{aligned} \tag{10}$$

where θ_p is the mean potential temperature, equal in the tunnel to the mean temperature.

	$-\rho\overline{u'w'}$ (kg m s^{-1})/(m^{2}s)				$\rho\, C_p\, \overline{w'\theta'}$ cal/(m^{2}s)			
ME.MO.MI.	IV.1	IV.2	II.1	II.2	IV.1	IV.2	II.1	II.2
Cospectra	0.21	0.19	0.12	0.084	37.5	26.0	-1.9	-2.3
Automatic Covariances			0.17	0.21			-0.18	-0.81
Profiles	0.20	0.19	0.13	0.075	73.6	59.0	-2.2	-2.6
Dissipation	0.16	0.18	0.080	0.063	29.9	29.0	-1.8	-2.2

Table 2. Flux estimates

The dissipative method is derived from the relation obtained by the combination of expressions (4) and (1),

$$\frac{\kappa z \; g \; (\kappa z \chi_\theta)^{1/2}}{\sqrt{2}\; \theta_p \; (\kappa z \varepsilon)^{5/6}} = (z/L)\, \phi_\chi^{1/2} \phi_\varepsilon^{-5/6} = F(z/L). \tag{11}$$

Taking into account that the transport term in the budget equation of $\overline{\theta'^2}$ is usually small (Wyngaard & Coté, 1971), one can replace $\phi_\chi(z/L)$ by $\phi_h(z/L)$ in relation (11). Using the semi-empirical relations proposed by Businger et al. (1972) for ϕ_h and ϕ_ε, Champagne et al. (1977) have computed the function F(z/L) for z/L < 0. Using Businger et al.'s relations for the stable cases, after the corrections for the value of Von Karman's constant suggested by Yaglom (1977),

$$\phi_\varepsilon(z/L) = 1.14 \left(1 + 2.31 \; (z/L)^{3/5}\right)^{3/2} \tag{12}$$

$$\phi_h(z/L) = 1.14 \; (0.74 + 4.1 \; z/L), \tag{13}$$

it comes

$$F_s(z/L) = (1.14)^{-1/3} (z/L)(0.74 + 4.1 \; z/L)^{1/2} (1 + 2.3 \; (z/L)^{3/5})^{-5/4} \tag{14}$$

This function, given by Mestayer (1980), allows to estimate z/L from z, θ_p, ε and χ_θ. The relations (5), (4), (12) and (13) can then be used to compute the fluxes.

We shall consider the cospectral estimates as the reference values in the discussion of Table 2 results. The dissipative method leads to under-estimations in all stable cases ; this could be due to the fact that relations (12) and (13) lead to the values 1.14 and 0.84 for $\phi_\varepsilon(o)$ and $\phi_h(o)$, instead of 1 and 0.74. The profile method leads to relatively good estimates except for the heat flux in the tunnel. The temperature profiles give an over-estimation by a factor around 2 which is not clearly explained. Selva (1979) has measured the stress, heat transfer and water vapour transfer coefficients by the profile and integral methods. His results indicate that the ratios of the estimates by the two methods increase along the tunnel length : their values remain smaller than 1.3 for the stress and water vapour coefficients but can be greater than two for the heat transfer coefficient, especially for the most unstable conditions: in this case tri-dimensional convectives effects are supposed to induce important lateral heat fluxes.

As for the atmospheric data the relatively good estimations by the profile methods appear satisfactory. They seem to indicate that this very commonly used method is still valid locally when the profiles do not perfectly follow the logarithmic law. The automatic covariance method appears to give poor estimations, especially for the heat fluxes. This can be due to the poor quality of the temperature signal from the sonic anemometer when the temperature fluctuation level is very low and the velocity fluctuation level relatively high. This certainly also results from the automatic procedure with a fixed time constant which cannot be well adapted to all conditions.

4.2. Non-Dimensional Functions

The table 3 presents the values of the non-dimensional functions defined by the relations (2), (3) and (4), which are the normalized production and dissipation rates of kinetic energy and temperature variance. For the neutral stability (z/L = 0) the classical values of ϕ_m and ϕ_h are A and 0.74 and the dissipation and production rates are expected to be about in balance. The laboratory data ME.MO.MI. IV.1 and 2 for very small z/L are rather close to the expected values, with the exceptions of the ϕ_h functions which have values more than twice greater than 0.74 because of the temperature gradient anomaly. In the atmosphere, the kinetic energy functions have values around 1, with a large dispersion of ϕ_ε. As for the temperature functions, the non-dimensional rates of dissipation largely over-balance the production rates and the latter have values neatly greater than 0.74, which is in agreement with the experimental determinations reported by Yaglom (1977).

	z/L	$\phi_m = \frac{\kappa z}{u_*}\frac{\partial \bar{U}}{\partial z}$	$\phi_\varepsilon = \frac{\kappa z \varepsilon}{u_*^3}$	$\phi_h = \frac{\kappa z}{\theta_*}\frac{\partial \theta_p}{\partial z}$	$\phi_\chi = \frac{\kappa z \chi_\theta}{2u_*\theta_*}$
ME.MO.MI IV.1	0.0057	0.98	0.92	1.55	0.69
IV.2	-0.0073	1.00	1.05	1.45	0.88
ME.MO.MI II.1	0.021	1.04	0.80	0.84	0.95
II.2	0.043	0.95	1.27	0.90	1.24

Table 3. Normalized production and dissipation rates

4.3. Spectra and Cospectra

Kaimal et al. (1972) have shown that the Kansas experiment spectra and cospectra converge when adequately normalized. We compare here our laboratory and atmosphere results to the spectral "empirical" curves proposed by Kaimal et al. for the neutral stratification, in the appropriate non-dimensional variables, i.e. the reduced frequency $nz/\bar{U}$ and the spectra and cospectra normalized by u_*, θ_*, ϕ_ε and ϕ_χ and multiplied by the frequency n.

The velocity spectra are actually normalized by the only corresponding dissipation rate , as

$$u_*^2\ \phi_\varepsilon^{2/3} = (\kappa z \varepsilon)^{2/3}. \tag{15}$$

All our u spectra were computed from the signals of the straight hot-wire in the multi-wire probe which were also used to estimate ε. The laboratory w spectra were measured with the X-wire probe. The w spectra in the atmosphere come from the sonic anemometer except for their high frequency parts which are the X-wire w spectra adjusted to match the levels of the sonic spectra. The first evidence is that the inertial sub-range ends at much lower reduced frequencies in the laboratory than in the atmosphere, as it can be expected at the view of the turbulence Reynolds numbers (Table 1). The u spectra (Figure 4) show a good agreement between the laboratory and Kansas results in the inertial sub-range but smaller energy levels at the lower frequencies in the laboratory. The atmospheric data agree very well with the shape of the Kansas curve on the whole range of frequencies but with slightly lower levels, leading to lower estimates of the Kolmogorov constant α_1 in the inertial sub-range

$$E_1\ (k_1) = \alpha_1\ \varepsilon^{2/3}\ k_1^{-5/3} \tag{16}$$

Our values of α_1 are 0.41 and 0.44 instead of 0.50, the Kansas average value.

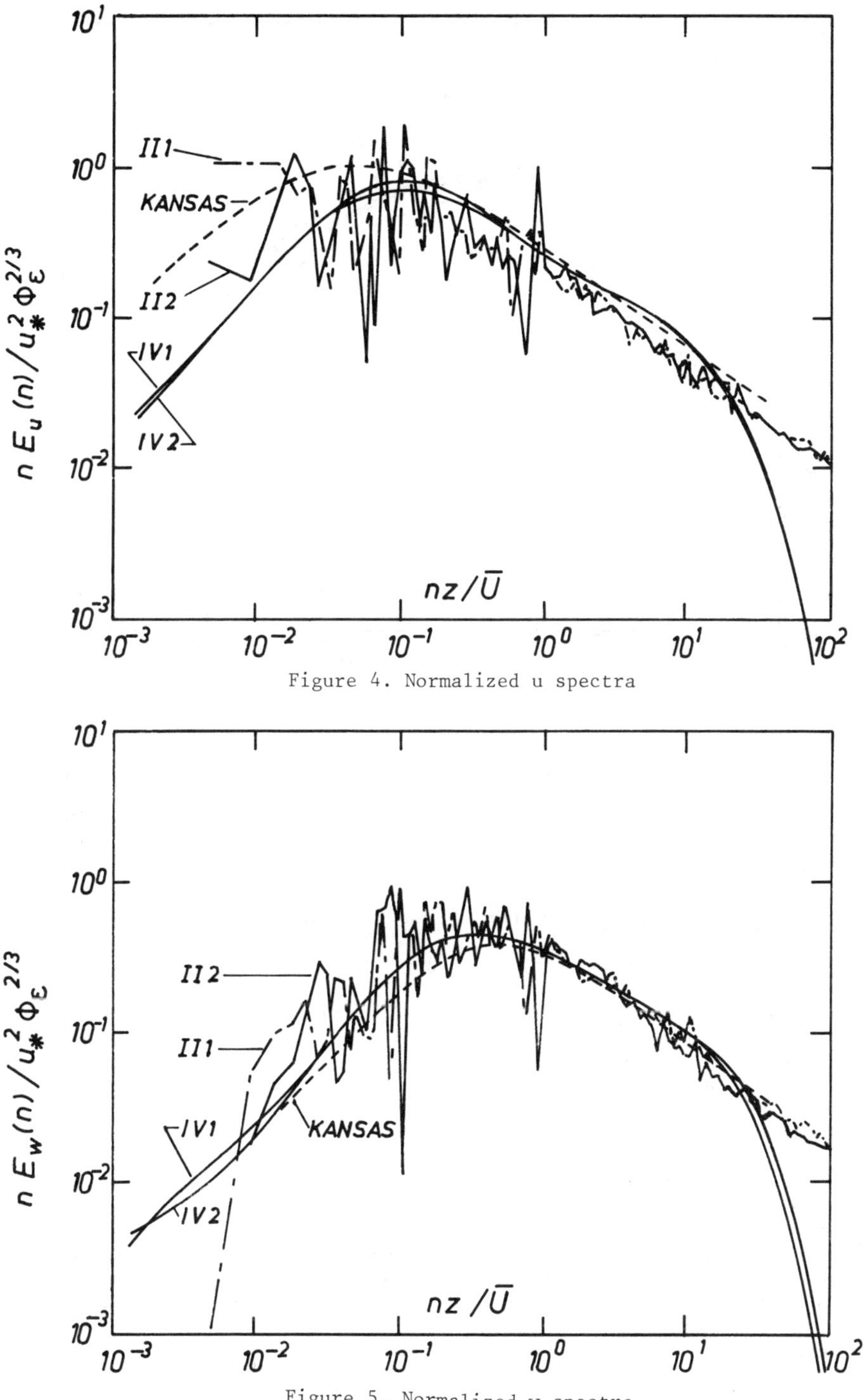

Figure 4. Normalized u spectra

Figure 5. Normalized w spectra

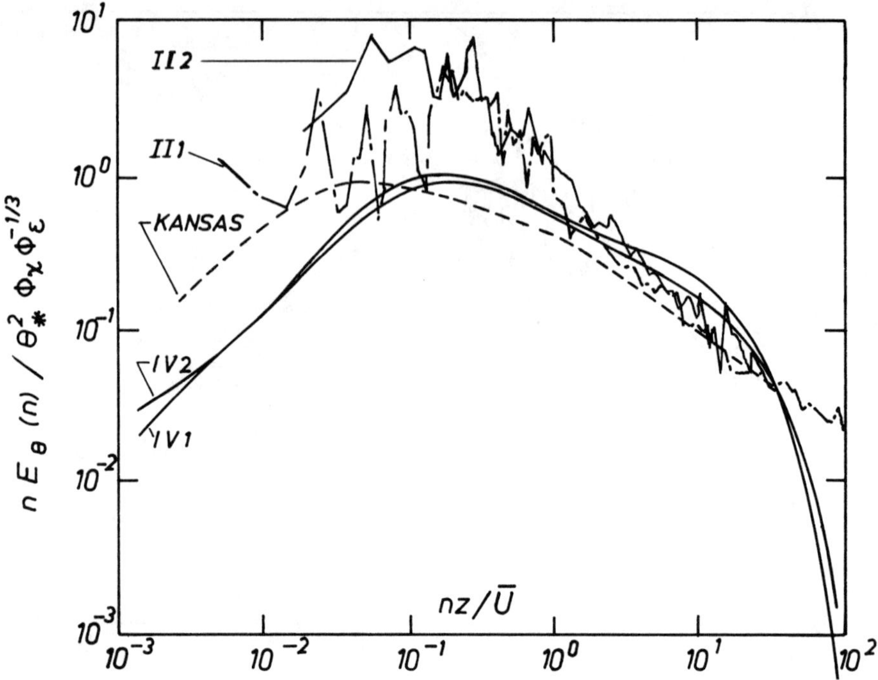

Figure 6. Normalized temperature spectra

On Figure 5, all the w spectra present a rather good agreement on the whole range of reduced frequencies, except naturally for the laboratory highest frequencies. This especially shows that the laboratory simulation is satisfactory in the range of frequencies around the maximum of turbulent kinetic energy production ($0.1 < nz/\overline{U} < 1$).

The temperature spectra are normalized by ε and χ_θ, as

$$\theta_*^{\,2}\,\phi_\chi\,\phi_\varepsilon^{-1/3} = (\kappa z \chi_\theta)(\kappa z \varepsilon)^{-1/3} / 2 \,. \qquad (17)$$

Actually, the Kansas curve is normalized by ϕ_h with the assumption that ϕ_h and ϕ_χ are equal. Our spectra (Figure 6) exhibit large departures from the Kansas curve. The laboratory spectra fall down more rapidly than the atmospheric curves not only on the high frequency side but also at the lower frequencies. The behaviour is also different in the inertial sub-range, with a slope slightly less than -2/3 and a level higher than that of the Kansas curve by a factor of nearly two. The position of the maximum is also different, close to $nz/\overline{U} = 0.2$, when the Kansas maximum occurs around $nz/\overline{U} = 0.05$. The atmospheric results of ME.MO.MI. II also differ significantly from the Kansas curve. In the frequencies around the maximum they seem to behave rather like the unstable spectra observed by Kaimal et al. (1972). These shapes could be attributed to those remnants of unstable structures which have been traced in the cospectrum positive lobe (Figure 3). In the inertial sub-range, the ME.MO.MI. II curves have more usual behaviours with slopes close to -2/3 and levels midway between our laboratory results and the Kansas curve.

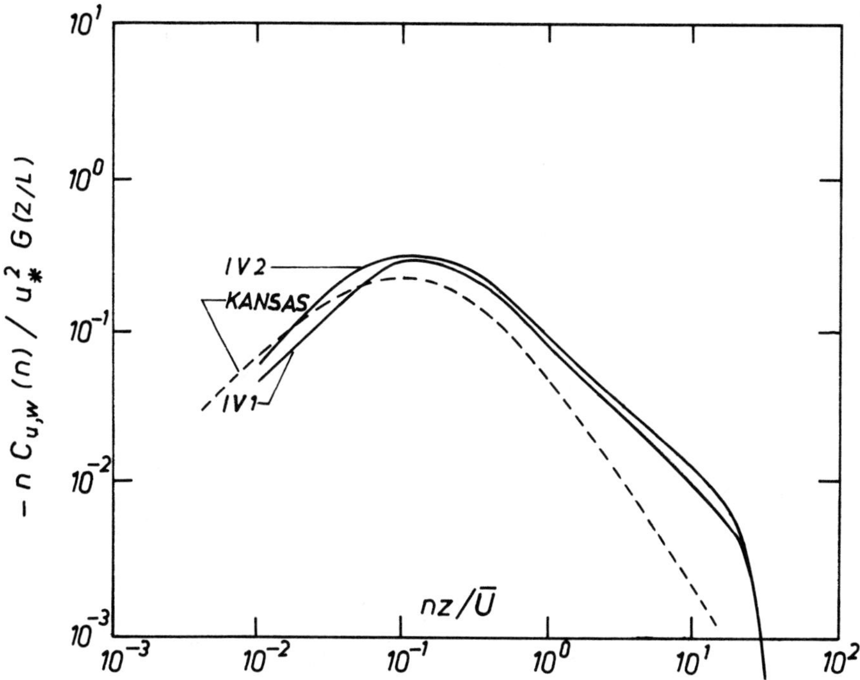

Figure 7. Normalized u, w cospectra

The cospectra are normalized by u_* and θ_* and by Kaimal et al.'s "regrouping" functions G, H and K, equal to 1 for $z/L < 0$ and, for $z/L > 0$, defined by the relations

$$G(z/L) = 1 + 7.9\ z/L\ ,\quad H(z/L) = 1 + 6.4\ z/L\ ,\quad K(z/L) = 1 + 17.4\ z/L.$$

Since, in the laboratory experiment the values of z/L are very small, the cospectra $C_{u,w}(n)$ (Figure 7) and $C_{w,\theta}(n)$ (Figure 8) are nearly normalized by their own integrals. So it is not clear why the ME.MO.MI. IV and Kansas curves have such different levels except if we consider that the Kansas data were normalized by the values of u_* estimated from the drag plate measurements. As for the slopes of the cospectra, our laboratory results indicate less steep power laws than expected from the theoretical considerations of Wyngaard & Coté (1972) and from the "empirical" Kansas curves : -0.94 and -0.88 for ME.MO.MI IV.2 u, w cospectra instead of -1.33 (Figure 7), also -0.88 for the w,θ cospectra instead of -1.33 (Figure 8) and -1.25 for one u, θ cospectrum (Figure 9) instead of -1.5 for the Kansas curve and -2 for Wyngaard & Coté (1972), although the unstable u,θ cospectrum indicates a rapid high frequency roll-off. We must note that Wyngaard & Clifford (1978) estimate that Taylor's hypothesis could be a cause of large errors in the determination of the high frequency parts of the cospectra. As these errors depend on the turbulence intensity, a closer examination of the cospectral behaviours would certainly need first an improvement

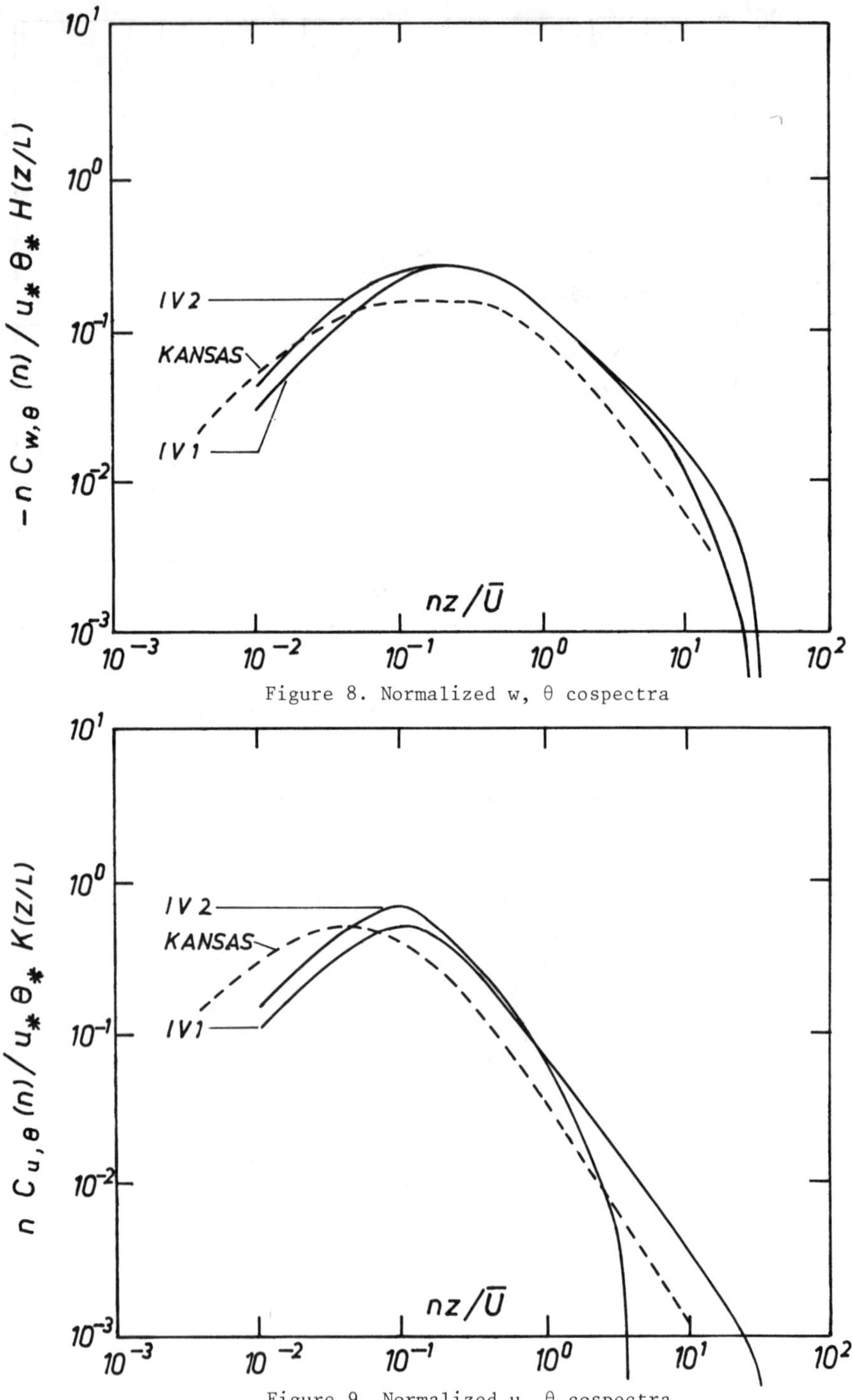

Figure 8. Normalized w, θ cospectra

Figure 9. Normalized u, θ cospectra

of the time-space relation.

4.4. Reynolds Spectral Analogy

The assumption that the mechanisms of turbulent transport are the same for momentum and heat is the basis of Reynolds's analogy which has been used to compute the mean temperature distribution from the mean velocity distribution in a large number of flows. Fulachier & Dumas (1976) have shown that this analogy can be extended to the spectral repartitions of temperature and velocity vector fluctuations in a low Reynolds number boundary layer over a slightly heated plate. Fulachier (1978) has shown that this extension of the analogy is also valid in various turbulent boundary layers where it can be assume that the temperature fluctuation is a passive scalar contaminant, especially in moderately stable atmospheric surface layers. We tested the spectral analogy for the weakly unstable tunnel flow where the three velocity component spectra have been measured. The comparison of Figure 10 between these spectra and that of the temperature, normalized by their variances, indicate that they neatly differ except for the highest frequencies for which there seems to be an agreement between the temperature and the transverse component spectra. Yet, there is a very good agreement between the temperature spectrum and the velocity vector frequency spectrum $E_q(n)$ (Fulachier & Dumas, 1976) obtained by addition of the three contributions to the turbulent kinetic energy at each frequency

$$E_q(n) = E_u(n) + E_v(n) + E_w(n) \tag{18}$$

normalized by the squared velocity modulus, $\overline{q'^2} = \overline{u'^2} + \overline{v'^2} + \overline{w'^2}$.

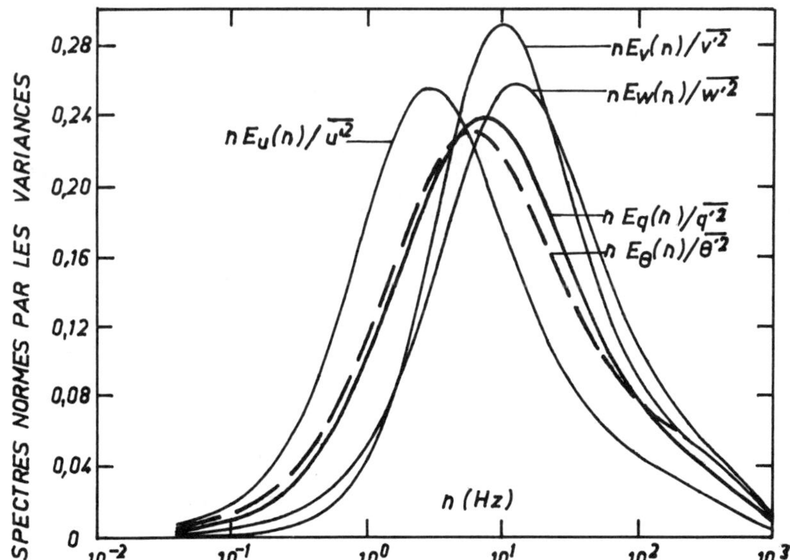

Figure 10. Velocity and temperature spectra normalized by the variances.
$E_q(n) = E_u(n) + E_v(n) + E_w(n)$, $\overline{q'^2} = \overline{u'^2} + \overline{v'^2} + \overline{w'^2}$

5. CONCLUSIONS

This ensemble of experimental results obtained with the same probes in the I.M.S.T. tunnel and on the Grau-de-Vendres site has allowed a comparison of the turbulence structure in the real and simulated surface layers. The dynamic field characteristics have been shown to be very similar and in good agreement with the results from the Kansas experiment, but large differences in the thermal structures have been put in evidence. The computation of the fluxes of momentum and sensible heat by direct and indirect methods have shown the necessity to compute the direct estimates from the cospectra, the validity of the estimations from the profiles even when they do not present "perfect" logarithmic repartitions and the limits of the dissipative method involving Businger et al.'s (1972) relations. The values of the non-dimensional production and dissipation functions have been estimated : the dynamic ϕ_m and ϕ_ε have values close to 1 ; the ϕ_h values are greater than 0.74 as already noted by Yaglom (1977) and the dissipation of temperature variance overbalance the production. A curious characteristic of the fully developed boundary layer of the tunnel has been put in evidence with vertical mean gradients two times larger than expected from the direct estimations of the heat flux and destruction rate . Finally the spectral extension of Reynolds analogy proposed by Fulachier & Dumas (1976) has been verified in the weakly unstable tunnel boundary layer.

ACKNOWLEDGEMENTS

We are grateful to P. Chambaud, M. Braun and A.M. Rugiero for their technical cooperation, and to all I.M.S.T. scientific and technical staffs. These works were mainly supported by the C.N.R.S., especially through the A.T.P. "Recherches Atmosphériques" of I.N.A.G.

REFERENCES

Businger J.A., Wyngaard J.C., Izumi Y. and Bradley E.F., 1971. J. Atmos. Sci. 28, 181.

Champagne F.H., 1978. J. Fluid Mech., 86, 67.

Champagne F.H., Friehe C.A., La Rue J.C. and Wyngaard J.C., 1977. J. Atmos. Sci. 34, 515.

Coantic M., Bonmarin P., Pouchain B. and Favre A., 1969. AGARD Conf. Proceed. CP 48, pp.17-0 , 17-14.

Fulachier L., 1978. Proc. of AIAA ASME Thermophysics & Heat Transfer Conference Palo Alto, Calif., 24-26 May 1978. The American Soc. of Mech. Engin., New York, N.Y. 10017.

Fulachier L. and Fumas R., 1976. J. Fluid Mech., 77, 257.

Haugen D.A., Kaimal J.C. and Bradley E.F., 1971. Quart. J. Roy. Meteorol. Soc. 97, 168.

Kaimal J.C. and Haugen D.A., 1969. J. Applied Meteorol., 8, 460.

Kaimal J.C., Wyngaard J.C., Izumi Y. and Coté O.R., 1972. Quart. J. Roy. Meteorol. Soc., 98, 563.

Lumley J.L., 1965. Phys. Fluids, 8, 1056.

Lynch R.A. and Bradley E.F., 1974. J. Applied Meteorol., 13, 588.

Mestayer P.G., Gibson C.H., Coantic M.F. and Patel A., 1976. Physics of Fluids, 19, 1279.

Mestayer P.G., Champagne F.H., Friehe C.A., La Rue J.C. and Gibson C.H., 1978. Turbulent fluxes through the sea surface, wave dynamics and prediction (Favre & Hasselmann, Ed.) Plenum Press, New-York.

Mestayer P. and Chambaud P., 1979. Boundary Layer Meteor., 16, 311.

Mestayer P., 1980. Thèse de Doctorat d'Etat, I.M.S.T., Université d'Aix-Marseille II, Marseille, France.

Monin A.S. and Obukhov A.M., 1953. Dokl. Akad. Nauk. S.S.S.R., 93, 223.

Monin A.S. and Obukhov A.M., 1954. Trudy Geofiz. Inst. Akad. Nauk. S.S.S.R. n°24, 163.

Monin A.S. and Yaglom A.M., 1965. Statistical Fluid Mechanics (Nauka Press, Moscow). Trad. anglaise (The M.I.T. Press, Cambridge, Massachusetts and London, England), T.1, 1971 ; T.2, 1975.

Obukhov A.M., 1946. Akad. Nauk. S.S.S.R. Institut. Teoret. Geofiz. Trudy 1, 95.

Selva J.P., 1979. E.D.F. Bulletin de la Direction des Etudes et Recherches, n°3, Ed. La Houille Blanche, Grenoble.

Wieringa J., 1980. Boundary-Layer Meteorol., 18, 411.

Wyngaard J.C., 1968. J. Scientific Instrum., Series 2, 1, 1105.

Wyngaard J.C. and Coté O.R., 1971. Journal of Atmos. Sci., 28, 190.

Wyngaard J.C. and Coté O.R., 1972. Quart. J. Roy. Meteorol. Soc., 98, 590.

Wyngaard J.C. and Clifford S.F., 1977. Journal of Atmos. Sci., 34, 922.

Yaglom A.M., 1977. Boundary Layer Meteorol., 11, 89.

Dilatation Effects in Wind Tunnel Simulation of Non Adiabatic Atmospheres

CLAUDE REY
Laboratory of Fluid Mechanics
Ecole Centrale de Lyon
B.P. 163
69130 Ecully, France

ABSTRACT

Wind tunnel simulation of atmospheric boundary layer is discussed from non dimensional equation for heat and momentum which display the different similitude parameters to be used. Exact similitude cannot be achieve for Reynolds number that involves serious difficulties when exact similitude for Richardson number is required. Peculiarly, non negligible dilatation effect upon inertia may appears in wind tunnel heated boundary layers, then the large turbulent structures are not only sensitive to buoyancy in density variations. This study make possible the distinction between buoyancy and dilatation effect on inertia, and corrections can be made for a well comparison of turbulent scales, with the atmosphere.

1. INTRODUCTION

The problems of atmospheric boundary layers simulation are complicated by supplementary mechanisms, when buoyancy interfere. This very frequent phenomenon in nature which is due to a non adiabatic evolution of the temperature (or density) with respect to the altitude, is not easy to occurately observe in situ, where turbulent fields are often in evolution ; that is significant for the results of Kansas experiments (1) on spectral data, which give a large discrepancy of the turbulent energy at the level of the large scales of atmospheric turbulence.

An experimental study in laboratory is well adapted to this case and the particular influence of a given parameter on the turbulent field can be easily pointed out. For modelling the buoyancy effect the chosen parameters will be the Richardson number or the Froude number.

Several works have been developed during this last years for stable or unstable boundary layer flows (2, 3, 4), with change of the stability (5, 6) or wall roughness downstream (7, 8), but the validation of these results for atmospheric situation has not be well-precised.

Concerning the mean velocity, temperature or centration field results it can be shown (9, 10) that a direct comparison between the last values and these of atmospheric flows keep good. Presently our purpose is to emphasize the difficulties which appear when turbulent fields are considered, for the whole turbulence mechanisms are not necessarily the same in wind tunnel as in the atmosphere. This is evident in the case of an atmospheric plume evolution from a heated or cooled islet, where the situation is neither bidimensional nor homogeneous in horizontal plane, so that the equation cannot be easily simplified. Peculiarly the Boussinesq approximations, which are valid in atmospheric flows, cannot be used anywhere in wind tunnel boundary layer and plume. This analyse becomes too much complicated when boundary layer approximation is not justified. In order to display the non negligible effects of the dilatation upon inertia, contrary to buoyancy, we only considere a see mingly simple case concerning the simulation of a non adiabatic quasi horizontal atmosphere flow, in a wind tunnel heated boundary layer. In this restrictive situation our purpose is to distinguish between the turbulent buoyancy driven structure and the turbulent structure evoluing with dilatation effect on inertia, in order to well transpose the wind tunnel data to the atmosphere. On the other hand, few information can be found about dilatation effect on inertia for a dilated flow by heat, and even less on the consequence for turbulent structures ; though we study a relatively simple situation, the results may allow us to precise the fondamental parameters required for the description of more complex dilated flows.

2. BASIC EQUATION AND SIMILITUDE

In the case of atmosphere boundary layer all variables are taken with respect to a standard equilibrium state defined using the temperature $T_o(z)$, the density $\rho_o(z)$ and the pressure $P_o(z)$. These quantities are related to each other by adiabatic relations, and we will state

$$\frac{\delta}{\delta t} P_o = 0$$
$$\frac{\delta}{\delta x_i} P_o = -\rho_o g \delta_{i3} \qquad (1)$$
$$\rho_o C_p \frac{\delta}{\delta x_i} T_o = \frac{\delta}{\delta x} P_o$$
$$P_o = \rho_o r T_o$$

taking into account the deviation with respect to the standard equilibrium state, we will write the following quantities as :

$\rho_a = \rho_o + \Delta\rho$	for the density
U_i	for the velocity component in the x_i direction
$T = T_o + \Delta T$	for the temperature
$P = P_o + \Delta P$	for the pressure

These notations allow us to write the set of deviation equations as :

. for the momentum equation :

$$\rho_a \frac{d}{dt} U_i \simeq -\frac{\delta}{\delta x_i} \Delta P - \Delta\rho\, g \delta_{i3} - 2\rho_a \varepsilon_{ijk} \Omega_j U_k + \frac{\delta}{\delta x_k}\left(\mu \frac{\delta}{\delta x_k} U_i\right) \qquad (2)$$

. for the continuity :

$$\frac{d}{dt}\Delta\rho + \rho_a \frac{\delta}{\delta x_k} U_k = \frac{\rho_o g U_3}{T_o(\gamma-1)C_p} \qquad (3)$$

. for the energy equation :

$$\rho_a C_p \frac{d}{dt}\Delta T = \Delta\rho\, g U_3 + \frac{d}{dt}\Delta P + \frac{\delta}{\delta x_k}\left(k \frac{\delta}{\delta x_k}\Delta T\right) \qquad (4)$$

. and for the equation of state :

$$\frac{\Delta P}{P_0}\frac{\rho_0}{\rho_a} = \frac{\Delta T}{T_0} + \frac{\Delta \rho}{\rho_a} \quad (5)$$

where Ω stands for the angular velocity of the earth's rotation ε_{ijk} is the permutation tensor, η_i is the component of the unity vector of the earth's axis, μ is the dynamic viscosity, and K is the thermal conductivity.

Using four reference scales, such as U_r, L_r, ΔT_r or $\Delta \rho_r$ for reference velocity, length, temperature or density, the previous equations can be given in dimensionless form and the following parameters appear :

The Reynolds number $Re = \frac{\rho_a U_r L_r}{\mu}$

The Richardson number $Ri = \frac{g\,\delta\Delta T/\delta z}{T_0(\delta U/\delta z)^2}$ or the Archimede number $Ar = \frac{-\Delta\rho_r\, g\, L_r}{\rho_a\, U_r^2}$

The Rossby number $Ro = \frac{U_r}{L_r \Omega}$

The Eckert number $E = \frac{U_r^2}{C_p\,\Delta T}$

The Prandtl number $Pr = \frac{\mu C_p}{k}$

Exact similarity of atmospheric and wind tunnel flows requires that each of these numbers have the same values.Coriolis effects are not easily simulated in wind tunnel but they can be neglected in the lower part of the atmosphere. On the other hand exact similitude cannot be achieved for Reynolds number in wind tunnel. Then when buoyancy interfere the similitude of Richardson number requires high value of the wind tunnel boundary layer temperature. Comparison between dilatation effect on inertia and buoyancy leads us to consider the following terms of the momentum equations :

$$\Delta\rho \frac{d}{dt} U_i \qquad \text{and} \qquad \Delta\rho\, g\, \delta_{i3}$$

or the dimensionless square Froude number :

$$\frac{1}{g}\frac{d}{dt}U_i$$

In the atmosphere this number is very small, then Boussinesq approximations can be done. In wind tunnel the value of this quantity depends on the Reynolds number used and on the scale of the model, generaly this number is generaly not negligible by comparison with unity. Then the momentum equation taking into account dilatation effect on inertia show a particular mechanism due to the dilatation and different from the buoyancy one. Same consideration can be made on the equation of energy where the term :

$$\Delta\rho\, C_p \frac{d}{dt}\Delta T$$

has to be studied.

3. THE EQUATIONS OF THE MEAN QUANTITIES

3.1. The mean quantities equations of the velocity field

The analyse of turbulent phenomena requires a statistical decomposition of quantities between mean values and fluctuations.

We will write :

$$\begin{aligned} \rho_a &= \rho_0 + \overline{\Delta\rho} + \rho \\ U_i &= \overline{U}_i + u_i \\ T &= T_0 + \overline{\Delta T} + \theta \\ P &= P_0 + \overline{\Delta P} + p \end{aligned} \qquad (6)$$

For a simplification, let us suppose that correlations with molecular properties fluctuations are negligible, and that the Richardson number is not very large, such as $\bar{W} = \bar{V} = 0$. If the flow is bidimensional, in the case of stationarity then the boundary layer assumption can be used. Then the momentum equation of the longitudinal component of the velocity U gives :

$$\frac{\partial}{\partial z}(\overline{\rho_a uw} + \overline{\rho w}\overline{U}) + \frac{\partial}{\partial x}(\bar{\rho}_a\overline{U} + \overline{\rho_a u^2} + 2\,\overline{\rho u}\,\overline{U}) \simeq -\frac{\partial}{\partial x}\overline{\Delta P} + \frac{\partial}{\partial z}\left(\mu\frac{\partial \overline{U}}{\partial z}\right) \qquad (7)$$

The concept of turbulent viscosity give the general form of equation (7) :

$$\frac{\partial}{\partial t}\overline{\rho_a U_i} + \frac{\partial}{\partial x_j}\overline{\rho_a U_i}\,\overline{U_j} \simeq -\frac{\partial \overline{\Delta P}}{\partial x_i} + \frac{\partial}{\partial x_j}(\mu_t + \mu)\frac{\partial \overline{U}}{\partial x_j} \qquad (8)$$

then we will write

$$\overline{\rho_a uw}\left(1 + \frac{\overline{\rho w}\,\overline{U}}{\overline{\rho_a uw}}\right) = -\mu_t\frac{\partial \overline{U}}{\partial z} \qquad (9)$$

the flux Richardson number is :

$$R_f = -\frac{\overline{\rho w}\,g}{\overline{\rho_a uw}\frac{\partial \overline{U}}{\partial z}}$$

and the Froude number :

$$Fr = \left(\frac{\overline{U}}{g}\frac{\partial}{\partial z}\overline{U}\right)^{1/2}$$

give us the Reynolds stress in the following form :

$$\overline{\rho_a uw} = \frac{-\mu_t}{1 - R_f Fr^2} \frac{\partial \overline{U}}{\partial z} \tag{10}$$

3.2. Consequence for atmospheric simulation.

If the evolution of mean quantitie along the x direction is quite negligible, then the contribution of dilatation $\Delta\rho$ in the equation (9) gives very negligible effects upon the turbulent viscosity μ_t. When the Reynolds number of the wind tunnel is correct for atmospheric simulation, we can write :

$$\left[\frac{1}{K z^+} \cdot \frac{\mu_t}{\mu}\right]_{\text{atmosphere}} \simeq \left[\frac{1}{K z^+} \cdot \frac{\mu_t}{\mu}\right]_{\text{wind tunnel}}$$

In the atmosphere the Froude number is very small compared to unity, then $R_f\ Fr^2 << 1$, $\rho_a \simeq \rho_o$ and

$$\rho_o \overline{uw} \simeq -\mu_t \frac{\partial \overline{U}}{\partial z}$$

the wind tunnel simulation of buoyant atmosphere lead to :

$$\left[\frac{\overline{uw}}{U_*^2}\right]_{\text{atmosphere}} \simeq \left[\frac{\overline{\rho_a uw}}{\rho_p U_*^2}(1 - R_f . Fr^2)\right]_{\text{wind tunnel}}$$

where U_* is the friction velocity and ρ_p is the value of ρ_a close to the wall.

Generally we can use the following approximation :

$$\overline{\rho_a uw} = \rho_o \left(1 + \frac{\overline{\Delta\rho}}{\rho_o}\right) \overline{uw}$$

then the last relation can be written as :

$$\left[\frac{\overline{uw}}{U_*^2}\right]_{\text{atmosphere}} \simeq \left[\frac{\overline{uw}}{U_*^2} \cdot \frac{\rho_o + \overline{\Delta\rho}}{\rho_o + (\overline{\Delta\rho})_{max}}(1 - R_f\ Fr^2)\right]_{\text{wind tunnel}} \tag{11}$$

3.3. Application to the temperature field.

The same analysis and hypothesis lead us to the relation :

$$\overline{\rho_a w\theta}\left(1 + \frac{\overline{\rho w}\ \overline{\Delta T}}{\overline{\rho_a w\theta}}\right) = -\frac{k_t}{C_p} \frac{\partial \overline{\Delta T}}{\partial z} \tag{12}$$

it can be seen here, that dilatation effect on the turbulent temperature field, concern the equation of kinetic energy through $\overline{\rho w}\, g$, the equation of horizontal heat flux through $\rho_a \overline{w\theta}\, \partial\overline{U}/\partial z$, and the analogy between temperature and velocity distribution through their respective gradient. So we can write :

$$\frac{\overline{\rho w}\,\overline{\Delta T}}{\rho_a \overline{w\theta}} = \frac{\overline{\rho w}\, g}{\rho_a \overline{uw}\frac{\partial \overline{U}}{\partial z}} \cdot \frac{\rho_a \overline{uw}\frac{\partial}{\partial z}\overline{\Delta T}}{\rho_a \overline{w\theta}\frac{\partial}{\partial z}\overline{U}} \cdot \frac{T_0\left(\frac{\partial}{\partial z}\overline{U}\right)^2}{g\frac{\partial}{\partial z}\overline{\Delta T}} \cdot \frac{\overline{\Delta T}}{T_0}$$

$$= - R_f \,.\, Pr_t \cdot \frac{1}{Ri} \cdot \frac{\overline{\Delta T}}{T}$$

Concerning the wind tunnel simulation of atmospheric flow the conclusions are the same as those obtained about velocity, because the Froude number is very small in the atmosphere, then $(-R_f/Ri).Pr_t.(\overline{\Delta T}/T) << 1$. Comparison of wind tunnel data and atmosphere gives, if is the friction temperature,

$$\left[\frac{\overline{w\theta}}{U_* T_*}\right]_{\text{atmosphere}} \simeq \left[\frac{\overline{w\theta}}{U_* T_*}\,\frac{\rho_0+\overline{\Delta\rho}}{\rho_0+\Delta\rho_m}\left(1-\frac{R_f}{Ri}Pr_t\frac{\Delta T}{T}\right)\right]_{\text{wind tunnel}} \qquad (13)$$

This last approximed relation put forward the complexity of the equations interaction when dilatation interfere.

4. THE EQUATION OF THE TURBULENCE

4.1. Application to the turbulent energy

General discussions concerning the Boussinesq approximations have been previously made (11,12, 13) but the conclusions are not always evident concerning the turbulent mechanisms and the tridimensional flows.

The use of a density decomposition between an adiabatic value, a mean deviation and a fluctuation gives very complicate equationsfor the turbulent kinetic energy and temperature variance. Complete equations taking into account all these density variations have been writtenby Favre (14) for the compressible flows. Here we will state these equations using the instantaneous equation of momentum and temperature divided by density. When Richardson numbers are not very large, we will suppose for simplification that the mean quantities gradients along the x direction are negligible by comparison with the others in the z direction. The flow is supposed horizontal, then the contribution of the mean pressure deviation $\overline{\Delta P}$ can be neglected, but no argument allows us to neglect the contribution of the pressure fluctuations even in the temperature equation.

With these considerations the equation of the kinetic energy be written as :

$$2\,\overline{uw}\frac{\partial \bar{U}}{\partial z} + \frac{\partial}{\partial z}\overline{wq^2} + 2g\left(-\frac{\overline{w\theta}}{T_0} + \frac{\overline{\rho w}}{\rho_0} + \frac{\overline{\rho w\theta}}{\rho_0 T_0}\right) + \frac{2}{\rho_0 T_0}\left(\bar{T}\,\overline{u_i\frac{\partial p}{\partial x_i}} + \overline{\theta u_i\frac{\partial p}{\partial x_i}}\right)$$

$$-\frac{2}{\rho_0 T_0}\left(\overline{u\theta}\frac{\partial}{\partial z}\left(\mu\frac{\partial \bar{U}}{\partial z}\right) - \bar{T}\,\overline{u_i\frac{\partial}{\partial x_k}\left(\mu\frac{\partial u_i}{\partial x_k}\right)} - \overline{u_i\theta\frac{\partial}{\partial x_k}\left(\mu\frac{\partial u_i}{\partial x_k}\right)}\right) \simeq 0 \qquad (14)$$

and similarly the equation of the temperature fluctuations variance $\overline{\theta^2}$ is :

$$2\,\overline{w\theta}\frac{\partial}{\partial z}\overline{\Delta T} + \frac{\partial}{\partial z}\overline{w\theta^2} - \frac{2}{\rho_0 C_p T_0}\left(\bar{T}\bar{U}\,\overline{\theta\frac{\partial p}{\partial x}} + \bar{T}\,\overline{u_i\theta\frac{\partial p}{\partial x_i}} + \bar{U}\,\overline{\theta^2\frac{\partial p}{\partial x}} + \overline{\theta^2 u_i\frac{\partial p}{\partial x_i}}\right)$$

$$-\frac{2}{\rho_0 C_p T_0}\left(\overline{\theta^2}\frac{\partial}{\partial x_k}\left(k\frac{\partial \overline{\Delta T}}{\partial x_k}\right) + \bar{T}\,\overline{\theta\frac{\partial}{\partial x_k}\left(k\frac{\partial \theta}{\partial x_k}\right)} + \overline{\theta^2\frac{\partial}{\partial x_k}\left(k\frac{\partial \theta}{\partial x_k}\right)}\right) \doteq 0 \qquad (15)$$

These relations are very different from the usual ones, generaly written in the atmosphere or not buoyant wind tunnel boundary layer. Then it is necessary to analyse each terms of these equations through the application to the wind tunnel simulation of the lower layers of the atmosphere.

The ratio between the integral and the dissipation turbulent scales $l\,/\,\lambda$ cannot be simulated in wind tunnel, then at the same time it is not possible to analyse all the size of the turbulent structure. Generally, the transfer phenomena are the interesting matter in the atmosphere, then we will only considere the simulation problems of the large turbulent scales, and we will not use the molecular properties in the non dimensional representation of the data. Effectively, the density variation effects on the molecular turbulent terms in equation 14 and 15 are not to be taken into account, only because its change the non simulated turbulent Reynolds, or Peclet number, that considerably simplifies our present problem.

4.2. Application to the turbulent scales

It is difficult to precisely well understand the integral scales evolution in non homogeneous fields. Then we generally use an estimation of this scale through the relations :

$$l = \frac{u'^3}{P} \qquad (16)$$

for the velocity integral turbulent scale and

$$l_\theta = \frac{u'\theta'^2}{P_\theta} \qquad (17)$$

for the temperature integral turbulent scale, where P and P_θ are respectively the turbulent kinetic energy and temperature variance production. Let us suppose

$$\frac{\overline{pw}}{\rho_0} \ll \frac{\overline{w\theta}}{T_0}$$

then, in our not very large Richardson number situation that allow us to write the turbulent production terms of equation 14 as :

$$P = -\overline{uw}\,\partial\overline{U}/\partial z\,(1 - R_f)$$

and similarly, with the use of equation 15 :

$$P_\theta = -\overline{w\theta}\,\partial\overline{\Delta T}/\partial z$$

If inertia is quite important by comparison with buoyancy ($z/L < 0.4$ for our experiments), then $\overline{q^2}$ and $\overline{\theta^2}$ are supposed not significantly influenced by the dilatation and the velocity gradient and temperature gradient $\partial\overline{U}/\partial z$ and $\partial\overline{\Delta T}/\partial z$ are supposed given by (15) :

$$\frac{\partial\overline{U}}{\partial z} \simeq \frac{U_*}{Kz}(1 - 15\,z/L)^{-1/4}$$

$$\frac{\partial\overline{\Delta T}}{\partial z} \simeq \frac{T_*}{Kz}\,0.74(1 - 9\,z/L)^{-1/2} \qquad (18)$$

where L is the Monin-Obukhov length.

These previous approximations and assumptions lead us to the relations for the dilatation effect on turbulent scales in wind tunnel. We can write the velocity integral scale with the use of 10 and 16 as :

$$l \simeq \frac{u'^3}{\frac{\mu_t}{\rho_0 + \overline{\Delta\rho}}\left(\frac{\partial\overline{U}}{\partial z}\right)^2} \cdot \frac{(1 - R_f Fr^2)}{(1 - R_f)} \qquad (19)$$

then when z/L is the same in the atmosphere and wind tunnel, 18 and 19 give :

$$\left[\frac{l}{Kz}\right]_{\text{atmosphere}} = \left[\frac{l}{Kz}\cdot\frac{1}{(1 + \frac{\overline{\Delta\rho}}{\rho_0})(1 - R_f Fr^2)}\right]_{\text{wind tunnel}}$$

Equivalent assumption and the relation 13 and 17 give the temperature integral scale as :

$$l_\theta = \frac{u'\theta'^2\,(\rho_0 + \overline{\Delta\rho})}{k_t(\partial\overline{\Delta T}/\partial z)^2}\left(1 - \frac{R_f}{R_i}\,Pr_t\,\frac{\overline{\Delta T}}{T_0}\right)$$

then for comparison with the atmosphere we will write :

$$\left[\frac{l_\theta}{kz}\right]_{\text{atmosphere}} \simeq \left[\frac{l_\theta}{kz}\,\frac{1}{(1+\frac{\Delta\bar{\rho}}{\rho_0})(1-\frac{R_f}{R_i}Pr_t\frac{\Delta\bar{T}}{T_0})}\right]_{\text{wind tunnel}}$$

generally, $R_f\,Pr_t/Ri \sim 1$ and the dilatation on inertia corrections are pratically the same for the velocity and temperature integral scales.

5. EXPERIMENTAL RESULTS

5.1. Experimental arrangement.

The wind tunnel used for these experiments is of the open circuit type with a 1.0 m x 1.2 m x 10 m test section. The flow develop a boundary layer along a heated smooth plate which can be made rough by transverse loophole bars 7.5 mm high. The wall temperature varies from the ambiant to 370 K and the velocity at the center of the channel is adjustable from 1 m/sec to 20 m/sec while the turbulence intensity at the same initial station is about 3 %. The thickness of the boundary layer can be change by the use of air injection through a restricted area, a short distance before the beginning of the test plate.

The turbulent quantities are measured by means of a three wire probe made at the laboratory;this probe is made up of both an X probe and a single wire sensitive only to θ. The wires of the X probe are 5 μm diameter, about 1 mm along and are operated by two DISA 55 M01 anemometers. The wire sensitive to θ is 1 μm diameter and is fed with a current constant of 0.15 mA.

The X wires being sensitive to both velocity and temperature, an adequate calibration (16) make possible to distinguish between velocity and temperature informations. This calibration and data processing can be only used when the dilatation effects on mean and turbulent quantities are not very important or can be easily taked into account with a correction method.

A Schlumberger 1860 integrating voltmeter gives the mean and r.m.s. quantities. The correlations are obtained through analog multipliers. For these correlations in the unstable case the average time is more than 100 sec. The turbulence scales are obtained by spectra measurements through a Bruel and Kjaer 2010 heterodyne analyser. Most of the data are taken at a 8.5 m distance from the beginning of the test plate where the boundary layer thickness δ is about 0.18 m in the neutral case and 0.24 m in the unstable case.

5.2. The mean quantities and the characteristic parameters

The data are taken from Schon (3) and Rey (4) thesis. In a previous paper (17) the effect of buoyancy into the boundary layer was shown with a distinction from the Reynolds number effects, the contribution of the density variation to inertia was supposed included in the Reynolds number. Here we want to precise the complexity of atmospheric flows studies using a wind tunnel simulation, and show the necessity to separate between the mechanisms due to density variation into inertia from those due to buoyancy.

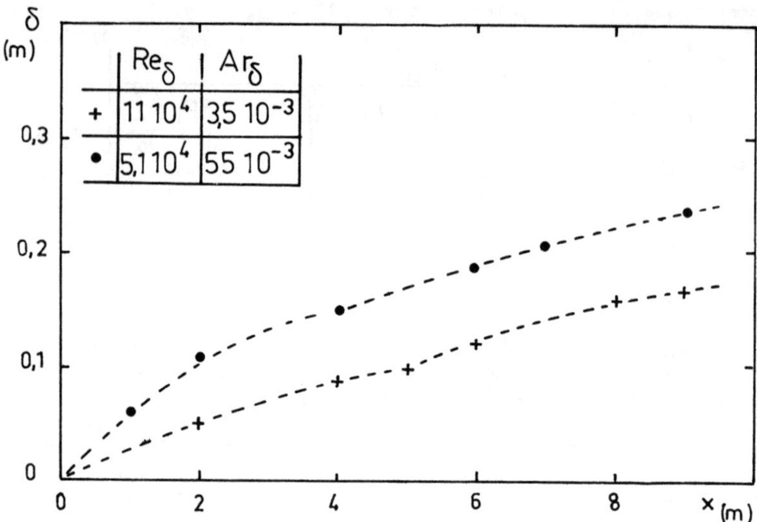

FIG. 1 - Evolution of the boundary layer thickness

The stability influence on the boundary layer development is given on the figure 1, which show the interaction between Reynolds number effects and the stratification in the wind tunnel ; then it is not possible to completly separate between these two effects. Nevertheless the turbulent mechanisms can be change with density variation in inertia (a fortiori with gravity) and this contribution to the Reynolds number is an information for our present problem.

The effect of the stratification is very clear on the mean velocity and temperature profiles given on figure 2 and 3. The shape of the curve shows that the turbulent flow is better developped in the unstable case contrary to the relatively small Reynolds number. Same experiments made over a rough plate (4,18) pratically show no stratification effect on the shape of the mean velocity profile.

Then, it seems that this influence on the turbulent flow development would be less important in the case of very large wind tunnel for which the test section can be choosed far downstream the beginning of the turbulent layer.

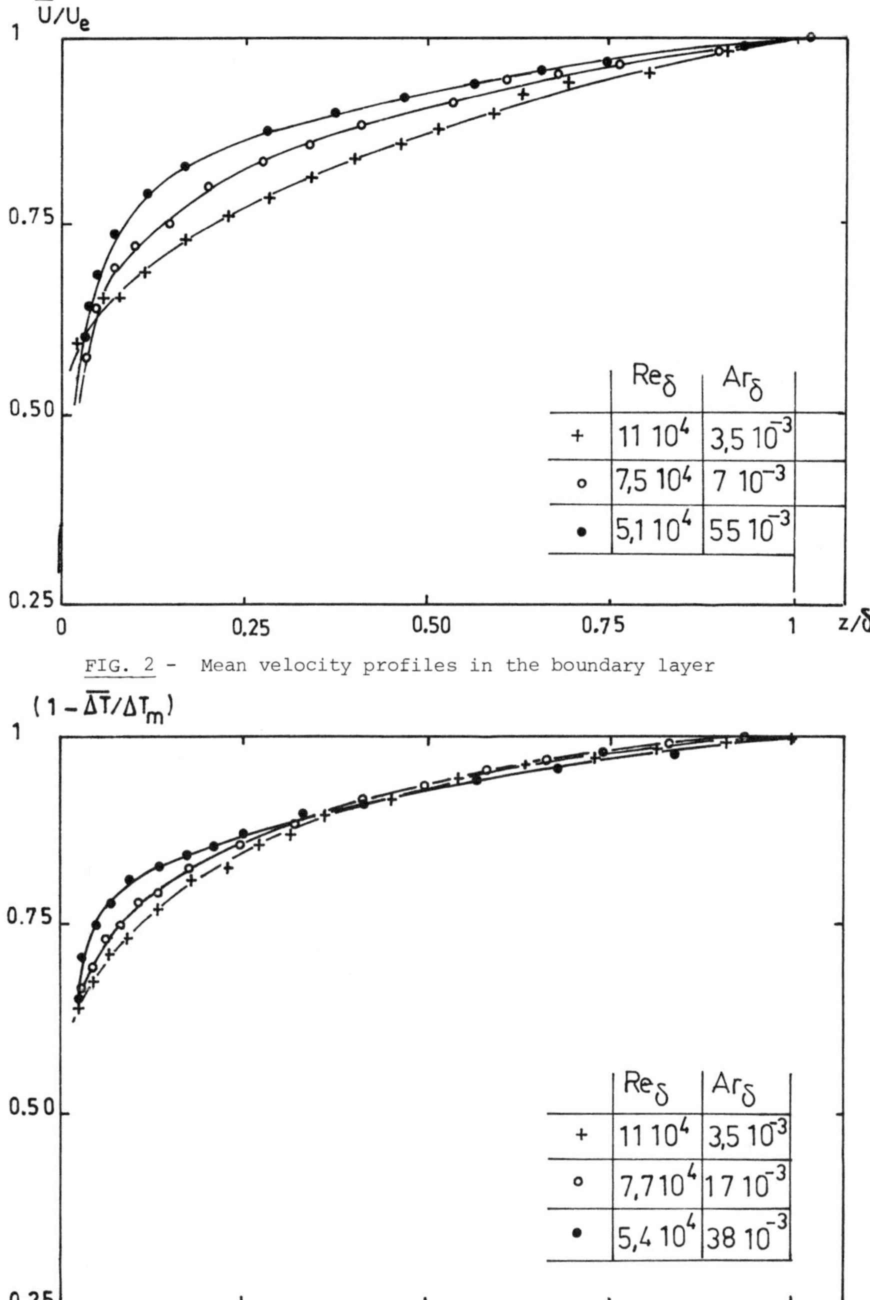

FIG. 2 - Mean velocity profiles in the boundary layer

FIG. 3 - Mean temperature profiles in the boundary layer

The data obtained in our experiments give the various parameters presently introduced ; the mean velocity, the Reynolds stress and the vertical heat flux carry out the first information concerning the real importance of density variation upon inertia, through the term

$$1 - R_f \, Fr^2$$

Figure 4 and figure 5 separately give the evolution of the flux Richardson number and Froude number. The Froude number is greater near the wall but at the same place the flux Richardson number is small ; then the maximum of the dilatation effect is not located near the wall contrary to our just feeling (4). The curve draw on figure 6 show with evidence that Boussinesq approximations are not valid in the unstable boundary layer. On the other hand a relatively small Archimède number ($Ar_\delta = 3{,}5\ 10^{-3}$) and flux Richardson number ($R_f \simeq 10^{-2}$) give a 7 % contamination due to density variation by heat when in this case, the temperature was usually considered as a nearly passive scalar.

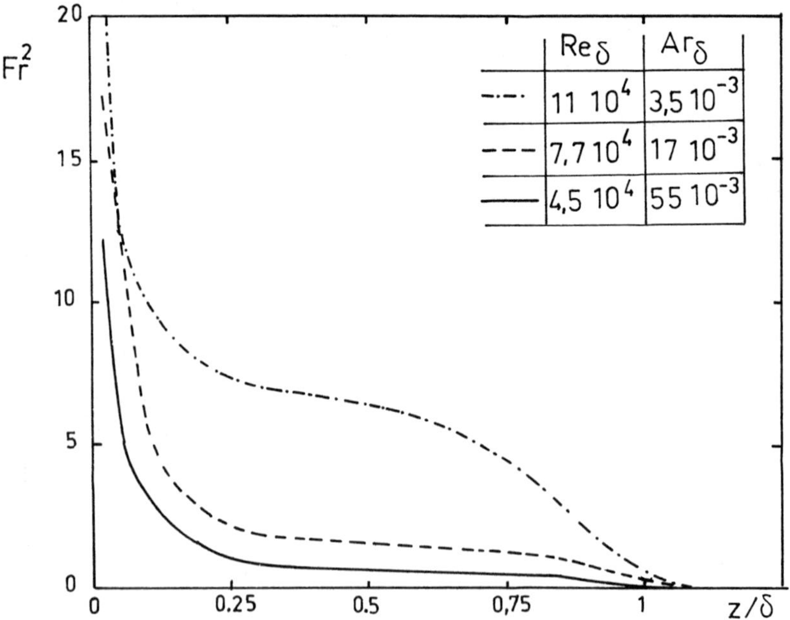

FIG. 4 - Evolution of the Froude number in the boundary layer

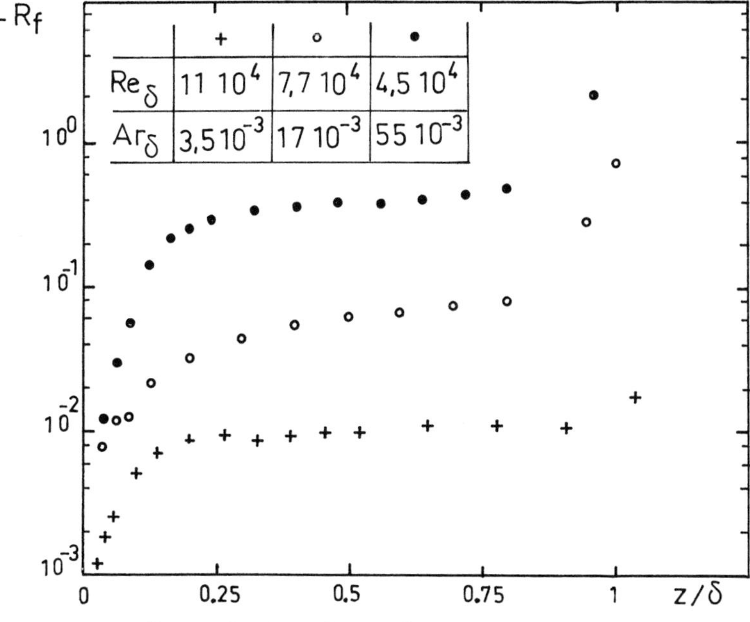

FIG. 5 - Evolution of the flux Richardson number

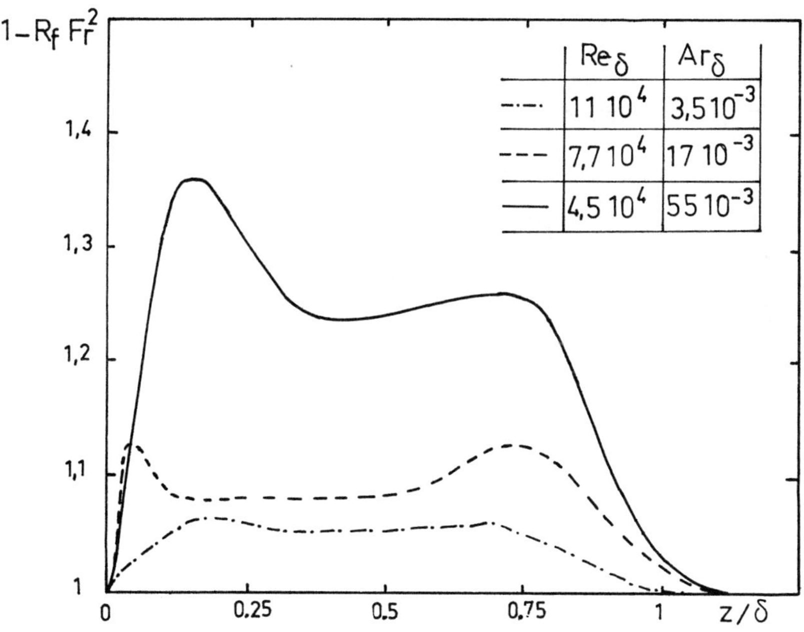

FIG. 6 - Influence of the dilatation on inertia in the boundary layer

Same observation can be made concerning the dilatation effect on the turbulent fields. The correspondant characteristic parameters are the gradient Richardson number, the turbulent Prandtl number and the temperature deviation.

Figure 7 shows the Richardson number evolution across the boundary layer (in the external part of the layer where $z/\delta \sim 1$ the values are not significant on account of the great discrepancy in velocity and temperature gradient determination);the turbulent Prandtl number given in figure 8 is clearly influenced by the thermal stratification and the same conclusions as those stated for the mean velocity and temperature curves can be made here.

The quantitative resultsconcerning the dilatation effect for the temperature field are given in figure 9. Contrary to the velocity field, these effect are quite negligible for the unstable flow, and for the case where the temperature was usually considered as a passive scalar, these effects are very weak. Here the effects are maximum near the wall which is different from the results of the velocity field.

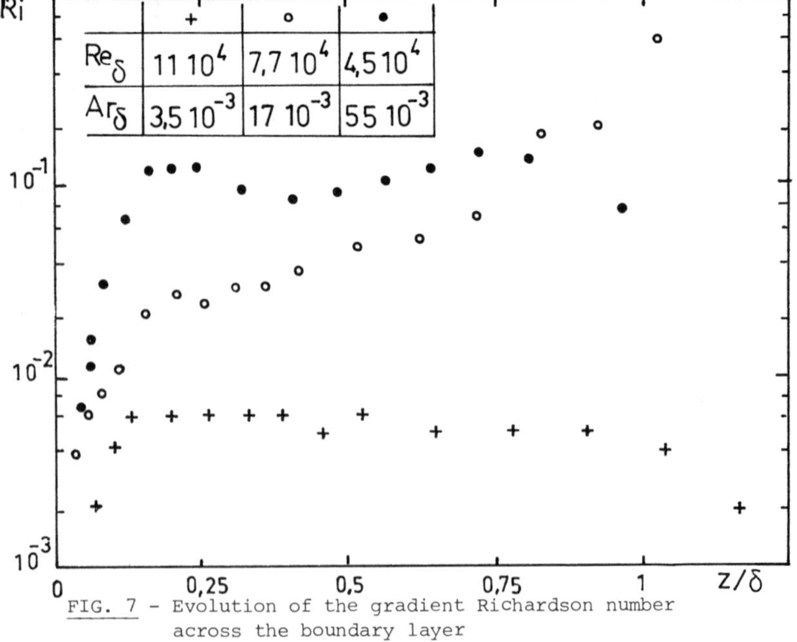

FIG. 7 - Evolution of the gradient Richardson number across the boundary layer

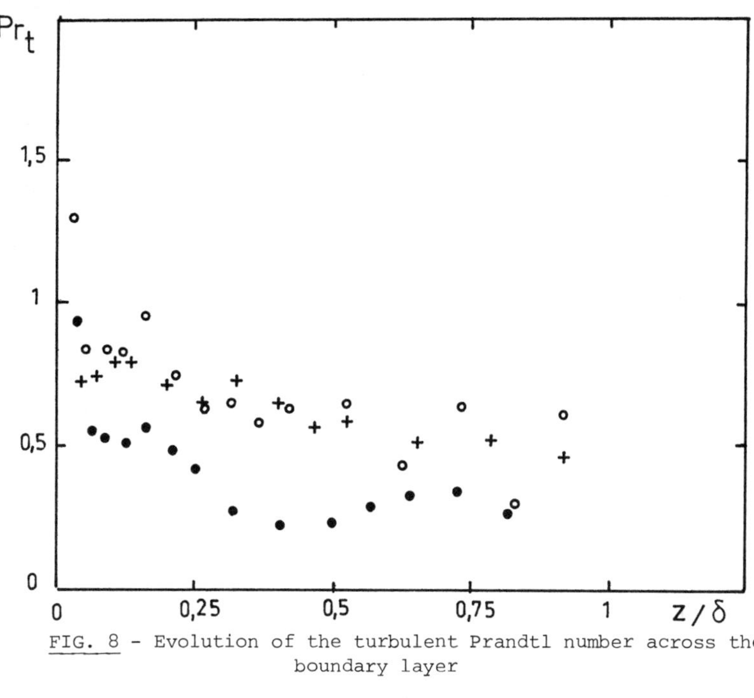

FIG. 8 - Evolution of the turbulent Prandtl number across the boundary layer

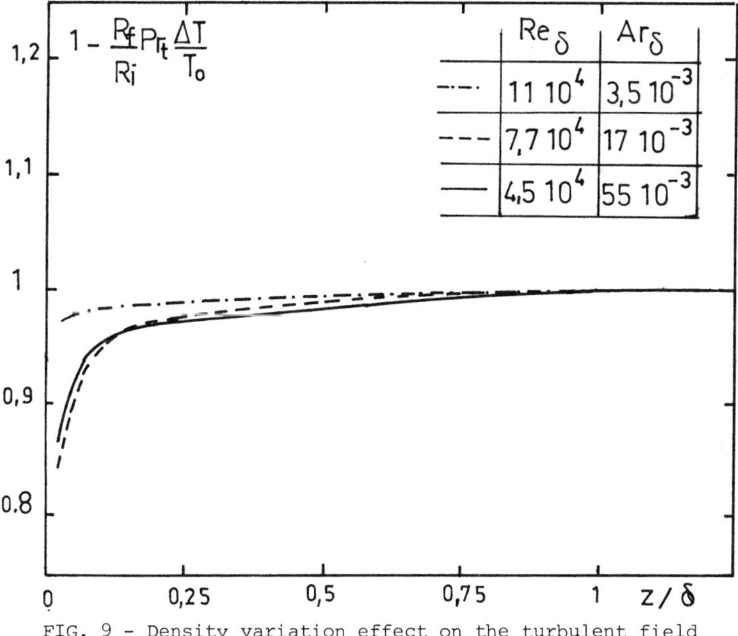

FIG. 9 - Density variation effect on the turbulent field

5.3. Consequence for the turbulent fluxes and scales.

The different quantities needed for correction are now calculated, and we are able to built an idealized correlation $\overline{uw}$ ($\overline{w\theta}$) for which the dilatation effect upon inertia would be absolutely negligible (even for non negligible Archimede number).

On the other hand our wind tunnel unstable situation, make difficult to determine a correct value for the friction velocity U_* or the friction temperature T_* . Then the non dimensional form of the correlation are given with the use of the external velocity U_e and the maximum variation of the temperature ΔT_m. With this representation Fulachier (19) observe the same decrease of the correlation $\overline{uw}$ with the unstake stratification increasing, which is partially due to density variation upon inertia.

The effect of the correction is endent on figure 10 showing the evolution of the non dimensional correlation $\overline{uw}/U_e^2$. The corrected curves are better comparable one another as for the non dimensional heat flux $\overline{w\theta}/U_e \Delta T_m$ given on figure 11. For the temperature field the correction is practilly unnecessary.

The effect on integral turbulent scales is given by figure 12. with the evolution of the length scale l defined by :

$$l = \frac{u^2}{q^2} l_u + \frac{v^2}{q^2} l_v + \frac{w^2}{q^2} l_w$$

where l_u, l_v and l_w are respectively the integral scales determined from the u , v and w spectra. If we compare the evolution of both Ar_δ = 3,5 10^{-3} and Ar_δ = 5 5 10^{-3} velocity integral scales with the corresponding temperature scales given on figure 13, we find that the comparison between the two fields scale evolution is better with the correction of the density variation upon inertia effect.

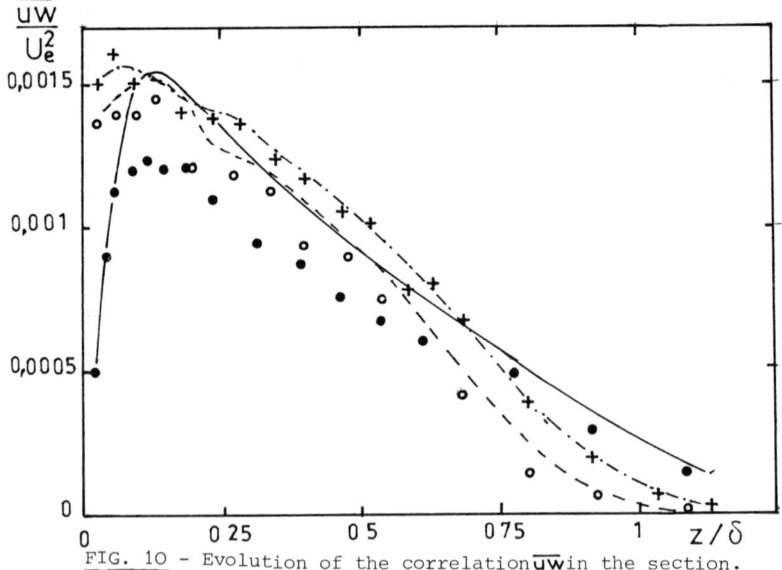

FIG. 10 - Evolution of the correlation $\overline{uw}$ in the section.
+, o, • direct measurement, — · —, - - - , — corrected curves.

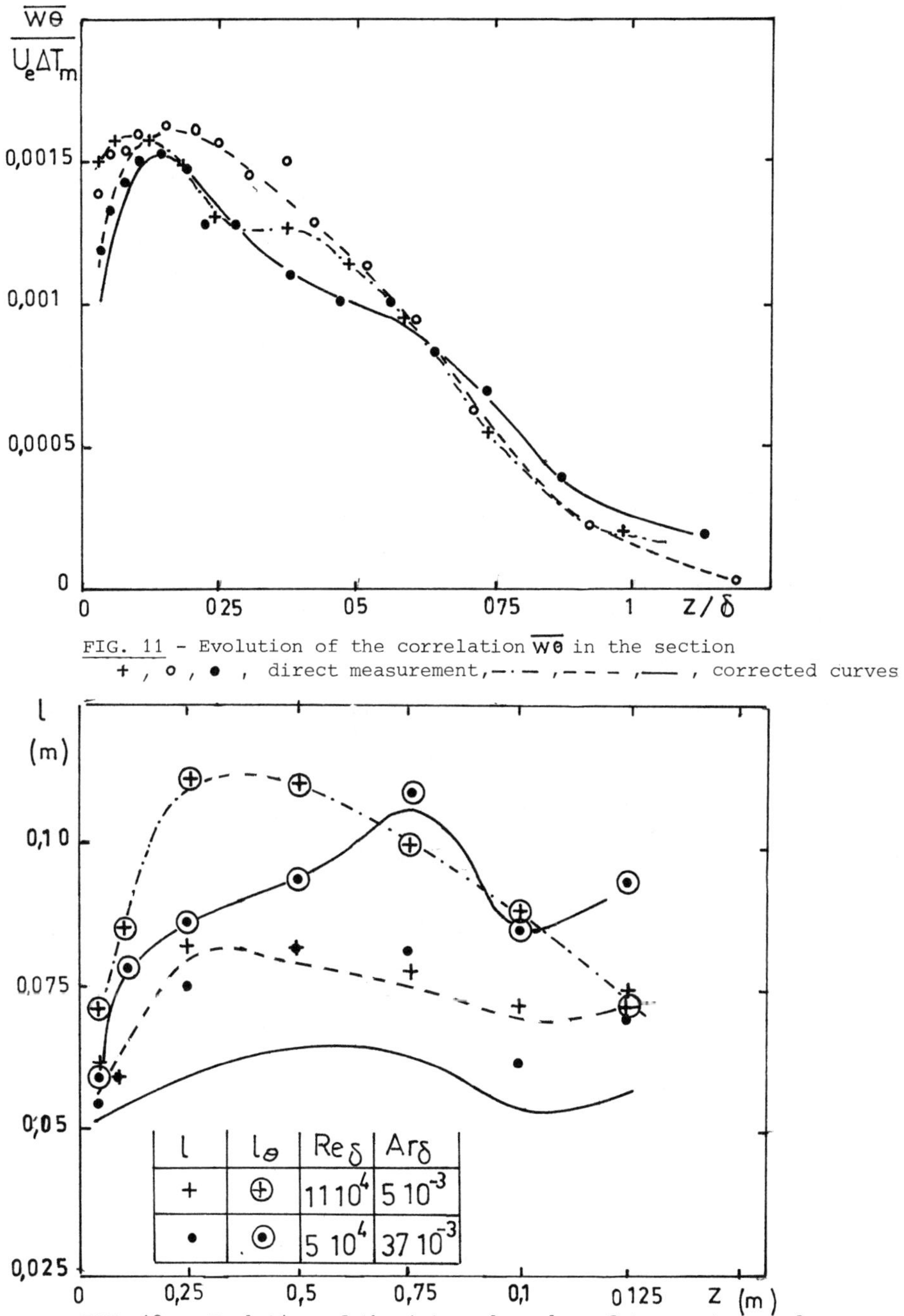

FIG. 11 - Evolution of the correlation $\overline{w\theta}$ in the section
+ , o , • , direct measurement, —·— , – – – , —— , corrected curves

FIG. 12 - Evolution of the integral scales of temperature and velocity. The line are the corrected curves.

6. CONCLUSION

Detailled study of the turbulent mechanisms in atmospheric flow need the use of wind tunnel modelling. The problem concerning the simulation of unstable atmospheric layers are specified here. Effectively the Reynolds stress are very sensitive to density variation in the inertia term and then the mechanisms of turbulence production are not the same, though the production term is formally unchanged. The measurement of the various parameters of stratified boundary layer allow us to precise the importance of these effect which are directly linked to the validity of Boussinesq approximations.

For the velocity field it is shown that Boussinesq approximation are not valid when the Richardson number is more than 10^{-2}. Even for the case when the temperature was usually considered as a passive scalar, the error due to Boussinesq approximation is about 6 %. On the other hand the temperature field is few sensitive to this effect of dilation and the correction carried out is this papers are unnecessary an the usual cases.

Concerning the turbulent scales, we have made various measurement of u, v, w and θ spectra for many case of the flow unstability. We especially present here the evolution of the integral scales l and l_θ caracteristical of turbulent velocity and temperature fields, in the case of two unstability. Same conclusion as those for the heat flux show, that the effect of dilatation on the temperature integral scale is relatively weak, but the difference between the scales l and l_θ become nearly the same for the two unstability, when dilatation corrections are made.

The turbulent structures seem to be differently influenced by the dilatation effect upon inertia : effectively these effect decrease the Reynolds stress but lightly increase the vertical turbulent heat flux. Then the effect are opposite for the two fields when the size of the turbulent structure are well increase by this dilatation effect.

Finally, it can be said that the density variation upon inertia give a kind of compensation from the buoyancy effect which can be one explication for the analogy (3, 19) between the two velocity and temperature field for unstable layer in wind tunnel.

REFERENCES

(1) KAIMAL J.C., WYNGAARD J.C., IZUMI Y and COTE O.R. Quart. J. Roy. Met. Soc. 1972, vol 98, N° 417 pp. 563-589

(2) ARYA S.P.S. J. Fluid. Mech. 1975, vol 68, part 2, pp. 321-343

(3) REY C., Thèse (1977) doctorat es Sciences. Univ. C. Bernard Lyon, France

(4) SCHON J.P., thèse (1974) doctorat es Sciences, Univ. C. Bernard, Lyon, France

(5) CHARNAY G., SCHON J.P., ALCARAZ E et MATHIEU J. (1977) 1er Symposium of turbulent shear flows, Univ. Park, Pennsylvania

(6) ANTONIA R.A., DANH H.G. and PRABHU A. J. Fluid Mech (1977) vol 80, part 1 pp. 153-177

(7) ANTONIA R.A. and LUXTON R.E. J. Fluid Mech. (1971) vol 48, part 4, pp.

(8) ANTONIA R.A. and LUXTON R.E. J. Fluid Mech. (1971) vol 53, part 4, pp.

(9) REY C., SCHON J.P., MOREL R et RIBON M. Comptes-rendus des XIVèmes Journées de l'Hydraulique, Paris, Septembre 1980

(10) REY C., KREISS J. et RIBON M. Comptes rendus du Colloque Météorologie et Environnement 6, 7, 8 Octobre 1980, Paris

(11) SPIEGEL E.A. and VERONIS G. Astrophys. J (1960) p. 442-447

(12) MIHALJAN J.M. Astrophys. J. (1962) vol 136, pp. 1126-1133.

(13) GENCE J.N., SCHON J.P. et MATHIEU J . J. de Mécanique (1978) vol 17, N° 2, pp. 299-321

(14) FAVRE A., J. de Mécanique (1965) vol 4, N°s3 et 4, pp. 361-421

(15) BUSINGER J.A., WYNGAARD Y., IZUMI Y and BRADLEY E.F., J. Atms. Sci. (1971), vol 28, N° 2, pp. 181-189

(16) SCHON J.P et BAILLE A. C.R. Acad. Sc. Paris (1972), t. 274, p. 116

(17) REY C., SCHON J.P. and MATHIEU J. Phys. Fluids (1979), vol 22, N° 6 pp. 1020-1028

(18) SCHON J.P., REY C., MERY P and MATHIEU J. (1976) ICHMT/IUTAM Symposium Dubrovnik, Yougoslavia

(19) FULACHIER L. thèse (1972) Université de Provence, Marseille, France

Buoyancy Flux Limited Mixed Layer Depth in Oscillating Grid Turbulence

E. J. HOPFINGER*
Department of Applied Mathematics and Theoretical Physics
University of Cambridge

ABSTRACT

The effect of a buoyancy flux on turbulent mixing has been studied in a water column in which turbulence without mean shear was produced by a vertically oscillating grid. On top of the water column a buoyancy flux B_o' *was imposed, either by creating a mass flux or by heating the surface layer. The depth to which buoyancy is mixed down is predicted by a simple turbulence closure model and is in good agreement with laboratory observations. It is found that the mixed layer depth* D *scales on the third power of the r.m.s. turbulent velocity* u *at the base of the mixed layer and on the local buoyancy flux* B *. When turbulence is generated near the bottom in a water column of depth* D *, onset of surface stratification occurs at :*

$$B_o D/u^3 \simeq 1$$

These "local" energy balance laws are discussed in the context of observation in natural flow systems.

1. INTRODUCTION

A stabilizing heat flux at the boundary, as for example cooling of the ground in the atmospheric boundary layer or heating of the ocean surface, can have drastic effects on the turbulence structure in the boundary layer or the surface layer. Quite often, the stable temperature gradient necessary for the transfer of the heat is quite weak and has no direct effect on the turbulence. The flux itself drains kinetic energy as a result of the potential energy increase of the fluid layer. Simple global energy arguments suggest that for a given turbulence generation at the boundary, charaterised by u_*^3 *, the depth* D *to which the heat input can be transported is finite and is proportional to the Monin-Obukov length*

$$D = a\, u_*^3 / B_o \quad , \qquad (1)$$

where B_o *is the buoyancy flux at the boundary. This relation was suggested by Kitaigorodskii (1960) (reported by Turner 1973, p. 301), who also obtained from an examination of selected ocean data a value of about 2 for the constant 'a'. At the depth* D *the turbulent velocity goes to zero and a density interface is*

\+ *On leave from the University of Grenoble.*

expected to develop at this depth. The depth D is therefore an indication of the nocturnal inversion layer height and the seasonal thermocline depth. In shallow waters we are interested in the conditions for the existence of a uniformly mixed water column when the top is heated by solar radiation or lighter water (thermal outflow for example) is intruding.

Although some observations seem to support global energy arguments, there is no existing evidence as to their general validity for predicting the mixed layer depth or the onset of surface stratification. In fact it is most likely that "local" energy balance is governing the depth to which buoyancy can be mixed, with other words, viscous dissipation must be included in any meaningful model. What it amounts to is that the velocity near the base of the mixed layer is not a constant fraction of the characteristic velocity u_* at the boundary.

A simple fluid system to study buoyancy flux effects on turbulence is oscillating grid turbulence which is a turbulence field without mean flow and is homogeneous in a plane. In the absence of a buoyancy flux the decay of the r.m.s. turbulent velocity is inversely proportional to the distance from the generating plane and the integral scale increases linearely with distance (Hopfinger and Toly 1976). When a stabilising buoyancy flux is imposed at the boundary the turbulent velocity decays more rapidly and goes to zero at a depth D.

A model is presented in section 3 which predicts this mixed layer depth in terms of the turbulent characteristics and the buoyancy flux at the boundary. The agreement with experiments is remarkably good. Implications of these results are discussed in section 6.

2. CHARACTERISTICS OF OSCILLATING GRID TURBULENCE

The turbulent velocity field produced by an oscillating grid was measured by Thompson and Turner (1975) and by Hopfinger and Toly (1976). Hopfinger and Toly's measurements were made in a much larger tank and they also investigated the dependence of the turbulent velocity on mesh size M and on the stroke S of the grid oscillations. They suggested the following empirical formula for the r.m.s. horizontal velocity at distance z from the "virtual origin"

$$u/fS = C(SM)^{1/2}\, z^{-1} \quad , \tag{2}$$

where f is the frequency of grid oscillation. The "virtual origin" almost coincides with the midplane of the grid, more precisely it falls 0.5 cm behind the midplane. The formula holds for grids made of square bars and a ratio of mesh to bar-size of 5. The coefficient of proportionality is 0.2 for Thompson and Turner's velocities and 0.26 for Hopfinger and Toly's measurements.

Recently, McDougall (1979) measured the r.m.s. horizontal and vertical velocity with a laser-Doppler anemometer at twenty position in a horizontal plane, 100 mm from the midplane of the grid. Conditions were otherwise identical to those of Thompson and Turner. These measurements indicated a considerable horizontal non-uniformity in the r.m.s. turbulent velocities. The average of the twenty values gave a r.m.s. horizontal velocity closely described by formula (2) with $C = 0.26$. It was therefore tempting to use the some coefficient for all the experiments, but it turned out (see section 5), that the present data were better correlated when using $C = 0.2$ for $S = 1$ cm, consistent with Thompson and Turner's results. For $S = 2.4$ cm the some value was chosen whereas when $S = 4$ cm a value of $C = 0.26$ was used. McDougall also found a r.m.s. vertical velocity w to be about 40% larger than the horizontal component. Hopfinger and Toly's measurements on the other hand indicated a ratio $w/u \simeq 1.25$ and these measurements showed that this ratio was independent of distance for $z \gtrsim M$. The value of 1.25 will be used for the present correlation of the data.

The integral scale ℓ was found to increase linearly with z. The coefficient β in $\ell = \beta z$ depends on the stroke for a given grid geometry. For the grid used $\beta = 0.1$ when $S = 1$cm and 0.25 when $S = 4$ cm. A linear interpolation was used for $S = 2.4$ cm. An important point in the interpretation of the results is the spatial decay of the r.m.s. velocity. The $u \propto z^{-1}$ behaviour expressed by equation (2) is in good agreement with the experimental results obtained by Hopfinger and Toly. Thompson and Turner's results on the other hand show a z^{-1} dependence only for $z \lesssim 13$ (see Thompson and Turner, figure 6a). Beyond $z \simeq 13$ the decay is much more rapid. The present results, as will be seen in section 5.1, clearly show the effect of this change in decay rate.

A power law decay is readily obtained from the turbulent kinetic energy equation which is for the homogeneous fluid system

$$\frac{d}{dz}\left(\frac{1}{\rho}\overline{w'p'} + \frac{1}{2}\overline{w'q^2}\right) + \varepsilon = 0 \quad , \tag{3}$$

where w' is the instantaneous vertical turbulent velocity component, p' the pressure fluctuation and ε the rate of dissipation. With $\left(\frac{1}{\rho}\overline{w'p'} + \frac{1}{2}\overline{w'q^2}\right) = r u^3$ and $\varepsilon = A u^3/\ell$, the equation governing the decay rate is :

$$du^3/dz = -\frac{b}{\beta} u^3$$

where $b = A/r$. The solution is

$$u = u_0 (z/z_0)^{-b/3\beta} \tag{4}$$

This indicates that $b/\beta = 3$ when $u \propto z^{-1}$. The correlation coefficient for $\overline{w'q^2}$ and the relative importance of pressure and diffusive transports of kinetic energy is not known. In general these depend on the flow in consideration. The pressure transport could take any proportion, whereas a correlation coefficient of 1/2 that is $\overline{w'q^2} = \frac{1}{2} wq^2$ is a reasonable assumption in regions of strong energy gradients. Writing $\frac{1}{\rho}\overline{w'p'} = \frac{k}{2}\overline{w'q^2}$ and using a correlation coefficient of 1/2 the coefficient r is $r = 1.1(1+k)$. This gives $A = 3.3(1+k)\beta$ which is of order 1 when k is taken equal to 1. From this indirect argument we deduce that the two transport terms are of equal order. For if the pressure transport were neglected a value of $A = 0.33$ would be obtained when $\beta = 0.1$. This value is clearly on the low side and a value of 0.66 when $k = 1$ is more acceptable. In a wake Townsend (see Hinze, 1959) measured also an approximate equipartition between the two transport terms. In the local energy balance model (see section 4) a value of $r = 2.2$ will be used.

3. MODIFICATION OF THE DECAY LAW BY A BUOYANCY FLUX

The energy balance equation (3) in the presence of a buoyancy flux takes the form :

$$\frac{d}{dz}\left(\frac{1}{\rho}\overline{w'p'} + \frac{1}{2}\overline{w'q^2}\right) = -\varepsilon + g\,\overline{w'\rho'}/\rho \quad , \tag{5}$$

where ρ *and* ρ' *are the mean and fluctuating densities. Making the same assumptions as in section 2, equation (5) can be written :*

$$r\frac{du^3}{dz} = -\frac{A}{\beta}\frac{u^3}{z} + g\,\overline{w'\rho'}/\rho \quad . \tag{6}$$

The buoyancy flux term can be expressed in terms of the known buoyancy flux per unit area across the plane $z = 0$ *(virtual origin),*

$$g\frac{\overline{\rho'w'}}{\rho} = \frac{B_0}{D}(z - D) \quad . \tag{7}$$

The relation between B_o *and the imposed surface buoyancy flux* B'_o *depends on the experimental arrangement .*

In the case of an imposed mass flux at velocity V , $B'_o = gV\Delta\rho_0/\rho$ *and when a heat flux* q_o *per unit area is imposed* $B'_o = g\alpha q_0/c_p\rho$ *, where* α *is the coefficient of thermal expansion,* C_p *the specific heat and* ρ *is the fluid density. In obtaining equation (7) it was assumed that the change in density in the mixed layer is independent of* z *. Because of steady state conditions this assumption is valid. The equation governing the local energy flux is then :*

$$\frac{du^3}{dz} = -\frac{b}{\beta}\frac{u^3}{z} + \frac{1}{r}B_0\left(\frac{z - D}{D}\right) \quad , \tag{8}$$

which has the solution, subject to the initial conditions $u = u_o$ *when* $z = z_o$*,*

$$(u/u_0)^3 = (z/z_0)^{-b/\beta}\left\{1 - \frac{1}{r}(z/z_0)^{b/\beta}\frac{B_0}{D}\frac{1}{u_0^3}\left[\frac{Dz}{1 + b/\beta}\left(1 - (z_0/z)^{(1+b/\beta)}\right) - z^2/(2 + b/\beta)\left(1 - (z_0/z)^{(2+b/\beta)}\right)\right]\right\} \tag{9}$$

For values of $z > z_o$ *such that* $(z_o/z)^4 << 1$ *(taking* $b/\beta = 3$*), equation (9) reduces to :*

$$(u/u_0)^3 = (z/z_0)^{-3}\left[1 - \frac{1}{r}\frac{B_0}{(u_0 z_0)^3}\frac{z^4}{D}\left(\frac{D}{4} - z/5\right)\right] \quad . \tag{10}$$

Mixing stops at $z = D$ *and equation (10) shows that* $u = 0$ *at* $z = D$ *when*

$$D = (20r)^{1/4}\left[(u_0 z_0)^3/B_0\right]^{1/4} \quad . \tag{11}$$

Equation (11) indicates that the mixed layer depth scales on $(u_o^3/B_o)^{1/4}$ *(keeping* z_o *constant) rather than on* u_o^3/B_o *as would suggest a Monin-Obukov length scaling. The 1/4th power is a consequence of a neutral decay of the form* $u \propto z^{-1}$. *A decay law of* $u \propto z^{-3/2}$, *as was suggested by Thompson and Turner (1975), would lead to* $D \propto (u_o^3/B_o)^{2/11}$.

4. THE APPARATUS

The experiments were carried out in the mixing box used by McDougall (1979) which is essentially the same as described by Turner (1968). It is 25.4 x 25.4 cm in crossection and 45 cm deep. The frequency of the grid oscillation was adjusted between 1 Hz to 6 Hz and the stroke was set at 1cm., 2.4cm and 4 cm.

A stabilizing buoyancy flux was introduced either by adding continously fresh water on top of a column of salt water or by heating the top of a fresh water column. The arrangements used are shown schematically in figures 1a to 1c.

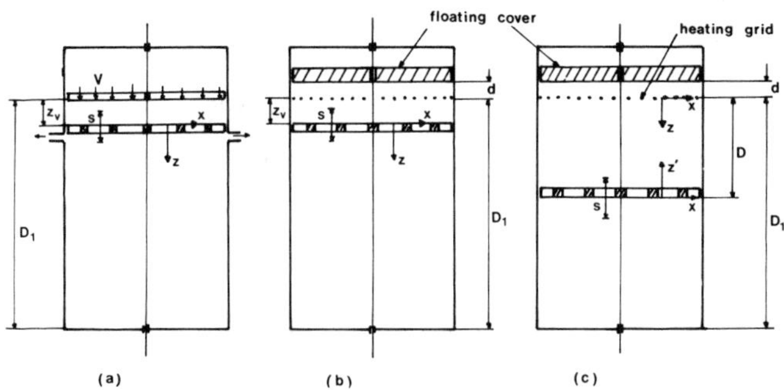

Fig.1. Schematic of the apparatus. (a) mass flux arrangement ; (b) and (c) heat flux arrangement.

The fresh water (mass flux) was introduced uniformely through a 1cm thick foam rubber placed on the surface of the salt water column (figure 1a). Mixed fluid was withdrawn through four holes in the side walls of the mixing box to keep the total water column at constant depth.

Heating was achieved by means of a 110 ohms heating grid, located at a depth d below the water surface (see figure 1b and 1c). This grid consisted of a 0.4 mm diameter enamelled wire and had a 1.4 cm mesh. It was connected to an AC variable power supply. Thermal insulation of the side walls of the box and in particular the floating polystyrene cover shown in figures 1b and 1c reduced thermal losses.

The experiments were conducted by filling the box with salt water in the case of mass flux, or fresh water when a heat flux was imposed. The turbulence generating grid was then started and after some time (a few minutes), when the turbulence was fully established, the mass flux or heat flux was suddenly switched on. Generally, the volume flux or the heat flux, rather than the buoyancy flux, were kept constant but this did not appreciably differ from a constant buoyancy flux during the time the experiments were conducted.

With the turbulence generating grid placed near the buoyancy input plane, as shown in figure 1a and 1b, a top mixed layer formed, whereas when the generating grid was located well below the heating grid (figure 1c), stratification in a thin surface layer occured (when the heating rate was sufficiently high) with a mixed layer below.

The mixed layer depth was determined from shadowgraphs. Dye was also sometimes added at the injection plane. The depth to which this dye was mixed down corresponded to the depth determined from the shadowgraphs. Salinity profiles were determined from samples withdrawn at different depths and the salt concentration measured with a refractometer. Temperature profiles were measured with a thermistor probe.

5. EXPERIMENTAL RESULTS

5.1. Mixed layer depth

The experimentally observed mixed layer depth D *, measured from the virtual origin, located distance* z_v *below the injection or heating plane (see figure 1a and 1b), is plotted logarithmically as a function of* $(u_o z_o)^3/B_o$ *in figure 2. For the experimental configuration shown in figures 1a and 1b the buoyancy flux* B_o *is related with the surface buoyancy flux* $B_o = B_o' D/(D+Z_v)$ *for mass flux (figure 1a) and* $B_o = B_o' D_1 D/(D_1+d)(D+Z_v)$ *for heat flux (figure 1b).*

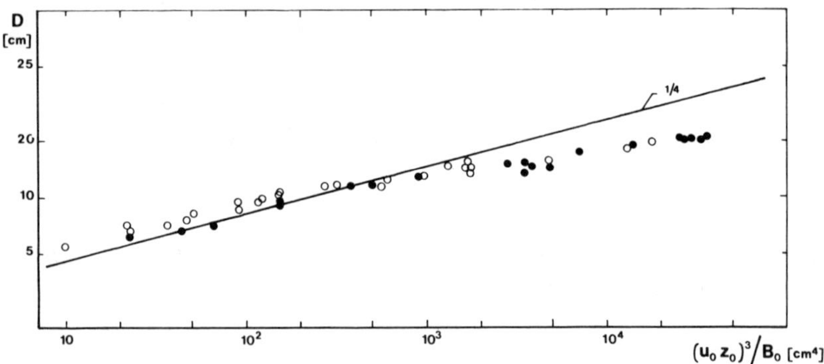

Fig.2. Mixed layer depth D *as a function of* $(u_o z_o)^3/B_o$. *o, mass flux ; ● heat flux. —— calculated from equations (11) with* $r = 2.2$.

The solid line of slope $1/4$ *was calculated from equation (11) using a value of* $r = 2.2$. *Two remarks can be made concerning figure 2 : one is that a correlation according to equation (11) collapses the experimental results on one curve, and two, the actual numerical value of the mixed layer depth is well predicted by equation (11) when* $D \lesssim 13$ *cm. Beyond* $D \approx 13$ *cm a slope 2/11 or less, corresponding to a more rapid neutral decay of the r.m.s. turbulent velocity, would be more adequate. This is in agreement with the two decay zones found by Thompson and Turner and is a consequence of the limited size of the apparatus (see section 2).*

A mixed layer became visible on the shadowgraph shortly after the buoyancy flux was established and in the layer $z > D$ the turbulent velocity decayed rapidly and ceased to exist. In the interfacial layer between the mixed layer and the non-turbulent water column a strong temperature or density gradient developed in time. Temperature profiles which show this behaviour are shown in figure 3a. This figure also indicates that the mixed layer depth remains constant in time. Runs carried out over 20 min. and longer showed no noticeable change in depth of the mixed layer. The base of the mixed layer, determined from the shadowgraphs, is located in about the middle of the interfacial layer. The turbulent front appeared very intermittent and D corresponds to its mean position. As the density gradient increased in time, internal waves trapped in the interfacial layer also became visible on the shadowgraph.

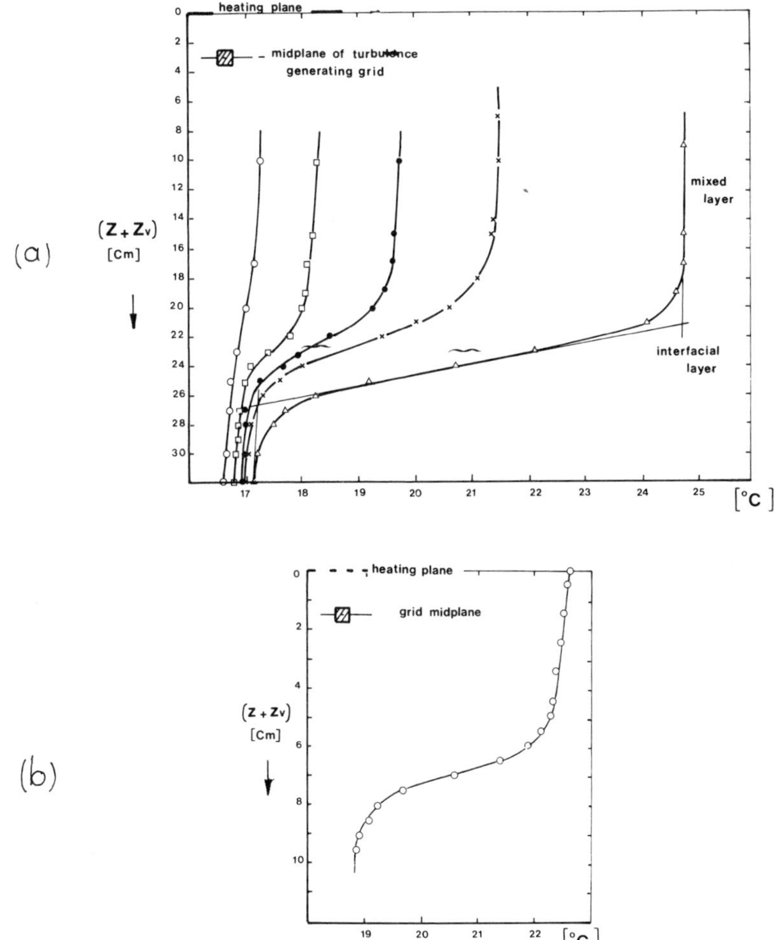

Fig.3. Typical temperature profiles. (a), $(u_o z_o)^3/B_o = 3.3 \times 10^4$; o, 1 min after creation of heat flux ; □, 3 min ; ●, 6 min ; x, 10 min ; △, 27 min ; ~, mixed layer depth determined from shadowgraphs. (b), $(u_o z_o)^3/B_o = 5 \times 10^2$, indicating a weak gradient in the mixed layer.

In the mixed layer the gradient required to carry the heat or buoyancy down to the base is very weak. The profiles shown in figure 3a, which correspond to $(u_o z_o)^3/B_o = 3.3 \times 10^4$ have no apparent gradient in the mixed layer. For smaller values of this parameter a measurable gradient exists as is seen from fig. 3b, corresponding to $(u_o z_o)^3/B_o = 5 \times 10^2$. The gradient effect on the dynamics of the turbulence remains in all cases negligible. Since in a large portion of the mixed layer the turbulence is also only weakly affected by the buoyancy flux the eddy viscosity $u\ell$ (and also uz since $\ell = \beta z$) is approximately constant. In figure 4 the observed average mean temperature gradient is plotted as a function of q/uz, which is the heat flux per unit area, referred to the plane of virtual origin, divided by the eddy viscosity characteristic of eddies of order z. It is noticed that $q \approx \rho c_p u z \, dT/dz$ and this indicates that heat or buoyancy is transported by eddies of size z rather than ℓ. Energetically these large scale eddies are relatively weak but they are very efficient in transporting a scalar quantity.

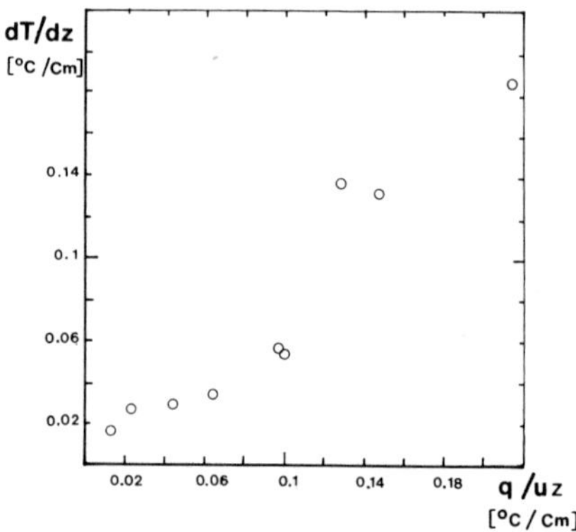

Fig.4. Measured temperature gradient in the mixed layer as a function of heat flux divided by eddy viscosity.

Since the heat flux vanishes at the base of the mixed layer the temperature gradient should go to zero there when a constant eddy viscosity is assumed. Observations indicate however that the gradient persists down to the base of the mixed layer (figure 3b) which implies the existence of a decreasing eddy viscosity as the base is approched. This is in fact consistent with a vanishing turbulent velocity at $z \simeq D$.

5.2. Onset of surface stratification

When turbulence is generated at a depth D below the heating plane, as shown in figure 1c, the surface temperature increases more rapidly than the bulk temperature when, for a given turbulence production, the surface buoyancy flux exceeds a certain value. The solution of equation (6) for the configuration shown in figure 1c indicates that the r.m.s. velocity vanishes at $z' = D$ when

$$B_o = \frac{5r}{\left(D/D_1 + 5\,\frac{D_1 - D}{4D_1}\right)} \left(u_o z_o'\right)^3 / D^4 \qquad (12)$$

The relation between B_o *and* B'_o *is* $B_o = B'_o \dfrac{D_1}{D_1+d}$.

The parameter for correlating the results for rate of surface heating is thus

$$H = \gamma \left[D^4 B_o / (u_o z_o')^3\right] \text{ where } \gamma \equiv D/D_1 + \frac{5}{4}\frac{D_1 - D}{D_1} \quad (\gamma = 1 \text{ when } D = D_1).$$

In figure 5 the rate of change in surface temperature $\Delta T_s = T_s - T_i$, *normalised by* $q_o / \rho C_p (D_1 + d)$, *is plotted logarithmically as a function of H. In these experiments, D was 12,5cm with* D_1 = *24 cm and d = 3 cm. The choice of* $d \gtrsim 3$ *cm, corresponding to* $d/\ell \gtrsim 1.4$ *for the stroke S = 2.4 cm, assures that the turbulence structure at the depth of the heating grid was not affected by the upper boundary where w goes to zero. The influence of a boundary extends to about one integral scale (Thomas and Hancock, 1977).*

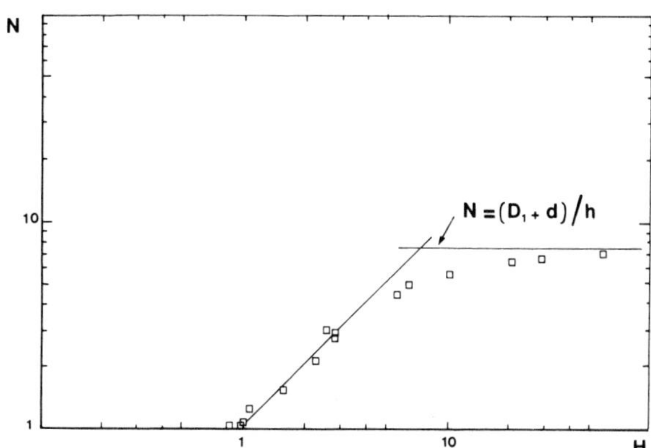

Fig.5. Rate of surface temperature increase $N = \Delta T_s / [q_o / \rho C_p (D_1 + d)]$ *as a function of* $\gamma D^4 / (u_o z_o)^3 / Bo$. *The heating grid was placed 3 cm below the free surface corresponding to 1.4 integral length scales. h is the surface layer thickness, practically equal to d.* $D_1 + d$ *is total depth of the water column.*

Figure 5 shows that the value of N remains equal to unity until a critical value of H is reached which is close to 1. For values of H greater than 1, surface stratification develops and N increases practically linearly with H. Further increase in H leads to a "runaway" stratification of a thin surface layer of thickness h. In this state the surface heating rate reaches assymptotically a value of $N = (D_1 + d)/h$. *The temperature profile given in figure 6 shows that the surface layer is separated by a relatively thin interfacial layer of thickness 0.15 D from the lower turbulent layer. The center of this interfacial layer is located somewhat below the heating grid position, giving* $h \simeq (d + 0.05\ D) = 3.6$ *cm.*

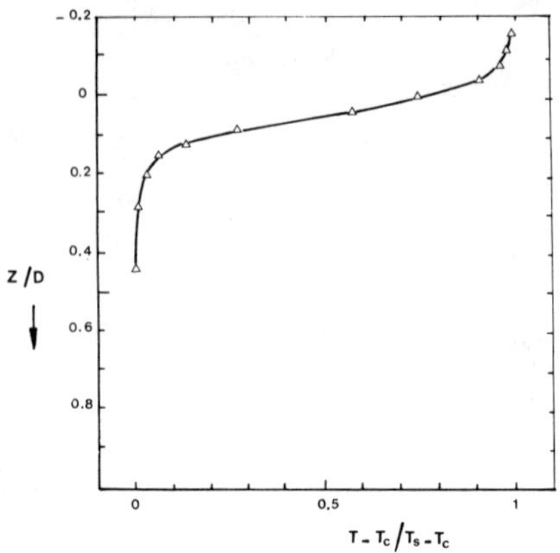

Fig.6. Temperature profile in the surface layer for $H = 21$, taken 5 min. after the heat flux was switched on.

When in Equation (12) $r = 2.2$ is taken, a value of $H \approx 11$ is obtained. At this value, almost all the heat supplied at the surface serves to heat up a thin surface layer only.

6. DISCUSSIONS AND CONCLUSIONS

The results indicate that the mixed layer depth is governed by local energy flux balance which explicitly takes into account the viscous decay of turbulent kinetic energy. It should be emphasised again that the law $D \propto [(U_0 z_0)^3/B_0]^{1/4}$ only applies to the present situation and any other turbulent structure would result in a different scaling.

A more universal expression for D can be obtained when rewriting equation (11) in terms of local velocity rather than the eddy coefficient $u_o z_o$ near the generating plane. Since in equation (11) $u_o z_o = uz$, where u_o is the r.m.s. turbulent velocity without buoyancy flux effects, we can write :

$$D = 44\, u^3 / B_o$$

If, furthermore, a local buoyancy flux $B = B_o \ell/D = \beta B_o$ is taken, that is the flux across a plane distance ℓ away from the base of the mixed layer, the expression for D is :

$$D = 44\, \beta\, u^3 / B \tag{13}$$

This result is the local equivalent to Kitaogorodskii's (1960) global argument (1). In relation (13) the coefficient of proportionality, equivalent to 'a' in (1), depends on the integral scale relative to the mixed layer depth. In the present exp riments, its value varies between 4.4 and 11, which is by a factor of 2 to 5 larger than the value proposed by Kitaigorodskii.

Recent observations by Kondo, Kanechika and Yasuda (1978) in the lower atmosphere give a mixed layer depth of $D \approx 9.5$ m for $u_* \approx 4.8$ cm/sec and $B_o \approx 0.8\ cm^2/sec^3$. This gives a coefficient of proportionality of about 7. Since D is within the constant stress layer, ω remains roughly constant across D and ℓ is of order D, in which case global and local energy arguments are equivalent. The agreement of Kando et al's observations with the present laboratory study is an interesting aspect but it is possible that the mixed layer depth is also affected by an existing interface and the result is fortitious.

When the criterion for onset of surface stratification $H \simeq 1$ is expressed in terms of local velocity u we get

$$B_o D/u^3 \simeq 1 \quad . \qquad (14)$$

In tidal regions, changes from a well mixed zone to zones with surface stratification follow approximately contours of $D/|U|^3$ (Simpson and Hunter, 1974), where $|U|$ is the average velocity over a tidal cycle. If u can be taken proportional to $|U|$ this behaviour agrees with relation (14). Global energy arguments used by Simpson and Hunter give

$$\epsilon C_D \overline{|U|^3} = \frac{1}{2} B_o D \, ,$$

where C_D is a friction coefficient in a quadratic bottom friction law and ϵ expresses the amount of kinetic energy input used for mixing. Observations in tidal regions seem to indicate that $\epsilon \sim 10^{-3}$. When using the present results for transition (equation (14)), we get :

$$\epsilon = \frac{1}{2} \frac{u^3}{C_D \overline{|U|^3}} \qquad (15)$$

A standard friction coefficient of $C_D \approx 2 \times 10^{-3}$ requires $u/|U| \approx 2 \times 10^{-2}$ in order for $\epsilon \approx 2 \times 10^{-3}$.

The criterion for onset of surface stratification is valid only when the initial water column is homogeneous. In general the evolution of the state of a water column would depend on its history.

For large values of H, when surface stratification increases rapidly, the density gradient in the interfacial layer, which gets stronger in time, has an increasing blocking effect on the transfer to the lower layer. Practically all the heat will then go into a thin surface layer only and this processus can be referred to as "runaway" stratification.

Many enlightning discussions with Dr. P.F. Linden during the course of this work are acknowledged.

REFERENCES

Hinze, J.O., 1959. Turbulence, McGraw Hill, p. 400

Hopfinger, E.J. and Toly, J.A., 1976. Spatially decaying turbulence and its relation to mixing across density interfaces. J. Fluid Mech., 78, pp. 155-175.

Kitaigorodskii, S.A., 1960. On the computation of the thickness of the wind-mixing layer in the ocean. Izv. Adak. Nauk, S.S.S.R. Geophys., Ser. 3, 425-431.

Kondo, J., Kanechika, O., and Yasuda, N., 1978. Heat and momentum transfers under strong stability in the atmospheric surface layer. J. Atmos. Sci., 35 1012-1021.

McDougall, T.J., 1979. Measurements of turbulence in a zero-mean-shear mixed layer. J. Fluid Mech., 94, pp. 409-431.

Simpson, J.H. and Hunter, J.R., 1974. Fronts in the Irish Sea. Nature, 250 p. 404.

Thomas, N.H. and Hancock, P.E., 1977. Grid turbulence near a moving wall. J. Fluid Mech., 82, pp. 481-496.

Thompson, S.M. and J.S. Turner, 1975, Mixing across an interface due to turbulence generated by an oscillating grid. J. Fluid Mech., 67, pp. 349-368.

Turner, J.S., 1968. The influence of molecular diffusivity on turbulent entrainment across a density interface. J. Fluid Mech., 23, pp. 639-656.

Turner, J.S., 1973. Buoyancy effects in fluids. Cambridge University Press.

Coherent Structures in Open-Channel Flow

HIROJI NAKAGAWA and IEHISA NEZU
Hydraulics Laboratory
Department of Civil Engineering
Kyoto University
Kyoto 606, Japan

ABSTRACT

In order to reveal the interrelation between the bursting phenomenon occurring originally in the wall region and the macroturbulence such as kolk-boils and cellular secondary current which are observed in the outer region of an actual river, the present study has examined the spanwise spatial structures of open-channel flows with and without the cellular secondary current which was created artificially. New conditional averaging technique that the signals of sampling probe are weighted-averaged by the sorted instantaneous Reynolds-stress signals of detection probe, has been used to evaluate the mean spanwise streaky spacings throughout the flow depth. The direct effect of the cellular secondary current on the bursting motion has not been recognized. It was suggested that the coherent structures of open-channel flow would indicate the double-structure of turbulence in which both bursting and macroturbulence coexist self-consistently.

1. INTRODUCTION

The boundary layer, pipe and open-channel flows are classified as the wall turbulence, which is characterized by the strong shear stress near the wall. The modern turbulence research has revealed that the wall turbulence field consists of both inner and outer layers. The inner layer is characterized by the viscous stress represented by the kinematic viscosity ν, and also the Reynolds stress represented by the friction velocity U_*. The outer layer is the region farther from the wall, in which the strong Reynolds shear stress is fairly relaxed and the effects of outer (mainstream) boundary conditions appear. An open-channel flow field can be also divided into the wall region (inner layer) and the outer region, as shown in Fig. 1. The wall region is controlled by the inner variables of ν/U_*, while the outer region is controlled by the outer variables of the flow depth h and the free-surface velocity U_{max}. Only in the outer region, the effect of the free surface peculiar to the open channel flow appears.

We consider a fully developed, two-dimensional and normal open-channel flow in Fig. 1. The x coordinate is aligned with the mainstream velocity. y is the vertical coordinate to the wall, and z is in the spanwise direction. The velocity fluctuations in the x, y and z directions are u, v and w, and also their r.m.s. values, i.e. the turbulence intensities, are u', v' and w', respectively. And, the edge y_w of the wall region is given by $y_w=100\nu/U_*$. When the Reynolds number $R_*\equiv hU_*/\nu$ is sufficiently large, $y_w/h=100/R_*$ becomes very small and thus the flow field is almost occupied by the outer region (e.g. Nezu 1977). The large-scale turbulence in an actual river observed hitherto by many river engineers corresponds to the fluid motions in the outer region, which were

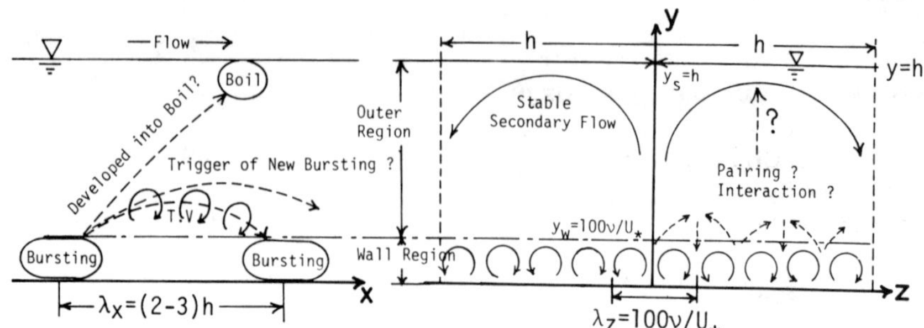

Fig. 1. Illustration of coherent structures in an open-channel flow.

generally termed 'macroturbulence' by Matthes(1947). He proposed the noticeable suggestions that there might exist a strong upward and intermittent current like a tornado near the river bed (he called it 'kolk') and its vortex developed up to the free surface and then became a so-called 'boil' which was observed as a vortex-ring-like swollen water surface of river. The hydraulic behaviours of boils were then examined through aerial photo-survey by Kinoshita(1967) and Coleman(1969), as described in Table 1. They found that a boil formed a low-speed zone and, at both spanwise sides with the spacing of twice the flow depth h, high-speed zones were formed, as shown in Fig. 1. They also suggested that there might exist cellular secondary currents in a straight river which had a pair of counter-rotating streamwise vortices with vortex diameter of about h *(see Figs. 1 and 2)*. Such the macroturbulence in the outer region has been regarded as very important phenomena in the hydraulic engineering, because it might be closely related to formation of various bed configurations and also it might cause sediment transportation and three-dimensional flow patterns in an actual river. However, the origin of the macroturbulence and its physical mechanism are quite unknown at present.

Table 1. Subdivision of wall turbulence field and its typical research.

	Wall Region ($y^+ \leq 100$)	Outer Region ($y^+ \geq 100$)
Phenomena	Bursting Process of ejections and sweeps appears in common in boundary Layer, Pipe and Open Channel Flows.	Bulges Motion (Boundary Layer) Core Motion (Pipe) Kolk-Boil & Cellular Secondary Currents (Open-channel)
Flow Visualization Hydrogen-bubble, Aerial Photosurvey Dye Tracer	Kline et al.(1967, 71, 74, 75) :B.L. Flow Brodkey et al.(1969, 73, 78) : Pipe Flow Grass(1971), Nakagawa & Nezu(1977) : Open-channel Flow	Bulges = Many studies(e.g. Falco 1977) since 1950's (NACA) Boils = Observation of water surface of river (Matthes 1947, Kinoshita 1967, Coleman 1969 Jackson 1976, etc.)
Probe Measurements Hot-wires Hot-films Laser D.V. Conditional Analysis	Laufer et al.(1971) Eckelmann et al.(1974, 77, 79) Hanratty et al.(1974, 77) Nishioka et al.(1977)	Almost in viscous sublayer Streaky Wall Structure
	Willmarth et al.(1972, 73, 75) Blackwelder et al.(1972, 76, 78)	Up to the edge of boundary layer ($y \leq \delta$)
	Sabot et al.(1976)	Up to the center axis ($y \leq D/2$) of pipe flow
	Nakagawa & Nezu(1977, 78, 79, 80)	Up to the water surface ($y \leq h$) of open-channel

On the other hand, many flow visualization studies as well as probe measurements oncs of wall turbulent shear flows in the last decade have found that the wall region is dominated by a sequence of events such as ejections and sweeps which has been termed the bursting phenomenon. This sequence of events is first noticed by the occurrence of low-speed streaks ($u<0$) in the streamwise direction. A part of these streaks lifts up away from the wall, i.e. $v>0$, and then breaks down in an ejection phase. This is followed by a sweep of high-speed fluid ($u>0$) that originates in the outer region and moves at small angle toward the wall, i.e. $v<0$. Then, the most important feature of the bursting phenomenon is to be accompanied with most of the production of turbulent energy and Reynolds stress, i.e. $-uv>0$, and also its occurrence is fairly random at one place or time, while a sequence of the bursting motions has a quasi-ordered process. For further details, Hinze(1975), Laufer(1975) and Willmarth(1975) have written excellent reviews which describe the recent status of research on solid-wall turbulence, especially on the bursting phenomenon of boundary layers.

As shown in Table 1, a close interrelation between the inner and outer layers, i.e. between the bursting and bulge motions in boundary layer, has been suggested through the visual observations and the hot-wire measurements by Laufer & Narayanan(1971), Blackwelder & Kovasznay(1972), Offen & Kline(1975), Falco(1977), Brown & Thomas(1977) and Praturi & Brodkey(1978). A close interrelation between the wall and free-surface regions, i.e. the bursting and boil phenomena in open channel, has been also suggested by Grass(1971), Jackson(1976) and the authors(1978). Fig. 1 shows an illustration of the coherent structures in an open-channel flow, on the basis of previous visual observations. The most notable characteristics are that (1) the bursting period T_B and the streamwise spacing λ_x of the bursting motions in an open-channel flow are controlled by the outer variables of h and U_{max}, while its spanwise spacing λ_z is controlled by the inner variables of ν/U_* (Nakagawa & Nezu 1978), (2) the boil or cellular secondary current appears at a nearly fixed position on the free surface and it is controlled by the outer variables (Kinoshita 1967, Jackson 1976). From these visual results, the following essential and important questions arise.

(1) Does a bursting motion grow into a transverse vortex (T.V.) in the outer region, and in turn does its vortex generate a new bursting motion at the downstream in the wall region ?

(2) Does a strong bursting motion develop up to the free surface and then become a boil or cellular secondary current ?

Therefore, an investigation of the space-time structure of the bursting phenomenon, which occurs originally near the buffer layer and then shows a coherent and organized behaviour during its convection process, will be contributive not only to the elucidation of these questions, but also to the refinement of the physical description of the bursting phenomenon.

In the light of the above, the authors(1980) have recently investigated the structure of space-time correlations of bursting events, such as ejections and sweeps, in the x-y center plane of a smooth open-channel flow *(see Fig. 1)*, by the conditional sampling of the instantaneous Reynolds-stress signals measured simultaneously at two different points. One probe was fixed near the edge of the buffer layer and used as a detection probe of the bursting motions, and the other probe was moved in both streamwise and vertical directions. The temporal and spatial scales of the ejection and sweep motions and their convective process were investigated in detail experimentally.

In the present study, the spanwise spatial structure in the y-z plane will be examined further, in order to reveal some interrelation between the high- and low-speed streaks of the bursting motions and the cellular secondary currents , as shown in Fig. 1.

2. NEW DETECTION SCHEME OF BURSTING PROCESS AND CONDITIONAL AVERAGE

Although the bursting phenomenon occurs fairly random at one place and time, a sequence of its bursting events has a quasi-ordered process. It seems, therefore, that its significant information cannot be obtained by the conventional 'long-time' averaging technique, as pointed out by many researchers. Hence, a conditional sampling method which detects the significant information from the fluctuating velocity signals has been explored by the authors(1980).

In general, the conditional average of an arbitrary velocity signal $u(x_1, y_1, z_1, t+\tau)$ can be defined as follows:

$$\langle u\rangle(x_1, y_1, z_1, \tau) = \frac{\int_T u(x_1, y_1, z_1, t+\tau)\cdot I(x_0, y_0, z_0, t)\, dt}{\int_T I(x_0, y_0, z_0, t)\, dt} \qquad (1)$$

where, $\int_T \cdots\cdots dt$ means $\lim_{T\to\infty} T^{-1}\int_0^T \cdots\cdots dt$. $I(x_0, y_0, z_0, t)$ is the detection function of the turbulence phenomenon in question which occurs at the point of (x_0, y_0, z_0, t). For example, the detection function I of the bulges phenomena in the boundary layer can be definitely given by the intermittency function.

However, the detection function I of the bursting phenomenon near the wall is not sufficiently established yet, although several detection functions have been proposed by Lu & Willmarth(1973), Brodkey et al.(1974), Blackwelder & Kaplan (1976), Wallace et al.(1977), Nakagawa & Nezu(1978) and others. As already pointed out in our previous paper(1978), the instantaneous Reynolds-stress signals $u(t)v(t)$ can be reasonably used as discriminating information, since they are directly related to the bursting phenomenon or the mechanism of turbulence production. Then, because the ejection and sweep events contribute mostly to the turbulence production, the sorting functions of ejection, $I_e(t)$, and sweep, $I_s(t)$, are defined here, respectively, as follows:

$$I_e(t) = \begin{cases} 1 & \text{for } u(t)<0 \text{ and } v(t)>0 \\ 0 & \text{otherwise} \end{cases}, \quad I_s(t) = \begin{cases} 1 & \text{for } u(t)>0 \text{ and } v(t)<0 \\ 0 & \text{otherwise} \end{cases} \qquad (2)$$

Unfortunately, $I_e(t)$ or $I_s(t)$ should not be regarded directly as a detection function $I(t)$ for the bursting motions, because the ejection and sweep events in the sorted Reynolds-stress signals still contain the interaction-like motions. So, Lu & Willmarth(1973), Sabot & Comte-Bellot(1976) and the authors(1977 a, 78, 79) introduced a threshold level H in the Reynolds stress signals, and assumed that the bursting motions occurred only when $|u(t)\cdot v(t)| \geq H$. However, the determination of the threshold level H is more or less arbitrary, because stationary values of the conditionally averaged patterns against the variation of H do not exist definitely (Nakagawa & Nezu 1978). Especially in the water flow measurements, such a threshold method has further the following shortcomings. Since the bursting period T_B is of the order of one second in ordinary laboratory experiments *(also see Fig.20)*, continuous velocity signals longer than 500 seconds are needed, in order to obtain an accurate ensemble average by the threshold method. However, the accurate signal data cannot be expected by long-time measurements, because the sensitivity of hot-film anemometers will decay by coating with the impurities in water flow.

In order to overcome the above shortcomings of the threshold method, the authors(1980) have recently proposed a new detection function weighted by the instantaneous Reynolds-stress signal itself, i.e. $I(t) \equiv u(t)v(t)\cdot I_e(t)$ for ejection phase and $I(t) \equiv u(t)v(t)\cdot I_s(t)$ for sweep phase. Then, the conditional average of the space-time correlations of ejection and sweep motions can be definitely given by, respectively,

$$<u>_e(x_1, y_1, z_1, \tau) = \frac{\int_T u(x_1, y_1, z_1, t+\tau)\cdot\{u(t)v(t)I_e(t)\}_0\, dt}{\int_T \{u(t)v(t)I_e(t)\}_0\, dt} \quad (3)$$

$$<u>_s(x_1, y_1, z_1, \tau) = \frac{\int_T u(x_1, y_1, z_1, t+\tau)\cdot\{u(t)v(t)I_s(t)\}_0\, dt}{\int_T \{u(t)v(t)I_s(t)\}_0\, dt} \quad (4)$$

where, the suffix $\{\ \}_0$ denotes the position of the detection probe.

Eqs.(3) and (4) imply that the stronger instantaneous Reynolds-stress is more contributive to the average structure of the bursting phenomenon. Although the above method cannot detect each instantaneous bursting event separately, it is preferable to the threshold method, because the structure of bursting motions can be evaluated more objectively without introducing any arbitrary parameter and the random components such as background turbulence are to be cancelled on averaging larger amounts of data.

The validity of Eqs.(3) and (4) was confirmed by the authors(1980) through comparison with the results obtained by previous threshold methods, e.g. method of $|u(t)v(t)| \geq H$, VITA-method of Blackwelder & Kaplan(1976) and pattern recognition of Wallace et al.(1977).

3. EXPERIMENTAL TECHNIQUES AND DATA PROCEDURES

3.1 Experimental Set-up

Three groups of experiments on a two-dimensional, fully developed and normal turbulent flow in a smooth open-channel, whose hydraulic conditions are described in Table 2, were conducted in a tilting flume 15m long, 50cm wide and 30cm deep at the Hydraulics Laboratory in Civil Engineering, Kyoto University. This experimental set-up was the same as that previously used by the authors(1977 a, 78, 79, 80). A settling chamber and screens to prevent the occurrence of large-scale disturbances and the distortion of main surface flow were set up at the entrance of the channel, and consequently a fully developed turbulent shear flow was obtained at the test section 9.5m downstream from the entrance.

By the way, most of the previous researches on the bursting phenomenon dealt with that over a smooth flat bed. The bursting phenomena over a rough bed, however, are little known at present since the existence of the wall region itself becomes unclear, although a burst-like behaviour has been discovered even over a completely rough bed by Grass(1971) and Nakagawa & Nezu(1977 a). On the other hand, all of the macroturbulences in the outer region of river *(see Table 1)* were observed over the rough flat or wavy sand beds. And, it is suspicious whether they appear over a smooth flat bed. Thus, a reasonable device is needed by which a macroturbulence such as cellular secondary currents is created artificially over a smooth flat bed. So, in the present study, three kinds of experiments were carried out as follows.

Uniform flow over a smooth bed (Cases R-1 and R-2). This flow is the most fundamental one in open-channels, since the bed or wall is uniformly smooth flat and solid one. Two kinds of hydraulic conditions were set, as described in Table 2. Case R-1 had the nearly same hydraulic condition as in our previous experiments that have examined the bursting structure in the x-y plane (Nakagawa

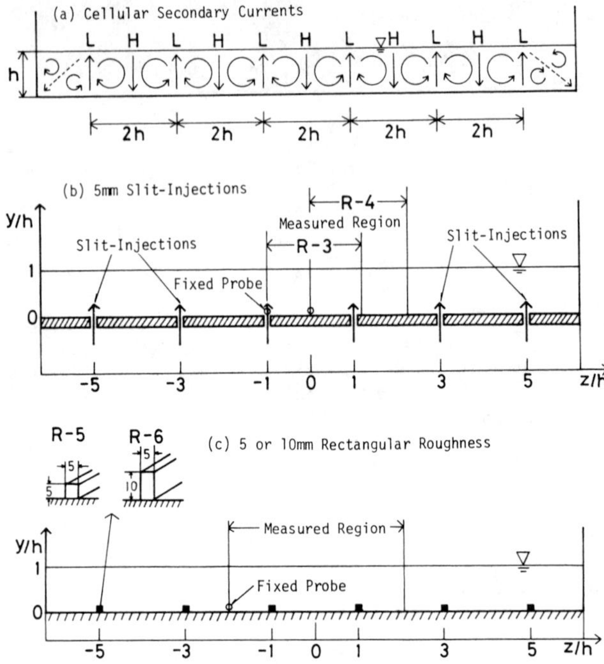

Fig. 2. Illustration of cellular secondary currents and the experimental set-up.

Table 2. Hydraulic conditions for the experiments.

Series of Experiments	Case R-1	Case R-2	Case R-3	Case R-4	Case R-5	Case R-6
Flow Conditions	Throughout Solid Bed		With Interval Injections		With Interval Roughness	
Flow depth h (cm)	7.7	7.9	4.0	4.0	4.0	4.0
Discharge Q (l/sec)	5.5	2.2	2.8	2.8	2.6	2.6
Free-surface velocity U_{max} (cm/sec)	20.4	8.4	21.0	21.0	22.2	19.9
Mean velocity U_m (cm/sec)	14.1	5.5	13.9	13.9	13.0	13.0
Friction velocity U_*(cm/sec)	0.94	0.42	0.96	0.96	1.05	0.91
Injection velocity v_0(cm/sec)	0.0	0.0	0.47	0.53	0.0	0.0
Reynolds number $R_e \equiv U_m h/\nu$	1.1×10^4	0.42×10^4	0.54×10^4	0.54×10^4	0.52×10^4	0.52×10^4
$R_* \equiv U_* h/\nu$	696.	318.	375.	376.	408.	358.
Froude number $F_r \equiv U_m/\sqrt{gh}$	0.16	0.063	0.22	0.22	0.21	0.21
Water temperature T_w(°C)	18.7	19.6	19.2	19.7	19.1	19.9
Roughness Element					5x5mm	5x10mm

et al. 1980). The Reynolds number in Case R-2 was equal to about a half of that in Case R-1, and it became then easier to compare the results of Case R-2 with visual data.

Flow with 5mm-slit injections with uniform spanwise spacings(Cases R-3, R-4) If the cellular secondary currents exist in an open-channel flow, they must have the same configuration as described in Fig. 2(a). In order to artificially create significant cellular secondary currents in the open channel, 5mm-slit injection flows were given through the porous plate, at spanwise intervals of 2h, as shown in Fig. 2(b). The appatatus of uniform flow of injections through the porous plate of 1040mm long, 490mm wide and 15mm thick, is described in detail in our previous paper(1979). On the basis of this experience, the test section was chosen at the position of 840mm downstream from the upstream edge of the porous plate, and the injection velocity v_0 was equal to about a half of the friction velocity U_*. In order to examine the effect of injection on the bursting detection probe, the probe was set at the injection point for Case R-3, and at the intermediate point between the adjacent injections for Case R-4, respectively , although the hydraulic conditions were nearly equal in both cases *(see Fig.2)*.

Flow with rectangular longitudinal roughness elements (Cases R-5 and R-6). Utami & Ueno(1977) pointed out that stable cellular secondary currents appeared in open-channel flows when the rectangular roughness elements were aligned in the mean flow direction at spanwise intervals of 2h, as described in Fig. 2(c). Similar effects will appear in the current lineations which are spanwise alternating ridges and broad hollows (Jackson 1976). So, in the present study, the rectangular and polished brazen roughness elements of 5x5mm cross-section for Case R-5 and 5x10mm cross-section for Case R-6, respectively, were aligned parallel to the mean flow *(see Fig. 2(c))*. These longitudinal roughness element with tapered upstream edge was 6000mm long, and the position of the test section was chosen 5000mm downstream from its tapered edge. The hydraulic conditions of Cases R-5 and R-6 were nearly the same as those of Cases R-3 and R-4.

3.2 Probe Measurements and Data Procedures

A V-type dual sensor (DISA 55A89) and an I-type single sensor (DISA 55A85) hot-films were used as the detection(fixed) probe and the sampling(movable) probe , respectively. As shown in Fig. 2, the position z_0 of the detection probe was fixed at the center of the channel, so that $z_0/h=0$ for Cases R-1 and R-2, $z_0/h=-1$ for Case R-3, $z_0/h=0$ for Case R-4 and $z_0/h=-2$ for Cases R-5 and R-6, respectively. The position z_1 of the sampling probe was systematically moved over the distances longer than 2h. It was set at about 90 different points in number for each run, in the y-z plane at $x_0 \equiv x_1 \equiv 0$. The vertical positions of the detection and sampling probes were set equally to each other, in which $y \equiv y_0 \equiv y_1 =$ 4, 7, 13, 24, 40 and 60mm for R-1 and R-2, y=4, 10, 20 and 35mm for R-3 and R-4, and y=4, 7, 10, 20 and 35mm for R-5 and R-6, in order to obtain the representative measured points in both wall and outer regions. Of course, the probe was arranged so as to avoid any effect of the wake by another probe.

The output signals of the anemometers were recorded in analog form by using an FM tape recorder for about two minutes at each measured point. The effects of impurities and bubbles in the water on the hot-films were almost negligible in this duration, and thus a stable operation of the anemometers was obtained. The output signals were then reproduced for converting into digital form, with the sample size N=10,000 and the sampling frequency f=100Hz, that is, the average evaluation time T=N/f=100 seconds. This sampling frequency f was chosen so that the spectral analysis could be performed at least to the extent of the inertial subrange (Nezu 1977, also see Fig. 18).

The temperature difference between the water (T_W) and the hot-film sensor (T_S) was set about 20°C through all tests. The maximum frequency response f_{max} of the hot-films is 30 kHz in the technical data of DISA, and it was confirmed by the spectral analysis (Nezu 1977) that f_{max} > 1000Hz in ordinary hydraulic conditions, which was sufficiently high for the present study. The hot-film anemometers were calibrated by towing the probe over a fixed length of about 2m in still water, by making use of a movable carriage driven by a variable motor before and after each run of measurements.

Consequently, the output voltage signals could be easily converted into the velocity signals through these calibration curves. Then, some statistical analyses described previously were carried out by a large digital computer, the FACOM M-200, at the Data Processing Center, Kyoto University. Most of the results were then plotted by an on-line X-Y plotter.

4. EXPERIMENTAL RESULTS AND DISCUSSION

4.1 Mean Velocity and Turbulence Intensities

Fig. 3 shows the mean velocity distributions $U^+ \equiv U/U_*$ measured by the detection probe at each test section of z_0 described in 3.2. The solid line is the well-known log-law distribution, which is given by the van Driest type for $y^+ \leq 30$, where the Karman constant $\kappa=0.4$ and the van Driest damping constant A=27 are used. The friction velocity U_* could be reasonably determined so that the experimental values of U, u', v' and $-\overline{uv}$ might coincide with the theoretical curves as accurately as possible. Obviously, the experimental values in Fig. 3 agree very well with the log-law profile. Fig. 4 shows the turbulence intensities u'/U_* and v'/U_* measured by the detection probe. Their experimental values are in good agreement with the theoretical curves which were given by Nezu(1977). Fig. 5 shows the correlation coefficient $R \equiv -\overline{uv}/u'v'$ of the Reynolds stress. The scatter of the present data is almost within that of the previous data. Further, the contribution rates of each bursting event to the Reynolds stress coincided well with the previous data of Nakagawa & Nezu(1977 a). From the above examination, it can be concluded that even in the flows with interval injections(Cases R-3 and R-4) and longitudinal roughness elements (R-5 and R-6),

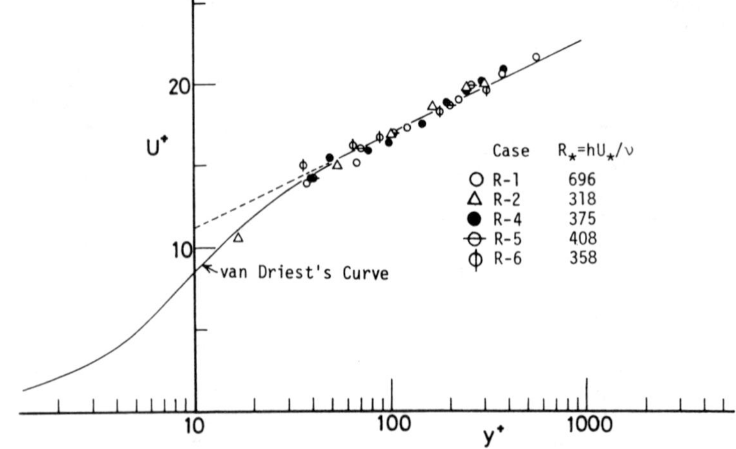

Fig. 3. Mean velocity distributions at the detection probe.

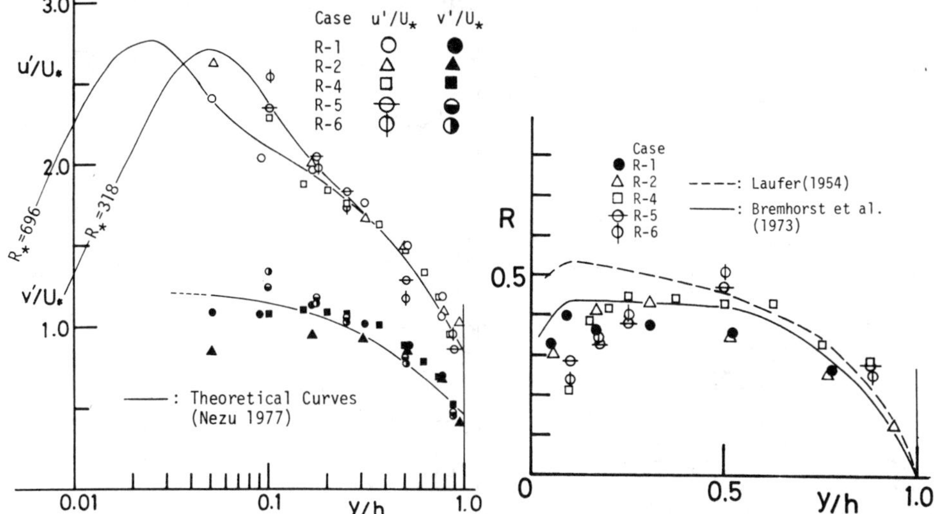

Fig. 4. Turbulence intensities at the detection probe.

Fig. 5. Correlation coefficient of the Reynolds stress at the detection probe.

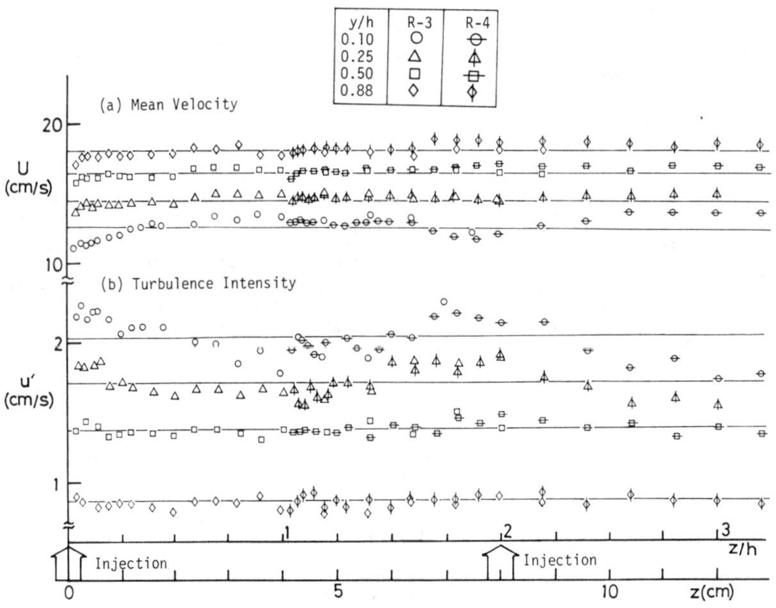

Fig. 6. Variations of mean velocity U(z) and turbulence intensity u'(z) against the spanwise direction, when 5mm slit-injections are given at spanwise intervals of 2h.

the turbulence characteristics of two-dimensional and fully developed open-channel are satisfactorily established at the position of the detection probe, and thus the bursting phenomenon can be accurately detected by the present probe measurements.

Next, Fig. 6 shows the variation of the mean velocity U(y,z) and the turbulence intensity u'(y,z) obtained by the sampling probe measurements at the different elevations, when the 5mm-slit injections were given spanwise at intervals of 2h. Obviously, a region of the low mean velocity and high turbulence intensity appears at intervals of 2h in the vicinity of the wall. This feature is consistent well with that of a secondary current in Fig. 1 (e.g. Müller 1977). However, the above feature becomes more indiscernible as the water surface is approached, probably due to a strong diffusion of the injections. Consequently, a large-scale cellular secondary current across the whole flow depth was not created here in Cases R-3 and R-4.

In the same manner, Fig.7 plots the spanwise variation of U(z) and u'(z) for Case R-5, where the long longitudinal roughness elements of k=5mm high were set spanwise at intervals of 2h. The measurement zone Δz was up to h at the elevation of y/h=0.1 which was under the top of the roughness elements, i.e. k/h=0.125. At the other elevations of y/h, Δz was up to 4h. Now, the spanwise average values of U(z) and u'(z) are denoted as $\overline{U}$ and $\overline{u}'$, respectively. Then, the low mean velocity ($U<\overline{U}$) and high turbulence intensity ($u'>\overline{u}'$) appear near the position of Δz/h =1 and 3, while the high mean velocity ($U>\overline{U}$) and low turbulence intensity ($u'<\overline{u}'$) appear at the neighbourhood of Δz/h=0, 2 and 4. These cyclic variations of U(z) and u'(z), whose wave length

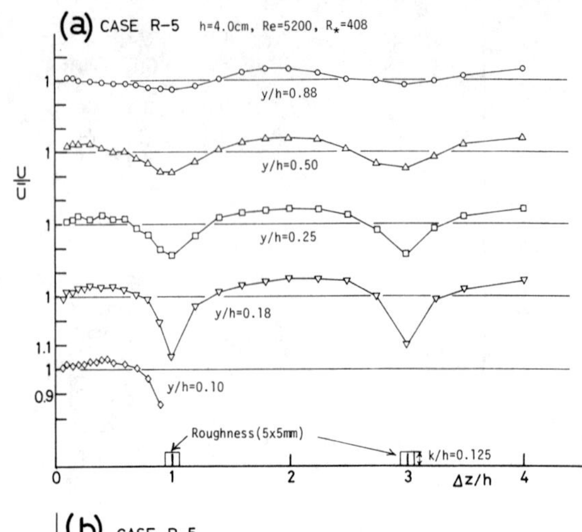

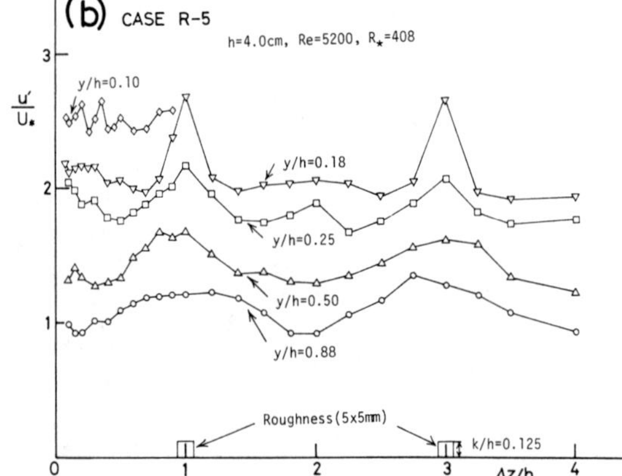

Fig. 7. Variations of U(z) and u'(z) against Δz/h, when the longitudinal roughness elements are set at spanwise intervals of 2h.

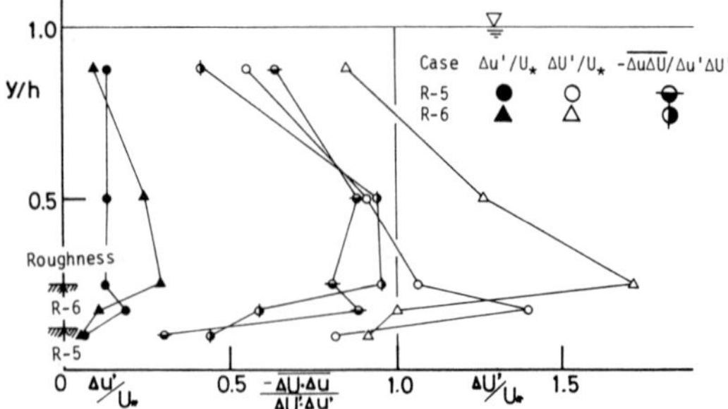

Fig. 8. Correlation coefficient between U(z) and u'(z).

is equal to 2h, are seen clearly throughout the water depth. The same results were also obtained for Case R-6, where the roughness height was $k/h=0.25$. The r.m.s. values $\Delta U'$ and $\Delta u'$ of the mean velocity difference $\Delta U \equiv U-\overline{U}$ and the turbulence intensity difference $\Delta u \equiv u'-\overline{u}'$, respectively, in respect to the spanwise direction, z, are shown in Fig. 8, together with the correlation coefficient $R_\Delta \equiv -\overline{\Delta U \Delta u}/\Delta U' \Delta u'$ between the two. By comparison between Figs. 3, 4 and Fig. 8, these values of $\Delta U'$ and $\Delta u'$ are less than 10 percent of $\overline{U}$ and $\overline{u}'$, respectively. It should be, however, noticed that R_Δ has positive values against the whole variation of y/h, and especially it shows a large positive correlation between ΔU and $-\Delta u$ at the middle of the flow depth, that is, $R_\Delta \simeq 0.9$ at $y/h \simeq 0.5$. The typical cellular secondary currents in Fig. 2 produce the largest ascending current, which probably corresponds to $\Delta u>0$, in the zone of the low mean velocity ($\Delta U<0$) and the largest descending current, which probably corresponds to $\Delta u<0$, in the zone of the high mean velocity ($\Delta U>0$), at the middle position of the flow depth. Hence, the present experimental values of R_Δ may satisfactorily explain these features of the secondary currents (cf. Müller 1977). In other words, the cellular secondary currents must have been created throughout the flow depth by setting the longitudinal roughness elements at spanwise intervals of 2h.

4.2 Spanwise Streaky Structure of Bursting Motions

It is well recognized by many visual observations that the spanwise spatial structure in the wall region consists of the high- and low-speed streaks, whose mean spacing $\overline{\lambda}$ is nearly equal to constant, i.e. $\overline{\lambda} \simeq 100\nu/U_*$. Our earlier visual study (1977 b) of an open-channel flow has also found that, although there existed clearly a streaky structure of $\overline{\lambda}^+ \simeq 100$ in the wall region, it became gradually indistinct outside the wall region and its spacing $\overline{\lambda}^+$ tended to increase with y/h. In order to obtain more quantitative evaluation, the spatial structures of $\langle u \rangle_e(y,z,\tau)$ and $\langle u \rangle_s(y,z,\tau)$ have been analyzed by making use of the conditional average of Eqs.(3) and (4), where $\Delta x \equiv 0$, $y \equiv y_0 \equiv y_1$ and $z \equiv z_1 \equiv \Delta z$.

Fig. 9 shows an example of the conditionally averaged spanwise structure $\langle u \rangle(y,z,0)$ obtained at $Re=1.1 \times 10^4$ (Case R-1), in which $\langle u \rangle_e(y,z,0)$ of the ejections is normalized by $\langle u \rangle_e(y,0,0)$. Similar results were also obtained for the sweeps. Evidently, the spanwise spatial scale increases

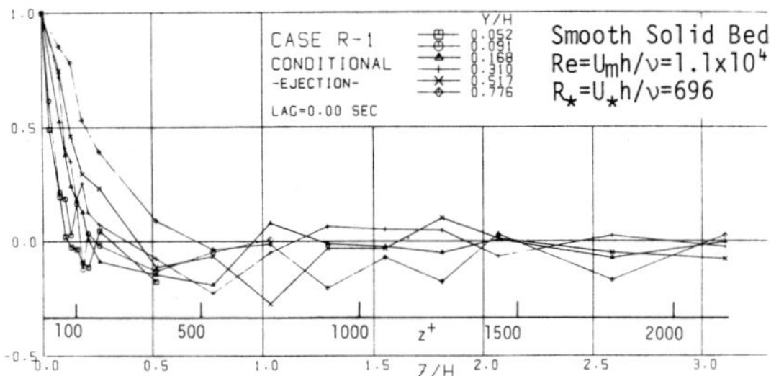

Fig. 9. Spanwise spatial structure of the ejections (Case R-1).

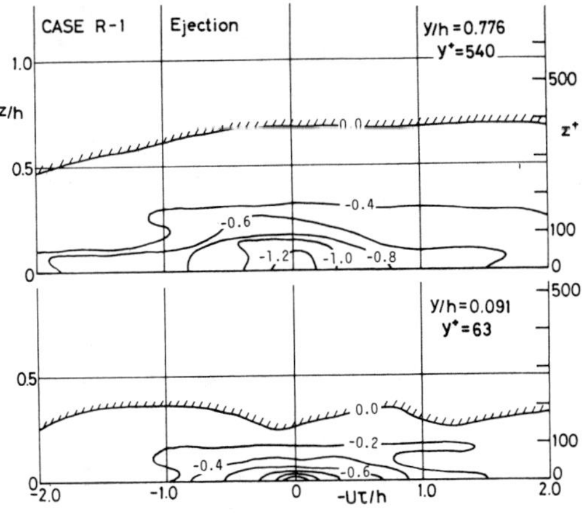

Fig. 10. Space-time iso-patterns of <u/u'> in the ejection phase.

with an increase of y/h. However, an existence of the stable cellular secondary current shown in Fig. 1 is not clear, because $<u>(z)$ decays fairly rapidly up to z/h=2. Surely, some periodicity of $<u>(z)$ is recognized near the wall, and it perhaps corresponds to the high- and low-speed streaks observed visually. Its intensity, however, is not so large, as compared with the results of Gupta et al. (1971) who analyzed the spatial structure in the viscous sublayer by VITA-method of the ten-hot-wire-rake signals, and Nishioka et al.(1977) who detected the bursting motions at $y^+=10$ by a three-hot-wire rake probe *(see Figs. 11 and 12)*. This difference can be explained in the following respects. (1) If multi-point-rake probe more than two sensors is used as a detection one, the detection criterion of the bursting motions will be able to be set more precisely and reasonably. (2) In the viscous sublayer($y^+\leq 10$), the coherent streaky structure appears, in itself, more clearly (Nakagawa & Nezu 1977 b). It is, however, difficult at present to use the multi-point-rake probe in water flow. Moreover, it has been found that the wall streaks became gradually intricate and indistinct farther from the wall ($y^+\geq 10$). Hence, in order to examine the streaky structure in both wall and outer regions, a new analytic method will be necessary as proposed in the next section.

Next, Fig. 10 shows an example of space-time structure of the ejections, in which the contour lines of $<u/u'>(z,\tau)$ are described at intervals of 0.2. Similar figures were also obtained for the sweeps. Because the convection velocity of the ejection motions is nearly equal to about 0.8 times the mean velocity U (Nakagawa et al. 1980), the value of $U\times(-\tau)$ cannot be precisely regarded as the convection distance Δx, but it is only a rough approximation of Δx. It should be noticed that the bursting structure enlarges similarly as it goes farther from the wall, and that it indicates a longitudinally elliptic shape extended in the streamwise direction. This shape is similar to the structure of the conventional space-time correlations evaluated by a long-time average, and thus it is recognized that the bursting motion is a kind of large-scale eddy structure, as the authors(1980) pointed out about the x-y plane structure.

4.3 Evaluation Method of Mean Streaky Spacing

Lee et al.(1974) observed the footprint of the coherent streaky structure on the wall by making use of electrochemical techniques, and they reported that the instantaneous velocity pattern of the wall streaks was approximately expressed by a cyclic function such as $\cos(2\pi z/\lambda)$, where λ is the wave length of the spanwise wall streaks. That is,

$$u(z,\ t)= u(0,\ t)\cdot\cos(2\pi z/\lambda) \sim \lambda\cdot\cos(2\pi z/\lambda) \qquad (5)$$

By ensemble averaging u(z, t) in respect to its wave length λ, we can obtain

$$\frac{<u>(z)}{<u>(z=0)} = \frac{\int_0^\infty \lambda\cdot\cos(2\pi z/\lambda)\cdot P(\lambda)\ d\lambda}{\int_0^\infty \lambda\cdot P(\lambda)\ d\lambda} \qquad (6)$$

where, $P(\lambda)$ is the probability of the occurrence of λ. Eq.(6) is equivalent to $<u>(z)$ of Eq.(1), whose average is weighted by λ itself. Our visual observation (1977 b) has found that $P(\lambda)$ was expressed well by the log-normal distribution;

$$P(\lambda)= \frac{\log e}{(2\pi)^{1/2}\sigma_0\lambda}\cdot\exp\{-\frac{1}{2}(\frac{1}{\sigma_0}\log\frac{\lambda}{\lambda_0})^2\} \qquad (7)$$

where, $\sigma_0\equiv(\log e)(\ln(1+\psi^2))^{1/2}$, $\lambda_0\equiv(1+\psi^2)^{-1/2}\cdot\bar{\lambda}$ and e=2.71828.
The calculated curves of Eq.(6) are shown versus $z/\bar{\lambda}$ in Figs. 11 and 12, as a parameter of $\psi\equiv\sigma/\bar{\lambda}$ (where, $\bar{\lambda}$ is the mean value of λ, and σ is the standard

deviation of λ). Even the curve of Eq.(6) with ψ=0.5† evaluated from our visual data does not show any noticeable periodicity. This means that the high- and low-speed streaks are fairly smoothed even by means of conditional average of two probe signals.

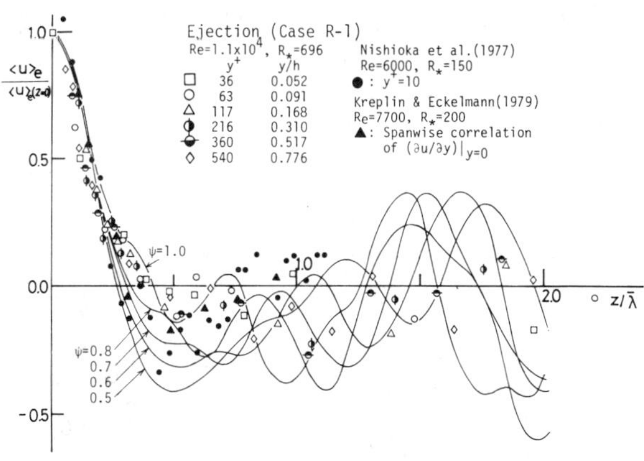

Fig. 11. Spanwise spatial structure of ejections normalized by the mean spacing $\bar{\lambda}_e$.

The effect of ψ on Eq.(6), however, is fairly small up to the first zero-crossing point z_c. Then, by using ψ=0.7†, the following equation can be obtained from Eq.(6).

$$\int_0^{z_c} \frac{\langle u\rangle(z)}{\langle u\rangle(z=0)}\, dz = 0.16\bar{\lambda} \quad (8)$$

The mean spacing $\bar{\lambda}$, therefore, can be determined tentatively to satisfy Eq.(8) with the experimental values in Fig. 9. The experimental values of <u> are plotted in Fig. 11, versus $z/\bar{\lambda}$, i.e. the spanwise separation distance normalized by the mean spacing $\bar{\lambda}$ evaluated from Eq.(8). Fig. 11 also shows the data obtained in an air pipe by Nishioka et al. (1977), and the recent data of Kreplin & Eckelmann(1979) who measured the spanwise correlations of the velocity gradient at the wall by flush-sensor hot-films in an oil channel. A good agreement between the experimental and calculated values is recognized up to $z/\bar{\lambda} \simeq 0.5$, which corresponds to the spacing between the neighbouring high- and low-speed streaks. Therefore, the present evaluation method of $\bar{\lambda}$, which corresponds to a kind of integral scales *(see Eq.(8))*, seems to be reasonable. Also, these spatial features are true of the sweeps, as shown in Fig. 12.

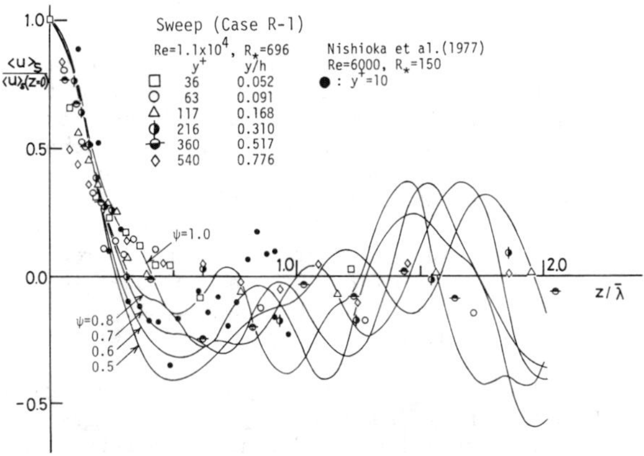

Fig. 12. Spanwise spatial structure of sweeps normalized by the mean spacing $\bar{\lambda}_s$.

4.4 Spanwise Streaky Structure in Uniform Flow over a Smooth Bed

Fig. 13 shows the mean streaky spacings $\bar{\lambda}^+ \equiv \bar{\lambda} U_*/\nu$ versus y^+ for the ejections and sweeps in the uniform flow over a smooth bed (Cases R-1 and R-2), which were systematically evaluated from Eq.(8). For a comparison, in Fig. 13 are replotted the typical visual data of Nakagawa & Nezu(1977 b) who used hydrogen-bubble techniques, and the previous anemometry data of Morrison et al.(1971), Gupta

---

† The value of the coefficient ψ of variation of the bursting period is nearly equal to 0.5-1.5 (Nakagawa & Nezu 1978).

et al.(1971) and Nishioka et al.(1977). Although there is some scattering among these experimental results, a good agreement between the previous and present data is recognized. This means, therefore, that the present evaluation method of the mean streaky spacing $\overline{\lambda}$ is quite reasonable.

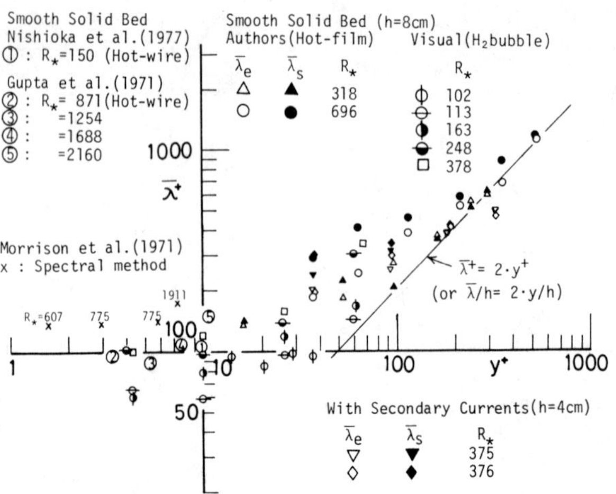

Fig. 13. Variations of the mean spacing $\overline{\lambda}^+$ between the high- and low-speed streaks, as a function of y^+.

$\overline{\lambda}^+$ is nearly constant, i.e. $\overline{\lambda}^+ \simeq 100$, in the buffer layer ($y^+ \leq 30$). It is surprising that Morrison et al. (1971) could obtain the streaky spacing $\overline{\lambda}^+=135$ in the viscous sublayer ($y^+ \leq 5$) by the long-time averaging spectral method, in contrast to the remarks of Gupta et al. Lee et al. (1974) and Kreplin & Eckelmann(1979) could also obtain the streaky spacing $\overline{\lambda}^+ \simeq 100$ of the velocity gradient at the wall by the long-time averaging correlation method. These results indicate that the streaky structure of the viscous sublayer can be detected even by the conventional long-time averaging methods. In other words, this suggests that the viscous sublayer may be regarded as a stable 'footprint' of the bursting motions.

Next, $\overline{\lambda}^+$ increases gradually in the region of $y^+ \geq 30$, and it is approximated by a relation of $\overline{\lambda}^+ = 2y^+$ in the outer region. This suggests that there may exist an interaction or pairing (coalescence) between the high- and low-speed streaks and that its spacing on the water surface becomes nearly equal to 2h, as shown in Fig. 1. As seen in the successive motion pictures by the authors(1977), the wall streaks coalesce and break up themselves repeatedly in space and time, and consequently their mean spacing $\overline{\lambda}^+$ is kept to be constant. The coalescence or pairing mechanism is recognized in common in the coherent turbulent structures such as mixing layer, boundary layer, jets and wakes (e.g. Laufer 1975). On the other hand, the break-up mechanism is caused by an inrush motion of accelerated fluids, i.e. sweeps, in the Offen-Kline model(1975). Consequently, the coalescence of the streaky structure seems to be predominant over its break-up in the region of $y^+ \geq 30$, because the inrush motion becomes weaker farther from the wall (Grass 1971 and Nakagawa & Nezu 1977 a). This can reasonably explain the above most important feature of the mean streaky spacing outside the buffer layer.

4.5 Spanwise Streaky Structure in Open-channel Flows with Cellular Secondary Currents

As an example, we consider Case R-5, where the essential features of the secondary currents appear throughout the flow depth, as mentioned before, and the spanwise structure $\langle u \rangle_e(z)/\langle u \rangle_e(z=0)$ of ejections

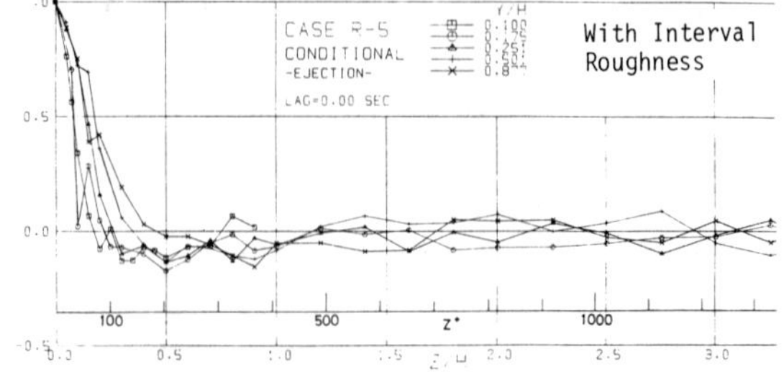

Fig. 14. Spanwise spatial structure of the ejections (Case R-5).

obtained by the measurements is shown in Fig. 14. As compared with Fig. 9 of Case R-1 where the secondary currents did not appear, a remarkable difference between the two cannot be recognized. This was also true of the other cases.

By the same evaluation method of Eq.(8) as in Cases R-1 and R-2, the mean spanwise streaky spacings $\overline{\lambda}_e$ and $\overline{\lambda}_s$ of the ejections and sweeps, respectively, were obtained for open-channel flows with cellular secondary currents. Figs.15 and 16 are an inner variable description ($\overline{\lambda}^+$ *vs.* y^+) and an outer variable one ($\overline{\lambda}/h$ *vs.* y/h), respectively, of all data in the present experiments. The most striking result is that the spanwise spacings $\overline{\lambda}_e$ and $\overline{\lambda}_s$ near the wall in Cases R-3, R-4, R-5 and R-6 coincide well with those obtained over the uniform solid bed of Cases R-1 and R-2, no matter how definite feature of the secondary currents appears in this region, as shown in Figs. 6 and 7. $\overline{\lambda}/h$ increases monotoneously with an increase of y/h, and tends to approach a relation of $\overline{\lambda}/h = 2y/h$. However, near the water surface, i.e. $y/h=0.88$, the values of $\overline{\lambda}/h$ for R-3 ~ R-6 are somewhat less than those for R-1 and R-2. Considering that the definite feature of the secondary currents could not be detected near the water surface in the case of R-3 and R-4, as mentioned earlier in Fig. 6, this difference between the two is probably caused not by the secondary currents, but the other effects such as small water waves. (The flow depth h was nearly equal to 8cm for R-1 and R-2, while it was nearly equal to 4cm for R-3 ~ R-6.)

Fig. 17 shows the ratio, $\overline{\lambda}_s/\overline{\lambda}_e$, of the sweep scale to ejection scale, against y/h. Although some scattering is seen among the data, a systematic variation affected by the secondary currents cannot be recognized. That is to say, irrespective of the existence of secondary currents, the spanwise scale of the sweeps

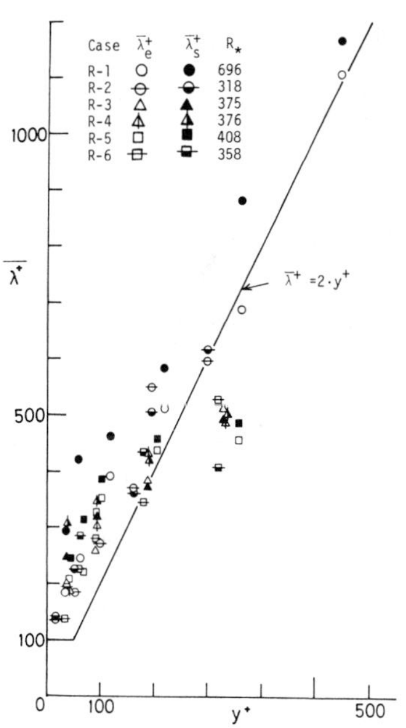

Fig. 15. $\overline{\lambda}^+$ *vs.* y^+ (inner variable expression).

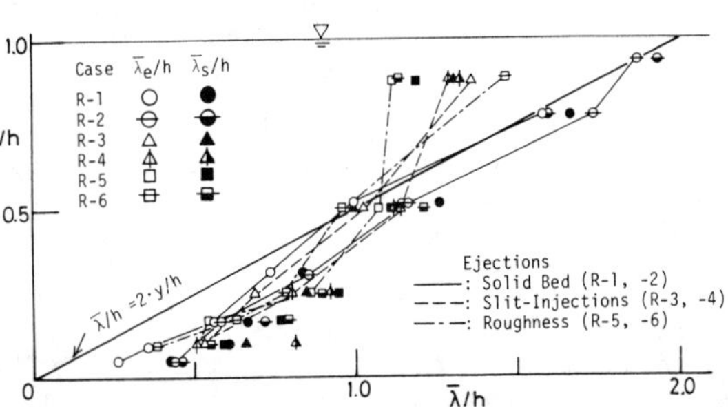

Fig. 16. $\overline{\lambda}/h$ *vs.* y/h (outer variable expression).

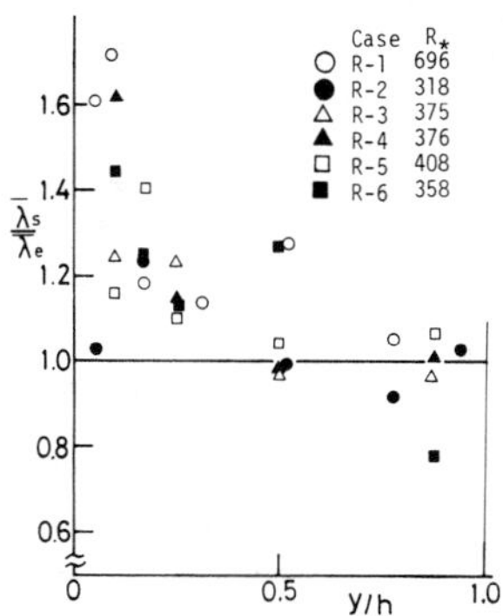

Fig. 17. Ratio of $\overline{\lambda}_s/\overline{\lambda}_e$ of the sweep scale to ejection one.

is larger than that of the ejections, i.e. $\overline{\lambda}_s > \overline{\lambda}_e$, near the wall where these bursting motions occur most violently. This feature has been also found by many visual observations and conditional probe measurements (e.g. Nakagawa et al. 1980). On the other hand, $\overline{\lambda}_s$ becomes equal to $\overline{\lambda}_e$ as the water surface is approached. This suggests strongly that the bursting motions tend to have isotropic property near the water surface, which has been also suggested by Nakagawa & Nezu (1977 a) in the examination of the contribution rates of bursting events to the turbulence production.

The above experimental results indicate that the high- and low-speed streaks of the bursting motions may not have a quite direct interrelation with the cellular secondary currents which scale with the outer variables. That is to say, the former cannot be identified with the latter.

4.6 Interrelation between the Bursting and Cellular Secondary Currents in the Productive Subrange of Spectrum

As mentioned in the Introduction, the large-scale eddies, such as kolk-boils and cellular secondary currents, in the outer region have been termed 'macroturbulence', in consideration of their kinematic behaviours. On the other hand, because the bursting phenomenon can be regarded as the mechanism by which the turbulent energy is produced from the mean flow, its phenomenon belongs almost to the productive subrange in the energy spectral space, which is controlled by a large-scale eddy represented by the integral scale L_x. It is, therefore, considered that the macroturbulence and bursting motion both belong to the wave number range lower than the inertial subrange in which the well-known -5/3 power law is satisfied. In this section, we specially examine an interrelation between the bursting and secondary currents in the productive subrange.

Fig. 18 is a typical example of the spectral distribution S(k) in Case R-1, where k is the wave number. These spectral analyses were carried out by making use of F.F.T. method with the sampling data size of 32,768. Also, similar results were obtained in the outer experimental cases. Nezu(1977) proposed a theoretical spectral distribution, as follows. In the range lower than the inertial subrange, the following Karman's interpolated formula is adopted;

$$S(k)= (2/\pi)\{ 1 + (k/k_0)^2 \}^{-5/6} \qquad (9)$$

In the range higher than the inertial subrange, the Heisenberg's interpolated formula is adopted. Then, if the Reynolds number $R_L \equiv u'L_x/\nu$ is given, both of the formulae can be further interpolated so that the dissipation spectrum may satisfy the relation of isotropic turbulence. Although the value of R_L is smaller as the wall is approached, it becomes nearly constant in the outer region, i.e. $R_L=(u'/U_*)(L_x/h)R_* \simeq R_*$. The solid curve described in Fig. 18 is a calculated one by the above theory, by using R_L=700, which is the average value in the outer region. The deviation between the experimental and calculated values in the viscous subrange is probably due to the limitations of hot-film measurements. However, a very good agreement between the two

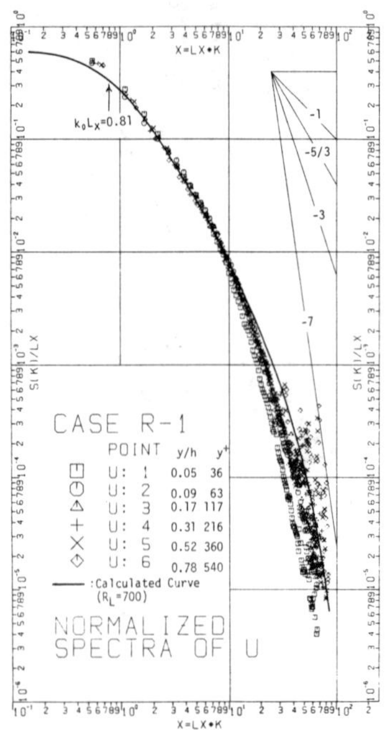

Fig. 18. Spectral distribution S(k) at the detection probe (Case R-1).

is recognized in the range lower than the inertial subrange, and thus this confirms the validity of Eq.(9). Then, the lower limit k_0 of the inertial subrange can be given by Eq.(9), and

$$k \leq k_0 \quad \text{(i.e. } f \leq f_0 \text{ in terms of frequency)} \qquad (10)$$

corresponds to the productive subrange. In the present study, the value of k_0 was nearly equal to $0.8L_x^{-1}$ in all experimental cases.

By the way, the present weighted average technique cannot detect each bursting motion separately, as mentioned in Chapter 2. So, by making use of the pattern-recognized techniques of Wallace et al.(1977) where an arbitrary parameter can be determined comparatively easily, the bursting period T_B was analyzed. Fig. 19 is a typical illustration of the pattern-recognized structure of u/u' which was measured by the single sampling probe at y/h≃ 0.1 in Cases R-1 and R-5. Similar patterns were also obtained at the other measured points and in the other cases. It should be noticed that the effects of the secondary currents on the bursting patterns of Case R-5 can hardly be seen, as compared with those of R-1.

Fig. 20 shows the mean bursting period T_B which was evaluated by the pattern recognized techniques, against the spanwise direction z/h. There is some scattering among these data, since the total number of occurrence of each bursting is only about 120 during the measurement time, i.e. 100 seconds. However, the spanwise distribution of T_B is nearly uniform in all cases. In other words, the effects of the secondary currents on the bursting period cannot be recognized. This result coincides with the notable result of Blackwelder & Woo(1974) that the mean bursting period in the wall region during the disturbance is independent of the pressure perturbation frequency given artificially in the outer region.

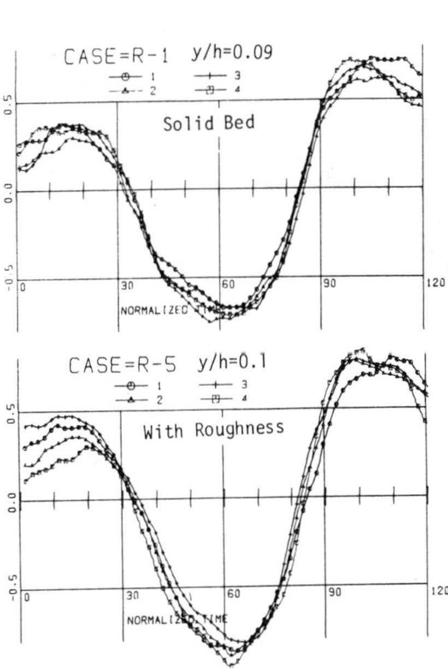

Fig. 19. Some examples of pattern-recognized structure of u/u'.

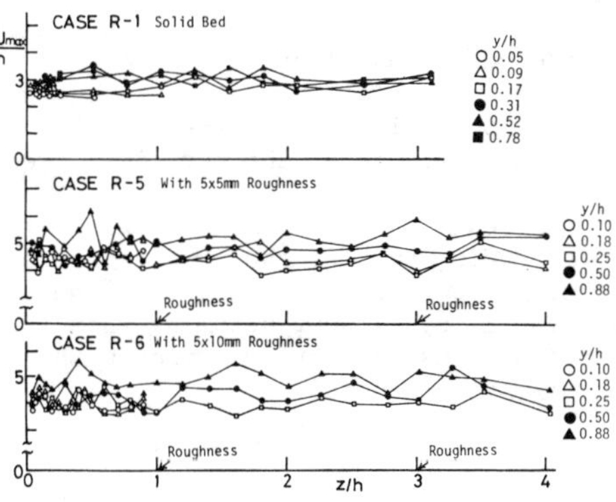

Fig. 20. Variations of bursting period T_B against z/h.

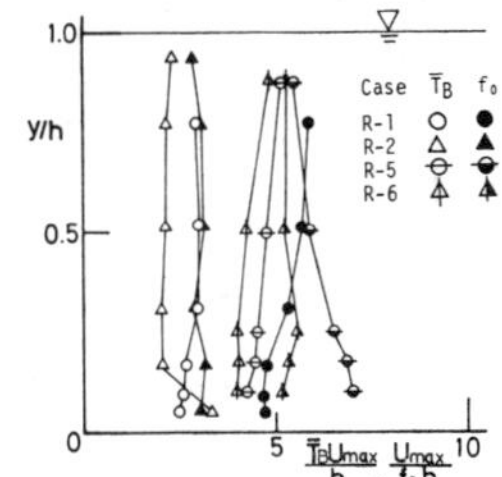

Fig. 21. Mean bursting period $\overline{T}_B$ and upper-limit frequency f_0 of the productive subrange.

Fig. 21 shows the average value $\overline{T}_B$ of the mean bursting period in respect of the spanwise direction, and the upper limit frequency f_0 of the productive subrange, i.e. Eq.(10). Although $\overline{T}_B$ is somewhat smaller than $1/f_0$, both of them are nearly constant against y/h. That is,

$$\overline{T}_B U_{max}/h = (3 - 5) \quad (11)$$

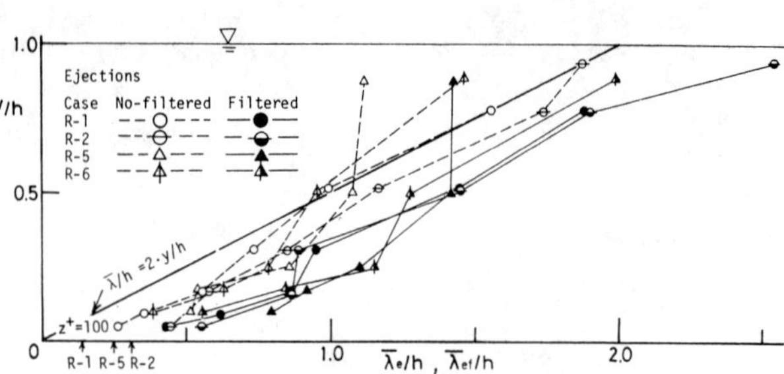

Fig. 22. Mean spacing $\overline{\lambda}_{ef}$ of ejections in the filtered signals.

Eq.(11) is also consistent with many previous data obtained in boundary layer, pipe and open-channel flows.

From the above examinations, it is expected that, if the higher frequency components of $f>f_0$ are filtered out of the raw velocity signals, an interrelation between the bursting and secondary currents will be revealed more clearly. The low-pass filter adopted here was a linear digital filter technique; The raw signals u(t) were converted into Fourier components by F.F.T. method. Next, after the Fourier components of $f>f_0$ were only reduced to zero, they were converted into the filtered signals $u_f(t)$ by inverse F.F.T. method. In the same manner as the raw signals u(t), the conditional average $\langle u_f \rangle(z)$ of the filtered signals $u_f(t)$ was obtained from Eq.(1). Then, the mean spanwise streaky spacings $\overline{\lambda}_{ef}$ and $\overline{\lambda}_{sf}$ of the ejections and sweeps, respectively, were evaluated by using Eq.(8) , and they are shown in Fig. 22. Similar figure was also obtained for the sweeps. $\overline{\lambda}_{ef}$ indicates a uniformly increasing deviation from $\overline{\lambda}_e$, irrespective of y/h and the experimental cases. Above all, an outstanding feature is that the values of $\overline{\lambda}_f$ in the flow with the secondary currents (R-5 and R-6) coincide well with those in the uniform flow (R-1 and R-2), and consequently the effects of the secondary currents on the bursting streaks cannot be recognized again.

Although there are not, at present, any measurement data of the velocity of the ascending and descending currents of the cellular secondary currents, their velocity seems to be at most the order of the vertical turbulence intensity v' (Jackson 1976). Now, assuming that the cellular secondary current in Fig. 2 is represented as a circular motion with the diameter of h and the rotating frequency f_s, we can obtain

$$U_{max}/(f_s h) \geq \pi(U_{max}/v') \quad (12)$$

Since $U_{max}/v' \geq 20$ from Figs. 3 and 4, Eq.(12) yields to $U_{max}/(f_s h) \geq 60$, which is much larger by one order than the bursting period $\overline{T}_B$ of Eq.(11). This suggests strongly that the bursting and cellular secondary currents are the different phenomena each other.

5. CONCLUDING REMARKS

We summarize here the interrelation between the bursting phenomenon in the wall region and the macroturbulence in the outer region, on the basis of many fragmentary information obtained from the previous and present studies.

(1) Evidences which suggest the existence of interrelation between the two phenomena

(a) The mean period of occurrence of boils is coincident with the mean bursting period, i.e. Eq.(11). The probability distributions of their occurrences are

both log-normal ones (Jackson 1976, Nakagawa & Nezu 1978, Fukuoka et al. 1980).
(b) The mean spanwise streaky spacing $\bar{\lambda}$ gradually increases in the vertical direction from $\bar{\lambda} \simeq 100\nu/U_*$ to 2h on the water surface (the present study).
(c) The transverse vortex in the outer region seems to give a trigger of occurrence of a new bursting motion in the wall region (Offen & Kline 1975, Praturi & Brodkey 1978, Nakagawa et al. 1980).
(d) The bursting process is explained well by a horseshoe vortex model (Offen & Kline 1975, Praturi & Brodkey 1978, Nakagawa et al. 1978 & 1980). The macroturbulence of kolk-boils is also explained by a horseshoe vortex model (Utami & Ueno 1977, Fukuoka & Fukushima 1980).

(2) Evidences which deny the direct interrelation between the two phenomena

(e) When the Reynolds number is sufficiently large, the thickness of the wall region becomes very small, and thus the scales of the bursting and macroturbulence are too different each other.
(f) The bursting motions occur randomly in space and time, while the boils and secondary currents occur at nearly fixed positions which are controlled by the flow depth (Kinoshita 1967, Jackson 1976).
(g) The effects of the cellular secondary currents on the bursting wall streaks cannot be recognized (the present study).
(h) The bursting phenomenon produces the turbulent energy from the mean flow, while the macroturbulence does not necessarily produce the turbulence and Reynolds stress.

Among the above items, (h) will be the most important hydrodynamic feature to make clear whether there exists any interrelation between the two phenomena or not. The most essential mechanism of turbulent open-channel flow is the production of turbulent energy from its mean flow. Its energy is transfered into smaller eddies through the cascade process and finally it is dissipated into heat by the viscosity. Thus, the bursting phenomenon is the most essential one in the wall turbulence. In other words, the flow over the wall without the bursting phenomenon is not turbulent any longer. On the other hand, the turbulent flows in which the macroturbulence cannot be detected must exist in the nature. Actually, the macroturbulence cannot be observed usually in the laboratory straight smooth flumes, similar to that used in the present study. That is to say, the macroturbulence is not directly caused by the bursting motions, but it is probably caused by the other factors such as various bedforms. For example, the large-scale eddies due to flow separation are produced on the crest of the sand wave bed.

We cannot conclude, however, that both of the bursting and macroturbulence are quite independent phenomena, because some kinematic analogy of the items *(1)* exists between the two. In this discussion, we should recall the hypothesis of double-structure of turbulence, which has been proposed by Falco(1977) and Nezu (1977), that the wall turbulence consists of both bursting motion controlled by the inner variables and macroturbulence controlled by the outer variables. According to this hypothesis, the former is the active component that contributes to the production of turbulence and Reynolds stress and indicates the universal characteristics, and the latter is the inactive component that has the much lower frequency relating to the flow history from the upstream. This conception of active-inactive components is consistent with that which had been already proposed by Bradshaw(1967).

The bursting motions occurring in the wall region will be convected outward from the wall and downstream *(see Fig. 1)* as if they might ride on the macroturbulence. Then, they will decay and at last disappear in the outer region. Because the composite vortex of the macroturbulence and bursting motions, that is, a kind of transverse vortex, probably pull the trigger on a birth of new bursting motion, its mean bursting period $\overline{T}_B$ becomes nearly constant throughout the flow depth as shown in Fig. 21. Both of the bursting and macroturbulence must coexist self-consistently even when the Reynolds number becomes sufficiently large. Therefore, more detailed investigations will be further needed whether or not the macroturbulence of kolk-boils and cellular secondary currents in an actual river corresponds to the inactive component and then the above hypothesis of double-structure of turbulence can explain well the coherent structures in an open-channel flow.

REFERENCES

1) Blackwelder, R.F. & Kaplan, R.E. 1976 J. Fluid Mech., **76**, 89.
2) Blackwelder, R.F. & Kovasznay, L.S.G. 1972 Phys. Fluids, **15**, 1545.
3) Blackwelder, R.F. & Woo, H.H.W. 1974 Phys. Fluids, **17**, 515.
4) Bradshaw, P. 1967 J. Fluid Mech., **30**, 241.
5) Bremhorst, K. & Walker, T.B. 1973 J. Fluid Mech., **61**, 173.
6) Brodkey, R.S., Wallace, J.M. & Eckelmann, H. 1974 J. Fluid Mech., **63**, 209.
7) Brown, G.L. & Thomas, A.S.W. 1977 Phys. Fluids, **20**, S243.
8) Coleman, J.M. 1969 Sediment. Geol., **3**, 129.
9) Corino, E.R. & Brodkey, R.S. 1969 J. Fluid Mech., **37**, 1.
10) Falco, R.E. 1977 Phys. Fluids, **20**, S124.
11) Fukuoka, S. & Fukushima, Y. 1980 3-rd Int. Symp. Stochastic Hydr., p. 609.
12) Grass, A.J. 1971 J. Fluid Mech., **50**, 233.
13) Gupta, A.K., Laufer, J. & Kaplan, R.E. 1971 J. Fluid Mech., **50**, 493.
14) Hinze, J.O. 1975 *'Turbulence'* 2-nd ed., McGraw-Hill, pp.586-770.
15) Jackson, R.E. 1976 J. Fluid Mech., **77**, 531.
16) Kim, H.T., Kline, S.J. & Reynolds, W.C. 1971 J. Fluid Mech., **50**, 133.
17) Kinoshita, R. 1967 Photographic Surveying, **6**, 1 *(in Japanese)*.
18) Kline, S.J., Reynolds, W.C., Schraub, R.A. & Runstadler, P.W. 1967 J. Fluid Mech., **30**, 741.
19) Kreplin, H.P. & Eckelmann, H. 1979 J. Fluid Mech., **95**, 305.
20) Laufer, J. 1954 NACA, Technical Report TR-1174.
21) Laufer, J. 1975 Ann. Rev. Fluid Mech., **7**, 307.
22) Laufer, J. & Narayanan, M.A.B. 1971 Phys. Fluids, **14**, 182.
23) Lee, M.K., Eckelman, L.D. & Hanratty, T.J. 1974 J. Fluid Mech., **66**, 17.
24) Lu, S.S. & Willmarth, W.W. 1973 J. Fluid Mech., **60**, 481.
25) Matthes, G.H. 1947 Amer. Geoph. Union, **28**, 255.
26) Morrison,W.R.B., Bullock,K.J. & Kronauer,R.E. 1971 J. Fluid Mech., **47**, 639.
27) Müller, A. 1977 2-nd Int. Symp. Stochastic Hydraulics, No.3, pp.1-22.
28) Nakagawa, H. & Nezu, I. 1977a J. Fluid Mech., **80**, 99.
29) Nakagawa, H. & Nezu, I. 1977b 5-th Symp. Flow Visualization, ISAS, Univ. of Tokyo, pp.47-50 *(in Japanese)*.
30) Nakagawa, H. & Nezu, I. 1978 Memoirs, Fac. Eng., Kyoto Univ., **40**, 213.
31) Nakagawa, H. & Nezu, I. 1979 Memoirs, Fac. Eng., Kyoto Univ., **41**, 240.
32) Nakagawa, H., Nezu, I. & Matsumoto, N. 1980 Memoirs, Fac. Eng., Kyoto Univ., **42**, 85. Or, 1980 3-rd Int. Symp. Stochastic Hydraulics, Tokyo, pp.577-588.
33) Nezu, I. 1977 *'Turbulent Structure in Open-channel Flows'*, Ph. D. thesis, Kyoto University.
34) Nishioka, M., Iida, S. & Shimizu, K. 1977 9-th Symp. Turbulence, ISAS, Univ. of Tokyo, pp.129-136 *(in Japanese)*.
35) Offen, G.R. & Kline, S.J. 1975 J. Fluid Mech., **70**, 209.
36) Praturi, A.K. & Brodkey, R.S. 1978 J. Fluid Mech., **89**, 251.
37) Sabot, J. & Comte-Bellot, G. 1976 J. Fluid Mech., **74**, 767.
38) Theodorsen, T. 1955 50-Jahre Grenzshichtforschung, (ed. H. Görtler & W. Tollmien) Frier., Vieweg & Sohn, pp.55-62.
39) Utami, T. & Ueno, T. 1977 Annuals, D.P.R.I. Kyoto Univ., **20**,331 *(in Japanese)*.
40) Wallace, J.M., Brodkey, R.S. & Eckelmann, H. 1977 J. Fluid Mech., **83**, 673.
41) Willmarth, W.W. 1975 Adv. Appl. Mech., **15**, 159.

Secondary Flow Vortices: A Structure in Turbulent Open Channel Flow

ANDREAS MÜLLER
Swiss Federal Institute of Technology
Zürich, Switzerland

ABSTRACT

Secondary flow vortices can be observed in all open channel flows. Their mean axial vorticity is produced by moments of the turbulent shear stress gradients.

To simulate secondary flow vortices as they occur in flows with sediment transport, four longitudinal stripes of roughness were fixed 16 cm apart on the bed of a 24 m long and .6 m wide laboratory flume. LDA and hot-film anemometry were used to measure the components of the instantaneous velocity vector.

The mean secondary flow is directed towards the wall above the rough part and away from the wall above the smooth part. Above the stripes, smaller vortices can be detected. In the balance of mean axial forces, the dominant terms are the vertical advection and the vertical turbulent shear stress gradient. Near the bed, both forces are large compared with the driving pressure force gS (S = Slope) and they change their sign from the rough to the smooth part. The influence of the secondary flow cells on details of the momentum transfer is investigated.

The stability of the secondary flow is studied by means of moving averages which split the velocity in the mean, in slow fluctuations and fast fluctuations. The presence of strong slow fluctuations indicates that the secondary flow cells are not stable.

1. INTRODUCTION

Open channel flows are characterized by features not common to normal boundary layer flows. Their free surface forms a moving upper boundary which limits the length scales in the vertical direction. The flow has normally a large aspect ratio. It is driven by gravity which is in equilibrium with the wall shear stress, so that the flow is uniform and fully developed. The corners between side walls and bottom induce a secondary flow which can induce further secondary flow cells towards the axis of the channel. In alluvial channels with a movable sand bed, several types of bed roughness can form under different flow regimes. The starting point of this study was the observation of stripes of rough and fine sand in flow direction which were produced by secondary flow vortices.

Secondary flow cells can be detected in the time mean. Their mean axial vorticity is produced by moments of the shear stress gradients in the cross plane. They also show marked influence on the mean turbulence quantities including properties of the burst cycle. In this respect, they can be regarded as a structure in turbulent flow. However, this does not answer the question whether they are really a stable flow structure or more the effect of an intermittent process which fluctuates in space and in time.

Secondary flow cells were extensively studied in duct flows (Gessner, 1973, Gerard, 1978). Hinze (1973) reports an experiment in a rectangular duct with roughness induced secondary flow. But cells are also found in the boundary layers of a wind tunnel (Eichelbrenner Preston, 1971). Müller (1977) has studied secondary flow cells above an alluvial bed in an open channel.

The present paper describes in a first part an experiment in which secondary flow cells were induced by longitudinal stripes of roughness (Müller, Studerus, 1979). A second part shows the influence of the mean lateral motion on the turbulence and finally the stability of the cells is investigated.

2. EXPERIMENT

To simulate the rough and smooth stripes observed in alluvial channels, four 6 cm wide longitudinal stripes of sand roughness ($d = 0.25$ cm) were fixed $\lambda = 16$ cm apart on the bed of a 60 cm wide and 24 m long channel (Fig. 1). Due to these stripes, the wall friction was periodic in the lateral direction, so that the three-dimensional flow above it could also be assumed to be periodic in z. The stripes were placed at such a location that the roughness induced secondary flow cells counteracted the two corner cells.

Instantaneous longitudinal and vertical components u and v were measured by LDA. With a horizontally crossed hot film fiber probe, u and w were measured at the same location.

Two drawbacks of this method had to be taken into account. The stagnation zone in front of the probe induced a downward motion of the order of 1 cm/sec, depending on the depth of submergence. The vertical velocities, therefore, had to be corrected by assuming that the mean mass flux through a horizontal plane of one period λ width was zero. The second problem was the response of the hot film probe to gradients of u. As the two wires were .1 cm apart, a gradient of u produced a difference in the readings of the upper and the lower wire which could not be distinguished from a w component. While $\overline{w}$ could be corrected on the basis of the mean $\overline{u}$ gradient, the

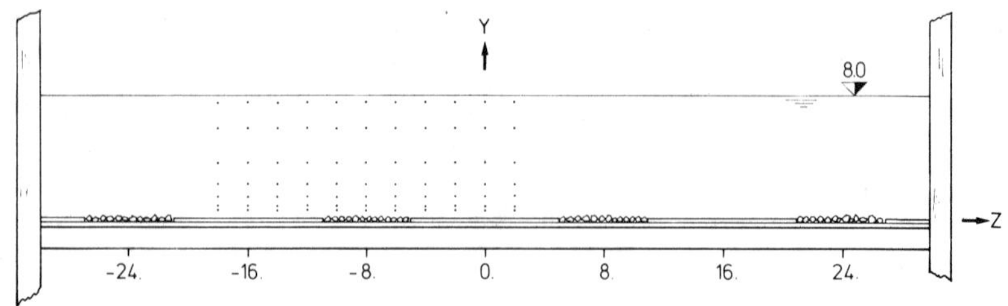

Fig. 1 Cross section of the flume with stripes of roughness and grid of measuring points

fluctuating part w' could not be corrected. Only the correlation $\overline{v'w'}$ could be adjusted by assuming that $\overline{v'w'}$ (y,z) is periodic in z with a mean value of zero.

Combined three components measurements were taken for a grid of 77 points across one period of the stripes. The grid was logarithmic in y and linear in z (Fig. 1). In every point, a record of 50,000 samples was taken at a rate of 420 Hz. The low pass filter was set at 100 Hz. The flow parameters were: Flow depth h = 8cm, slope S = 0.00152, discharge Q = .026 m^3/sec giving a mean velocity u_o = 55 cm/sec, a mean shear stress $\tau_o/\rho = g\,R\,S = 9.4 (cm/sec)^2$ and a mean shear velocity u_{*o} = 3.1 cm/sec. The Re-number was $44 \cdot 10^3$ and the Froude-number .6.

3. EFFECTS OF THE MEAN SECONDARY FLOW

The mean secondary flow velocity is directed towards the rough part of the wall and away from the wall over the smooth part (Fig. 2). Due to these cells, mean axial momentum is transported into the region of high shear above the rough stripes. The contourlines of $\bar{u}$ in Fig. 3 show that this high shear is combined with a maximum of $\bar{u}$ above the stripes, while the lower shear over the smooth part reduces the velocity in the upper part.

Profiles $\bar{u}(y)$ are logarithmic in both regions. The wall shear stress, which is estimated from the gradient of these profiles assuming the von Karman constant $\varkappa$ = .4, is twice the mean shear stress on the roughness (τ_R/τ_o = 2.35) and about one half on the smooth part of the wall (τ_S/τ_o = 0.56). Therefore, the velocity scales are different in the two regions.

Within the Reynolds equation for x-momentum, the vertical advection $-\bar{v}\frac{\partial \bar{u}}{\partial y}$ and the shear stress gradient $-\frac{\partial}{\partial y}\overline{u'v'}$ are the dominant forces. Contourlines of these forces normalized with the driving pressure force gS (g = acceleration due to gravity) are represented in Fig. 4 and 5. Above the roughness, the vertical advection tends to accelerate, the shear stress gradient tends to decelerate the flow. Over the smooth part, vertical advection decelerates and the shear stress gradient accelerates, just as in flows with an adverse pressure gradient. Both forces exceed the driving pressure force by a factor of two to five. The profiles of $\overline{u'v'}$(y) in Fig. 6 visualize the same fact and show also the variation of the wall shear stress which can explain the selective sediment transport mentioned in the introduction.

Fig. 7 represents detailed measurements of mean vertical velocity close to the rough stripes. In 165 grid points, velocity records of 60 sec length were taken. Secondary flow cells with diameters in the order of one cm can be recognized near the bed. These small

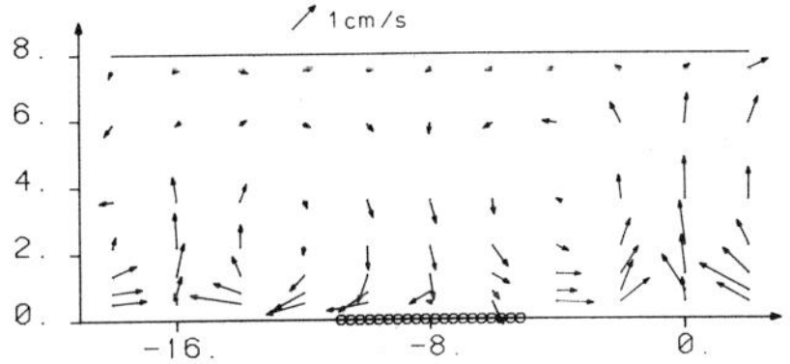

Fig. 2 Secondary flow velocity vectors ($\bar{v}$,$\bar{w}$)

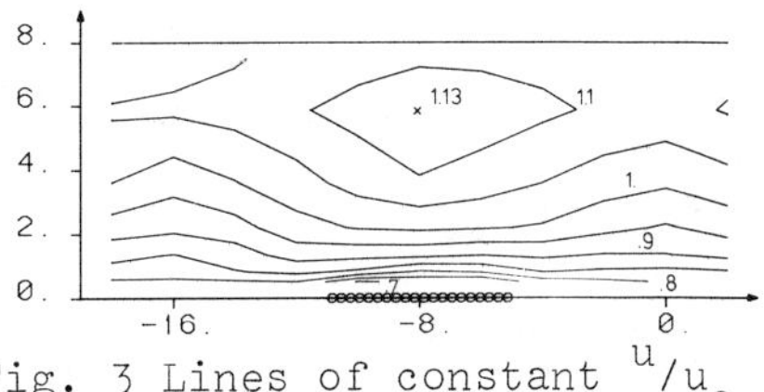

Fig. 3 Lines of constant u/u_o (u_o = 55 cm/sec)

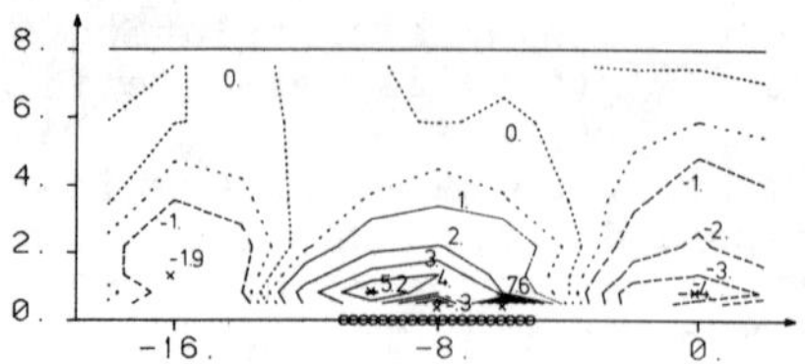

Fig. 4 Lines of constant $-\nabla \frac{\partial \overline{u}}{\partial y}$

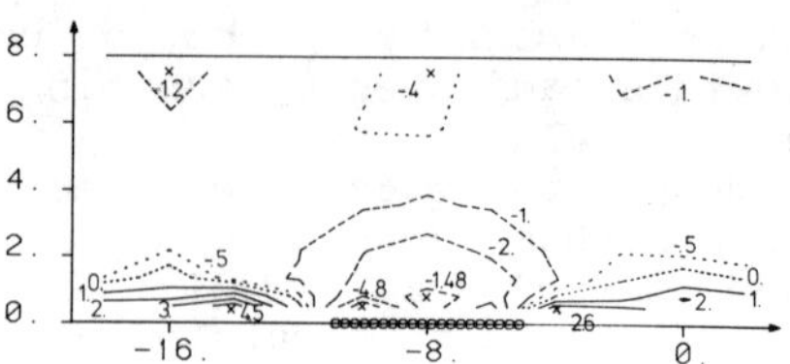

Fig. 5 Lines of constant $-\frac{\partial}{\partial y} \overline{u'v'}$

cells were not as stable as the main cell and could change their location and strength from run to run. They were not further investigated, as only two point measurements can give better insight in such a structure.

To study the influence of the main cells on the turbulent momentum exchange in more details, the joint probability distribution p(u',v') was calculated. Fig. 8 shows p(u',v') multiplied with u'·v', i.e. the distribution of the turbulent momentum flux in the u',v'-plane, at y = 0.04, 0.27, 0.45 for the profiles at z = -8 cm and 0 cm. In Fig. 9, the integral of u'·v'·p(u',v') in each of the four quadrants are represented, i.e. the classified Reynolds stresses, as defined by Wallace et al (1972). The downward flow (z = -8 cm) and the upward flow (z = 0 cm) of the secondary flow cells have a distinct influence on the momentum exchange and change also the relative importance of sweep and ejection events.

4. STABILITY OF SECONDARY FLOW CELLS

In the preceding sections, it is shown how the organized mean velocities in the cross plane (Fig. 2) control flow properties. However, the question is not yet answered whether these cells are

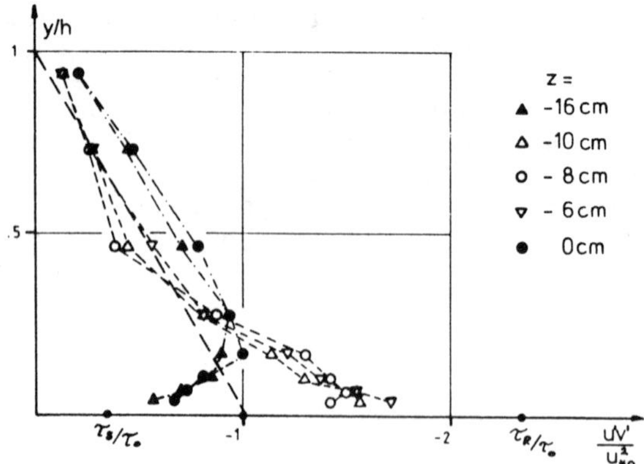

Fig. 6 Profiles of shear stress $u'v'(y,z)/u_{*o}^2$ normalized with the mean wall shear stress

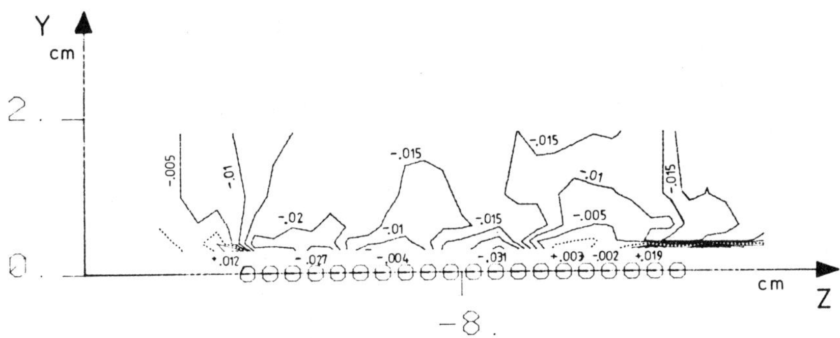

Fig. 7 Contourlines of $\bar{v}/u_o$ showing the influence of small secondary flow cells close to the stripe of roughness

really a flow structure, i.e. a steady cylindrical motion, or just the statistical property of an intermittent large scale structure.

In order for the secondary flow cells to be quasi steady, the time scale T_s of these cells must be much larger than the time scales of the burst cycle T_B. A moving average with an integration time T large compared with T_B and small compared with T_s should smooth out "burst turbulence" and reveal the stability of the secondary flow cells.

With such a procedure, the velocity signal is split up in three parts.

$$u(t) = \bar{u} + \hat{u}(t) + u''(t)$$
$$v(t) = \bar{v} + \hat{v}(t) + v''(t)$$

were

$$\bar{u} = \frac{1}{T_o} \int_o^{T_o} u(t)\, dt$$

is the mean over the total record length T_o = 119 sec. $\hat{u}(t)$ are the slow velocity fluctuations, which are not filtered out by the moving average

$$\hat{u}(t) = \frac{1}{T} \int_{t-T/2}^{t+T/2} u(t)dt - \bar{u}$$

and u" are the fast fluctuations.

The mean square of the turbulent fluctuations gets divided into three parts.

$$\overline{u'^2} = \overline{\hat{u}^2} + 2\,\overline{\hat{u}\,u''} + \overline{u''^2}$$
$$\overline{v'^2} = \overline{\hat{v}^2} + 2\,\overline{\hat{v}\,v''} + \overline{v''^2}$$

and the turbulent shear stress is composed of four terms

$$\overline{u'v'} = \overline{\hat{u}\,\hat{v}} + \overline{\hat{u}\,v''} + \overline{u''\,\hat{v}} + \overline{u''\,v''}$$

Obviously, this decomposition depends on the integration time T which is chosen equal to $10 \cdot h/u_o$ = 1.45 sec for the present analysis. Fig. 10a - c shows the time functions a) u' v'(t) b) $v(t) = \bar{v} + \hat{v}(t) + v''(t)$, c) $u(t)\bar{u} + \hat{u}(t) + u''(t)$ for an interval of 2 seconds at the location y/h = 0.27, z = -8 cm.

They illustrate that the burst activity is filtered out in the moving average with the chosen integration time.

Fig. 11 shows that the slow fluctuations contribute only a small fraction to the Reynolds stress. However, the slow fluctuations $\hat{u}$ and $\hat{v}$ are much better correlated as u" v". The correspon-

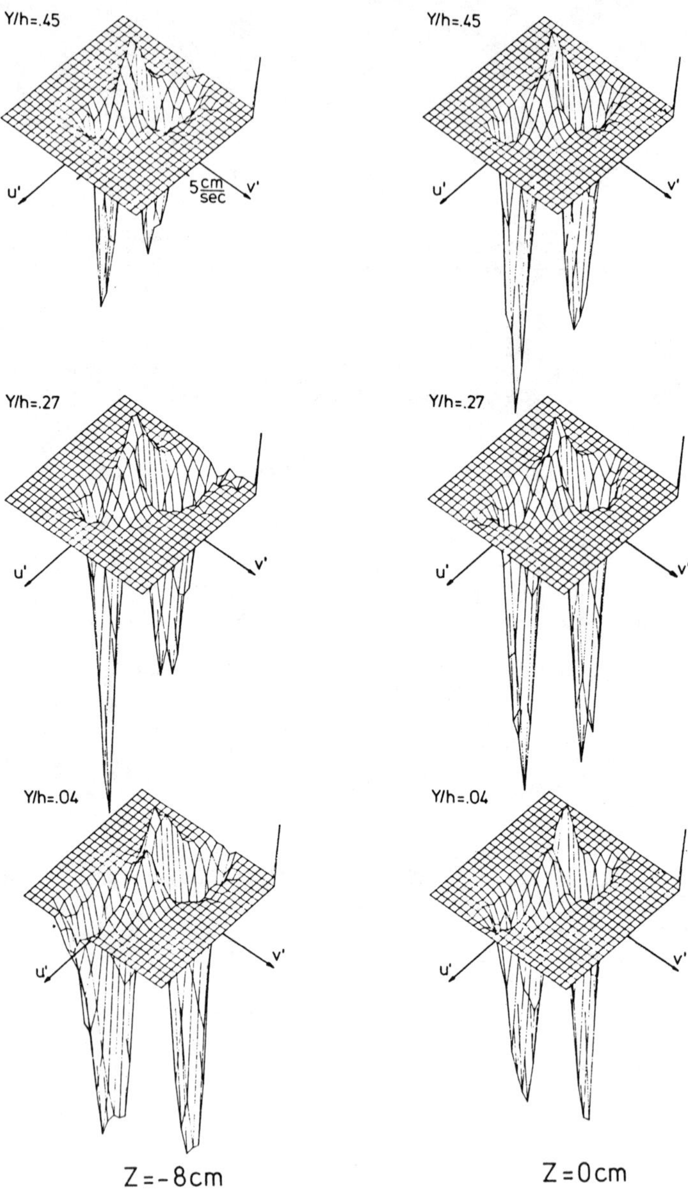

Fig. 8a - f Distributions of the momentum flux in the u'v'-plane for two profiles with mean downward (z = -8 cm) and mean upward motion (z = 0 cm)

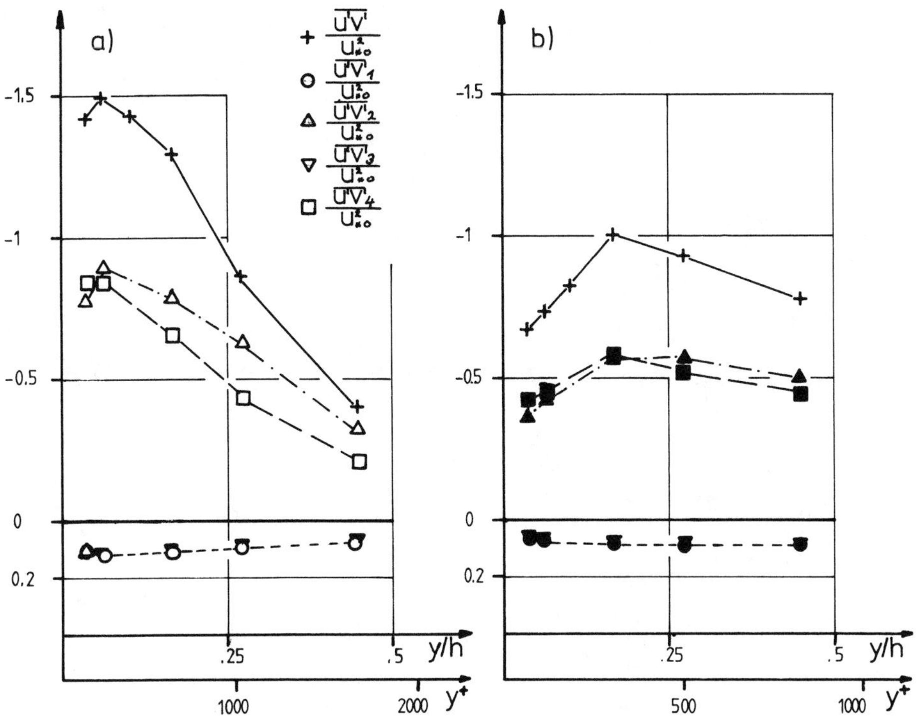

Fig. 9 a) b)
Classified Reynolds stress contributions of quadrant 1 to 4 of the u'v'-plane

a) mean downward motion b) mean upward motion

ding correlation coefficients $R_S = \frac{\overline{\hat{u}\,\hat{v}}}{\sqrt{\overline{\hat{u}^2}\cdot\overline{\hat{v}^2}}}$ $R_F = \frac{\overline{u''v''}}{\sqrt{\overline{u''^2}\cdot\overline{v''^2}}}$

are represented in Fig. 12. The cross terms $\overline{\hat{u}\ v''}$ and $\overline{u''\ \hat{v}}$ are negligible.

In Fig. 13a - c, time functions $\bar{v} + \hat{v}(t)$ are represented for a period of 100 sec in three points representative for a) downward motion (z = -8 cm, y/h = 0.27), b) the center of the flow cell (z = -4 cm, y/h = 0.45) and c) for upward motion (z = 0, y/h = 0.27).

At all three points, the vertical velocities fluctuate with amplitudes larger than the mean and zero crossings up to 150 h/u_o = 22 sec. They make clear that the secondary flow cell is not a structure which is stable in space and time. The fluctuations at the center of the cell z = -4 cm cannot be explained without assuming spatial fluctuations. The fluctuations above the roughness (z = -8 cm) and above the smooth wall (z = 0 cm) may be caused by a vortex which turns continuously, but fluctuates in space or by a structure with temporal and spatial fluctuations. The nature of such structures which make up the stable secondary flow cell, as

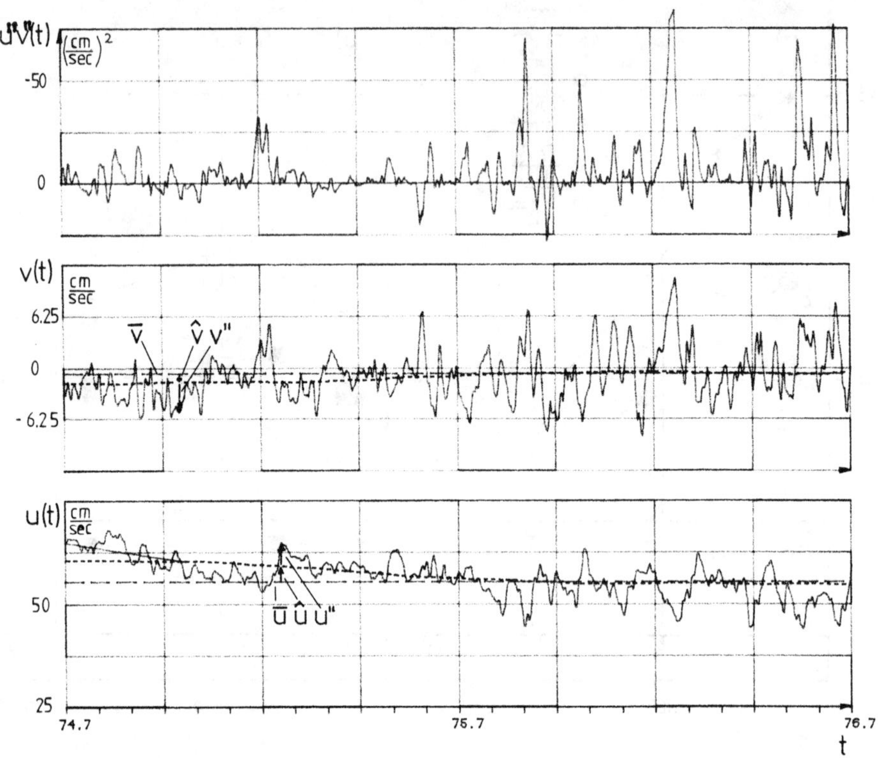

Fig. 10 Filteringeffect of the moving average with an integration time $t = 10\ ^h/u_o = 1.45$ sec at $y/h = 0.27$, $z = -8$ cm

a) $u''v''(t)$
b) $v(t) = \bar{v} + \hat{v}(t) + v''(t)$
c) $u(t) = \bar{u} + \hat{u}(t) + u''(t)$

represented in Fig. 2, cannot be understood without two point correlations.

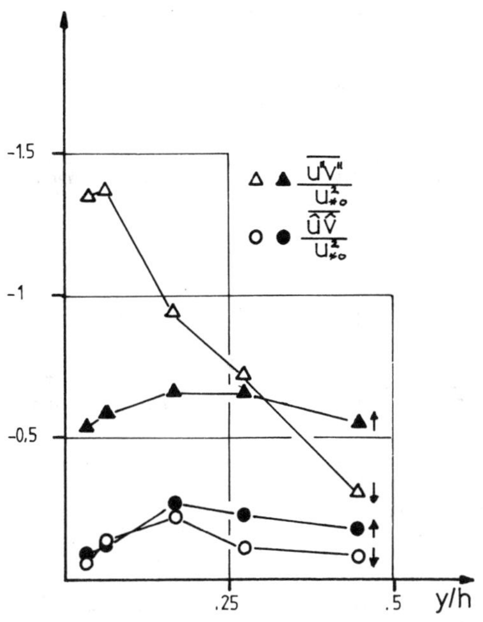

Fig. 11 Contribution of the slow fluctuations $\frac{\overline{\hat{u}\ \hat{v}}}{u_{*o}^2}$ and the fast fluctuations $\frac{\overline{u''\ v''}}{u_{*o}^2}$ to the Reynolds stress

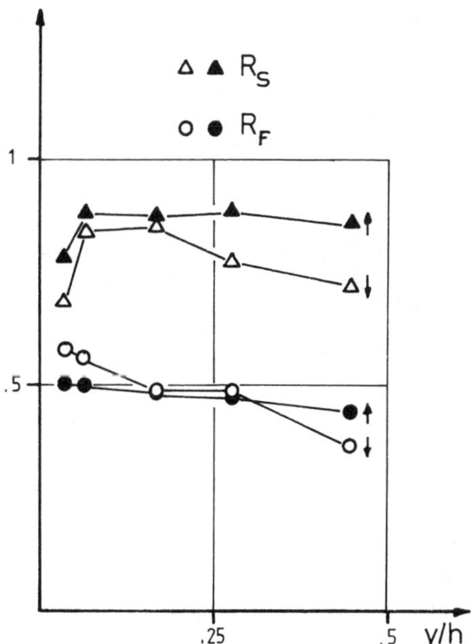

Fig. 12 Correlation coefficients R_S and R_F for the slow and the fast fluctuations

Figs. 11 and 12 Open symbols: Downward motion
Full symbols: Upward motion

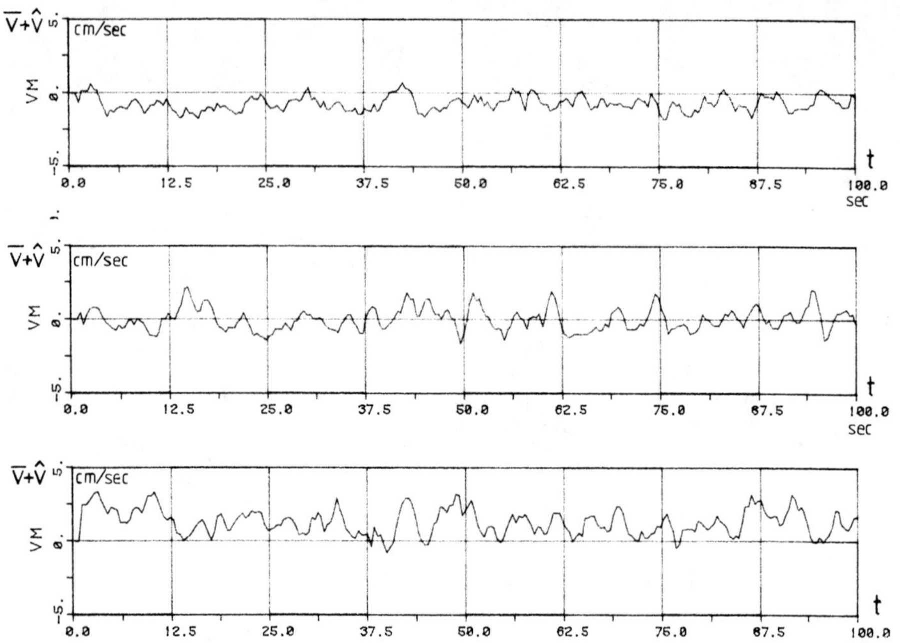

Fig. 13 Moving averages $v_M = \bar{v} + \hat{v}(t)$ showing instabilities of the vertical motion

a) y/h = 0.27, z = -8 cm: mean downward motion
b) y/h = 0.45, z = -4 cm: center of cell
c) y/h = 0.27 z = 0 cm: mean upward motion

REFERENCES

Eichelbrenner, E.A., Preston, J.H. (1971),"On the role of secondary flow in turbulent boundary layers in corners (and salients)", Journal de Mécanique 10, 91 - 112

Gerard, R. (1978),"Secondary flow in non-circular conduits", ASCE J. Hydraulic Div. 105, 755 - 773

Gessner, F.B. (1973),"The origin of secondary flow in turbulent flow along a corner", JFM 58, 1 - 25

Hinze, J.O. (1973),"Experimental investigation on secondary currents in the turbulent flow through a straight conduit", Appl. Sci. Res. 28, 453 - 465

Müller, A. (1977),"Effect of secondary flow on turbulence in an open channel flow", in "Hydraulic Problems solved by stochastic Methods", Water Resources Publ., Fort Collins

Müller, A., Studerus, X. (1979),"Secondary flow in an open channel", Proc. XVIII IAHR-Congress, Cagliari, Vol. 3, 19 - 24

Wallace, J.M., Eckelmann, H., Brodkey, R.S. (1972),"The wall region in turbulent shear flow, JFM 54, 39 - 48

Experiments on Particle Motions near the Bottom in Turbulent Open Channel Flow

B. MUTLU SUMER
Technical University of Istanbul
Faculty of Civil Engineering
Taksim, Istanbul, Turkey

ABSTRACT

Paths of suspended heavy particles were recorded in three dimensions and time. Accord between the particle motion and the bursting process provided further support for the "suspension mechanism" proposed in Sumer and Oğuz [5]. The present study suggested that the "suspension mechanism" could be extended to the rough boundary case. Employing the present path data, the lift force on a particle was determined. Making use of the lift force information, a criterion was obtained for the limiting stage of suspension.

1. INTRODUCTION

If the correct conditions exist, a small heavy particle released into a channel flow is prevented from settling on the bottom ; it travels most of the time close to the bottom of the channel. The mechanism which prevents the particle from settling should be closely associated with the turbulence structure near the bottom, namely the bursting process.

To the author's knowledge, Sutherland [1] was the first investigator who tried to explain particle entrainment by turbulent flows by making use of turbulence observations near the wall. Grass [2] filmed the process of suspension of sand particles in a turbulent boundary layer on a flat plate ; Grass' work directly established the link between the particle suspension and the bursting process. Jackson [3] and Lumley [4] noted that a probable candidate for the suspension mechanism is the bursting process. Sumer and Oğuz [5] made observations of the motion of heavy suspended particles near the smooth bottom in turbulent flow in an open channel, which led the authors to make an attempt to explain the mechanism of particle suspension close to the wall in turbulent flows in terms of the bursting process.

The objective of the present study is to study further the motions of heavy suspended particles near the bottom in turbulent flow in an open channel when the bottom is both smooth and rough.

2. EXPERIMENTAL FACILITY

A 0.30 x 0.30 x 10 m flume was used. The fluid was tap water. The flow properties are given in Table : 1.

Table : 1 - Flow properties

Bottom	Flow depth h(cm)	Mean flow velocity U_m(cm/s)	$Re = \frac{hU_m}{\nu}$	Shear velocity u_*(cm/s)	Mean roughness height k (mm)	$\frac{ku_*}{\nu}$
Smooth	6.8	29.8	21000	1.44	0	0
Rough	7.0	29.1	21000	2.19	3.6 mm pebbles	81

Plastic particles were employed. Three types of particles were used in the smooth bottom experiments. One was slightly heavier than water (Particle A) the other was moderately heavier than water (Particle B), and the third was considerably heavier than water (Particle C). All had almost the same size. Although the Particle C did not quite satisfy the requirement that the particles should stay in suspension, it was purposely chosen so as to get an insight into what is occuring in the case of considerably-heavier-than-water particles. As to the rough bottom experiments, only one type of particle was used since the specific object of these experiments was to get an insight into the mechanics of particle motions in the case of rough boundary flow. The particle used in the tests was moderately heavier than water. The particle is designated with the letter D. With Particle D, two runs were conducted : Run D1 and Run D2. The particle properties are given in Table 2.

Table : 2 - Particle properties

Bottom	Particle	Particle size d(mm)	du_*/ν	Settling velocity w(cm/s)	Water temp. T(^{0}C)	Specific gravity s	Settling velocity parameter $w/(\kappa u_*)$
Smooth	A	2.9	44	0.64	23.4	1.0029	1.0(5)
	B	3.0	46	1.33	23.3	1.0075	2.2
	C	3.1	47	3.06	22.3	1.0258	5.1
Rough	D(RunD1)	3.0	67	1.49	21.4	1.0090	1.6
	D(RunD2)	3.0	67	1.41	22.2	1.0081	1.5

In the experiments, a stereophotogrammetric method was employed to record the particle paths in three-dimensions and time, the latter being achieved with the aid of a stroboscope. For more information about the photogrammetric method, see Jacobi [6], which was reproduced in Sumer and Deigaard [7] .

3. SMOOTH BOTTOM RESULTS

A particle path consists of an alternation between upward and downward paths as it travels close to the bottom (see Fig.1 for a schematic description) In Table : 3 are shown the main results accompanied with the relevant data for corresponding actual bursting events.

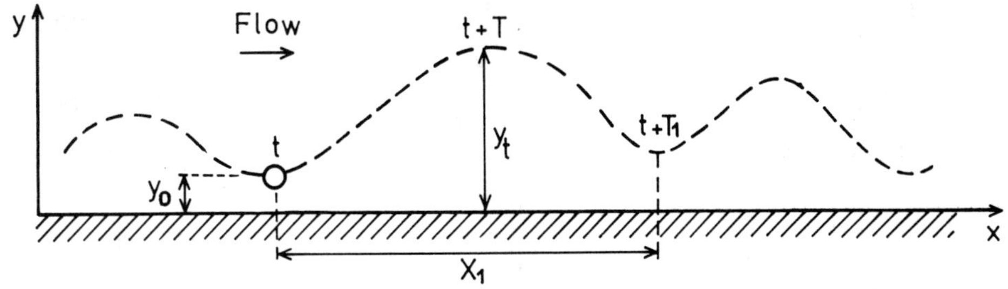

Fig.1. Schematic description of particle path.

Table : 3 - Path data and data for actual bursting events

	Present "particle" data Particle A	Particle B	Data for corresponding actual bursting events
Mean y_o; $y_o u_*/\nu$	40	38	Fluid ejections originate in $5<y^+<50$ (Ref. [8]).
Mean y_t ; $y_t u_*/\nu$	125	103	Ejected fluid elements reach a y^+ of 100 or a litte more (Ref. [9]).
Mean "rise" time ; Tu_∞/h	3	2.4	Mean "life time of a burst" $Tu_\infty/h = 2.3$ (Ref. [3]).
Mean periodicity ; $T_1\ u_\infty/h$	5.3	4.4	Mean periodicity of burst $T_1\ u_\infty/h = 5$ (Ref. [3]).
Mean wave length ; X_1/h	3.5	2.8	Characteristic wave length of burst, $X_1/h = 4$ (Ref. [10]).
Mean transverse displacement within the zone $y^+<50$; Zu_*/ν	56		½ x Mean spacing of lowspeed wall streaks ; ½ $\lambda u_*/\nu = 50$ (Ref. [3]).

In Figs. 2a and 2b respectively are plotted the data for vertical component of particle velocity corresponding to upward and downward paths. In Fig.2, also the velocity data for corresponding actual bursting events (Ref. [11]) are plotted for comparison. Of the latter data, those corresponding to $y^+=135$ and $y^+=195$ seem to be dubious as these y^+ positions were located in the central part of the oil channel used by Brodkey et al. (in fact, $y^+=195$ corresponds to the centre of the channel, i.e. $y/b = 1$, where b is the half -width of the channel); thus any motion detected at these positions might have been influenced by the bursting process in the other half of the channel. For more information about the path and velocity statistics, the reader is referred to Ref. [7].

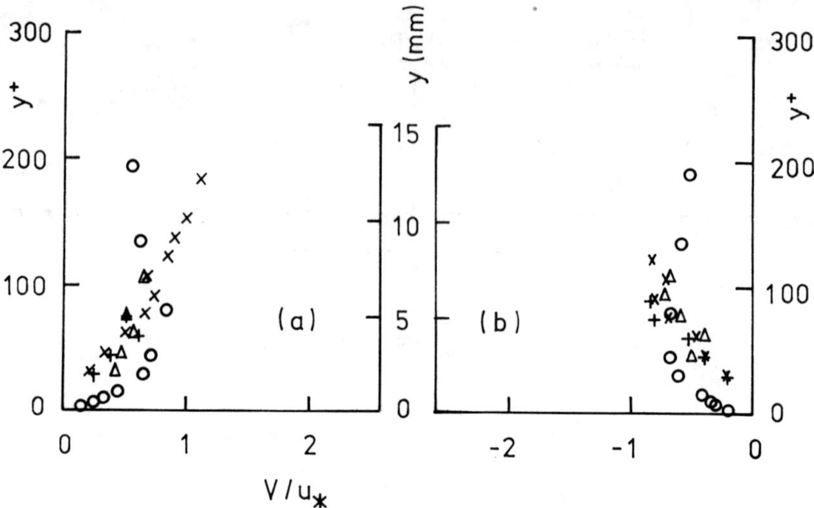

Fig.2. Vertical velocity data for smooth bottom. (a)Upward path. (b)Downward path. Δ, Particle A ; x, Particle B; +,Particle C ; o, Data for corresponding actual bursting events [11].

The new evidences obtained in the present study provide further support for the mechanism of particle suspension close to the bottom proposed in Ref. [5]. This mechanism views the particle "rise" from the wall region as the upward travel of the particle within the main body of the lifted fluid (carrier fluid) of a burst. The particle "rise" terminates as the burst breaks up, and then the particle starts to fall. The immediate neighbourhood of the particle in its "fall" is an high-speed fluid zone penetrating towards the bottom which hits the bottom and spreads out sideways, that sweeps the particle into a low -speed wall streak. The latter is actually the onset of another burst which enters the main body of the flow immediately. Thus the particle, being in the body of the lifted fluid of this new burst, rises again. It is this process which makes it possible for the particle to stay in suspension continuously.

Relatively heavier particles. To get an insight into the mechanics of motions of relatively heavier particles near the bottom, the main results obtained for Particle C are compared with those for Particle A in Table : 4.

The data summarized in Table : 4 as well as others in [7] suggest an overshooting effect in the particle's upward path : relatively heavier particles ejected from the wall region cannot be maintained in the lifted fluid of the accompanying burst but leave the main body of the lifted fluid due to gravity before the accompanying burst breaks up. Note that the latter coincides with what is conventionally known as the "crossing-trajectories effect"concept, i.e.,a particle falls out of an eddy before the eddy loses its identity.

Table : 4- Data for relatively heavier particles

	Slightly-heavier-than -water particle (Particle A)	Considerably-heavier than-water particle (Particle C)
Mean y_o; $y_o u_*/\nu$	40	33
Mean y_t; $y_t u_*/\nu$	125	54
Mean "rise" time ; $Tu\infty/h$	3	1.1
Mean periodicity ; $T_1 u\infty/h$	5.3	2.0
Mean wave length ; X_1/h	3.5	1.1

4. ROUGH BOTTOM RESULTS

In contrast to the case of smooth wall, particle position in the vertical in the rough bottom case will not be expressed in terms of $y^+(= yu_*/\nu)$ units as the roughness Reynolds number $ku_*/\nu(= 81)$ falls into the rough boundary category $ku_*/\nu>70$, thus the viscosity has no effect on the flow.

To facilitate comparison, any result of the rough bottom case will be accompanied by its counterpart obtained in the smooth bottom case. Such a comparison is possible, because a) the flow velocity and the flow depth (thus Re number) were kept almost unchanged in both the smooth and rough wall tests, and b) the properties of the corresponding particles were also kept almost the same in these two experiments ; Particle B of the smooth wall case and the particle used in the rough wall case (Particle D) have almost the same characteristics (see Table 2).

In Table : 5 are given the main results obtained for rough bottom case along with the corresponding smooth bottom data. Fig.3 presents vertical velocity data for both smooth and rough boundary cases. For more information about the path and velocity statistics, the reader is referred to Ref. [7].

The present findings show that particle motions close to the rough bottom are similar to those in the case of smooth bottom, however with more violent appearence. Qualitative, even quantitative (partially), accord between the smooth and rough boundary results suggests that the suspension mechanism given for the smooth boundary flow could be extended to the rough boundary case.

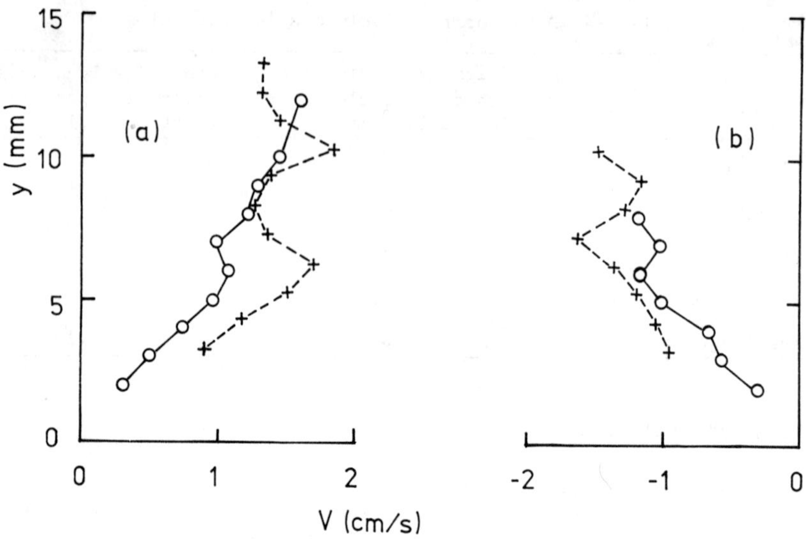

Fig.3. Vertical velocity data. (a) Upward path. (b) Downward path. o, Smooth bottom; +, Rough bottom.

Table : 5 - Data for rough boundary

	Smooth bottom (Particle B)	Rough bottom (Particle D) Run D1	Run D2
Mean y_o; in mm	3.9	5.3*	6.5*
Mean y_t; in mm	8.1	11.6*	14.4*
Mean "rise" time ; $Tu\infty/h$	2.3	2.0	2.3
Mean periodicity ; $T_1u\infty/h$	4.6	4.0	4.6
Mean wave length ; X_1/h	3.1	2.5	3.0

*The y distances are from the base bottom.

5. LIFT FORCE

Experiments show that, if the correct conditions exist, a small heavy particle goes into suspension from near the bottom. This implies that the particle experiences a lift force, which acts against gravity plus other possible resisting agents. In sediment transport, the determination of lift forces is of considerably importance for the entrainment and (or) suspension of particles.

Consider a spherical heavy particle migrating upwards. The vertical component of the equation of motion of such a particle is :

$$\underset{1}{\frac{\pi d^3}{6}\rho_p \frac{dv_p}{dt}} = -\underset{2}{\frac{\pi d^3}{6}(\gamma_p-\gamma)}+\underset{3}{F_D}-\underset{4}{\frac{\pi d^3}{6}\frac{\partial p}{\partial y}}+\underset{5}{\frac{1}{2}\frac{\pi d^3}{6}\rho\frac{d}{dt}(v-v_p)} \qquad (1)$$

In this equation, the term 1 is the force to accelerate the particle, the term 2 is the submerged weight of the particle, the term 3 is the drag and the term 5 is the force to accelerate the apparent mass of the particle relative to the ambient fluid. Herein, p is the instantaneous deviation from the hydrostatic pressure. The term 4 represents the force due to the gradient of the instantaneous extra pressure which builds up in the fluid environment surrounding the particle. It appears that this is the only force in equation (1) which would drive the particle upwards. To the author's knowledge, Engelund [12] was the first investigator who recognized its significance in the mechanics of sediment suspension.

The "suspension mechanism" given in the above paragraphs suggests that a particle moving towards the bottom in a "high-speed fluid" environment eventually finds itself swept into a temporarily formed "low-speed fluid" lump with which it is instantly lifted up into the main body of the flow. At the instant of lift-up, one has a negligible drag on the particle, as the relative flow between the particle and the surrounding lifted fluid is too weak :

$$F_D \simeq 0 \tag{2}$$

On the other hand, using u_* and d as the characteristic velocity and characteristic length respectively, one can write the Navier-Stokes equation in the nondimensionalized form as in the following

$$\frac{dv}{dt} = -\frac{1}{\rho}\frac{\partial p}{\partial y} + \left(\frac{du_*}{\nu}\right)^{-1} \nabla^2 v \tag{3}$$

Since $du_*/\nu >> 1$ in our case, one has

$$\frac{dv}{dt} \simeq -\frac{1}{\rho}\frac{\partial p}{\partial y} \tag{4}$$

From equations (1), (2) and (4), the "pressure gradient" force at the instant of lift up is

$$-\frac{1}{\rho_p}\frac{\partial p}{\partial y} = \frac{2}{3}\left(1+\frac{1}{2s}\right)\frac{dv_p}{dt} + \frac{2}{3}\, g\left(1-\frac{1}{s}\right) \tag{5}$$

which is the lift force (per unit mass of particle) by definition. Herein, s is the specific gravity of particle, $s = \rho_p/\rho$.

Using equation (5), the lift force was determined employing the path records of the present study. In Fig.4 are shown the mean lift force data plotted against the distance from the wall. From this figure, the following points should be noted : a) A thin layer exists near the wall where the lift force has appreciable values ; b) The positions where the lift force has appreciable values coincide with those where most of the ejections originate ; c) The lift force increases with increasing bed roughness ; d) The effect of rough wall upon the lift force tends to disappear with the distance from the wall.

Other possible lift forces have been discussed elsewhere [7] ; the latter showed that the present case has no dealings with these additional lift forces.

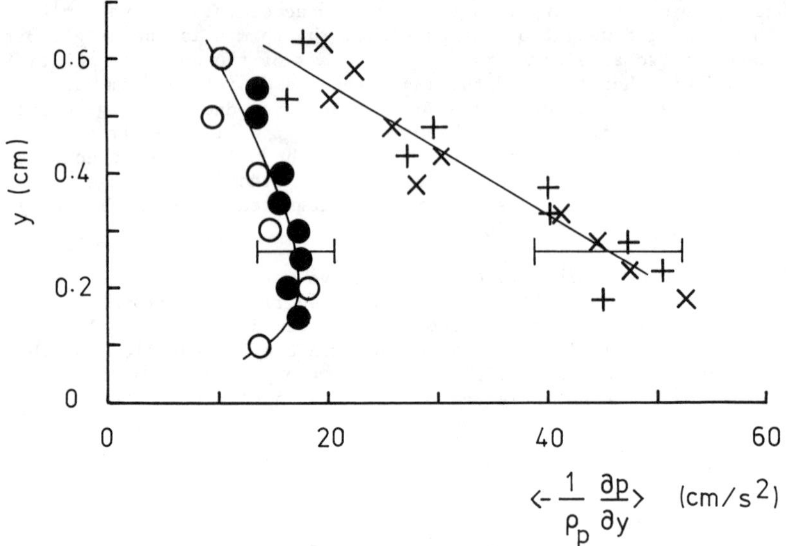

Fig.4. Mean lift force. o, Smooth bottom (Particle A); •, Smooth bottom (Particle B); x, Rough bottom (Particle D, Run D1); +, Rough bottom (Particle D, Run D2).

Einstein and El-Samni [13] determined the lift force on an individual element of an immobile bed by measuring the dynamic pressure difference between the top of the element and the base plate on which the elements sit. Einstein and El-Samni's result, when converted to the following form (see Ref. [14]) reads $\Delta p/(\rho u_*^2) = 2.5$. This latter quantity $\Delta p/(\rho u_*^2)$ can be written in the present case as

$$\frac{\Delta p}{\rho u_*^2} = \frac{2}{3}\frac{sd}{u_*^2}\left(-\frac{1}{\rho_p}\frac{\partial p}{\partial y}\right) \tag{6}$$

Picking up the values of $\left(-\frac{1}{\rho_p}\frac{\partial p}{\partial y}\right)$ (y locations of which are the nearest the bottom) from Fig.4, equation (6) gives $\Delta p/(\rho u_*^2) = 2.2$ for Particle D (Run D1) and 1.9 for Particle D (Run D2). These values appear to be in accord with the Einstein and El-Samni's value 2.5 .

6. THE LIMITING STAGE OF SUSPENSION

The definition of suspension adopted in this work is as follows. A particle is considered to go into suspension if it is lifted up from near or from the bottom. This definition implies that we have suspension if $dv_p/dt > 0$ at the instant of lift-up and we have the limiting stage of suspension if $dv_p/dt = 0$. Thus one can write the following criterion

$$\text{Suspension occurs when} \quad \frac{dv_p}{dt} > 0 \tag{7}$$

or, from equation (5)

$$\text{Suspension occurs when} \quad \frac{3}{2}\left(-\frac{1}{\rho_p}\frac{\partial p}{\partial y}\right) > g\left(1-\frac{1}{s}\right) \qquad (8)$$

Taking $< \frac{\partial p}{\partial y} > \pm \sigma$ as the representative value for the "pressure gradient" term in (8), and considering that the statistical properties of the "pressure gradient" near the bottom is expected to scale with the wall variables (that is, u_* and ν in the case of smooth bottom and u_* and k in the case of rough bottom), the criterion in (8) becomes

$$\text{Suspension occurs when} \begin{cases} \dfrac{(s-1)g\nu}{u_*^3} < -\dfrac{3}{2}\left[< \dfrac{\partial p^+}{\partial y^+} > \pm\sigma\right] \\ \text{(Smooth bottom)} \\ \dfrac{(s-1)gk}{u_*^2} < -\dfrac{3}{2}\left[< \dfrac{\partial p^+}{\partial y'} > \pm\sigma\right] \\ \text{(Rough bottom)} \end{cases} \qquad (9)$$

Herein, $p^+ = p/(\rho u_*^2)$, $y^+ = yu_*/\nu$ and $y' = y/k$. Picking up the maximum mean value and the standard deviation of "pressure gradient" from Fig.4, one has

$$\text{Suspension occurs when} \begin{cases} \dfrac{(s-1)g\nu}{u_*^3} < 0.084 \pm 0.034 \\ \text{(Smooth bottom)} \\ \dfrac{(s-1)gk}{u_*^2} < 5.9 \pm 1.56 \\ \text{(Rough bottom)} \end{cases} \qquad (10)$$

Thus, a particle will go into suspension at any rate if it meets the requirement in (10). Relevance of the preceding result to the other works has been discussed in [7]. In Ref. [7] has also been given the Shields representation of the criterion. This gave a "suspension region" in the usual Shields diagram as is shown in Fig.5, where Shields' [15, p.20] relevant flume data and Middleton's [16, Table:3] river data determining the limiting stage of suspension are plotted for comparison. In addition, Shields' curve determining the initiation of motion at the bed is also plotted in this figure.

7. CONCLUSIONS

Using a stereo-photogrammetric technique, observations of the three-dimensional motions of small heavy particles near the bottom of a turbulent open channel were made. The latter showed that the kinematical quantities in connection with particle motions are in accord with those of the bursting process.

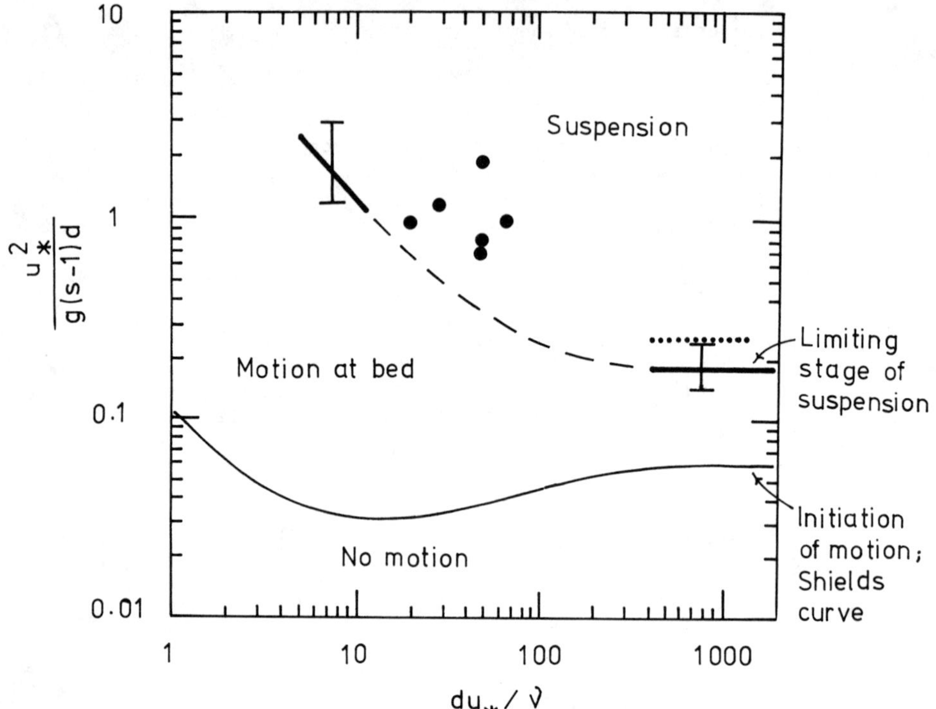

Fig.5. Shields representation of the limiting stage of suspension. —I—, Present study; ......, Shields' flume data; •,Middleton's river data.

The new evidences obtained in the present study provided further support for the mechanism of particle suspension close to the bottom proposed in Ref. [5].

As to the rough bottom case, under the light of present findings a similar mechanism seems to occur with more violent appearence.

As to the motion of relatively heavier particles, the present findings suggest that the particle ejected from the wall region falls out of the lifted fluid of the accompanying burst before the latter breaks up.

The information on the mechanics of suspended particle motion gained in the present study led us to determine the lift force on a particle at the instant the particle is lifted up from near the bottom. This force appeared to be that due to a local, temporary pressure gradient in the vertical.

A small heavy particle in an open channel flow will move in suspension if it is light enough ; if not, it will settle on the bottom. Hence there exists a stage after which no suspension of particles is observed, the limiting stage of suspension. Making use of the lift force information, a criterion was obtained for the latter.

A report covering the whole aspects of the present study is available as a publication of the Inst. of Hydrodyn. and Hyd. Engrg., Tech.Univ. of Denmark [7] .

ACKNOWLEDGEMENTS

The auther wishes to express his gratitude to Prof. F.A.Engelund and Drs. J.Fredsøe and R.Deigaard for their continuous interest and generous help in the course of investigation. The National Council for Scientific Research (Denmark), the Institute of Hydrodyn. and Hyd. Engrg. of the Tech.Univ. of Denmark, the Scientific and Technical Research Council of Turkey and the Faculty of Civil Engrg. of the Tech.Univ. of Istanbul are all gratefully acknowledged.

REFERENCES

1. Sutherland, A.J. 1967. Proposed mechanism for sediment entrainment by turbulent flow. J.Geophys. Res., Vol.72, No.24, p.6183.

2. Grass, A.J. 1974. Transport of fine sand on a flat bed : turbulence and suspension mechanics. Euromech 48, Inst. Hydrodyn. Hyd. Engrg. Tech.Univ. Denmark, p.33.

3. Jackson, R.G. 1976. Sedimentological and fluid-dynamic implications of the turbulent bursting phenomenon in geophysical flows. J.Fluid Mech., Vol.77, p.531.

4. Lumley, J.L. 1978. Two-phase and non-Newtonian flows. In : Turbulence ; Topics in Appl. Phys. (ed. P.Bradshaw), p.289, 2nd ed., Springer-Verlag.

5. Sumer, B.M. and Oğuz, B. 1978. Particle motions near the bottom in turbulent flow in an open channel. J.Fluid Mech., Vol.86, p.109.

6. Jacobi, O. 1979. Photogrammetric tracking of a particle in running water. Euromech 120, Fluid Mech. Dept. Tech.Univ. Denmark, Paper 3.3 .

7. Sumer, B.M. and Deigaard, R. 1979. Experimental investigation of motions of suspended heavy particles and the bursting process. Inst. Hydrodyn. and Hyd. Engrg. Tech.Univ. Denmark, Series Paper 23.

8. Nychas, S.G. et al. 1973. A visual study of turbulent shear flow. J.Fluid Mech., Vol.61, p.513.

9. Pratury, A.K. and Brodkey, R.S. 1978. A stereoscopic visual study of coherent structures in turbulent shear flow. J.Fluid Mech., Vol.89, p.251.

10. Hinze, J.O. 1975. Turbulence. McGraw-Hill.

11. Brodkey, R.S. et al. 1974. Some properties of truncated turbulence signals in bounded shear flows. J.Fluid Mech., Vol.63, p.209.

12. Engelund, F. 1970. A note on the mechanics of sediment suspension. Basic Res. Prog. Rep. Hydraulic Lab. Tech.Univ. Denmark, Vol.21, p.7.

13. Einstein, H.A. and El-Samni, E.A. 1949. Hydrodynamic forces on a rough wall. Rev. of Modern Phys., Vol.21, No.3, p.520.

14. Vanoni, V.A. 1975. In: Sedimentation Engineering (ed. V.A.Vanoni). A.S.C.E. Manual and Reports on Engineering Practice, No.54.

15. Shields, A. 1936. Anvendung der Aehlichkeitsmechanik und der Turbulenzforschung auf die Geschiebebewegung. Mitteilungen der Preussichen Versuchsanstalt für Wasserbau und Schiffbau, No.6.

16. Middleton, G.V. 1976. Hydraulic interpretation of sand size distributions. J.Geology, Vol.84, p.405.

Environmental Flows

E. J. HOPFINGER
Institut de Mécanique
Université de Grenoble

Environmental flows encompass flow situations which range from air and water polution to wind effects on buildings. Various conferences have been held recently which were entirely devoted to some topics of environmental flows : "Dispersion in natural flow systems" at Delft (1978 , Euromech 109), "Flow over Hills and Mountains" in Munich (1979 , Euromech 113), "Stratified Flows" in Trondheim (1980 , I.A.H.R. Symposium). The present Symposium intended to discuss observations of coherent structures in environmental flows and the way in which these structural processes are modified by, for instance, stratification. Many environmental flows exhibit coherent structures, made visual for example by cloud patterns. These are generally a result of convection or Coriolis effects. Evidence of coherent structures in the atmospheric boundary layer is harder to come by butthere is little doubt that processes are similar to those in laboratory boundary layers. Detailled measurements carried out in the benthic,tidal boundary layer (Gordon and Witting, 1977) indicate the existence of repetitive structures and a highly intermittent momentum transport. Structures characteristic of free shear flows can be observed in the cean and in lakes, and laboratory studies of the stratified shear layer are therefore relevant to a great number of environmental flow problems. Although there has been no paper presented at this Symposium on environmentally relevant free turbulent shear flows , it is instructive to start the discussion with the stratified turbulent mixing layer.

Free shear flows

Any mixing in a stably stratified fluid results in an increase in potential energy, which is necessarily gained at the expense of kinetic energy. For a mixing layer of shear thickness δ_U and density thickness δ_ρ the change in kinetic and potential energies are :

$$\Delta KE = -\frac{1}{24}\rho(\Delta U)^2(\delta_{U_f} - \delta_{U_i})$$

$$\Delta PE = \frac{1}{24} g \Delta\rho (\delta_{\rho_f}^2 - \delta_{\rho_i}^2)$$

where ΔU and $\Delta\rho$ are the mean velocity and mean density changes across the layer and a constant gradient was assumed. If all the KE would show up as potential energy then (neglecting the initial thickness)

$$\delta_U/\delta_\rho = Ri$$

where $Ri = g\Delta\rho\,\delta_\rho/(\Delta U)^2\rho$. With $\delta_U \approx \delta_\rho$ the upper bound on Ri is about 1. Observations indicate (Thorpe 1973, Koop and Browand 1979) that the growth of the layer stops when Ri $\gtrsim$ 1/3 or even less depending on the initial stability. Two dimensional calculations by Patnaik, Sherman and Corcos(1976) give the same result. This shows that mixing is inefficient and most of the kinetic energy (70% or more) is dissipated by viscosity either directly or in part by way of internal waves. The effect of the initial stability of the fluid system on the amount of mixing which will be effected, is best illustrated by the following diagram, initially proposed by Posmentier (1977)

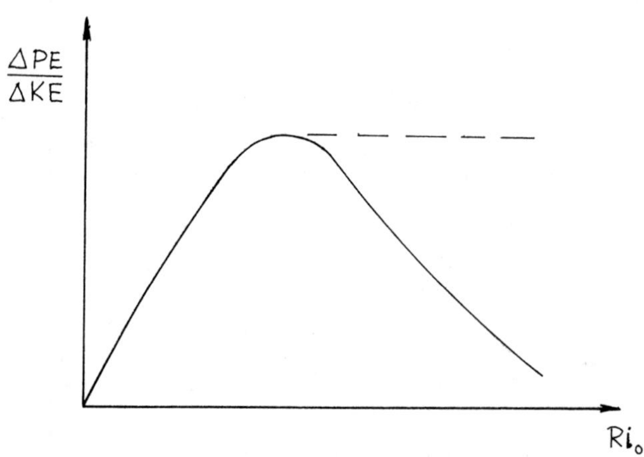

The experimental results of Thorpe and Koop and Browand are well represented by the solid curve. Wakes, and mixing caused by grid turbulence, seem to exhibit a similar behaviour, as has been shown by Linden (1979). The solid line implies that in a very stably stratified fluid practically no transfer of mass or heat is possible. The gain in potential energy relative to the kinectic energy is a global flux Richardson number (space and time averaged) $\overline{R}_f$. The diagram then indicates that no critical value of $\overline{R}_f$ exists, in contrast with older models (see for instance Turner 1973) and beliefs suggesting a critical value of $\overline{R}_f$. Only an upper limit of $\overline{R}_f$ would exist. This is one of the reasons why measured values of $\overline{R}_f$ range from about 0.05 to 0.3. When local flux Richardson numbers are measured a wide variation is also possible because of diffusion of energy which is not included in the usual definition of a local flux Richardson number.

How are these observations tied up with the dynamics of coherent structures ? The observations by Koop and Browand show clearly that the pairing events decrease with increasing stability and pairing is completely inhibited when Ri local reaches a certain value (Ri $\approx$ 1/3 when Ri_o is small) and consequently the growth of the mixing layer stops. Since at this stage the energy production ceases, the vorticity is redistributed in space and turbulence decays by viscosity.

The observation of Ri $\approx$ 1/3, which is close to the stability criterion of 1/4 of a stratified laminar shear layer (assuming equal scales), would suggest that the stratification acts on the production mechanism of turbulence which is then viewed as an instability (Corcos and Hopfinger, 1976). This is only a conjecture and a purely energetic reasoning which does not evoke any specific mechanism can explain the stop in growth rate at the values of Ri observed. Jets and wakes are also inhibited in their growth when $Ri \approx 1/3$ (Townsend, 1957, Hopfinger, 1972, Prych, Harty and Kennedy, 1964) although these differ in structure from shear layers. Future research will perhaps clarify the detailed mechanisms concerning the effect of stable stratification on processes related with coherent structures.

Wall Shear Flows

Stratified wall shear flows are understood to a much lesser extent than free shear flows. Measurements in the atmospheric boundary layer have mainly been interpreted in the light of similarity theories in which the mean flow and turbulent quantities are functions of L/Z (L is the Monin-Obukov length $L = -u_*^3/kB$; k is the von Karman constant and B the buoyancy flux at the boundary). These theories give generally good results in weakly stratified situations. It must be relized, however, that the stability dependence of turbulent transport processes is not the same in the inner and outer layer and Mizushina, Ueda, Mitsumoto, Komori and Fukui* have investigated the whole region of an unstably stratified boundary layer, showing the difference in stability dependence of the fluxes. Visual observations showed a longitudinal vortex flow. This is to be expected when convection cells are subjected to a mean shear. Longitudinal vortex rolls are also a commun feature in open channel flow (Müller*). Spectra and Cospectra obtained in a weakly stratified laboratory boundary layer were found to differ from those obtained in the atmospheric boundary layer (Mestayer, Pages, Coantic and Saissac*). Whether or not these departures are intrinsic to a difference in flow structure or are just a result of measurement anomalies has not been made clear.

In a strongly stably stratified atmosphere similarity theories are less satisfying. An attempt to study the stably stratified boundary layer in more detail has hitherto only been made by Arya (1975) and Nicholl (1979). The results of Arya indicate that the boundary layer tends to a more laminar state when the stratification is increased but whether or not the boundary layer can be relaminarized and at which Richardson number is not known. Work on stably stratified channel flow like plane Poiseuille flow is particularly sparce (Britter, 1974). This would be one of the few stratified flow situations where the Richardson number could be set at any value and kept constant.

Interfacial layers

In environmental flow situations, transfer of heat and mass is sometimes observed even when the overall Richardson number is high (of order 10). These observations are not in contradiction with the relatively low values observed in laboratory shear flows or with the energy constraint of $Ri \lesssim 1$ discussed above. When the condition of equal length scales and a constant gradient is relaxed, high values of an overall Richardson number are not excluded. This leads to the concept of interfacial layers or in short interfaces. Such interfaces are well documented in the interior of the ocean. Near the surface of the ocean and in the atmospheric boundary layer, interfaces (thermoclines and inversions) are a result of strong mixing caused in a stably stratified fluid by turbulence generated at the boundary.

Numerous models of the rate of deepening of the mixed layer in the presence of mean shear have been proposed and the activity centered around this problem is still considerable (see for instance Trondheim Symposium, section vertical exchange). However, at present no unambiguous experimental results are known.

---

* Indicates that the paper has been presented at the ICHMT/IUTAM Symposium, Dubrovnik.

The experiments of Kato and Phillips (1969) suffer from side wall, and possibly rotation effects. The controversy is to decide on whether in a linearely stratified environment the mixed layer deepens at a rate $D \propto t^{1/3}$ (Kato and Phillips) or $D \propto t^{1/2}$, indicated by most models. Measurements carried out recently in a boundary layer which penetrates into a density interface (Piat and Hopfinger, 1980) tend to support $D \propto t^{1/2}$. The latter results suffer however from low Reynolds number and a limited range of Richardson number $Ri_* = g \frac{\Delta\rho}{\rho} \delta / u_*^2 \lesssim 70$. It is interesting to note that in these shear experiments the shear generated across the interface can give rise to a shear instability and Kelvin-Helmholtz billows in the interface. Occasional observations of coherent structures in the mixed layer may be a result of the instability in the interface.

When no mean shear exists like in penetrative convection or in laboratory studies of mixing across an interface, driven by turbulence generated by an oscillating grid, the state of knowledge is also fragmentary. Mechanically generated turbulence results in $dD/dt \propto u\, Ri^{-3/2}$ (Turner, 1968, Hopfinger and Toly, 1976), whereas penetrative convection could be made to agree with $dD/dt \propto u\, Ri^{-1}$ (Deardorff, 1980). An interesting aspect is the interface thickness which seems to scale on the Taylor microscale λ (work in progress in Grenoble). The scale λ has therefore physical significance in mixing phenomena. A suggestion of its physical significance in shear flows has been advocated by Long (1980). How a stable stratification, which permits the radiation of energy by internal waves, affects the mixing rate is still obscure. Linden (1975) has observed a considerable decrease in the rate of deepening, whereas in other situations no noticeable effect seems to be manifest (Deardorff, 1980). The interaction between the turbulence arriving at the interface and the wave field is poorly understood.

Buoyancy flux effects on an initially unstratified turbulent field have received little attention and the paper by Hopfinger* is concerned with this problem. Experiments performed in a turbulent water column without mean shear showed that heat, put in at the surface of the water column, was mixed down to a depth proportional to a Monin-Obukhov length constructed with "local" variables.

Probably the most debated question in laboratory modeling of environmental flows concerns possible Reynolds and Peclet number effects. There is no clear answer to this question and it has to be examined in each flow situation. The answer will also depend on the aim of the research. High Richardson and Reynolds numbers are difficult to obtain simultaneously in laboratory flows. Richardson number (or Froude number) similarity at high Reynolds numbers may be achieved with large density differences. Inertial effects may then become of importance as was pointed out by Rey*.

During the discussion which followed it become clear that in the future more emphasis should be placed on mechanisms in environmental flows which are a consequence of coherent structures. Laboratory studies of the change in coherent processes in stratified shear flows would not only be of interest in the context of environmental flows but could help in understanding turbulent shear flows in general.

ARYA, S.P.S., 1975, Buoyancy effects in a horizontal flat plate boundary layer, J. Fluid Mech., 68, 321.

BRITTER, R.E., 1974, An experiment on turbulence in a density-stratified fluid, Ph. D. Thesis, Monash University, Victoria, Australia.

CORCOS, G.M. and HOPFINGER, E.J., 1976, L'instabilité motrice de la turbulence en écoulements libres, J. de Physique, tome 37, C1-95.

DEARDORFF, J.W., WILLIS, G.E. and STOCKTON, B.H., 1980, Laboratory studies of the entrainement zone of a convectively mixed layer (to appear in J. Fluid Mech.).

GORDON, C.M. and WITTING, J., 1977, Turbulent structure in a benthic boundary layer. In Bottom Turbulence, Elsevier Oceanographic, Series 19.

HOPFINGER, E.J., 1972, Development of a stratified shear flow. Proceedings of stratified shear flow conference, Novosibirsk (A.S.C.E. Publication, 1973, New-York).

HOPFINGER E.J. and TOLY, J.A., 1976, Spatially decaying turbulence and its relation to mixing across density interfaces. J. Fluid Mech., 78, 155.

KATO, H. and PHILLIPS, O.M., 1969, On the penetration of a turbulent layer into a stratified fluid. J. Fluid Mech., 37, 643.

KOOP, C.G. and BROWAND, F.K., 1979, Instability and turbulence in a stratified fluid with shear. J. Fluid Mech., 93, p. 135.

LINDEN, P.F., 1975, The deepening of a mixed layer in a stratified fluid. J. Fluid Mech., 71, 385.

LINDEN, P.F., 1979, Mixing in stratified fluids. Geophys. Astrophys. Fluid Dynamics, 13, 3.

NICHOL, C.I.H., 1970, Some dynamical effects of heat on a turbulent boundary layer. J. Fluid Mech., 40, 361.

PATNAIK, P.C., SHERMAN, F.S. and CORCOS, G.M., 1976, A numerical simulation of Kelvin-Helmholtz waves of finite amplitude. J. Fluid Mech., 73, 215.

PIAT, J.F. and HOPFINGER, E.J., 1980, A boundary layer topped by a density interface, (submitted to J. Fluid Mech.).

POSMENTIER, E.S., 1977, The generation of salinity fine structure by vertical diffusion, J. Phys. Ocean, 7, 298.

PRYCH,E.A., HARTY, F.R. and KENNEDY, J.F., Turbulent wakes in density stratified fluids of finite extent, MIT Hydrodynamics Laboratory, Report 65, 1964.

THORPE, S.A., 1973, Experiments on instability and turbulence in a stratified shear flow, J. Fluid Mech., 61, p. 731.

TOWNSEND, A.A., 1957, Turbulent flow in a stably stratified atmosphere. J. Fluid Mech., 3, 361.

TURNER, J.S., 1973, Buoyancy effects in fluids, Cambridge, University Press.

TURNER, J.S., 1968, The influence of molecular diffusivity on turbulent entrainment across a density interface, J. Fluid Mech., 33, 639.

TURBULENCE MODELING

A Theoretical Model for Coherent Structures in Wall Turbulence

M. T. LANDAHL
Royal Institute of Technology
Stockholm, Sweden
and Massachusetts Institute of Technology
Cambridge, Massachusetts, USA

ABSTRACT

The paper formulates a theory for coherent structures in the wall region of a turbulent boundary layer in an attempt to model such structures that come out from experiments employing conditional sampling techniques. The starting point is the Navier-Stokes equations written as perturbations (finite) on a parallel flow. These are then conditionally sampled leading to a system of nonhomogeneous equations describing an initial-value problem. The nonhomogeneous driving terms consist of Reynolds stress terms due to the small-scale noncoherent velocity fluctuations.

For more detailed analysis some further assumptions regarding the small-scale flow, partly supported by observations, are made. Thus, the horizontal extent of the region of non-zero sampled small-scale Reynolds stresses is assumed to be much greater than the vertical one. This is justified by the observations that the streamwise size of typical coherent structures is of the order several hundred viscous wall units, whereas the vertical one is only of the order a few tens. Furthermore, the small-scale motion is assumed to develop on a much shorter time scale than the coherent one, as would be the situation if the small-scale motion were the result of localized secondary instability of inflexional type. On this basis the evolution of the coherent structure is analyzed using a Lagrangian formulation. The assumption of a large horizontal scale allows one to introduce a boundary-layer type analysis. The effect of induced pressure, viscosity, and nonlinearity are found to play an important role only for long times after the initiation of the structure, when it has been convected several times its own length. For moderate times, when structure has traveled a distance of the order its own streamwise length, the evolution is controlled primarily by continuity and horizontal inertia. The streamwise perturbation velocity may then be expressed very simply in terms of a "mixing length" measuring the vertical displacement of the fluid element. Calculations based on the approximation for moderate times give results in qualitative agreement with the conditionally sampled data of Blackwelder & Kaplan [7]. Thus, it shows the tendency towards formation of thin shear layers, a velocity defect near the wall for short times being replaced by a slowly decaying velocity excess for large times, and a tendency towards inflow ("sweep") of high-speed fluid from the outer region.

1. INTRODUCTION

One of the most active areas of research in the field of turbulence during recent years has been the study of the turbulence structure near a wall. The discovery by Kline et al. [1] that the turbulence in the near wall region exhibits recognizable and repeatable patterns ("coherent structures") in the intermittent turbulent bursting cycle has spawned a vigorous research, primarily experimental, to try to explore and explain their main features. For reviews of the developments in this area we may refer to the article by Willmarth [2] (for research before 1975) and to proceedings [3], [4], [5] of a number of symposia and workshops devoted to this subject. As has become more and more apparent over the years, the experimental difficulties in the study of such structures are very great, indeed, and especially those involved in trying to relate characteristic spatial-temporal patterns seen in visualization experiments to results from point measurements by velocity probes in the flow. Such difficulties are discussed by Offen & Kline [6]. Nevertheless, despite considerable differences of opinion among experimentalists concerning observational details and the merits of the different conditional sampling methods used to educe coherent features, there is considerable agreement on some major observational findings such as the formation of low-speed streaks and their lift-up, and the high contribution to Reynolds stresses by the ejection and sweep phases during bursting. Also, the appearance during this cycle of local regions of high shear in form of highly inclined thin shear layers has been firmly established. These manifest themselves as rapid velocity increases, "fronts" [7, 8] which in measurements employing hot-wire rakes through the boundary appear to move towards the wall because of their inclination in the downstream direction.

The interpretation of the experimental data in terms of mechanistic cause and effect relationships for the observed motion has been severely hampered by the lack of adequate theoretical models. Because of the extreme theoretical difficulties encountered in an analysis of a highly vortical, unsteady, and three-dimensional viscous flow, in which nonlinearities are an essential feature, one is forced to resort to several simplifying assumptions to make the problem tractable. To choose what assumptions can be made without leaving out any essential part of the mechanisms involved is of course quite difficult and requires careful consideration of what effects are most essential. In view of the strong three-dimensionality of the observed features, two-dimensional models are unlikely to capture the essential features of the flow. On the other hand, the rapidity with which the structures evolve indicates that they are controlled primarily by inertial effects, so that viscosity may not be an essental ingredient. Another observational feature of the structures is their large horizontal dimensions - particularly in the streamwise direction - compared to their thickness, which should allow one to make use of boundary-layer type approximations. Such ideas underlie the model proposed earlier by the author and explored in earlier publications [9]-[11]. The present paper represents a further development of the model including additional comparisons with experiments.

3. ANALYSIS

We consider turbulent fluctuations of velocities $u_1, u_2, u_3 = u, v, w$ and pressure p superimposed on a mean shear flow, which will be assumed to be parallel and of velocity $\widetilde{U}_i = u(y)\delta_{1i}$. These are governed by the equations

$$\frac{\partial u_i}{\partial t} + U \frac{\partial u_i}{\partial x_j} + u_2 \frac{dU}{dy} \delta_{1i} = -\frac{1}{\rho}\frac{\partial p}{\partial x_i} + \nu\nabla^2 u_i + \frac{\partial}{\partial x_i}\tau_{ij} \tag{1}$$

$$\frac{\partial u_i}{\partial x_i} = 0 \tag{2}$$

where $x_1 = x$, $x_2 = y$, $x_3 = z$, and τ_{ij} are the "fluctuating Reynolds stresses" defined by

$$\tau_{ij} = -u_i u_j + \overline{u_i u_j} \tag{3}$$

overbar denoting mean values. It is also assumed that the mean flow is in balance with the mean Reynold stresses which for boundary layers would generally require the flow to be non-parallel, but the effects of non-parallelism on the fluctuating field shall be ignored. We shall study the behavior of a large-scale "coherent" structure educed by some conditional sampling criterion. The associated ensemble average is denoted by tilde (~). The criterion is assumed to be such as to single out flow features which are of large horizontal (streamwise and spanwise) scale compared to their thickness, i.e., to be of a "pancake" shape. This may be expressed in terms of the smallness of the parameter

$$\varepsilon = (\delta/\ell) \ll 1 \tag{4}$$

where δ is the thickness (in direction $x_2 = y$ normal to the wall) and ℓ the smallest of the streamwise (ℓ_1) and spanwise (ℓ_3) dimension of the structure. This assumption is well justified as far as the streamwise dimensions are concerned by the visual observations [1], which indicate lengths of up to 1000 in viscous wall units whereas the thickness of the structures are a few tens in y^+. Less justified is the assumption of large spanwise scale, a typical scale being the observed average distance between low-and high-speed streaks, which is about 50 in viscous wall lengths. However, in the interest of constructing the simplest possible model which will incorporate some of the basic effects of three-dimensionality, this assumption is adopted anyway, but will of course require justification by a separate analysis. The second basic assumption in the theory is that the large-scale structures are initiated by the collective action of small-scale Reynolds stresses produced by localized regions of secondary instability. This is supported by the visual observations [1] which reveal the existence of a rapid and intensive break-up stage during bursting. Such regions may result from breakdown of thin shear layers evolving from older large-scale eddies. The time scale of evolution of the small-scale secondary instability is typically of the order $\tau_s = (dU/dy)^{-1}$, whereas the large-scale eddy evolves on a time scale $\tau_\ell = \ell/U$. A correlary of the assumption of a large horizontal scale is that the ratio of these scales is small, i.e., that

$$\tau_s/\tau_\ell = U/(\ell U') = O(\varepsilon) \ll 1$$

Consequently, in the statistical analysis of how the large-scale structures evolve in the time scale τ_ℓ one may assume that each mumber of the ensemble making up the coherent structure is created instantaneously in that time scale. We introduce in the system (1), (2)

$$u_i = \tilde{u}_i + u'_i \tag{5}$$

where $\tilde{u}_i$ is the conditional average and u'_i the small-scale fluctuations defined in such a manner that

$$\tilde{u}'_i = 0 \tag{6}$$

tilde denoting the conditional average. Application of conditional averaging to (1), (2) gives

$$\frac{\tilde{D}(\tilde{u}_i + U\delta_{1i})}{Dt} = -\frac{1}{\rho}\frac{\partial \tilde{p}}{\partial x_i} + \nu\nabla^2\tilde{u}_i + \frac{\partial \tilde{\tau}'_{ij}}{\partial x_j} \tag{7}$$

$$\frac{\partial \tilde{u}_i}{\partial x_i} = 0 \tag{8}$$

where

$$\frac{\tilde{D}}{Dt} = \frac{\partial}{\partial t} + (\tilde{u}_j + U\delta_{1j})\frac{\partial}{\partial x_j} \tag{9}$$

and

$$\tilde{\tau}'_{ij} = \overline{u_i u_j} - \widetilde{u'_i u'_j} \tag{10}$$

are the conditionally averaged Reynolds stress fluctuations due to the small scales. To solve the system (7) (8) approximately we introduce Lagrangian coordinates ξ_i ($\xi_1 = \xi$, $\xi_2 = \eta$, $\xi_3 = \zeta$) where ξ_i denotes (pseudo-)material coordinates for the conditionally averaged large-scale motion. From the component equations for the streamwise ($\tilde{u}_1 = \tilde{u}$) and spanwise ($\tilde{u}_3 = \tilde{w}$) conditionally averaged velocities one then finds from (7) formally, upon integration with respect to time

$$\tilde{u} + U = \tilde{u}_o(\xi,\eta,\zeta) + U(\eta) + \int_0^t \left[-\frac{1}{\rho}\frac{\partial \tilde{p}}{\partial x} + \nu\nabla^2\tilde{u} + \frac{\partial \tilde{\tau}'_{1j}}{\partial x_j}\right]\tilde{D}t_1 \tag{11}$$

$$\tilde{w} = \tilde{w}_o(\xi,\eta,\zeta) + \int_0^t \left[-\frac{1}{\rho}\frac{\partial \tilde{p}}{\partial z} + \nu\nabla^2\tilde{w} + \frac{\partial \tilde{\tau}'_{3j}}{\partial x_j}\right]\tilde{D}t_1 \tag{12}$$

where index o refers to values at t = 0 and $\int \sim \tilde{D}t_1$ denotes integration with respect to time holding ξ,η,ζ fixed. In keeping with the basic assumptions of the theory, the terms $\tilde{\tau}'_{ij}$ are to be regarded as the driving terms. Furthermore, they act on a much shorter time scale than the other stress terms in the integrals, so that on the time scale of evolution of the large scales they may be treated as delta functions in time acting at some (average) initial time $t = t_i$. We may therefore write (11) and (12) as

$$\tilde{u} + U = \tilde{u}_i(\xi,\eta,\zeta) + U(\eta) + \int_{t_i}^t \left[-\frac{1}{\rho}\frac{\partial \tilde{p}}{\partial x} + \nu\nabla^2\tilde{u}\right]\tilde{D}t_1 \tag{13}$$

$$\tilde{w} = \tilde{w}_i(\xi,\eta,\zeta) + \int_{t_i}^t \left[-\frac{1}{\rho}\frac{\partial \tilde{p}}{\partial z} + \nu\nabla^2\tilde{w}\right]\tilde{D}t_1 \tag{14}$$

where

$$\tilde{u}_i = \tilde{u}_o + \tilde{F}_{1\eta} \qquad (15)$$

$$\tilde{w}_i = \tilde{w}_o + \tilde{F}_{3\eta} \qquad (16)$$

and

$$\tilde{F}_1 = \int_0^{t_i} \int_0^{\eta} \frac{\partial \tilde{\tau}'_{1j}}{\partial x_j} \, dy \, \tilde{D}t_1 \simeq \int_0^{t_i} \tilde{\tau}'_{12} \, \tilde{D}t_1 \qquad (17)$$

$$\tilde{F}_3 = \int_0^{t_i} \int_0^{\eta} \frac{\partial \tilde{\tau}'_{3j}}{\partial x_j} \, dy \, \tilde{D}t_1 \simeq \int_0^{t_i} \tilde{\tau}'_{23} \, \tilde{D}t_1 \qquad (18)$$

The approximations in (17, (18) involves neglect of the terms containing horizontal (x,z-)derivatives compared to those with vertical (y-)derivatives, in consistency with the assumption of large horizontal scale. In (13)-(14) the original problem has thus been reformulated [11] as a new initial-value problem with initial data given by (15)-(18) in terms of the small-scale stresses and the velocity field existing before their onset. An approximate method of solution of the initial-value problem, valid for short times, was developed in Ref. [9]. This is based on neglecting the integrals in (13), (14). To assess the range of validity of such an approximation one needs an estimate of the pressure, which may be obtained from a boundary layer approximation, which gives that the induced pressure is approximately constant through the layer and given by its value at the outer edge of the region of strong shear, where it may be calculated from potential-flow theory [9]. One finds that

$$\frac{\tilde{p}}{\rho} = 0(\tilde{v}_\delta U_\delta) \qquad (19)$$

where index δ denotes values at the outer edge. For small times

$$\tilde{v}_\delta = 0(\tilde{u}_{i\delta}\,\delta/\ell) \qquad (20)$$

so that the contribution from the pressure to the integral (setting $\partial/\partial x = 0(1/\ell)$) becomes of order

$$\frac{(t-t_i)\, U_\delta \, \tilde{u}_{i\delta} \, \delta}{\ell^2} \qquad (21)$$

which is small compared to the initial velocity $\tilde{u}_i$, whenever

$$\frac{(t-t_i)\, U_\delta \, \delta}{\ell^2} \ll 1 \qquad (22)$$

Thus, the pressure gradient term may be neglected in the determination of $\tilde{u}$

for times which are short compared to the time it takes for the large-scale structure to be convected a distance of the order (ℓ/δ) times its own length. Application of the estimate (19) to the equation for $\tilde{w}$ leads to the same condition if one assumes that $\tilde{w}$ and $\tilde{u}$ are of the same order as are the spanwise and streamwise scales.

An estimate of the viscous term shows that this may be neglected whenever

$$\frac{\nu(t-t_i)}{\delta^2} \ll 1 \tag{23}$$

(For a convected eddy a slightly more refined estimate may be obtained, see [10].) As an example of the restrictions in time imposed by (22) and (23) we may take as typical values $\delta^+ = 20$, $\ell^+ = 200$ in viscous wall units which give $t^+ - t_i^+ \ll 200$ from (22) and $t^+ - t_i^+ \ll 400$ from (23), using $U_\delta^+ = 10$.

Thus, neglecting the integral terms in (13) and (14) we find

$$\tilde{u} + U = \tilde{u}_i(\xi,\eta,\zeta) + U(\eta) \tag{24}$$

$$\tilde{w} = \tilde{w}_i(\xi,\eta,\zeta) \tag{25}$$

and

$$x = \xi + (t-t_i)(U+\tilde{u}_i) \tag{26}$$

$$z = \zeta + (t-t_i)\,\tilde{w}_i \tag{27}$$

Here, the origin of the material coordinates has been defined such that ξ,η,ζ are the values of x,y,z at $t = t_i$.

The coordinate y is determined through application of continuity. Because of the conservation of volume of fluid elements in a constant-density flow, one can show (Lamb. [12] art.14) that (8) is equivalent to

$$J = \frac{\partial(x,y,z)}{\partial(\xi,\eta,\zeta)} = 1 \tag{28}$$

which gives for y the following partial differential equation [9]

$$A_1 \frac{\partial y}{\partial \xi} + A_2 \frac{\partial y}{\partial \eta} + A_3 \frac{\partial y}{\partial \zeta} = 1 \tag{29}$$

where

$$A_1 = x_\zeta z_\eta - x_\eta z_\zeta \tag{30}$$

$$A_2 = x_\xi z_\zeta - x_\zeta z_\xi \tag{31}$$

$$A_3 = x_\eta z_\xi - x_\xi z_\eta \tag{32}$$

which may be solved by integration along the characteristics x,z const., given by

$$\frac{\partial \xi}{\partial \eta} = \frac{A_1}{A_2} \quad , \quad \frac{\partial \zeta}{\partial \eta} = \frac{A_3}{A_2} \tag{33}$$

yielding

$$y = \int_0^{\eta} \frac{d\eta_1}{(A_2)_{x,z}} \tag{34}$$

$$A_2 = 1 + (t-t_i)(\tilde{u}_{i\xi} + \tilde{w}_{i\zeta}) - (t-t_i)^2(\tilde{u}_{i\zeta}\tilde{w}_{i\xi} - \tilde{u}_{i\xi}\tilde{w}_{i\zeta}) \tag{35}$$

In consistency with the short-time approximation we shall assume that only linear terms need be retained, i.e., that the initial velocities are small, so that

$$|(t-t_i)\,\tilde{u}_i\,/\ell| << 1 \tag{36}$$

which is consistent with the assumption (22), provided

$$\tilde{u}_i/U_\delta = 0(\varepsilon) \tag{37}$$

and similarly for $\tilde{w}_i$. The result (34) may then be linearized to yield for the displacement of the fluid element $\ell_m \equiv y - \eta$, the local "mixing length",

$$\ell_m = (t-t_i) \int_0^{\eta} \tilde{v}_{i\eta}(\xi_1, \zeta, \eta_1)\, d\eta_1 \tag{38}$$

where the integral is to be evaluated holding

$$x = \xi + (t-t_i)\, U(\eta) \tag{39}$$

$$z = \zeta \tag{40}$$

fixed. To within the linearized approximation, the velocity components are then given by

$$\tilde{u} = \tilde{u}_i(\xi,\eta,\zeta) - \ell_m U' \tag{41}$$

$$\tilde{v} = \left(\frac{\partial}{\partial t} + U\frac{\partial}{\partial x}\right) \ell_m \tag{42}$$

$$\tilde{w} = \tilde{w}_i(\xi,\eta,\zeta) \tag{43}$$

The evaluation of the velocity components within the initial time period restricted by (22) and (39) thus becomes a simple matter of evaluating the integral (38), which is readily accomplished by numerical quadrature.

Certain interesting propoerties of the solution may be derived by expanding the solution for large values of $t-t_i$. One can demonstrate [9] that $\tilde{v}$ tends to zero as $(t-t_i)^{-1}$ for $t-t_i \rightarrow \infty$ but that $\tilde{u}$ tends to a non-zero value, independent of time, in a convected frame of reference. The solution for $\tilde{w}$ represents a purely convective motion; in the absence of spanwise pressure gradients, viscous stresses, and interaction with the mean shear, the fluid elements retain their initial spanwise velocity. In reality, of course, there will be spanwise pressure gradients induced by the three-dimensional motion, but for structures of large horizontal scales they will act over a longer time scale. In Ref. [9] some initial efforts were made to include them, but the theory then becomes rather complicated.

3. ASSUMPTIONS REGARDING THE SMALL-SCALE STRESSES

In order to carry out calculations based on the approximate theory presented it is necessary to make additional assumptions regarding the small-scale stresses and the flow conditions preceding the bursting. For the small-scale turbulent stresses it shall be assumed that the streamwise shear-stress component, $\tilde{\tau}'_{12}$, accounts for the dominating effects so that the spanwise shear stress τ'_{23} may be neglected. At present, there is no theoretical or experimental basis on which to judge whether this is justifiable, but it will be adopted in the interest of simplification. Since $\tilde{\tau}'_{12}$ is an average of stresses produced by events which are likely to be largely statistically independent, it is reasonable to assume for $\tilde{F}_1$ a Gaussian distribution in the streamwise coordinate of the form

$$\tilde{F}_1 = \tilde{G}(y,z)\exp\left[-\left(\frac{x-x_o}{\ell_1}\right)^2\right] \tag{44}$$

where x_o is the streamwise center of gravity of the distribution, and ℓ_1 its streamwise length. Conceivably, x_0 as well as ℓ_1 could be functions of y and z, but for simplicity they shall be assumed to be constants. Similarly, the distribution with z could also be taken to be Gaussian, but it is not needed for the present calculations.

For the variation of $\tilde{F}_1$ with y no such simple distribution can be employed since the fluctuation field is not homogeneous in this coordinate and both the scales and the intensity of the Reynolds stresses depend on the distance to the wall. Fortunately, some experimental data are abailable from Kim et al. [13] for the variation with y of the Reynolds stress distribution during bursting which may be used as a guide. The results of their measurements are shown in Figure 1. Also shown is a fit to the data by the curve

$$C'\left(\frac{y}{\ell_2}\right)^3 e^{-(y/\ell_2)^2} \tag{45}$$

for which $C' = 6.5$ and $\ell_2^+ \equiv \ell_2 u_*/\nu = 16$ have been chosen. This form was proposed by Bark [14] and used with some success in his calculations of wave-number frequency spectra on basis of the wave-guide model [15]. The values of C' and ℓ_2^+ selected here are slightly different from the values $C' = 6.36$ and $\ell_2^+ = 15$ used by Bark; with the present choice a someqhat better fit is obtained in the region $y^+ < 20$, which is considered the most interesting one in the present context. The data of Kim et al. [13] represent averages during bursting and, presumably, include contributions from both large-and small-scale motions. Nevertheless, it is likely that a major contribution comes from the small scales in view of the observed rapid onset of and the intensive small-scale activity during the breakup stage, and that in any case the y-distribution of the small-scale stresses is of a similar form and adequately represented by the expression (45). Accordingly, we shall choose for $\tilde{F}_1$ the functional form

$$\tilde{F}_1 = C\left(\frac{y}{\ell_2}\right)^3 \exp\left[-\left(\frac{x-x_o}{\ell_1}\right)^2 - \left(\frac{y}{\ell_2}\right)^2\right] \tag{46}$$

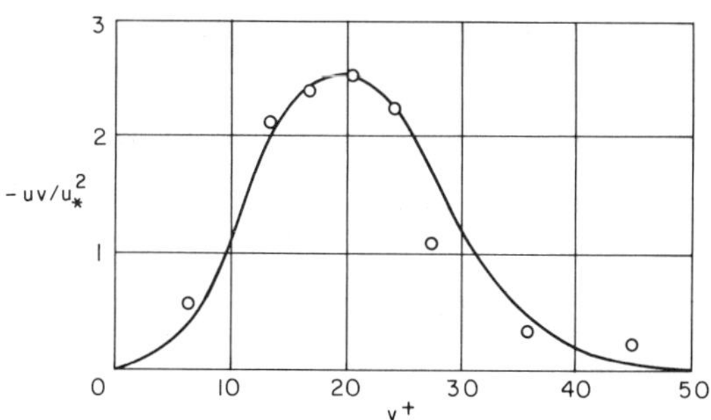

Figure 1. The uy component of the Reynolds stress tensor as a function of y^+ during bursting. o measured points from Kim et al. [13]; —— fitted curve according to formula (45).

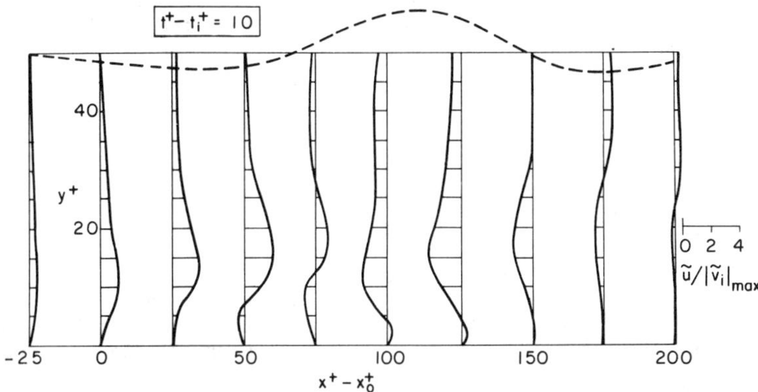

Figure 2. Streamwise perturbation velocity field for model coherent structure at $t^+ - t_i^+ = 10$. – – – location of fluid elements originally located at $y^+ = 50$ for $|\tilde{v}_i|_{max} = 0.5\ u_*$.

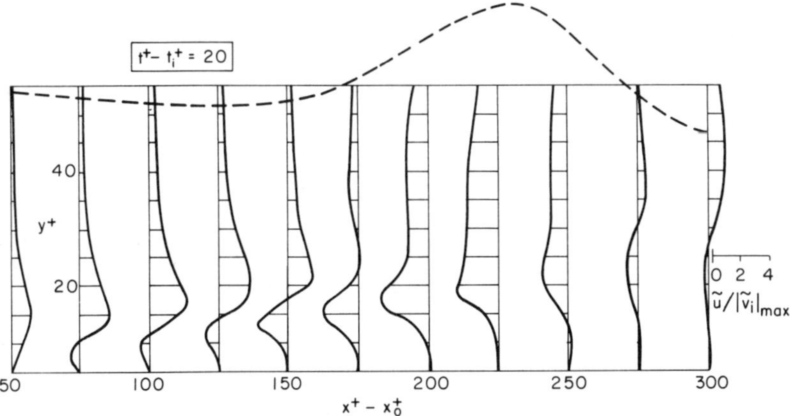

Figure 3. Streamwise perturbation velocity field for model coherent structure at $t^+ - t_i^+ = 20$. – – – location of fluid elements originally located along $y^+ = 50$ for $|\tilde{v}_i|_{max} = 0.5\ u_*$.

The coefficient C is not directly comparable to the coefficient C' above since $\tilde{F}_1$ represents an integral over time of the stresses. However, for the present purpose it is not needed, since we will only consider a linear version of the theory in which all perturbation quantities become proportional to C, and only relative and not absolute values are of interest at this stage, especially the qualitative behavior of the streamwise component. Also, C would generally be a function of z, but its variation is not needed in (38)

From (15) it follows that $\tilde{F}_1$ is directly related to the change in streamwise velocity due to the small-scale stresses, namely

$$\tilde{F}_{1\eta} = \tilde{u}_i - \tilde{u}_o \tag{47}$$

Kim et al. [13] observed that during the chaotic breakup stage, which in the present model is hypothesized to correspond to the period of the action of the small-scale stresses, the instantaneous velocity profile returns very nearly to the mean profile. Accordingly, we shall assume that the small-scale mixing is such as to completely wipe out the inflectional profile preceding the breakup, so that one may set

$$\tilde{u}_i = 0 \tag{48}$$

in (47) and hence

$$\tilde{F}_{1\eta} = - \tilde{u}_o \tag{49}$$

The linear theory for disturbances in an inviscid shear flow predicts that the vertical velocity component decays with time (but not the horizontal ones, see [16]). Thus, if the bursting regions are well separated in time and space, one may in addition set

$$\tilde{v}_o = 0 \tag{50}$$

From (48) and (49) together with continuity one hence finds that

$$\tilde{v}_i(\xi,\eta,\zeta) = - \tilde{F}_{1\xi} \tag{50}$$

which is to be used together with (46) and (48) in (38)-(43) in the calculations which follow.

4. NUMERICAL EXAMPLE

For numerical calculations one needs, in addition to the sampled small-scale Reynolds stresses and the initial perturbation velocity components, also the mean velocity distribution. For this the expression proposed by Reichardt [17] was used

$$U^+ \equiv \frac{U}{u_*} = \frac{1}{\varkappa}\,\ell n\,(1+\varkappa y^+) + c\left[1-e^{-(y^+/d^+)} - \frac{y^+}{d^+}\,e^{-0.33\,y^+}\right] \tag{51}$$

which with $\varkappa = 0.4$, $d^+ = 11$ and $c = 7.4$ gives an excellent fit to the experimental data for all values of $y^+ = \frac{yu_*}{\nu}$ in the wall and buffer regions. In the expression (46) for $\tilde{F}_1$, $\ell_2^+ = 16$ was chosen, as discussed in the previous section. For the streamwise scale factor, the value $\ell_1^+ = 50$ was selected corresponding to an overall length for the initial structure of about 200 viscous wall units. What value is actually chosen for ℓ_1 will not make any essential qualitative difference in the final results since $\tilde{u}$, as well as t and x, may be readily rescaled to correspond to any other chosen value of ℓ_1. The final results are presented as ratios of $\tilde{u}$ to the maximum in the absolute value of the initial vertical velocity, $|\tilde{v}_i|_{max}$, so that the results become independent of C. The computation involves the evaluation of a single integral (38) which is readily accomplished by numerical quadrature. The simple trapezoidal rule was used with a step size of $\Delta y^+ = 1$. The results from the calculations for the "model" coherent structure thus obtained are presented in Figures 2-4. In Figure 2 and 3, the perturbation velocity distribution $\tilde{u}(y^+)$ is shown for two different values of time, $t^+ - t_i^+ = 10$ and $t^+ - t_i^+ = 20$, and for a number of values of $x^+ - x_o^+$. The dotted line shows, for $|\tilde{v}_i|_{max} = 0.5_*$, the position of fluid elements originally located along $y^+ = 50$. For small times the streamwise length of the perturbed region is of the order $\Delta x^+ \simeq 200$, but it grows with t^+, and for $t^+ = 20$ the perturbed region occupies a streamwise length of more than $\Delta x^+ \simeq 300$. Of special interest is the formation of a thin layer of strong perturbation shear highly inclined to the vertical which is seen to extend between about $x^+ - x_o^+ \cong 75$ and $x^+ - x_o^+ \cong 175$. As time increases, the thickness of this internal shear layer decreases so that the shear increases [10].

In the variable-interval time-averaging (VITA) technique employed by Blackwelder & Kaplan [7] sampled perturbation velocity distributions $\tilde{u}(y^+)$ and $\tilde{v}(y^+)$ were obtained as function of time relative to detection using as a detection criterion

$$D(t) = \begin{cases} 1 \text{ if } \widehat{var} > k\, u^2_{r.m.s} \\ 0 \text{ otherwise} \end{cases} \tag{52}$$

where k is a threshold value (taken as 1.2) and where $\widehat{var}$ (the localized variance) is defined by

$$\widehat{var}(x_i, t, T) = \widehat{u^2}(x_i, t, T) - \left[\hat{u}(x_i, t, T)\right]^2 \tag{53}$$

Here the short-time average $\hat{Q}$ of any quantity $Q(x_i, t)$ is defined as

$$\hat{Q}(x_i, t, T) = \frac{1}{T}\int_{t-\frac{1}{2}T}^{t+\frac{1}{2}T} Q(x_i, t_1)\,dt_1 \tag{54}$$

The variance will be largest if $\hat{u}$ is as small as possible at the same time as $\widehat{u^2}$ is as large as possible. This would generally tend to occur when u goes through a zero with $|\partial u/\partial t|$ being as large as possible. Thus, the detection criterion (52) would tend to single out motions which give large accelerations or decelerations at instances when u passes zero. For the typical coherent structure considered we therefore select its time origin t_i^+ and streamwise center x_o^+ such that $|\partial\tilde{u}/\partial t^+|$ has a maximum at $t^+ = 0$, $x^+ = 0$ when $\tilde{u} = 0$. After some numerical searching it was found that with the model considered this criterion is satisfied for $t_i^+ = -20$ and $x_o^+ = -170$. Of significance is that $\partial\tilde{u}/\partial t^+$ was always found to be positive when $\tilde{u} = 0$.

The values of $\tilde{u}$ determined with this choice of x_o^+, t_o^+ at $x^+ = 0$, and for several values y^+ are presented as function of time in Figure 4.

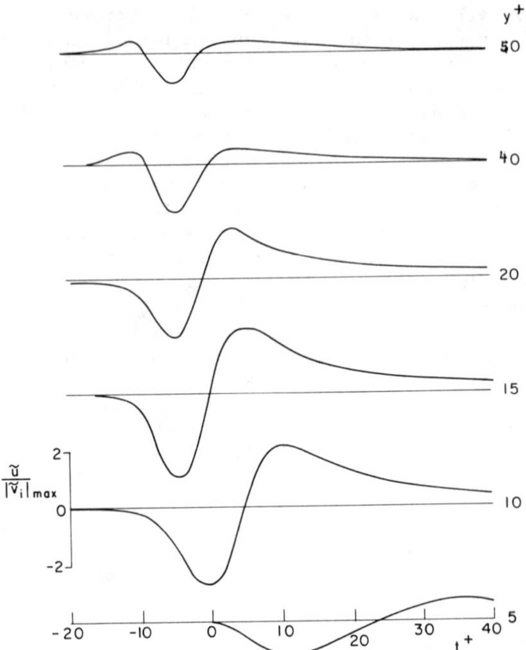

Figure 4. Streamwise perturbation velocities at various distances from the wall as function of time relative to initiation of model coherent structure. $x_o^+ = -170$, $t_i^+ = -20$.

5. COMPARISONS WITH EXPERIMENTS

A comparison with available experimental data shows that the simple theory reproduces qualitatively correctly many of the observed properties of coherent structures. Perhaps most interesting is the appearance of the thin shear layer, (Figure 3) which is formed as the structure is being convected downstream. Local regions of strongly inflectional velocity profiles indicating the appearance of such layers were observed already in the experiments by Kline et al. [1]. According to the theory, the shear layer intensifies as it moves downstream and is continuously tilted over by the mean shear towards the horizontal. The high-speed highly inclined "fronts" seen in the experiments by Blackwelder & Kaplan [7], Chen & Blackwelder [8], Kreplin & Eckelmann [18] and others are clearly manifestations of such shear layers. The layers are of great interest from a dynamical point of view since they would be likely to become inflectionally unstable and eventually break down as proposed by Kline et al. [1]. According to the theory presented such a breakdown could initiate a new coherent structure further downstream, thus sustaining the process.

It is also interesting that the theory shows the shear layer to appear in the "sweep" region of the coherent structure, i.e., underneath the region where there has been an inrush of high-speed fluid towards the wall from the outer region. For a fixed streamwise position of observation the theory predicts an outflow of low-speed fluid to occur first followed by an inflow of high-speed fluid with the shear layer forming between the high- and low-speed flows. This is in qualitative accord with observations in many experiments such as those by Kline et al. [1], Corino & Brodkey [19] and Offen & Kline [6]. Downstream of the internal shear layer the theory shows that a bulge forms at the outer edge (taken as $y^+ = 50$) of the wall region. This is highly reminiscent of the large scale flow (on the scale of the boundary layer thickness) proposed by Falco [20], Brown & Thomas [21], and others. Falco [20] observed typical eddies of scales of the order 100 viscous wall lengths along the upstream face of the bulges. These may be manifestations of local shear layer instability. Brown & Thomas [21] found high-frequency wall shear stress fluctuations to occur just downstream of the sweep region, which they proposed arise from Taylor-Görtler instability but more likely have their origin in shear layer instability. Although these observations concern mainly the outer flow region, away from the viscous wall region, the present theoretical model is likely to apply qualitatively also to the flow in the outer region which should be even more dominated by inviscid mechanisms, although the mean shear is considerably smaller there than in the wall region, so that effects arising from the interaction with the mean shear should be less.

The theoretical results presented in Figure 4 are directly comparable with the conditionally sampled velocities obtained by Blackwelder & Kaplan [7] using their VITA technique. The theoretical $\tilde{u}$-velocities show a behavior quite similar in character to the conditionally sampled experimental ones. Thus, a deceleration occurs prior to the rapid acceleration at all distances from the wall. As a consequence of the shearing by the mean flow the acceleration arrives earlier at the outer positions than nearer the wall, as is also seen in the experimental data (Figure 10 in Blackwelder & Kaplan [7]). The theory predicts the strongest accelerations to occur at $y^+ = 15$-20 in accordance with the experiments, but the behavior far from the wall is somewhat different in that the experiments give a clear excess velocity for $y^+ > 40$, whereas the theory predicts a period of negative velocity for all values of y^+ before the acceleration takes place. (This acceleration is considerably less than in the region $y^+ = 10$-20, however.) The most striking difference between theory and experiments is seen in the velocity at $y^+ = 5$ for which the experiments indicate a velocity signature very much like the one for $y^+ = 10$ (although occurring somewhat later), but for which the theory gives a much lower value with the acceleration occurring considerably later. This discrepancy between theory and experiments could be due to shortcomings of the Reynolds stress model (46) chosen, or a consequence of the neglect of pressure gradients.

6. CONCLUSIONS

The inviscid model for a large-scale coherent structure of boundary layer turbulence presented here is based on the hypothesis that the large-scale motion is initiated by the collective effects of short-lived small-scale Reynolds stresses produced by local secondary instability. The interaction of the large-scale motion with the mean shear is found to produce a local internal shear layer which intensifies and extends in the streamwise direction as it moves downstream. The mechanism involved in the shear intensification may be identified as stretching of the mean vorticity by the spanwise perturbation velocities [9]. Contrary to suggestions by Willmarth [2], and by Offen & Kline [6], streamwise pressure gradients need not be invoked to explain the appearance of local inflectional velocity profiles since they are predicted also by a theory in which horizontal pressure gradients are neglected altogether. The neglect of pressures can be shown to be permitted for short to moderate times [9] before the structure has been convected downstream several times its own streamwise length. Eventually, however, pressure effects do become important but to include them will make the theory considerably more complicated [9]. The effects of nonlinearity and viscosity also need be incorporated for large times, but to include both of these would probably require a direct numerical attack on the Navier-Stokes equations.

The statistical model equations for the coherent structure was arrived at by application of conditional sampling of the inviscid equations assuming the sampled small-scale Reynolds stresses to be short-lived and have a Gaussian distribution in the streamwise direction over a distance which is large compared to the thickness of the stress-producing region. Such assumption are consistent with the basic idea in the model that these stresses are produced by the breakdown of thin shear layers highly elongated in the streamwise direction. The assumption of a large spanwise scale, as well, is more difficult to justify in view of the experimental data which show a spanwise scale of typically $\Delta z^+ = 50$ or less; presumably spanwise pressure gradients would be more important than streamwise ones, some initial efforts to account for pressure effects and spanwise gradients are presented in [9] and further results will appear in [22].

Depending on what criterion is used for the conditional sampling one would expect to see different dominant features in the coherent structures. Here, the VITA criterion used in Blackwelder & Kaplan [7] was chosen for a closer study. This appears to produce a rather distinct pattern; in particular it singles out the rapid acceleration following the zero crossing of the streamwise velocity. By searching for the largest acceleration occurring at a zero crossing for the calculated model structure a reference distance and time relative to its initiation corresponding approximately to the satisfaction of the VITA criterion could be determined.

Application of other conditional sampling techniques, such as the pattern recognition technique by Wallace et al.[23], for example, would in principle be possible. In their technique they search for a typical pattern between two consecutive maxima in u characterized by a gradual deceleration from the first maximum followed by a rapid acceleration. From the calculations based on the present model coherent structure, however, no pattern of such a character can be seen. Instead, the characteristic pattern consists of first a deceleration to a minimum in u followed by a rapid acceleration, in turn followed by a slow deceleration (see Figure 4). A possible interpretation of the patterns found by Wallace et al. [23] is that they describe the time interval between two consecutive bursts.

ACKNOWLEDGEMENTS

This research was supported in part by the U.S. Air Force Office of Scientific Research under Grant No. AFOSR 79-0006. The author is indebted to Mrs. Gunnel Nordenfelt for preparing this manuscript on short notice.

REFERENCES

1. Kline, S.J. and Reynolds, W.C. and Schraub, F.A. and Runstadler, P.W. 1967. The structure of turbulent boundary layers. J. Fluid Mech. Vol. 30, pp. 741-775

2. Willmarth, W.W. 1975. Structure of turbulence in boundary layers. Adv. in Applied Mechanics, Vol. 15, pp. 159-254. New York: Academic Press.

3. Frenkiel, F.N. and Landahl, M.T. and Lumley, J.L., editors 1977. Structure of turbulence and drag reduction. Proceedings of IUTAM Symposium, Washington D.C. 1976. Phys. Fluids, Vol. 20, No. 10, Part II.

4. Fiedler, H., editor 1977. Structure and mechanisms of turbulence I and II. Proceedings Berlin 1977. Lecture Notes in Physics, Vol. 75 and 76. Berlin, Heidelberg, New York: Springer-Verlag.

5. Smith, C. R. and Abbot, D. E. (editors) 1978. AFOSR/Lehigh Workshop on Coherent structure of turbulent boundary layers. Proceedings, Lehigh University 1978. Bethlehem, Pa.: Lehigh University

6. Offen, G.R. and Kline, S.J. 1975. A proposed model of the bursting process in turbulent boundary layers. J. Fluid Mech. Vol. 70, pp. 209-228.

7. Blackwelder, R.F. and Kaplan, R.E. 1976. On the wall structure of the turbulent boundary layer. J. Fluid Mech. Vol. 76, pp. 89-112.

8. Chen, C.H.P. and Blackwelder, R.F. 1978. Large-scale motion in a turbulent boundary layer: a study using temperature contamination. J. Fluid Mech. 89, pp, 1-31.

9. Landahl, M.T. 1978. Modeling of coherent structure in boundary layer turbulence. Proceedings, AFOSR/Lehigh Workshop on Coherent structure of turbulent boundary layers, pp. 340-371, Lehigh University.

Landahl, M.T. 1977. Dynamics of boundary layer turbulence and the mechanism of drag reduction. Phys. Fluids, Vol. 20, No. 10, Part II, pp. S55-63.

11. Landahl, M.T. 1975. Wave breakdown and turbulence. SIAMJ. Appl. Math. Vol. 28. pp. 735-356.

12 Lamb, H. 1932. Hydrodynamics, Sixth edition. New York: Dover publications.

13 Kim, H.T. and Kline, S.J. and Reynolds, W.C. 1971. The production of turbulence near a smooth wall in a turbulent boundary layer. J. Fluid Mech. Vol. 50, pp. 133-160.

14. Bark, F.H. 1975. On the wave structure of the wall region of a turbulent boundary layer. J. Fluid Mech. Vol. 70, pp. 229-250.

15. Landahl, M.T. 1967. A wave guide model for turbulent shear flow. J. Fluid Mech. Vol. 29, pp. 441-459.

16. Landahl, M.T. 1980. A note on an algebraic instability of inviscid parallel shear slows. J. Fluid Mech. Vol. 98, pp. 243-251.

17. Reichhardt, H. 1951. Vollständige Darstellung der turbulenten Geschwindigkeitsverteilung in glatten Leitungen. Z.A.M.M. Vol. 31, pp. 208-219.

18. Kreplin, H.-P. and Eckelmann, H. 1979. Propagation of perturbations in the viscous sublayer and adjacent wall region. J. Fluid Mech. Vol. 95, pp. 305-322.

19. Corino, E.R. and Brodkey, R.S. 1969. A visual investigation of the wall region in turbulent flow. J. Fluid Mech. Vol. 37, pp. 1-30.

20. Falco, R.E. 1977. Coherent motions in the outer region of turbulent boundary layers. Phys. Fluids, Vol. 20, No. 10, Part II, pp. S124-S132.

21. Brown, G.L. and Thomas, A.S.W. 1977. Large structure in a turbulent boundary layer. Phys. Fluids, Vol. 20, No. 10, Part II, pp. S243-S252.

22. Russell, J. and Landahl, M.T. 1980. The evolution of an eddy of large horizontal scale in an inviscid shear flow. To be published.

23. Wallace, J.M. and Eckelmann, H. and Brodkey, R.S. 1972. The wall region in turbulent shear flow. J. Fluid Mech. Vol. 54, pp. 39-48.

A Phenomenological Model of the Organized Vortex in the Two-dimensional Mixing Layer

R. TAKAKI and H. FIEDLER
Hermann-Föttinger-Institut
Technische Universität Berlin

ABSTRACT

A phenomenological model is proposed to predict values of a deterministic and a stochastic part of the Reynolds stress in the two-dimensional mixing layer. In the model an organized large scale vortex in the layer is composed of a vortex sheet with a finite thickness rolled up to a spiral shape. This rolling motion is assumed to give the deterministic part. Turbulence within the sheet is analysed by use of the Nee-Kovasznay model. This turbulence and a mean velocity profile within the sheet are assumed to give the stochastic part of the Reynolds stress. The geometrical shape of the sheet is assumed on the basis of flow visualization. The two parts of the Reynolds stress, averaged along a line parallel to the mean flow direction, are calculated. Results agree well with experiments particularly for the deterministic part.

1. INTRODUCTION

The turbulent two-dimensional mixing layer shows a typical coherent structure and has been investigated by many workers under various experimental conditions. In a recent experiment by Mensing [1] the air was injected into still air at the same temperature and the mixing layer was made turbulent by a roughness at the nozzle exit. An excitation was introduced by a loud speaker. Two or three elliptical vortices were observed, which were formed through a stretching and a rolling up of the turbulent shear sheet produced at the nozzle. The mixing layer seemed to loose its deterministic structutre again further downstream. These vortices were deformed, and as a result the major axis of the ellipse was inclined towards the main stream direction by about 40 degrees (Fig. 1). The sense of inclination changed from fowards (vortex A in Fig.1) to backwards (vortex B). On the other hand, Fiedler and Rösgen [2] measured several correlations in an excited turbulent mixing layer using a similar wind tunnel. Measured quantities were decomposed into two parts, a deterministic and a stochastic one, by a computer. The most remarkable result from their measurements was that the deterministic part of the correlation $\overline{u'v'}$ became negative at a certain distance from the nozzle exit (about 400 mm for the maximum flow velocity of 10m/s and excitation frequency of 20 Hz). The stochastic part was always positive and the sum of the constituents was also positive. Since the mean flow profile had a negative gradient in their flow configuration, the negative value of $\overline{u'v'}$ implies a diffusion of turbulence against the gradient or a negative production of turbulence.

In general the mechanism of occurrence of the negative value is not under-

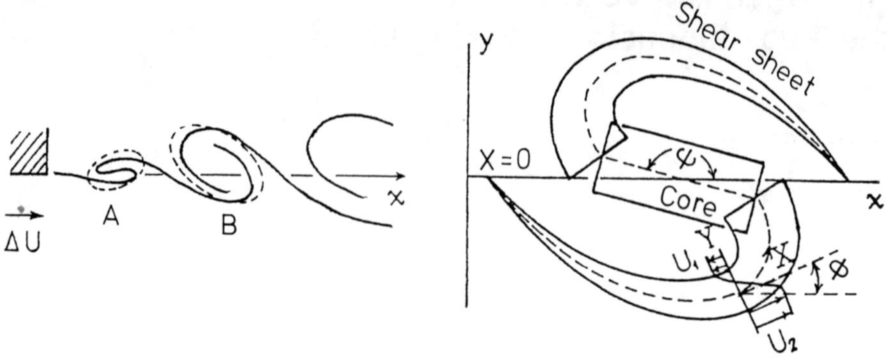

Fig.1. Sketch of the vortices

Fig.2. Assumed structure of a vortex

stood well, and probably differs depending on experimental conditions. In the present case, however, the smoke photograph by Mensing [1] suggests strongly that this phenomenon is closely related to the deformation of the organized vortex. Therefore, it is worthwhile to try to predict the value of the Reynolds stress by constructing a suitable model which includes the geometrical shape and other essential features of the organized vortex. This trial is the purpose of this paper. The phenomenon of vortex merger is not considered because it was not observed in the experiments [1,2].

Recently a computer experiment was made by Knight [3] about an amalgamation of two vortices in the two-dimensional mixing layer. He decomposed the velocity fluctuations into coherent and random parts and obtained a similar negative production of turbulence for the coherent part, which occured just after the amalgamation. The mechanism of the negative production may be different from that in the present case, but his vorticity contour after the amalgamation is analogous to the deformed vortex assumed here. Therefore, an explanation of the negative production of turbulence as a consequence of a single deformed vortex is expected to have a universal meaning to a considerable degree.

2. BASIC ASSUMPTIONS

The large scale vortex is assumed to be composed of two branches of spiral-shaped vortex sheet and a core region connecting these two branches as shown in Fig.2. The fluid in the sheets and the core is turbulent. The sheets are stretched owing to a velocity field induced by a row of large scale vortices. Further assumptions about the core region will be made in the following section. Secondly, stretching and rolling up of the shear sheet is responsible for the deterministic part of the Reynolds stress. On the other hand, both a mean velocity distribution ($u(Y)$ in Fig.2) and the small scale turbulence within the sheet are assumed to be responsible for the stochastic part. The former may in principle be deterministic. But, when the large scale vortex suffers a jitterring or when the shape of the rolling sheet fluctuates, quantities with a medium scale structure, such as the mean velocity profile within the sheet, are looked upon as stochastic. Contributions to the Reynolds stress are assumed to come only from the fluid inside the sheet. According to the smoke photograph [1], diffusion of turbulence is strong and a potential region in the large scale vortex was very narrow, hence a contribution from the potential region is negligible. For quantitative predictions a velocity of rolling motion and some characteristics of the turbulent shear sheet, such as a turbulent viscosity, a width of the sheet and the mean velocity distribution, are necessary. In order to estimate a velocity field which yields the stretching of the sheet, Stuart's theory [4] is useful. Characteristics of the turbulence in the sheet are analyzed by use of the Nee-Kovasznay model [5,6]. This model is employed because it is simple yet sufficient for the present purpose. The third assumption is concerned with an effect of the rolling motion on the turbulence in the sheet. This turbulence is assumed to be strongly effected by the stretching of the sheet, but little by its curvature. Castro and Bradshaw[7] made a measurement about a curved mixing layer bounding an impinging jet, and reported a decrease of Reynolds stress followed by an overshoot at a maxmum curvature. However, in the impinging jet, both the curvature and the outer velocity field vary along the layer, and it is difficult to extract a pure effect of curvature from the measured data. On the other hand the stretching of the sheet has without doubt a large effect on the turbulence.

The procedure of the analysis is then as follows: First, a mean velocity profile, a turbulent viscosity and a spacial growth rate of a plane turbulent shear layer are obtained by use of the Nee-Kovasznay model. The rate of stretching of the layer, which is a given parameter in the model equation, is estimated from the Stuart's theory [4]. Then, a geometrical shape of the rolling sheet is assumed so that it is consistent with the observations by Mensing [1]. Finally, by integrating contributions from the sheet along a line parallel to the center line of the mixing layer, one obtains both the deterministic and the stochastic part of the Reynold stress.

3. ANALYSIS OF A STRETCHED TURBULENT SHEAR SHEET

Let a size of a large scale vortex be denoted by L (Fig.2), a velocity difference across the mixing layer by ΔU. Two coordinate systems are introduced. A coordinate system (x,y) is the conventional one, the x-axis being parallel to the main stream direction. Another system (X,Y) is defined as follows: X denotes a length along the sheet, where X=0 at the junction with a neighboring large scale vortex. Y is the coordinate perpendicular to the sheet. Velocities on both sides of the sheet, their difference and thickness of the sheet are U_1, U_2 $\sigma = U_2 - U_1$ and δ, respectively. The mean velocity field is expressed by a superposition

$$(u,v) = (U,V) + (\tilde{u},\tilde{v}), \tag{1}$$

where (U,V) is a potential part and $(\tilde{u},\tilde{v})$ is a rotational part coming from the sheet. One can put $U=(U_1+U_2)/2$ without loss of generality. U is looked upon as a stretching velocity. Since a part of the sheet at X=0 does not move and the sheet is stretched everywhere, the following linear functions are assumed:

$$U = KX, \quad V = -KY. \tag{2}$$

(U,V) and $(\tilde{u},\tilde{v})$ must satisfy the continuity equation separately. The Nee-Kovasznay model with boundry layer approximation consists of equations governing the mean velocity (u,v) and the turbulent viscosity n as follows:

$$\frac{\partial u}{\partial X} + \frac{\partial v}{\partial Y} = 0, \tag{3}$$

$$\frac{Du}{Dt} = -\frac{1}{\rho}\frac{\partial P}{\partial X} + \frac{\partial}{\partial Y} n \frac{\partial u}{\partial Y}, \tag{4}$$

$$\frac{Dn}{Dt} = \frac{\partial}{\partial Y} n \frac{\partial n}{\partial Y} + A(n-\nu)\left|\frac{\partial u}{\partial Y}\right| - B\frac{n(n-\nu)}{l^2} - C\frac{n(n-\nu)}{\sigma^2}\frac{\partial U}{\partial X}\left|\frac{\partial u}{\partial Y}\right|, \tag{5}$$

where D/Dt, P and ν denote the Lagrangian derivative, the mean pressure field and the molecular viscosity, respectively. A,B and C are nondimensional universal constants and l is a certain length scale concerned with a dissipation of turbulence. It is appropriate to assume $l = \delta$. The last term of (5) stands for a dissipation due to the stretching. Originally this model is proposed for a turbulent boundary layer with an outer velocity U_∞, and there this term has a form $C\frac{n(n-\nu)}{U_\infty^2}\frac{dU_\infty}{dX}\left|\frac{\partial u}{\partial Y}\right|$. But in the present case we have two outer velocities U_1 and U_2. Here , from a physical consideration on the effects of the outer flow, the present form is employed. Note that dU_∞/dX is the stretching rate of the boundary layer and that the denominater U_∞^2 is a square of an important velocity scale, which is σ^2 in the present case. Values of A,B and C were given by Nee and Kovasznay [5,6], but those of A and B are improved by Takaki [8]. They are

$$A = 0.20, \quad B = 2.3, \quad C = 1. \tag{6}$$

The effect of the molecular viscosity is neglected here. $(\tilde{u},\tilde{v})$ satisfies the following boundary conditions:

$$\tilde{u} \longrightarrow \mp\frac{1}{2}\sigma, \quad \tilde{v} \longrightarrow \pm\int_0^Y \frac{1}{2}\frac{d\sigma}{dX}dY + v_0, \quad \text{for } Y \longrightarrow \pm\infty, \tag{7}$$

where v_0 does not depend on X or Y since the flow is potential outside of the sheet. The function $\sigma(X)$ must be known to solve for $\tilde{u}$ and n.

Here a similar solution is sought, where all quantities have a form of $X^m \cdot f(\xi)$, where $\xi = Y/\delta$. This type of similarity is not confirmed by experiments, but it is necessary to simplify the analysis. Some trials show that the similar solution exists if the following assumptions are made:

$$\sigma = \bar{\sigma} X, \quad \delta = aX, \tag{8}$$

where $\bar{\sigma}$ and a are positive constants, which still may depend on time. A comment will be given later about these assumptions. In the similar solution quantities depend on the time t through σ or a. From the conservation of vorticity in the two-dimensional flow, the former must satisfy the following equation:

$$\frac{\partial \sigma}{\partial t} + \frac{\partial}{\partial X}(U\sigma) = 0. \tag{9}$$

Substituting (2) and (8) into this equation, one obtains

$$\frac{\partial \bar{\sigma}}{\partial t} = -2K\bar{\sigma}, \tag{10}$$

which gives the time dependence of $\bar{\sigma}$. Although σ (and $\bar{\sigma}$) varies with time, the turbulence within the sheet is assumed quasi-stationary. From (2), (5) and (8) a potential flow field $\phi_\pm$ outside of the sheet and the pressure field P are obtained, where ± show the regions above and below the sheet respectively.

$$\phi_{\pm} = \frac{1}{2} K (X^2-Y^2) \mp \frac{1}{4} \bar{\sigma} (X^2-Y^2) + v_o Y, \tag{11}$$

$$\frac{1}{\rho} P = -\frac{1}{2} KX^2 - \frac{1}{8} \bar{\sigma}^2 X^2 + O(Y^2) + \text{func}(t), \tag{12}$$

where the Bernoulli's theorem is used in deriving (12). In the boundary layer approximation this pressure field is valid also within the sheet.

Now, based on the assumptions (8), similarity solutions are expressed as

$$\tilde{u} = \bar{\sigma} X \cdot F(\xi), \tag{13}$$

$$\tilde{v} = \bar{\sigma} X \cdot G(\xi), \tag{14}$$

$$n = \bar{\sigma} X \cdot \delta \cdot N(\xi) = \bar{\sigma} aX^2 \cdot N(\xi), \tag{15}$$

where

$$\xi = Y/aX \quad (a = \text{const}), \tag{16}$$

and F,G and N are nondimensional functions. Substituting (1),(2) and (10)-(16) into (3)-(5), one has the following set of equations:

$$F - \xi F' + G' = 0, \tag{17}$$

$$-2\varkappa\xi F' + F(F-\xi F') + GF' - \frac{1}{4} = \frac{1}{a}\frac{d}{d\xi}(NF'), \tag{18}$$

$$-2\varkappa\xi N' + F(2N-\xi N') + GN' = \frac{1}{a}\frac{d}{d\xi} NN' + \frac{A}{a} N|F'| - \frac{B}{a} N^2 - C\varkappa N^2|F'|, \tag{19}$$

where the prime denotes a derivative with respect to ξ and

$$\varkappa = K/\bar{\sigma}. \tag{20}$$

$\varkappa$ should be a given parameter and is estimated in the next section. Boundary conditions are

$$F \rightarrow \mp\frac{1}{2}, \quad N \rightarrow 0 \quad \text{for } \xi \rightarrow \mp\infty. \tag{21}$$

This set of equations (17)-(19) seems to be very complicated, but its approximate solution is obtained easily in the following manner. The first two equations (17) and (18) are satisfied exactly by

$$F = -\xi, \tag{22}$$

$$G = -\frac{1}{4}, \tag{23}$$

$$N = N_o(1-4\xi^2), \tag{24}$$

for the condition $a\varkappa/N = 4$. (25)

This solution is not valid for $|\xi| > 1/2$, but this region is not important since $N(\pm 1/2) = 0$ and the fluid is laminar for $|\xi| > 1/2$. The third equation (19) is satisfied exactly in two regions near to $\xi = \pm 1/2$, but not in the inner region. Hence another condition for the parameters a, $\varkappa$ and N_o is necessary so that

(19) is satisfied approximately in the whole region. This condition is obtained by substituting (13)-(15) into (19) and integrating both sides from -1/2 to 1/2, the result of which is

$$\frac{4}{3}\kappa N_o = \frac{A}{a}\frac{2}{3}N_o - \frac{B}{a}\frac{8}{15}N_o{}^2 - C\kappa\frac{8}{15}N_o{}^2. \tag{26}$$

κ and a can be eliminated from this equation by use of (25), and a second order equation for N_o is obtained, with a positive root

$$N_o = -\frac{1}{2C}\left(\frac{B}{4}+\frac{5}{2}\right) + \sqrt{\frac{1}{2C}\left(\frac{B}{4}+\frac{5}{2}\right)^2 + \frac{5A}{16C}} = 0.020, \tag{27}$$

where (6) is used. Although C=1 is only a rough estimation, N_o does not depend sensitively on C (N_o=0.0197 for C=5, N_o=0.0203 for C=0). In order to check on the accuracy of the solution (22)-(24), each term of (19) and an error, i.e. a difference of both sides of (19), were calculated numerically after an estimation of κ (explained in the next section) was made. The result showed that the error was always less than 5% of the largest term in (19).

From the solutions for $\tilde{u}(Y)$ and n(Y) obtained above, a Reynolds stress in the coordinate system (X,Y) is calculated from

$$\overline{u''v''} = -n\frac{\partial\tilde{u}}{\partial Y} = \bar{\sigma}^2 X^2 N_o(1-4\xi^2), \tag{28}$$

where (u'',v'') denote small scale fluctuations in the system (X,Y). Since this system is rotated against the system (x,y), a transformation accompanied by a coordinate rotation is necessary in order to obtain the Reynolds stress in the conventional sense. Let an angle of rotation be ϕ (Fig.2), then the conventional Reynolds stress is

$$-\overline{u'v'} = -\overline{u''v''}(\cos^2\phi - \sin^2\phi) - (\overline{u''^2} - \overline{v''^2})\cos\phi\,\sin\phi\ . \tag{29}$$

Since ($\overline{u''^2}-\overline{v''^2}$) can not be derived from the resent model, a rough estimation is made based on some experimental data. In many measurements the profile of $\overline{u'v'}$ was similar to those of $\overline{u'^2}$ and $\overline{v'^2}$, and the peak value of ($\overline{u'^2}-\overline{v'^2}$) differs from that of $\overline{u'v'}$ by a factor of O(1). This factor was about 1.3 in the experiment by Wygnanski and Fiedler [9], 1.5 by Spencer and Jones [10], and it varied from -0.33 to 0.96 by Pui and Gastshore [11] depending on the velocity ratio on both sides of the layer. Since these values are for a non-stretched mixing layer, it is meaningless to seek for a precise value. Here the following simple formula is assumed:

$$\overline{u''^2}-\overline{v''^2} = \overline{u''v''}. \tag{30}$$

4. GEOMETRICAL SHAPE OF THE SHEET AND ESTIMATION OF κ

The following expressions for the shape of center line of the sheet is assumed:

$$x(\theta) = r^*\cos\theta + \beta y(\theta),\quad y(\theta) = r^*\sin\theta,\quad r^*(\theta) = \frac{L}{2}\exp(-\gamma\theta), \tag{31}$$

where the parameter θ varies from 0 to θ_c , and β signifies the strength of the vortex distortion. Two examples of the spiral are shown in Fig.3. These spirals closely resemble the large scale vortices as observed by Mensing [1], where β=1 corresponds to the first vortex (A in Fig.1) and β=-1 to that further downstream.

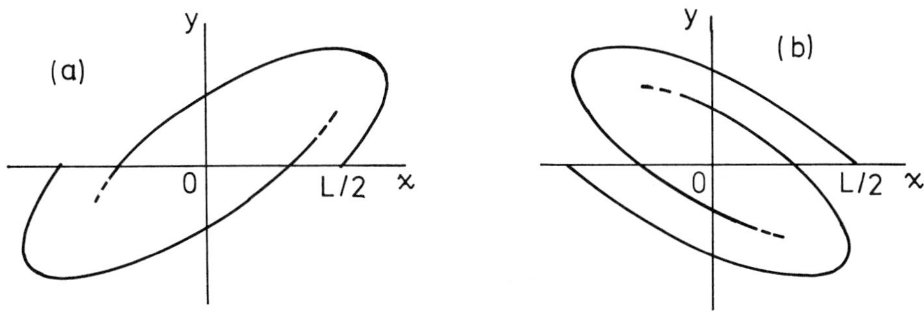

Fig.3. Examples of assumed shape of spiral expressed by (31).
(a) ß=1, (b) ß= -1.

Hence the parameter ß is considered to vary from 1 to -1 as the vortex is convected downstream. θ_c increases with development of rolling up. The large scale vortex is constructed by connecting two ends of spirals ($\theta=\theta_c$) by a straight core sheet which has the same width and the same velocity difference as those at the ends of the spirals. The core is assumed to be unstretched. Once γ, ß and θ_c are given, $\bar{\sigma}$ is determined from conservation of circulation. Let the lengths of one branch of the spiral and the core sheet be X_c and l_c respectively, then the total circulations for one branch and the core are $1/2\cdot\bar{\sigma} X_c{}^2$ and $\bar{\sigma} X_c l_c$ respectively, and the conservation law is

$$2\cdot \frac{1}{2}\bar{\sigma} X_c{}^2 + \bar{\sigma} X_c l_c = \Delta UL, \tag{33}$$

from which $\bar{\sigma}$ is obtained.

The rate of stretching, K, is estimated in the following way. A flow field induced by a periodic distribution of vorticity, which is obtained by Stuart, is expressed by a stream function

$$\psi = -\frac{\Delta UL}{4}\ln\left(C^*\cosh\frac{2\pi y}{L} + A^*\cos\frac{2\pi x}{L}\right), \tag{34}$$

where notations are changed to the present ones. C* and A* are constants and satisfy

$$A^* = (C^{*2}-1)^{1/2}. \tag{35}$$

A set of streamlines according to (34) looks like a cat's eye, but the vorticity distribution depends on A*. For $A^* \to \infty$ it is a row of point vortices, for A*=0 the vorticity is uniform in x-direction and for a finite value of A* it is a row of vortices with a finite size. According to the measurement by Browand [12], a ratio of the peak value of the vorticity in a large scale vortex to an average value on the center line is about 1.6. This ratio is found from Stuart's solution if A*=0.5. Hence, assuming this value of A*, one can obtain a velocity V_c at any point in the large scale vortex. The end of the spiral sheet is considered to be pulled by this velocity V_c, and the stretching rate is given by

$$K = V_c/X_c. \tag{36}$$

The parameter a is also determined from (25) and (27).

Now the Reynolds stress is calculated by drawing a line y=const, and integrating contributions from intersections of the line with the sheet. For a part of the sheet with X and an angle of inclination ϕ, one has a stretching velocity vector $(u_D,v_D) = (KX\cos\phi, KX\sin\phi)$, a width of the sheet aX and finally a contribution to the deterministic part

$$\int_{\text{intersection}} dx \quad \bar{u}_D\bar{v}_D = KX^2\cos\phi\ \sin\phi \frac{aX}{|\sin\phi|}, \tag{37}$$

where the denominator $|\sin\phi|^{-1}$ comes from the fact that the intersection of the line y=const with the sheet is oblique. Similarly the mean velocity profile within the sheet has the following contribution to the stochastic part:

$$\int_{\text{intersection}} dx \quad \overline{u_S v_S} = \frac{\bar{\sigma}^2}{12} X^2\cos\phi\ \sin\phi \frac{aX}{|\sin\phi|}. \tag{38}$$

The contributions of the small scale turbulence at an intersection is, as obtained from (28)-(30),

$$\int_{\text{intersection}} dx \quad \overline{u_S v_S} = \bar{\sigma}^2 X^2 N_o^2 \frac{2}{3} \frac{aX}{|\sin\phi|}(\cos^2\phi - \sin^2\phi + \cos\phi\sin\phi). \tag{39}$$

The intersection with the core sheet contributes only to the stochastic part, because the core has no stretching as assumed. An intersection at the end of the spiral is considered to give a contribution which is half of that for the full width of the sheet, because the length of intersection is also a half. But it is of no use to discuss its physical meaning. It is looked upon as a convention to obtain a continuous profile of the Reynolds stress.

5. RESULTS AND DISCUSSIONS

Calculations are made for values of parameters, $\gamma = 1/2\pi$, $\beta=1$, 0 and -1, and $\pi < \theta_c < 6.7$. These values are chosen so that the shape of the spiral resembles the smoke observation as mentioned above. Predicted values normalized by ΔU^2 on the center line y=0 are shown in Fig.4, where the abscissa ψ denotes the angle of rotation of the core sheet. Since ψ increases with time, it is closely related to the variable x in the experiment, i.e. a distance from the nozzle exit. From comparisons with observations, it is appropriate to put $\beta=1$ for vortices with $\psi \sim \pi$ and $\beta=-1$ for $\psi \sim 2\pi$ (twice folding of the sheet) and $\beta=0$ for intermediate states. It is meaningless to give more refined values to β, because it is difficult to extract precise data about the vortex distortion from experiment. The result shows that the deterministic part and the total value have peaks at an earlier stage and then become negative. Kinks for curves with $\beta=-1$ come from a situation where two ends of the spiral meet and cross the line y=0. This situation has no physical meaning, because the core part of an actual vortex is much more turbulent in the later stage and the organized structure in the core has already decayed. Hence it is worthwhile to modify the calculation taking this decay into consideration. It is assumed that a part of the

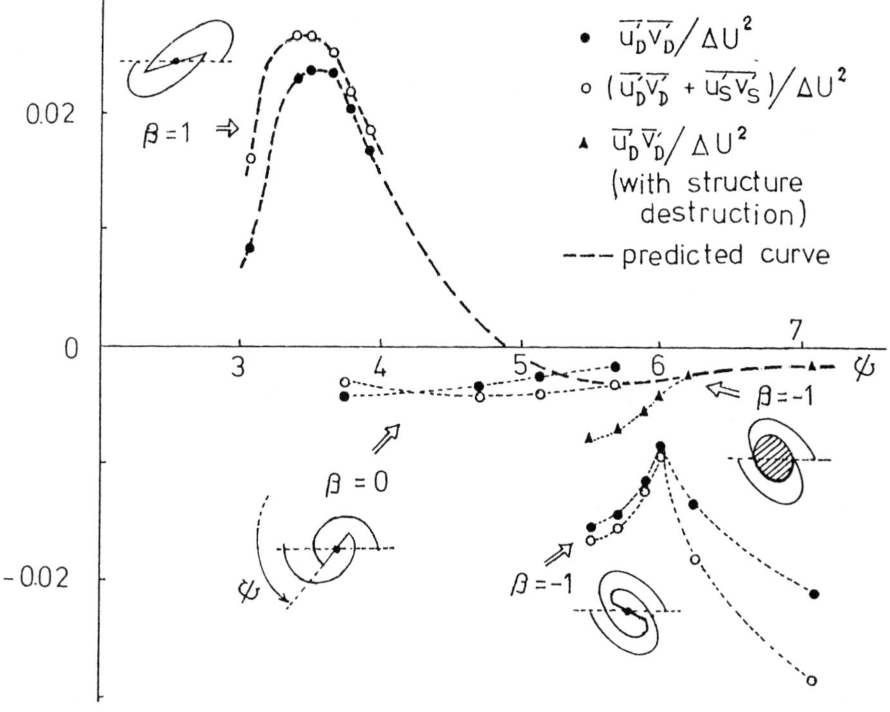

Fig.4. Theoretical results

spiral for $\theta \leq \pi$ remains and contributes to the deterministic part and that the other part is mixed with the core sheet to form a large turbulent core without any deterministic structure. The new results from this assumption are shown by black triangles in Fig.4. Modification of the stochastic part is difficult in the present theory. After all the predicted curves are shown by dashed lines in Fig.4.

These dashed lines agree qualitatively well with the experimental ones by Fiedler and Rösgen [2]. In their presentation the Reynolds stress is normalized by $(d\bar{u}/dy)x/\Delta U$, whose experimental value is approximately 3. If this value is assumed, agreement is also quantitative, especially for the deterministic part. However, the measured total values are larger than the theoretical ones (the peak value is 0.1 in the experiment and 0.07 in the theory after multiplication by 3). On the other hand, the total tends to become negative at $\psi \sim 5$ in the theory. Probably, this quantitative disagreement comes from the fact that the contribution from the core part is not estimated properly, and that the contribution from the part of the sheet at X=0 is neglected in the theory. Actually, the sheet has a finite thickness there and the assumption (8) is inaccurate.

Profiles of the predicted deterministic part at several stages of rolling up are shown in Fig.5, where only the result with the assumption of structural

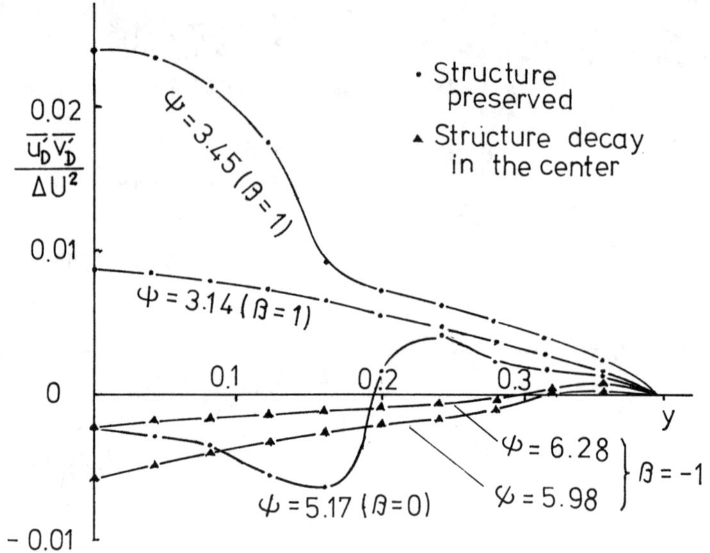

Fig.5. Profiles of the deterministic part

decay are shown in Fig.5. Since the theoretical curves are symmetric with respect to y=0, only the part with $y \geq 0$ is shown. For earlier stages ($\psi \sim \pi$) the profiles have a positive peak at the center (y=0) and then decrease monotonically to 0. For later stages ($\psi \sim 2\pi$) it has a negative peak at the center and a small positive part at y>0.3. These predicted curves show good qualitative agreement with the experiments [2].

In conclusion, appropriate assumptions about the geometrical shape of vortex, along with a model equation and an assumption of the structural decay in the later stage, may lead to a good prediction.

ACKNOWLEDGEMENT

The authors would like to express their thanks to Alexander von Humboldt Foundation for a finantial support of this work.

REFERENCES

1. Mensing, P. 1979. DGLR Symp. "Strömungen mit Ablösungen",München.

2. Fiedler, H.E. and Rösgen, Th. 1979. 32nd Meeting of APS.

3. Knight, D.D. 1979. Proc. Bienial Symp. on Turbulence, Univ. Missouri-Rolla.

4. Stuart, J.T. 1967. J. Fluid Mech. Vol.29, pp 417-440.

5. Nee, V.W. and Kovasznay, L.S.G. 1969. Phys. fluids, Vol. 12, pp 473-484.

6. Nee, V.W. and Kovasznay, L.S.G. 1968. Proc. Comp. Turb. Boundary Layers, Vol.1, pp330-319.

7. Castro, I.F. and Bradshaw, P. 1976. J. Fluid Mech. Vol.73, pp 265-304.

8. Takaki, R. 1979. unpublished.

9. Wygnanski, I. and Fiedler, H.E. 1970. J. fluid Mech., Vol. 41, pp 327-361.

10. Spencer, B.W. and Jones, B.G. 1971. A.I.A.A. Paper, no. 71-613.

11. Pui, N.K. and Gastshore, I.S. 1979. J. Fluid Mech., Vol. 91, pp 111-130.

12. Browand, F.K. 1975. in Turbulent Mixing in Nonreactive and Reactive Flows, ed. by Murthy, S.N.B., Plenum Press, N.Y. and London, pp 317-321.

Some Problems of Second Order Modelling of Mass Transfer in a Turbulent Gas Mixture

P. CHASSAING and H. HA MINH
Institut de Mécanique des Fluides de l'Institut National Polytechnique
Laboratoire associé au CNRS 005
2, rue Charles Camichel
31071 Toulouse Cedex, France

ABSTRACT

In this paper, some of the problems encountered in the study of density effects in a turbulent flow are investigated both theoretically and experimentally. The situation is restricted to the mixing of two non reactive gases in a free subsonic round jet. The theoretical part is devoted to the question of the second order modelling of such a flow when using mass-weighted averages. Taking into consideration the peculiarities of the problem, an eight transport equations model is proposed. The experimental contribution consists in the measurements of the following properties in 21 sections of a CO_2-Air jet issuing from a long pipe (Re_o = 54100) : the mean velocity and concentration, the four significant Reynolds stresses. The conclusions are two-fold :

- Partial valuation, from a direct comparison between numerical predictions and experimental data, of a model based on rather crude extensions of schemes previously developed in buoyant flows. The result is that, if qualitative agreement is generally obtained, the quantitative one is still in progress.

- Analysis of the difficulties of a turbulence modelling in a variable density flow in order to find out the most critical points that are mainly to improve.

1. INTRODUCTION

The second order turbulence models have been extensively improve during the last few years. Even though new trends seem to be very attractive, such as for instance, large eddy simulations [1], [2], the numerical predictions based on the former are nevertheless able to give satisfactory results in several different flows [3], [4], including some situations in which buoyancy effects are important [5].

In most cases, however, the fluid is considered to be strictly incompressible, otherwise, the Boussinesq's approximations are most often introduced. In this paper, on the contrary, the attention will be focussed on some problems directly linked with a particular density variation. The peculiarities of the problem are presented hereafter (§ 2). The theoretical analysis and numerical considerations are given respectively in sections 3 and 4. Some aspects of the experimental contribution are detailed in section 5 and the last one (§ 6) is devoted to the comparison of predictions and measurements.

2. THE PECULIARITIES OF THE PROBLEM

The density variation of a moving fluid may occur from several origins : high speed flows (compressibility effects), heated fluid flows (thermal effects) multi-component flows (inhomogeneous effects)...

In practical engineering problems, for example combustion or rocket-jet propulsion, these different effects are coupled. Owing to the resulting complexity, it seems reasonable to restrict attention to some specific and more tractable situation.

The problem we consider here is one such situation. Briefly, it can be described as the mixing of two non reactive gases that takes place in a free turbulent round jet when the exhausting species is different from the atmosphere. Regarding the exit conditions, the jet issues from a long pipe so that the dynamic characteristics correspond to a fully developed duct flow. Finally, the temperatures of the atmosphere and the issuing gas are the same, and the exit velocity lies in the subsonic incompressible range.

As a direct consequence of the following conditions, it may be seen that the density of the mixture (ρ) and the mass fraction of the exhausting species (C) are linked with, according to the relation :

$$\rho = a\,\rho\,C + b \tag{1}$$

where a and b are two constants, the values of which only depend on the densities of the pure components.

Before dealing with the theoretical analysis, it must be noticed that, if some consequences of Eq.(1) will be very helpful, such an equation obviously restricts the domain for practical applications.

3. THE THEORETICAL ANALYSIS

3.1. The equations of the instantaneous motion

In addition to Eq.(1), the instantaneous values of the six unknown functions ρ, C, U_k (velocity) and P (pressure) are governed by the following equations :

$$\frac{\partial \rho}{\partial t} + \frac{\partial}{\partial x_k}(\rho U_k) = 0 \tag{2}$$

$$\frac{\partial \rho C}{\partial t} + \frac{\partial}{\partial x_k}(\rho C U_k) = \frac{\partial}{\partial x_k}\left(\rho \mathcal{D} \frac{\partial C}{\partial x_k}\right) \tag{3}$$

$$\frac{\partial \rho U_i}{\partial t} + \frac{\partial}{\partial x_k}(\rho U_i U_k) = \rho g_i - \frac{\partial P}{\partial x_i} + \frac{\partial}{\partial x_k}\left[\mu\left(\frac{\partial U_i}{\partial x_k} + \frac{\partial U_k}{\partial x_i}\right)\right] - \frac{2}{3}\frac{\partial}{\partial x_i}\left[\mu \frac{\partial U_k}{\partial x_k}\right] \tag{4}$$

These equations respectively express the continuity of the mixture and the conservation of mass fraction and momentum, according to classical assumptions in fluid dynamics (e.g Fick diffusion law and Newton-Stokes constitutive equation). The only point to be noticed is that the diffusivity ($\mathcal{D}$) and the viscosity (μ) are then depending on the density.

3.2. Averaging the equations

Dealing with turbulent motions and following a classical approach, the first question to discuss is that of averaging the previous equations. In most papers in fact, this problem is restricted to a choice between the classical (Reynolds) and mass weighted (Favre) averages. We shall comment this point later on and only make here the three following remarks :

i) As shown by Favre [5] the expressions of the mean and fluctuating parts of the L.H.S. of the transport equations are both simpler when combining the Reynolds averages (for ρ and P) and the mass weighted averages for all other quantities.

ii) As a consequence of using two different types of averages, the Reynolds mean value of a Favrian fluctuation is different from zero.

iii) Owing to the peculiarities of the problem, the mean value of any Favrian fluctuation can be derived from a suitable second order moment. This directly results from Eq.(1).

Let us consider, for instance, any property - say F - which can be mass averaged :

$$F = \tilde{F} + \phi \tag{5}$$

with $\tilde{F} = \overline{\rho F}/\overline{\rho}$ (6)

and $\overline{\phi} = - \overline{\rho'\phi}/\overline{\rho}$ (7)

In Eq.(7), ρ' denotes the fluctuating part of the density with respect to the Reynolds mean value $\overline{\rho}$.

Averaging Eq.(1), one readily obtains :

$$\overline{\rho} = a\,\overline{\rho}\,\tilde{C} + b \tag{8}$$

which, by subtraction from Eq.(1) and after some simple algebraic manipulations leads to :

$$\rho' = \frac{a}{b}\,\overline{\rho}\,.\,\rho\gamma \tag{9}$$

where γ stands for the mass fraction fluctuation with respect to the Favrian mean value $\ddot{C}$.

By substituing Eq.(9) into Eq.(7), one obtains the final result :

$$\overline{\phi} = - \frac{a}{b}\,\overline{\rho\gamma\phi} \tag{10}$$

Thus one can conclude that, for the problem under consideration, the difficult question of determining the mean values of the Favrian fluctuations needs no specific treatment (see, e.g.[6]) and is in fact a part of the general problem of second order modelling.

3.3. The open set of equations

Averaging Eqs.(2),(3) and (4), one obtains the equations for the mean motion that can be written (discarding the molecular diffusion terms) :

$$\frac{\partial\bar{\rho}}{\partial t} + \frac{\partial}{\partial x_k}(\bar{\rho}\tilde{U}_k) = 0 \tag{11}$$

$$\frac{\partial(\bar{\rho}\tilde{C})}{\partial t} + \tilde{U}_k \frac{\partial(\bar{\rho}\tilde{C})}{\partial x_k} = -\bar{\rho}\,\tilde{C}\,\frac{\partial\tilde{U}_k}{\partial x_k} - \frac{\partial}{\partial x_k}(\overline{\rho\gamma u_k}) \tag{12}$$

$$\frac{\partial}{\partial t}(\bar{\rho}\tilde{U}_i) + \tilde{U}_k\frac{\partial}{\partial x_k}(\bar{\rho}\tilde{U}_i) = -\bar{\rho}\tilde{U}_i\frac{\partial\tilde{U}_k}{\partial x_k} - \bar{\rho}g_i - \frac{\partial\bar{P}}{\partial x_i} - \frac{\partial}{\partial x_k}(\overline{\rho u_i u_k})$$
$$+ \frac{\partial}{\partial x_k}(\overline{\mu S_{ik}}) - \frac{2}{3}\frac{\partial}{\partial x_i}(\overline{\mu S}) \tag{13}$$

In Eq.(13), S_{ik} stands for the tensor $(\frac{\partial U_i}{\partial x_k} + \frac{\partial U_k}{\partial x_i})$ and $S = \frac{1}{2}S_{ii}$.

Exact (open) transport equations can also be derived for the second order moments $\overline{\rho u_i u_k}$, $\overline{\rho u_i\gamma}$ and $\overline{\rho\gamma^2}$. The mathematical procedure is well known and detailled in ref.[7]. Discarding again the molecular effects, the resulting equations are :

$$\frac{\partial}{\partial t}(\overline{\rho u_i u_j}) + \tilde{U}_k\frac{\partial}{\partial x_k}(\overline{\rho u_i u_j}) = \overset{*}{G}_{ij} + \tilde{G}_{ij} + \overset{*}{R}_{ij} + \tilde{R}_{ij} - \frac{\partial}{\partial x_k}(\tilde{I}_{ijk}) - \bar{\varepsilon}_{ij} \tag{14}$$

$$\frac{\partial}{\partial t}(\overline{\rho u_i\gamma}) + \tilde{U}_k\frac{\partial}{\partial x_k}(\overline{\rho u_i\gamma}) = \overset{*}{H}_i + \tilde{H}_i + \overset{*}{S}_i + \tilde{S}_i - \frac{\partial}{\partial x_k}(\tilde{J}_{ik}) - \bar{\phi}_i \tag{15}$$

$$\frac{\partial}{\partial t}(\overline{\rho\gamma^2}) + \tilde{U}_k\frac{\partial}{\partial x_k}(\overline{\rho\gamma^2}) = \overset{*}{L} + \tilde{L} - \frac{\partial}{\partial x_k}(\tilde{K}_k) - 2\bar{\phi} \tag{16}$$

The detailled expressions of the R.H.S. terms of Eqs.(14),(15) and (16) are given in Table 1.

RHS Terms \ Quantity	$\overline{\rho u_i u_j}$	$\overline{\rho u_i \gamma}$	$\overline{\rho\gamma^2}$
"Bulk" Production	$\overset{*}{G}_{ij} = -\overline{\rho u_i u_j}\frac{\partial\tilde{U}_k}{\partial x_k}$	$\overset{*}{H}_i = -\overline{\rho u_i\gamma}\frac{\partial\tilde{U}_k}{\partial x_k}$	$\overset{*}{L} = -\overline{\rho\gamma^2}\frac{\partial\tilde{U}_k}{\partial x_k}$
Gradient Production	$\tilde{G}_{ij} = -(\overline{\rho u_i u_k}\frac{\partial\tilde{U}_j}{\partial x_k} + \overline{\rho u_j u_k}\frac{\partial\tilde{U}_i}{\partial x_k})$	$\tilde{H}_i = -(\overline{\rho\gamma u_k}\frac{\partial\tilde{U}_i}{\partial x_k} + \overline{\rho u_i u_k}\frac{\partial\tilde{C}}{\partial x_k})$	$\tilde{L} = -2\,\overline{\rho\gamma u_k}\frac{\partial\tilde{C}}{\partial x_k}$
Mean pressure interaction	$\overset{*}{R}_{ij} = -(\bar{u}_i\frac{\partial\bar{P}}{\partial x_j} + \bar{u}_j\frac{\partial\bar{P}}{\partial x_i})$	$\overset{*}{S}_i = -\bar{\gamma}\frac{\partial\bar{P}}{\partial x_i}$	0
Pressure correlation	$\tilde{R}_{ij} = \overline{p'(\frac{\partial u_i}{\partial x_j} + \frac{\partial u_j}{\partial x_i})}$	$\tilde{S}_i = \overline{p'\frac{\partial\gamma}{\partial x_i}}$	0
Diffusion	$\tilde{I}_{ijk} = \overline{\rho u_i u_j u_k} + \overline{p'u_i}\delta_{jk} + \overline{p'u_j}\delta_{ik}$	$\tilde{J}_{ik} = \overline{\rho\gamma u_i u_k} + \overline{p'\gamma}\delta_{ik}$	$\tilde{K}_k = \overline{\rho\gamma^2 u_k}$
Destruction	$\bar{\varepsilon}_{ij} = \overline{\mu S_{ik}\frac{\partial u_j}{\partial x_k}} + \overline{\mu S_{jk}\frac{\partial u_i}{\partial x_k}} - \frac{2}{3}\overline{\mu s_{ij}\frac{\partial U_k}{\partial x_k}}$	$\bar{\phi}_i = \rho\,\mathcal{D}\,\overline{\frac{\partial C}{\partial x_k}\frac{\partial u_i}{\partial x_k}} - \frac{2}{3}\overline{\mu\frac{\partial\gamma}{\partial x_i}\frac{\partial U_k}{\partial x_k}} + \overline{\mu S_{ik}\frac{\partial\gamma}{\partial x_k}}$	$\bar{\phi} = \rho\,\mathcal{D}\,\overline{\frac{\partial C}{\partial x_k}\frac{\partial\gamma}{\partial x_k}}$

Table 1. R.H.S. terms of the transport equations of the second order moments.

Before discussing the closure problem, some specific features must be first emphasized :

i) As a consequence of the "transport" form which is adopted for the convection terms, "bulk production" terms are added to the classical gradient ones that are only present in the constant density case. A similar remark also holds for Eqs.(12) and (13).

ii) There is no gravity linked terms in the second ordre moment equations. This point directly results from using Favre's averages. In fact, it can be easily seen [7] that, according to the Boussinesq's approximations, the gravity terms can be explicitely deduced from the so called "mean pressure interaction" terms.

iii) The terms involving the pressure fluctuation have been splitted into two parts, one of which being included in the diffusion terms. It must be recalled that, as for the constant density case, this can be achieved in a different way, as pointed out by Lumley [8].

3.4. The closure problems

Let us turn now to the core of the theoretical analysis, that is the closure of the previous set of equations. Considering a second ordre closure level, only the pressure correlation, diffusion and destruction therms are to be modeled.

Pressure correlation. As an aid for modeling the pressure strain correlation, one often uses the Poisson's equation for the pressure fluctuation. When the density changes, this equation can be written as follows [7] :

$$- \Delta p' = \underbrace{(\rho\tilde{U}_i u_k + \rho\tilde{U}_k u_i + \rho'\tilde{U}_i\tilde{U}_k)}_{(I)}{}_{,ik} + \underbrace{(\rho u_i u_k - \overline{\rho u_i u_k})}_{(II)}{}_{,ik} - \underbrace{(\rho' g_i)_{,i}}_{(III)}$$

$$+ \underbrace{(\rho'\tilde{U}_i + \rho u_i)_{,it}}_{(IV)} - \underbrace{(\mu s_{ik})_{,ik} + \frac{2}{3}\Delta(\mu s)}_{(V)} \qquad (17)$$

When the fluid is incompressible, the R.H.S. of Eq.(17) only includes the two first terms which reduce respectively to the classical linear $- \rho(\bar{U}_i u_k + \bar{U}_k u_i)_{,ik}$ - and quadratic $- \rho(u_i u_k - \overline{u_i u_k})_{,ik}$ - contributions. In buoyant flows a gravity term, equivalent to (III) is added. But when the density changes, two more contributions must be considered due to the non sationarity and viscosity effects.

Eq.(17) suggests to model both $\tilde{R}_{ij}$ and $\tilde{S}_i$ as a sum of five different terms. However, because of the lack of information about the two last contributions only the three first ones have been considered in the present study. Thus, the models that have been used are :

$$\tilde{R}_{ij} = \underbrace{-b_o K \frac{\overline{\nu}\,\overline{\varepsilon^{\star}}}{k^{\star}} (\overline{\rho u_i u_j} - \frac{2}{3}\overline{k^{\star}}\delta_{ij})}_{(a)} \underbrace{- b_1 \overline{k^{\star}}\tilde{S}_{ij} - b_2(\tilde{G}_{ij} - \frac{2}{3}\tilde{G}\delta_{ij}) - b_3(\tilde{\Gamma}_{ij} - \frac{2}{3}\tilde{G}\delta_{ij})}_{(b)}$$

$$\underbrace{- b_4(\overset{\star}{R}_{ij} - \frac{2}{3}\overset{\star}{R}\,\delta_{ij})}_{(c)} \qquad (18)$$

and :

$$\tilde{S}_i = \underbrace{-c_o K \frac{\overline{\nu}\,\overline{\varepsilon^*}}{k^*} \overline{\rho u_i \gamma}}_{(a')} + \underbrace{c_2 \overline{\rho \gamma u_k} \frac{\partial \tilde{U}_i}{\partial x_k} - c_3 \overline{\rho \gamma u_k} \frac{\partial \tilde{U}_k}{\partial x_i}}_{(b')} - \underbrace{c_4 \tilde{S}_i^*}_{(c')} \qquad (19)$$

Considering first groups (a) and (b) in Eq.(18), one readily recognizes a direct extension of the rather classical proposals developed in constant density situations and reviewed by Reynolds [9]. k^* is the turbulent kinetic energy $(\frac{1}{2}\overline{\rho u_i u_i})$, $\overline{\varepsilon^*}$ the pseudo-dissipation rate defined later on and $\tilde{\Gamma}_{ij}$ stands for the quantity :

$$\tilde{\Gamma}_{ij} = -\left[\overline{\rho u_i u_k} \frac{\partial \tilde{U}_k}{\partial x_j} + \overline{\rho u_j u_k} \frac{\partial \tilde{U}_k}{\partial x_i}\right]$$

The last term (c) in this equation is equivalent to the one considered in buoyant flows, as significant in a second order expansion [10].

In Eq.(19), the first term is taken according to Monin proposal, the second (b') and third (c') groups are respectively equivalent to the schemes developed in buoyant motions by Zeman et al [11] and Gibson et al [12].

The values of the constants K, b_α and c_α are prescribed later on.

Diffusion. The turbulent diffusion is modelled according to a gradient type law. Following Daly et al [13], the diffusivity is taken as proportional to the Reynolds stress :

$$\tilde{I}_{ijk} = -\frac{b_5}{K} \frac{\overline{k^*}}{\overline{\rho}\,\overline{\nu}\,\overline{\varepsilon^*}} \overline{\rho u_k u_\ell} \frac{\partial \overline{\rho u_i u_j}}{\partial x_\ell} \qquad (20)$$

$$\tilde{J}_{ik} = -\frac{c_5}{K} \frac{\overline{k^*}}{\overline{\rho}\,\overline{\nu}\,\overline{\varepsilon^*}} \overline{\rho u_k u_\ell} \frac{\partial \overline{\rho u_i \gamma}}{\partial x_\ell} \qquad (21)$$

$$\tilde{K}_k = -\frac{d_5}{K} \frac{\overline{k^*}}{\overline{\rho}\,\overline{\nu}\,\overline{\varepsilon^*}} \overline{\rho u_k u_\ell} \frac{\partial \overline{\rho \gamma^2}}{\partial x_\ell} \qquad (22)$$

In buoyant flows, schemes similar to (21) and (22) have been used by Wyngaard et al [14], [15], for instance.

Destruction. Assuming local isotropy, the destruction term ϕ_i is zero and $\overline{\varepsilon}_{ij}$ can be taken as :

$$\overline{\varepsilon}_{ij} = \overline{\varepsilon}\, \delta_{ij}$$

with :

$$\overline{\varepsilon} = \frac{1}{2} \overline{\mu\, s_{ij}\, s_{ij}} - \frac{2}{3} \overline{\mu\, ss} \qquad (23)$$

denoting by $s_{ij}(s)$ the fluctuating part of $S_{ij}(S)$.

When expliciting the expression (23) using Favre averages, at least two drawbacks can be pointed out. In Eq.(23), the viscosity -and no the density- fluctuations are involved, and by taking the Favrian mean value of S_{ij}, one introduces density effects into a solely kinematic property of the flow field. Consequently, defining the pseudo-dissipation rate ε^* :

$$\overline{\varepsilon^*} = \frac{1}{2} \overline{\rho s_{ij} s_{ij}} - \frac{2}{3} \overline{\rho s^2} \qquad (24)$$

it can be seen that :

$$\bar{\varepsilon} = K\, \bar{\nu}\, \overline{\varepsilon^{*}} \qquad (25)$$

where nothing can be said about the ratio $K(\sim \overline{\nu\rho s^2}/\bar{\nu}\cdot\overline{\rho s^2})$ except that $K = 1$ for incompressible motions.

Contrasting with the quantity $\bar{\varepsilon}$, an exact transport equation can be derived for the pseudo-dissipation rate $\overline{\varepsilon^{*}}$. The result, which does not look very simple, is given in ref. [7] and clearly reveals that specific terms are added to the classical equation due to the density variation. These terms are not taken into consideration in the present study in which an extension of the proposals of Wyngaard et al [14], [15], Lumley [16], André et al [17], is adopted. It consists in adding the mean pressure term $(\overline{u_i}\ \partial\bar{P}/\partial x_i)$ to the classical incompressible closure scheme and introduces a new constant d_4.

A similar treatment also holds for the "mass dissipation rate" $\bar{\phi}$. In the present study, a rather simpler closure scheme is adopted since the "pseudo mass dissipation rate $\overline{\phi^{*}}$ is taken as :

$$\overline{\phi^{*}} = \overline{\rho\ \frac{\partial\gamma}{\partial x_k}\frac{\partial\gamma}{\partial x_k}} = K_m . C_6 . \frac{\overline{\varepsilon^{*}}}{k^{*}}\ \overline{\rho\gamma^2} \qquad (26)$$

in which only two new constants K_m and C_6 are introduced.

4. NUMERICAL CONSIDERATIONS

To sum up, besides the mean motion equations (Eqs. 11-12-13), eight transport equations are to be solved for $\overline{\rho u^2}$, $\overline{\rho v^2}$, $\overline{\rho w^2}$, $\overline{\rho uv}$, $\overline{\rho u\gamma}$, $\overline{\rho v\gamma}$, $\overline{\rho\gamma^2}$ and $\overline{\varepsilon^{*}}$.

Owing to the boundary layer approximations, the mean pressure gradient can be easily explicited and considering a steady, axisymmetrical mean flow, the common form of these equations is :

$$\tilde{U}\frac{\partial\bar{F}}{\partial X} + \tilde{V}\frac{\partial\bar{F}}{\partial r} = \frac{1}{r}\frac{\partial}{\partial r}\left(\bar{\Delta}_F\frac{\partial\bar{F}}{\partial r}\right) + \bar{S}_F$$

Convection Diffusion Source

In order to solve this set of parabolical equations, the Patankar and Spalding procedure [18] has been adopted. However several dispositions have been introduced in the general method, the details of which are given in ref.[7].

5. THE EXPERIMENTAL CONTRIBUTION

The experimental set up consists in a 3,70 m long, seamless bronze duct of 3,6 cm in diameter. The gas mixture (80 % CO_2, 20 % air) issues vertically downwards at an exit velocity on the center line of 13.0 m/s. The Reynolds number based on the diameter is 54100.

The volume fraction χ of the carbon dioxide is measured by a 2 mm, constant rate sampling probe, connected to a Schlumberger "ANIR 1211" infrared analyzer.

The conventional mean velocity and Reynolds stresses are measured by constant temperature hot wire anemometers. In CO_2-Air mixtures, it has been previously pointed out [19] that the sensitivity to concentration vanishes when the overheat ratio is correctly adjusted. This result is illustrated in Fig.1 for a linearized signal.

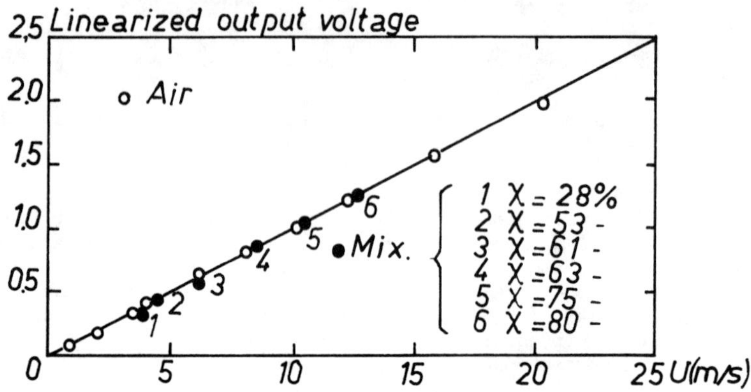

Fig.1. Linearized hot wire anemometer calibration in CO_2-Air mixture

The significant components of the turbulent stresses are deduced from the R.M.S. values of the signals given by normal and slanted single wire probes - DISA 55 P 11 and 55 P 12 -. Corrections for directional characteristics have been introduced according to Hinze proposal [20], the value of the yaw factor being 0.275.

6. THE COMPARISON OF THE RESULTS

Before beginning the discussion, the following points must be emphasized :

i) The inlet conditions, that are prescribed in the computation, are directly taken from the experiments except for the turbulent mass fluxes and the dissipation rates. Consequently, the predicted values may depend on the chosen profiles. Numerical trials have shown that the computed values of any transport quantity were no longer sensitive to the inlet values of $\rho u\gamma$, $\rho v\gamma$ and $\rho\gamma^2$ beyond X/D_o = 25, but no similar conclusion was reached when changing the dissipation rate profiles.

ii) The predictions are, of course, modified by changing the values of the constants. However, it must be noticed that it is not possible to adjust the constants in order to fit, simultaneously, all the experimental results. Moreover, using the values of the incompressible situation does not lead to satisfactory results. The values adopted in the present study are given in Table 2 with $K = K_m = 1.$, and are not to be considered as definitive ones.

b_0 = 1.6	c_0 = 2.10	d_1 = 1.44
b_1 = 0.01		d_4 = 0.50
b_2 = 0.37	c_2 = 0.80	d_5 = 0.15
b_3 = 0.01	c_3 = 0.20	d_6 = 1.92
b_4 = 0.30	c_4 = 0.50	
b_5 = 0.25	c_5 = 0.30	
	c_6 = 0.70	

Table 2. Values of the constants.

iii) The measured mean values are not the Favre averages used throughout the theoretical treatment. In fact, it may be easily seen that they are linked since :

$$\tilde{U} = \overline{U} + \frac{a}{b}\,\overline{\rho u \gamma} \tag{27}$$

and $$\tilde{C} = \overline{C} + \frac{a}{b}\,\overline{\rho \gamma^2} \tag{28}$$

The differences are thus only function of the second order moments and can be easily estimated.

We shall turn now to the comparison of experimental and computed results, considering both streamwise variations and cross distributions.

6.1. Streamwise variations

▪ The first characteristic to be examined is the expansion of the jet given in Fig.2. Both the total radius, R, and the conventional half velocity width, δ, are plotted. The variations of these parameters are closely linear beyond X = 45 Ro. The predicted slope for δ (0,133) is greater than the one corresponding to the constant density case (0,086) which is in rather good agreement with the experimental data of the CO_2-Air jet. Turning now to the total radius R, the discrepancy is reduced, a fact that leads to conclude that the expansion is rather well predicted while the shape of the velocity profiles is probably inexact.

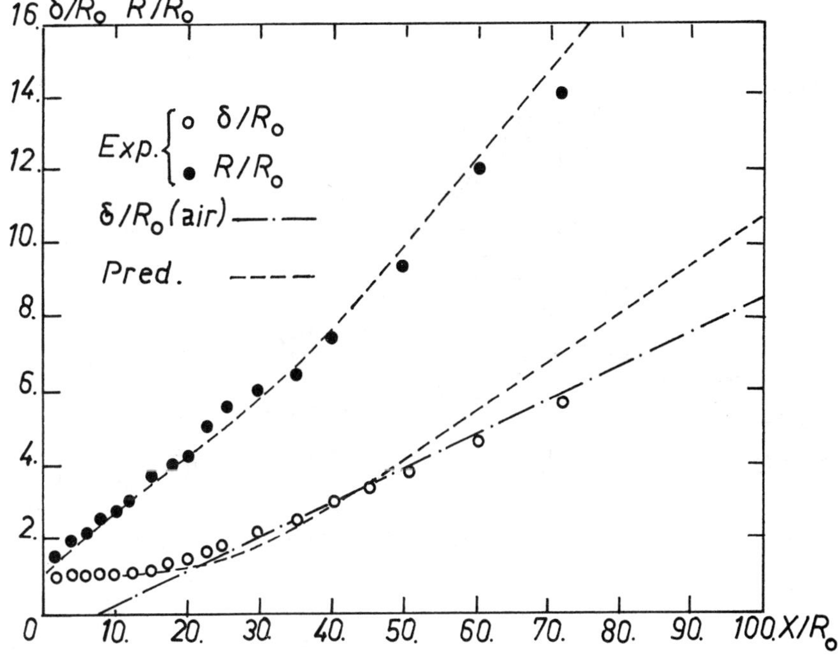

Fig.2. The CO_2 - Air jet expansion

▪ Let us consider now the variations of the velocity - Fig.3.- and the volume fraction - Fig.4.- along the axis.

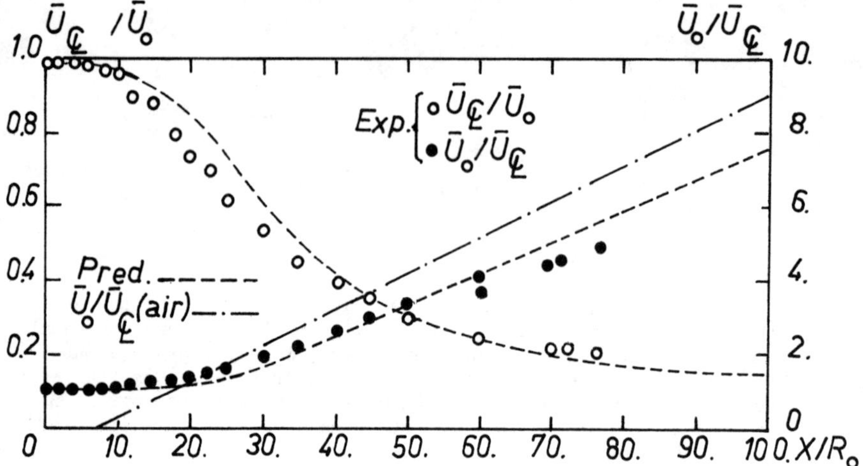

Fig.3. Velocity variation along the axis

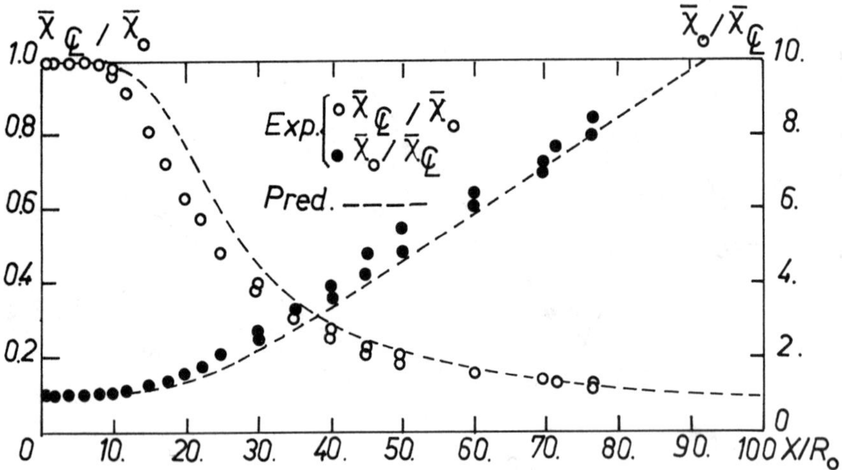

Fig.4. Volume fraction variation along the axis

From about the abscissa X = 45 Ro, these two quantities decrease as X^{-1}. More precisely the following laws can be adopted :

$$\frac{\bar{U}_{\mathbb{C}}}{\bar{U}_o} = A \frac{D_e}{X - X_u} \tag{29}$$

and

$$\frac{\bar{X}_{\mathbb{C}}}{\bar{X}_o} = B \frac{D_e}{X - X_c} \tag{30}$$

In Eqs.(29) and (30), X_u and X_c stand for the abscissa of the virtual origins for the velocity and concentration respectively, and D_e is the equivalent diameter :

$$D_e = D_o\sqrt{\rho_o/\rho_a} \qquad (31)$$

The experimental values of the coefficients A,B,X_u and X_c are given in table 3 :

A	B	X_u/D_o	X_c/D_o
5.45	3.45	4	4.5

Table 3 : Experimental values of decrease law coefficients for the velocity and concentration

The coefficients A and B are slightly lower than those reviewed in the litterature [7], but the ratio B/A is quite correct.

Regarding the prediction, it can be observed that the computed values are always greater than the experimental data near the exit ($X/R_o < 40$). This fact is all the more important that, considering for instance the velocity variation, the slope of the linear part - when plotting $\bar{U}_o/\bar{U}_{\mathcal{L}}$ versus X/R_o - is rather close to that of the incompressible case (resp. 0.084 and 0.087) while the two curves largely differ in the early stage of the decrease.

- The streamwise variations of the turbulent intensities are given in Fig.5. Using a fixed reference $\bar{U}_o$, the turbulent intensities go through a maximum at an abscissa of about 20 R_o, as it is also observed for the constant density pipe jet. Using a local reference, $U_{\mathcal{L}}$ for instance, the tendancy towards the self

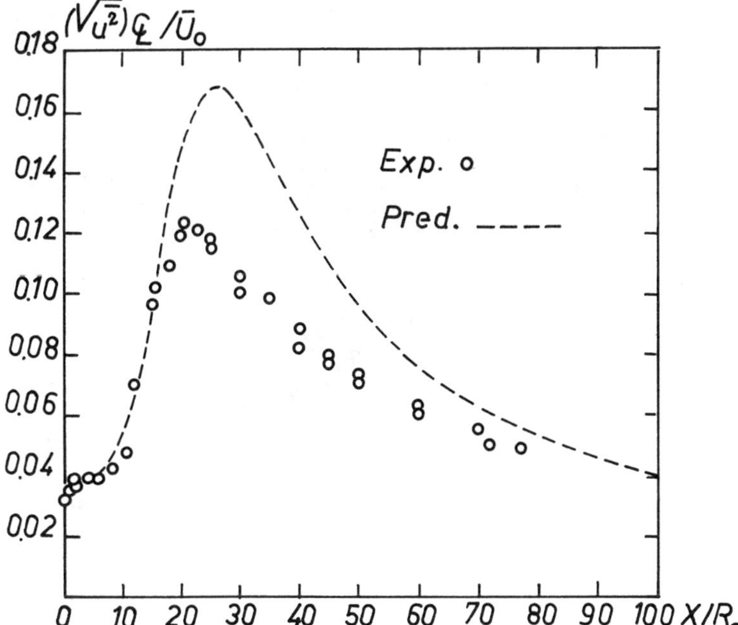

Fig.5. : Turbulent intensities along the axis. a) I_u component.

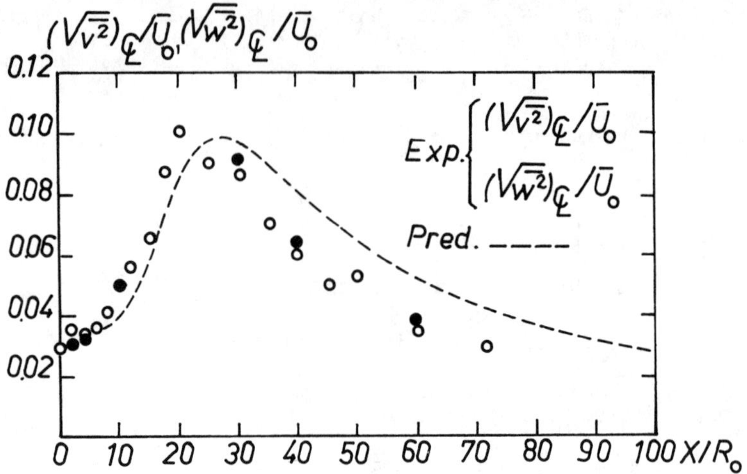

Fig.5. : Turbulent intensities along the axis. b) I_v and I_w components.

preserving conditions clearly emerges, since both the measured and predicted values reach a constant level beyond X = 50 R_o. In this region, the agreement between experiment and prediction is rather good (about 7% for $\sqrt{\overline{u^2}}$). On the contrary, in the peak region, the discrepancy is important (about 30% for $\sqrt{\overline{u^2}}$). However it seems difficult to find out the exact reason since, as noticed by Wygnanski et al [21] the experimental data are likely under-estimated due to the frequency cut-off of the R.M.S. voltmeter.

As a partial conclusion, it can be said that most of the discrepancies observed between the measured and predicted streamwise variations are located in an intermediate region ($10R_o \leqslant X \leqslant 50R_o$). As it was checked, they are not due to differences in the definition of mean quantities ($\overline{U}$ and $\tilde{U}$ for instance) and are slightly greater than those found in the incompressible case [7].

6.2. Profile

The transverse distributions are given in two sections : X = $20R_o$ and X=$50R_o$. The first one corresponds roughly to the peaks of turbulent intensities whereas the second one is located in the restructurating part of the jet.

In Figs. 6a and 6b the velocity profiles ($\overline{U}/\overline{U}_{\mathfrak{L}}$) are plotted versus r/X.

The result in Fig. 6a confirms the previous observation about the discrepancy in the streamwise variation of δ. In this section, the model gives an underpredicted value of about 17%. In the section X = 50Ro, however, the agreement is better, the computed profile being very close to the experimental one found by Wygnanski et al [21] in an incompressible jet. This point seems to conflict with the measured profiles since a slightly different self preserving form is reached beyond X = 40Ro. Here agin, it has been checked that the differences between $\overline{U}$ and $\tilde{U}$ could not entirely explain the observed discrepancies.

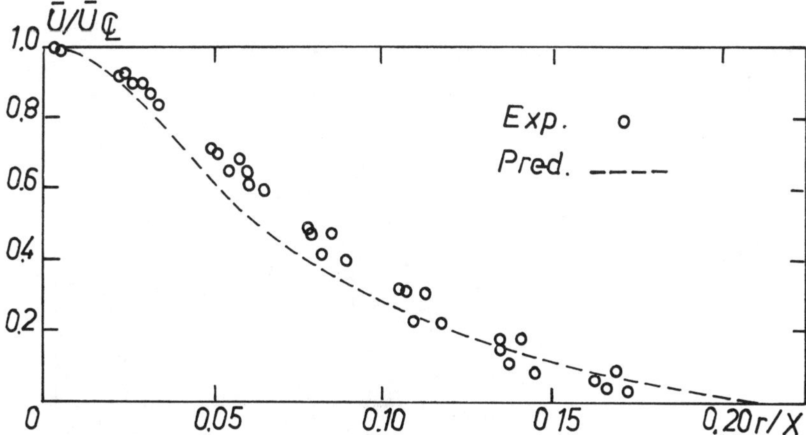

Fig. 6a. : Velocity profile in the section X = $20R_o$

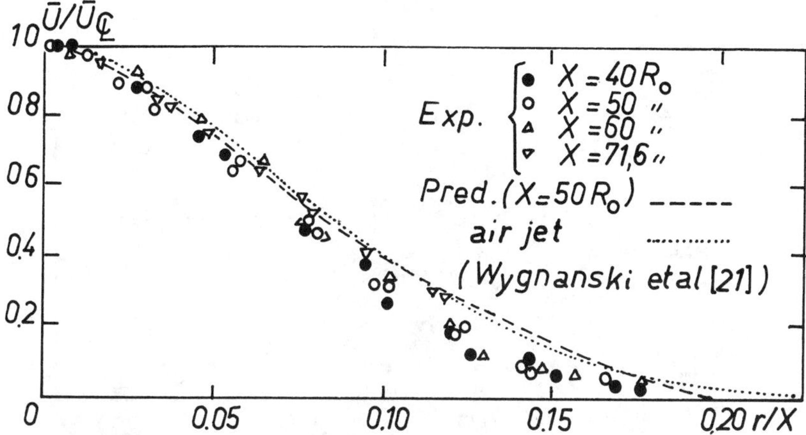

Fig. 6b. : Velocity profile in the section X = $50R_o$

The concentration profiles are given in Figs. 7a and 7b.

Owing to experimental errors, the only conclusion that emerges from Fig. 7a is the rather good agreement between experiment and prediction. In Fig. 7b however, the same observation cannot be made. The computation leads to overpredicting the profile in a way that contradicts the experimental tendancy, as observed by Corrsin et al [22] for instance.

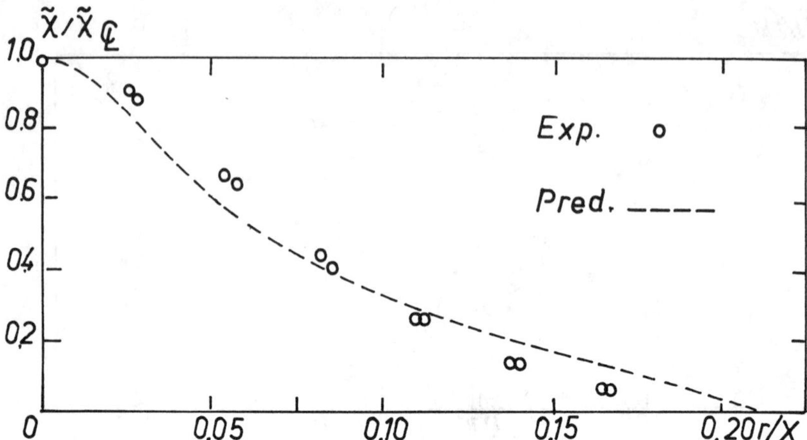

Fig. 7a. : Concentration profile in the section X = $20R_o$

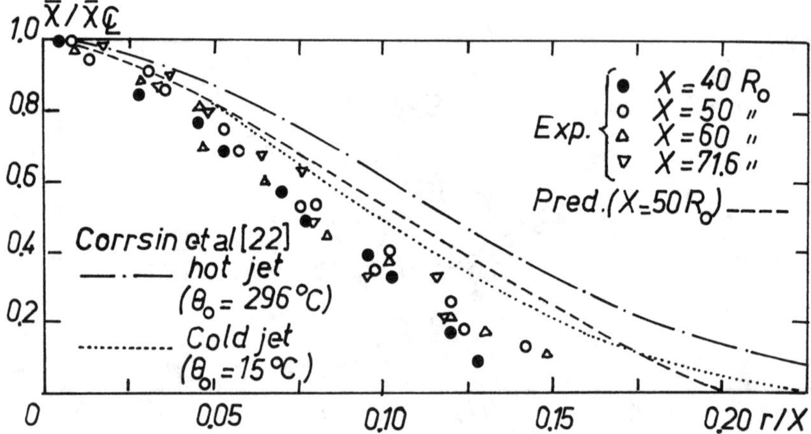

Fig. 7b. : Concentration profile in the section X = $50R_o$

The Reynolds stresses are given in Figs. 8a and 8b.

All the curves are plotted using the maximum value as scaling factor in order to avoid to make the errors corresponding to different properties interact (as for example $\overline{u^2}/\overline{U}_{\mathbb{C}}^2$). The overall observation is a rather good agreement between the experimental data and the computed results (in particular the radial positions of the peaks are well predicted). Of course this fact cannot be taken as a conclusion before having checked that the maximum values are correctly calculated. The comparison is given in Table 4.

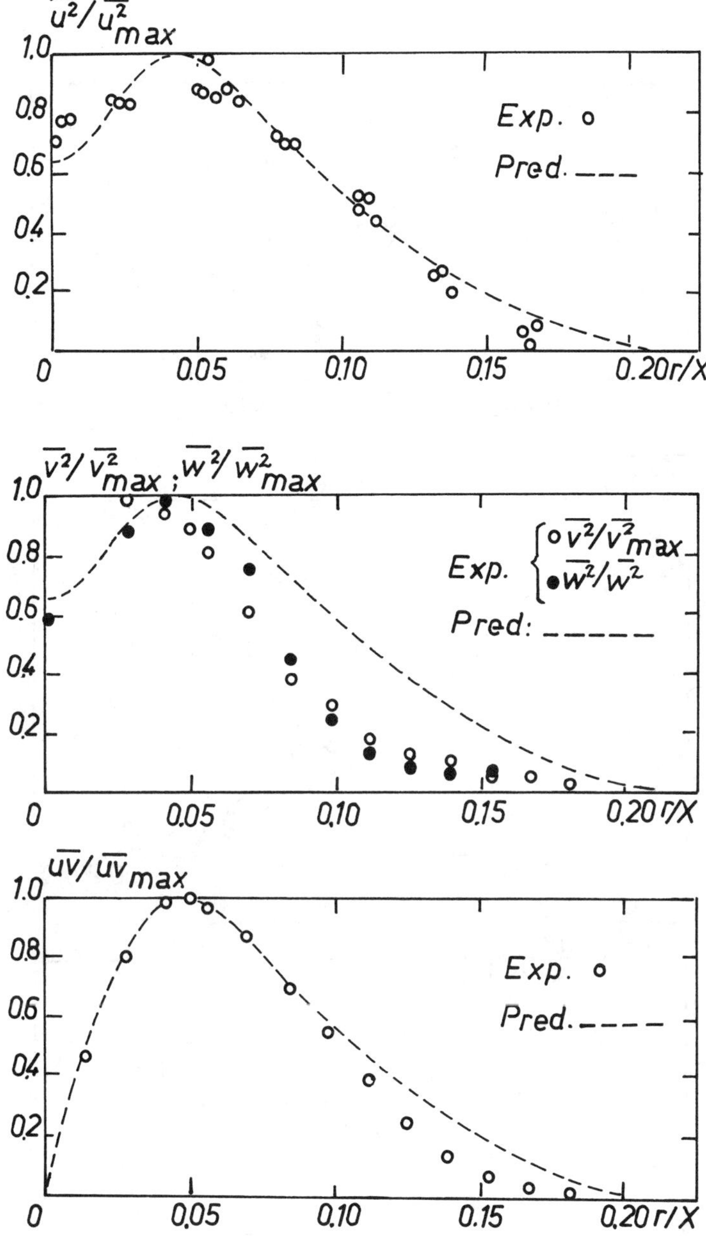

Fig. 8a. : Reynolds stresses in the section X = 20 R_o

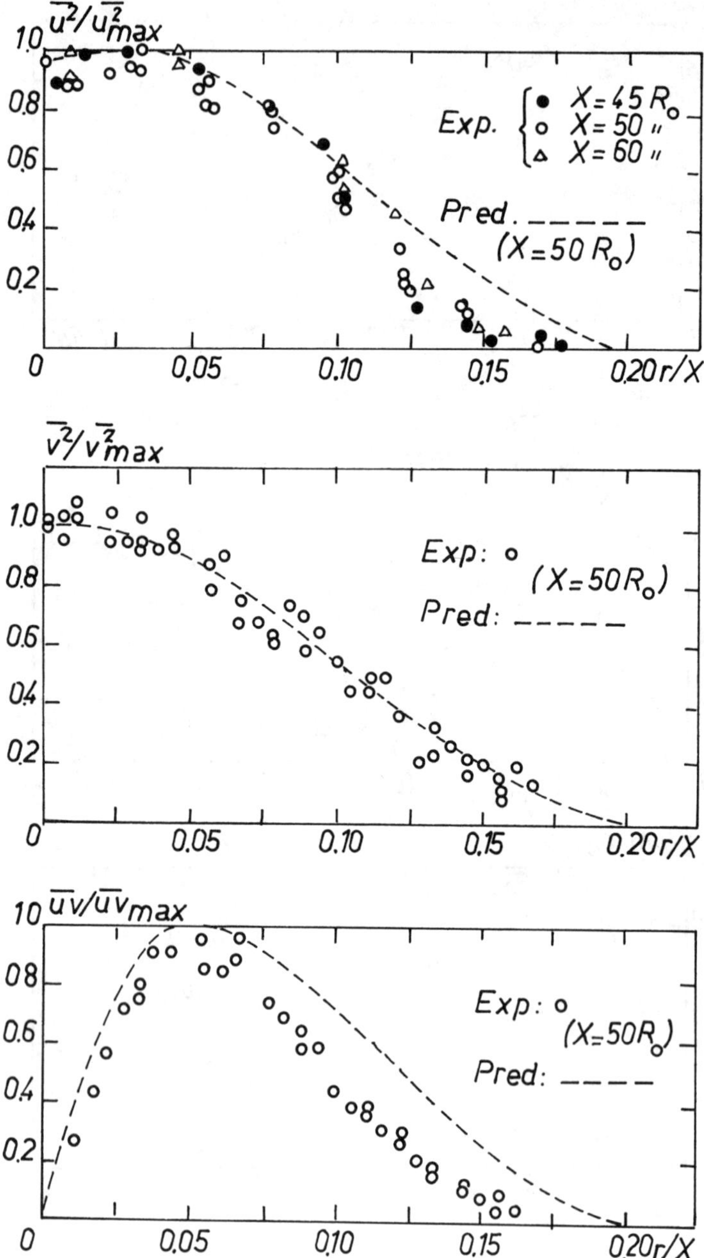

Fig. 8b. : Reynolds stresses in the section X = 50 R_o

		Exp.	Pred.
$\frac{\overline{uv}_{max}}{\overline{U}^2_{\mathfrak{L}}}$	X = 20 R_o	1.1 10^{-2}	1.6 10^{-2}
	X = 50 R_o	1.6 10^{-2}	2.9 10^{-2}
$\frac{\overline{u^2}_{max}}{\overline{U}^2_{\mathfrak{L}}}$	X = 20 R_o	3.5 10^{-2}	4.9 10^{-2}
	X = 50 R_o	6.0 10^{-2}	9.8 10^{-2}
$\frac{\overline{v^2}}{\overline{U}^2_{\mathfrak{L}}}$	X = 20 R_o	1.5 10^{-2}	1.6 10^{-2}
	X = 50 R_o	3.0 10^{-2}	4.3 10^{-2}

Table 4. Comparison of the measured and predicted maximum values of the Reynolds stresses.

Considering the results in Table 4, two observations immediately appear :

- the computed values are systematically greater than the experimental ones
- the differences depend on the quantity and the section.

Consequently, it seems difficult to explain the discrepancies only by "averaging type differences" ($\overline{uv}$ and $\overline{\rho uv}/\overline{\rho}$ for instance), even though it would be worth to check this point. Finally it is fair to say that, it is possible to fit the experiments by adjusting some constants, but that is always done in prejudice of other characteristics previously presented.

7. CONCLUSION

This study must not be considered as a completed work but only as one of the first attempts to predict a variable density turbulent flow in such a situation as a direct comparison between experimental data and numerical results is possible. Consequently, most of the conclusions are largely depending on the peculiarities of the problem that was considered. For sake of clearness, they are grouped into two parts :

▪ Considering first how the closure problem has been solved in the present study, it appears that the model, using Favre's averages, is based on rather straight extensions of the schemes previously developed in incompressible and buoyant motions. Nevertheless, without having checked if realizability conditions were satisfied, the computation proved to be possible and gave qualitatively good results. It must be add that, if the detailed comparison is restricted in this paper to a heavy jet (ρ_e/ρ_a = 1.44), the inverse case can be easily computed. As an example, the centre-line values of the concentrations corresponding to three different density ratios are given in Fig. 9.

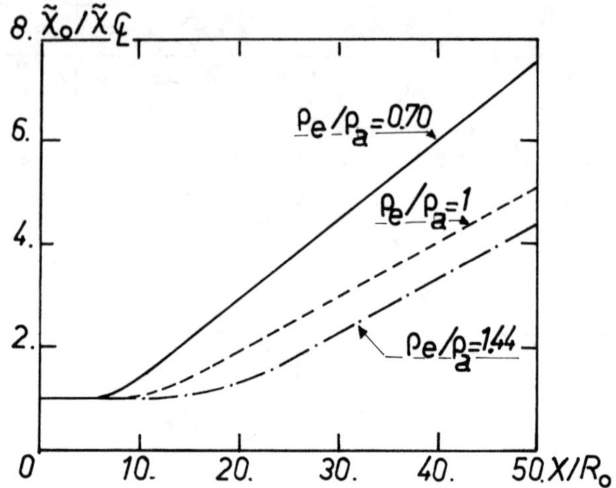

Fig. 9 : Computed values of the centre-line concentration for three density ratios.

It can be observed that, without changing any constant, the model leads to qualitatively correct predictions. However, as pointed out in section 6, some quantitative discrepancies do exist with the CO_2-Air jet experiments. Without ignoring that errors are also introduced in the measurements, there is a lot of reasons to think that the predictions are mostly to be improved, a point that will be discussed in the second part of the conclusions.

▪ The numerical predictions can be quantitatively wrong for several reasons :

. The choice of the constants, a rather refractory problem at the moment, owing to the relative paucity of experimental data with respect to the incompressible situation.

. The prescription of the exit conditions. It must be recall that, in the present study, neither the mass fraction correlations nor the dissipation rates profiles have been prescribed according to measured values.

. Defects of the closure schemes. It is the authors' opinion that important progress is to be expected from improvements of the model. This can be made in different ways : taking into account invariant conditions; introducting new arguments in the schemes owing to specific effects of the density variations ; modelling new terms since, as pointed out in the theoretical approach, many of them have been discarded in the present study ; refining the treatments of the dissipation rates, in particular the mass dissipation one.

ACKNOWLEDGEMENTS

This work has been supported by the C.N.R.S. under grants n° ATP 2914 and 3962. The authors are most indebted to Professors B.E. LAUNDER and J.L. LUMLEY for the fruitful discussions they had with them and would like to take this opportunity to express there sincere gratitude. They wish also to thank Miss P. HERAULT, MM. J. BONNEFONT and J.C. PONS for their skilful technical contribution.

REFERENCES

1. Mansour, N.N. and Moin, P. Reynolds, W.C. Ferziger, J.H. 1977. Improved Methods for Large-Eddy simulations of Turbulence. Proc. Turb. Shear Flows. Univ. Park. Penn. Vol.1, pp.14.21-14.29

2. Ferziger, J.H. and Mehta, U.B., Reynolds,W.C. 1977. Large Eddy simulation of Homogeneous Isotropic Turbulence. Proc. Turb. Shear Flows. Univ. Park. Penn. Vol.1, pp.14.31-14.39

3. Launder, B.E. and Morse, A. 1977. Numerical Prediction of Axisymmetric Free Shear Flows with a Second-Order Reynolds Stress Closure. Proc. Turb. Shear Flows. Univ. Park. Penn. Vol.1, pp.4.21-4.30

4. Ha Minh, H. and Chassaing P. 1977. Perturbations of Turbulent Pipe Flow. Proc. Turb. Shear Flows. Univ. Park. Penn. Vol.1, pp.13.9-13.17

5. Favre, A. 1965. Méthode des vitesses moyennes, méthode des vitesses macroscopiques pondérées par la masse volumique. Jl. de Méc. Vol.4, pp. 390-421.

6. Jones, W.P. 1979. Models for turbulent flows with variable density and combustion. V.K.I. Lec. Ser. 1979-2. Prediction Methods for Turbulent Flows

7. Chassaing,P. 1979. Mélange turbulent de gaz inertes dans un jet de tube libre. Thèse I.N.P.T. n°42.

8. Lumley, J.L. 1975. Pressure strain correlation. The Physics of Fluids. Vol.18, (6), pp.750

9. Reynolds, W.C. 1976. Computation of Turbulent Flows. Ann. Rev. Fluid. Mech. Vol.8. pp.183-208

10. Siess, J. 1975. Etude d'après la méthode de J.L. Lumley de la pénétration de la turbulence dans un milieu stratifié. Thèse Doc-Ing. Université Aix-Marseille II

11. Zeman, O. and Lumley, J.L. 1977. Buoyancy effects in entraining turbulent boundary layers : A second order closure study. Priv. Comm.

12. Gibson, M.M. and Launder, B.E. 1978. Ground effects on pressure fluctuations in the atmospheric boundary layer. Jl. of Fluid Mech. Vol.86-3, pp.491-511.

13. Daly, B.J. and Harlow, F.H. 1970. Transport equations in turbulence. The Physics of Fluids, vol.13-11, pp.1634-1649

14. Wyngaard, J.C. and Cotté, O.R. 1974. The evolution of a convectice planetary boundary layer. A higher-order-closure model study. Bound. Lay. Meteo. Vol.7, pp. 289-308.

15. Wyngaard, J.C. 1975. Modeling the planetary boundary layer. Extension to the stable case. Bound. Lay. Meteo. Vol.9, pp.441-460.

16. Lumley, J.L. 1975. Prediction methods for turbulent flows. Introduction. V.K.I. Lec. Ser. 1975

17. André, J.C. and De Moor, G. Lacarrère, P. Therry, G. Du Vachat, R. 1977. The clipping approximation for Inhomogeneous Turbulence. Proc. Turb. Shear Flows. Univ. Park. Penn. Vol.1, pp.4.39-4.46.

18. Patankar, S.V. and Spalding, D.B. 1970. Heat and Mass transfer in Boundary layers. Intertext. Books. Lond. 2nd. Ed.

19. Chassaing, P. 1977. Heat transfer from cylindrical anemometer probes in CO_2-Air mixtures. The Physics of Fluids. Vol.20-8, pp.1260-1262.

20. Hinze, J.O. 1959. Turbulence - Mc Graw Hill Ed.

21. Wygnanski, I. and Fiedler, H. 1969. Some measurements in the self preserving jet. Jl. of Fluid Mech. Vol.38-3, pp.577-612.

22. Corrsin, S. and Uberoi, M.S. 1949. Further experiments on the flow and heat transfer in a heated turbulent air jet. N.A.C.A.-T.N. 1865.

Buoyancy Effects and Mutual Interaction of Suspended Particles and Velocity Field in Turbulent Shear Flows

K. HANJALIĆ and M. IVANOVIĆ
Institute of Process, Power and Environmental Engineering
Mašinski fakultet, Sarajevo

V. VUJOVIĆ
Institute of Thermal and Nuclear Engineering
ITEN-Energoinvest, Sarajevo

V. MILIŠIĆ
Institute of Hydraulic Research "Jaroslav Černi", Beograd

ABSTRACT

A second order, single point closure scheme has been applied to modelling of the velocity and concentration field and their mutual interaction in the case of turbulent shear flow laden with a dilute concentration of suspended, heavy, inertialess, solid particles. The predicted results for some selected flows indicate noticeable modifications of the hydrodynamic field and turbulence structure due to the presence of the particulate phase.

1. INTRODUCTION

The progress in turbulence research during the past few years has brought some new prospects for better understanding and prediction of at least some of the complex phenomena encountered in multiphase and multicomponent turbulent flows. Increasing activities are evident particularly in experimental research where the laser and some other advanced techniques bring new possibilities for the investigation of the interaction between the components of the flow mixture. The same could be said for analytical and computational methods, where attempts are being made to correlate the dominant physical processes and to incorporate these into general calculation schemes, capable of handling various specific aspects of the multiphase flows, encountered in nature or in engineering systems. Within the latter approach, the practice of closure modelling of turbulent flows, which claims a number of successes in predicting the important features of various cases of singlephase turbulent flows, seems to offer promising grounds for the reasonably faithful mathematical imitation and calculation of the basic turbulent interactions and flow characteristics of multiphase and multicomponent flows.

The research reported here falls into the last category; it is an attempt to extend a general second order modelling technique aimed at predicting some particular cases of turbulent flows laden with a small amount of continuously distributed solid particles.

An extensive coverage of various aspects of multiphase and multicomponent flows can be found in literature, but relatively few papers are concerned with turbulent cases. Among these, the majority deal with discrete particles or droplets and their response to turbulent fluctuations. Systematic surveys of the topic of two-phase flow related to the solid particles diluted in turbulent flow can be found in review articles by Batchelor [1] and more recently by Lumley [2], and Durst and Lee [3]. An equally abundant amount of literature can be found on the topic of modelling the turbulent phenomena, but only sporadic papers report the attempts to model the twophase turbulent flows, and in most cases the particulate phase is treated as passive with respect to the turbulence structure, e.g. [4].

2. THE OBJECTIVE OF THE PRESENT STUDY AND THE APPROACH ADOPTED

A moving mixture of fluid and particles of a different phase should generally be treated as a twophase flow and its mathematical description requires a separate application of the conservation laws to each phase with a proper account of their mutual interactions. Because of a number of additional phenomena involved, which are otherwise absent in single phase flows, the mathematical treatment becomes increasingly complex. This is particularly evident in the case of turbulent flows where the interaction of the basic fluid and the particulate phase is strongly reflected in the turbulence structure.

There are, however, many circumstances of both practical and theoretical interest, corresponding to certain ranges of particle sizes and concentration, where some of the effects can be ignored, permitting substantial simplifications and the subsequent solution of the governing equations. A particular advantage is achieved if the flow mixture can be regarded as a continuum; the assumption is valid only if a macroscopic point contains a sufficient number of continuously distributed particles to permit the definition of the local average properties of the flow mixture, with particle sizes sufficiently small in comparison to their distance from each other to allow the neglect of their mutual interaction.

In the case of turbulent flows, there is an additional requirement that the particle size be smaller than the important dynamic scales of turbulence. Although these requirements sound very stringent, many cases of twophase flows of engineering interest seem to satisfy reasonably well the continuum criteria, to mention the dust particles in the atmosphere and the sedimet particles in natural water flows, as two typical examples.

The transport of suspended sediment particles in liquid flows is a particular case of twophase, twocomponent flow with the particle material density of about 2 to 3 times larger than the fluid density, but with volume concentration of the order of magnitude of 10^{-3}. Although the size of particles ranges from a few microns to several milimeters, those having diameters smaller than about 0.1 milimeter seem to satisfy the continuum criteria in typical engineering situations. In such cases the particles may also be regarded as inertialess so that the slip velocity in the flow direction can be neglected. The Kolmogorov length scale, regarded as the smallest important turbulence scale is, in most cases, considerably larger than the particle size and one may also presume a full response of the particles to the fluctuating motion.

Excluding the narrow region close to the bottom of the liquid flow, where complex interactions occur between the deposited material and particles suspended in the fluid, causing the downstream rolling motion of the particles along the solid boundary with occasional upward movements and reintrainment in the flow, (e.g. [5]),the particles in the main fluid body are transported with the fluid under the combined action of the convective motion, the turbulent diffusion and the particle weight. The turbulence fluctuations in the vertical direction balance the gravitational force on particles, hence the turbulence acts as an essential agent which keeps the particles suspended. However, presence of the particles and their movement have a countereffect on the fluctuating fluid motion. If this effect is large enough, subject to the particle size and concentration, an ultimate modification of the turbulence field occurs, so that in a general case there can be no fully developed situation.

This consi deration is the focal point of the present paper; most of the studies found in literature, which are concerned with the prediction of the concentration field of suspended sediment transport regard the suspension field as passive with respect to the fluid velocity field. Based on this assumption, many authors presume a prescribed, fully developed mean velocity field and analyse only the concentration field. A common aproach to this problem is to solve analytically or numerically the equation for the conservation of particle concentration taking into account the particle weight, but prescribing turbulent diffusion coefficients in different coordinate directions, such as in ref [4].

It is known, however, that the presence of even a small quantity of solid particles in a turbulent shear flow imposes a noticeable effect on the velocity field and turbulence structure, the extent of the influence depending primarily on the particle concentration and their size relative to the dynamic scale of turbulence. In the case of heavy particles, the major effect comes from the negative buoyancy which is known to cause a damping of the flow fluctuation. In order to study these phenomena, a particular case of turbulent flow, laden with heavy, but inertia-free, non-reacting solid particles has been considerd with an attempt to simulate mathematically the velocity and concentration field and their mutual interaction. The consideration was restricted to the particle sizes and the values of bulk concentrations that permit the flow mixture to be regarded as a continuum and treated as a simple fluid. The modelling of the concentration field was based on the more established approach to the modelling of the thermal buoyant field. However, several important features, by which the suspension field differs from the thermal one had to be accounted for. The major effect originating from the particle weight was introduced through the nonzero relative velocity of the particles in the vertical direction in the convective terms of allequations. In addition, the buoyancy effects due to the substantial difference between the densities of the particle material and fluid are included in the consideration.

3. BASIC EQUATIONS

We start by considering a general case of twophase flow mixture where the particulate phase has a volume concentration denoted by S, so that the mixture density can be expressed in the form

$$\rho = \rho_f(1-S) + \rho_p S \qquad (1)$$

Here ρ_f and ρ_p represent the densities of fluid and particle material, respectively. In the present investigation, restricted to the incompressible fluid, both densities will be regarded as constant, while the mixture density depends on the concentration S. However, due to the turbulent fluctuations of the volume concentration, the mixture density will also fluctuate. Adopting the Reynolds decomposition of the instantaneous quantities into the mean and fluctuating part, we may define the fluctuating density in terms of the fluctuating concentration as follows

$$\rho' = (\rho_p-\rho_f)S = \alpha \rho s \qquad (2)$$

where $\alpha = (\rho_p-\rho_f)/\rho$ is the buoyancy parameter.

To obtain the equation governing the concentration of the particles we consider the mass conservation of the particulate phase. Because we treat particles as heavy, but without inertia, the absolute particle velocity in the i-direction can be expressed as

$$U_i^p = U_i^f + V_m \delta_{im} \qquad (3)$$

where U_i^f and U_i^p stand for the mean velocities of the fluid and particluate phase respectively, V_m represents the particle relative velocity and m denotes the direction of the gravitational vector.

If the consideration is restricted to the cases of small volume concentration ($S \ll 1$), the fluid velocity in the convective term can be replaced by the mixture velocity U_i, so that the mean mass conservation equation for the particulate phase for the steady flow can be written in the form:

$$(U_j + V_m \delta_{jm}) \frac{\partial S}{\partial x_j} = \frac{\partial}{\partial x_j} (- \overline{su_j}) \qquad (4)$$

The turbulent flux of the particle concentration su_j is related to the turbulent mass flux of the particulate species $\overline{\rho' u_j}$ through the equation (2). To solve the equation (4), either $\overline{su_j}$ or $\overline{\rho' u_j}$ has to be specified, in addition to the mean flow velocity and particle relative velocity V_m.

Generally, the particle relative velocity depends on many parameters, but for small volume concentrations it can be regarded as independent of S and equal to the particle settling velocity V^s, i.e. the velocity by which a single particle descends in a quiescent fluid. As such, V^s can be treated as an input parameter and for the particular fluid and particle size and density, it can be determined from the appropriate drag law. A number of correction formula can be found in the literature for modifying V^s, so as to account for the particle-to-particle interactions, movement of the fluid, turbulence intensities, etc. However, we are concerned here primarily with the overall effect of the twophase interactions. Therefore, we have used the particle settling velocity as an empirical input, so that the determination of V^s is for the present investigation irrelevant, as long as the chosen value corresponds to the particle size that falls within the scope of the adopted assumptions and limitations.

In the case of a steady regime of a stable suspension, in which the particle concentration profile does not change in the flow direction and with negligible fluid velocity component in the vertical direction, the mean concentration equation (4) could be reduced to the simple form (with x_1 denoting the flow direction and the x_2 coordinate oriented vertically upwards, so that $V_m=(0,-V^s,0)$):

$$V^s S = \overline{su_2} \tag{5}$$

expressing the fact that there is no net transfer of particles in the vertical direction. In other words, the downward gravitational sedimentation is balanced with the upward turbulent flux of particles. If the turbulent flux su_2 is expressed in terms of the mean concentration gradient and an eddy diffusivity coefficient, with the latter one specified e.g. in terms of mixing length, the equation (5) can be integrated to yield the well known exponential expression for the vertical distribution of the mean concentration in a stable suspension.

Our aim was, however, to consider a more general case when the concentration field does not satisfy the stable conditions either due to the variable convective transport or simply due to the imbalance between the gravitational fallout and the vertical turbulent transport. If the concentration profile does not remain steady in the flow direction, one will expect that the turbulence structure will also change, affecting the mean velocity profile with feedback consequences on the concentration profile itself. To account for the convective transport one has to solve the mean momentum equation together with the mean concentration equation.

The mean momentum equation for the flow mixture can be written in the general form

$$\frac{\partial}{\partial x_j}\left[\varrho_f(1-S)\,U_i^f U_j^f + \varrho_p S U_i^p U_j^p\right] = -\frac{\partial p}{\partial x_i} + \frac{\partial}{\partial x_j}(\tau_{ij}) + \varrho g_m \delta_{im} \tag{6}$$

where τ_{ij} represents the total shear stress tensor and other symbols have standard meanings. Following the assumptions stated earlier for the case of small concentrations ($S \ll 1$) and neglecting the density variation in inertial terms in the analogous manner as implied by the Boussinesq approximation for the thermal fluctuation field, one may write the mean momentum equation in a simplified form as follows:

$$U_j \frac{\partial U_i}{\partial x_j} = \frac{\partial}{\partial x_j}(-\overline{u_i u_j}) - \frac{1}{\varrho}\frac{\partial p}{\partial x_i} + (1+\mathcal{L})\, g_m \delta_{im} \tag{6a}$$

where U_i is the mean velocity of the flow mixture and $\mathcal{L} = S(\varrho_p - \varrho_f)/\varrho$ is the "loading" parameter.

It should be noted that for small S the density diference term $\mathcal{L} g_m \delta_{im}$ has a significant value only in the case of large particle-to-fluid density ratios, e.g. dust particles in the air, while in the case of fine sediment particles in water flow it can be neglected.

Equations (4) and (6a) form the basic set of the mean flow equations which require a specification of the turbulent fluxes in order to achieve their closure.

4. THE MATHEMATICAL MODEL OF TURBULENT FLUXES

4.1. General form

In order to close the mean momentum and concentration equations we adopt here the general approach of the second order, single point closure modelling for high Re-number flows as proposed by Hanjalić and Launder [6] and subsequently refined by Launder et al. [7]. The model entails the solution of the transport equations for turbulent stresses $\overline{u_i u_j}$, which in modelled form, with the inclusion of buoyancy terms [8,9] may be written in the form

$$U_1 \frac{\delta \overline{u_i u_j}}{\delta x_1} = \mathcal{D}_{\overline{u_i u_j}} + P_{ij} + G_{ij} - \left(\frac{2}{3}\delta_{ij} - C_1 a_{ij}\right)\varepsilon - C_2\left(P_{ij} + G_{ij} - \frac{2}{3}\delta_{ij}(P+G)\right) \qquad (7)$$

Without going into further details, which are described in the mentioned references it will be recalled here that P_{ij} and G_{ij} represent the stress production terms. These quantities are defined as follows, with $k = 0.5(\overline{u_i u_i})$ - turbulent kinetic energy.

$$P_{ij} = -\left(\overline{u_i u_1}\frac{\delta U_j}{\delta x_1} + \overline{u_j u_1}\frac{\delta U_i}{\delta x_1}\right); \qquad G_{ij} = -\alpha g_m\left(\delta_{mi}\overline{u_j s} + \delta_{mj}\overline{u_i s}\right);$$

$$a_{ij} = \left(\frac{\overline{u_i u_j}}{k} - \frac{2}{3}\delta_{ij}\right); \qquad P = P_{ii}/2; \qquad G = G_{ii}/2 \qquad (8)$$

It remains to be said that $\mathcal{D}_{\overline{u_i u_j}}$ - the turbulent diffusion of the stresses is modelled here in the form of the invariant gradient expression:

$$\mathcal{D}_{u_i u_j} = -\frac{\delta}{\delta x_1}\left[C\frac{k}{\varepsilon}\left(\overline{u_1 u_k}\frac{\delta \overline{u_i u_j}}{\delta x_k} + \overline{u_j u_k}\frac{\delta \overline{u_1 u_i}}{\delta x_k} + \overline{u_i u_k}\frac{\delta \overline{u_j u_1}}{\delta x_k}\right)\right] \qquad (9)$$

with C - diffusion coefficient. In the equation (7), C_1 and C_2 are the empirical coefficients which appear as a consequence of modelling of the pressure redistribution term, and these have been determined within the narrow range by testing the model in number of different non-buoyant flows. The novelty is in the appearance of the gravitational production G_{ij} which, according to the practice of Launder [8] appeares together with P_{ij} in the modelled form of the pressure-strain term. It should be pointed out that the gravitational production term contains the buoyancy parameter as well as the turbulent flux of the concentration su_j, which has to be specified.

To close the hydrodynamic model, the turbulence energy dissipation rate was supplied from the transport equation which in standard modelled form reads:

$$U_j \frac{\delta \varepsilon}{\delta x_j} = \frac{\delta}{\delta x_1}\left(C_\varepsilon \frac{k}{\varepsilon}\overline{u_k u_1}\frac{\delta \varepsilon}{\delta x_1}\right) + \frac{\varepsilon}{k}\left(C_{\varepsilon 1}P + C_{\varepsilon 3}G - C_{\varepsilon 2}\ \right) \qquad (10)$$

where C_ε, $C_{\varepsilon 1}$, $C_{\varepsilon 2}$ and $C_{\varepsilon 3}$ are further empirical constants. The first three have widely accepted values, while $C_{\varepsilon 3}$ is still uncertain in spite of a number of proposals, mainly in connection with modelling the bouyant thermal field. As could be seen from the Table 1. where the currently used values of all empirical coefficients are summarised, $C_{\varepsilon 3}=C_{\varepsilon 1}/2$ was used in the present case, although further investigations and comparison with experimental data is required before the value of $C_{\varepsilon 3}$ could be finally adopted.

We turn now to the consideration of the turbulent concentration flux. Following the standard procedure, the exact form of the transport equation for this quantity can be derived yielding

$$\frac{D\overline{su_i}}{Dt} = \underbrace{-\left(\overline{u_i u_l}\frac{\partial S}{\partial x_l} + \overline{su_l}\frac{\partial U_i}{\partial x_l}\right)}_{(i)} + \underbrace{\frac{\overline{\rho' s}}{\rho} g_m \delta_{im}}_{(ii)} \underbrace{- (\gamma+\nu)\overline{\frac{\partial s}{\partial x_l}\frac{\partial u_i}{\partial x_l}}}_{(iii)} + \underbrace{\overline{\frac{p'}{\rho}\frac{\partial s}{\partial x_i}}}_{(iv)} \underbrace{- \frac{\partial}{\partial x_l}\left(\overline{u_l u_i s} + \frac{\overline{p' s}}{\rho}\delta_{il}\right)}_{(v)} \tag{11}$$

If the turbulent stresses and concentration fluxes are to be provided from their transport equations, as implied by the second order closure scheme, the generation term (i) arising from the mean velocity and mean concentration gradient can be treated in the exact form. The gravitatinal generation term (ii) can be expressed by means of relationship (2) in terms of the concentration variance $\overline{s^2}$, implying also the necessity for solving an equation for this quantity, as adopted by the practice of modelling the turbulent thermal field.

All other terms in the equation (11) require modelling, and we follow here the practice proposed by Launder [8] and Hossain and Rodi [10], which rests on basically analogous arguments as in the pr ocedure of modelling the stress equation. Without elaborating the details, which may be found in the review article by Launder [9], only the finally adopted expression of various terms will be presented here.

The pressure-concentration gradient correlation is expressed in the form:

$$\overline{p'\frac{\partial s}{\partial x_i}} = -C_{s1}\,\varepsilon\frac{\overline{su_i}}{k} + C_{s2}\left(\overline{su_j}\,\frac{\partial U_i}{\partial x_j} + \alpha g_m \delta_{mi}\overline{s^2}\right) \tag{12}$$

where C_{s1} and C_{s2} are the empirical coefficients. The first term represents the "return to isotropy" (in fact the diminuation) of the turbulent concentration flux, while the second term can be interpreted as a tendency towards isotropizing the production process of su_i, all due to the action of the fluctuating pressure.

Invariant gradient approximation is again applied to model the turbulent diffusion of su_i, yielding the modelled form of term (v) in the equation (11):

$$\mathcal{D}_{\overline{su_i}} = -\frac{\partial}{\partial x_l}\; C\frac{k}{\varepsilon}\left(\overline{u_i u_l}\,\frac{\partial\overline{su_j}}{\partial x_l} + \overline{u_j u_l}\,\frac{\partial\overline{su_i}}{\partial x_l}\right) \tag{13}$$

Finally it remains to say that the molecular destruction term (iii) can be neglected if the turbulence Re number is sufficiently high - the condition assumed to prevail in the considered case.

Introducing the particle relative velocity into the convective term and neglecting higher order terms, the modelled equation for the concentration flux can be written in the form:

$$(V_m \delta_{1m} + U_1) \frac{\delta \overline{su_i}}{\delta x_1} = \mathcal{D}_{\overline{su_i}} + P_{is} + G_{is} - C_{s1} \varepsilon \frac{\overline{su_i}}{k} - C_{s2}(P_{is} + G_{is} - \overline{u_i u_j} \frac{\delta S}{\delta x_j}) \quad (14)$$

where, for the sake of clarity, the abbreviated notations are introduced, analogous to those adopted in the stress equation:

$$P_{is} = -(\overline{u_i u_j} \frac{\delta S}{\delta x_j} + \overline{u_j s} \frac{\delta U_i}{\delta x_j}) \qquad G_{is} = -\alpha g_m \delta_{mi} \overline{s^2} \quad (15)$$

Following the same arguments, the modelled form of the equation for the concentration variance $\overline{s^2}$ can also be written in the form:

$$(V_m \delta_m + U_1) \frac{\delta \overline{s^2}}{\delta x_1} = \mathcal{D}_{s^2} + P_s - C_s \varepsilon \frac{\overline{s^2}}{k} \quad (16)$$

where $P_s = 2\overline{su_j}\,(\delta S/\delta x_j)$ is the exact generation term. The diffusion term is expressed in the form of gradient of $\overline{s^2}$, analogous to (13) with diffusion constant equal to C, while the form of the molecular destruction term reflects the assumption that the ratio of the "mechanical" to "concentration" time scales is constant (e.g. ref. [9]).

Equations (4),(6a),(7),(10),(11) and (14), together with the associated definitions of various terms now form a closed set which can be solved numerically, provided the empirical coefficients are specified. Expecting that for the considered case of small volume concentrations and small particle sizes the modelled turbulent interactions will not differ substantially from those encountered in the turbulent thermal field, we adopt here the standard values of the coefficients as summarised in Table [1]:

Table 1.

C	C_1	C_2	C_s	C_{s1}	C_{s2}	C_ε	$C_{\varepsilon 1}$	$C_{\varepsilon 2}$	$C_{\varepsilon 3}$
0.08	2.2	0.55	0.7	3.2	0.5	0.15	1.44	1.92	0.7

4.2. A simplified version of the model for nonrecirculating flows

Because of the tensorial character of the stress equation (7) and the vector character of the concentration flux equation (11) the total number of the modelled differencial transport equations to be solved in a general three-dimensional flow, numbers 11, in addition to the mean flow equations. It is known, however, that in thin, nonrecirculating flows only one component of the shear stress affects the mean flow. Furthermore, since all three components of the normal stresses behave in a similar manner, they can be expressed approximately in terms of their sum, i.e. the turbulent kinetic energy, thus giving a possibility to solve only the transport equation for k instead of for each component of $\overline{u_i u_i}$, as was done in ref [6]. Similarly, only the lateral component of the concentration flux, $\overline{su_2}$, dominates the turbulent transport in thin shear flows. In these situations one may approximate algebraically the longitudinal flux $\overline{su_1}$ and thereby reduce further the number of differential equations and the numerical task, without substantially affecting the accuracy of the solution. Hence for the twodimensional nonrecirculating flows we may recommend the simplified model which entails the solution of differential transport equations for $\overline{u_1u_2}$, k, ε, $\overline{su_2}$ and $\overline{s^2}$.

The remaining turbulence quantities that appear in the model are the lateral normal stress component $\overline{u_2^2}$ and the longitudinal concentration flux $\overline{su_1}$, but because their role is of secondary importance, we may neglect the combined effect of the convective and diffusive terms in their transport equations, yielding the explicit algebraic expressions, as in ref [10].

5. ILLUSTRATION OF THE MODEL APPLICATION

5.1. Specification of the selected test flow

Because the emphasis of the present work was placed on the study of the possibilities of modelling the interaction between the velocity and concentration field in a turbulent shear flow, we have chosen a simple case of plane open channel flow, laden with fine sediment particles. This has allowed us to eliminate other effects that may be encountered in complex recirculating flows, but also to use the simplified from of the turbulence model and to reduce the computing effort. To approach more closely a real engineering situation, the selected flow depth was 1 m and the bulk velocity 0.5. m/s yielding the mean flow Reynolds number of 326 000., with particle sizes and concentrations that corresponded to the values encountered in typical natural water flows with suspended sediment transport.

Two values of the initial bulk volume concentrations were considered, the basic case with $S = 3.75\text{x}10^{-4}$ (m^3/m^3) (corresponding to the mass concentration of 1 kg of particle material having density of 2670 kg/m^3, per 1 m^3 of the mixture volume), and the second case with a concentration twice as small, $S = 1.87\text{x}10^{-4}$. These values, associated with the density difference yield the loading factor $\mathcal{L} = 6\text{x}10^{-4}$ and $3\text{x}10^{-4}$ respectively.

In order to illustrate the effect of the particle settling velocity, the results obtained with three different values will be presented, with V^S=0.001, 0.002 and 0.003 m/s. If the Stokes drag law is pressumed, then the selected values of V^S correspond roughly to the particle sizes of 70, 140 and 210 microns.

The considered cases are summarized in Table 2, where the reference case, corresponding to the flow with the passive concentration field, is denoted by "1".

Table 2

Cases	1	2	3	4	5
Settling velocity (m/s)	0.001	0.001	0.002	0.003	0.001
Initial bulk concentration S	$3.75\text{x}10^{-4}$				$1.75\text{x}10^{-4}$
Loading factor $\mathcal{L}$	$6\text{x}10^{-4}$				$3\text{x}10^{-4}$

5.2. The solution procedure and the initial and boundary conditions

The equation set was solved numerically by finite-difference procedure of Patankar and Spalding, [11], marching downstream from the channel entrance.

The initial profiles of all variables were presumed to be uniform, so that the process of flow development can be followed.

The boundary conditions were specified in accord with the practice reported in ref. [6]. To avoid the need for solving the equations within the viscous sublayer close to the wall, where the applied turbulence model does not hold, the first mesh node was placed just outside the viscosity affected region. Here the mean velocity was determined by means of the logarithmic wall function. The turbulence quantities k, $\overline{u_1u_2}$, $\overline{su_2}$ and $\overline{s^2}$ were determined from the condition of zero diffusive flux of these quantities towards the wall, while for ε the condition of local energy equilibrium was utilised. The free surface was treated as a symmetry plane for all even turbulence correlations except for ε, for which the length scale damping correction of Rodi [12] was applied. The odd turbulence correlations were given zero values at the free surface.

For the mean concentration equation, the condition of no net transport of particles was used for both boundaries, i.e.:

$$V^S S - \overline{su_2} = 0 \qquad \text{at} \qquad y = 0, H$$

5.3. Discussion of results

As indicated earlier, due to the continuous fallout of the particles, the concentration field, and, in the case of active interaction, all hydrodynamic quantities will undergo a continuous modification along the flow. We chose, therefore, to present the profiles of various quantities at the specific position of 265 flow depths downstream, which proved to be sufficiently far from the channel entrance for the particle-to-fluid interaction to show up in all considered cases. The development of some of the characteristic parameters along the flow will also be presented.

The profiles of mean velocity for the considered cases are shown in Fig. 1,

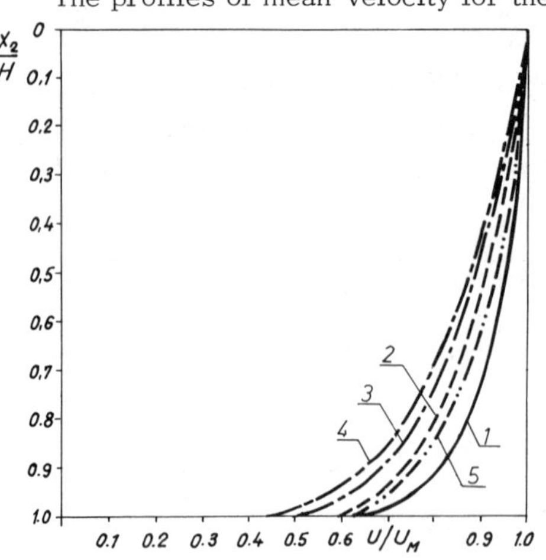

Fig. 1 - Mean velocity profiles

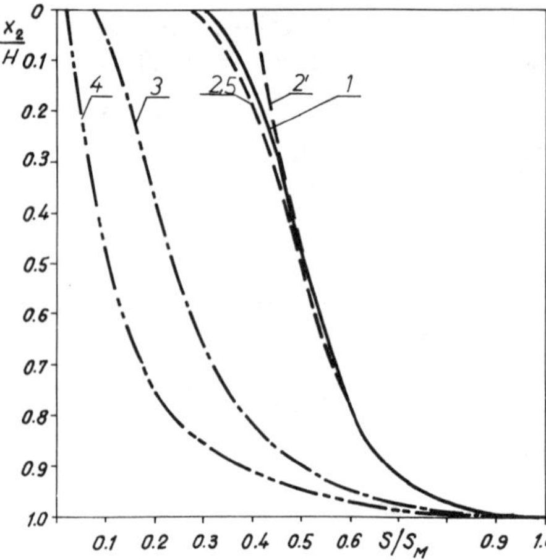

Fig.2. - Mean concentration profiles normalized with maximum value

where the profile numbered "1" corresponds to the case of fully developed non--buoyant plane channel flow. The earlier application of the full-stress as well as of the three-equation model yielded the predictions in excellent agreement with experimental results, e.g. [6], and here no further comparison will be shown. The buoyancy influence due to the particle presence is clearly shown by another four profiles. These should be analysed in parallel with the mean concentration profiles, which are presented in Figures 2 and 3. In Fig. 2 the mean concentration was normalized with the maximum value, reached at the bottom, while in Fig. 3 the initial bulk value was used for normalization in order to show more clearly the differences between the absolute values of S in various regions of the channel cross-section for the considered cases.

The profiles 2,3 and 4 reflect the influence of the settling velocity and, indirectly, the effect of the particle size, for the same value of the initial bulk concentration. As expected, the higher values of the settling velocity will cause the faster fallout of the particles and, hence, the larger concentration in the region close to the flow bed, as seen in Fig. 3. Higher concentrations cause in turn the more intensive damping of the turbulent shear stress, which in turn increases the mean velocity gradient, yielding more laminar-like profiles.

A similar trend is indicated by the increase of the initial concentration, as shown by curves 2 and 5, which correspond to the same settling velocity.

In Fig. 2 the influence of the free surface boundary condition is also demonstrated; the curve 2' corresponds to the same case as the curve 2, but without the length scale correction at the free surface - the result that would be obtained if the free surface were regarded fully as a symmetry plane. As shown here, the length scale damping modifies the profile only in the region close to the free surface, not affecting substantially the rest of the profile.

The damping effect of the particle presence on the turbulence is best illustrated by the profiles of turbulent quantities for different cases, listed in Table 2. Firstly, the turbulent kinetic energy profiles are shown in Fig. 4. In order to isolate the particle effect on turbulence energy, the profiles have been normalized with the bulk flow velocity instead of the friction velocity, as usual, to avoid the effect of the continuous change of the wall shear stress in the downstream direction. The curves 2,3 and 4 clearly indicate the damping due to the buoyancy terms in the enrgy equation; the effect being particularly visible and systematic in the region close to the bottom.

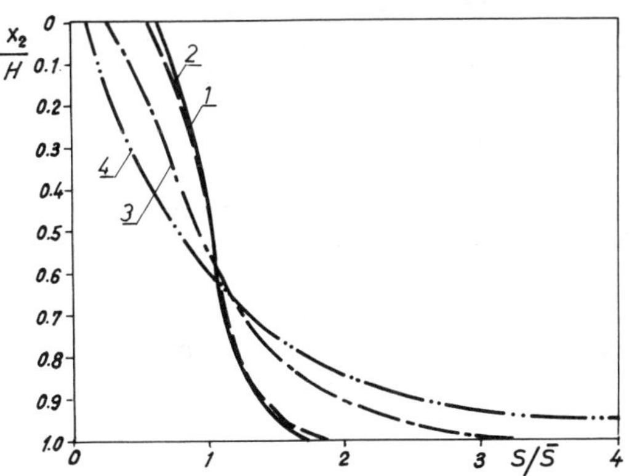

Fig.3. - Mean concentration profiles normalized with bulk value

The crossing of the curves, in particular those denoting cases 2 and 3 requires further discussion.

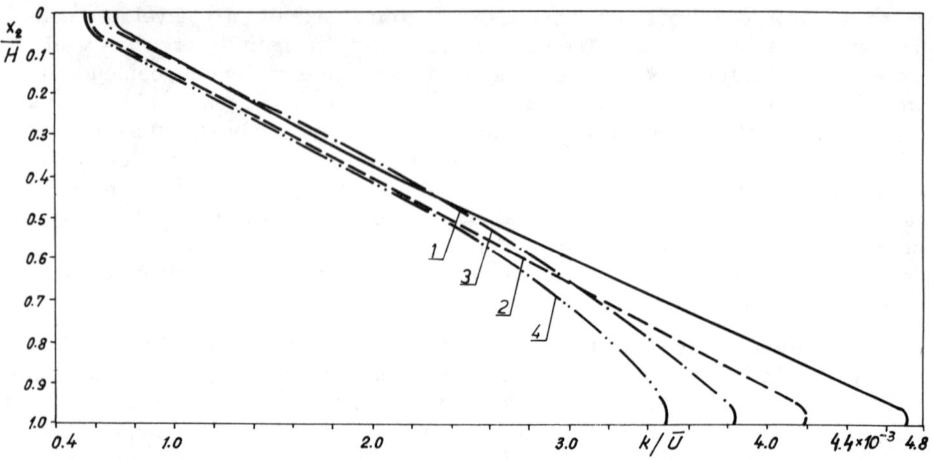

Fig. 4. - Turbulent kinetic energy profiles

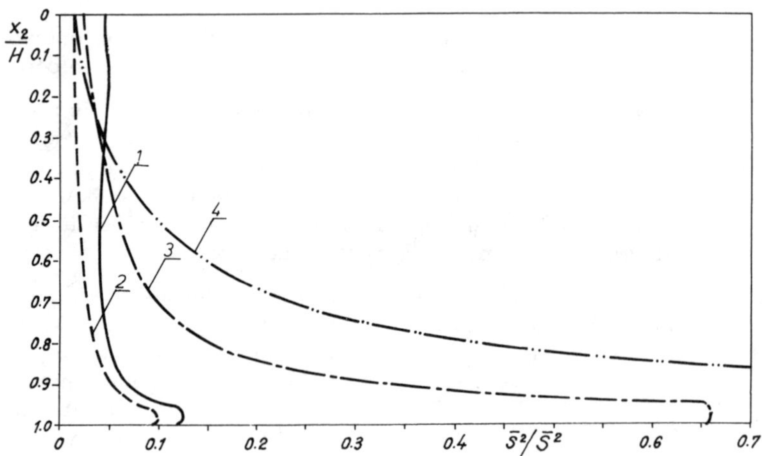

Fig. 5. - Concentration variance profiles

As mentioned earlier, higher settling velocity causes a higher damping of the turbulence energy in the bottom region and a smaller effect in the upper region. If the turbulent transport of energy in the vertical direction remains of the same importance, there will be a crossing of the kinetic energy profiles, as shown by curves 2 and 3. However, if the damping in the bottom region is large enough to supress the high energy generation process, the upward turbulent transport, which represents the major supply of energy in the region close to the free surface, will be diminished and the energy profile will show smaller values all acrosse the flow depth, as demonstrated by curve 4 relative to curve 3. It should be pointed out, that the described situation corresponds only to a particular position in the channel downstream from the inlet. Different shapes and mutual dispositions of the curves for the same settling velocities may be expected at different stations in the flow.

Similar, but steeper and more pronounced variation is shown by the concentration variance $\overline{s^2}$, shown in Fig. 5. Again, for clarity, the profiles have been normalized with the initial bulk volume concentration. Very large values, indicated by curve 4 in the bottom region, correspond to the large mean concentration in this area, but it should be mentioned that the relative values of $\overline{s^2}$, normalized with the local values of the concentration, also show larger values for higher settling velocities.

Next, in Fig. 6 the profiles of the vertical turbulent flux of concentration are shown, normalized with the product of the initial bulk velocity and bulk concentration. It is interesting to note the large difference between the curves 1 and 2,

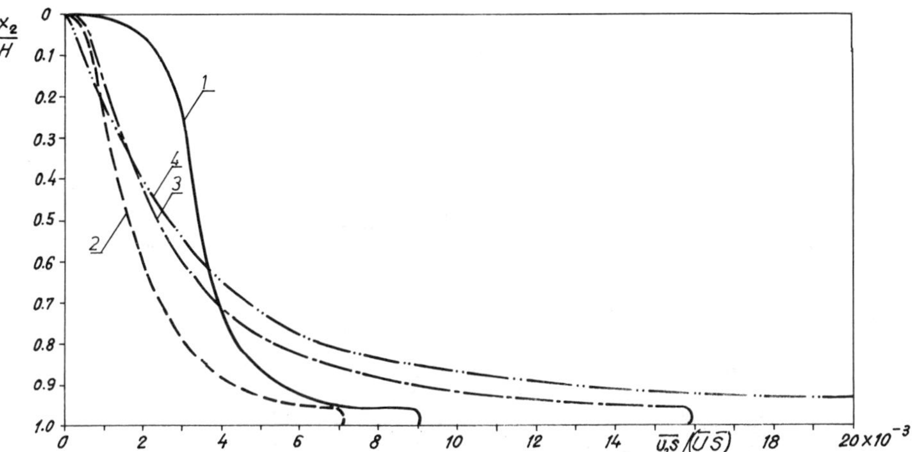

Fig. 6. - Profiles of turbulent concentration flux

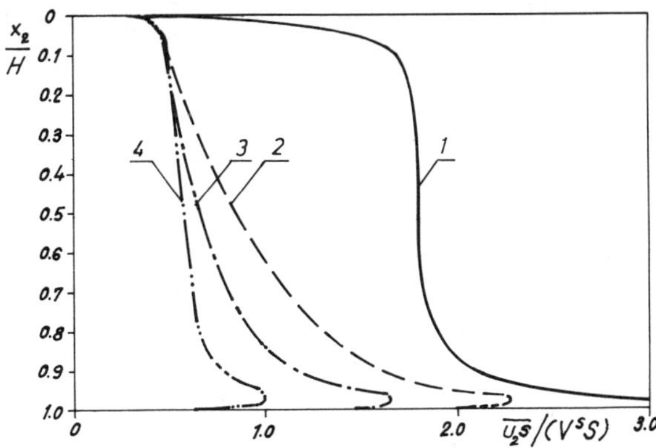

Fig. 7. - Turbulent concentration flux normalized with local mean transport in vertical direction

corresponding to the same situation, but the first one treating the concentration field as passive and the second one with an active influence upon the turbulence. In order to illustrate the convective influence and the departure from the stable situation, implied by relationship (5), the ratio of $\overline{su_2}/V^S S$ is plotted in Fig. 7. Curve 1 and 4, although for different reasons, show constant ratio over the majority of the channel depth indicating reasonably close conditions pressumed for the case of stable suspension, while curves 2 and 3 show a different behaviour.

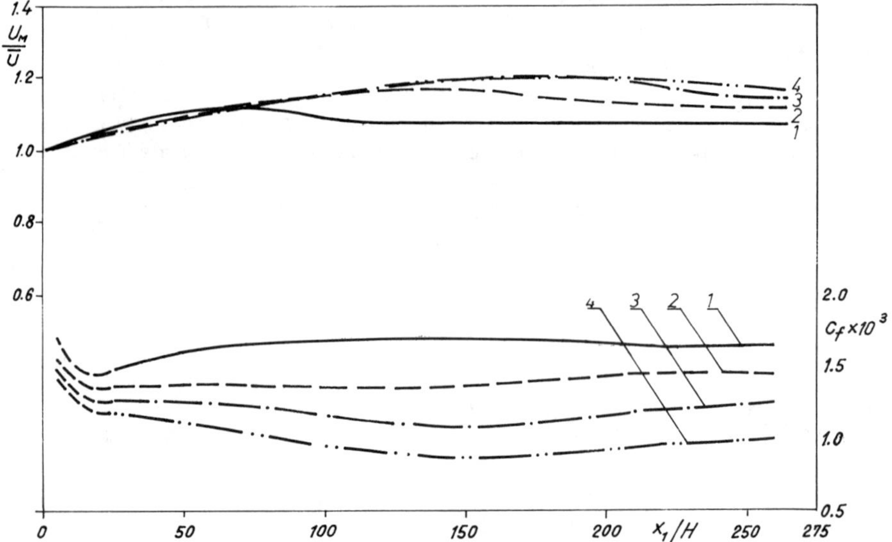

Fig. 8. - Development of free surface velocity and wall friction factor

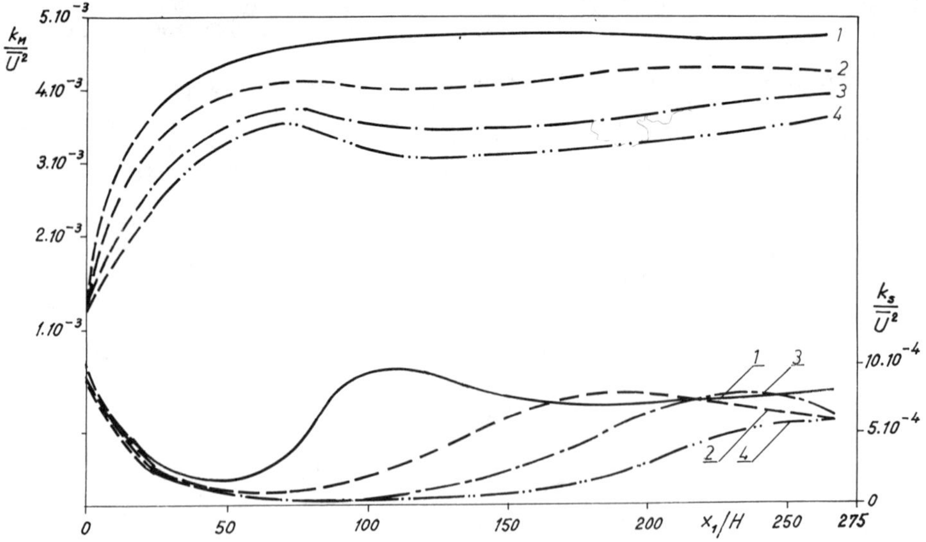

Fig. 9. - Development of free surface- and maximum turbulent kinetic energy

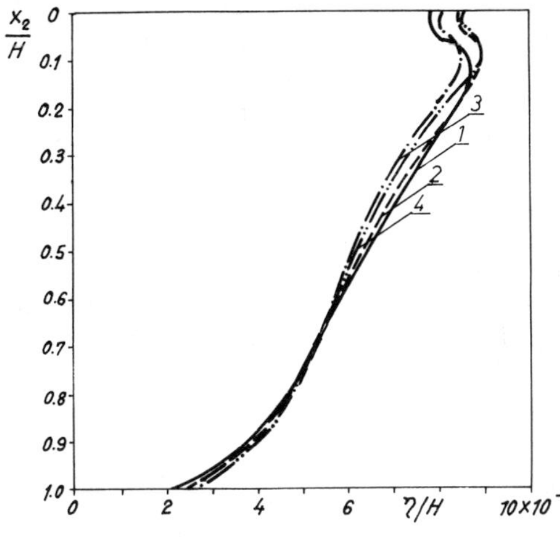

Fig. 10. - Kolmogorov length scale

To illustrate the effect of the particle presence on the mean flow development, Fig. 8 presents the variation of the free surface mean velocity U_M and the wall friction factor c_f along the flow. Curve 1 corresponds again to the non-buoyant case showing the familiar increase and overshooting of U_M, reaching its maximum at the position of about 65 flow depths downstream, with a gradual fall off to the fully developed value at about 150 to 200 channel depths downstream from the inlet position. The presence of the particles obviously slows down the flow development with the free surface velocity reaching its maximum further downstream as the settling velocity increases. Similar behaviour is exhibited by friction factor, in which case the influence of the concentration field is shown to affect not only its development, but also to decrease the wall shear stress, in accord with the modification of the mean velocity profiles, shown in Fig. 1. It should be mentioned here that the initial portion of the channel for $x_1 < 30$ H should not be regarded as reliable since the adopted numerical mesh in this region was too coarse for the proper consideration of the thin boundary layer, developing along the wall.

Fig. 9. shows the development of the free surface and maximum value of the turbulent kinetic energy k_S and k_M. Ignoring again the initial region of the channel, where the adjustment of the kinetic energy takes place, the profiles show the damping of the kinetic energy, particularly the maximum value in the wall region, and a corresponding effect on the turbulence development.

Finally, to support the assumption of the turbulent diffusion controled transport of particles, we present the normalized profiles of the Kolmogorov length scale in Fig. 10. As can be seen, for H = 1m the considered particle sizes up to 200 microns are all smaller than the smallest dynamic scale of turbulence across the whole flow depth.

6. CONCLUSIONS

It has been demonstrated that the extension of the full-stress-flux model of turbulence, and its simplified version, based on the solution of the differential transport equations for all important single point turbulence correlations, can produce the plausible predictions of the major mean and turbulence quantities in turbulent shear flow laden with a dilute solution of fine solid particles. The model proves to be capable of accounting properly for the buoyancy effects and mutual interactions between the hydrodynamic and concentration fields, and the consequent effects on turbulent structure.

The modelled equations contain the empirical coefficients, proved in earlier works of these and other authors, to yield the acceptable agreemants with experimental data in a number of singlephase isothermal flows, as well as in several cases of thermal buoyant flows. Only one coefficient, $C_{\varepsilon 3}$, appearing in the buoyancy production term in the ε - equation, contains a degee of uncertainty and requires further testing with reference to reliable laboratory experiments;a task that was not fully accomplished in the present investigation either,due to the serious lack of experimental data for the considered flow situation. On the whole, however, the model seems generally applicable to the prediction of the problem of suspended sediment transport, even in more complex flow configurations, and, with some refinements, may serve as a basis for the study of other phenomena associated with the solid particles-fluid mixture.

ACKNOWLEDGEMENTS

The authors gratefully acknowledge the valuable discussions with Dr S. Bruk of the Institute of Hydraulic Research "Jaroslav Černi" in Beograd.

The research was supported in part by the Selfgovernig Community of Interests for the Sciences in Bosnia and Herzegovina, and the Institute of Hydraulic Research "Jaroslav Černi". The numerical calculations were carried out on the UNIVAC 1110 computer of ERC - Energoinvest, Sarajevo.

REFERENCES

1. Batchelor, G.K., Proc. 2nd Austr. Conf. on Hydraul. and Fluid Mech., 19, p.p. 019-041, University of Auckland, 1965.
2. Lumley, J.L., Ch. 7., "Topics in Applied Physics; Vol. 12 - Turbulence" (ed. Bradshaw, P.), Springer, Berlin, 1976.
3. Lee, S.L., Durst, F., Von Karman Inst. for Fluid Dynamics-Lect.Ser.1980-3.
4. Sumer, B.M., J. Hydr. Division, HY 11, pp. 1323-1337. 1977.
5. Sumer. B.M., Oguz B., J. Fluid Mech., 86, part 1, pp. 109-127, 1978.
6. Hanjalić, K., Launder, B.E., J. Fluid Mech., 52, part 4. pp 609-638, 1972.
7. Launder, B.E., Reece, G.J., Rodi, W., J.Fluid Mech, 68, p. 537, 1975.
8. Launder, B.E., J. Fluid Mech., 67, p 569, 1975.
9. Launder, B.E., Ch.6, "Topics in Appl. Phys., 12 (ed.Bradshow P.) Springer, 1976.
10. Hossain,M.S., Rodi, W., in "Turbulent Buoyant Convection" (ed.Spalding D.B.) Hemisphere Publishing Corp., Washington D.C., 1976.
11. Patankar, S.V., Spalding, D.B., "Heat and Mass Tranfer in Boundery Layers", Intertext Books, London, 2 nd Ed., 1970.
12. Rodi, W., SFB 80/T127, Sonderforschungsbereich 80, Univ. Karlsruhe, 1978.

FLOW VISUALIZATION

Turbulent Boundary Layer Structure

M. R. HEAD and P. BANDYOPADHYAY
University Engineering Department
Trumpington Street
Cambridge, England

1. INTRODUCTION

The flow visualisation film to be presented provides evidence in favour of the hairpin-vortex hypothesis outlined below. A complete account of the investigation leading to this hypothesis is to be published [1] and a briefer description has already been given in ref. [2] . The purpose of the present paper is to provide a background against which the film can be viewed.

2. HAIRPIN VORTEX HYPOTHESIS

The hypothesis is that, at least up to values of Re_θ in the region of 10,000, the zero-pressure-gradient turbulent boundary layer consists very largely of horseshoe or hairpin vortices generated in the immediate vicinity of the wall and extending in some cases right through the boundary layer. The horseshoes or hairpins are substantially straight and inclined downstream at a characteristic angle of 40° - 50° to the surface. Because they are generated in the immediate wall region, the lateral scales of the hairpins and their longitudinal spacings are initially set by the wall variables U_τ and ν. Their length, however, is ultimately limited only by the boundary layer thickness, so that there is a large Reynolds number effect on boundary layer structure, and at high Reynolds numbers at least a minority of the hairpins will become greatly elongated, as indicated in Fig. 1.

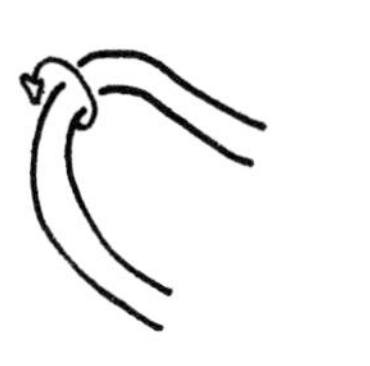

very low Re
(loops)

low-moderate Re
(horseshoes)

moderate-high Re
(hairpins)

Fig. 1. Effect of increasing Reynolds number

There are several factors that complicate the simple picture just described. First there is the presence of a wide variety of scales in the wall region, since it is only in the mean that the scaling with U_τ and ν will be observed (see e.g. Van Maanen [3]). Again, some measure of mutual induction and entanglement between vortex loops, horseshoes or hairpins would seem inevitable, and it is only occasionally we should expect regular arrays of these features to be observed [4] . Dissipation and vortex stretching will evidently have important effects which can scarcely be discussed here. We may however remark that the observed characteristic hairpin angle is very similar to the angle where vortex stretching is a maximum, as Theodorsen [5] originally proposed; also that the joint effects of dissipation, mutual induction, and the existence of a wide variety of scales may well account for the existence of a universal law of the wall at high Reynolds numbers, with only a minority of the hairpins surviving into the outer region of the layer.

3. IMPLICATIONS OF THE HYPOTHESIS AND COMPARISON WITH THE RESULTS OF FLOW VISUALISATION

On the basis of the hypothesis we should expect the low Reynolds number boundary layer to be characterised by a relatively small number of low aspect ratio vortex loops or horseshoes, while the high Reynolds number layer should consist of much more closely packed narrow features inclined at the characteristic angles to the surface, much as indicated in the sketch below (Fig. 2). This shows up quite clearly in the film, where a longitudinal light plane is used to illuminate the smoke-filled boundary layer at low and high Reynolds numbers.

With hot wires staggered at approximately the characteristic hairpin angle to the surface we should expect to obtain very similar signals, with zero time delay, for at least part of the time, and several examples of similar oscilloscope traces are shown in the film.

Finally, we should expect to see very different views with transverse light planes inclined 45° downstream and 45° upstream, as indicated in Fig. 3 (top of next page). With the light plane inclined 45° downstream we should see rapidly changing loop-like structures, while with it inclined 45° upstream we should see large numbers of vortex pairs, apparently in continuous motion towards the wall.

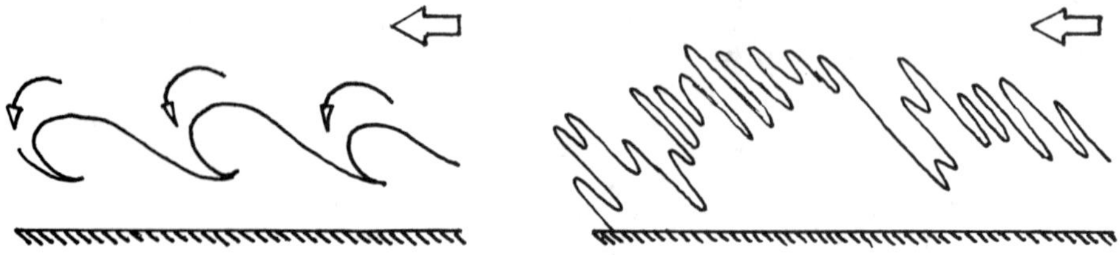

low Reynolds number ($Re_\theta \simeq 500$)
sections of vortex loops

high Reynolds number ($Re_\theta \simeq 10{,}000$)
hairpins

Fig. 2. Contrast between boundary layers at low and high Reynolds numbers

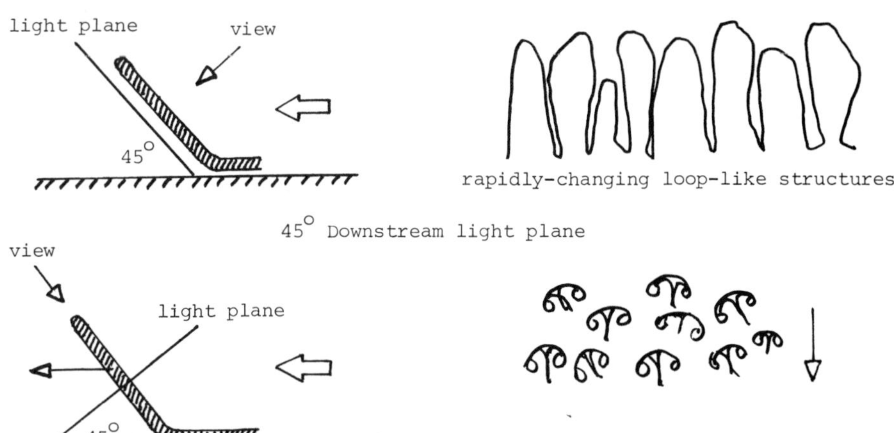

Fig. 3. Contrast between views with downstream and upstream light planes

Sequences showing views with downstream and upstream inclined light planes, at three widely different Reynolds numbers, show precisely the expected behaviour, with very large reductions in lateral scales as the Reynolds number is increased. This provides, we believe, fairly convincing evidence in favour of the hairpin vortex hypothesis.

REFERENCES

1. Head, M.R. and Bandyopadhyay, P. New aspects of turbulent boundary layer structure. J. Fluid Mech. to be published.

2. Head, M.R. and Bandyopadhyay, P. 1979. Flow visualisation of turbulent boundary layer structure. AGARD Conference Proceedings No. 271, pp. 25-1 to 25-12.

3. Van Maanen, H.R.E. 1979. Experimental studies of coherent structures in the turbulent boundary layer of pipe flow using laser-doppler anemometry. AGARD Conference Proceedings No. 271, pp. 3-1 to 3-20.

4. Bandyopadhyay, P. 1977. Combined flow visualisation and hot-wire anemometry in turbulent boundary layers. Lecture Notes in Physics No. 75, pp. 205-216, Springer Verlag.

5. Theodorsen, T. 1952. Mechanism of turbulence. Proc. 2nd Midwestern Conference on Fluid Mechanics, Ohio State University.

Visualizations of Turbulent Structures of Wakes and Boundary Layers

RÉGIS DUMAS, PIERRE BONMARIN, and LOUIS FULACHIER
Institut de Mécanique Statistique de la Turbulence
Laboratoire Associé au C.N.R.S. No. 130
12, Avenue Général Leclerc - 13003 Marseille

ABSTRACT

Through two examples concerning turbulent flows, a wake and a boundary layer, the usefulness of visualizations for a better understanding of the structure of these flows is underlined. We use a vertical hydrodynamical tunnel continuous running.

When two parallel rods are different in diameters, the mean velocity profile in the downstream flow becomes slightly dissymetrical and a counter-gradient turbulent stress appears in the central part. The film shows that the flow pattern is strongly dissymetrical; it discloses that the counter-gradient stress is due to the fluid entrainement by the big vortex issued from the larger rod.

Besides the turbulent boundary layer on one of the test-section wall has been filmed. The filament lines outside the boundary layer bear out the three-dimensional behaviour of the intermittency region. In the vicinity of the wall($y^+>15$), the generation of hairpins vortex are observed. These observations support the accepted hypothesis concerning the generation of the turbulence at the wall.

1. INTRODUCTION

The visualizations are a suitable tool to display the coherent vortices which are present in most turbulent flows, at least at their beginning. They are also Lagrangian pictures of the flow which are very useful to the physical understanding of the Eulerian measurements by the classic velocimeters.

In this mind, a vertical hydrodynamic tunnel for visualizations has been built. The first experiments concern the flow pattern behind two identical parallel cylinders, perpendicular to the mean flow. Various modes of linkage between the wakes are disclosed. If the rods are different in diameters, a counter-gradient turbulent stress [1] appears in the central part of wake. The photographs give an explanation of the phenomenon which appears to be due to the coherent vortices. Besides a turbulent boundary layer has been visualized specially in order to study the sublayer and the intermittent region. The pictures provide supplementary useful materials concerning the structure. The films which have been also undertaken are even more clear for the understanding of the structure of the abovementioned flows.

2. HYDRONAMIC TUNNEL.

The vertical hydronamic tunnel is of gravity type[2,3,4,], it is continuous working, a water pump lifting the fluid from lower tanks to an upper tank above the collector ot the test section. The dimensions of the square test section are 20 x 20 x 120 cm. The pressure head remains at constant value owing to the water outfall set on the upper and the receiving lower tanks.

Fig. 1. Hydrodynamic tunnel.

The stability of the bulk velocity of better than .1 % in the range of working 5 cms^{-1} to 46 cms^{-1}. The collector and the test-section are entirely transparent made of plexiglass. This appears to be very useful to insert the dye nozzles without trouble to the potential flow inside the collector. The main difficulty has been to damp the vortices inside the upper tank generated by the pipe inflows from the lifting pump. Various arrangements has been undertaken. So the preturbulence level in the test sections gives a signal negligible against the noise background of the laser-Doppler velocimeter. The streak lines are also perfectly straight (see [3]). Finally the smallest observable features in the photos were of the same size than the Kolmogorov microscale.

3. VISUALIZATION FACILITIES - LASER VELOCIMETRY.

The dye nozzles may be inserted inside the flow particularly in the potential flow of the collector and in the recirculating flow behind the rods. Also dye injections may be made through thin slits or holes at the surface of the body. The hole diameters are .5mm . The dyes were made of rhodorsil (milky silicone oil emulsion), fluoresceine (green) and rhodamine B (red). The total amount of water is roughly $5m^3$. This is a sufficient quantity to provide a long time about 25 minutes until the dyes make the water cloudy. The illuminations are supplied by classic flash light synchronised on the camera rate(max. 500 photo. sec^{-1}) for the film [5]. The thickness of the plane light beam was about 1 cm, not too thin because the vortices to be investigated are fluctuating in the third direction perpendiculary. Eulerian measurements of velocity are made with a Laser-Doppler operating in fringe mode with a tracker frequency detection (laser neon 15 mW, Disa system). The optic is equiped with a frequency shift, Bragg cell. The noise background of the velocimeter may be estimated by putting the beams intersection in the indisturbed flow, it is roughly equivalent to a relative intensity of turbulence of .5 %. Theoretically it is irrelevant for the values of the tengential turbulent stress $\overline{u'v'}$ [4].

4. TURBULENT WAKES BEHIND TWO PARALLEL CYLINDERS.

The main purpose of the research was to explain the mechanism of the counter-gradient turbulent which occurs in the wake of two parallel rods of different diameters. Before these studies it was necessary to investigate the pattern flow behind two rods of same diameter. The photo 2a and 2b present typical pictures of the two modes of vortice structure in the case of moderate linkage between the wake of cylinders ($G/\emptyset = 2$, G distance between the rods, $\emptyset$ diameter, $R_\emptyset = 1800$). Figure 2a shows the expected linkage with symmetrical pattern, but the figure 2b discloses a disymmetrical mode which had never been clearly visualized, although it was assumed. The change of modes occurs randomly in the time. Figures 3a and 3b show the patterns relative to a strong linkage ($G/\emptyset = 0.44$).

The previous symmetrical mode disappears, only coherent vortices issued from one of the rods remain with random sequences of the phases on the left side (3a) or on the right (3b).

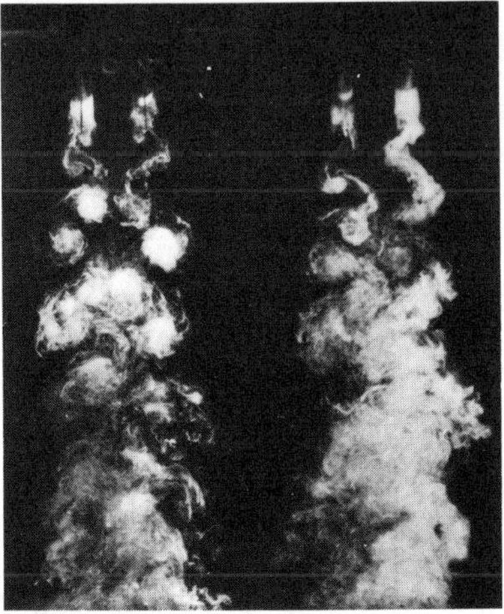

(a) (b)

Fig. 2. Turbulent wakes behind two cylinders of same diameter : moderate linkage.

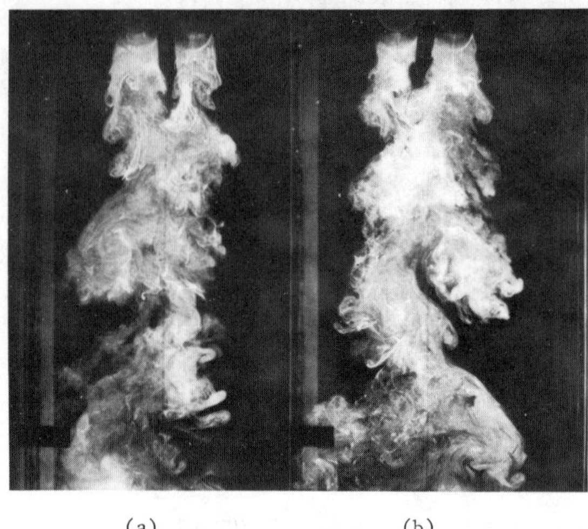

(a) (b)

Fig. 3. Turbulent wakes behind two cylinders of same diameter : strong linkage.

Now figure 4 presents the wake corresponding to parallel rods different in diameter (ϕ_1 = 20 mm, ϕ_2 = 12 mm, G/ϕ_1 = .35, $R_{(\phi_1+\phi_2)/2}$ = 2800).The picture looks a little like on figure 2 but here the coherent vortices are always due to the larger rod. The instabilities from the smaller one do not shed coherent vortices. In this situation PALMER and KEFFER [1] have disclosed that the counter-gradient turbulent diffusion is not shifted towards the side opposite to the maximum ot the turbulent kinetic energy, as it is usual in the disymmetrical flows with extremum of mean velocity.

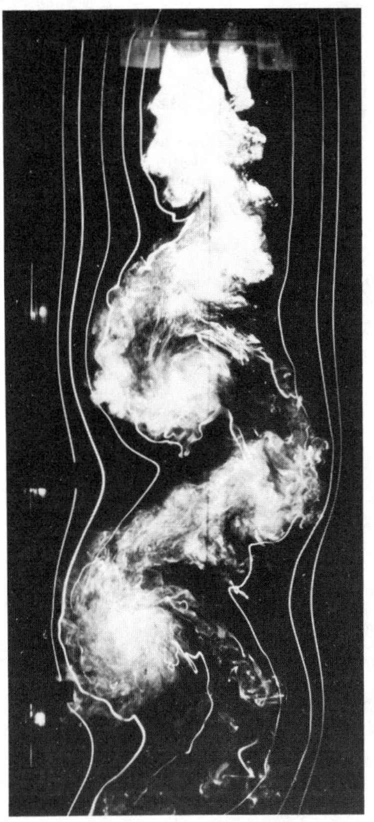

(a) Wake and streakline

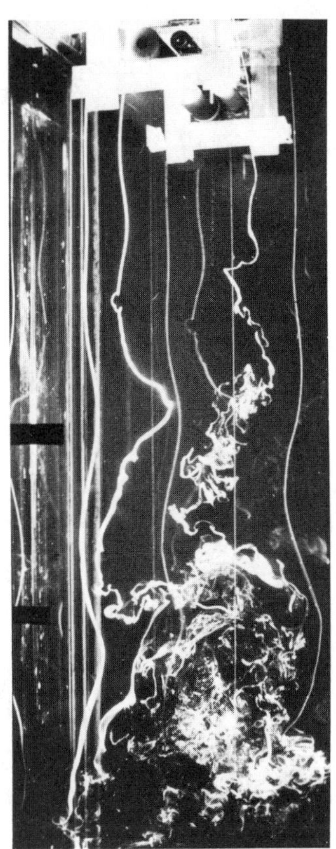

(b) Streakline only

Fig.4. Turbulent wakes behind two cylinders different in diameter : strong source interaction.

This result was verified as it is shown on figures 5 and 6. Figure 5 shows the mean longitudinal velocity U/U max, U_{max} being the upstream general velocity, and the turbulent tangentiel stress $\overline{u'v'}/U^2$ max. The half wide of the wake corresponding to U = ΔU/2 is L and y_M corresponds to $\partial U/\partial y = 0$ (with ΔU = U max - U min).

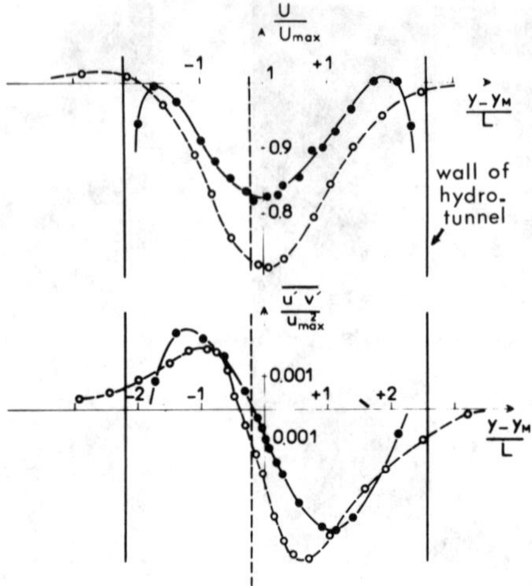

—●— Hydrodynamic tunnel, —○— Wind-tunnel (Fraunié [6])

Fig.5. Mean longitudinal velocity and Reynolds stresses behind two cylinders different in diameter.

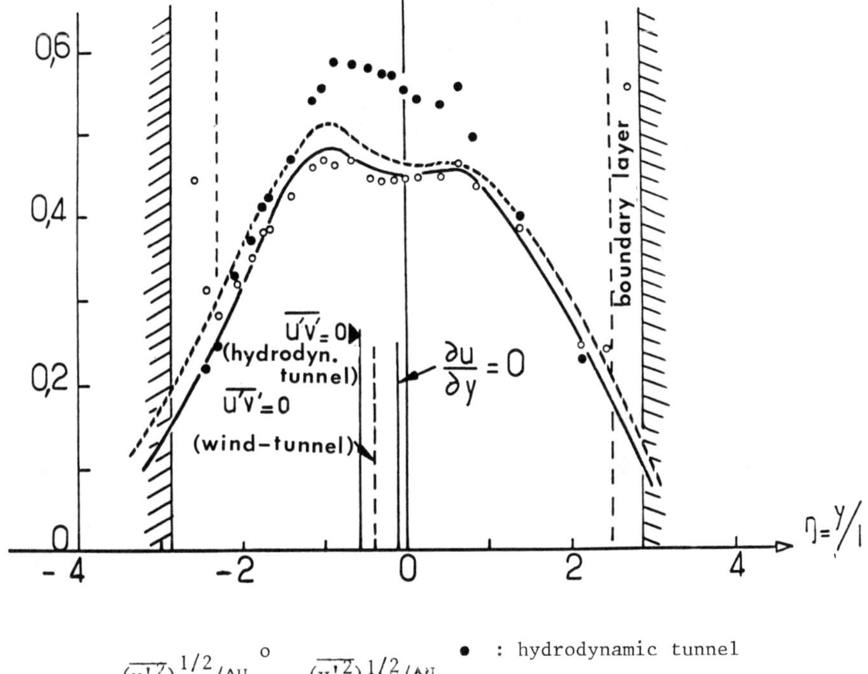

$(\overline{u'^2})^{1/2}/\Delta U$ ○ , $(\overline{v'^2})^{1/2}/\Delta U$ ● : hydrodynamic tunnel

---- : wind-tunnel, (Fraunié [6])

Fig.6. Longitudinal and transverse turbulent stresses.

— The following figure 6 concerns the longitudinal ,$\overline{u'^2}$, and transverse ,$\overline{v'^2}$, turbulent normal stress (The half wide of the wake is here 1).

The velocities U,u',v', have been measured with a laser-Doppler velocimeter. On these figures 5 and 6, it is displayed also the results obtained by Fraunié [6] in a wind tunnel with hot wires anemometers. The geometrical configuration was quite similar, but in the hydrodynamical tunnel the mirror effect of the wall is not at all negligible. Yet the results exhibit the same general trend for the flows in the two experiments. In particular the zero value of the turbulent stress $\overline{u'v'}$ is shifted towards the higher kinetic turbulent energy side. Now the visualization display that the coherent vortices, seen on figure 4, induce very strong entrainment of fluid in the central part of the wake corresponding to very large negative values, in our system of coordinates, of instantaneous products u'v'. These values explain the shift of Reynolds stress $\overline{u'v'}$. Besides the film underlines much better these entrainments of fluid. Also, figure 4b, which shows only the streaklines, points out clearly these entrainments transverly to the mean flow.

We recall also that according to the classical modeling originated by A.A.Townsend, it is assumed :

$$\overline{u'v'} = -\nu_t \frac{\partial U}{\partial y} + \overline{u'V}$$

ν_t is the eddy viscosity. The additional term $\overline{u'V}$ takes into account the diffusion due to the large turbulent eddies which are not related to the local mean velocity gradient as the smaller eddies. According to the model of P. Bradshaw ($|\overline{u'v'}|/\overline{q'^2} \simeq \cdot 15$ close the maximum of $|u'v'|$, with $\overline{q'^2} = \overline{u'^2} + \overline{v'^2} + \overline{w'^2}$), the maximum for the absolute value $|\overline{u'v'}|$ is following the turbulent kinetic energy $\overline{q'^2}$, that is in agreement with the usual shift of the zero value of $\overline{u'V}$. This shift corresponds, in a wake flow, to a positive signe for $\overline{u'V}$ in the vicinity of $\partial U/\partial y = 0$ point. But in the present situation the additional stress $\overline{u'v'}$ is negative related to the entrainment flow by the coherent vortices. Then the earlier cross experiments carried out by Daien [4] in the wake behind two airfoils of different sizes - where the vortices are much less coherent than downstream the cylinders - point out that the sign of the additional term becomes positive as it is expected in the dissymetrical wake.

5. TURBULENT BOUNDARY LAYER.

The mechanism in the depth of the turbulent boundary layer is still questionable. So we have performed some visualizations as it is pointed out on the scheme (Fig. 7).

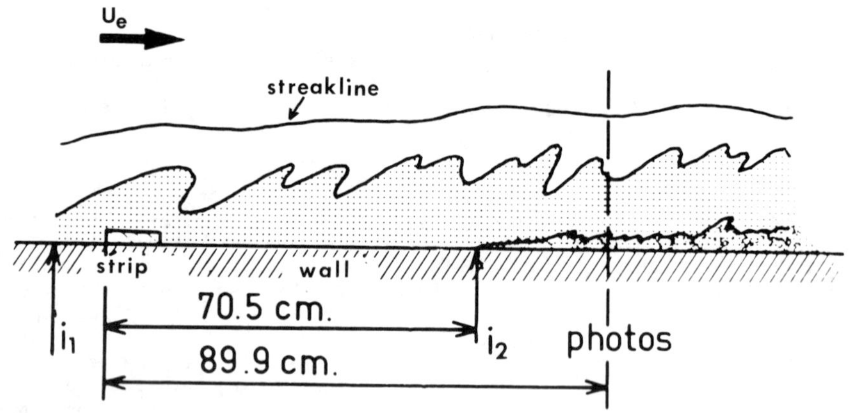

Fig. 7. Scheme of visualization.

An upstream hole injection dyes all the boundary layer with a spanwise of the order of its thickness δ. A second hole injection dyes at the origin only the viscous layer on a thickness of the order $y^+ = u_\tau y/\nu = 15$, u_τ friction velocity. The inflow is adjusted to the smaller rate compatible with the visualization. Nevertheless the inflow makes locally unstable the viscous layer, but we can assume that the process is signifiant for the natural instabilities which occur randomly close to the wall. Besides two streaklines are injected in the potential flow close to the boundary layer.

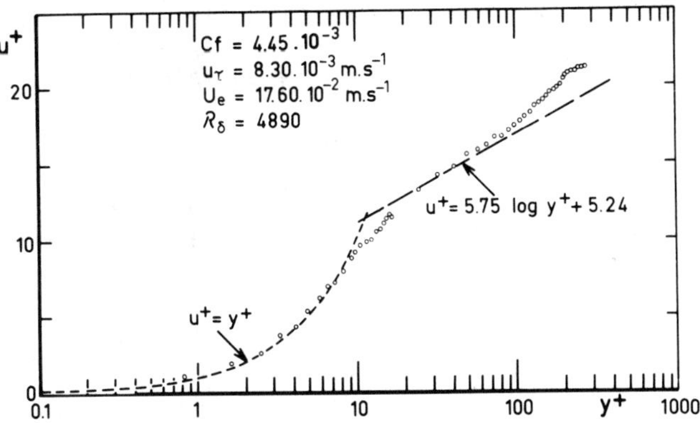

Fig. 8. Mean velocity profile (R_δ= 4 890).

Figures 8 and 9 give the mean velocity profiles obtained at $R_\delta = U_e \delta/\nu$ = 4 890 and R_δ = 13 270 (U_e = outer mean velocity, δ = boundary layer thickness at U = .99 U_e and $u^+ \triangleq U/u_\tau$).

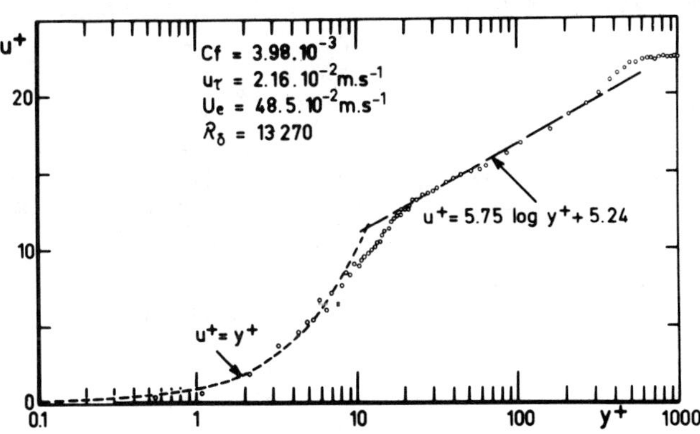

Fig. 9. Mean velocity profile (R_δ = 13 270).

The wall friction velocity u_τ is measured from the mean velocity slope at the wall. Even for the lower Reynolds number, the mean velocity profile exhibits an inertial region in agreement with the classic values of the Van Driest profile :

$$u+ = 5.75 \log y^+ + 5.24$$

A typical photograph is given (Fig.10), but this black and white photo is obviously much less instructive that the couloured original one and than especially the coloured film.

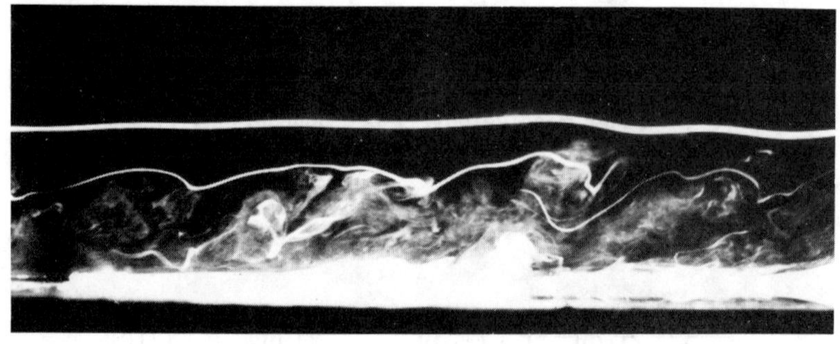

Fig. 10. Visualization of the boundary layer.

The intermittency region looks to be strongly threedimensional. The streak-line co:nes and goes around the turbulent bulges. The outer potential flow enters very little into the turbulent one.

Close to the wall the ejections of fluid towards the central regions are discoutinuous and look often like hair-pins or sheets streched in threedimensional way.

On these visualizations and the film, we did not disclose the intermittency would drive the ejections as it is sometimes assumed. But one may argue that the bursting process is threedimensional (see for example[7]). Indeed it appears that the fluid ejected from the wall region is enclosed in spanwise direction by inflows coming from the outer region (see for example figure 11 [8]). So one needs to use true threedimensional visualizations in order to display a possible effect of the outer flow on the ejection process. It may be that in the present pictures the linkage does not appear.

ACKNOWLEDGEMENTS

The authors write to express their gratitude to M. Astier who has carried up all the experimental facilities.

REFERENCES

1. Béguier, C. and Giralt, F., Fulachier, L. and Keffer, J.F. - 1978 - Negative production in turbulent shear flow. Proc. Symp. on Turbulence - Springer Verlag, Berlin.

2. Werle, H. - 1979 - Le tunnel hydrodynamique au service de la recherche aérospatiale. Onera Publ. 156.

3. Domptail, C. - 1979 - Sillages turbulents en aval d'un et de deux barreaux parallèles en tunnel hydrodynamique : visualisation et vélocimétrie laser. Th. Doct. Spécialité I.M.S.T. - Université Aix-Marseille 2.

4. Daien, E. - 1980 - Contribution à l'étude d'écoulements turbulents par visualisations. - Th. Doct. Spécialité I.M.S.T. - Université Aix-Marseille 2.

5. Bonmarin, P. - 1978 - Une méthode de visualisation appliquée à l'étude des écoulements en tunnel hydrodynamique et en soufflerie air-eau. - 7ème Coll. Nat. Cinéma rapide, Paris.

6. Fraunié, P. - 1979 - Contribution à l'étude des champs dynamiques et thermiques dans le sillage de deux cylindres de section circulaire. - Th. Doct. Spécialité I.M.S.T. - Université Aix-Marseille 2.

7. Offen, G.R. and Kline, S.J. - 1975 - A proposed model of bursting process in turbulent boundary layers. Journal of Fluid Mech. 70, 109.

8. Elena, M. Fulachier, L. and Dumas R. - 1979 - Etude expérimentale des apports et des éjections de fluide dans une couche limite turbulente. Agard Conf. Proc. n° 271.

Visualization Study of the Axisymmetric Mixing Layer of a High Reynolds Number Jet

A. K. M. F. HUSSAIN
University of Houston
Texas 77004

ABSTRACT

The near-field of a 18cm air jet has been visualized in the azimuthal plane via a thin sheet of laser light. At low jet speeds, the initially laminar axisymmetric shear layer rolls up into an essentially periodic train of toroidal vortices which undergo breakdoen through pairing. At higher jet speeds, say at the jet Reynolds number of 3.6×10^5, the layer becomes turbulent soon after separation and all its mean and turbulence measures confirm its self-preservation for the range $0.5 \leq x/D \leq 4.5$. Contrary to suggestions otherwise, visualization reveals that the mixing layer in the self-preservation region is mostly in a state of disorganization. Large-scale coherent structures are formed only intermittently and typically survive for a distance comparable to their streamwise length, their turnover time being about the survival time. The structures undergo a variety of changes including tearing and complete, partial, and fractional pairings. These interactions and slippage of the high-speed side structures past the ones on the low-speed side can explain the heretofore unexplained but perplexing data that the passage frequency of the large-scale structures has been found to vary considerably across the width of the mixing layer; these also explain the failures of prior efforts in the eduction of coherent structures of the layer. When the upstream boundary layer is tripped to make it fully turbulent before the separation point, the mixing layer is also intermittently organized typically into a group of comparatively more compact coherent structures. Based on visualization and hot-wire data, it appears that with increasing Reynolds number, the mixing layer becomes progressively more infrequently and intermittently organized but its coherent structure becomes comparatively more energetic and more efficient in the production of coherent Reynolds stress.

1. Introduction

Accumulated evidences suggest that large-scale coherent structures are inherent to all turbulent shear flows [1-3]. This realization would have meant nothing more than a passing curiosity and transient excitement in the continuing "Rise and Fall of Ideas in Turbulence", except for two unique but related factors. There is a growing belief among many serious researchers that these structures dominate the transport phenomena and noise production in turbulent shear flows. The quasi-deteriministic nature of these structures also provide some hope that a viable theory of turbulence is a likely possibility provided that these structures are directly incorporated as the building blocks in the formulation of a new theory [4].

For a viable theory to germinate, there should be adequate physical understanding of these structures. Progress in this front is so rudimentary that there is no general agreement on the definition of these sturctures [4]. In the absence of exciting and continuing advances in turbulence, researchers are unduly overcautious and tend to view new suggestions with considerable suspicion. Consequently, an effort to formulate a definition of the structures [4] evoked more protests than support. This first effort needs both serious scrutiny and improvement.

The most important questions facing the "chasers" of coherent structures is the dynamics of these structures. How significant are these structures? How much of the transports of momentum, heat and mass and production of aerodynamic noise are due to these structures? Are these structures essentially universal or do they vary from one flow class to another? Do they depend on the Reynolds number and the initial condition? And so on. These questions have been more elaborately discussed elsewhere [4,5].

The role of these structures has technological significance because control or modification of these structures may help control transport phenomena and noise production in turbulent shear flows. The large dispersion in the shape, size, orientation, strength, convection velocity and "age" of the structures in each flow would make detailed eduction of structure properties prohibitive. Even if a large number of sensors in a three-dimensional array were used, the signals from the sensors must be recorded, and the structures identified from the signals through some well-verified detection scheme. The structures then will have to be classified on the basis of their characterizing parameters, viz., shape, size, orientation, strength, convection velocity, etc. That is, the multi-dimensional probability density function in these parameters must be determined. In order to educe the spatial distributions of the structure properties, the necessary spatial filtering must be done by ensemble averaging only structures of identical properties. Clearly, the experimentation time and computation capability required will be prohibitive [4]. In order to simplify the detection, coherent structures have been organized through controlled excitation [6-9] so that the eduction can be facilitated by triggering on the easily available phase reference. Even though arguments have been put forward that these structures are merely the natural structures induced at controlled phases [4,6], conclusive evidence is still lacking. Furthermore, while impulsive excitation [7] gives induced structures separated by natural structures, periodic excitation generated only the natural structures at regular intervals [8]. Clearly, a direct investigation of the natural structures, though prohivitive, is unavoidable. Investigations of the natural structures in the near and far fields of jets and mixing layers via a rake of vorticity and X-wires are in progress in our laboratory.

Visual study of the natural structures is clearly desirable in order to obtain direct insight into the flow physics as well as effectively design the quantitative study via hot-wire arrays. It is worth mentioning here that flow visualization, which was extensively used in the early investigations in fluid mechanics, is finding a rebirth via coherent structure studies where researchers seem to have agreed to make it an indispensable adjunct to quantitative studies [9]. With this in mind, it was decided to visualize the mixing layer of a large subsonic jet.

There were four specific motivations for this study. First: extent of mixing layer organization, structure interaction and evolution. Second: Reynolds number effect. Third: initial condition effect. Fourth: to test bifurcation hypothesis of Lau.

While the roll-up of the initially laminar shear layer into distinct toroidal vortex rings is well known, how the structures evolved downstream was not known. The evolution of the structures may involve interactions like pairing or tearing or transitions or breakdown and decay. Since vortex pairing is of major dynamic significance both in entrainment and noise production, it was felt necessary to understand the details of the pairing process and the extent to which it is preponderant in the undisturbed flow. Parenthetically, considerable details fo the physics of pairing of induced vortices in a jet have been reported separately [8,10]. The prevalence of the structures and their interactions at different Reynolds numbers and initial conditions is a logical inquiry. In spite of the suggestions by some researchers to the contrary, we believed that the dependece should be significant. In order to address this question with minimal effort, we performed the visualization study at two different jet Reynolds numbers (one at four times the other) and for the two asymptotic initial conditions: laminar and fully turbulent.

Consistent with the prevalent notions about the large-scale coherent structure in an axisymmetric mixing layer [11-14], the structure passage frequency should be the same on the high-speed and low-speed sides of a mixing layer. However, experiments [15,16] show that the passage count on the high-speed side is about at least three times that on the low-speed side. This is clearly in contrast with the notion of a single large-scale structure constituting the entire mixing layer width; pairing or tearing cannot explain this variation, either. This prompted Lau [16] to propound the hypothesis that the initially rolled up toroidal vortices in an axisymmetric jet bifurcate at $x/D \simeq 1.5$ alternately into two streams of vortex rings: one stream accelerating and converging on the jet axis at $x/D \simeq 8$, and the other stream decelerating and moving away from the jet axis. This provocative suggestion, though claimed on the basis of extensive experimental data and visualization by Lau [16], faced vehement objection by us because our experiences indicated that his suggestion was a very unlikely possibility. We thus were obligated to definitively test the Lau hypothesis. Visualization appeared to us to be the most direct avenue available to test Lau's hypothesis.

2. Apparatus and Procedure

The visualization study was carried out in an 18cm axisymmetric air jet discharging into a large laboratory space so that the influences of recirculation and free-stream turbulence [17,18] were not likely to be significant. The facility is operated by a full wave rectifier-controlled dc motor so that the speed can be varied continuously up to 100% but held constant at any value for a long period of time. Flow from a centrifugal blower passes through acoustic mufflers and a diffuser before entering the 76cm diameter settling chamber fitted with a number of honeycombs and screens. The nozzle has a contour which is a compromise of the cubic and the Batchelor-Shaw profiles and terminates into a tip with a 20^{0} bevel so that the primary and entrained air streams merge at the separation point essentially parallel. A number of tips with identical bevel angles can be attached in order to obtain different initial momentum thicknesses. Rings with various serrations can be inserted between the nozzle and the tip in order to obtain turbulent initial conditions. The blower-motor assembly is isolated from the floor. When the jet is in operation, the nozzle is free from any perceptible vibration. Details of the flow facility has been described elsewhere [19].

The mixing layer was visualized with titanium oxide smoke after the smoke was bubbled through a water bath to remove both heavier particles and Hcl acid. The smoke is released via a tygon tube held pressed on the tip on the low-speed side of the jet about 0.5 cm upstream from the separation point. The smoke comes out at a negligible speed and is entrained to mark the mixing layer. This method of injection was found to be far superior to injection into the nozzle as the latter method would disturb the flow. Small changes of the location of the smoke release point had no effect on the flow patterns.

Most discussions relate to observations for the exit speed of 30 m/s when the centerline (free stream) longitudinal turbulence intensity is about 0.25%. The study has been carried out for two asymptotic initial conditions: laminar and fully turbulent. For the laminar case the exit mean velocity profile agreed very closely with the Blasius profile with a shape factor of 2.54. The fluctuations consisted of frequencies considerably lower than the instability frequencies of the exit shear layer. The turbulent initial condition was achieved with a trip ring inserted upstream of a long tip. The trip ring consisted of a spanwise array of teeth whose height, width, depth, length and spacing each was equal to the local displacement thickness. The mean velocity profile in the universal (U^+,y^+) coordinates had distinct logarithmic and wake regions typical of the flat plate equilibrium boundary layer: both the slope of the logarithmic region and the wake strength agreed with those expected of the flat plate equilibrium turbulent boundary layer at the corresponding value of the momemtum thickness Reynolds number. The longitudinal fluctuation intensity increased monotonically from the free-stream value to its peak value at $y^+ \simeq 13$. The u-spectrum showed continuous broadband distribution up to 10kHz withouth any sepctral peak. All these measures of the exit boundary layer confirmed that it was fully turbulent. For a discussion of the criteria for characterization of the initial condition for both laminar and turbulent cases see ref. [4]. The exit boundary layer and the near-field was checked to be axisymmetric.

The flow was illuminated with a 6W Ar-Ion laser beam located about 8m downstream from the exit. The beam was flared into a sheet via a cylindrical lens so that it formed an illumination plane of about 2mm thickness passing through the jet axis and the tygon tube center. It should be recognized that this kind of illumination technique has two constraints. First, azimuthal motion of smoke-marked fluid may be confused with streamwise fluid motion. However, experimentation with other visualization techniques including spot and flood lighting convinced us that the illumination in the thin sheet is the best method for observing the details of the structure dynamics and entrainment. It may be worthwhile to review in this connection the visualization technique used in a few other similar studies. The technique used by Brown & Roshko [1] for depicting the end view of the plane mixing layer structure illuminated the entire flow and thus integrated the structure over the entire span. The illumination of the entire jet used by Crow & Champagne [20], Browand & Laufer [11], Davies & Baxter [21], Yule [14] and Moore [22] hardly revealed any details over the cross-section of the structure. Careful examination of our picture sequences obtained from the high-speed movie allows one to differentiate the effects of azimuthal motion from the structure interactions. The visualization technique with a scanning laser beam employed by Fiedler [32] is comparable to ours except that it suffers from the constraint that the structure photographed is distorted as different parts are exposed at different times.

The second constraint of our visualization technique is that the smoke intensity progressively decreases in the downstream direction due to diffusion. This is the inherent limitation of smoke visualization employed in any research. Visualization with smoke wire in air or hydrogen bubble technique in water has a potential advantage in this respect because markers are introduced locally but suffers from the constraint that description in a plane is not available. The second constraint is vortex shedding unavoidable even at moderate Reynolds

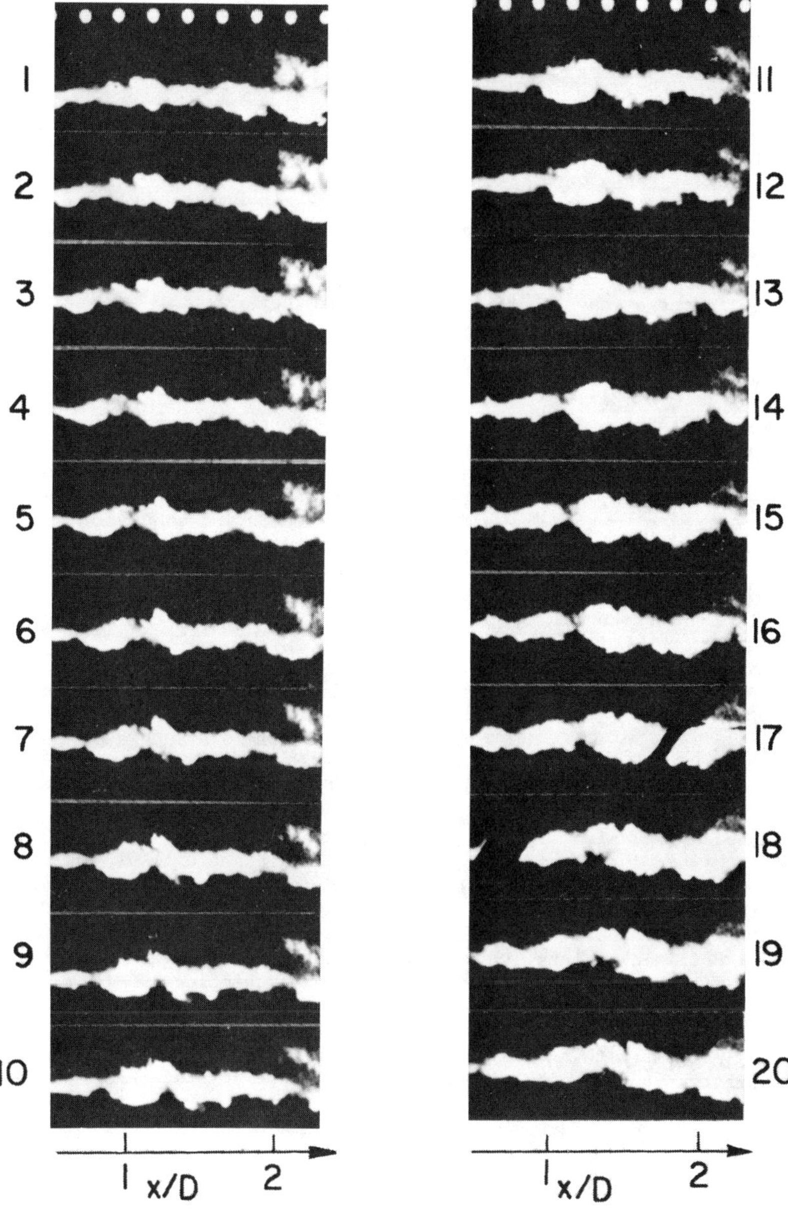

Figure 1. Picture sequence of the sectional view of the mixing layer; $\delta t = 0.56ms$.

number flows. Attempts to visualize our flow with the smoke-wire technique failed because of the constraint, in addition to the two just mentioned, of large temperature differences of the wire across the jet boundary. This prevented the oil droplets from having a steady drift down the wire.

The flow in the illuminated plane was photographed on a high speed film with a movie camera located normal to the illuminated plane. Incandescent red lights at 5.08 cm spacing was placed in view in order to provide spatial references. These lights were placed sufficiently away from the flow so that these have no influence on the flow. The movies were taken with a Fastax movie camera at the framing rate of 2000/sec. Since the film has to start from rest, the interval between successive frames depended on the location of a frame with respect to the start of the film. the time interval between two successive frames at any location can be determined from the timing marks placed on the film.

3. Observations

The instantaneous details of fluid motion in the mixing layer, its state of organization or disorganization, the motion of the turbulent-nonturbulent interface, and the evolution and interactions of the large-scale coherent structures are vividly clear from the high-speed movie, especially when viewed at very low framing rate. The smoke is entrained during the roll-up of the shear layer. During transition it marks the entire vortical fluid. It can be shown that during the structure transit time, the smoke and vorticity boundaries cannot be noticeably different. Thus, the smoke boundary is indeed the boundary of the turbulent mixing layer.

From repeated observations of the movies (available on request) a large number of conclusions can be drawn unambiguously. In order to transmit the observations to the research community, the usual procedure is to describe the events in words or at best use some schematic diagrams. Such a process suffers from the limitations that the reader has no way to verify the description other than observing the films himself/herself. It is generally agreed that what one finds in a turbulent flow visualization film depends on the "eye of the beholder ." Also, it is perhaps true that one could find what one is looking for (this author would suggest that such is not unlikely in the film of, for example, Corino & Brodkey). Thus, in order to supplement our observations with some hard evidences as well as document some limited aspects of the film in the printed form of the literature, we have undertaken the painstaking process of obtaining prints of successive frames of the movie at some selected events and mounted these in sequence. A few such examples are included in this paper.

It is necessary to emphasize that the mixing layer captured visually was first characterized *via* conventional hot-wire measurements. Profiles of $U(y)$, $V(y)$, $u'(y)$, $v'(y)$, $\overline{uv}(y)$ showed similarity for $0.5 \leqslant x/D \leqslant 4.5$. The integral measures of the width of the layer namely the momentum thickness

$$\theta_{0\cdot 1} = \int_{y_{0\cdot 1}}^{\infty} \frac{U}{U_e}\left(1 - \frac{U}{U_e}\right) dy,$$

width $B = y_{0.95} - y_{0.1}$ and vorticity thickness $\delta_\omega = (1/U_e)/(\partial U/\partial y)_{max}$ all show linear variation over this x-range. The time average boundaries of the mixing layer also show linear growth. The spectra of $u(t)$, $v(t)$, and $uv(t)$ showed full development. All these measures confirmed that the mixing layer was self-preserving over $0.5 \leqslant x/D \leqslant 4.5$

Figures 1-4 show four sequences of pictures obtained from the high-speed

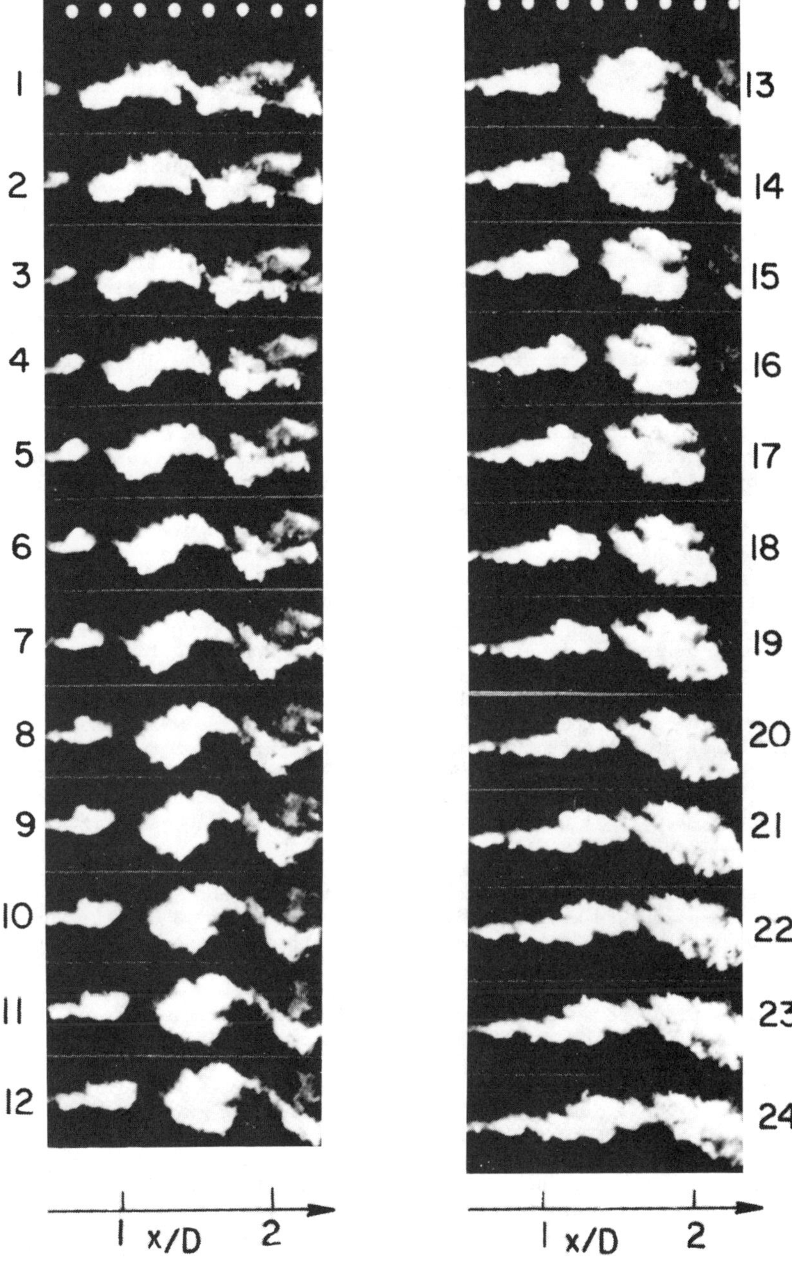

Figure 2. Picture sequence of the sectional view of the mixing layer; $\delta t = 0.50$ms.

movie of the axisymmetric mixing layer. Figures 1 and 2 cover the streamwise range $0.4 \leq x/D \leq 2.3$ and Figure 3 covers the range $0.5 \leq x/D \leq 3.5$. Figure 4 shows details of the structure over the shorter streamwise range $1.2 \leq x/D \leq 3.2$. These scales have been corrected for parallax effect. The succession of picture frames in time is identified by the numbers on the margin and the time difference δt between two adjacent frames is noted in the caption. Note the spatial reference scales identified by dots 5.08 cm apart at the top of each column. Even though the mixing layer is mostly in a state of disorganization, these sequences are chosen to capture the infrequent instants of organization. Figure 1, however, gives some perception of disorganization as denoted by, say, frames 1 and 20. Note the sharp difference of the cross-section with that depicted in those of Brown & Roshko [1], Winant & Browand [24], Ashurst [25], etc. Note that the gaps in frames 17 and 18 are due to accidental scratching of the negative during development. Starting from the left end of frame 2, there is an incipient roll-up of the layer into a large-scale structure up to frame 12 but this structure fails to evolve further downstream. The disorganization as revealed by frames 1 and 20 is the typical state of the mixing layer.

An example of organization, which occurs randomly, is captured in Figure 2. A pairing process in progress is appartent in frame 1. The pairing is complete by frame 12, resulting into a structure with a rounded cross-section. This structure does not proceed onto another stage of pairing but appears to undergo breakdown into finer scales as it is leaving the view. The extent and rate of rotation of the structure is apparent between frames 7 and 12. The structure on the left hand side (say frame 6) progressively increases in cross-section presumably due to azimuthal motion of the structure.

A number of interesting features of coherent structure interaction is captured in Figure 3. Pairing starting from the left hand end of frame 16 is complete in frame 23. However, a high-speed part of the paired structure is torn away and carried downstream by the high-speed core fluid. Note that the apparent merger of two structures starting in the middle of frame 7 never materializes. A low-speed part of the downstream structure is snatched away by the trailing structure leaving the rest of the leading structure intact. This is an example of partial pairing. Note the spec of the low-speed fluid remaining essentially stationary with respect to the laboratory. In fact, this spec of fluid even moves upstream in the laboratory frame, this flow reversal having been studied in our laboratory with a cold-wire-hot-wire pair [10,26]. Note that the structure rolled up in the middle of frame 16 does not undergo pairing but is stretched out by the straining effect of the stronger, paired structure trailing it.

Figure 4 provides an interesting example of structure slippage. Note that the two smaller structures on the left hand side of frame 1 approach each other but moves past each other (frames 9-15). The low-speed side structure is eventually captured by the relatively advancing trailing structure. This is an example of fractional pairing. Still another structure approaches the fast moving structure (starting in the middle of frame 15) but slips over it without merger (frame 26). Figure 4 also shows examples of two other events. The braid connecting the large-scale structure on the right hand side in frame 1 is gradually depleted as the braid fluid is drawn into the adjacent structures through their induced motions. This depletion of braid was experimentally measured in [10] as well as analytically predicted in [27]. The large-scale structure between frames 1 and 6 on the right hand side also gives an indication of the rate of rotation of the structure.

Measuring the rotation of a structure as identified by a front or an edge

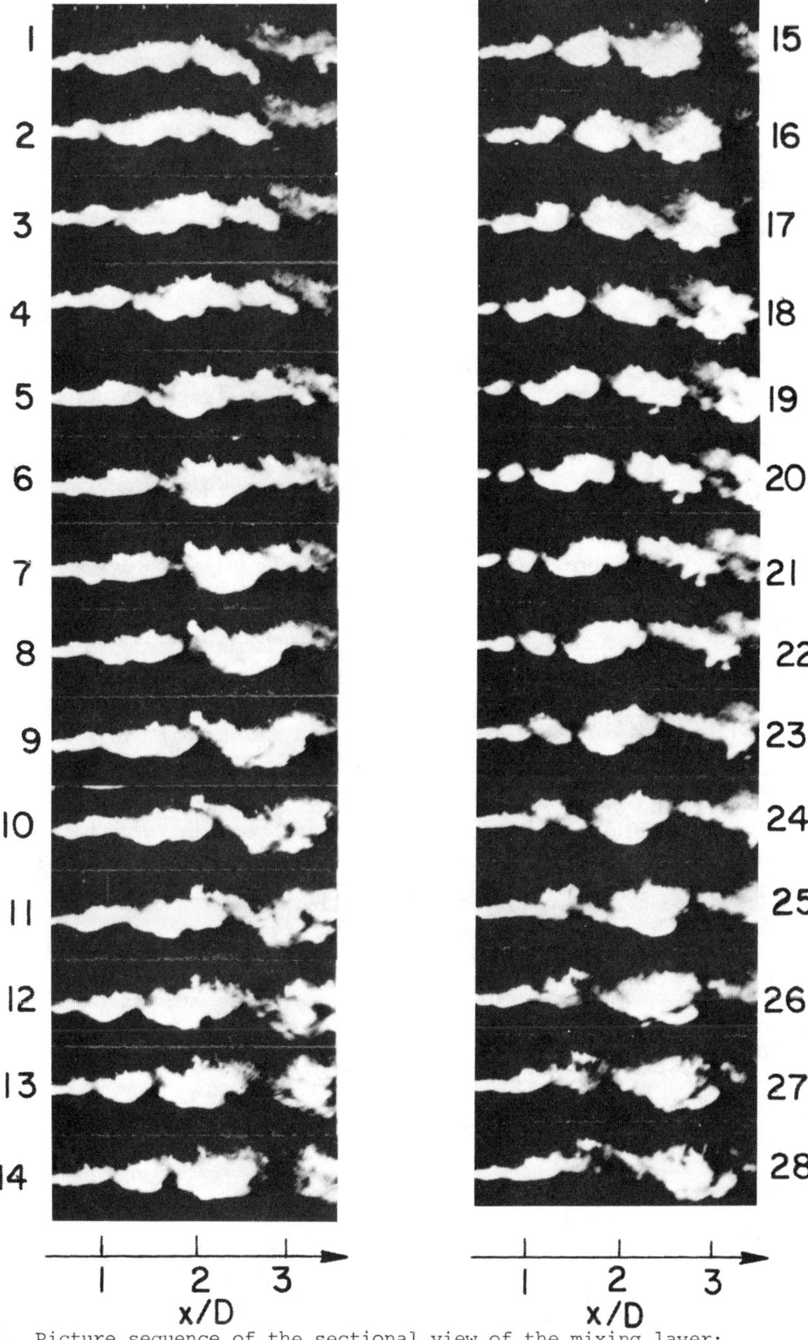

Figure 3. Picture sequence of the sectional view of the mixing layer; δt = 1.43ms.

as well as noting the time lapse one can determine the rotation rate. An average over a number of such events gives the turnover time of a structure.

By tracing the space-time trajectory of an identified fluid element one can measure the time-history of its convection velocity. From this the dispersion of convection velocity at each transverse location in the mixing layer can be determined. Also by tracking the large-scale structure in space-time one can also determine the structure convection velocity.

A number of such statistical measurements have been made and only the results will be summarized. Other details can be found in [15]. Similar results have also been obtained for initially fully turbulent mixing layer.

4. Summary

Based on flow visualization of a high Reynolds number axisymmetric mixing layer and studies of the high-speed movie at slower speeds, a number of observations can be made. Some of these are qualitative and would be convincing if the movies are viewed by the readers themselves.

While at a low Reynolds number ($Re_D \simeq 9 \times 10^4$) the mixing layer is nearly periodically organized into rolled up structures, at a higher Reynolds number ($Re_D \simeq 3.6 \times 10^5$) the mixing layer is mostly in a disorganized state. Only occasionally and intermittently the layer is organized into large-scale coherent structures. This appears to be in contrast with the state of organization revealed in the studies of Brown & Roshko [1], Winant & Browand [24], Lau & Fisher [12] and appears to be more in line with the observations of Yule [14], Chandrsuda et al. [28], Bruun [13] etc.

Even when organized, there is a large dispersion in the shape, size, orientation, strength, convection velocity and interaction of these structures. These structures are quite short-lived, typically surviving for a distance about equal to the structure length. The structure turnover time equals about one survival period. The structure either disintegrates through breakdown into finer-scale structures or undergo interactions like tearing or pairing. Tearing, when a high-speed segment of a large-scale structure is torn and carried downstream by the high-speed core fluid leaving the low-speed segment behind, appears to be at least as prevalent as pairing. Pairing is not typically of the complete type when two large-scale structures spanning the width of the mixing layer pair up. But typically pairing is of the fractional type when substructures torn from earlier larger structures merge, or of the partial type when a structure captures only a part of an adjoining structure leaving the rest alone to continue as a structure. No complete pairing is observed beyond about two diameters downstream from the exit. This is also consistent with the observations of Bruun [13] and Yule [14]. The dominant structure size and convection velocity agree with those determined from double Fourier-transformation of the longitudinal velocity space-time correlation [29]. The convection velocity of the large-scale structures is about 60% of the core fluid velocity. This closely agrees with the data of Kleis et al [30], Zaman & Hussain [8], Petersen [31], Bradshaw et al. [32], Lau et al. [33].

The two-street bifurcation model propounded by Lau is not borne out by the flow visualization movies. The large variation of the structure passage frequency across the width of a mixing layer, which prompted Lau to propose the structure bifurcation hypothesis, appears to be explainable by tearing, fractional and partial pairings and relative slippage of the high-speed side structures past the slower ones on the low-speed side. It appears that these factors were the reasons why Bruun [13] educed different structures on the high-speed

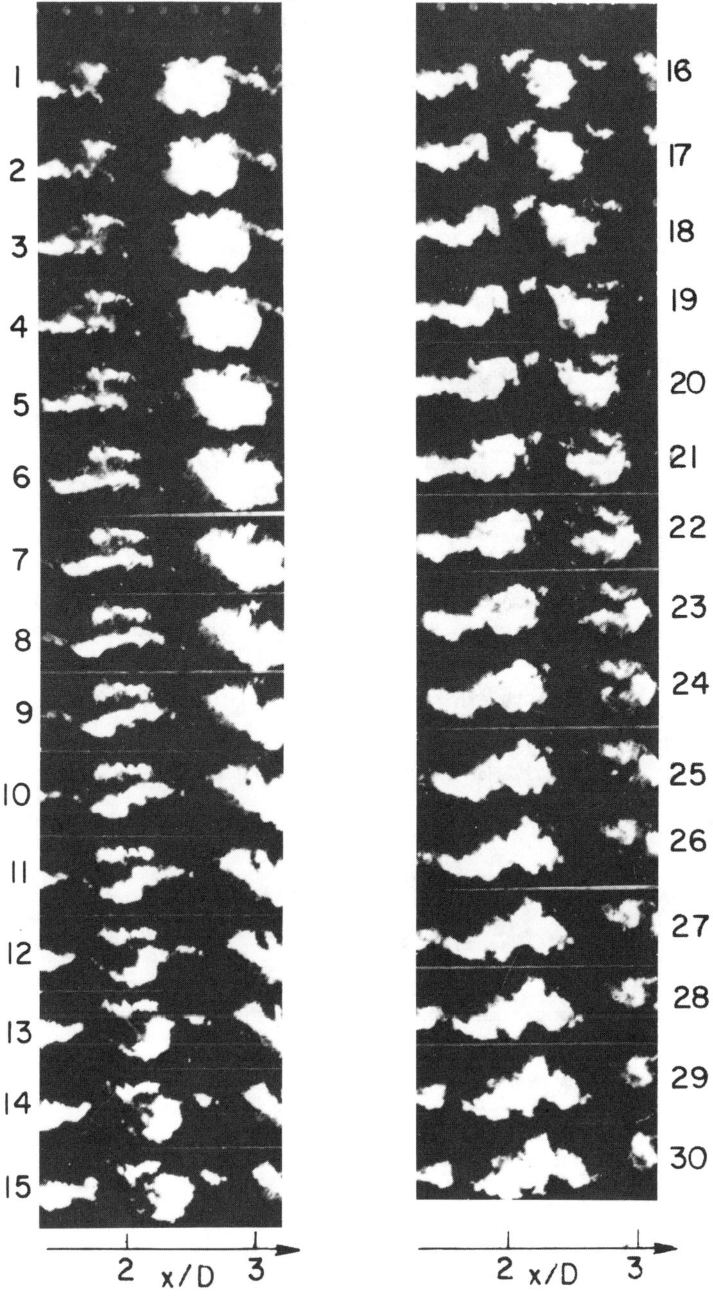

Figure 4. Picture sequence of the sectional view of the mixing layer; $\delta t = 0.56ms$.

and low-speed sides of the layer and why Yule [14] was unable to educe any structure on the low-speed side while triggering the eduction on the high-speed side velocity signal.

With increasing distances from the locations of structure roll-up, the large-scale structures appear progressively weaker and diffuse. It appears that smaller-scale structures are as important in transport and mixing as the large-scale coherent structures. This is consistent with other quantitative studies of phase-average Reynolds stress and production over the spatial extents of coherent structures [4].

Acknowledgement

This research was supported by the National Science Foundation under Grant ENG-7822110.

References

1. Brown, G. L. & Roshko, A. 1974 *J. Fluid Mech. 64*, 775.

2. Roshko, A. 1976 *A.I.A.A. J. 10*, 1349.

3. Liepmann, H. W. 1979 *American Scientist 67*, 221.

4. Hussain, A. K. M. F. 1980, "Coherent Structures and Studies of Perturbed and Unperturbed Jets" *Lecture Notes in Physics*, Proceedings of Conference on Coherent Structures, Madrid, June 25-27, 1980 (in press).

5. Hussain, A. K. M. F. 1979 *6th Symposium on Turbulence*, U. of Missouri-Rolla (in press).

6. Hussain, A. K. M. F. & Zaman, K. B. M. Q. 1980a *J. Fluid Mech.* (to appear).

7. Sokolov, M., Hussain, A. K. M. F., Kleis, S. J. & Husain, Z. D. 1980 *J. Fluid Mech, 98*, 65.

8. Zaman, K. B. M. Q. & Hussain, A. K. M. F. 1980 *J. Fluid Mech.* (to appear).

9. Hussain, A. K. M. F. & Zaman, K. B. M. Q. 1977 *Lecture Notes in Physics 75*, 31.

10. Hussain, A. K. M. F. & Zaman, K. B. M. Q. 1980b *J. Fluid Mech.* (to appear).

11. Browand, F. K. & Laufer, J. 1975 *Turb. Liquids*, U. of Missouri-Rolla, 333.

12. Lau, J. C. & Fisher, M. J. 1975 *J. Fluid Mech. 67*, 229.

13. Bruun, H. H 1977 *J. Fluid Mech. 64*, 775.

14. Yule, A. J. 1978 *J. Fluid Mech. 89*, 413.

15. Hussain, A. K. M. F. & Clark, A. R. 1980 *J. Fluid Mech.* (to appear).

16. Lau, J. C. 1978 (submitted to *J. Fluid Mech.*)

17. Kotsovinos, N. E. 1976 *J. Fluid Mech. 77*, 305.

18. Bradshaw, P. 1977 *J. Fluid Mech. 80*, 795.

19. Husain, Z. D. & Hussain, A. K. M. F. 1979 A.I.A.A. J. 17, 48.

20. Crow, S. C. & Champagne, F. H. 1971 J. Fluid Mech. 48, 547.

21. Davies, P. O. A. L. & Baxter, D. R. J. 1977 Lecture Notes in Physics, 75, 125.

22. Moore, C. J. 1977 J. Fluid Mech. 80, 321.

23. Fiedler, H. 1980 Lecture Notes in Physics (to appear).

24. Winant, C. D. & Browand, F. K. 1974 J. Fluid Mech. 63, 237.

25. Ashurst, W. T. 1977 Turb. Shear Flows, Penn State U., 11.43.

26. Antonia, R. A., Chambers, A. J. & Hussain, A. K. M. F. 1980 Phys. Fluids 23, 695.

27. Corcos, G. M. & Sherman, F. S. 1976 J. Fluid Mech. 73, 241.

28. Chandrsuda, C., Mehta, R. D., Weir, A. D. & Bradshaw, P. 1977 J. Fluid Mech. 85, 693.

29. Clark, A. R. & Hussain, A. K. M. F. 1979 Turb. Shear Flows, London, 2.30.

30. Kleis, S. J., Hussain, A. K. M. F. & Sokolov, M. 1980 J. Fluid Mech. (to appear).

31. Petersen, R. A. 1978 J. Fluid Mech. 89, 469.

32. Bradshaw, P., Ferriss, D. H. & Johnson, R. F. 1964 J. Fluid Mech. 19, 591.

33. Lau, J. C., Fisher, M. J. & Fuchs, H. V. 1972 J. Sound Vib. 22, 379.

Participants

N. Afgan, Boris Kidrič Institute of Nuclear Sciences, P.O. Box 522, 11 001 Beograd, Yugoslavia

P. André, Universite d'Orleans, Ecole Superieure des Transports de l'Energie et de la Propulsion, B.P. 6057, 45 017 Orleans Cedex, France

R. A. Antonia, The University of New Castle, New South Wales 2308, Australia

J.-P. Bonnet, Universite de Poitiers, Centre d'Etudes Aerodynamiques et Thermiques, Route de l'Aerodrome, 86 000 Poitiers, France

P. Chassaing, Institut National Polytechnique, Institut de Mécanique des Fluides, 2 rue Camichel, 31 071 Toulouse Cedex, France

R. Creff, Universite d'Orleans, Ecole Superieure des Transports de l'Energie et de la Propulsion, B.P. 6057, 45 017 Orleans Cedex, France

J. F. Debieve, Institut de Mécanique Statistique de la Turbulence, 12, Avenue Général Leclerc, 13 003 Marseille, France

A. Dinckelacker, Max-Planck-Institut für Strömungforschung, Postfach 867, 3400 Göttingen, FRG

F. Durst, Sonderforschungsbereich 80, Universität Karlsruhe, Kaiserstrasse 12, 7500 Karlsruhe 1, FRG

R. E. Falco, Michigan State University, Dept. of Mechanical Engineering, East Lansing, Michigan 48824, USA

M. Favre-Marinet, Institut de Mecanique I.M.G., B.P. 53 x, 38 041 Grenoble Cedex, France

N. Frossling, Chalmers University of Technology, Dept. of Applied Thermo and Fluid Dynamics, Hörsalsvägen 7, 412 96 Göteberg, Sweden

L. Fulachier, Universite d'Aix-Marseille, Inst. de Mécanique Statistique de la Turbulence, 12 Avenue Général Leclerc, 13 003 Marseille, France

F. Giralt, Universidad de Barcelona, Facultad de Quimica, Tarragona, Spain

I. Gokalp, Aerothermique—CNRS, 4 Ter, Route des Gardes, 92190 Meudon, France

H. Gouin, Université d'Aix Marseille III, rue Henri Poincare, 13 397 Marseille, France

K. Graichen, Academy Scie. GDR, Mohrenstrasse 39, Berlin, GDR

K. Hanjalić, Mašinski fakultet, Omladinsko šetalište b. b., 71 000 Sarajevo, Yugoslavia

M. R. Head, University of Cambridge, Dept. of Engineering, Cambridge CB16QR, England

E. J. Hopfinger, Université de Grenoble, Institut de Mécanique, B. P. 53X, 38041 Grenoble, Cedex

A. K. M. F. Hussain, Aerodynamics & Turbulence Lab., Cullen College of Engineering, University of Houston, Houston, Texas 77004, USA

M. Ichijo, Faculty of Engineering, Hokkaido University, N-13, W-8, Sapporo, Japan

J. J. Jovanović, Boris Kidrić Institute of Nuclear Sciences, P.O. Box 522, 11 001 Beograd, Yugoslavia

T. Keck, Sonderforschungsbereich 80, University of Karlsruhe, Kaiserstrasse 12, 7500 Karlsruhe, FRG

J. F. Keffer, University of Toronto, Dept. of Mechanical Eng., 5 King's College Rd., Toronto, Canada M5S 1A4

S. Klebanoff, National Bureau of Standards, FM 105, Washington, D.C. 20234, USA

Y. Kobashi, Faculty of Engineering, Hokkaido University, N-13, W-8, Sapporo, Japan

M. Landahl, The Royal Institute of Technology, Dept. of Mechanics, 100 44 Stockholm 70, Sweden

J. Laufer, University of Southern California, School of Engineering, Dept. of Aerospace Engineering, University Park, Los Angeles, CA 90007, USA

P. A. Libby, University of California, Dept. of Applied Mechanics and Engineering Sciences, Mail Code B-010, La Jolla, California 92093, USA

C. Maksimović, Gradevinski fakultet, Bulevar revolucije 73, 11 000 Beograd

P. Mestayer, Université d'Aix Marseille, Inst. de la Turbulence, 12, Av. Général Leclerc, 13 003 Marseille, France

T. Mizushina, Kyoto University, Chemical Engineering Dept., Kyoto 606, Japan

A. Muller, IHW, ETH Zürich, ETH Hönggerberg, 8093 Zürich, Switzerland

I. Nezu, Dept. of Civil Engineering, Kyoto University, Kyoto 606, Japan

C. Rey, Ecole Centrale de Lyon, Lab. de Mécanique des Fluides, 36, Route de Dardially, B. P. 163, 69 130 Ecully, France

A. de Roquefort, Univ. de Poitiers, Poitiers, France

A. Rosengard, Commissariat a l'Energie Atomique, B. P. 2, 91 190 Gif sur Yvette, France

K. Suzuki, Dept. of Mechanical Engineering, Kyoto University, Kyoto 606, Japan

R. Takaki, Hermann-Föttinger- Institut, TU Berlin, 1000 Berlin 12, FRG

N. van Thinh, Univ. de Paris VI, Mecanique Experimentale des Fluides, Campus Universitaire, Batiment 502, 91 405 Orsay Cedex, France

K. Torii, Dept. of Mechanical Engineering, Jokohama National University, Hodogaya-ku, Jokohama 232, Japan

H. Ueda, National Institute for Environmental Studies, Japan Environment Agency, P.O. Yatabe, Taukuba, Ibaraki 300-21, Japan

P. Vukoslavčević, University Titograd, Titograd, Yugoslavia

J. M. Wallace, Dept. of Mechanical Eng., The Univ. of Maryland, College Park, Maryland 20742, USA

Z. Zarić, Boris Kidrič Institute of Nuclear Sciences, P.O. Box 522, 11 001 Beograd, Yugoslavia

Index